The Emergence of Modern Humans

An Archaeological Perspective

The Emergence of Modern Humans

An Archaeological Perspective

edited by

PAUL MELLARS

EDINBURGH UNIVERSITY PRESS

22 George Square, Edinburgh

Transferred to Digital Print 2009

Set in Linotron Plantin
by Koinonia Ltd, Bury,

Printed and bound in Great Britain by
CPI Antony Rowe, Chippenham and Eastbourne

British Library Cataloguing
in publication data
The Emergence of modern humans.
1. Man. Evolution
I. Mellars, Paul
573.2

ISBN 0 7486 0130 9

Contents

Preface vii
Paul Mellars
List of contributors ix

A. REGIONAL STUDIES

1. Social and Ecological Models for the Middle Stone Age in Southern Africa 3
Stanley H. Ambrose and Karl G. Lorenz

2. A Critique of the Consensus View on the Age of Howieson's Poort Assemblages in South Africa 34
John Parkingon

3. The Middle and Upper Palaeolithic of the Near East and the Nile Valley: the Problem of Cultural Transformations 56
Anthony E. Marks

4. The Amudian in the Context of the Mugharan Tradition at the Tabun Cave (Mount Carmel), Israel 81
Arthur J. Jelinek

5. A Technological analysis of the Upper Palaeolithic Levels (XXV-VI) of Ksar Akil, Lebanon 91
K Ohnuma and C.A. Bergman

6. Middle to Upper Palaeolithic Transition: the Evidence for the Nile Valley 139
Philip Van Peer and Pierre M. Vermeersch

7. The Szeletian and the Stratigraphic Succession in Central Europe and Adjacent Areas: Main Trends, Recent Results, and Problems for Resolution 160
P. Allsworth-Jones

8. Chronological Change in Périgord Lithic Assemblage Diversity 243
Jan F. Simek and Heather A. Price

9. The Middle-Upper Palaeolithic Transition in Southwestern France: Interpreting the Lithic Evidence 262
Tim Reynolds

10. The Early Upper Palaeolithic of Southwest Europe Cro-Magnon Adaptations in the Iberian Peripheries, 40 000—20 000 BP 276
Lawrence Guy Straus

11. The Transition from Middle to Upper Palaeolithic at Arcy-sur-Cure (Yonne, France): Technological, Economic and Social Aspects — 303
Catherine Farizy

12. Peopling Australasia: the 'Coastal Colonization' Hypothesis re-examined — 327
Sandra Bowdler

B. GENERAL STUDIES

13. Middle Palaeolithic Socio-Economic Formations in Western Eurasia: an Exploratory Survey — 347
Nicolas Rolland

14. Aspects of Behaviour in the Middle Palaeolithic: Functional Analysis of Stone Tools from Southwest France — 389
Patricia Anderson-Gerfaud

15. A Multiaspectual Approach to the Origins of the Upper Palaeolithic in Europe — 419
Janusz K Kozlowski

16. From the Middle to the Upper Palaeolithic: the Nature of the Transition — 438
Marcel Otte

17. Early Hominid Symbol and Evolution of the Human Capacity — 457
Alexander Marshack

18. Human Cognitive Changes at the Middle to Upper Palaeolithic Transition: the Evidence of Boker Tachtit — 499
Sheldon Klein

19. Symbolic Origins and Transitions in the Palaeolithic — 517
Mary LeCron Foster

Author Index — 541
Site and Localities Index — 549

Preface

The question of the biological and behavioural origins of fully 'modern' human populations is not only one of the most important issues in palaeoanthropology, but also one of the most difficult and controversial. Throughout almost the whole of the present century opinions on this question have tended to polarise between two sharply opposed schools of thought: on the one hand are those who see the transition from anatomically 'archaic' to fully 'modern' populations as an essentially gradual, continuous process, reflecting a gradual working out of long-term evolutionary processs in all regions of the world; on the other hand are those who see this as a much more dramatic event, involving some form of large-scale population dispersal – and eventually replacement – deriving ultimately from one major geographical centre. Intermediate positions involving varying degrees of local evolution, combined with equally local episodes of population displacement and interaction, can of course be added to this range of options. As a generalisation, however, it is probably fair to say that the tendency in most of the recent literature has been to opt fairly firmly for one of these two sharply opposed schools of thought.

The Cambridge conference on the 'Origins and Dispersal of Modern Humans' was planned to bring together specialists on all aspects of the biological and behavioural origins of modern human populations, at a time when many crucial developments in molecular biology, relative and absolute dating methods, novel approaches to the analysis and interpretation of the archaeological data etc. were bringing an impressive array of new and highly important evidence to bear on these long debated problems. The success of the conference depended party on the wide range of specialists present at the meeting (fifty-five in all, from areas as far afield as Australasia, southern Africa and North America) and partly on the wide range of complementary disciplines represented – including not only the long-established fields of biological anthropology and archaeology, but also the newly emerging fields of molecular biology, human ethology and palaeo-linguistics. The opportunity to confront the crucial issues of modern human origins and dispersals from such a wide range of geographical and theoretical perspectives was an experience which most of the contributors seemed to find extremely stimulating and which certainly led to some very lively exchanges and discussions at the meeting itself.

The task of collecting together the wide range of papers presented at the Cambridge conference for publication has itelf presented something of a challenge. Since it was clearly impossible to include all of the fifty-five papers given at the conference in a single volume, there was the difficult task of dividing these in some meaningful way into two separate volumes. Eventually – and after careful discussions with the publishers – it was decided to include all of the strictly 'biological' papers, together with some of the more general 'behavioural' and archaeological papers in the first volume, and to reserve the second volume primarily for the more detailed and specific archaeological case studies. Any division of this kind is inevitably arbitrary to some degree, and there are clearly many papers which could have fitted equally well into either volume. Nevertheless, the emphasis in the present volume is placed very firmly on the archaeological aspects of the transition from archaic to modern humans, as a way of illustrating not only the general patterns of human behaviour and cultural development across this crucial transition, but also some of the interesting and highly signficant contrasts in the patterns of behavioural and technological development which can be observed in different regions of the world.

The editorial task of preparing the papers for publication has, once again, been dependent on help from many colleagues and students at Cambridge. Above all, none of this editorial work could have been completed successfully without the computing expertise of Nigel Holman, who successfully arranged for the transcription of all the papers onto computer files and played a major role in the detailed editing of the text and bibliographies, and the distribution of computerised galley-proofs to authors. Similar help in the computerisation of texts was provided by Mrs Doreen Simpson of the University Computing service, while many of the illustrations for the volume were redrawn by Sarah Skinner and John Rodford. The whole of the editorial work was carried out in the Department of Archaeology at Cambridge, for which the cooperation of the Head of Department, Professor Colin Renfrew, and the Departmental Secretary, Mrs José John, is warmly acknowledged. Needless to say, none of the organisation of the Cambrige conference would have been possible without the generous financial support of several organizations, including the L.S.B. Leakey Foundation for Anthropological Research, The British Academy, the Royal Society, the Boise Fund of Oxford University, and the Association for Cultural Exchange. It was of course this support which made it possible to bring together so many colleagues for the original meeting in Cambridge, and which ultimately contibuted to the highly diverse and stimulating range of view-points reflected in the published papers.

Lastly, I owe a special debt of thanks to my co-organiser in the Cambridge meeting, Chris Stringer, and to the authors of the various papers included in the two volumes of the conference proceedings. Even if these papers have not succeeded in resolving more than a fraction of the immensely difficult problems involved in the study of modern human origins, they have at least served to bring these issues into much sharper focus and – I hope – to point the way to some of the exciting lines of research to be explored over the course of the next decade.

PAUL MELLARS
CAMBRIDGE
June 1990

List of contributors

P. ALLSWORTH JONES, Department of Archaeology, University of Ibadan, Ibadan, Nigeria.

S. H. AMBROSE, Department of Anthropology University of Illinois at Urbana-Champaign, 109 Davenport Hall, 607 South Mathews Ave., Urbana, Illinois 61801, USA.

P. ANDERSON-GERFAUD, Institut de Préhistoire Orientale, Jalés, Berrias, 07460 Saint-Paul-Le-Jeune, France.

C. A. BERGMAN, 2110 Anderson Ferry Road, Cincinnati, Ohio 45238, USA.

S. BOWDLER, Centre for Prehistory, University of Western Australia, Nedlands, Western Australia 6009, Australia.

C. FARIZY, Laboratoire d'Ethnologie Préhistorique, 44 Rue de l'Amiral Mouchez, 75014 Paris, France.

M. LECRON FOSTER, Department of Anthropology, University of California, Berkely, California 94720, USA.

A. JELINEK, Department of Anthropology, University of Arizona, Tucson, Arizona 85721, USA.

S. KLEIN, Computer Sciences Department, University of Wisconsin-Madison, 1210 West Dayton Street, Madison, Wisconsin 53706, USA.

J. K. KOZLOWSKi, Institute of Archaeology, Jagellonian University, 11 Ul. Golebia, P-31007, Kracow, Poland.

A. E. MARKS, Department of Anthropology, Southern Methodist University, Dallas, Texas 75275, USA.

A. MARSHACK, 4 Washington Square Village, New York, New York 10012, USA.

M. OTTE, Service de Préhistoire, Université de Liège, 7 Place du XX Août, B-4000 Liège, Belgium.

K. OHNUMA, The Institute for Cultural Studies of Ancient Iraq, Kokushikan University, Machida, Tokyo, 194-01, Japan.

J. E. PARKINGTON, Department of Archaeology, University of Cape Town, Private Bag, Rondesbosch, Cape 7700, South Africa.

T. E. G. REYNOLDS, Department of Archaeology, University of Cambridge, Downing Street, Cambridge CB2 3DZ.

N. C. ROLLAND, Department of Anthropology, University of Victoria, PO Box 1700, Victoria, British Columbia, Canada V8W 2Y2.

J. F. SIMEK, Department of Anthropology, University of Tennessee-Knoxville, 252 South Stadium Hall, Knoxville, Tennessee 37996-0720, USA.

L. G. STRAUS, Department of Anthropology, University of New Mexico, Albuquerque, New Mexico 87131, USA.

P. VAN PEER, Laboratorium voor Prehistorie, Instituut voor Aardwetenschappen, Katholieke Universiteit Te Leuven, Redingenstraat 16bis, 3000 Leuven, Belgium.

P. M. VERMEERSCH, Laboratorium voor Prehistorie, Instituut voor Aardwetenschappen, Katholieke Universiteit Te Leuven, Redingenstraat 16bis, 3000 Leuven, Belgium.

A
Regional Studies

1. Social and Ecological Models for the Middle Stone Age in Southern Africa

STANLEY H. AMBROSE AND KARL G. LORENZ

INTRODUCTION

The Howieson's Poort lithic tool tradition of Southern Africa has a suite of unusual features that has led researchers to characterize it as "a remarkably precocious lithic entity" (Butzer 1982: 42), and to consider it transitional between Middle Stone Age (MSA) and Later Stone Age (LSA) industries, occupying a culture-stratigraphic and technological position analogous to that of the Magosian further north in subsaharan Africa (Sampson 1974). The transitional interpretation may be untenable because of its antiquity and stratigraphic position midway through the sequence of MSA industries.

Singer and Wymer (1982: 107-9) interpreted its abrupt appearance as the result of the "intrusion of people from outside the region with different cultural and possibly physical heritage" over one of "the reaction of the indigenous population to a change of circumstances." Population replacement hypotheses are notoriously unfashionable and are now usually dismissed *a priori* as explanations of last resort by those studying the prehistory of hunter-gatherers (e.g. Deacon 1978: 102; Parkington 1980: 80; but see Bettinger and Baumhoff [1982] for an ecological approach to population replacement). While not slaves to current fashions, we here present an ecological model and review archaeological evidence which suggests that the appearance and disappearance of the Howieson's Poort *was* the result of changes in the adaptation of an indigenous population in response to environmental change.

In this paper we will briefly review the evidence for assemblage composition, chronology, subsistence and environmental change in the MSA of Southern Africa, and evaluate the archaeological data in terms of predictive ecological models of hunter-gatherer social and territorial organization proposed by Dyson-Hudson and Smith (1978) and Wilmsen (1973). These models are based on a cost-benefit analysis of subsistence and settlement strategies analogous to those applied by ecologists to other mobile animal species (Brown 1964; Brown and Orians 1970; Orians and Pearson 1979; Horn 1968). We conclude that the transition seen between the MSA 2 and the Howieson's Poort is similar to that seen between the Albany and Wilton LSA industries in Southern Africa during the early to middle Holocene, and represents a similar transformation in resource procurement strategies and social and territorial organization in response to environmental change. Deacon and Thackeray (1984) have made passing reference to this parallelism, and we here explore this theme in some detail. However, the most important conclusion arising from this analysis is identical to that of Klein (1975, 1978a, 1979): MSA humans apparently did not have the same capacity for adaptation to resources as terminal Pleistocene humans because they failed to respond to similar resource structures in analogous ways. Our conclusion contradicts that of Deacon (1989), who considers their behaviour to be effectively modern.

THE MIDDLE STONE AGE OF SOUTHERN AFRICA

Lithic assemblage composition and industrial affiliations

In all sites in which Howieson's Poort lithic assemblages are found, they are stratified above MSA industries typically assigned to MSA 1 or 2 (Pietersburg Industrial Complex), or stratified below MSA 3 or 4 (Bambatan Industrial Complex). Fine-grained non-local lithic raw materials (mainly crystal quartz and cryptocrystalline silicas such as chert and silcrete) dominate Howieson's Poort assemblages, while other MSA assemblages typically have low frequencies of fine-grained lithic raw materials from distant sources, and high frequencies of coarse-grained, locally available lithics. MSA 3 and 4 assemblages typically have slightly higher percentages of non-local raw materials than MSA 1 and 2 assemblages. These differences in lithic raw material use through time in one place are best seen in the long stratified sequence at Klasies River Mouth, which contains a fairly complete sequence of MSA industries (Singer and Wymer 1982).

The Howieson's Poort lithic tool tradition is characterized by artifact classes that typify other MSA industries, including flakes with facetted platforms, side-scrapers, and occasionally unifacial and bifacial points. It is combined with notched blades, end-scrapers, relatively large, thin backed blades, trapezoidal, triangular and crescentic backed blade segments (microliths) made on fine-grained siliceous raw materials (Volman 1984). The latter suite of artifact types is characteristic of Later Stone Age (LSA) Holocene lithic industries. Volman (1984) suggests these artifacts were hafted in composite tools in ways similar to those of LSA tool types, but notes that direct evidence for hafting techniques has not yet been obtained. Singer and Wymer (1982) concur, and further suggest that the blades upon which these tools were made were produced by a punched blade technique. However, it is difficult to differentiate punch-struck from soft hammer blades (Bordes and Crabtree 1969). Hence Howieson's Poort stone tool functions and production techniques remain uncertain.

This combination of MSA and LSA typological features has led many researchers to conclude that this was a late MSA or transitional MSA/LSA (Second Intermediate) industry related to the Magosian of lower latitudes in subsaharan Africa (Clark 1959; Sampson 1974). The 'modern' or 'advanced' aspect of this lithic industry led to the conflation of technology with age, and with human biological and cultural evolutionary stages. This has been a significant impediment to an understanding of the cultural, technological and stratigraphic succession of Southern Africa.

Stratigraphy and chronology

Sites where Howieson's Poort assemblages have been found stratified below or between more typical MSA industries include Border Cave, Boomplaas, Apollo 11, Rose Cottage Cave, Border Cave, Sehonghong, Moshebi's Shelter, Montagu Cave, Peers Cave, Boomplaas and Klasies River Mouth (hereafter KRM) (Volman 1984; Carter 1969). Binford (1984) has reinterpreted the stratigraphic position of the Howieson's Poort at KRM, placing it later in time. Binford bases his reinterpretation of the cultural stratigraphy on a faunal seriation. The logical foundation for such an exercise requires careful explication to say the least. His methodological procedure and logic have been challenged by Marean (1986) and others.

Nonetheless, there is continued uncertainty about the age and stratigraphic position of the Howieson's Poort. Parkington's discussion in this volume is mainly predicated on the assumption that finite radiocarbon dates less than 40 000 years old on these levels are meaningful. However, there are many ways in which carbon can be contaminated to appear younger, but very few ways (e.g. contamination with coal [Tankersley *et al.* 1987]) to make charcoal spuriously older. Thus the presence of infinite radiometric dates together with finite ones, in our opinion, indicate that the finite dates rather than the infinite ones are in error and should be rejected. Parkington notes that Howieson's Poort assemblages rarely have overlying MSA 3/4 horizons, implying there may be more than one Howieson's Poort-like industry within the MSA. If Howieson's Poort-like industries appeared at several points in time then one would expect them to occur more than once in a single stratigraphic sequence; this situation has not as yet been documented.

During the 1960s Howieson's Poort occurrences were dated to 18 740 BP at the type site (Sampson, 1974), 19 100 BP and 23 200 BP at Montagu Cave (Keller 1973), and >18 400 BP and 28 100 BP at KRM Cave 1A (Singer and Wymer 1982). A reassessment of radiocarbon dating of Southern African sites by Vogel and Beaumont (1972) suggests that all finite dates (e.g. less than 40 000 BP) on Howieson's Poort and other MSA industries are probably unreliable. Using improved techniques of pre-treatment for radiocarbon samples Vogel and Beaumont (1972) have shown that the entire MSA is probably outside of the range of radiocarbon dating. Dates of greater than 40 000 and 50 000 BP have been obtained from Montagu Cave (Keller 1973) and redating of the Howieson's Poort layers at Rose Cottage Cave to >48 000 BP (Vogel and Beaumont 1972) and those from KRM to >50 000 BP (Singer and Wymer 1982) show that the Howieson's Poort is greater than 50 000 years old. Dates younger than 35 000 BP on post-Howieson's Poort MSA occurrences from Boomplaas and Apollo 11 Cave (Deacon and Thackeray 1984; Deacon *et al.* 1989;

Wendt 1976) are considered reliable. It is thus possible that the MSA ended as late as 26 000 BP.

Regrettably, reliable chronometric dating techniques are presently unavailable for the MSA. Agreement between different dating techniques applicable to this time period is often an elusive goal (Colman *et al.* 1986). In Southern Africa heavy reliance is placed on correlations with marine oxygen-isotope stages for dating. However, the details of climatic changes during the last 125 000 years remain poorly resolved in terrestrial sequences, and mis-correlation is possible. Correlation by oxygen-isotope ratios of shells in coastal sites is risky. Shells of different ages can have similar ^{18}O values. ^{18}O values of one sea shell in the Howieson's Poort levels of the KRM sequence have been interpreted by Shackleton (1982) as correlating with oxygen-isotope stage 3, in the range of 30 000 to 50 000 years ago, or possibly a cooler part of the older stage 5. Butzer (1982) correlated these levels with the cryoclastic deposits at Nelson Bay Cave, and interpreted the oxygen-isotope data as correlating with stage 5b. Correlation of cryoclastic sediments at Border Cave by Butzer *et al.* (1978) and Nelson Bay Cave (Butzer 1973) with oxygen-isotope stage 5a or 5b suggest an age near the end of the last interglacial period, between roughly 80 000 and 95 000 BP. Aspartic acid racemization 'dates' on bones from an MSA 4 level at Klasies River Mouth (Singer and Wymer 1982) suggest an age older than 65 000 BP. Amino acid racemization rates are controlled by environmental factors and are thus useful only for relative dating within sequences.

Although there is an emerging consensus of an age of around 70 000 BP for the Howieson's Poort (Deacon 1989), confirmation of its age by more reliable dating techniques is needed. Electron spin resonance and other dating methods are being applied to the KRM sequence (Deacon *et al.* 1986) and ionium dates of 98 000 and 110 000 BP have been obtained for MSA 2 levels (Hendey and Volman 1986). Thermoluminescence of burned cryptocrystalline silica artifacts would provide additional evidence for absolute dating, and has provided interesting insights into the chronology of the European Mousterian (Mellars 1986). Accelerator mass spectrometry of carbon and other elements (Partridge *et al.* 1984) may offer additional avenues for absolute dating of the MSA.

Palaeoenvironmental and economic reconstructions

The Howieson's Poort appears to date to the boundary between interglacial and glacial conditions. The climate of the time has been characterized as cool and humid on the basis of sedimentological evidence from Border Cave, Boomplaas, Montagu Cave, Nelson Bay Cave and KRM (Butzer 1973, 1982, 1984; Butzer *et al.* 1978). Cryoclastic or *éboulis sec* zones are present in the first four sites. Ferruginous crusts overlying the Howieson's Poort horizons in Nelson Bay Cave and Boomplaas suggest cooler, moist conditions followed this period. Butzer (1982) interprets Shackleton's (1982) ^{18}O values for shells from the KRM sequence as consistent with this interpretation. The absence of beach sands in the

Howieson's Poort layers in KRM (Butzer 1982) suggests that there was a slight decline in sea level at this time, which is also consistent with the oxygen-isotope evidence for global cooling. Sea level obviously did not decline far at this time, because marine shellfish and seals were still brought to the site. The underlying MSA 2 layers were deposited during the warmer, moist, last interglacial period, when sea level was at least as high as at present. Overlying MSA 3 layers are thought to have been deposited partly during a warmer humid climate corresponding to oxygen-isotope stage 5a, but the bulk of MSA 3 and 4 levels were apparently formed in a cooler climate, coeval with oxygen-isotope stages 3 and 4 (between 75 000 and 40 000 BP [Volman 1984]), when sea level may have been lower than at present. Recent work suggests that the KRM sequence terminates during the marine transgression at the oxygen isotope stage 4-3 boundary when precipitation was higher (Deacon 1989).

Faunal evidence for environmental conditions during Howieson's Poort times is available from Border Cave, Boomplaas Cave, and KRM (Klein 1978b, 1977, 1976). Interpretation of environmental conditions from faunal remains is complex when there are relatively small numbers of individuals from a diverse range of species with sometimes subtly different habitat preferences. Grayson (1981) suggests that presence or absence of species characteristic of specific habitats are more reliable indicators of prehistoric environmental conditions than are proportions of species. Grayson's caveat should be considered even when comparing faunal assemblages from a single sequence. In the KRM sequence (Klein 1976) within the MSA 1 levels there was a shift from large nomadic grazing ungulates that prefer open habitats (equids and alcelaphines) to higher proportions of species that prefer closed habitats (namely steenbok [*Raphicerus campestris*] and kudu [*Tragelaphus strepsiceros*]) and a corresponding decrease in open country forms. Klein (1976) considers this transition to be analogous to that which occurred at the Late Pleistocene/Holocene boundary at Nelson Bay Cave. The Howieson's Poort fauna at KRM also emphasized smaller, non-gregarious species that prefer closed habitats, but species such as wildebeest, hartebeest and zebra (*alcelaphini* and *equidae*), which prefer open grassland environments, began to appear in small numbers. The reversal in proportions of equids plus alcelaphines *versus Tragelaphus* plus *Raphicerus* was completed by MSA 3 and 4 times. Klein (1976) has, however, by lumping the Howieson's Poort with MSA 3 and 4 to enlarge sample size, made it appear that the transition to open grassland habitats was completed by Howieson's Poort times. This contrasts with sedimentological and isotopic evidence, and his own raw data. The faunal assemblages from Howieson's Poort levels at Border Cave and Boomplaas are also dominated by smaller, non-gregarious species characteristic of closed habitats (Beaumont *et al.* 1978; Klein 1977, 1978b; Deacon 1979). Although the Boomplaas fauna is far too small for reliable habitat reconstruction, concordant trends in changes in faunal assemblage composition in each of these sequences suggest that the inferred climate changes are not an artifact of faunal sampling error.

The MSA peoples of Southern Africa were apparently less effective predators than LSA peoples. Klein (1975) notes that in coastal MSA middens fish and flying birds were rare in comparison to shellfish and other sessile prey, or those that may have been scavenged (penguins and seals), while they were abundant in coastal LSA ones. Comparative analyses of ungulate species composition and mortality patterns (based on dental crown height measurements) of MSA and LSA faunal assemblages performed by Klein (1975, 1977, 1978a, 1979, 1980, 1981, 1982) led him to conclude that the former were less effective hunters than the latter: large dangerous species (eg. suids, elephants, rhino and giant buffalo) were underrepresented. These species were represented mainly by age classes that were more susceptible to disease and predation, namely the old and very young, rather than prime-aged adults in good condition. An 'attritional' mortality pattern for larger bovids is characteristic of larger social carnivores who select more vulnerable prey. This mortality pattern occurs in the MSA and contrasts remarkably with that of the typically 'catastrophic' mortality pattern found in LSA faunal assemblages.

HUNTER-GATHERER SUBSISTENCE AND SETTLEMENT

Theoretical considerations

Hunter-gatherers do not go about their daily business consciously assessing risk the caloric value of resources, resource encounter rates, energy expenditures and cost-benefit ratios of different subsistence strategies and resource mixes (unlike modern anthropological observers of their behaviour). However, when making decisions about when, where and which resources should be exploited or ignored, which areas should be defended, shared or trespassed, how large a camp and foraging or hunting party should be, and how often and how far the group should move residential locations, their behaviour often seems to be consistent with strategies of minimal energy expenditure, maximum energy capture, or foraging trip distance minimization.

Nowadays such adaptive strategies are referred to under the rubric of 'optimization', or 'optimal foraging' strategies. Optimization models include a family of predictive models of hunter-gatherer socioterritorial organization first proposed by Julian Steward (1938), and later formalized by Wilmsen (1973), Harpending and Davis (1977), Smith (1981), Gamble (1978) and Dyson-Hudson and Smith (1978). A functional relationship is posited between the abundance and predictability in space and time of subsistence resources (hereafter referred to as resource structure [Figure 1.1]) and hunter-gatherer residential group size, residential and logistical mobility, territory size, territorial defence strategy, intergroup relationships and diet breadth (Table 1.1).

This family of models posits four possible extremes of hunter-gatherer socioterritorial organization. It thus has more flexibility than dichotomous classifications of hunter-gatherer subsistence and settlement systems, for example, into 'broad' *versus* 'narrow' spectrum (Flannery 1969), 'processor' *versus* 'traveller' (Bettinger and Baumhoff 1982), 'forager' *versus* 'collec-

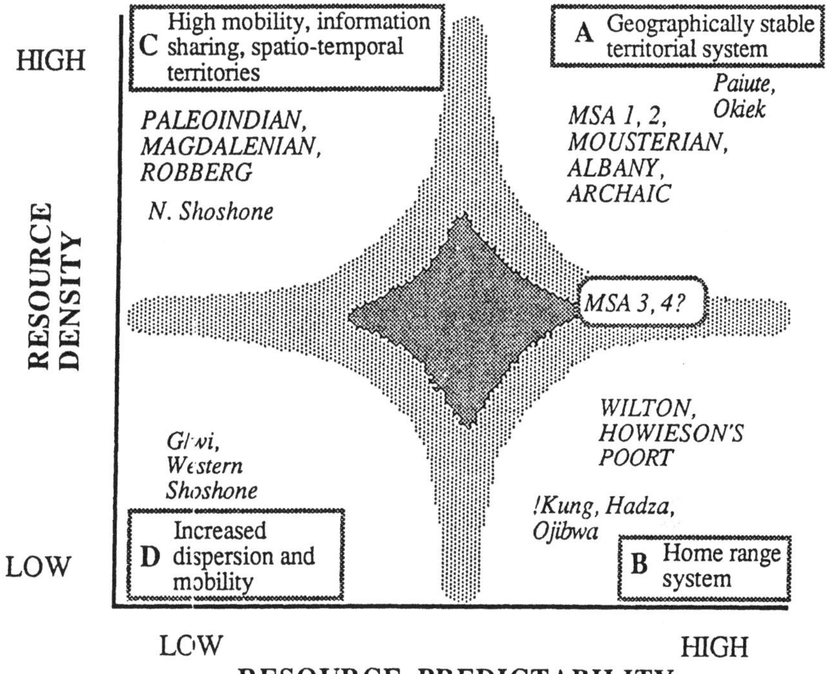

Figure 1.1. The relationship between resource structure and hunter-gatherer socioterritorial organization strategies (redrawn from Dyson-Hudson and Smith 1978). The names of selected archaeological cultures (capitalized italics) and modern hunter-gatherer groups (lower case italics) referred to in the text are placed in their approximate positions on the resource structure

tor' (Binford 1980), or 'patch foraging' *versus* 'mobile mixed' (Deacon and Thackeray 1984). Some general models are based on variables that are partly correlated with resource structure -for example the relationship between plant food consumption on the one hand, and latitude (Lee 1968), or effective temperature (Binford 1980; Kelly 1983) on the other. People do not forage for temperature, rainfall or latitude – they forage for food. At any given latitude or effective temperature the structure of the subsistence resource base can vary widely in terms of predictability, patchiness and density in space in time.

The following observations and descriptions of the responses of hunter-gatherers, and non-human animals to changes in the structural characteristics of the resources they exploit, illustrate the relationships between resource structure on the one hand and subsistence, settlement and socioterritorial organization, and the diversity in adaptations that can occur at a single latitude on the other. Harpending and Davis (1977)

Table 1.1. The relationships between resource structure and socioterritorial organization of hunter-gatherers (upper), and the long term regional and intra-site archaeological patterning (lower) this is expected to produce. Optimal and sub-optimal settlement location, settlement mobility, group size and territorial strategies for these four poles of a continuum of resource structure are illustrated in Figures 1.1 and 1.2.

BEHAVIOURAL CORRELATES

Resource Structure	Territorial Strategy	Information Exchange	Residential Mobility	Group Size	Population Density	Diet Breadth
A Predictable and dense	Territorial defence	Low	Low scheduled	Small	High	Moderate
B Predictable and scarce	Home range, semi-permeable	Medium	Medium, scheduled	Small	Medium	High
C Unpredictable and dense	Undefended very permeable	Very high	High, opportunistic	Large	Medium	Very low
D Unpredictable and scarce	Undefended, very permeable	High	Very high, opportunistic	Very small	Very low	Very high

ARCHAEOLOGICAL CORRELATES

Resource Structure	Macro-regional Assemblage Variability	Raw Material Sources	Occupation Site Intensity	Intra-site Spatial Organization	Faunal & Floral Diversity
A Predictable and dense	High stylistic variability	All local, embedded procurement	High at home base	Very structured activity and discard locations	Moderately high
B Predictable and scarce	Low stylistic variability	Mostly local	Moderate at home base	Moderately structured	High
C Unpredictable and dense	High stylistic uniformity	Diverse, many distant exotics	Very low at home base	Poorly structured	Very low, large game
D Unpredictable and scarce	High stylistic uniformity	Local, and distant exotics	Very low at home base	Poorly structured	Very high, mostly plant

observed that !Kung San bands had contrasting systems of settlement, mobility and territoriality in wetter, stable *versus* drier, unpredictable environments. They found that resources were highly concentrated on forested dunes spaced at intervals of several kilometres in the northern part of their study area. As a result of the stability, predictability and abundance of resources they found "a pattern of sedentary !Kung villages on the dune edges, with low inter-village mobility of individuals and families" (Harpending and Davis 1977: 275). In the arid southern Kalahari the primary resources – mobile ungulates and scarce and scattered plant foods – resulted in a pattern where 'there is little coherent occupation of the land for families constantly move over thousands of square kilometers.' These observations on diversity in mobility, interaction and settlement among !Kung foragers led them to propose that:

> In an environment where resources are distributed evenly and are of many different species, a hunter-gatherer group minimizes energy expenditure by being sedentary and possibly territorial; alternatively, if resources are widely scattered and only locally abundant, then group mobility will be necessary to exploit the full range of available resources. (Harpending and Davis 1977: 275).

Bicchieri (1969) performed a comparative study of the subsistence and settlement systems of BaMbuti pygmies, Hadza and southern Kalahari San. The desert San and the Hadza were characterized as having migratory settlement patterns in response to a seasonally variable, patchy distribution of scarce to sufficient resources. The Ituri forest BaMbuti had a resource base characterized as abundant and ubiquitous. Bicchieri suggested this stable resource base was correlated with demarcated territories and more systematic, predictable settlement locations. From these observations he proposed that: 'As the habitat varies from restrictive to permissive in obtaining the means of livelihood... opportunistic migration [is] replaced by cyclic reuse of campsites'(Bicchieri 1969: chart).

Dyson-Hudson and Smith (1978) have summarized Julian Steward's observations on three Great Basin and Plateau hunter-gatherer groups in order to illustrate the relationship between resource abundance and predictability in space and time, and subsistence, settlement and socio-territorial organization in a restricted region. Their choice of this region is important because it shows how much variability can occur within a small range of latitude.

The Western Shoshone lived in an arid region where resources were neither abundant nor predictable. Plant resources were collected by individual dispersed nuclear families, and hunting was a solitary endeavour. Pinion pine groves occasionally produced an erratic but superabundant nut crop, allowing temporary congregation of dispersed families at different places each year. Occasional cooperative hunts, in the form of antelope drives and rabbit net drives, also resulted in a temporary glut of resources, but again in unpredictable locations. Defended territories did not exist. Rather, there was a system of overlapping home ranges, with a high degree of information sharing, in order to increase the reliability of access to unpredictable and scarce resources.

The Owens Valley Paiute occupied a far richer environment than the Western Shoshone. A more secure water supply from the adjacent mountains, and elevation-stratified vegetation zones, increased the abundance, predictability and diversity of resources. All subsistence requirements could be fulfilled within a 3.2 km radius of the settlement. Consequently, the Paiute were sedentary, had smaller, physically defended territories, and lived in larger, closely and regularly spaced settlements, with a population density ten times higher than that of the Western Shoshone. Information sharing between settlements was presumably unnecessary.

The Northern Shoshone depended for a large part of their subsistence during the summer months on migratory bison herds on the plains. Such herds can be characterized as abundant and concentrated, but spatially and temporally unpredictable. In response, the Northern Shoshone formed large migratory bands for cooperative hunting. During the fall, when the hunt was over, bands dispersed into small family units, and scattered, unpredictable plant and small mammal resources were gathered in the valleys. During the winter the bands regrouped in large villages where they lived off of surplus meat cached from the summer kills. Neither buffalo herds nor plant resources were spatio-temporally predictable enough to be territorially owned. Consequently, group ranges overlapped extensively.

The seasonal variation in group size and foraging strategies noted among the Northern Shoshone is not unique. Among the Efe pygmies of the Ituri forest Bailey (1986) has observed that foraging for animal resources with nets is a group endeavour. However, when honey becomes available they disperse and forage alone to exploit this evenly distributed, predictable resource.

Intraspecific diversity in subsistence, settlement and socioterritorial organization is not restricted to human hunter-gatherers. This diversity has been demonstrated to correlate with differences in resource structure among a wide range of animal species. For example, Itani and Suzuki (1967: 356) observed that chimpanzees in a dry savanna woodland environment tend to live in larger groups, while forest-dwelling groups tend to have smaller social units. They also noted that chimpanzee groups are more evenly distributed in the forest than in the savanna woodland, and that open-country groups tend to migrate longer distances, moving as a cohesive unit, as is popularly envisioned for savanna-dwelling baboons. Territory size also varies with resource density and predictability. Forest chimpanzee territories of 7.5 and 35 km^2 have been reported for the Budongo forest and Kabogo mountains, respectively (Itani and Suzuki 1967: 372). Savanna woodland chimpanzee groups have ranges of from 100 to 200 km^2 (Izawa 1970; Suzuki 1969). Similar contrasts in group structure and size, and territory size have been noted by Hamilton *et al.* (1976) and Rowell (1966) between savanna and swamp-dwelling baboons. Finally, analogous contrasts in zebra, lion and spotted hyena socioterritorial organization have been observed by Kruuk (1975) between the stable environment of Ngorongoro Crater and the unstable one of the Serengeti Plain.

It is important to note that the same habitat may be perceived as having different resource structures depending on the morphological and behavioural attributes of the species, or the technological capacities of the human population. For example, the Kabale forest of Uganda presents a spatio-temporally abundant, predictable resource base for black and white colobus monkeys who possess the digestive adaptations for leaf-eating, but presents a relatively patchy and unpredictable resource base for fruit-dependent red colobus monkeys (Clutton-Brock 1974). Hence the effective resource structure for these two species differs, and this has a great influence on their respective socioterritorial organizations: black and white colobus monkeys live in small groups with well-defined defended territories, while red colobus monkeys live in larger, more mobile groups and have overlapping home ranges.

For human populations a change in technology – as, for example, the adoption of traps and snares – would turn small game from a scarce and relatively unpredictable resource base to a more predictable one, and permit greater sedentism (Hayden, 1981, 1982; Torrence 1983), as among the Northern Ojibwa (Dyson-Hudson and Smith 1978: 31-3). The adoption of projectile weapons – spear throwers and bows, especially with poisoned projectile tips must have increased the success rate for large game hunting in the late Upper Pleistocene. Large game would have then become an attractive resource, and previously-exploited smaller game and plant foods would have been passed over in preference for pursuit of more mobile resources. The result would have been an increase in mobility and range size.

RESOURCE STRUCTURE, SOCIOTERRITORIAL ORGANIZATION AND THE ARCHAEOLOGICAL RECORD

Humans and other animals thus have highly flexible mobility and socio-territorial organization patterns. Different configurations of resource structure appear to select for different organizational responses in hunter-gatherer adaptations. The structural characteristics of the resources that dominate the subsistence base may grade from one basic configuration to another in different habitats, or in the same habitat at different times within a group's range (as among the Northern Shoshone and Efe). One can thus model variation in subsistence and settlement within and between populations. This approach is therefore not deterministic but is flexible and suitable for modelling functional, seasonal and longer term (e.g. inter-annual, glacial/interglacial) changes in human foraging strategies and socio-economic organization.

Resource structure has two significant parameters: predictability and abundance. Resource predictability itself has two important attributes: predictability in space, and predictability in time. Resource abundance has three major attributes: average abundance or density, patchiness or clustering, and temporal abundance. For example, streams where salmon run every year represent resource locations that are patchy, high density and predictably located in space and time. Herds of large, nomadic wild herbivores such as caribou, bison or wildebeest are high density resource patches whose precise location changes daily, and the timing of annual migrations may vary, rendering them unpredictable in space and time. Pinon pine groves are predictably located in space, but mast abundantly at irregular intervals, averaging once every 3-4 years, rendering them temporally unpredictable and territorially undefendable. The nut trees of the eastern woodlands and California oaks mast on a more predictable annual schedule and are more evenly distributed throughout their habitat. They can thus be characterized as relatively abundant and predictable in space and time, as can coastal marine resources such as shellfish. In comparison to open grassland mammalian herbivores, those preferring forest, woodland and bush tend to be sedentary, non-gregarious and live in evenly spaced home ranges or territories. They are thus predictably and evenly distributed in space and time, and moderately abundant. In semi-arid and arid regions both plant and animal abundance and predictability in space and time declines, as in the American Great Basin, parts of the Kalahari desert in Africa, and the Western Desert of Australia, where resources comprise mainly small mammals, reptiles and plants.

We shall now present some generalizations about the relationship between resource structure and socioterritorial organization. The four basic extremes of configurations of resource structure described by Dyson-Hudson and Smith (1978), recent non-human and human adaptations to them, and the patterns in the archaeological record they are expected to produce are briefly summarized below and in Table 1.1, and illustrated in Figures 1.1 and 1.2. For each configuration of resource structure examples from non-human animals, recent hunter-gatherers and the archaeological record are suggested. These predictions are based on a qualitative cost/benefit analysis of different strategies of territorial organization, using energy as the currency to be optimized. Dyson-Hudson and Smith (1978) focus on variation in territorial defence strategies. Territoriality, defined as defence and exclusive use of a section of the landscape and the resources it contains, is expected when the costs of defence are outweighed by the benefits gained from exclusive access to the resources contained in the area. Since territory perimeter length increases at a rate of 3.142 times the diameter, the costs of perimeter defence escalates disproportionately with territory size increase. Hence only small territories (which must have dense and predictable resources) can be effectively defended (Figure 1.1). Wilmsen's (1973) treatment of this relationship, based on Horn's (1968) model of avian territoriality, focuses on the settlement location, group size and intergroup boundary relations that minimize mean round-trip distances between resources and consumers given different resource structures (Figure 1.2).

Type A: Resources are predictable in space and time, and abundant, as in uniform, well-watered woodland and forest environments, where the

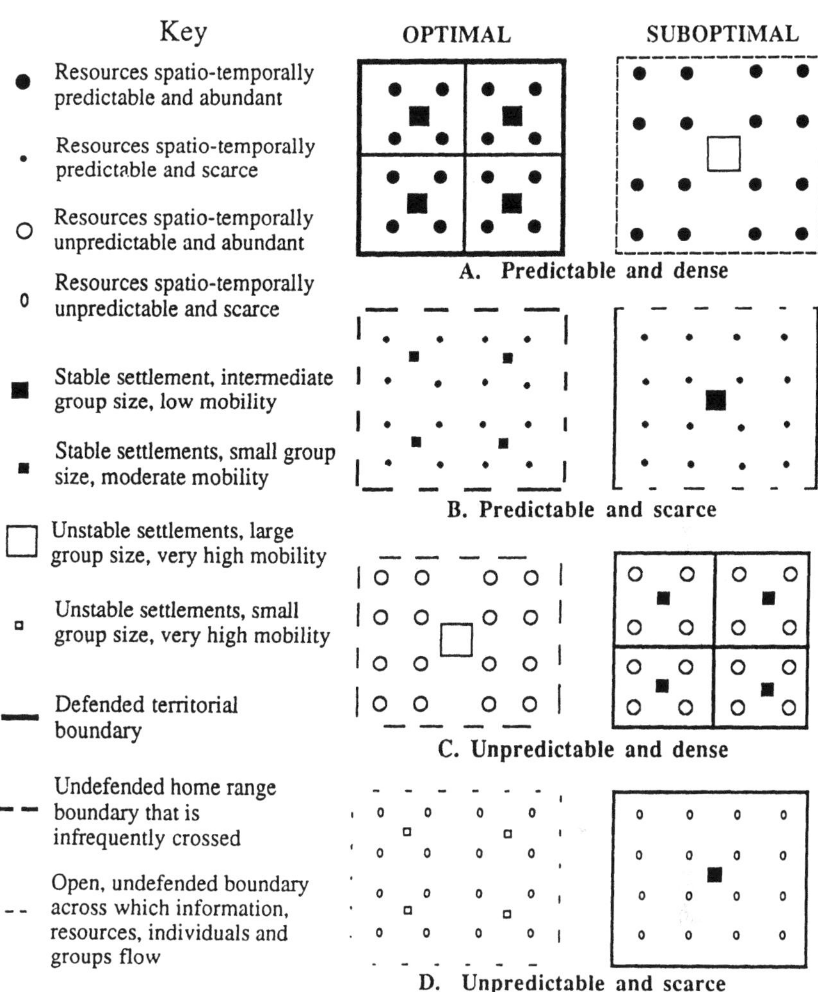

Figure 1.2. Optimal and suboptimal socio-territorial organization strategies for four extremes of resource structures (based loosely on Wilmsen 1973). The optimal settlement location is defined as that which provides the mean minimum round-trip distance between resources and consumers (Wilmsen 1973). Territorial boundaries should be closed and defended when the benefits derived from exclusive use of the territory exceed the costs of defence (Dyson-Hudson and Smith 1978).

dominant animal prey species are generally small, sedentary, non-gregarious and evenly spaced, and plant foods are evenly distributed as well. A high density, predictable resource base would fulfil a minimal band's needs within a small area. As the perimeter of such an area is small, territories could be defended because the benefits derived from exclusive access to the resources would outweigh the costs of defence. Though regional population densities may be high, residential group size may be small in order to minimize the mean round-trip distance between consumers and resources (Figure 1.2). Within territorial systems, group size is probably a function of resource density (agricultural populations probably all fall somewhere in this part of the resource structure spectrum). Base camps could be relocated infrequently, at regularly scheduled times, to predictable locations. Cooperative foraging and sharing of information about resource distributions would be unnecessary in such stable environments, and combined with territorial boundary defence, could lead to a high incidence of endogamy and cultural/linguistic differentiation.

This adaptation would generate an archaeological record characterized by small, intensively occupied sites containing a moderately high diversity of faunal and floral resources and predominantly local lithic raw material sources obtained through an 'embedded procurement' strategy (Binford 1979). Intrasite spatial patterning should be highly organized (Yellen 1977; Schiffer 1978). Such sites should have high archaeological visibility and be closely spaced. In rock-shelters and caves, thickly stratified, rapidly accumulated deposits should be found. Intentional burial of the dead should occur, because it may in many cases function as a sign of territorial ownership (Brandt 1988; Chapman 1981; Charles and Buikstra 1983).

Arboreal folivorous primates, gorillas and baboons in enriched closed habitats, whose resource bases can be characterized as spatio-temporally predictable and abundant, characteristically have low mobility, small defended territories and small group sizes (Eisenberg *et al.* 1972) as do many forest-dwelling ungulates (Geist 1974; Jarman 1974). Ethnographic examples include the Owens Valley Paiute (Dyson-Hudson and Smith 1978), the Mbuti of Central Africa (Harako 1976) and the Okiek of western Kenya (Blackburn 1971, 1982). Prehistoric examples may include the Western European Mesolithic (Gamble 1978), the eastern North American Archaic (Hayden 1982; Caldwell 1958) and the early Holocene Eburran and Albany Industries of Kenya and Southern Africa, respectively (Ambrose 1984, 1985, 1986; H. Deacon 1976; J. Deacon 1978).

Type B: Resources are spatio-temporally predictable and sparse, as in a semi-arid woodland or scrub environment where overall productivity is low but dominant resources come in small, evenly dispersed packages. Territory size should be larger than in Type A due to decreased resource encounter rates. When the costs of defending a larger territory perimeter exceed the benefits of exclusive use, territorial defence should be abandoned. Groups should still remain within a defined home range because of the uncertainties of locating resources in unfamiliar environments that may be depleted by adjacent resident groups (Moore 1981). Resource reliability could be increased by maintaining formal reciprocal arrangements with adjacent groups (Yellen and Harpending 1972). Residential group size should remain small, with higher frequencies of site relocation than in Type A.

The archaeological record of hunter-gatherer adaptations to this type of resource structure would include small, less intensively occupied home base sites in a larger variety of microhabitats. Resource diversity in home base sites should be higher than in Type A. Extra-territorial relationships should be evinced by the presence of moderate proportions of exotic raw materials (Gould 1978). Intrasite spatial organization should be less structured due to shorter occupation spans (Yellen 1977; Schiffer 1978, 1983).

Examples of non-human adaptations to this type of resource structure include many frugivorous arboreal primates (because ripe fruits are less abundant and predictable than leaves) and many asocial carnivores (leopards, jackals, brown hyenas). Ethnographic analogues may include the 18th Century Ojibwa (Dyson-Hudson and Smith 1978), the Hadza (Woodburn 1968) and the !Kung San (Lee 1979; Yellen 1977). An archaeological example of this kind of adaptation may be the South African Wilton Industry (H. Deacon 1976; J. Deacon 1978).

Type C: Resources are spatio-temporally unpredictable and dense, as in open savanna grasslands seasonally populated by large herds of nomadic ungulates in Eastern and Southern Africa, and the prairies of North America populated by bison herds. Territories should be very large and undefended because of the patchiness and high mobility of the highest ranked resources. Cooperation and sharing of information between groups would increase the likelihood of obtaining resources. Group size should be large in order to provide large, cooperative hunting parties. In addition, large groups can be supported because of the superabundance of food that such a strategy provides. Residential mobility should be high, and site relocation opportunistic, due to unpredictable prey locations. For a given standing exploitable biomass, carrying capacity would be lower in this configuration of resource structure than in Type A because energy expended in travel and pursuit of resources is greater for Type C, and the high mobility would undoubtedly decrease female fertility and increase infant mortality.

Archaeological evidence for this type of adaptation would include ephemeral occupation sites used for brief periods, and thus have little intrasite patterning of debris discard and activity area differentiation (Schiffer 1978, 1983). Floral and faunal resource diversity should be low, comprising mainly large, nomadic prey, and the lithic raw materials should include a high proportion of exotic raw materials (Wilmsen 1973). Habitation site visibility should be very low. Intentional burial of the dead should not be expected if it functions as a territorial marker.

Prehistoric examples of this type of adaptation may include the Robberg Industry of South Africa (H. Deacon 1976; J. Deacon 1978), the Western European Upper Palaeolithic (Gamble 1978) and Palaeoindians of North America (Wilmsen 1973; Hayden 1982). An ethnographic example is the Northern Shoshone buffalo hunters during the summer (Dyson-Hudson and Smith 1978) and the Nunamiut during the Caribou migrations (Laughlin 1975).

Type D: Resources are spatio-temporally unpredictable and sparse, as in the American Great Basin, the Western Desert of Australia, and the central Kalahari Desert, where resource encounter rates, particularly for game, are highly variable both seasonally and inter-annually. Territories should be very large and undefended, because the costs of defence of such a scarce resource base within such a large perimeter would far outweigh the benefits. Reciprocal exchange relationships should extend far beyond the annual home range to maximize access to territories and resources. Residential group size should be minimal during most seasons, and residential mobility should be extremely high. Occupation sites should be opportunistically relocated and be found in a wide range of microhabitats, except where water availability constrains their placement.

The archaeological record of such an adaptation would be characterized by ephemeral short-term occupation sites in a wide variety of environments, with an extremely high diversity of mainly low-ranked (small sized) plant and animal resources. Lithic raw materials should include proportions of exotic raw materials similar to those of Type B above. Intrasite spatial patterning of debris discard and activity area differentiation should be minimal because of the short duration of site occupation and small group size (Yellen 1977; Schiffer 1978, 1983). The archaeological visibility of such an adaptation would be extremely poor. A prehistoric example of this pattern may be the post-2000 BP LSA hunter-gatherers of the Western Cape, South Africa (Parkington 1984). Ethnographic examples include the Western Shoshone (Dyson-Hudson and Smith 1978), and the G/wi of the central Kalahari (Silberbauer 1981).

ARCHAEOLOGICAL CORRELATES OF SOCIOTERRITORIAL ORGANIZATION STRATEGIES

Placing prehistoric populations within this model of hunter-gatherer socioterritorial organization requires the integration of several lines of evidence, including the socio-territorial characteristics of animal prey (Foley 1983), environmental reconstruction for estimating the abundance and density of animal and plant resources (Dyson-Hudson and Smith 1978), lithic source utilization (Wilmsen 1973; Hayden 1982), the distribution of occupation sites on the prehistoric landscape, the intensity of site use, and the spatial organization of activities within sites (Yellen 1977; Schiffer 1978; 1983). Detailed justification of the predictions made above, and discussion of the methods of testing the predicted archaeological correlates of hunter-gatherer adaptations described above and listed in Table 1.1B, are beyond the scope of this paper. We will, however, focus on how lithic resource procurement is influenced by ecological factors, since this is the easiest data to extract from published reports and presents an obvious pattern in Southern African sites.

The working hypothesis of this paper is that the abundance of exotic lithic raw materials should be inversely correlated with resource abundance and predictability. Functional considerations must first be ruled out, as in the case of the Western Desert of Australia, where extremely hard cherts were preferred to work mulga wood (Gould and Saggers 1985). Changes in raw material frequencies through time, and the replacement of one technological tradition by another using a different suite of raw materials, can then be interpreted as evidence for alteration in foraging range size, degree of information sharing and intergroup boundary maintenance behaviour through time, in response to changes in the abundance and predictability of resources. From this perspective, lithic raw material procurement is hypothesized to be determined primarily by ecological factors and the human socioterritorial organization and mobility pattern rather than by the requirements of a particular kind of technology. Ecology may thus ultimately dictate the kind of lithic technology adopted.

The physical properties of raw materials may have a marked effect on the the basic approach to lithic tool manufacture (Clark 1980; Jones 1979). Coarse-grained low quality rocks are more abundant than fine-grained ones in most parts of the world, but do not lend themselves to the production of thin, regular blades. When groups increase their access to fine-grained, easily worked lithic resources, either through exchange or increased residential and logistic mobility, one would expect them to be exploited. This may permit the production of delicate forms not permitted on coarse-grained rocks and result in a fundamental shift in the character and composition of the lithic assemblage. The reverse case could occur with a permanent shift from high mobility and information/resource exchange to low mobility, endogamy and territorial defence. In this ecological model of change in lithic resource procurement, industrial change is perceived as an opportunistic response to changes in the flaking properties of lithic raw materials by a single time-transgressive population rather than assimilation or replacement of one culture by another bearing a different lithic technology.

J.D. Clark (1980: 54) notes that "... raw material can be used to help understand the extent of the territorial range of a prehistoric group and the significance of technological changes present in a stratified cultural sequence...the identification of non-local raw materials can give some indication of distances travelled or of exchange relationships." Lithic resources may thus enter a site as a result of embedded procurement, specific quarrying trips, or through exchange with adjacent groups (Gould and Saggers 1985). The source locations of raw materials found in an archaeological site may map distances travelled and/or intergroup interactions and thus regional social networks and/or territory size.

As an example of intergroup exchange, Gould (1978) found that among stone tool-using Western Desert Aborigines most raw materials came from sources within 32 km of the settlements. There was, however, a notable proportion of adzes and circumcision knives made on exotic raw materials from sources up to several hundred kilometres away – far beyond the limits of the annual ranges of the present owners of the tools. These exotic lithics were not obtained directly from the source areas by the owners, but were acquired through a totemic social network of exchange. In most cases the stone originated from a quarry associated with the present owners dream-time totem. In the Western Desert food resources are scarce and unpredictable. Information-sharing and pooling of resources would thus be highly adaptive. Gould thus understandably came to the conclusion that exotic lithic raw material exchange along totemic affiliation lines functioned to maintain a broad base of contacts and obligations with distant affines and relatives who could be relied upon in times of resource stress. In other regions this may take the form of ritualized gift exchange, as among the Kalahari San. We assume that where resources are more predictable and abundant, territorial groups, especially endogamous ones like the Northern Kalahari San, would find lithic resource exchange unnecessary.

PREHISTORIC EXAMPLES OF CHANGES IN LITHIC RESOURCE PROCUREMENT

North America. Perhaps the most well-documented example of a change in lithic resource procurement is that between the terminal Pleistocene and early Holocene in North America. Hayden (1982) has documented the available evidence and uses an explanatory framework basically similar to that used here. The only major difference between his mode of explanation and ours is that Hayden (1981, 1982) assumes that the transition to a broad spectrum economy was a matter of choice, rather than a necessary response to a climate-induced change in resource structure.

North American terminal Pleistocene Palaeoindian big game hunters were adapted to an abundant, dense, but spatio-temporally unpredictable resource base: the large gregarious, nomadic Late Pleistocene megafauna (Wilmsen 1973). Exploitation of such a resource probably required very high mobility, opportunistic site relocation and frequent travel over long distances. Such an adaptation would account for the relative abundance of kill sites and the extreme rarity of repeatedly-occupied habitation sites. The latter are rarely deeply stratified, but cover large areas, suggesting large group sizes. Lithic raw material use is congruent with the resource structure they exploited (Hayden 1982). For example, some raw materials at the Lindenmier site were obtained from sources within a 500 km radius of the site – from Utah to Texas (Wilmsen 1973). Many artifacts came from sources within a 3 km to 375 km radius of the site. The strict uniformity of projectile point style over the eastern two thirds of North America suggests large, overlapping, undefended territories, and together with exotic raw material patterns, institutionalized information, materials and perhaps mate exchange.

The Palaeoindian pattern stands in marked contrast to the succeeding Archaic phase: during the early Holocene, climatic amelioration favoured the spread of forests, creating a more predictable, dense and uniform resource structure. This period witnessed the rise of broad spectrum economies, more reliance on local lithic raw materials, higher site densities, and micro-regional differentiation of material culture (Hayden 1982). This pattern probably reflects decreased mobility, increased population density and the appearance of defended territorial boundaries.

Western Europe. Gamble (1978), also using an approach basically similar to that espoused in this paper, combined settlement pattern, climatic and environmental reconstruction, faunal and lithic raw material source-use data to model the changes in adaptation from Late Pleistocene to early Holocene adaptations in the Swabian Alb of southern Germany. Late Pleistocene Magdalenian peoples concentrated on horse and reindeer, a specialized strategy that resulted in large annual territories and uniformity in material culture over large areas. "Exchange, as evinced by the introduction of exotic raw materials into the annual territories of German groups, is also pronounced" (Gamble 1978: 181). In contrast, Mesolithic hunter-gatherers had an extremely broad spectrum adaptation to a diverse range of smaller game that preferred closed habitats. Fish, waterfowl and plants – stable, predictable and abundant resources – became a significant factor in the economy. Annual ranges for Mesolithic groups were estimated to be much smaller. Breakdown of the previous system of information and resource exchange and the rise of defended territories is evinced by the total absence of exotic raw materials, the increase in regional variability in artifact assemblage composition, and the appearance of local idiosyncratic tool forms.

Southern Africa. During the last 25 000 years in Southern Africa three major changes in lithic technology and adaptations are represented by the sequence of industries called Robberg, Albany and Wilton. The changes between these technological modes is considered to reflect changes in adaptations in response to changes in resource structure rather than population replacement (H. Deacon 1976; J. Deacon 1978).

Late Upper Pleistocene (25 000-12 000 BP) Robberg occurrences are characterized by nearly exclusive use of fine-grained non-local raw materials (crystal quartz, 'silcrete' and other cryptocrystalline raw materials) on which were made tiny bladelets and virtually no formal tools. The faunal assemblages are dominated by large, gregarious nomadic herbivore prey (equids and alcelaphines). Robberg occurrences are so rare and ephemeral that this industry remained unnamed until the mid 1970s. No open sites have been recognized and burials of this age are unknown. The Deacons suggest that a pattern of large migratory game hunting by large mobile groups with low population densities, living in large territories without fixed boundaries, could account for the patterning seen in the Robberg Industry (Deacon 1978: 107).

In contrast, during the warmer, moister terminal Pleistocene and early Holocene, Albany occurrences (12 000-8000 BP) are extremely abundant, deeply stratified and relatively highly visible on the landscape. A greater diversity of game species was exploited, many of which were characteristic of more closed habitats, and in coastal sites marine resources are well represented. The lithic industry comprises large scrapers and flakes made on local, often coarse-grained raw materials. Intentional burials are common in this period. This pattern is consistent with decreased mobility and perhaps the rise of smaller, defended territorial boundaries in environments with more abundant and predictable resources.

In the middle Holocene a trend toward greater aridity and decreased primary productivity is apparent. In the Wilton Industry (8000-2000 BP) the diversity of smaller animal prey resources characteristic of closed habitats and bush increases. Small occupation sites are found in a greater diversity of microhabitats (Parkington 1984). The proportions of fine-grained, non-local lithic raw materials increase significantly compared to the Albany but do not approach those in the Robberg Industry. The pattern of hunting of smaller, solitary, sedentary prey, consistent scheduling of plant collection and repeated use of occupation sites and dominance of local raw materials suggest that:

> By contrast [with the makers of the Robberg Industry] the hunting and collecting subsistence strategy evinced for the Wilton would correlate with smaller group organization, a higher population density, and more restricted territorial boundaries if we accept the interdependence of social organization, population density and subsistence strategy (Deacon 1978: 109).

If a change in lithic industries is caused by a change in lithic resource procurement patterns rather than by population replacement, then one would predict that major changes in the organization of lithic technology would be preceded by changes in the proportions of local coarse-grained *versus* fine-grained exotic rocks in regional cultural sequences. Moreover, such changes should be correlated with environmental changes that would have altered the spatio-temporal abundance and predictability of food resources and influenced subsistence, mobility and territorial organization. The temporal patterns in the detailed, well-documented sequence at Nelson Bay Cave shows that the transitions between Robberg, Albany and Wilton industries can be interpreted as cases of relatively abrupt changes in lithic technology in response to gradual changes in lithic resource procurement patterns. From Figure 1.3b (based on Table 2 of J. Deacon 1978) it is clear that within the Robberg levels there is a gradual increase through time in quartzite (the ubiquitous, local, low-quality rock) and a corresponding decline in non-local quartz, silcrete and chalcedony. With the increase of quartzite to 86% of the assemblage at 12 000 BP the technology based on tiny blades is replaced by the large flake-based Albany Industry. Quartzite continues to increase gradually to 98% of the total

---

Figure 1.3. (A) The percentages of local versus non-local raw materials in the whole flake assemblage from the Middle Stone Age (MSA) sequence at Klasies River Mouth (from Singer and Wymer 1977: 110); (B) The percentages of local versus non-local raw materials in the total stone artifact assemblage from the Later Stone Age (LSA) sequence at Nelson Bay Cave (from J. Deacon 1978: 90); (C) The percentages of local versus non-local raw materials in the total artifact assemblage from layers 6 to 16 of Shelter 1A at Klasies River Mouth (from Singer and Wymer 1982: 111). Levels 1-5 and 17-20 have been excluded because of small sample sizes. In both archaeological sequences it is apparent that abrupt changes in lithic industries are imposed on more gradual changes in raw material percentages.

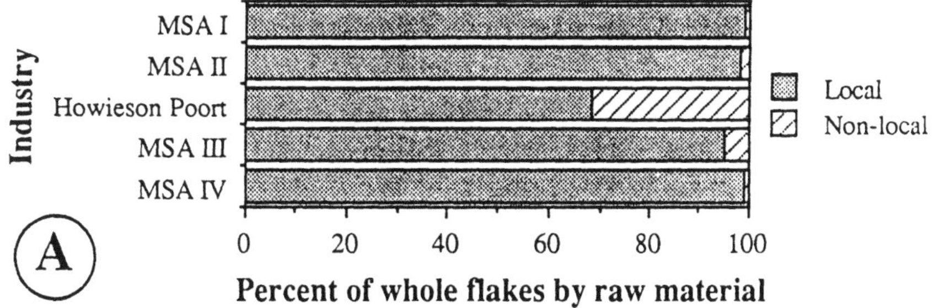
Klasies River Mouth MSA raw materials
Industry
MSA I
MSA II
Howieson Poort
MSA III
MSA IV
Local
Non-local
0
20
40
60
80
100
Percent of whole flakes by raw material
A

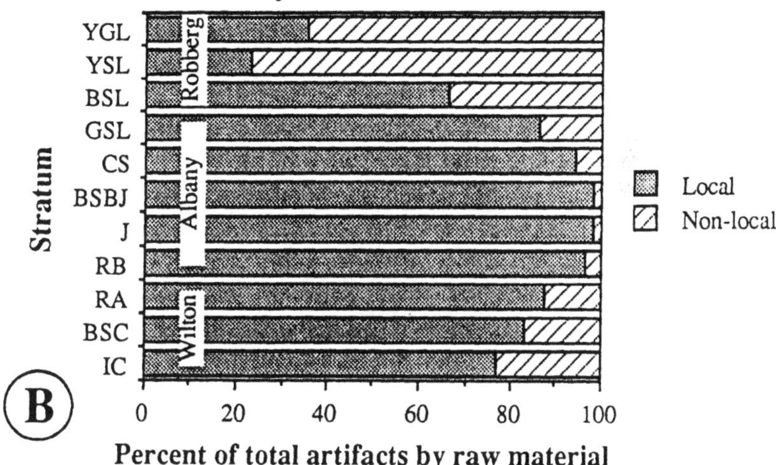
Nelson Bay Cave LSA raw materials
Stratum
YGL
YSL
BSL
GSL
CS
BSBJ
J
RB
RA
BSC
IC
Robberg
Albany
Wilton
Local
Non-local
0
20
40
60
80
100
Percent of total artifacts by raw material
B

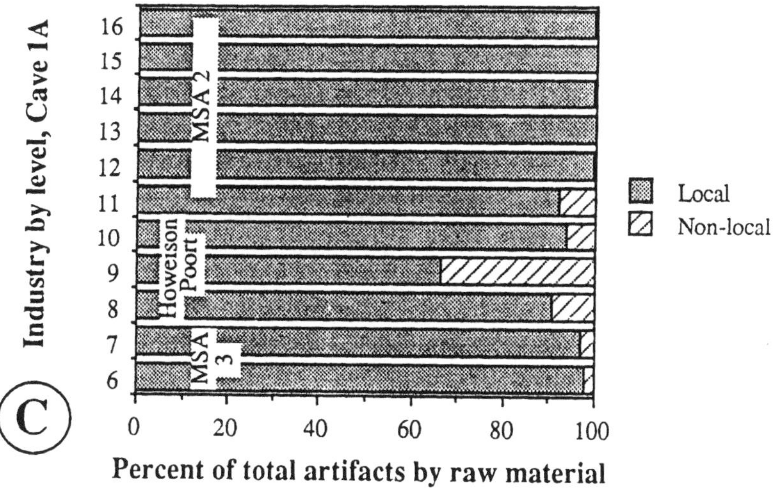
Klasies River Mouth MSA raw materials
Industry by level, Cave 1A
16
15
14
13
12
11
10
9
8
7
6
MSA 2
Howeison Poort
MSA 3
Local
Non-local
0
20
40
60
80
100
Percent of total artifacts by raw material
C

assemblage. A decline in quartzite abundance beginning at 8000 BP is followed by the transition to the Wilton Industry. After this transition quartzite declines from 82% to 77% of the lithic assemblage. J. Deacon's (1978: 90) Table 3, showing raw materials as percentages of formal tools at Nelson Bay Cave, and her Figure 4, showing quartz *versus* quartzite as percentages of raw materials, also indicate that changes in lithic technology are abruptly imposed on gradual changes in raw material frequencies.

The species composition of the faunal assemblages also shows transitional patterns of resource exploitation (Deacon 1978: 105): the decline in equids and alcelaphines begins in the latest Robberg levels, but is not complete until the earliest Albany ones. Grysbok is the most abundant small terrestrial mammal in the Wilton layers, but its frequency first rises dramatically in the latest Albany levels. The latest Robberg levels are also typologically similar to the earliest Albany ones. They have similar percentages of large scrapers, heavy edge-flaked pieces and bladelet cores even though the large tools and cores are generally considered to be mutually exclusive types restricted to Albany and Robberg industries, respectively.

We interpret the Robberg-Albany transition at Nelson Bay Cave as indicating that the non-local lithic resource procurement system of the Robberg broke down and became uneconomical as mobility and exchange decreased, and possibly as territoriality increased in response to resource-base enrichment at the Pleistocene/Holocene boundary. The response of the makers of the Robberg Industry was to totally reorganize their approach to flaked stone tool manufacture to more effectively accommodate the poor mechanical properties of locally available quartzite. Another reorganization occurred during the middle Holocene, and is reflected by the Albany/Wilton transition. The increase in availability of finer-grained raw materials on which small crescents and end-scrapers were made was probably a result of increased range size and the re-establishment of small-scale intergroup information and resource exchange networks in the face of declining resource abundance and predictability. A home range system was probably established at this time.

THE SOUTHERN AFRICAN MSA

Where then do the various stages of the MSA fit into the theoretical framework of adaptations outlined above?

MSA 1 and 2. Later MSA 1 and MSA 2 assemblages dating to the last interglacial probably formed under equable conditions in relatively moist, warm closed habitats. Associated faunal evidence suggests reliance on generally small prey that are characteristic of closed habitats, as at KRM (Klein 1976). This resource structure would have permitted reduced mobility and more intensive use of fewer sites in the seasonal round. This settlement pattern may account for the comparatively high archaeological visibility (Deacon and Thackeray 1984) and high densities of artifacts in MSA 1/2 occurrences relative to other kinds of MSA occurrences (Beaumont *et al.* 1978). For example, Carter (1976) notes that in the

Orange River Valley MSA occupations in caves are substantial and continuous, and open sites are abundant. Lithic assemblage are wholly dominated by local, often coarse-grained lithic raw materials (Figure 1.3a), as at KRM (Singer and Wymer 1982) Cave of Hearths bed 4, Border Cave, Rose Cottage Cave and the Southern Cape sites (Sampson 1974). The resource structure of MSA 1/2 times could be characterized as spatio-temporally abundant and predictable, and thus falling into area A of Figure 1.1, and the archaeological evidence is consistent with its predicted socioterritorial system (Table 1.1).

Howieson's Poort. Change toward a cooler climate at the end of the last interglacial period may have reduced exploitable game biomass, and thus reduced the spatio-temporal abundance and predictability of faunal and floral resources. The paucity of data and uncertainties regarding dating and correlation with marine climate sequences precludes an accurate reconstruction of resource structure for this time period. Nonetheless, the faunal assemblages from Border Cave, Boomplaas and KRM all suggest a trend towards more open environments, but a mosaic of open and closed habitats may have been available.

Lithic assemblages are characteristically enriched in non-local lithic resources. At Border Cave, a chalcedony whose known source is over 40 km away comprises 6% of the lithics in the MSA 1/2 levels, and 46% in the Howieson's Poort levels (Beaumont 1973: 43). As was argued for the Robberg/Albany/Wilton sequence it is likely that the change from MSA 2 to Howieson's Poort, and Howieson's Poort to MSA 3, can be viewed as a technological response to shifting patterns of raw material resource procurement. In both Nelson Bay Cave (Volman 1979) and KRM (Singer and Wymer 1982) the latest MSA 2 layers have elevated percentages of exotic raw materials. In Nelson Bay Cave the percentages of local raw materials are highest in the initial (98-99%) and final stages (96-99%) of the Howieson's Poort horizon, and lowest in the middle of this horizon (83-86%). In KRM there is an increase in local quartz midway through this occupation (Figure 1.3). Figure 1.3c shows that abrupt changes in lithic technology are imposed on more gradual changes in raw material frequencies in a way similar to that of the LSA at Nelson Bay Cave.

Settlement and site-use patterns are difficult to determine for the Howieson's Poort, a fact which in itself may be revealing. Open sites are poorly documented outside of the Cape coastal region (Sampson 1974: 250). Virtually no open sites are known in more arid regions. Shelter and cave sites rarely contain thick deposits. Overall, this phase of the MSA is less well represented in the archaeological record than the previous one, which may suggest relatively lower population densities. The only site where relative artifact densities can be assessed is KRM (Singer and Wymer 1982). The density of finds is highest for this industry, suggesting high intensity occupation.

Our first impression of the 'best fit' for the Howieson's Poort in the theoretical continuum of resource structure and socioterritorial organization was with a Type C mode, analogous to that suggested for the

Palaeoindian, Swabian Magdalenian and Robberg (Wilmsen 1973; Hayden 1982; Gamble 1978; J. Deacon 1978; H. Deacon 1976). This assessment was based on the abundance of exotic raw materials, and on the apparent presence of open grassland nomadic prey species at KRM (Klein 1976). However, since this faunal assemblage is clearly not representative of an open grassland environment, and exotic raw materials do not reach the high frequencies of the Robberg industry, the best fit for the Howieson's Poort seems to be with a Type B mode of socio-territorial organization, and thus most closely similar to that of the makers of the middle Holocene Wilton Industry, as suggested by Deacon and Thackeray (1984). The makers of the Wilton Industry also preferred non-local lithic materials for shaped tools (Humphreys 1972; Deacon 1978), and at Montagu Cave the percentages of nonlocal raw materials in the Wilton and Howieson's Poort are nearly identical (Keller 1973). Nonetheless, it cannot be taken to be a precise analogue for the Wilton for one very important reason: Howieson's Poort environments contained a suite of now extinct megafaunal prey that were unavailable to Wilton peoples. These resources would, in theory, have been a prime attraction for effective MSA hunters, and should have selected for greater mobility and less dependence on gathering. However, Klein's studies on ungulate mortality profiles suggests that they were not effective predators, and thus could not respond to the increased abundance of dense, spatio-temporally unpredictable resources at the end of the last interglacial period in the same ways as terminal Pleistocene humans.

MSA 3 and 4. The MSA 3 and 4 in Southern Africa should also fit into area C of the continuum of adaptations because these occurrences at KRM have faunas dominated by large, gregarious, nomadic ungulates such as alcelaphines and buffalo, and date to a time when more open, cooler, drier climates prevailed during the middle Upper Pleistocene. However, the emphasis on exotic lithic raw materials, though greater than in the MSA 1/2, is much less than in the Howieson's Poort (Figure 1.3a). The early MSA 1 from KRM amplifies this anomaly: despite the presence of open grassland species in the faunal assemblage there was nearly exclusive use of local lithic raw materials. This suggests continued low logistic and residential mobility, and combined with the relative rarity of sites dating to the isotope stages 3, 4 and 6, suggests very low population densities, perhaps restricted to a few favourable environments.

DISCUSSION AND CONCLUSIONS

The Howieson's Poort stands in stark contrast to the other MSA lithic industries in Southern Africa because of its use of non-local, fine-grained raw materials and the apparently advanced tool types that were made. In this paper we have argued that the contrast in lithic raw material use resulted from an expansion of local group foraging ranges and/or the development of intergroup exchange networks. The alternative hypothesis – that change in lithic raw material procurement strategies resulted from the need to use fine-grained raw materials to satisfy the needs of the microlithic technology – is rejected because a close examination of raw material frequencies within MSA sequences indicates that changes in raw material procurement patterns preceded changes in lithic technology. The published data base is remarkably sparse despite the ease with which raw material frequencies can be recorded, hence the hypothesis we favour is only weakly supported. More MSA and LSA sequences need to be examined in order to determine if this pattern is spurious.

If our hypothesis is correct, then the Howieson's Poort marks the first time in human history when there was a significant change in human territorial organization. In spite of resource-structure changes associated with twenty glacial/interglacial cycles during the previous two million years, this kind of adaptive response is not evident. This change in interaction patterns has obvious implications for patterns of gene flow and the transmission of cultural information across the African landscape.

This discontinuity between expectations and observations for MSA 1 and MSA 3/4 resource-procurement patterns raises serious questions about the relevance of the theoretical predictive framework presented above. Although the model appears to account for the diversity of terminal Pleistocene and Holocene human hunter-gatherer adaptations, it fails to account for the nearly exclusive use of local raw materials by early MSA 1 and MSA 3/4 peoples. This lithic source-use pattern presumably reflects low mobility, small home ranges, and little information and resource sharing, despite the presence of a presumably dense, patchy, spatio-temporally unpredictable resource base, in the form of large mammals in open environments. Perhaps faunal densities were so high that residential and logistic mobility remained comparatively low. Alternatively, MSA peoples may not have been able to fully exploit the abundant, gregarious, nomadic large herd species with the same effectiveness as terminal Pleistocene and historic hunter-gatherers, and thus could not respond in the same fashion. This explanation is unexpected: We have assumed that MSA and even Early Stone Age populations could take full advantage of such hunting opportunities. However, the conclusion seems inescapable and is consistent with Klein's (1978a) conclusion that MSA peoples were less effective hunters than LSA peoples, based on ungulate mortality profiles constructed from dental attrition stages. Our conclusion stands in marked contrast to that of Deacon, who views their subsistence and social behaviour as effectively identical to those of Later Stone Age hunter-gatherers in Southern Africa

The kind of adaptations represented by terminal Late Pleistocene *Homo sapiens sapiens* may therefore have no counterpart in the MSA in Southern Africa. Modern human adaptations thus cannot be used to model those of MSA peoples in Southern Africa, but the obvious discrepancies that arise from this exercise have served to illuminate the differences. Middle Stone Age humans, even if they were fully modern in anatomy, were not modern in their capacity for lithic and faunal resource exploitation. Olga Soffer (1989) has come to a similar conclusion for the

Eastern European Middle Palaeolithic. The inability to effectively exploit large, unpredictable, dangerous mobile prey may have forced pre-LSA and pre-Upper Palaeolithic humans to rely on a broad spectrum of locally available resources. This exploitation pattern probably restricted mobility and led to the embedded procurement of local lithic resources. It would also account for the consistently low diversity of lithic source use throughout most of human history.

The ecological model, though failing to account for the monotony of MSA adaptations to changing environments when modern humans serve as the standard for comparison, cannot be abandoned without a little *post-hoc* accommodative speculation. Before we write off MSA peoples as unskilled mental midgets or foraging automatons (Binford 1984) one must first ask: if MSA humans were given an LSA/Upper Palaeolithic technology, one which may have included spear throwers, bows and arrows, and deadly poisons, and taught how to use it, would they have have been able to respond to the availability of large game as effectively as terminal Pleistocene humans? An analogy can be drawn between the brain of *Homo sapiens* and the present state of computer technology. Modern computer hardware development has advanced so far over that of software that our computer programs, as remarkable as they may seem, do not exploit the full potential of the available technology. Perhaps early Upper Pleistocene man had modern biological hardware, but simply lacked the software – the cumulative body of knowledge and tradition required to make effective use of a technology not yet invented.

This analysis of earlier Upper Pleistocene resource exploitation patterns has been predicated on the assumption that a diversity of environmental conditions have existed in the past, that resource structure varies in complex ways on two dimensions, and that modern man has adapted to this diversity in equally diverse ways. Few authors have considered the diversity of climatic conditions and resource structures to which Middle Stone Age and Middle Palaeolithic humans have adapted. Rather, like Whallon (1989), a simple contrast between harsh glacial and favourable interglacial environments is made. Our analytical approach, so elegantly presented by Dyson-Hudson and Smith (1978), shows that such a dichotomy is overly simplistic at best. Nonetheless, with the exception of the Howieson's Poort, there is little evidence of diversity in human adaptations prior to 40 000 BP.

REFERENCES

Ambrose, S.H. 1984. *Holocene Environments and Human Adaptations in the Central Rift Valley, Kenya*. Unpublished Ph.D. Dissertation, University of California at Berkeley.

Ambrose, S.H. 1985. Excavations at Masai Gorge Rockshelter, Naivasha, Kenya. *Azania* 20: 29-67.

Ambrose, S.H. 1986. Hunter-gatherer adaptations to non-marginal environments: an ecological and archaeological assessment of the Dorobo model. *Sprache und Geschichte in Afrika* 7: 11-42.

Bailey, R.C. 1986. Time allocation and association patterns among

Efe pygmy men in northeast Zaire. *American Journal of Physical Anthropology* 69: 172.

Beaumont, P. B. 1973. Border Cave: a progress report. *South African Journal of Science* 69: 41-46.

Beaumont, P.B., De Villiers, H. and Vogel, J.C. 1978. Modern man in sub-Saharan Africa prior to 49,000 year BP: a review and evaluation with particular reference to Border Cave. *South African Journal of Science* 74: 409-419.

Bettinger, R.L. and Baumhoff, M. 1982. The Numic spread: Great Basin cultures in competition. *American Antiquity* 47: 485-503.

Bicchieri, M.G. 1969. A cultural ecological comparative study of three African foraging societies. *National Museums of Canada Bulletin* 228: 172-179.

Binford, L.R. 1979. Organization and formation process: looking at curated technologies. *Journal of Anthropological Research* 35: 255-273.

Binford, L.R. 1980. Willow smoke and dogs' tails: hunter-gatherer settlement systems and archaeological site formation. *American Antiquity* 45: 4-20.

Binford, L.R. 1984. *Faunal Remains at Klasies River Mouth*. New York: Academic Press.

Blackburn, R.H. 1971. *Honey in Okiek Personality, Culture, and Society*. Unpublished Ph.D. Thesis, Michigan State University.

Blackburn, R.H.1982. In the land of milk and honey: Okiek adaptations to their forests and their neighbors. In R.B. Lee and E. Leacock (eds) *Politics and History in Band Societies*. Cambridge: Cambridge University Press: 283-305.

Bordes, F. and Crabtree, D. 1969. The Corbiac blade technique and other experiments. *Tebiwa* 12: 1-21.

Brandt, S.A. 1988. Early Holocene mortuary practices and hunter-gatherer adaptations in southern Somalia. *World Archaeology* 20: 40-56.

Brown, J.L. 1964. The evolution of diversity in avian territorial systems. *Wilson Bulletin* 76: 160-169.

Brown, J.L. and Orians, G.H. 1970. Spacing patterns in mobile animals. *Annual Review of Ecology and Systematics* 1: 239-262.

Butzer, K.W. 1973. A provisional interpretation of the sedimentary sequence from Montagu Cave (Cape Province), South Africa. *University of California Anthropological Records* 28: 89-92.

Butzer, K.W. 1982. Geomorphology and sediment stratigraphy. In R. Singer and J. Wymer (eds) *The Middle Stone Age at Klasies River Mouth, South Africa*. Chicago: University of Chicago Press: 33-42.

Butzer, K.W. 1984. Late Quaternary environments in South Africa. In J.C. Vogel (ed.) *Late Cainozoic Palaeoclimates of the Southern Hemisphere*. Rotterdam: Balkema: 235-264.

Butzer, K.W., Beaumont, P.B. and Vogel, J.C. 1978. Lithostratigraphy of Border Cave, Kwazulu, South Africa: a Middle Stone Age sequence beginning c. 195,000 BP. *Journal of Archaeological Science* 5: 317-341.

Caldwell, J.R. 1958. *Trend and Tradition in the Prehistory of the Eastern United States*. American Anthropological Association Memoirs 88.

Carter, P.L. 1969. Moshebi's Shelter: excavation and exploitation in eastern Lesotho. *Lesotho* 8: 1-11.

Carter, P.L. 1976. The effects of climate change on settlement in eastern Lesotho during the Middle and Later Stone Age. *World Archaeology* 8: 197-206.

Chapman, R.W. 1981. The emergence of formal disposal areas and the 'problem' of megalithic tombs in prehistoric Europe. In R.W. Chapman, I.A. Kinnes and K. Randsborg (eds) *The Archaeology of Death*. Cambridge: Cambridge University Press: 71-81.

Charles, D.K. and Buikstra, J.E. 1983. Archaic mortuary sites in the central Mississippi drainage: distribution, structure and behavioral implications. In J.L. Phillips and J.A. Brown (eds) *Archaic Hunters and Gatherers in the American Midwest*. New York: Academic Press: 117-145.

Clark, J.D. 1959. *The Prehistory of Southern Africa*. Harmondsworth: Penguin.

Clark, J.D. 1980. Raw material and African lithic technology. *Man and Environment* 4: 44-55.

Clutton-Brock, T.H. 1974. Primate social organization and ecology. *Nature* 250: 539-542.

Colman, S.M., Choquette, A.F., Rosholt, J.N., Miller, G.H. and D.J. Huntley. 1986. Dating the upper Cenozoic sediments in Fisher Valley, southeastern Utah. *Geological Society of America Bulletin* 97: 1422-1431.

Deacon, H.J. 1976. *Where Hunters Gathered: a Study of Holocene Stone Age People in the Eastern Cape*. South African Archaeological Society Monograph Series 1.

Deacon, H.J. 1989. Late Pleistocene palaeoecology and archaeology in the southern Cape, South Africa. In P. Mellars and C. Stringer (eds) *The Human Revolution: Behavioural and Biological Perspectives on the Origins of Modern Humans*. Edinburgh: Edinburgh University Press: 547-564.

Deacon, H.J., Deacon, J., Scholtz, A., Thackeray, J.F., Brink J.S. and J.C. Vogel. 1984. Correlation of palaeoenvironmental data from the Late Pleistocene and Holocene deposits at Boomplaas cave, southern Cape. In J.C. Vogel (ed.) *Late Cainozoic Palaeoclimates of the Southern Hemisphere*. Rotterdam: Balkema: 339-352.

Deacon, H.J., Geleijnse, V.B., Thackeray, A.I., Thackeray, J.F, Tusenius, M.L. and Vogel, J.C. 1986. Late Pleistocene cave deposits in the southern Cape: current research at Klasies River. *Paleoecology of Africa* 17: 31-37.

Deacon, H.J. and Thackeray, J.F. 1984. Late Pleistocene environmental changes and implications for the archaeological record in Southern Africa. In J.C. Vogel (ed.) *Late Cainozoic Palaeoclimates of the Southern Hemisphere*. Rotterdam: Balkema: 375-390.

Deacon, J. 1978. Changing patterns in the Late Pleistocene/early Holocene prehistory of Southern Africa as seen from the Nelson Bay Cave stone artifact sequence. *Quaternary Research* 10: 84-111.

Deacon, J. 1979. *Guide to Archaeological Sites in the Southern Cape*. Occasional Publications of the Department of Archaeology, University of Stellenbosch 1: 1-149.

Dyson-Hudson, R. and Smith, E.A. 1978. Human territoriality: an ecological reassessment. *American Anthropologist* 80: 21-41.

Eisenberg, J.F., Muckenhirn, N.A. and Rudran, R. 1972. The relation between ecology and social structure in primates. *Science* 176: 863-874.

Flannery, K.V. 1969. Origins and ecological effects of domestication in Iran and the Near East. In P.J. Ucko and G.W. Dimbleby (eds) *The Domestication and Exploitation of Plants and Animals*. Chicago: Aldine: 73-100.

Foley, R. 1983. Modelling hunting strategies and inferring predator characteristics from prey attributes. In J. Clutton-Brock and C. Grigson (eds) *Animals and Archaeology*. Vol. 1: *Hunters and their*

Prey. Oxford: British Archaeological Reports International Series S163: 63-75.

Gamble, C. 1978. Resource exploitation and the spatial patterning of hunter-gatherers: a case study. In D. Green, C. Haselgrove and M. Spriggs (eds) *Social Organization and Settlement: Contributions from Anthropology, Archaeology and Geography*: Vol. 1. Oxford: British Archaeological Reports International Series S47: 153-185.

Geist, V. 1974. On the relationship of ecology and behavior in the evolution of ungulates: theoretical considerations. *American Zoologist* 14: 205-220.

Gould, R.A. 1978. The anthropology of human residues. *American Anthropologist* 80: 815-835.

Gould R.A. and Saggers, S. 1985. Lithic procurement in central Australia: a closer look at Binford's idea of embeddedness in archaeology. *American Antiquity* 50: 117-136.

Grayson, D.K. 1981. A critical view of the use of archaeological vertebrates in paleoenvironmental reconstruction. *Journal of Ethnobiology* 1: 28-38.

Hamilton, W.J., III, Buskirk, R.E and Buskirk, W.H. 1976. Defense of space and resources by Chacma (*Papio ursinus*) baboon troops in an African desert and swamp. *Ecology* 57: 1264-1272.

Harako, R. 1976. The Mbuti as hunters: a study of ecological anthropology of the Mbuti pygmies. *Kyoto University African Studies* 10: 37-99.

Harpending, H. and Davis, H. 1977. Some implications for hunter-gatherer ecology derived from the spatial structure of resources. *World Archaeology* 8: 275-283.

Hayden, B. 1981. Research and development in the Stone Age: technological transitions among hunter-gatherers. *Current Anthropology* 22: 519-548.

Hayden, B. 1982. Interaction parameters and the demise of Paleoindian craftsmanship. *Plains Anthropologist* 27: 109-123.

Hendey, Q.B. and Volman, T.P. 1986. Last interglacial sea levels and coastal caves in the Cape Province, South Africa. *Quaternary Research* 25: 189-198.

Horn, H.S. 1968. The adaptive significance of colonial nesting in the Brewer's blackbird *Euphagus cyanocephalus*. *Ecology* 49: 682-694.

Humphreys, A.B.J. 1972. Comments on aspects of raw material usage in the Later Stone Age of the Middle Orange River area. *South African Archaeological Society, Goodwin Series* 1: 46-53.

Itani, J. and Suzuki, A. 1967. The social unit of chimpanzees. *Primates* 8: 355-381.

Izawa, K. 1970. Unit groups of chimpanzees and their nomadism in the savanna woodland. *Primates* 11: 1-46.

Jarman, P.J. 1974. The social organization of antelope in relation to their ecology. *Behavior* 48: 215-267.

Jones, P.R. 1979. Effects of raw material on biface manufacture. *Science* 204: 835-836.

Keller, C.M. 1973. Montagu Cave in prehistory. *University of California Anthropological Records* 28: 1-150.

Kelly, R.L. 1983. Hunter-gatherer mobility strategies. *Journal of Anthropological Research* 39: 277-306.

Klein, R.G. 1975. Middle Stone Age man-animal relationships in Southern Africa: evidence from Die Kelders and Klasies River Mouth. *Science* 190: 265-267.

Klein, R.G. 1976. The mammalian fauna of the Klasies River Mouth sites. *South African Archaeological Bulletin* 31: 75-98.

Klein, R.G. 1977. The mammalian fauna from the Middle and Later Stone Age (later Pleistocene) levels of Border Cave, Natal Province, South Africa. *South African Archaeological Bulletin* 32:14-27.
Klein, R.G. 1978a. Stone age predation on large African bovids. *Journal of Archaeological Science* 5: 195-217.
Klein, R.G. 1978b. A preliminary report on the larger mammals from the Boomplaas Stone Age cave site, Cango Valley, Outdshoorn district, South Africa. *South African Archaeological Bulletin* 32: 127-145.
Klein, R.G. 1979. Stone age exploitation of animals in Southern Africa. *American Scientist* 67: 151-160.
Klein, R.G. 1980. Environmental and ecological implications of large mammals from Upper Pleistocene and Holocene sites in Southern Africa. *Annals of the South African Museum* 81: 223-283.
Klein, R.G. 1981. Stone age predation on small African bovids. *South African Archaeological Bulletin* 36: 55-65.
Klein, R.G. 1982. Age (mortality) profiles as a means of distinguishing hunted species from scavenged ones in Stone Age archaeological sites. *Paleobiology* 8: 151-158.
Kruuk, H. 1975. Functional aspects of social hunting by carnivores. In G. Barends, C. Beer and A. Manning (eds) *Function and Evolution in Behavior.* Oxford: Clarendon Press: 119-141.
Laughlin, W.S. 1975. Aleuts: ecosystem, Holocene history and Siberian origin. *Science* 189: 507-515.
Lee, R.B. 1968. What hunters do for a living or how to make out on scarce resources. In R.B. Lee and I. DeVore (eds) *Man the Hunter.* Chicago: Aldine: 30-48.
Lee, R.B. 1979. *The !Kung San: Men, Women and Work in a Foraging Society.* Cambridge: Cambridge University Press.
Marean, C.W. 1986. On the seal remains from Klasies River Mouth: an evaluation of Binford's interpretations. *Current Anthropology* 27: 365-367.
Mellars, P. 1986. A new chronology for the French Mousterian period. *Nature* 322: 410-411.
Moore, J.A. 1981. The effects of information in hunter-gatherer societies. In B. Winterhalder and E.A. Smith (eds) *Hunter-Gatherer Foraging Strategies.* Chicago: University of Chicago Press: 194-217.
Orians, G. and Pearson, N. 1979. On the theory of central place foraging. In D.J. Horn, R. Mitchell and G.R. Stairs (eds) *Analysis of Ecological Systems.* Columbus: Ohio University Press: 155-177.
Parkington, J.E. 1980. Time and place: some observations on spatial and temporal patterning in the Later Stone Age sequence in Southern Africa. *South African Archaeological Bulletin* 35: 73-83.
Parkington, J.E. 1984. Soaqua and Bushmen: hunters and robbers. In C. Schrire (ed.) *Past and Present in Hunter-Gatherer Studies.* Orlando: Academic Press: 151-174.
Partridge, T.C., Netterberg, F., Vogel, J.C. and Sellschop, J.P.F. 1984. Absolute dating methods for the South African Cainozoic. *South African Journal of Science* 80: 394-400.
Rowell, T.E. 1966. Forest living baboons in Uganda. *Journal of Zoology* 149: 344-364.
Sampson, C.G. 1974. *The Stone Age Archaeology of Southern Africa.* New York: Academic Press.
Schiffer, M.B. 1978. Methodological issues in ethnoarchaeology. In R.A. Gould (ed.) *Explorations in Ethnoarchaeology.* Albuquerque: University of New Mexico Press: 229-247.
Schiffer, M.B. 1983. Toward the identification of formation

processes. *American Antiquity* 48: 675-706.

Shackleton, N.J. 1982. Stratigraphy and chronology of the Klasies River Mouth deposits: oxygen isotope evidence. In R. Singer and J.J. Wymer (eds) *The Middle Stone Age at Klasies River Mouth in South Africa*. Chicago: University of Chicago Press: 194-199.

Silberbauer, G.B. 1981. *Hunter and Habitat in the Central Kalahari Desert.* Cambridge: Cambridge University Press.

Singer, R. and J. Wymer 1982. *The Middle Stone Age at Klasies River Mouth in South Africa.* Chicago: University of Chicago Press.

Smith, E.A. 1981. The application of optimal foraging theory to the analysis of hunter-gatherer group size. In B. Winterhalder and E.A. Smith. (eds) *Hunter-Gatherer Foraging Strategies*. Chicago: University of Chicago Press: 36-65.

Soffer, O. 1989. The Middle and Upper Palaeolithic transition on the Russian Plain. In P. Mellars and C. Stringer (eds) *The Human Revolution: Behavioural and Biological Perspectives on the Origins of Modern Humans.* Edinburgh: Edinburgh University Press: 714-742.

Steward, J.H. 1938. *Basin-Plateau Aboriginal Socio-Political Groups.* Bureau of American Indian Ethnology Bulletin 120. Washington (DC).

Suzuki, A. 1969. An ecological study of chimpanzees in a savanna woodland habitat. *Primates* 10: 103-148.

Tankersley, K.B., Munson, C.A. and Smith, D. 1987. Recognition of bituminous coal contaminants in radiocarbon samples. *American Antiquity* 52: 318-329.

Torrence, R. 1983. Time budgeting and hunter-gatherer technology. In G. Bailey (ed.) *Hunter-Gatherer Economy in Prehistory: a European Perspective.* Cambridge: Cambridge University Press: 11-22.

Vogel, J.C. and Beaumont, P.B. 1972. Revised radiocarbon chronology for the Stone Age in South Africa. *Nature* 237: 50-51.

Volman, T.P. 1979. Appendix 2: the Middle Stone Age sequence at Nelson Bay Cave. In J. Deacon *Guide to Archaeological Sites in the Southern Cape.* Occasional Publications of the Department of Archaeology, University of Stellenbosch 1: 65-67.

Volman, T.P. 1984. Early prehistory of Southern Africa. In R.G. Klein (ed.) *Southern African Prehistory and Paleoenvironments.* Rotterdam: Balkema: 169-220.

Wendt, W.E. 1976. 'Art Mobilier' from the Apollo 11 Cave, South-west Africa: Africa's oldest dated works of art. *South African Archaeological Bulletin* 31: 5-11.

Whallon, B. 1989. Elements of cultural change in the Later Palaeolithic. In P. Mellars and C. Stringer (eds) *The Human Revolution: Behavioural and Biological Perspectives on the Origins of Modern Humans*. Edinburgh: Edinburgh University Press: 433-454.

Wilmsen, E.N. 1973. Interaction, spacing behavior and the organization of hunting bands. *Journal of Anthropological Research* 29: 1-31.

Woodburn, J. 1968. An introduction to Hadza ecology. In R.B. Lee and I. DeVore (eds) *Man the Hunter*. Chicago: Aldine: 49-55.

Yellen, J. 1977. *Archaeological Approaches to the Present.* New York: Academic Press.

Yellen, J. and Harpending, H. 1972. Hunter—gatherer populations and archaeological inference. *World Archaeology* 4: 244-253.

2. A Critique of the Consensus View on the Age of Howieson's Poort Assemblages in South Africa

JOHN PARKINGTON

INTRODUCTION

Assemblages called Howiesons Poort (Figure 2.1) are usually described as essentially Middle Stone Age in character but with smaller flake products, more blades (perhaps punch struck), a wider range of stone raw materials (especially fine grained ones) and diagnostically high frequencies of usually non-microlithic backed elements – specifically segments, together with other geometrics and truncated blades. These technical and typological characteristics, along with the observations that Howiesons Poort (henceforth 'HP') assemblages are sometimes sandwiched between 'ordinary' Middle Stone Age ('MSA') assemblages, obviously prompts comparison with the long rock-shelter sequences of the Southern and Eastern Mediterranean coasts and Southwestern Europe. In this sense, then, the age of HP assemblages is of more than local significance. This is particularly so since a range of hominid skeletal fragments, some described as indisputably anatomically modern (Bräuer 1984; Rightmire 1984), are associated with HP and other MSA levels in cave sequences in South Africa. It is no exaggeration to argue that the ages of MSA hominids, the age of HP assemblages, and the stratigraphic relationships between various kinds of MSA assemblages are inextricably intertwined research issues. Models for the origins and dispersal of modern people thus require robust statements on the South African MSA chronology.

In this paper I challenge what I take to be the most widely promoted view of the age of HP assemblages, particularly at Border Cave (Beaumont *et al.* 1978; Butzer *et al.* 1978; Volman 1984; Deacon and Thackeray 1984) not because it can be shown to be wrong, but because we need to understand how constraining it is. We need to know how far 'sideways' from the consensus we can move before we encounter strong argumental barriers – "we cannot argue for X because of evidence Y". Part of my critique is that many of the arguments brought to bear on the age of HP assemblages, and thus related hominids, are neither independent nor strongly constraining.

More specifically I suggest that, in the absence of a solid absolute chronological framework, estimates of age are derived from a three-part argument that (1) identifies stratified patterns in biological or geological

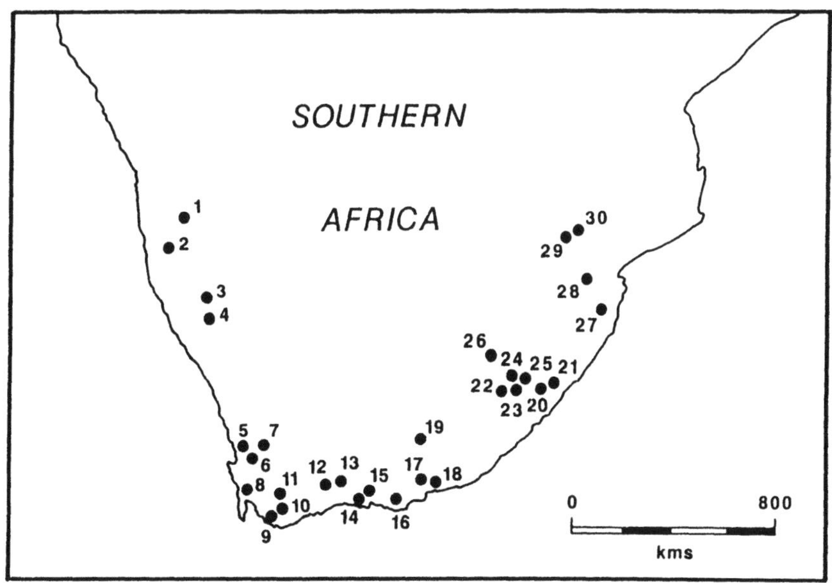

Figure 2.1. Southern African sites with Late Pleistocene stone tool assemblages. Not all of the sites have produced Howiesons Poort assemblages. 1. Zebrarivier; 2. Nos; 3. Pockenbank; 4. Apollo XI; 5. Elands Bay Cave; 6. Diepkloof; 7. Klipfonteinrand; 8. Peers Cave; 9. Die Kelders; 10. Byneskranskop; 11. Montagu Cave; 12. Buffelskloof; 13. Boomplaas; 14. Nelson Bay Cave; 15. Kangkara; 16. Klasies River Mouth; 17. Melkhoutboom; 18. Howiesons Poort; 19. Highlands; 20. Shongweni; 21. Umhlatuzana; 22. Melikane; 23. Moshebis; 24. Sehonghong; 25. Ha Soloja; 26. Rose Cottage; 27. Border Cave; 28. Lion Cavern; 29. Heuningneskrans; 30. Bushman Rock Shelter.

parameters; (2) interprets these in palaeoclimatic terms – wetter, drier, colder, warmer, grassier, bushier etc.; and then (3) relates these interpretations to the isotope chronology established from deep-sea cores. At each stage in the process there is the danger of generating arguments of convenience rather than arguments of necessity, and examples of interdependence of evidence are not hard to find. In the final resort the situation often resolves itself into a researcher attempting to *match* cycles or fluctuations in some proxy measure of environmental or climate change to the cycles of the isotope chronology. Given that the rates of deposition, scales of change in biological and geological materials and sensitivities of proxy measures are all unknown and presumably variable, the opportunity for mismatching is obvious.

What then *is* the consensus view? I should point out that I do not imply unanimity nor agreement in detail by this term. But on three central issues there is substantial consensus. Most MSA research workers would subscribe to the view that the HP assemblages are a coherent synchronic phenomenon, and thus by implication that dating some HP levels means dating that phenomenon as a whole. Put another way, this is to assume, not as yet to demonstrate, that the HP is a synchronic response made by widely dispersed groups in Southern Africa – an earlier version, it is said (Deacon *et al.* 1986), of the Wilton in the Holocene. This assumption naturally leads to a discussion of the meaning behind the synchronous response and of concepts of style or fashion. The fact that HP assemblages appear always to form consecutive sets of observations at sites, rather than appearing to interdigitate with other MSA horizons, is taken as supportive evidence of this view. Of course the contemporaneity of such sets between sites requires more substantiation.

A second, and obviously related, pillar of agreement is that the age of these penecontemporaneous HP assemblages lies in some part of oxygen-isotope stage 5, most often assumed to be 5b or 5d. Such an age would translate as approximately 95 000 or 105 000 years. The implication of a stage 5 age for the HP, taken with evidence referred to later placing earlier MSA assemblages in the earlier stage 5 (substage 5e) is that much, even most, of the MSA is of 'interglacial' age. This in turn then generates a discussion as to why periods of apparently high archaeological visibility are related to 'interglacial' or 'interstadial' stages, and leads to references to population fluctuations or 'local extinctions' during 'glacials' (Deacon and Thackeray 1984).

A third area of agreement relates to the nature of the subcontinental-wide MSA sequence. It has become accepted that subsequent to the HP 'episode' there was a universal return to conventional MSA technology for several tens of millennia prior to the appearance of Late Stone Age ('LSA') techniques and tool forms. Needless to say this pattern has stimulated much discussion and has moved attention away from the concept of the HP as 'intermediate', at least in a chronological sense.

My approach here is to isolate and critique arguments for a stage 5 age of specific HP assemblages and to illustrate 'play' in the system. I do this in the spirit of Karl Butzer's (1984) remark that we are at the stage of "first approximation" and in the interests of understanding how proximate we are. In a later section I flirt with the idea of a shorter chronology than the consensus allows. In doing so I raise the related questions of the nature of the MSA/LSA transition and the prevalence of MSA assemblages more recent in age than the HP.

ABSOLUTE DATES

Obviously the most constraining arguments would come from sets of absolutely dated HP assemblages. Janette Deacon (1979) has reviewed the radiocarbon dates apparently associated with HP levels at a wide range of sites and illustrated an unexpectedly variable situation. Since then no evidence has accumulated to gainsay her observation that ages range from the Holocene to a series of greater-than 50 000 year old dates. Taken at face value, of course, there is no support here for either of the first two assumptions of the consensus view, but this is 'solved' by rejecting all but the greater-than dates. The rejection of dates less than 20 000 years, on the grounds of recent contamination by younger carbon, is understandable in that we have increasingly coherent regional sequences for this period with nothing typologically or technologically resembling the HP. Apparently finite dates between 30 000 and 45 000 years, rejected on the same grounds, seem more controversial, as the expected sequence in this time period is not as easily derived. Janette Deacon's (1979) rejection of all apparently finite dates as "diverting", pointing to the existence of finite and infinite dates in the same stratigraphic level at some sites, meets with widespread, if tacit, agreement. Mechanisms for contamination by more recent materials are easily hypothesized, but it is worth noting here that the consensus view is salvaged only by rejecting a number of finite and in some cases stratigraphically consistent ages.

Bada and Deems (1975) did suggest ages for some of the Klasies River Mouth ('KRM') Cave 1 levels based on the racemization of aspartic acid in bones and calibrated by measurements on bones from Nelson Bay Cave of known (^{14}C) age. Their estimates of between 90 000 and 110 000 years for the MSA I layers (Figure 2.2) are consistent with a later Last Interglacial age for the earliest occupation in that cave. We should note, though, that the assumption that the temperature history of the NBC control bones has been the same as those from KRM is untested, and that a "1 degree difference in average temperature causes about a 20% change in the racemization rate constant" (Bada and Deems 1975: 218). Moreover, recent ^{14}C accelerator dates for North American skeletons previously dated by aspartic acid racemization has led to drastically revised ages, all of them more recent than claimed (Taylor *et al.* 1985). The authors here noted that "there is no clear relationship between the ^{14}C age and the extent of aspartic acid racemization in bone samples" (Taylor *et al.* 1985: 139) in many cases.

More likely sources of ages beyond 50 000 years are thermoluminescence, electron spin resonance and uranium series disequilibrium, although only the latter has as yet produced Southern African MSA-related results in press. Uranium series disequilibrium dates of 98 000 years overlie the earliest MSA levels in KRM Cave 1, adding support to the case for an MSA presence in Southern Africa prior to the end of oxygen-isotope stage 5 (Deacon *et al.* 1986; Hendey and Volman 1986; Kronfeld and Vogel 1980). Such an age provides a strong *post quem* date for the HP horizons which lie several metres higher up the complex stratigraphic pile of the KRM main site complex. Bada and Deems' (1975: 219) suggestion of about 65 000 years for the MSA IV of KRM I may or may not be reasonable, but in any case cannot yet stratigraphically be shown to predate the HP of Shelter 1A.

OXYGEN ISOTOPES, SEA LEVEL CORRELATIONS AND THE HOWIESONS POORT AT KLASIES RIVER MOUTH

I do not think it an exaggeration to point to the pivotal role of Nick Shackleton's research on the oxygen-isotope composition of Southern African fossil shells in the generation and popularization of the consensus view. Although prior to this there were infinite radiocarbon dates with MSA assemblages, it was the association with beach deposits at the base

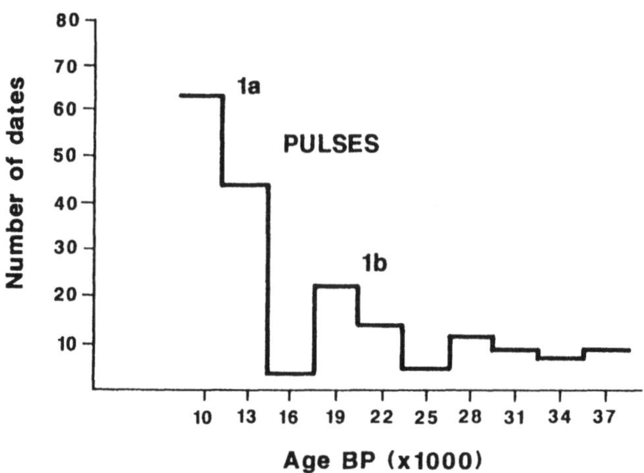

Figure 2.2. Distribution of radiocarbon dates for Late Pleistocene sites in Southern Africa, showing 'pulsing' of dates.

of Klasies River Mouth Cave I and with oxygen-isotope values suggestive of modern sea temperatures that led directly to a proposed stage 5e age for some MSA occupations. Shackleton's (1982) case for a stage 5e age for level 38 in KRM 1, on the grounds that such a substage was the last time when sea temperatures were as high as, or higher than, the Holocene, is strong, and is certainly independent of other biological or sedimentological inferences. Not surprisingly other interpretations tend to refer to the robusticity of this observation.

It is interesting to note the placing (Figure 2.3) by Shackleton of HP assemblages, admittedly on the basis of only a single shell from KRM 1A level 20. This shell showed values that "would be consistent with deposition within stage 3, although a cooler part of stage 5 cannot be excluded" (Shackleton 1982: 196). Not much constraint here, though we should remember that he actually placed the HP at the stage 2/3 interface – about 35 000 years ago. Equally interesting is Shackleton's reference to the isotope curve as predominantly a reflection of the locking up of isotopically-light ice in the continental ice caps, a reference which tempts us to see the curve as an approximation of sea level. Shackleton's (1982: 198) isotope curve can be read as indicating a sea level only half way down to the maximum recession of 20 000 years ago for much of the time between 100 000 and 35 000 years ago. Taken with Butzer's (1982) remark that a drop in sea level of up to 50 m would still leave the coast within 5 km of KRM, this would suggest that the coastline was only substantially and relatively permanently distant from the site in isotope stage 2. My point is that many have taken the presence of shells and the bones of marine animals through the whole of the MSA sequence at KRM as evidence for a stage 5 age. Much later dates may be possible even with the marine evidence.

If the base of the KRM MSA sequence is reliably of stage 5 age, are there

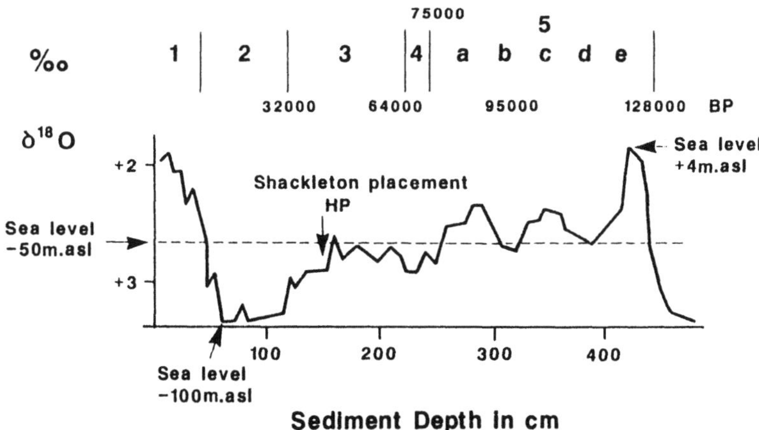

Figure 2.3. Oxygen isotope curve for the last 130 000 years, with the suggested placement of the Klasies River Mouth Howiesons Poort levels by Shackleton (redrawn after Shackleton, in Singer and Wymer 1982).

ways of extrapolating up to the HP levels? Deacon *et al.* (1986: 35) offer such an extrapolation in their observation that "there is a marked decline in the mass of shell per unit volume of excavated deposit upwards through the KRM lA sequence and this correlates with demonstrable changes in morphometric characteristics of the sands". Some of the upper HP levels in fact seem not to have any shell associated, linking the HP occupation to a major sea level regression which may be "isotope stage 4 but could be as early as stage 5b" (Deacon *et al.* 1986: 35). Thus the problem becomes one of knowing the local sea level history, the record of regressions and transgressions, the distances people would have been prepared to carry shells, and the continuity, or lack of it, in the shell midden sequence at KRM. If the Huon Peninsula record (Chappell and Shackleton 1986) is any guide there have been numerous regressions between 98 000 and about 45 000 years ago, most of which would have taken the shoreline more than 2 km from the site. Holocene shell middens may be found more distant than that from the source of shellfish but we know little as yet about MSA behaviour in this respect.

SEDIMENTS, PAST CLIMATES AND THE HOWIESONS POORT AT BORDER CAVE

Karl Butzer's work at Border Cave (Butzer *et al.* 1978; Butzer 1984) has been only marginally less influential than Nick Shackleton's at KRM in developing the consensus view. At the former site Peter Beaumont has excavated what he has called 'Epi-Pietersburg' or HP assemblages from levels with MSA above and below them (Beaumont 1973, 1978; Butzer *et al.* 1978; Beaumont *et al.* 1978). Butzer's assessment of the age of the HP levels (Figure 2.4) follows a two-pronged argument based partly on extrapolation of the radiocarbon dates from the upper levels and partly on his interpretation of the climatic sequence reflected in the sediments and his placement of these in the oxygen-isotope chronology. The HP is found in Butzer's levels 7 and 8 associated with an apparently finite date of 42 000 (but with a very large standard error) and three other dates all greater than about 48 000 years.

The extrapolation argument is quite straightforward, perhaps misleadingly so. "Assuming that the spacing and duration of depositional breaks remained broadly comparable" (Butzer *et al.* 1978: 332) through the sequence, he uses the cumulative mean depths of the top four levels (1a, 1b, 2, 3) plus the underlying breaks (apparently ± 12 000 years, ± 4000 years, 0 years and ± 10 000 years) as a means of deriving a calibration rate for the site as a whole. In an elaboration he adds estimates for compaction in the lower levels, and arrives at an approximation of between 186 000 and 208 000 for the base of the sequence. The HP levels 'date' to between 80 000 and 95 000 years (Butzer *et al.* 1978: 334), though I make the extrapolation rates given me more than 107 000 for the base of level 8, and 80 100 for the top of level 7.

I suggest there are serious problems with this argument. There is no reason why the spacing and duration of breaks should remain constant, otherwise we would have a lot less trouble than we do in dating archaeological sites. In fact there seems good reason to suspect very variable rates of deposition from the distribution of quite different kinds of discontinuities, depositional breaks, weathering horizons and hiatuses in the sequence. Rather we should see cave use in Southern Africa, as elsewhere, as wildly episodic – or not commit ourselves until we know. In either event, Butzer's extrapolation assumes that the >49 000 year date for the top of level 4 is really a finite 49 000 year date in order to calculate his calibration of 1.84 cm per thousand years. If it is 'really' 59 000 years, and if we take not cumulative *mean* deposit thickness but cumulative *maximum* thickness, we derive a rate of 2.82 cm per thousand years. I am not suggesting we should make this assumption, merely that it seems to me there is a lot of play in the calibration argument. Also, what is the justification for taking a >49 000 year date in level 4 as finite but rejecting similar ages for level 8 in favour of extrapolated ages of twice the size?

Some measure of the usefulness of the extrapolation exercise may be gained by using the sedimentation rate to interpolate ages in the upper units from which we do have finite radiocarbon ages to act as checks. It is very clear from Table 2.1 that the interpolated ages of the dated horizons are without exception all seriously in error, a pattern which derives from, and dramatically underlines, the episodic nature of cave infilling. If the sedimentation rate will not give accurate estimates where we can check them, what confidence can we possibly have lower down where we cannot? My suggestion is rather that much of the depositional column below the finite radiocarbon dates probably went into the cave in short bursts of rapid accumulation separated by episodes of relative non-deposition. By making an assumption of something we would really like to learn about at

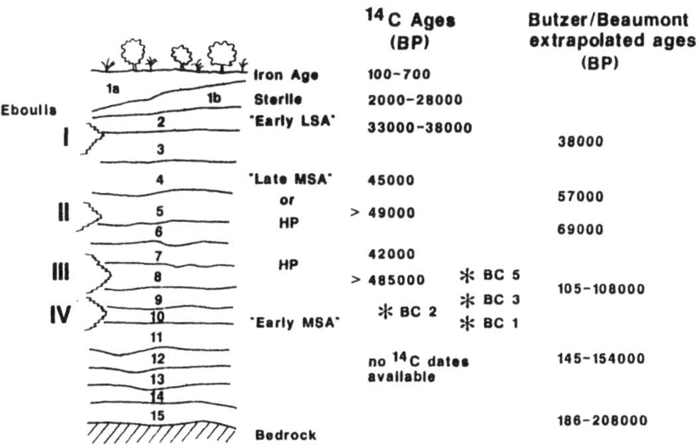

Figure 2.4. Sketch section through the Border Cave sequence, with radiocarbon dates and extrapolated ages suggested by Butzer *et al.* (1978).

Border Cave – the sedimentation rate – Butzer has effectively prevented himself from understanding the nature of the sedimentation process. We should remember that the Border Cave ('BC') sediment sequence is a series of alternating widespread ashes, which could form in hundreds of years, and brown sands with markedly fewer artifacts. It is important to try to establish how such cyclic stratigraphy was generated; it certainly differs from the massive sets of superimposed hearths at KRM and Diepkloof in HP times.

Beaumont's argument that the HP assemblages at BC are 80 000 to 100 000 years old is in my opinion based on a similar misreading of the pattern of depositional buildup (Beaumont *et al.* 1978). The origin and substance of his argument is that in the uppermost dated levels deposition can be shown to occur in mild periods and non deposition in harsher periods, specifically 0-13 000 years BP and 13-28 000 BP respectively. This is clearly wrong as Table 2.2 attempts to show. My reading of the radiocarbon stratigraphy is that there is indeed a burst of deposition about 300-500 years ago but that prior to that the next such episode is 33 000-38 000 years ago, assuming these dates to be finite. There is no evidence whatsoever for either natural or cultural deposition between 700 and 13 000 years ago, effectively the bulk of the 'mild' Holocene, nor is it clear that there was any less or more between 13 000 and 28 000 years ago. Beaumont does not help his argument by not clearly distinguishing between patterns of natural deposit accumulation and the build up of artifactual debris in the cave.

The climatic argument is based on a detailed interpretation of rubble and fine components, sedimentological discontinuities and palaeosol weathering. Central to this is the integration of the artifact-bearing levels with a sequence of rubble horizons which Butzer refers to as "*éboulis secs*" and interprets as the result of frost weathering (Butzer *et al.* 1978: 330).

Table 2.1. Interpolated dated for the sequence at Border Cave, using the sedimentation rates of Butzer, Beaumont and Vogel (1978).

Depth (cm)	Level	^{14}C Age B.P.	Age by Sediment Interpolation	Discrepancy
8-15	1a.	90 ± 105	4348-8152	4000-8000
8-15	1a.	170 ± 45	4348-8152	4000-8000
8-15	1a.	500 ± 45	4348-8152	4000-8000
30-38	1a.	500 ± 70	16,304-20,652	16,000-20,000
38-46	1a.	590 ± 70	20,652-25,000	20,000-25,000
30-38	1a.	440 ± 55	16,304-20,652	16,000-20,000
13	1a.	340 ± 45	7065	6500
13	1a.	480 ± 45	7065	6500
38-46	1bTop	2010 ± 50	20,652-25,000	18,000-23,000
46-53	1b Base	13,3000 ± 150	25,000-28,804	12,000-15,000
46-53	1b Top	650 ± 70	25,000-28,804	25,000-28,000
61-69	1b Base	28,500 ± 1800	33,152-37,500	5000-9000
69-76	2 Top	33,000 ± 2000	37,5000-41,304	4500-8000
69-Level 3	2 Base	38,600 ± 1500	37,500-48,913	1000-10,000

This appears to be arguable (Deacon, Lancaster and Scott 1984: 392, 401) but even if correct is not without its problems. Thus éboulis I at BC appears to grade laterally into levels 2 and 3 with six finite radiocarbon dates between 33 000 and 38 000 BP, a fairly good match with oxygen-isotope stage 3, an 'interstadial'. By contrast level 1b with radiocarbon dates of 13 300 and 28 500 BP, arguably including some of stage 2, has less effective frost weathering (Butzer *et al.* 1978: 330). Butzer is understandably cautious about when and by what mechanism conditions with lower temperature and enough available moisture would have prevailed. Again the purpose here is not to offer an alternative interpretation, merely to ask how we set about matching alternating episodes of wetter or drier, warmer or colder climates beyond the range of radiocarbon dating with the isotope record? Do we know what to expect in the vicinity of Border Cave during the oscillations of stage 3 and can we eliminate such oscillations from our interpretations of the sediment sequence? It seems to me there is 'play' in the climatic inference as well as in the reference to an appropriate undulation in the oxygen-isotope record. Few arguments of necessary relationships are available to us, and it is notable that in his most recent synthesis Butzer (1984) makes substantial use of inferences based on micromammalian remains to support his argument.

MICRO-ANIMALS, PALAEOENVIRONMENTAL RECONSTRUCTION AND THE HOWIESONS POORT AT BORDER CAVE

Like Butzer, Margaret Avery has two interlinked arguments in her attempt to use micromammal samples to date and add context to the artifactual assemblages from Border Cave (1982a). On the one hand she uses the relative frequencies of species of rodents and shrews to infer the local vegetational mosaic, and on the other she uses the Shannon index of

Table 2.2. Suggested pulsing of occupations at Border Cave, using the radiocarbon dates of Beaumont, de Villiers and Vogel (1978).

Level	^{14}C Ages B.P.	Association	Pulse
1a	90	Iron Age	
1a	170	Iron Age	
1a	500	Iron Age	
1a	500	Iron Age	Uppermost at Site
1a	590	Iron Age	Suggested Age: 300-500 B.P.
1a	440	Iron Age	
1a	340	Iron Age	
1a	480	Iron Age	
1b	2010	Sterile	?Iron Age contaminated
	650	Sterile	Iron Age
	13,300	Sterile	?
	28,500	Sterile	?Second Pulse at Site
2	33,000	Early LSA	
2	38,600	Early LSA	
3	33,000	Early LSA	
	34,800	Early LSA	
3	37,500	Early LSA	Second Pulse at Site
3	36,800	Early LSA	Suggested Age: 33,000-38,000
3	36,100	Early LSA	
3	35,700	Early LSA	
3	45,000	Early LSA	Third Pulse at Site
4	47,200	MSA	Third Pulse, Age
4	45,400	MSA	unknown by ^{14}C.

general diversity in stratified assemblages as a measure of the harshness or unpredictability of contemporary conditions (Avery 1982a: 189).

The first argument proceeds by dividing the species into groupings, one consisting of those best represented in levels 10-to 5a, including the HP, the second of those best represented in levels 4 to 1b, and the third of those best represented in the uppermost (1a) and lowermost (11b) levels of the cave. Although reasonable, this procedure seems to ascribe undue significance to species such as *Thamnomys dolichurus* which is represented by only 17 individuals spread through 16 stratigraphic levels, *Lemniscomys griselda* (only 7 individuals) or *Rhabdomys pumilio* (only 24 individuals). To use their ecological preferences in reconstructions of vegetation changes seems to be to put too much significance on the difference between one or no specimens per level. Particularly in the cases of *Thamnomys dolichurus* and *Rhabdomys pumilio* there is a tendency to be present in larger samples and absent in smaller ones. Nevertheless, Avery's overall reconstruction is adopted by Butzer (1984) and refers to open grassland conditions during the accumulation of levels 4 to 1b whereas from 10 to 5a "the vegetation was relatively thick with dense grass, perhaps fairly damp (*Otomys irroratus, Rhabdomys pumilio, Lemniscomys griselda*), and rather more extensive forest or thick scrub (*Thamnomys dolichurus*)" (Avery 1982a: 192).

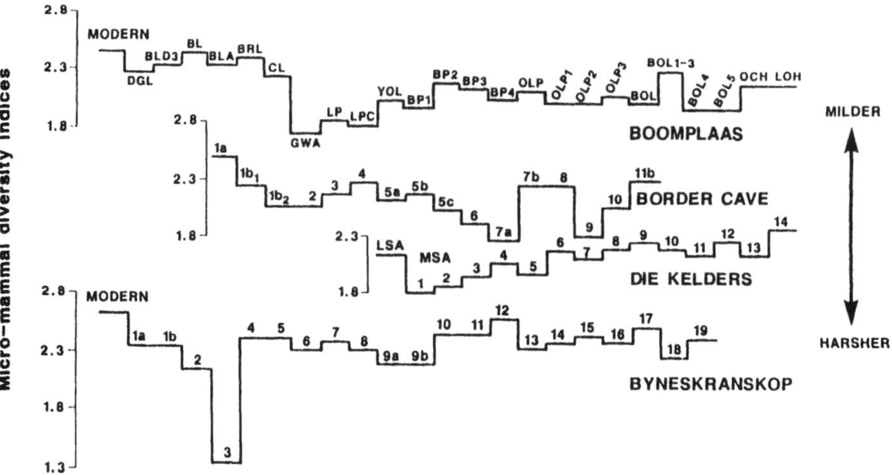

Figure 2.5. Micromammalian diversity indices for the stratigraphic sequences at Boomplaas, Border Cave, Die Kelders and Byneskranskop (redrawn after Avery 1982a, 1982b).

As levels 4 to 1b include everything from greater than 49 000 to 2010 years ago, this reconstruction seems to imply a much dampened amplitude of change as compared to the sediment pattern, or compared to intuitive expectations from a period subsuming oxygen-isotope stages 1, 2 and 3. Significantly, perhaps, and sensibly, Margaret Avery does not attempt to fit the bushier/open grassland dichotomy into the isotope chronology. She does, however, refer to some lack of fit between her inferences and those made by Richard Klein from his analysis of larger mammals from the same levels. Particularly problematic was Klein's (1977) suggestion that it was bushier during the accumulation of levels 3 and 2, whereas the micromammals seemed to reflect open grassland. In arguing strongly that agreement between different data sets need not necessarily be expected, Avery illustrates the problems of linking the results of different kinds of analyses. Consider her statements that "in fact none of the interpretations may be contradictory since the data may refer to different situations. It is quite possible, for instance, that the amount of grassland available was adequate for the large herbivores represented, even at time when the vegetation might seem relatively bushy from the small mammal evidence. It is equally possible that the large and small mammal evidence may refer to either or both the Lowveld and the Lebombos" (Border Cave is on a major escarpment) (Avery 1982a: 194). Whilst this is obviously true it does cast some doubt on our ability to infer and recognize the significance of environmental changes as reflected in faunal material.

In her second argument Margaret Avery does make an explicit attempt to relate changes in diversity through the BC sequence to the isotope record (Figure 2.5). Level 1a, the top of the sequence, has the highest index (2.52), an observation which is taken to justify its interpretation as "not harsh". Unfortunately the robusticity of this argument is lessened by her recording noted elsewhere (Avery 1982b: 309), of a very low diversity (1.75) from owl pellets in the reputedly mild vicinity of Glentyre cave, southern Cape, in the 1980s. Avery comments that "the index may be artificially low, as in the case of Glentyre, the situation apparently being caused by the predator's reliance on one major prey item" (Avery 1982b: 309). Some of the lower BC levels have indices as low as this, though most are between 2.0 and 2.3. It is hard to know what to make of them, especially as level 1b, which subsumes the peak of isotope stage 2, is only 2.25 whereas several are below 1.95 – presumably harsher than harsh! None of the indices approach the 'mildness' of la, the present, so we must presume that stage 5e is not reflected, but what is? There are no arguments of necessity linking the fluctuations in the diversity index with the undulations of the isotope chronology. We simply do not know what to make of level-to-level differences or the scale of undulations in the diversity curve. They don't look much like the isotope curve, but this is perhaps because we cannot assume a continuous, calibrated record.

My suggestion is that the arguments of Beaumont, Butzer and Avery in support of an age of 80 000 years or more for the HP levels at BC are tenuous and fragile. They may be right, but I see no strong argumental barrier to an age considerably less than this.

LARGE MAMMALS, PALAEOENVIRONMENTS AND THE HOWIESONS POORT AT BORDER CAVE

In the Border Cave faunal analysis, Richard Klein (1977) refers to the HP in levels 7 and 8 as 'MSA 2', which is stratified above MSA 1 in levels 9 and 10, and below MSA 3 (levels 4-6) and early LSA (levels 2-3). The fact that he was able to identify to species level only 313 out of some 139 000 bones from Beaumont's excavation (Butzer *et al.* 1978: 327) underlines the problem of extreme fragmentation and sample sizes at this site. Klein attempted to solve this problem partly by using various indices of species frequency as his observational framework, and partly by pooling the numbers of animals with similar ecological requirements so as to highlight patterns of change. One of his aims was to reconstruct the palaeoenvironmental sequence for Border Cave, match this with the regional scheme and thus estimate the age of the faunas.

Klein's initial remark is that apart from extinct forms, all the species identified were "historic inhabitants of the eastern Lowveld in which Border Cave is located" (1977: 18) and thus that in a general sense "the environment was broadly similar to the present one, with a persistent mosaic of dense, low thickets, particularly along watercourses, and large expanses of grassland and savannah" (1977: 18-19). It is worth comparing this statement with Margaret Avery's (1982a: 198) that "the majority of species [of micro-mammals] that occurred regularly at Border Cave in the past is still to be found in the area today": change appears to have been modest and related to shifts in the mosaic of habitat types. But, noting particularly that animals preferring bushier environments (bushpig, buffalo and tragelaphine antelopes) are more common in the early LSA levels, whereas those preferring more open vegetation (zebra, warthog and alcelaphines) are more common in MSA 3, Klein develops an argument for the age of the successive faunal assemblages. The first problem is to know what "bushier" implies. "It seems reasonable to hypothesize that the times with bushier conditions reflect the periods of lowered temperatures and changed atmospheric circulation patterns that characterized the Later Pleistocene everywhere, including southern Africa" (Klein 1977: 21). Note that levels 2 and 3 at Border Cave which have the early LSA and the "bushier" fauna have radiocarbon dates of 33 000 to 38 000 BP and thus appear radiometrically to relate to oxygen-isotope stage 3, and that there is no large mammal fauna at BC from stage 2. Without much supporting evidence we arrive at a bushier = colder guideline, which implies that MSA 3 was thus warmer, MSA 2 and the top of MSA 1 colder, but which does not tell us which intervals are involved. At this point Klein needs help and notes that "Butzer's (in preparation) analysis of the profile at the Klasies River Mouth MSA site in the southern Cape suggests that artifacts called 'Howiesons Poort' there may date from a marked cold oscillation within the later part of the Last Interglacial at about 95 000 BP. The Border Cave MSA 2 is very similar to the Klasies 'Howiesons Poort' and may well be the same age" (Klein 1977: 21). So, at Border Cave MSA 1 and 2 are stage 5b, MSA 3 is 5a and the early LSA is early stage 4, working up from the HP-related age established on artifact similarity. This seems to render the radiocarbon dates rather meaningless as the early LSA has dates which put it in stage 3, dates which incidentally were used to start the argument in the first place.

Very clearly Richard Klein never intended his study to be seen as independent of other associated research or as anything but a tentative interpretation of his observations. Small sample sizes and the absence of strong predictive arguments as to what kinds of environmental changes we could expect under what kinds of past climatic regimes make this inevitable. The result is that the study of the larger mammals at Border Cave is of little use in establishing the age of HP assemblages or associated hominid remains, as Richard Klein himself notes (personal communication *in litt.* 1986).

Since Klein's work there has been considerable research into the vegetation history of the Zululand lowveld which suggests that the colder = bushier equation may be in error (Watson and Macdonald 1983; Whateley and Porter 1983). Both Klein (1971) and Beaumont *et al.* (1978) used the faunal composition of the Umfolozi game reserve some 120 km south of BC as the best indication of the kinds of animals to be expected near the site in mild circumstances. The faunal composition appeared (Mentis 1970) to be dominated by grazers, thus supporting the bushier = colder argument. Ironically, had they chosen the fauna from the Hluhluwe reserve, which is in fact closer to BC, they would have been compelled to draw the opposite conclusion (Brooks and MacDonald 1983). In fact it appears that the high grazer numbers in the lowveld at the beginning of this century may be the result of 1500 years or so of Iron Age farming impact on the vegetation (Hall 1984). Once the reserves were proclaimed in 1895, effectively putting to an end clearing and cutting of woody vegetation for smelting and agriculture, open grasslands began to be replaced by more closed vegetation to the extent that the most recent censuses are heavily dominated by browsers or mixed feeders. It is not my purpose here to argue the reverse case to Klein, Beaumont, Butzer or Avery, merely to point to the fragile nature of both the palaeoenvironmental reconstructions *and* the chronological inferences drawn from the BC fauna.

IS THERE A CASE FOR A SHORTER CHRONOLOGY?

The burden of my argument so far is that biological and sedimentological observations do not constrain us to accept a stage 5 age for HP levels, particularly at Border Cave. The necessary linkages between global measures of ocean temperatures and ice volume, on the one hand, and regional changes in vegetation structure, on the other, have not been made. Until we know whether stage 5b should be bushy or grassy it will not help our dating endeavour to have a measure of bushiness. Much of the evidence could be re-marshalled in support of a new consensus, one that argued for an age of much less than 95 000 years or stage 5b for the HP. Any revision of that estimate must be viewed in association with the questions of the date of the MSA/LSA transition, the nature of post-HP MSA assemblages and indeed, in the long run, most significantly the *meaning* of the MSA/LSA transition.

A consensus could cohere around Deacon's suggestion of an age at the stage 4/5a interface for the HP at KRM (Deacon *et al.* 1986). Here I make two suggestions which would imply a more radical departure from the current consensus. First I ask whether some of the HP assemblages from Southern Africa might not be properly ^{14}C dated to between 35 000 and 50 000 years? And second, I ask whether we must envisage throughout Southern Africa a phase of post-HP MSA technology, effectively isolating the HP from the LSA which in some sense it presages? Before discussing these it is necessary to say something about problems of nomenclature and classification, because this is largely a literature-based survey only partly supported by personal investigation.

The case of Montagu Cave is instructive here. In Keller's original (1973) statement on the MSA levels he noted that had he found these assemblages on the surface he would have been tempted to try to separate out MSA and LSA components, assuming he had mixed assemblages. In the event he had found at least seven, probably more, superimposed assemblages all of which had the combination of MSA elements (in the form of medium to large quartzite flakes, some with facetted platforms) and LSA elements, (in the form of small bladelet and bipolar cores with small flakes and bladelets made of finer grained rocks). He referred all of this series to the Howiesons Poort. Volman, some years later (1981, 1984) and informed by the growing consensus, argued that some of Keller's upper HP levels were in fact post HP/MSA III assemblages. His main ground for saying this was the decrease in backed elements toward the top of the Montagu MSA levels. Recently Poggenpoel and I (manuscript in preparation) have re-examined samples from all of Keller's levels and conclude, along with Keller, that there are no discernible differences in assemblage composition among them. The fact that the uppermost levels have produced small samples almost certainly explains the scarcity of the (always fairly rare) backed pieces, and Volman's argument that one small segment there is derived from the LSA above is rendered fragile by the undoubted existence of quite tiny quartz segments elsewhere in the HP (at Umhlatuzana for example). My point is that these assemblages can easily be classified differently by different authors in the absence of definitional constraints. The 'facts' are just as malleable as the models.

The same kind of point could be made by considering the HP and overlying levels in the Lesotho sites excavated by Carter (1969, 1976; Carter and Vogel 1974). Here again Volman's expectation of finding post-HP MSA levels has guided his classification of the assemblage sequence here. My (in preparation) analysis of the levels above those with segments does not illustrate particularly strong MSA links there, but rather a transition toward LSA technology.

Bearing this in mind let us return to the age of HP levels. My personal interest in this derives from the dates we have obtained (through John Vogel of the CSIR in Pretoria) from the HP levels at Diepkloof. An already published date of 29 400 ± 675 BP (Pta-1051, Parkington and Poggenpoel 1987) has been complemented by new dates of 40 800 ± 1400 (Pta-4489) and 42 400 ± 1600 BP (Pta-4488), both of the latter being apparently finite but considered minimal by Vogel (*in litt.* 12.5.87). The temptation clearly exists for consensus believers to reject these as contaminated by overlying LSA charcoal (Butzer 1979) and to use a stratigraphically lower date of >45 270 (Pta-1054) as evidence that all of the HP at Diepkloof is much older. Similar kinds of arguments could be used to reject the notion, expressed some time ago that "in Eastern Lesotho a sophisticated blade industry with crescents and backed pieces occurs earlier than 30 900 BP at Sehonghong and around 40 000 BP at Ha Soloja" (Carter and Vogel 1974: 570). The as-yet unpublished observation of Jonathan Kaplan (personal communication) from the site of Umhlatuzana in Natal, where HP assemblages are associated with a stratigraphically consistent series of six radiocarbon dates ranging from 28 000 to 45 000 years, are likely to suffer the same fate.

And perhaps rightly so. I think we should make some points explicit in our decisions as to what these observations mean: (1) If we apply the principle that once one HP assemblage is associated with a >50 000 'date' then all such assemblages must be that age, we will never discover temporal variation in the age of the HP if it does exist; (2) If we systematically regard all radiocarbon dates over 30 000 years as too young, then we have produced a black hole in our chronology wherein we can date things that are <30 000 or >50 000 but not those in between. Presumably some Southern African assemblages fall into that time range, but which? (3) If in other geographic regions (Eastern Europe or the Near East, for example) the pattern is to accept resolution in the ^{14}C chronology between 30 000 and 50 000 years, then Southern Africa will necessarily appear early by comparison; (4) Lastly I think we should bear in mind that we have not yet demonstrated the unitary nature of the HP. It remains possible that there are several phenomena linked only by a superficial sharing of backed elements.

What of the nature of post HP assemblages throughout Southern Africa? Again beginning at Diepkloof, there are no overlying MSA levels at the site, although clearly it could be argued that assemblages of MSA character are simply missing here. Montagu Cave (Keller 1973), Peers Cave (Jolly 1948), other smaller Cape Peninsula sites (Goodwin 1953; Malan 1955) and Klipfonteinrand (Volman 1984) also document HP levels without any superimposed MSA, a pattern which leaves the western Cape totally deficient in this aspect of the consensus position. A similar case can be made for the Transvaal, with reference to Cave of Hearths (Mason 1962), for Natal, with reference to Umhlatuzana (Kaplan, personal communication) and, in my opinion, Border Cave (Beaumont 1978), and for Lesotho with reference to Melikane, Moshebis, Sehonghong and Ha Soloja (Carter and Vogel 1974). In fact the only published evidence for the existence of MSA horizons overlying HP ones comes from Boomplaas (Deacon 1979), Klasies River Mouth Shelter 1A (Singer and Wymer 1982) and Apollo XI cave (Freundlich *et al.* 1980). The consensus position thus requires us to see only these sites as preserving a particular portion of the regional MSA sequence wherein, above the HP, MSA blank forms and production modes reassert themselves in the absence of blades, bladelets and backed artifacts. Once again some of the implications should be made explicit: (1) If we assume there is a single regional-wide sequence of MSA change we will never discover contemporary variability; (2) In the absence of absolute dating how will we distinguish between, on the one hand, spatial variation in assemblage character brought about by raw material differences, regionally variable behaviour or needs and, on the other, temporally sequential patterns of tool making behaviour. Acceptance of the single sequence model will force us to try to give all or most variability some temporal status; (3) Some regions seem to reflect the existence of a post-HP informal and perhaps transitional MSA/LSA set of assemblages. These as yet find no place in the consensus model.

I believe there is a case for leaving open the possibility that some HP assemblages are correctly dated to isotope stage 3, that is between 50 000 and 35 000 years old, and that some of these are overlain not by assemblages with a resurgent MSA character but by transitional assemblages with increasing numbers of bladelet and bipolar cores and of small bladelets of LSA character. If we use the consensus not as something to emerge from our study but as a set of guiding principles then we will never discover whether it is right or wrong, merely that it can be applied.

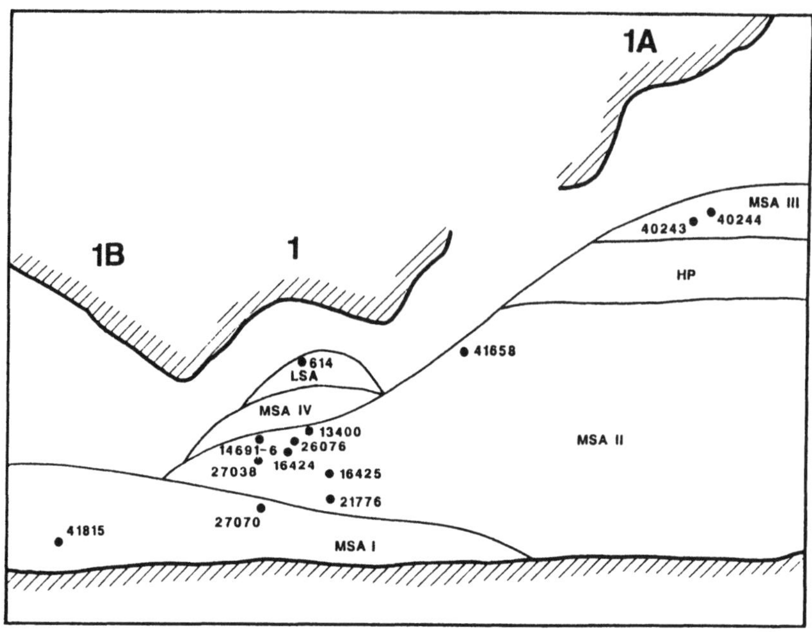

Figure 2.6. Hominid remains from Klasies River Mouth Main Site (redrawn after Singer and Wymer 1982).

THE AGE OF MSA HOMINIDS IN HOWIESONS POORT CONTEXTS AT BORDER CAVE AND KLASIES RIVER MOUTH

One reason for establishing the security of the MSA chronology for Southern Africa is the existence of anatomically modern hominid fragments at some sites. The possibility that anatomically modern humans were present in Southern Africa earlier than they were elsewhere has been referred to both by anatomists (Stringer 1986) and by biochemists (Cann *et al.* 1987; Wainscoat *et al.* 1986). The most dramatic case is obviously Border Cave, where we have claims for burial accompanied by items of personal decoration and the existence of anatomically modern people linked to the age of the HP levels (Beaumont *et al.* 1978). The hominids have been described in detail elsewhere (most recently Bräuer 1984; Rightmire 1984); here I refer only to their apparent age.

At Border Cave the most pertinent specimens are BC3 and BC5 (Rightmire 1984) for which Beaumont *et al.* report inferred ages of "105 ± 3 kyr BP" and "90 ± 5 kyr BP" (1978: 414) respectively, but BC1 and BC2 have also been claimed as MSA in age. The infant skeleton BC3 was apparently a burial, excavated by the 1941 research team (Cooke *et al.* 1945) and considered by Beaumont and Butzer to be equivalent in age to level 9, just below the HP. They believe there is sufficient evidence in the observations of the 1941 excavation to argue that the level containing the HP, a few centimetres above the burial, had not been truncated by the interment. A direct measurement of the age of the bone seems the only way to solve the problem.

The lower jaw BC5 was recovered from the edge of an excavation on a visit to the site in 1974 and is described as having come from "just above the base of the intact 3WA" (Beaumont 1978: 61), the HP level. Beaumont felt this might also have come from a grave. Again only a direct age estimate from the bone itself will remove lingering doubts as to its antiquity. BC1 and BC2 (a complete cranial vault and a partial mandible) were recovered from the dump left by a guano excavation in 1940, a hole that apparently did not penetrate beyond level 11a. Partly because of nitrogen and aspartic acid readings on BC1 and partly on the basis of soil adhesions in the interstices of the skull, Beaumont (1978: 108) believes BC1 to have come from level 10. For similar reasons he places BC2 in level 9 or 10. He thus believes that four of the BC hominids are between 90 000 and 105 000 years old. In the absence of direct dating of the fossils we have to fall back on impressions of relative preservation and fragmentation. Two things stand out. In the first place there is the semi-articulated or articulated state of the infant skeleton and the cranial fragments from the guano excavators dump, which is in very marked contrast to the extreme fragmentation of animal bones from the site. And beyond this there is the visible distinction between faunal and human remains in the state of preservation. These observations are consistent with the view that most human remains at BC are buried into older levels and are not likely to be older than 40 000 years.

Reliable age estimates for KRM hominids will depend on Deacon's reconstruction and dating of events surrounding the build up and subsequent erosion of a huge cone of deposit in front of Cave 1 and Shelter 1A. Much of the Cave 1 deposit, including several hominid fragments (Figure 2.6), has slid into the site from outside and the problem remains to integrate the stratigraphic sequences. Although the relationship with marine geomorphological and biological remains gives general assistance, the key obviously lies in absolute dating of sediments or faunal remains. What does seem clear is the contrast with BC. At KRM human remains are fragmentary, partly burnt and in a similar state of fragmentation and preservation to the animal bones. Here there seems no question of burial or non-contemporaneity, and the fact that KRM, unlike BC, includes some anatomically robust and perhaps archaic specimens is probably significant. They appear more likely to be older than 50 000 years in age.

CONCLUSION

The consensus view promoted for the age of HP assemblages and associated hominids may be in error. Arguments for HP assemblages at 95 000 years are not based on constraining observational sets but on extrapolations, climatic inferences and intuitive matchings. Radiocarbon ages for such assemblages, whilst admittedly at the limits of machine capability, suggest that more recent dates are likely. This in turn would prompt us to revise our models for the transition from HP assemblages to early LSA ones. Only with a sound chronology could we begin to compare the details of the Southern African sequence, technological as well as skeletal, with those from elsewhere.

ACKNOWLEDGEMENTS

I am very grateful to Pat Carter and Jonathan Kaplan for allowing me access to their unpublished excavated material from Lesotho and Umhlatuzana respectively. My preliminary comments in no way, of course, commit them to similar views. I also thank Pat Carter, Thomas Volman, Richard Klein, Peter Beaumont, John Vogel, Janette Deacon, Aron Mazel, Royden Yates, Philip Rightmire, Lyn Wadley, Margaret Avery and Francis Thackeray for sharing their ideas with me. Finally I am grateful to Glen Mills for drawings, and Dawn Fourie for typing and editing.

REFERENCES

Avery, D. M. 1982a. The micromammalian fauna from Border Cave, KwaZulu, South Africa. *Journal of Archaeological Science* 9: 187-204.

Avery, D. H. 1982b. Micromammals as palaeoenvironmental indicators and an interpretation of the late Quaternary in the southern Cape Province, South Africa. *Annals of the South African Museum* 85: 183-374.

Bada, J. L. and Deems, L. 1975. Accuracy of dates beyond the 14-C dating limit using the aspartic acid racemisation reaction. *Nature* 255: 218-219.

Beaumont, P. B. 1973. Border Cave: a progress report. *South African Journal of Science* 69: 41-46.

Beaumont, P. B. 1978. *Border Cave.* Unpublished MA Thesis. Department of Archaeology, University of Cape Town.

Beaumont, P. B., De Villiers, H. and Vogel, J. C. 1978. Modern man in sub-Saharan Africa prior to 49 000 years BP: a review and evaluation with particular reference to Border Cave. *South African Journal of Science* 74: 409-419.

Beaumont, P. B. and Vogel, J. C. 1972. On a new radiocarbon chronology for Africa south of the Equator, parts 1 and 2. *African Studies* 31: 66-89, 155-182.

Binford, L. R. 1984. *Faunal Remains from Klasies River Mouth.* New York: Academic Press.

Bräuer, G. 1984. A craniological approach to the origin of anatomically modern *Homo sapiens* in Africa and implications for the appearance of modern Europeans. In F. H. Smith and F. Spencer (eds) *The Origins of Modern Humans: a World Survey of the Fossil Evidence.* New York: Alan R. Liss: 327-410.

Brooks, P. M. and Macdonald, I. A. W. 1983. The Hluhluwe-Umfolozi Reserve: an ecological case history. In R. N. Owen-Smith (ed.) *Management of Large Mammals in African Conservation Areas.* Pretoria: HAUM Publishers: 187-200.

Butzer, K. W. 1978. Sediment stratigraphy of the Middle Stone Age sequence at Klasies River Mouth. *South African Archaeological Bulletin* 33: 141-151.

Butzer, K. W. 1979. Geomorphology and geo-archaeology at Elandsbaai, western Cape, South Africa. *Catena* 6: 157-166.

Butzer, K. W. 1985. Late Quaternary environments in South Africa. In J. C. Vogel (ed.) *Late Cainozoic Palaeoclimates of the Southern Hemisphere.* Rotterdam: Balkema: 235-265.

Butzer, K. W., Beaumont, P. B. and Vogel, J. C. 1978. Litho-stratigraphy of Border Cave, Kwazulu, South Africa: a Middle Stone Age sequence beginning c. 195 000 BP. *Journal of Archaeological Science* 5: 317-341.

Butzer, K. W. and Vogel, J. C. 1979. Archaeo-sedimentological sequences from the sub-montane interior of South Africa: Rose Cottage Cave, Heuningneskrans and Bushman Rock Shelter. Paper presented at a workshop entitled 'Towards a better understanding of the Upper Pleistocene in sub-Saharan Africa' organised by the Southern African Association of Archaeologists, Stellenbosch.

Cann, R. L., Stoneking, M. and Wilson, A. C. 1987. Mitochondrial DNA and human evolution. *Nature* 325: 31-36.

Carter, P. L. 1969. Moshebi's Shelter: excavation and exploitation in eastern Lesotho. *Lesotho* 8: 1-11.

Carter, P. L. 1976. The effects of climatic change on settlement in eastern Lesotho during the Middle and Later Stone Age. *World Archaeology* 8: 197-206.

Carter, P. L. and Vogel, J. C. 1974. The dating of industrial assemblages from stratified sites in eastern Lesotho. *Man* 9: 557-578.

Chappell, J. and Shackleton, N. J. 1986. Oxygen isotopes and sea level. *Nature* 324: 137-140.

Cooke, H. B. S., Malan, B. D. and Wells, L. H. 1945. Fossil Man in the Lebombo Mountains, South Africa: The 'Border Cave' Ingwavuma District, Zululand. *Man* 45: 6-13.

Deacon, H. J. 1979. Excavations at Boomplaas Cave: a sequence through the Upper Pleistocene and Holocene in South Africa. *World Archaeology* 10: 241-257

Deacon, H. J., Geleijnse, V. B., Thackeray, A. I., Thackeray, J. F., Tusenius, M. L. and Vogel, J. C. 1986. Late Pleistocene cave deposits in the southern Cape: current research at Klasies River. *Palaeoecology of Africa* 17: 31-37.

Deacon, H. J. and Thackeray, J. F. 1984. Late Pleistocene environmental changes and implications for the archaeological record in Southern Africa. In J. C. Vogel (ed.) *Late Cainozoic Palaeoclimates of the Southern Hemisphere*. Rotterdam: Balkema: 375-390.

Deacon, J. 1978. Changing patterns in the late Pleistocene/early Holocene prehistory of Southern Africa as seen from the Nelson Bay Cave stone artifact sequence. *Quaternary Research* 10: 84-111.

Deacon, J. 1979. The Howiesons Poort and related industries in Southern African with special reference to the name site collection. Paper presented at a workshop entitled 'Towards a better understanding of the Upper Pleistocene in sub-Saharan Africa' organised by the Southern African Association of Archaeologists, Stellenbosch.

Deacon, J., Lancaster, N. and Scott, L. 1984. Evidence for Late Quaternary climatic change in Southern Africa. In J.C. Vogel (ed.) *Late Cainozoic Palaeoclimates of the Southern Hemisphere*. Rotterdam: Balkema: 391-404.

Freundlich, J. C., Schwabedissen, H. and Wendt, W. E. 1980. Köln Radiocarbon Measurements II. *Radiocarbon* 22: 68-81.

Goodwin, A. J. H. 1953. Two Caves at Kalk Bay, Cape Peninsula. *South African Archaeological Bulletin* 8: 59-77.

Hall, M. 1984. Prehistoric farming in the Mfolozi and Hluhluwe Valleys of Southeast Africa: an archaeo-botanical survey. *Journal of Archaeological Science* 11: 223-235.

Hendey, Q. B. and Volman, T. P. 1986. Last Interglacial sea levels and coastal caves in the Cape Province, South Africa. *Quaternary Research* 25: 189-198.

Jolly, K. 1948. The development of the Cape Middle Stone Age in the Skildergat Cave, Fish Hoek. *South African Archaeological Bulletin* 3: 106-107.

Keller, C. M. 1973. Montagu Cave in Prehistory. *University of California Anthropological Records* 28: 1-150.

Klein, R. G. 1972. The late Quaternary mammalian fauna of Nelson Bay Cave (Cape Province, South Africa): its implications for megafaunal extinctions and environmental and cultural change. *Quaternary Research* 2: 135-142.

Klein, R. G. 1976. The mammalian fauna of the Klasies River Mouth sites, South Africa. *South African Archaeological Bulletin* 31: 35-98.

Klein, R. G. 1977. The mammalian fauna from the Middle and Later Stone Age (later Pleistocene) levels of Border Cave, Natal Province, South Africa. *South African Archaeological Bulletin* 32: 14-27.

Kronfeld, J. and Vogel, J. C. 1980. Fingerprints in groundwater. *Nuclear Active* 22: 30-32.

Malan, B. D. 1955. The archaeology of the Tunnel Cave and Skildergat Kop, Fish Hoek, Cape of Good Hope. *South African Archaeological Bulletin* 10: 3-9.

Mason, R. 1962. *Prehistory of the Transvaal.* Johannesburg: Witwatersrand University Press.

Mentis, M. T. 1970. Estimates of natural biomasses of large herbivores in the Umfolozi Game Reserve area. *Mammalia* 34: 363-393.

Parkington, J. E. and Poggenpoel, C. A. 1987. Diepkloof rockshelter. In J. E. Parkington and M. Hall (eds) *Papers in the Prehistory of the Western Cape South Africa.* Oxford: British Archaeological Reports International Series S352: 269-293.

Rightmire, G. P 1984. The fossil evidence for hominid evolution in Southern Africa. In R G. Klein (ed.) *Southern African Prehistory and Paleoenvironments*. Rotterdam: Balkema: 147-395.

Rightmire, G. P. 1984. *Homo sapiens* in Sub-Saharan Africa. In F. H. Smith and F. Spencer (eds) *The Origins of Modern Humans: a World Survey of Fossil Evidence.* New York: Alan R. Liss: 295-325.

Shackleton, N. J. 1982. Stratigraphy and chronology of the Klasies River Mouth deposits: oxygen isotopes evidence. In R. Singer and J. J. Wymer (eds) *The Middle Stone Age and Klasies River Mouth in South Africa.* Chicago: University of Chicago Press: 194-199.

Singer, R. and Wymer, J. J. 1982. *The Middle Stone at Age Klasies River Mouth in South Africa.* Chicago: University of Chicago Press.

Stringer, C. 1984. Human evolution and biological adaptation in the Pleistocene. In R. J. Foley (ed.) *Hominid Evolution and Community Ecology*. London: Academic Press: 55-83.

Taylor, R. E., Payen, L. A., Prior, C. A., Slota, P. J. Jr., Gillespie, R., Gowlett, J. A. J., Hedges, R. E. M., Jull, A. J. T., Zabel, T. H., Donahue, D. J. and Berger, R. 1985. Major revisions in the Pleistocene age assignments for North American human skeletons by C-14 accelerator mass spectrometry: none older than 11 000 C-14 years BP. *American Antiquity* 50: 136-140.

Volman, T. P. 1981. *The Middle Stone Age in the Southern Cape*. Unpublished Ph.D. Thesis, University of Chicago.

Volman, T. P. 1984. Early prehistory of Southern Africa. In R. G. Klein (ed.) *Southern African Prehistory and Paleoenvironments*. Rotterdam: Balkema: 169-220.

Wainscoat, J. S., Hill, A. V. S., Boyce, A. L., Flint, J., Hernandez, M., Thein, S. L., Old, J. M., Lynch, J. R., Falusi, A. G., Weatherall,

D. J. and Clegg, J. B. 1986. Evolutionary relationships of human populations from an analysis of nuclear DNA polymorphisms. *Nature* 319: 491-493.

Watson, H. K. and Macdonald, I. A. W. 1983. Vegetation changes in the Hluhluwe-Umfolozi Game Reserve Complex 1937 to 1975. *Bothalia* 14: 265-269.

Whateley, A. and Porter, R. N. 1983. The woody vegetation communities of the Hluhluwe-Corridor-Umfolozi Game Reserve Complex. *Bothalia* 14: 745-758.

3: The Middle and Upper Palaeolithic of the Near East and the Nile Valley: the Problem of Cultural Transformations

ANTHONY E. MARKS

INTRODUCTION

The change in the archaeological record from what is recognized as Middle Palaeolithic to what is called Upper Palaeolithic traditionally has been viewed as being, at least, dramatic and, at most, a profound and fundamental change in both behaviour and human potentials. If the change is viewed as a replacement of an archaic, culturally less complex folk (the Neanderthals) by people who were fully modern in physical type and cultural potentials, then there is little problem with this traditional view. However, if replacement was not the mechanism for change everywhere, then these views make it very difficult to deal methodologically and even conceptually with the potential development of the Middle Palaeolithic into the Upper Palaeolithic. The Near East and Northeast Africa are both areas where developmental transitions have been postulated and, so, where traditional views of replacement, mainly derived from and supported by Western European data, may have little relevance. With this in mind, this paper will examine some of the present issues which bear directly and indirectly on the potentially developmental nature of the change from the Middle to the Upper Palaeolithic in the Near East and the Nile Valley.

As in most Old World areas outside Europe, methods and concepts developed from specific European experiences were imported wholesale into the Near East and somewhat into the Nile Valley and were applied, uncritically, to data sets which, in a few cases, differed quite significantly from their European counterparts. The two most obvious examples were the wholesale adoption of French systematics for the study of lithic artifact assemblages and the many years during which questionably recognizable local geological features were crammed into the Alpine Pleistocene sequence. There are other, less pronounced examples and even more extreme ones (e.g. the transfer of European culture names, such as Magdalenian (Buzy 1929), to assemblages in the Sinai) which, combined, all had the effect of moulding Near Eastern prehistory in the image of Europe – if not exactly the same, then at least one in which then current European ideas could find comfort.

This is not to say that, for instance, French typological systematics in lithic assemblage studies did not represent a marked advance over what was common practice prior to their adoption. Without question, they did provide a more or less universal language for the description of assemblages but they also, in doing so, limited its vocabulary and permitted only a restricted structure for such studies. The present author knows this all too well, since he was one of the first to apply Bordian systematics to the Middle Palaeolithic of the Nile Valley (Marks 1968a). It was not that the systematics were flawed. In fact, used for the reasons for which they were developed in France – to deal with the then current Mousterian/Levalloisian question (Bordes 1953) – they were highly effective. In the Near East and the Nile Valley, however, this was not as relevant a problem and, therefore, the systematics simply were not particularly well suited to answer many questions which were inherent in Near Eastern and Nilotic African prehistory. The same is certainly true for the comparable systematics adopted for the Upper Palaeolithic; they are only partly relevant outside Europe. Thus, while temporarily improving communication in local prehistory, it is now all too clear that, without modification, they are unable to address effectively those questions and to incisively resolve those problems which face Near Eastern and Nilotic prehistorians today.

Perhaps, because of the obviously uncomfortable fit between the Alpine glacial sequence and the non-glacial climatic shifts in the Near East, only a modicum of confusion resulted from using them. In fact, it would seem that most problems of correlation are to be found in Middle and Early Pleistocene contexts (Horowitz 1979), rather than in the Upper Pleistocene where radiometric dating is more available. On a regional level, the complexity of even local correlations is acute (Farrand with comments 1982) and there seems to be a tendency, at times, to underestimate or overestimate the effects of intra-regional climatic differences as an explanatory model for synchronic cultural variation (Copeland 1986).

The situation in the Nile Valley is not so marked. There has been less interest in European/Nilotic climatic correlations, with emphasis placed on the correlation between Upper Pleistocene sites, the Nile sequence, and local climatic conditions (e.g. de Heinzelin 1968; Said 1983; Butzer and Hansen 1968). Even in methodology, workers in Northeast Africa tended to borrow relatively little from European systematics, except in Middle Palaeolithic studies (Marks 1968a; Guichard and Guichard 1968). Since European Upper Palaeolithic typologies failed to correspond with post-Mousterian tool assemblages in the Nile Valley, they were not adopted, although many of the other assumptions inherent in European systematics were modified and used (Schild *et al.* 1968).

In spite of these historic problems, the prehistory of the Near East and the Nile Valley is much better known today than it was only a few years ago. There are areas, however, such as the inland Near East and Nubia, where little work has been done for the past 20 years. Other areas, such as the marginal zones of the Levant (e.g. Bar Yosef and Phillips 1977; Marks 1976a, 1977, 1983a) and central Egypt (e.g. Wendorf and Schild 1976) have witnessed a great deal of recent work and, on balance, the data base has expanded enormously. In addition, new dating techniques have begun to provide a somewhat sounder footing for local and even intra-regional correlations, while traditional systematics are being modified to answer questions which are specifically derived from local data. Finally, but with difficulty, many of the older implicit assumptions regarding the meaning of assemblage variability are being explicitly reexamined and, in many cases, are being modified or abandoned. Thus, we are in a very exciting period in Near Eastern and Nilotic Upper Pleistocene studies; a period of flux both in the data base and in the ways it is being approached.

It is not surprising, therefore, that the past ten years have witnessed the emergence of quite different views of Upper Pleistocene cultural development, of different interpretations derived from different kinds of analyses of the same data sets, and quite different overall views based largely on familiarity of different workers with different parts of the region. While exciting, unfortunately truth has not yet raised her head and identified herself clearly, so we are faced with looking closely at the various competing interpretations now current concerning Upper Pleistocene cultures and their transformations. Such an examination may well point to new directions for research.

Since 1951, it has been accepted that there was a local developmental transition from the Middle Palaeolithic to the Upper Palaeolithic in the Levant (Garrod 1951, 1955, 1962; Copeland 1975; Azoury 1971). A later and certainly less well known position suggested that there might have been a separate local transition from the Shanidar-type Mousterian to the Upper Palaeolithic Baradostian in the inland Near East (Hole and Flannery 1967).

In the Nile Valley, continuity also was proposed from the Mousterian onward but, in lacking a traditional European Upper Palaeolithic technocomplex, this continuity was seen as involving less a transformation than an ongoing 'conservatism' (Caton-Thompson 1946; Huzayyin 1941). Thus, in all areas of the region, continuity rather than disjunction has been postulated between the Middle Palaeolithic and what followed.

It is only in the Levant, however, that a clear technological transition has been defined between the Levallois-based Middle Palaeolithic and the blade-based Upper Palaeolithic. For the other areas, continuity was postulated either on the perceived continuation of a limited number of tool types across the Mousterian/Upper Palaeolithic boundary (Hole and Flannery 1967: 152) or, even more generally, in the retention of basic flake technology, as opposed to a change to blade technology (Alimen 1957). Only in Nubia can one document a visible, detailed technological transition from a Levallois to a blade technology within a single industrial sequence (Marks 1968c). In this case of the Halfan, however, the transition takes place in a geographic cul-de-sac and at a date too late to relate to the broader issues of *Homo sapiens sapiens* biocultural evolution. Still, it is informative and provides an example that not only can such a transition take place within a technological system but that it actually did so in one case, at least.

Although a transition from the Levantine Middle Palaeolithic to the local Upper Palaeolithic has been postulated for many years and has been described in some detail more than once (Garrod 1955; Copeland 1970; Marks 1983b), it was only with the extensive core reconstructions at Boker Tachtit in the Negev (Volkman 1983; Marks and Volkman 1983), that it was possible to put this transition into a secure and detailed technological framework and to trace the various transformations in core reduction strategies involved.

With the recent advances in dating techniques, it has been possible to pin down this specific transition to a relatively short period, so that both its nature and its timing are now reasonably known (Marks 1981a, 1983b). However, it has become obvious that merely identifying, dating and describing a 'transitional phase' of lithic technology provides, in and of itself, only a little enlightenment. While the technological adjustments in stone flaking leading from the Middle Palaeolithic to the Upper Palaeolithic follow quite logical but somewhat unexpected trends, the concept of a distinct transitional phase (Garrod 1955), defined technologically and typologically, even with its own type-fossil (Volkman and Kaufman 1983), has the tendency to merely create a separate 'phase' between those of the Middle and Upper Palaeolithic. There has even been a type-list proposed for the phase (Hours 1974), to help distinguish it clearly from what came before and after. However, it does not seem to have been adopted and there seems to be a general movement away from compartmentalizing this largely demonstrable continuous cultural development.

In order to view the process of cultural transformation within the context of a continuum, one cannot merely study the transition itself, however it might be defined, but must look in detail at the continuum, both before and after the point of perceived transformation. It is truly unfortunate that the nature of the local archaeological evidence only permits us to recognize and define, in great detail, a transition in lithic reduction from a Levallois-based to a consistent blade-based strategy which surely must have been among the least significant aspects of the more general cultural transformations which took place during the Middle and Upper Palaeolithic in the Levant. At best, such a lithic, technological transition can only be one minor aspect of a changing adaptation and it is certainly questionable whether it, by itself, should be thought of as a phase, beyond the very narrow view of a sequence of lithic assemblages. Still, since lithic studies are central to Upper Pleistocene prehistory, the description of the process of change is a step forward.

At the moment, in fact, there is perhaps less diversity of opinion concerning the nature of the transitional technology than there is concerning the nature of either local Middle Palaeolithic or early Upper Palaeolithic development. This, of course, is recognized to be a function of an extremely limited data base pertaining to the transition, *per se*, and not to any extraordinary insight on the part of those working with that particular problem.

For numerous reasons, both historic and because of the nature of the data, the central focus of transitional Middle to Upper Palaeolithic studies has been in the Levant. Therefore, this paper will focus most attention there. This is not meant to slight the Nile Valley or the inland Near East but, as much as anything, it reflects the author's own knowledge and current interests.

THE MIDDLE PALAEOLITHIC

Until recently, the post last Interpluvial Middle Palaeolithic sequence in the Levant seemed quite straightforward and rather simple compared with that in Europe (Garrod 1962). Following the work of Garrod (Garrod and Bate 1937) at the cave of Tabun, near Haifa, a basic developmental, two phase Mousterian was recognized. The earliest phase, the Lower Levalloiso-Mousterian (now often referred to as 'Early Levantine Mousterian', 'Tabun D-type' Mousterian, or 'Phase 1' Mousterian) was largely characterized by elongated Levallois blanks, including many unretouched points, relatively abundant Upper Palaeolithic type tools, a fair number of non-Levallois blades, as well as by typical Mousterian side-scrapers. The Upper Levalloiso-Mousterian (now often referred to as 'Late Levantine Mousterian', 'Tabun C-B type' Mousterian, or 'Phase 2-3' Mousterian), on the other hand, was characterized by ovoid Levallois blanks, very few squat Levallois points, rare Upper Palaeolithic tools, and large numbers of side-scrapers and Mousterian points (Ronen 1979).

In 1975, Copeland redivided the Levalloiso-Mousterian as seen at Tabun into three phases (Copeland 1975). To her, the material from each layer was distinct and represented a separate developmental phase. Most importantly, the uppermost phase, seenin Tabun B, showed a mixture of traits of the other two phases, including a tendency back towards blank elongation. Although this assemblage type seems to be present at a few sites, there is as yet no agreement on its significance – whether it is a functional variant within the Late Levantine Mousterian (Ronen 1979) or whether it is, indeed, a final Mousterian developmental stage (Copeland 1986; Meignen *et al.* 1988).

The lack of agreement in this matter appears trivial and, in fact, would be so if the Levantine Mousterian, in general, was a developmental dead end. With a demonstrated transition to the Upper Palaeolithic out of some form of local Middle Palaeolithic, it is important to know what the terminal Mousterian was like – was there a single assemblage form or was it a period of assemblage, and presumably adaptive, heterogeneity? Important as it may be, the level of ambiguity and disagreement in this case, in fact, pales beside the other unresolved questions of both fact and interpretation over what ten years ago seemed known.

The stratigraphic sequence of assemblage types at Tabun has been confirmed at other cave sites where both the Early and Late Levantine Mousterian are found (e.g. Ksar Akil, Nahr Ibrahim, etc.). Recent re-excavations at Tabun (Jelinek *et al.* 1973; Jelinek 1982a) also have confirmed Garrod's basic interpretation of the Mousterian sequence.

Through detailed metric analyses, it has been argued, again, and in more detail, that the sequence is not only temporal but also developmental (Jelinek 1981a, 1982b).

A series of internal geologic events, as well as correlations with beach lines in front of the cave have produced a relatively sound temporal sequence (Jelinek *et al.* 1973; Jelinek 1982b). Of particular importance is the correlation of the sedimentary hiatus between Garrod's levels C and D with oxygen-isotope Stage 4 (Jelinek 1982b), which is firmly dated to about 70 000 BP. This places Early Levantine Mousterian prior to that date and the Late Levantine Mousterian after it. A series of radiocarbon dates suggests that Tabun C might date to about 50 000 BP or, perhaps, somewhat earlier. Thorium/Uranium dates from some other Early Levantine Mousterian assemblages confirm its pre-70 000 BP dating (Bar Yosef and Vandermeersch 1981; Schwarcz *et al.* 1979), while many ^{14}C dates on Late Levantine Mousterian sites suggest its presence as late as 40 000 BP (Henry and Servello 1974).

On the other hand, Thorium/Uranium dates from what seems to be above and below a quite typical Late Levantine Mousterian beach assemblage at Naa'me,in Lebanon, have produced dates in the order of 90 000 BP (Sanlaville 1981) – *at least* 25 000 years before the evolution of the Late Levantine Mousterian out of the Early Levantine Mousterian at Tabun, only about 120 km away (Copeland 1981).

A second discordant note has been sounded in the dating of the undescribed but apparently Late Levantine Mousterian levels at Qafzeh, only some 30 km from Tabun. Based on its microfaunal assemblage and recent TL dates, it has been concluded that the levels involved should date either just before or the same as the levels of Tabun D (Tchernov 1981, Bar Yosef, in press). This seems to support the Naáme date and throws into serious question the evolutionarily developmental aspects of the stratigraphic sequence at Tabun.

To make things even more complicated, in spite of considerable recent work in the southern marginal zones of the Levant – southern Jordan and the Central Negev – and in spite of hints to the contrary (Gilead 1984) only Early Levantine Mousterian has been found (Marks 1981a; Clark and Lindly, this volume). Data from high elevations in southern Jordan, moreover, suggest that this Mousterian occurs, at least, temporally late, during a period of climatic desiccation contemporaneous with the Late Levantine Mousterian at Tabun (Henry 1982; Lindly 1986).

Thus the situation is complex, indeed. While under certain circumstances, this would not be particularly relevant to the question of the Middle to Upper Palaeolithic transition and the emergence of modern humans (should there be any direct connection between the two), the variety of Levantine hominid fossils and their various archaeological associations make it of vital importance.

From early work, two fossils forms were recognized; a 'Neanderthal' found in Tabun C and a 'progressive Neanderthal' found at Skhul (McCown and Keith 1939) – the latter correlated stratigraphically and culturally with Tabun B (Garrod and Bate 1937). It was an acceptable relative sequence in that the later form was closer to modern man than the earlier one. Nothing is simple any more. Quite apart from the question of whether there ever were true Neanderthals (in the European sense) in the Near East (Stringer 1974), even locally, problems abound.

Recent work has uncovered additional fossil hominids. At Kebara a Neanderthal was found associated with a reported Late Levantine Mousterian occupation dated by TL to after 60 000 BP (Bar Yosef *et al.* 1986), while at Tabun a reevaluation of the stratigraphic position of the Tabun C Neanderthal tentatively suggests it belongs in Tabun D (discussion after Trinkaus 1982: 317). Thus, we may have Neanderthals in both the Early and Late Levantine Mousterian, and *Homo sapiens sapiens* also associated with the Late Levantine Mousterian. In addition, the excavations at Qafzeh have unearthed a series of *Homo sapiens sapiens* associated with a reported Late Levantine Mousterian, dated by microfauna and TL to before or the same as the Early Levantine Mousterian at Tabun (Bar Yosef, in press). Thus, on the face of it, there were both Neanderthals and *Homo sapiens sapiens* in the Levant from at least 80 000 BP onwards.

New ESR dates from Skhul (Stringer *et al.* 1989) reinforce the early Qafzeh dating for *Homo sapiens sapiens* but, at the same time, complicate this picture. Now it would appear that all early *Homo sapiens sapiens* in the Levant were there prior to 80 000 BP, to be replaced by Neanderthals sometime after 60 000 BP (provided, of course, that the Tabun Neanderthal, in fact, is associated with Tabun C and not Tabun D). According to Stringer *et al.* (1989) *Homo sapiens sapiens* were only episodic visitors to the Levant, while according to Bar Yosef (e.g. Vandermeersch and Bar Yosef 1988) Neanderthals probably only arrived in the Levant at *c.* 60 000 BP. While this follows presently available data and appears reasonable when only the fossils are considered, such an interpretation is at odds with what we think we know about the archaeology.

Should the above interpretation be accurate, then the Late Levantine Mousterian was made by early *Homo sapiens sapiens* before 80 000 BP but by Neanderthals after 60 000 BP. Certainly the association between apparently Late Levantine Mousterian assemblages and early *Homo sapiens sapiens* is well attested to at Skhul and Qafzeh, while the association between the same industry and a Neanderthal is equally well documented at Kebara. While this, in itself, raises some interesting questions as to the assumed relationship between adaptations and lithic industries, as historic fact, it poses no problems. Who, however, made the Early Levantine Mousterian? This is a problem because it formed the base from which developed the Levantine Upper Palaeolithic. Presumably, its makers in its terminal development were *Homo sapiens sapiens* and archaeologically, at least, there appears to be a continuous development in the Levant of it from before 80 000 BP to *c.* 45 000 BP when it evolves into the Initial Upper Palaeolithic. Could this continuity involve both *Homo sapiens sapiens* and Neanderthals, each coming from a different area with, presumably, different historic traditions and adaptations? How should we view all this? I would suggest we do so very carefully. In order to clarify the situation, a number of important points need verification, confirmation, or even rethinking:

1. The early date for Qafzeh is important for two reasons; the association of *Homo sapiens sapiens* and the reported presence of a Late Levantine Mousterian lithic assemblage. While there can be no doubt that the fossils are what they have been interpreted to be (Vandermeersch 1981), there must be less certainty regarding the lithic assemblage, since it has never been published. Even when excavated and reported by Neuville (1951), the Mousterian at Qafzeh seemed somewhat out of place and was never described in detail. This point, of course, can be resolved easily and should be, considering the present claims for its age.

2. The dating of the Naa'me beach lines seems acceptable (Sanlaville 1981: 29) but is it truly certain that the Late Mousterian beach assemblage is stratigraphically placed between the two dated deposits? Is there additional information which could be brought to bear?

3. The stratigraphic position of the Tabun cranium, now possibly placed in Garrod's Layer D, is the only Neanderthal so dated, although flake measurements suggest that the associated Neanderthal skull from Amud also may be in this time range (Jelinek 1982a: 99). The last Interpluvial fossil from Zuttiyeh appears to be an archaic *Homo sapiens sapiens* rather than an early Neanderthal (Vandermeersch 1982). The reconstruction of the Tabun D fossil's original position seems reasonable but is it accurate? If not, then Bar Yosef's contention that Neanderthals themselves or their genes entered relatively late into the Levant (Bar Yosef, in press) is not contradicted by the evidence. In fact, his estimate of that time corresponds nicely with the traditional date of the Late Levantine Mousterian appearance.

4. What is the significance of the curve derived by Jelinek (1982a: 84) for the variance on the width/thickness ratios for complete flakes throughout the Tabun sequence? It has been presented as evidence for a gradual improvement in human capabilities and, in doing so, it has been assumed to be linked to biocultural change. Yet, as he notes, the samples used were drawn from artifact assemblages produced under different conditions. Some samples seem to come from floors where basic core reduction took place and others from areas where it seems clear that only selected artifacts were carried into the cave (Jelinek 1982a). Could this have had an effect on the means and variances of the width/thickness ratios? Also, Jelinek sees a point of accelerated change just when the Late Levantine Mousterian begins (Jelinek 1982b). Is it possible that this point of acceleration is actually a developmental break and the curve is, in fact, two different developmentally unrelated curves?

As importantly, as yet there appears to be no theoretical basis for explaining the observed changes in terms of human evolution, much less in terms of progressive adaptive technological change. The curve is there; that is about all that can be said about it, except that the core reduction strategies of the Early Levantine Mousterian and those of the Late

Levantine Mousterian were quite different and seemed to have been used to produce differently shaped end-products. In that regard, there is a clear disjunction, not the continuity which the flake measurements may be interpreted to show.

Regardless of how these questions are resolved, a few facts will remain. Perhaps the most important is that both a Neanderthal (at Kebara) and *Homo sapiens sapiens* (at Skhul and Qafzeh) are directly associated with the Late Levantine Mousterian and, therefore, the Mousterian cannot be thought of as the material cultural of Neanderthal man (Oakley 1964: 144) as is still commonly done (White 1982). In fact, a slightly greater geographic perspective would remind us that two very 'Mousterian' looking industries, the MSA of East Africa and the Aterian, were associated with *Homo sapiens sapiens* (Clark 1981). On the other side, the presence of a non-Levallois blade technology dated to the Last Interpluvial at El Kowm (Hours 1982), as well as the better known Amúdian of the central Levant (Jelinek, this volume), should banish the idea that a dominant blade technology was somehow bioculturally excluded from pre-*Homo sapiens sapiens* potentials.

Even should the Tabun model of Middle Palaeolithic evolutionary development turn out to be consistent with all data for the Central and Northern Levant, we are still faced with a developmental evolution *within* the Early Levantine Mousterian which temporally parallels that seen at Tabun. Not only that, but geographically they were virtually cheek by jowl. The distance from Tabun to Wadi Hasa in southern Jordan is only 210 kilometres. How might this be accounted for; in fact, how can we account for the contemporaneous presence of both the Early and Late Levantine Mousterian either temporally early or late, or both, in such a small area as the Levant? It has been suggested implicitly that Early Levantine Mousterian adaptation was successful in the climatically marginal zones of the Levant and, therefore, there was no reason to evolve into the Late Levantine Mousterian (Jelinek 1981b). However, the Early Levantine Mousterian developed and first existed in the southern Levant under Mediterranean environmental conditions, not under semi-arid conditions. Therefore, while it may have been well adapted to semi-arid conditions, it also must have been well adapted to Mediterranean climatic conditions. Using the same reasoning, there would have been no reason to evolve into the Late Levantine Mousterian. A variation on this theme has been used to separate the prehistory of Lebanon from that of the southern, marginal zones (Copeland 1986). The reasoning is that one should not expect the two areas to be the same given the environmental differences and the significant distance between the areas. The latter is not realistic; it is only 290 km from Beirut to Beersheva – a trivial distance in hunter-gatherer territorial terms.

The environmental argument is also weak. While there are numerous microenvironmental zones in the Levant, hot/cold and wet/dry shifts are essentially clinal and, more often than not, oriented along north-south axes (Horowitz 1979). Therefore, while the extremes are marked, their boundaries are blurred and constantly shifting. Given the limited temporal control we have for the prehistory of much of the Upper Pleistocene, apparent synchronic variability cannot easily be attributed to direct environmental determinism.

One possible explanation for the apparent synchronic occurrence of both the Early and Late Levantine Mousterian, which is only secondarily environmentally deterministic, is that the differences in technology and typology which distinguish the two industries reflect diverging settlement systems out of the same base. From work in the Central Negev and southern Jordan, it appears that the temporally early Early Levantine Mousterian was associated with a radiating settlement system (Marks 1983b) but that the temporally late Early Levantine Mousterian of southern Jordan was associated with a geographically limited but elevationally marked circulating settlement system (Henry 1982; but see Coinman *et al.* 1986). On the other hand, further north, the Late Levantine Mousterian seems to have been associated with a radiating settlement system, like that seen much earlier in the Central Negev (Bar Yosef, in press). Since much of the typological difference between the two industries rests in the much greater numbers of side-scrapers and Mousterian points in the Late Levantine Mousterian as compared to the Early Levantine Mousterian, it is possible that this is tied directly to differential intensity of occupation and to the resharpening of tools (Dibble 1987), rather than to any innate cultural difference in what tools were considered appropriate for manufacture. If this is the case, it might suggest that Late Levantine Mousterian typological configurations reflect increasing occupational stability, while the decrease in the number of typical Mousterian retouched tools in the temporally late Early Levantine Mousterian of southern Jordan reflects an increasing occupational mobility. If so, then the Late Levantine Mousterian might well have developed out of the Early Levantine Mousterian in those areas where settlement stability could be maintained. The technological differences observed, which now do appear to be very marked, might be a response to increasing tool resharpening, so that ovoid blanks rather than elongated ones were produced to accommodate the expected increase in individual tool life.

At Ksar Akil, the increase in ovoid blank production is incremental from the lowest of the Late Levantine Mousterian levels to the uppermost, as is the increase in the percentage of retouched tools to the total assemblages (Marks and Volkman 1986). This latter, again, indicates an increasing intensity of artifact utilization beyond tool resharpening.

The above scenario might even be useful should Naa'me and Qafzeh prove to be as old as they now seem. If so, the two now contemporaneous industries might be viewed merely as facies of the same industry, each reflecting the relative differences in settlement stability or mobility in their areas. It is striking that Early Levantine Mousterian is not nearly so well represented in the northern core area of the Levant as is the Late Levantine Mousterian and, as well, that the Late Levantine Mousterian is undocumented in the extreme Southern Levant. While the environments may well have shifted through time, the Northern Levant would have always been either colder and/or wetter than the Southern Levant. Under the latter conditions, it would be optimal environmentally, and under the former, the relative cold might well have encouraged increased local habitation stability in caves protected from the inclement weather of the winter months.

In addition, this scenario seemingly can accommodate the postulated late arrival of Neanderthal 'influence' into the Levant. If only a matter of indirect gene flow, then it has no bearing on lithic assemblages, but if actual movements of Neanderthals were involved, their 'traditional' Mousterian patterns, at most, would merely have reinforced those tendencies already adaptive to occupational stability – provided, of course, that traditional Mousterian side-scrapers and points reflect relatively developed occupational intensity and, thus, occupational stability. If Bar Yosef is correct (Bar Yosef, in press; this volume), any such movement of Neanderthals into the Levant would have taken place during a relatively cold phase, which would be in accord with above scenario.

All of this, of course, is in direct opposition to the recent interpretation of the density of bone in Neanderthal legs as being a direct and clear indication of not only biologically determined almost continuous mobility but also, no less, of a relatively unpatterned mobility. If *Homo sapiens sapiens* can be linked to the Early Levantine Mousterian (which has yet to be done) its radiating settlement system should become acceptable to those who hold with this genetically compulsive wanderlust on the part of the Neanderthals but not so for the same settlement pattern for the Late Levantine Mousterian – or at least for those of its makers who were Neanderthals.

The situation in the Nile Valley is less complex but little more comprehensible. Problems abound which directly reflect mainly the paucity of data, rather than its interpretation (Vermeersch and Van Peer, this volume). Although a number of Mousterian industries have been recognized (perhaps more than need to be), none is dated with any security. At best, a general relative dating is possible which places the Khormusan as the youngest. Since a single ^{14}C date puts it at greater than 40 000 BP (Wendorf *et al.* 1979) and it alone of all the Mousterian industries occurs in Nile silts, the interpretation is obvious and tells us little that was not already known. Because of proximity to sub-Saharan Africa, the Nubian Middle Palaeolithic industries were potentially open to southern influences and, in fact, certain of the Nubian industries contain bifacial foliates (Chmielewski 1968; Guichard and Guichard 1968). It is certainly not clear whether these relate to the south or to the Aterian to the west or whether they were a local development out of the Late Acheulian. However, the presence of only one documented Aterian assemblage with a single clear pedunculate in the Nile Valley (Singleton and Close 1980) indicates that the Aterian certainly neither developed *in situ* in the Nile Valley out of the local Middle Palaeolithic nor exploited the Valley to any significant degree. Perhaps the only point of congruence between the

Aterian and the Nilotic Middle Palaeolithic rests in the extremely fine Levallois technique which characterizes both the late Aterian of the Western Desert (e.g. Hester and Hoebler 1970) and the Khormusan (Marks 1968b). Of course, this same characteristic typifies the Mousterian of Libya (McBurney 1967) and it may not have any generic significance.

In Nubia there is no Upper Palaeolithic in the sense used here, and whatever developments took place, if any, out of the Nubian Middle Palaeolithic can tell us nothing about the origins of any leptolithic tradition. In addition, the belief that there were no true Neanderthals in Africa (Stringer 1974), at any time, indicates that the problems of African cultural transformations are different than those of the Near East.

Further north in Egypt there is an early leptolithic Upper Palaeolithic and, so, the local Middle Palaeolithic is of some importance. Unfortunately, few such sites have been studied in central or Lower Egypt. To date, all appear to be extensive quarry/workshops with a marked paucity of retouched tools (Vermeersch and Van Peer, this volume). One can sympathize with Caton-Thompson who years ago characterized her Upper Levalloisian by an absence of characteristic elements (Caton-Thompson 1952). Yet, these sites have been defined technologically and, on this basis, there seem to be two kinds of assemblages, neither one of which seems to be (1) heading toward a leptolithic orientation, or (2) toward any technological connections with the Near East. In addition, these Middle Palaeolithic sites cannot be dated, but there does seem to be a marked hiatus between them and the earliest known Upper Palaeolithic. Thus, at the moment, it looks as if the Nile Valley played no role in the initial transition from the Middle to the Upper Palaeolithic which took place in the adjacent southern margins of the Levant (Vermeersch and Van Peer, this volume).

THE MIDDLE TO UPPER PALAEOLITHIC TRANSITION IN THE LEVANT

As noted in the introduction, some form of developmental transition between Middle and Upper Palaeolithic in the Levant has been recognized since before 1951. The original publications of Haller (1946) and Ewing (1947) of their Lebanese finds; the subsequent reevaluation by Garrod of her El Wad materials (Garrod 1951) resulting in the recognition of a transitional industry (the Emiran); its rejection on geological grounds (Bar Yosef and Vandermeersch 1972); and the transition's redefinition by Copeland (1970) based on the Ksar Akil sequence is well known, recently discussed in detail (Marks 1983b; Copeland 1986), and need not be repeated here. The same may be said for the discovery of the site of Boker Tachtit in the Central Negev, its dating, the extensive reconstruction of its cores and the resulting description and definition of the actual technological steps which led, in that case, from a Levallois-based technology to one with a consistent, single platform blade producing technology which I, at least, feel is non-Levallois (Marks 1983a, 1983b; Marks and Volkman

1983; Volkman 1983).

While the interpretations derived from the Boker Tachtit analyses have not been questioned – except for the Levallois *versus* non-Levallois conceptualizations (Copeland 1983) – there is still no real agreement on the significance of Boker Tachtit as a model for the Levantine Middle to Upper Palaeolithic transition. Clearly, it is accepted as one model, even more – a path actually taken but, perhaps, only one of many paths. Part of this uncertainty comes from the fact that Ksar Akil and Boker Tachtit are almost at opposite ends of the Levant, although the distance between them is less than 400 kilometres. Another problem comes from the somewhat different approaches used in the Boker Tachtit studies (Volkman 1983; Marks and Volkman 1983) and those traditional studies of the Ksar Akil material which seem to be seeing different things (Azoury 1986; Copeland 1975). Finally, the nature of the two sites is different. Boker Tachtit is an ephemerally reoccupied site, while during the transition Ksar Akil may have been an intensively occupied site or, as is almost as likely, an ephemerally occupied site which underwent very slow geological aggradation, resulting in dense artifact accumulations. Combined, these differences have led to some statements claiming that two different transitions took place in the Levant, one in the south and one in the north (Copeland 1986). This position, in fact, follows the opinion of Garrod (1951) who saw two facies in her Emiran, one Lebanese and one Palestinian.

To what extent is it possible to substantiate a single or multiple passage from the Middle to the Upper Palaeolithic within the Levant? Is there, as claimed, room enough in the Levant for everybody's interpretations? I think not, since a multitude or even two such paths implies at least two different, unrelated developments which produced essentially the same specific results in lithic technology. It seems unlikely, at best.

First, since there can be no direct correlation between the appearance in the Levant of *Homo sapiens sapiens* at 80 000 BP and the defined technological transition to a consistent leptolithic complex which took place at *c.* 45 000 BP, biocultural explanations are inappropriate. Thus, any explanatory model must relate to adaptive changes on the part of already existing *Homo sapiens sapiens*. Since the technological changes seen during the transition at Boker Tachtit can be documented, to a reasonable extent, as a culmination of some long term trends present only within the evolution of the Early Levantine Mousterian in the climatically marginal zones, it is only sensible to look for reasons for adaptive change within those areas and, particularly, in the Southern Levant.

The Middle Palaeolithic data from the north, whether it be from the temporal sequence at Tabun (Jelinek 1982b) or Ksar Akil, point to technological and even typological changes leading away from, not toward, a leptolithic technology and toward an intensification, not a reduction, in traditional Mousterian retouched tools (Marks and Volkman 1986). Thus, there is a prima facie case for an origin in the climatic marginal zones of the Levant, rather than in the core Mediterranean zone.

Based on the data from the Early Levantine Mousterian sites of the central Negev and Southern Jordan, it has been possible to present a model linking intensification of elongate blank production with increases in residential mobility, initially in the face of deteriorating climatic conditions, and then as an adaptive mechanism to increasing mobility during a period of renewed climatic amelioration which reopened low elevation areas, at least, to seasonal exploitation (Marks 1983; 1988). The emphasis here is on technological intensification of an already existing core reduction strategy and the elimination of less efficient core reduction strategies, particularly during the period of final Early Levantine Mousterian expansion into areas where raw materials sources were long forgotten and, therefore, unpredictable. This unpredictability was reinforced by the only marginal nature of the climatic amelioration which tied exploitation to a limited number of widely spaced sources of surface water. In many cases, these sources did not coincide with immediately available raw material, making increased efficiency in core reduction strategy adaptive.

Since the Early Levantine Mousterian of the Southern Levant had retained and intensified the production of elongated blanks from opposed platform and single platform cores during its period of limited elevational transhumance which was temporally coeval with the Late Levantine Mousterian in the north (the temporally late Late Mousterian, that is), the final change really involved nothing more than learning to produce, consistently, blanks of a desired size and shape. While the specialized opposed platform Levallois technique utilized during this period produced both elongated blades and Levallois points, the need to constantly rejuvenate the two platforms resulted in the rapid decrease in the length of the flaking surface (Volkman 1983). While the assemblage from Boker Tachtit, Level 2, clearly illustrates a number of attempted solutions to this problem, only with the adoption of a consistent single platform core reduction was it possible to produce both sizable elongated blanks *and* points which in all but method of manufacture look Levallois.

It is suggested that the environmental pressures in the climatically marginal zones of the Levant were missing from the core Mediterranean zone and that the long term technological trends toward blade production were also missing. Thus, if a separate transition took place in the north, it must have had a different genesis.

The argument in favour of a separate transition in the north seems to lie in two areas. The first is the assumption that stratigraphic sequences equal developmental sequences and that this is particularly true at Ksar Akil because of its deep and archaeologically rich deposits (Copeland 1970, 1986). Until the recent examination of the upper six excavation levels of Mousterian at Ksar Akil (Marks and Volkman 1986), a claim for continuity was, perhaps, valid, although by no means was that interpretation widely held. In fact, Azoury (1986: 92, Note 2: 234) who studied the transitional levels felt that there was no transition from the underlying Mousterian at Ksar Akil.

Second, in comparison with the tool assemblages of Boker Tachtit, the

Ksar Akil early transitional levels looked somewhat different (Marks 1983b). Not only did they contain a high percentage of chamfered pieces (Newcomer 1975) but they also exhibited some tendency toward carinated tool forms which was totally missing from Boker Tachtit (Marks and Ferring 1987). At first, I felt that this was sufficient evidence to postulate two different transitions (Marks 1983b; 1985: 134). Upon reflection, I no longer believe that is true. Not only reflection is involved, however. Two recent studies of the Ksar Akil transitional levels of 23 through 21 clearly document, better than I was able to do in a few days with the London Institute samples, the identity in technology between those levels and the technology of Boker Tachtit, Level 4 (Ohnuma 1986; Marks and Ferring 1987). What technological differences exist may be explained in the obviously more intensive core reduction at Ksar Akil as compared with Boker Tachtit. Since two different people did these studies on two different samples, the robustness of the results are impressive. At this point, I simply believe that two different, essentially contemporaneous sites with the same specific technology cannot represent simple chance convergence along two unrelated vectors of change.

There are still the typological differences to address. The Ksar Akil tool samples may be seen to represent the end products of relatively intensive activities, as compared with those at Boker Tachtit. In addition, large amounts of sizable nodules of fine flint were immediately available at Boker Tachtit, which was not the case at Ksar Akil. The two factors combined easily can account for the higher percentage of tools in the Ksar Akil assemblages (even assuming the discard of much debitage during excavations at Ksar Akil), since the transitional levels have considerably higher percentages of tools than do the underlying Mousterian levels, and there are the numbers of heavily retouched tools which are thick, relative to their length. Also, the high percentages of chamfered pieces indicate extensive tool rejuvenation. It appears that the chamfering blow is merely a way of resharpening the working edge of a scraper. Such blows were applied to scraper edges and scraper retouch was applied to chamfered edges to rejuvenate them (Newcomer 1970). In fact, when chamfered pieces disappear between levels 22 and 21, they are replaced almost exactly in level 21 by end-scrapers (Ohnuma 1986; Bergman and Ohnuma, this volume).

A recent but, as yet, unpublished study of a small sample of chamfered pieces and end-scrapers from the transitional levels at Ksar Akil indicate that all were used in hide processing, although some suggest dry and some wet hide working (C. Christopher, personal communication). In that regard, therefore, the transitional occupations at Ksar Akil may be viewed, all things considered, as a locus for processing hides and the subsidiary camp activities which went with it. Boker Tachtit, Level 4, on the other hand, is most reasonably interpreted as an ephemeral hunting camp, not necessarily specialized in the production of points, as may be the case for Level 1, but lacking the intensity of processing activities seen at Ksar Akil.

Therefore, on balance, I believe that there was only a single technological and adaptive transition out of the temporally late Early Levantine Mousterian and that the early transitional levels at Ksar Akil represent the stage in this transition seen at Boker Tachtit, Level 4. At Ksar Akil, however, not only is the context of occupation different but that part of the transition is seen throughout a considerable stratigraphic span, unlike at Boker Tachtit. Therefore, the Ksar Akil sequence, with all its problems, provides the best view we have of what I would consider Initial Upper Palaeolithic. After all, in any continuum the breaks are arbitrary and imposed by those studying it. Where better to begin the Upper Palaeolithic than with a technology where virtually all core reduction is pointed toward blade production? The Levallois-like points of Boker Tachtit, Level 4 and Ksar Akil, 25-21, are really no more than pointed, symmetrically converging blades. Given the early popularity of Upper Palaeolithic type tools in the southern Early Levantine Mousterian, not to mention in the Amúdian, it is better not to use them as a deciding criterion in any definition. Perhaps, the disappearance of typical Middle Palaeolithic retouched tools is more significant but, then, one yet may be forced to call a technologically Levallois-based assemblage Upper Palaeolithic. For the Levant, at least, it must be recognized that Upper Palaeolithic tools and consistent blade production did not evolve hand in hand.

Although it has been possible to document the technological change between the Middle and Upper Palaeolithic, unfortunately it tells us little about cultural transformations. In fact, given the long term technological trends which culminated in the Boker Tachtit, Level 4, technology, it is not reasonable to suggest that it reflects any significant cultural transformation at that time.

Using a different kind of data base, there is some indication of behavioural changes which may parallel the technological ones. Utilizing information from intra-site spatial analyses, it appears that during the Early Levantine Mousterian (at the site of Rosh Ein Mor and still present in the basal level of Boker Tachtit), there was marked spatial differentiation between certain groups of tool types, suggestive of clearly different areas for specific functions (Hietala and Stevens 1977; Hietala and Marks 1981). In addition, these clusters seem to be quite distinct, with little or no exchange of artifacts between them (Hietala 1983). In the uppermost level of Boker Tachtit, however, two quite distinct concentrations show considerable evidence for a two way exchange of both finished tools and cores (Hietala 1983). While the data are very limited and, at best, suggestive, it is possible that this apparent change in the use of space related either to a shift from highly specific activity loci to more general activities within delimited areas, or, more generally, to an increase in group interaction in the form of artifact sharing. A great deal more information will be needed before these observations, much less their interpretations, can be confirmed.

THE UPPER PALAEOLITHIC

Since it is held that there was only a single transition, then the early Levantine Upper Palaeolithic evolved out of the Initial Upper Palaeolithic of Boker Tachtit, Level 4, and Ksar Akil, Levels 25-21. Having said so, however, there is very little direct evidence to monitor the direction and effects of Upper Palaeolithic status. Only two sites appear to fall into this period, Boker D and Ksar Akil, Levels 20-18. Since Boker D was almost totally destroyed by fluviatile action (Marks 1983a), it provides little more than a small artifact assemblage and an amorphous burned area (Marks and Kaufman 1983). The artifacts are consistent with Boker Tachtit, Level 4, but the converging blades (Levallois-like points) or just plain blades with converging lateral edges, are missing. Yet, the small size of the sample precludes any firm interpretations.

The situation is somewhat different at Ksar Akil, where these levels are about one metre in thickness and contained a good density of artifacts. On the most general level, it appears that the development is one of increasing control in blade production, a shift to opposed platform core reduction, a shift from chamfered pieces to end-scrapers, and a growth in the tendency to back blades (Bergman and Ohnuma, this volume). None of this tells us very much except that there was continued adjustment in core reduction strategies away from earlier, Levallois-like blank production.

It is only with the appearance of Level 17 at Ksar Akil (Bergman 1987) and Boker A in the Central Negev (Jones *et al.* 1983) that we do have some reasonable information. A *Homo sapiens sapiens* was recovered from Level 17 (Bergman 1987), a not unexpected association. Both assemblages represent, perhaps, the finest blade/bladelet production during the Upper Palaeolithic in the Levant. Yet, based on ^{14}C dates, it is likely that these assemblages date to greater than 35 000 BC (Marks 1981b). Technologically, they are clearly within the Ahmarian industry, although at Ksar Akil scrapers are common and burins very rare (Bergman 1987), while at Boker A, the opposite is true (Jones *et al.* 1983). This is the first of numerous examples of extreme tool kit variability in the Ahmarian industry and strongly suggests that the proportional occurrence of tool classes in any given assemblage is probably of little significance beyond immediate functional considerations. The use of this kind of variability as a criterion for industry designation (Gilead 1981), something which is still current, seems ill-conceived. Too much depends upon spatial sampling and site location. For instance, at Boker B, Level III, a typical Ahmarian assemblage rich in blade and bladelet tools, el Wad points account for 14% of the tool assemblage over a 55 m^2 excavation. On the other hand, of the 74 el Wad points recovered, over 50 of them came from a single 5 m^2 area where they represent 80% of the tools (Marks 1976b). If this area had not been exposed, the proportions of blade/bladelet tools relative to scrapers and burins would have changed drastically. Comparable intersite retouched tool variability can be seen in the Ahmarian sites near Lagama in

Sinai (Bar Yosef and Phillips 1977). Therefore, it seems safer and more realistic to rely on technological criteria in assemblage groupings.

The Ahmarian, aside from a magnificent blade technology, shows few attributes traditionally associated with the Upper Palaeolithic. There is no art and only a hint of simple bone tool manufacture (Newcomer 1987; Newcomer and Watson 1984). Burials are unknown (unless the Ksar Akil 17 skeleton was a burial) and the few sites with faunal remains suggest no evidence for any specialized exploitation. All sites are small and rather ephemeral, which is consistent with all open Upper Palaeolithic sites in the Levant. The larger open sites, when examined closely, almost always show evidence for repeated occupations, rather than single occupations by large residential units: e.g. Ein Aqev East (Ferring 1977). These observations are valid not only for the early Ahmarian but for those Ahmarian assemblages dating to close to 20 000 BC, as well.

One of the more pressing problems in early Upper Palaeolithic studies in the Levant is the appearance at about 29 000 BP of assemblages which are very distinct from the Ahmarian and fall into what has traditionally been called the Levantine Aurignacian (Garrod and Bate 1937; Garrod 1954). Technologically, they are heavily oriented toward flake production and tools are mainly heavily retouched scrapers, often carinated or laterally carinated. Associated with these are some bladelet tools but these are mainly produced on twisted bladelets which are a by-product of carinated tool manufacture. It is associated with this lithic industry that bone tools occur in small but reasonable numbers (as at Ksar Akil, Levels 7 and 8; Newcomer 1987) – and, for the first time, where evidence is seen for personal ornamentation in the form of beads, perforated bones, etc. (Belfer-Cohen and Bar Yosef 1982). Although never as rich as in some European industries, they are sufficiently numerous that they must be accepted as a usual part of the industry. On the other hand, art seems absent and burials are unknown.

A third group of assemblages occurs in the caves of the central and Northern Levant – as for example at Ksar Akil. These seem to contain elements of both the Ahmarian and the Levantine Aurignacian (Ronen 1976; Bergman 1987). Unfortunately, all come from old excavations where stratigraphic units were gross, at best. So, the chance that they represent mechanical mixing is there, but they occur too many times to really believe in such mixing. Although there is no sure answer to this problem, it is possible that they are, in fact, Ahmarian sites evidencing a somewhat greater residential stability and occupational intensity than is seen at most open sites in the south. A single Negev site, Sde Divshon (Ferring 1976), shows a similar pattern, accentuated by slope wash which removed many of the smaller bladelets normally associated with the Ahmarian. Because of this, some have called it Levantine Aurignacian (Gilead 1981), while others refer it to Ahmarian (Marks 1981b).

Even with these seemingly intermediate assemblages, it seems unlikely that the Ahmarian and the Levantine Aurignacian are merely functional facies of the same general industry. Given the markedly distinct technology of each after c. 30 000 BP, they probably should be considered as separate entities. It is likely, however, that the Levantine Aurignacian developed out of the early Ahmarian or, perhaps, even out of the Initial Upper Palaeolithic. If the latter, one might view these assemblages with so-called mixed attributes as a form of Proto-Levantine Aurignacian. At Ksar Akil, at least, they fall stratigraphically between the clear Ahmarian of Level 17, and probably 16, and the equally clear Levantine Aurignacian of Levels 8 and 7 (Bergman 1987). While hardly proven, it is a construct worth consideration.

It is not being suggested that the difference between the Ahmarian and the Levantine Aurignacian can be explained solely in terms of differential residential mobility but that relatively different settlement patterns helped to establish variable patterns of blank production (elongated blade *versus* rather thick flake) which directly affected the final form of the tools. Once the lithic reduction patterns were adopted they remained even during situations where Levantine Aurignacian groups indulged in mobile settlement strategies or where Ahmarian groups showed more residential stability than was usual. After all, the time involved was not great and the differences in settlement patterning were probably relatively minor, as compared with what happened during the Middle Palaeolithic.

CONCLUSIONS

It may be said fairly that this is not the time for presenting definitive conclusions about almost any aspect of Near Eastern Upper Pleistocene studies. Everything is in flux. However, it is vital to point out to our European colleagues that the temporally comparable prehistory of Europe and the Near East are not the same. The association between the appearance of *Homo sapiens sapiens* and 'fully cultural behaviour' is not valid in the Near East and, therefore, is probably not valid anywhere else. Those attributes which are traditionally posited to separate the Neanderthals from us, place not only the early *Homo sapiens sapiens* of the Levant with the Neanderthals but also virtually *all* those folk who produced a true leptolithic tradition in the Near East from 47 000 BC until *c.* 30 000 BC.

If there was, in the general sense, a moment of 'becoming human', which I personally doubt, the line must be drawn well after the biological evolution of modern man. In fact, the archaeological data from the Old World suggests that we are probably not able to recognize the transition to such 'fully cultural behaviour' in those areas where preservation is poor. Even where preservation is good, such a criterion as a shift to specialized hunting of a limited range of species is tied to very special ecological conditions which do not seem to have ever existed in the Near East and, therefore, is useless for global generalities.

Much work still needs to be done in the Near East. There is still the question of the inland Near East and the origin of the Baradostian and its possible relationship to the Shanidar Mousterian. From where did the Early Upper Palaeolithic of central Egypt come, and is it connected to the early leptolithic development in Libya, the Dabban? We are far from having all the answers and even those things which we thought we knew only a short time ago, are no longer so clear.

REFERENCES

Alimen, H. 1957. *The Prehistory of Africa*. London: Hutchinson Scientific and Technical.

Azoury, I. 1971. *A Technological and Typological Analysis of the Transitional and Early Upper Paleolithic levels of Ksar Akil and Abu Halka*. Unpublished Ph.D. Dissertation, Institute of Archaeology, University of London.

Azoury, I. 1986. *Ksar Akil, Lebanon: a Technological and Typological Analysis of the Transitional and Early Upper Paleolithic Levels of Ksar Akil and Abu Halka*. Oxford: British Archaeological Reports International Series S289.

Bar Yosef, O. (in press). Upper Pleistocene Human Adaptations in Southwestern Asia. In E. Trinkaus (ed.) *Corridors, Cul-de-Sacs and Coalescence: the Biocultural Foundations of Modern People*. Cambridge: Cambridge University Press. In Press.

Bar Yosef, O. and Phillips, J. 1977. *Prehistoric Investigations in Gebel Maghara, Northern Sinai*. Qedem 7. Jerusalem: Institute of Archaeology.

Bar Yosef, O. and Vandermeersch, B. 1972. The stratigraphical and cultural problems of the passage from Middle to Upper Paleolithic in Palestinian caves. In F. Bordes (ed.) *The Origin of Homo Sapiens*. Paris: UNESCO: 221-226.

Bar Yosef, O. and Vandermeersch, B. 1981. Notes Concerning the possible age of the Mousterian layers in Qafzeh Cave. In P. Sanlaville and J. Cauvin (eds) *Préhistoire du Levant*. Paris: Centre National de la Recherche Scientifique: 281-286.

Bar Yosef, O., Vandermeersch, B., Goldberg, P., Laville, H., Meignen, L., Rak, Y., Tchernov E. and Tillier, A.M. 1986. New data on the origin of Modern Man in the Levant. *Current Anthropology* 27: 63-4.

Belfer-Cohen, A. and Bar Yosef, O. 1982. The Aurignacian at Hayonim Cave. *Paléorient* 7:19-42.

Bergman, C. 1987. *Ksar Akil, Lebanon: a Technological and Typological Analysis of the Later Paleolithic Levels of Ksar Akil*. Vol. 2: *Levels XIII-VI*. Oxford: British Archaeological Reports International Series S329.

Bordes, F. 1953. Levalloisien et Moustérien. *Bulletin de la Société Préhistorique Française* 50: 226-35.

Butzer, K. and Hansen, C. 1968. *Desert and River in Nubia*. Madison: University of Wisconsin Press.

Buzy, D. 1929. Une station Magdalénienne dans le Négeb (Ain el-Qedeirat). *Revue Biblique* 38: 364-381.

Caton-Thompson, G. 1946. The Levalloisian of Egypt. *Proceedings of the Prehistoric Society* 12: 57-120.

Caton-Thompson, G. 1952. *Kharga Oasis in Prehistory*. London: Athlone Press.

Chmielewski, V. 1968. Early and Middle Paleolithic Sites near Arkin, Sudan. In F. Wendorf (ed.) *The Prehistory of Nubia*: Vol. 1. Dallas: Fort Burgwin Research Center and Southern Methodist University Press: 110-147.

Clark, J.D. 1981. New men, strange faces, other minds. *Proceedings of the British Academy* 67: 163-192.

Coinman, N., Clark, G. and Lindly, J. 1986. Prehistoric hunter-gatherer settlement in the Wadi Hasa, west-central Jordan. In L. Straus (ed.) *The End of the Paleolithic in the Old World*. Oxford: British Archaeological Reports International Series S284: 129-170.

Copeland L. 1970. The Early Upper Paleolithic flint material from levels VII-V, Antelias Cave, Lebanon. *Berytus* 19: 99-143.

Copeland L. 1975. The Middle and Upper Paleolithic of Lebanon and Syria in the light of recent research. In F. Wendorf and A. Marks (eds) *Problems in Prehistory: North Africa and the Levant*. Dallas: Southern Methodist University Press: 317-350.

Copeland L. 1981. Chronology and distribution of the Middle Paleolithic as known in 1980, in Lebanon and Syria. In P. Sanlaville and J. Cauvin (eds) *Préhistoire du Levant*. Paris: Centre National de la Recherche Scientifique: 239-264.

Copeland L. 1983. Levallois/non-Levallois determinations in the early Levant Mousterian: problems and questions for 1983. *Paléorient* 9:15-28.

Copeland L. 1986. Introduction to volume 1. In I. Azoury *Ksar Akil, Lebanon: a Technological and Typological Analysis of the Transitional and Early Upper Paleolithic Levels of Ksar Akil and Abu Halka*. Oxford: British Archaeological Reports International Series S289. 1-19.

de Heinzelin, J. 1968. Geological History of the Nile Valley in Nubia. In F. Wendorf (ed.) *The Prehistory of Nubia*: Vol. 1. Dallas: Fort Burgwin Research Center and Southern Methodist University Press: 19-55.

Ewing, J.F. 1947. Preliminary note on the excavation at the Paleolithic site of Ksar Akil, Republic of Lebanon. *Antiquity* 21: 186-196.

Farrand, W. 1982. Environmental conditions during the Lower/ Middle Paleolithic transition in the Near East and the Bankins. In A. Ronen (ed.) *The Transition from the Lower to Middle Paleolithic and the Origins of Modern Man*. Oxford: British Archaeological Reports International Series 151: 105-112.

Ferring, C. 1976. Sde Divshon: an Upper Paleolithic site on the Divshon Plain. In A. Marks (ed.) *Prehistory and Paleoenvironments in the Central Negev, Israel*. Vol. 1: *The Avdat/Aqev Area (Part 1)*. Dallas: Southern Methodist University Press: 199-226.

Ferring, C. 1977. The Late Upper Paleolithic site of Ein Aqev East. In A. Marks (ed.) *Prehistory and Paleoenvironments in the Central Negev, Israel*. Vol. 2: *The Avdat/Agev Area (Part 2) and the Har Harif*. Dallas: Department of Anthropology, Southern Methodist University: 81-118.

Garrod, D.A.E. 1951. A transitional industry from the base of the Upper Paleolithic in Palestine and Syria. *Journal of the Royal Anthropological Institute* 81: 121-129.

Garrod, D.A.E. 1955. The Mugharet el-Emireh in lower Galilee: type station of the Emiran industry. *Journal of the Royal Anthropological Institute* 85: 141-162.

Garrod, D.A.E. 1962. An outline of Pleistocene prehistory in Palestine-Lebanon-Syria. *Quartinaria* 6: 541-546.

Garrod, D.A.E. and Bate, D.M.A. 1937. *The Stone Age of Mount Carmel:* Vol. 1. Oxford: Clarendon Press.

Gilead, I. 1981. Upper Paleolithic tool assemblages from the Negev and Sinai. In P. Sanlaville and J. Cauvin (eds) *Préhistoire du Levant*. Paris: Centre National de la Recherche Scientifique: 331-342.

Gilead, I. 1984. Paleolithic sites in northeastern Sinai. *Paléorient* 10: 135-142.

Guichard, J. and Guichard, G. 1968. Contributions to the study of the Early and Middle Paleolithic of Nubia. In F. Wendorf (ed.) *The Prehistory of Nubia*: Vol. 1. Dallas: Fort Burgwin Research Center and Southern Methodist University Press: 148-193.

Haller, J. 1946. Notes de préhistoire phenicienne: l'Abri d'Abou Halka (Tripoli). *Bulletin du Musée de Beyrouth* 6: 1-20.

Henry, D.O. and Servello, A.F. 1974. Compendium of C-14 determinations derived from Near Eastern prehistoric sites. *Paléorient* 2: 19-44.

Henry, D.O. 1982. The prehistory of Southern Jordan and relationships with the Levant. *Journal of Field Archaeology* 9: 417-444.

Hester, J. and Hoebler, P. 1970. *Prehistoric Settlement Patterns in the Libyan Desert.* University of Utah Papers in Anthropology 92 (Nubia Series 4). Salt Lake City: University of Utah Press.

Hietala, H. 1983. Boker Tachtit: intralevel and interlevel spatial analysis. In A. Marks (ed.) *Prehistory and Paleoenvironments in the Central Negev, Israel.* Vol. 3: *The Avdat/Aqev Area (Part 3)*. Dallas: Southern Methodist University Press: 217-282.

Hietala, H. and Stevens, D. 1977. Spatial analysis: multiple procedures in pattern recognition studies. *American Antiquity* 42: 535-559.

Hietala, H. and Marks, A. 1981. Changes in spatial organization at the Middle to Upper Paleolithic site of Boker Tachtit, Central Negev. In P. Sanlaville and J. Chauvin (eds) *Préhistoire du Levant.* Paris: Centre National de la Recherche Scientifique: 305-318.

Hole, F. and Flannery, K. 1967. The prehistory of southwestern Iran: a preliminary report. *Proceedings of the Prehistoric Society* 33: 147-170.

Horowitz, A. 1979. *The Quaternary of Israel.* New York: Academic Press.

Hours, F. 1974. Remarques sur l'utilisation des listes-types pour l'étude du Paléolithique Supérieur et de l'Epipaleolithique du Levant. *Paléorient* 2: 3-18.

Hours, F. 1982. Une nouvelle industrie en Syrie entre l'Acheuléen supérieur et le Levalloiso-Mousterien. In J. Starcky and F. Hours (eds) *Archéologie du Levant.* Lyon: Maison de l'Orient Méditerranéen: 33-46.

Huzayyin, S. 1941. *The Place of Egypt in Prehistory.* Cairo: Mémoires de l'Institut d'Egypte 49.

Jelinek, A.J. 1981a. The Middle Paleolithic in the Southern Levant from the perspective of the Tabun cave. In P. Sanlaville and J. Cauvin (eds) *Préhistoire du Levant.* Paris: Centre National de la Recherche Scientifique: 265-285.

Jelinek, A.J. 1981b. The Middle Paleolithic of the Levant: Synthesis. In P. Sanlaville and J. Cauvin (eds) *Préhistoire du Levant.* Paris: Centre National de la Recherche Scientifique: 299-302.

Jelinek, A.J. 1982a. The Middle Paleolithic in the Southern Levant, with comments on the appearance of modern *Homo Sapiens.* In A. Ronen (ed.) *The Transition from Lower to Middle Paleolithic and the Origin of Modern Man.* Oxford: British Archaeological Reports International Series S151: 57-104.

Jelinek, A.J. 1982b. The Tabun Cave and Paleolithic Man in the Levant. *Science* 216: 1369-1375.

Jones, M., Marks, A. and Kaufman, D. 1983. Boker: the artifacts. In A. Marks (ed.) *Prehistory and Paleoenvironments in the Central Negev, Israel.* Vol. 3: *The Avdat/Aqev Area (Part 3)*. Dallas: Southern Methodist University Press: 283-332.

Lindly, J. 1986. A preliminary lithic analysis of the Mousterian site 634 from west-central Jordan. Paper presented at the Annual Meeting of the Society of American Archaeology, New Orleans (La), 1986.

McBurney, C. 1967. *The Haua Fteah (Cyrenaica) and the Stone Age of the Southeast Mediterranean.* Cambridge: Cambridge University Press.

McCown, T. and A. Keith 1939. *The Stone Age of Mount Carmel:* Vol. 2. Oxford: Oxford University Press.

Marks, A. 1968a. The Mousterian industries of Nubia. In F. Wendorf (ed.) *The Prehistory of Nubia*: Vol. 1. Dallas: Fort Burgwin Research Center and Southern Methodist University Press: 194-314.

Marks, A. 1968b. The Khormusan: an Upper Pleistocene industry in Sudanese Nubia. In F. Wendorf (ed.) *The Prehistory of Nubia:* Vol. 1. Dallas: Fort Burgwin Research Center and Southern Methodist University Press: 315-391.

Marks, A. 1968c. The Halfan industry. In F. Wendorf (ed.) *The Prehistory of Nubia.* Dallas: Fort Burgwin Research Center and Southern Methodist University Press: 392-460.

Marks, A. 1976a. (ed.) *Prehistory and Paleoenvironments in the central Negev, Israel.* Vol. 1: *The Avdat/Aqev area (Part 1).* Dallas: Southern Methodist University Press.

Marks, A. 1976b. Terminology and chronology of the Levantine Upper Paleolithic as seen from the Central Negev, Israel. In F. Wendorf (ed.) *Deuxième Colloque sur la Terminologie de la Préhistoire du Proche-Orient.* UISPP, Nice: 49-76.

Marks, A. 1977. (ed.) *Prehistory and Paleoenvironments in the central Negev, Israel.* Vol. 2: *The Avdat/Aqev Area (Part 2) and the Har Harif.* Dallas: Southern Methodist University Press.

Marks, A. 1981a. The Middle Paleolithic of the Negev. In P. Sanlaville and J. Cauvin (eds) *Préhistoire du Levant.* Paris: Centre National de la Recherche Scientifique: 287-98.

Marks, A. 1981b. The Upper Paleolithic of the Negev. In P. Sanlaville and J. Cauvin (eds) *Préhistoire du Levant.* Paris: Centre National de la Recherche Scientifique: 299-304.

Marks, A. 1983a. (ed.) *Prehistory and Paleoenvironments in the central Negev, Israel.* Vol. 3: *The Avdat/Aqev Area (Part 3).* Dallas: Southern Methodist University Press.

Marks, A. 1983b. The Middle to Upper Paleolithic transition in the Levant. In F. Wendorf and A.E. Close (eds) *Advances in World Archaeology* Vol. 2. New York: Academic Press: 51-98.

Marks, A. 1985. The Levantine Middle to Upper Paleolithic transition: the past and present. In M. Liverani, A. Peroni and R. Peroni (eds) *Studi di Paletnologia in onore di Salvatore M. Puglisi.* Rome: Universita di Roma, La Sapienza. Departimento di Scienze, Storiche, Archeologiche e Antropologiche dell' Antichita: 123-136.

Marks, A. 1988. The Middle to Upper Paleolithic transition in the Southern Levant: technological change as an adaptation to increasing mobility. In M. Otte (ed.) *L'Homme du Neandertal.* Vol. 8: *La Mutation.* Liège: Etudes et Recherches Archéologiques de l'Université de Liège (ERAUL 35): 109-124.

Marks, A. and Ferring, R. 1987. The Early Upper Paleolithic of the Levant. Paper presented at the Annual Meeting of the Society for American Archaeology, Toronto, 1987.

Marks, A. and Kaufman, D. 1983. Boker Tachtit: the artifacts. In A. Marks (ed.) *Prehistory and Paleoenvironments in the Central Negev, Israel.* Vol 3: *The Avdat/Aqev Area.* Dallas: Southern Methodist

University Press: 69-126.

Marks, A. and Volkman, P. 1983. Changing core reduction strategies: a technological shift from the Middle to the Upper Paleolithic in the Southern Levant. In E. Trinkaus (ed.) *The Mousterian Legacy: Human Biocultural Change in the Upper Pleistocene*. Oxford: British Archaeological Reports International Series S164: 13-33.

Marks, A. and Volkman, P. 1986. The Mousterian of Ksar Akil: levels XXVIA through XXVIIIB. *Paléorient* 12: 5-20.

Meignen, L. and Bar Yosef, A. 1988. Kébara et le Paléolithique Moyen du Mont Carmel (Israël). *Paléorient* 14: 123-129.

Neuville, R. 1951. *Le Paléolithique et le Mésolithique du Desert de Judée*. Paris: Archives de l'Institut de Paléontologie Humaine, Mémoire 24.

Newcomer, M.H. 1970. The chamfered pieces from Ksar Akil (Lebanon). *Bulletin of the Institute of Archaeology (University of London)* 8-9: 177-191.

Newcomer, M.H. 1987. Study and replication of bone tools from Ksar Akil, Republic of Lebanon. Appendix 3. In C. Bergman *Ksar Akil, Lebanon: A Technological and Typological Analysis of the Later Paleolithic Levels of Ksar Akil*. Vol. 2: *Levels XIII-VI*. Oxford: British Archaeological Reports International Series S329: 284-307.

Newcomer, M. and Watson, J. 1984. Bone Artifacts from Ksar Aqil (Lebanon). *Paléorient* 10: 143-148.

Oakley, K.P. 1964. *Frameworks for Dating Fossil Man*. London: Weidenfeld and Nicolson.

Ohnuma, K. 1986. *A Technological Study of the Upper Paleolithic Material from Levels XXV-XIV from Ksar Akil*. Unpublished Ph.D. Thesis, University of London.

Ronen, A. 1976. The Upper Paleolithic in northern Israel: Mt. Carmel and Galilee. Colloque 3. *Deuxième Colloque sur la Terminologie de la Préhistoire du Proche-Orient*. UISPP, Nice.

Ronen, A. 1979. Paleolithic industries: Upper Acheulian, Early Paleolithic flake industries, Middle Paleolithic. In A. Horowitz (ed.) *The Quaternary of Israel*. New York: Academic Press: 300-305.

Said, R. 1983. Proposed classification of the Quaternary of Egypt. *Journal of African Earth Sciences* 1: 41-5.

Sanlaville P. 1981. Stratigraphie et chronologie du Quaternaire du Levant. In P. Sanlaville and J. Cauvin (eds) *Préhistoire du Levant*. Paris: Centre National de la Recherche Scientifique: 21-32.

Schild, R., Chmielewska, M. and Wieckowska, H. 1968. The Arkinian and Shamarkian industries. In F. Wendorf (ed.) *The Prehistory of Nubia*: Vol. 2. Dallas: Fort Burgwin Research Center and Southern Methodist University Press: 651-767.

Schwarcz, H., Blackwell, B., Goldberg, P. and Marks, A. 1979. Uranium series dating of travertine from archaeological sites, Nahal Zin, Israel. *Nature* 277: 558-60.

Singleton, W. and Close, A. 1980. Report on Site E-78-11. In A. Close (ed.) *Loaves and Fishes: the Prehistory of Wadi Kubbaniya*. Dallas: Department of Anthropology, Southern Methodist University: 229-238.

Stringer, C. 1974. Population relationships of later Pleistocene hominids: a multivariate study of available crania *Journal of Archaeological Science* 1: 317-342.

Stringer, C., Grün, R., Schwartz, H.P. and Goldberg, P. 1989. ESR dates for the hominid burial site of Es Skhúl in Israel. *Nature* 338: 756-758.

Tchernov, E. 1981. The biostratigraphy of the Middle East. In P.

Sanlaville and J. Cauvin (eds) *Préhistorie du Levant*. Paris: Centre National de la Recherche Scientifique: 67-98.

Trinkaus, E. 1982. Neanderthal postcrania and the adaptive shift to Modern Man. In E. Trinkaus (ed.) *The Mousterian Legacy: Human Biocultural Change in the Upper Pleistocene*. Oxford: British Archaeological Reports International Series S164: 165-200.

Vandermeersch, B. 1981. *Les Hommes Fossiles de Qafzeh (Israel)*. Paris: Centre National de la Recherche Scientifique.

Vandermeersch, B. 1982. The first *Homo sapiens sapiens* in the Near East. In A. Ronen (ed.) *The Transition from the Lower to the Middle Paleolithic and the Origin of Modern Man*. Oxford: British Archaeological Report International Series S151.

Volkman, P. 1983. Boker Tachtit: core reconstructions. In A. Marks (ed.) *Prehistory and Paleoenvironments in the Central Negev, Israel*. Vol. 3: *The Avdat/Agev Area (Part 3)*. Dallas: Southern Methodist University Press: 127-190.

Volkman, P. and Kaufman, D. 1982. A reassessment of the Emireh Point as a possible type fossil for the technological shift from the Middle to the Upper Paleolithic in Levant. In E. Trinkaus (ed.) *Human Biocultural Change in the Upper Pleistocene*. Oxford: British Archaeological Reports International Series S164: 35-52.

Wendorf, F. and Schild, R. 1976. *Prehistory of the Nile Valley*. New York: Academic Press.

Wendorf, F., Schild, R. and Hass, H. 1979. A new radiocarbon chronology for prehistoric sites in Nubia. *Journal of Field Archaeology* 6: 219-223.

White, R. 1982. Rethinking the Middle/Upper Paleolithic transition. *Current Anthropology* 23: 169-176.

4: The Amudian in the Context of the Mugharan Tradition at the Tabun Cave (Mount Carmel), Israel

ARTHUR J. JELINEK

BACKGROUND: THE DISCOVERY AND DESCRIPTION OF THE AMUDIAN

The earliest appearance in the Levant of industries whose typological and technological composition includes substantial numbers of prismatic blades and tools made on blades has been assigned to the early Late Pleistocene at the sites of Abri Zumoffen, Jabrud I, Tabun, and Zuttiyeh. A presumed association of blades and blade tools with bifaces and Mousterian-type scrapers was first mentioned by Turville-Petre (1927: 22) in his description of the collection of artifacts that he made from the deposits at the Zuttiyeh Cave near the Sea of Galilee. The discovery of human remains (the 'Galilee Skull') in association with Middle Palaeolithic artifacts in the basal levels of the site attracted the attention of prehistorians to the region and stimulated further work in the archaeology of this period in the Levant.

Within a few years after Turville-Petre's discoveries at Zuttiyeh, Professor D.A.E. Garrod initiated her excavations in the Wadi el-Mughara on the west edge of Mount Carmel and Alfred Rust began his work in the Jabrud rock-shelters in the Wadi Skifta, north of Damascus. By 1933 Rust had isolated and briefly defined an industry characterized primarily by end-scrapers and burins from Levels 15 and 13 at the Jabrud I shelter that he referred to as "Prä-Antelian (Prä-Aurignacien)" (Rust 1933: 211). These levels were described in this first report as interstratified between industries dominated by transverse and multiple-edged side-scrapers, to which he gave the name "Jabrudien" *op cit:* 209). Deeper in the section this Jabrudian industry was found interstratified with "Acheuléen" industries characterized by abundant bifaces (Note 1).

Soon afterwards Garrod (in Garrod and Bate 1937: 67) described the occurrence of considerable numbers of characteristic Upper Palaeolithic tools in her Layer E at the Tabuṇ Cave in a context that she designated as "Acheulean (Micoquian)". She described the situation of these artifacts as follows: "At 5.50 m below datum (that is, at the base of Ea) Châtelperron points and other forms recalling the Lower Aurignacian of Europe began to occur in much greater numbers and continued to about 6.50 m below datum, after which they were found only sporadically and in decreasing numbers. They were associated with the hand-axes and side-scrapers which were normally characteristic of this level, but differed from them very markedly in technique."

Subsequent to these initial descriptions, both Rust and Garrod modified the terminology used to refer to these industries. Rust, in his final report on the Jabrud shelters, settled on the term "Prä-Aurignacien" for the Level 13 and 15 artifacts from Jabrud I (Rust 1950: 13). He accommodated to the presence of both bifaces and the Jabrudian scraper assemblage in some other levels by creating the term "Acheuleo-Jabrudien" (*op cit:* 16). In his final sequence, the Pre-Aurignacian levels were placed in the following succession:

Level	*Industry*
11	Acheuleo-Jabrudian
12	Final Acheulian (Pre-Mousterian)
13	Pre-Aurignacian
14	Late Jabrudian
15	Pre-Aurignacian
16	Jabrudian

In a detailed discussion of the context of the Layer E blade tools at Tabun, Garrod (1956) followed Rust's usage and employed the term "Pré-Aurignacien" for these artifacts. Inasmuch as a re-examination of the overall distribution of artifacts in Layer E at Tabun failed to distinguish the kind of segregation of levels rich in scrapers from those rich in hand-axes that characterized Rust's Jabrudian and Acheulian industries at Jabrud I, she referred to the industries of Layer E as "Acheuléo-Jabroudien". Subsequently, following a preliminary analysis of the artifacts from the Abri Zumoffen, near Adlun on the coast of southern Lebanon, Garrod noted that the early blade tool industries from Tabun, Zumoffen, and Zuttiyeh differed typologically from those at Jabrud I in the dominance of backed elements in the former as opposed to burins in the latter. She proposed the name "Amudian" for the industries dominated by backed elements, a name derived from the Wadi Amud in which Zuttiyeh (where the early blade industry had first been noted) is located (Garrod and Kirkbride 1961a: 11). This name is now widely accepted for the early blade industries of the Tabun Cave and the Abri Zumoffen.

In her last treatment of the relationship between the Amudian and the Pre-Aurignacian of Jabrud I, Garrod (1970) considered them to be essentially the same industry, with some local variation. She agreed with Rust's earlier views that this blade-using technology was the result of a distinct intrusive population. This is made explicit in her interpretation of the contexts that contained a mixture of Yabrudian and Amudian elements at Zumoffen: "I think there is therefore good evidence that the makers of the flake tools and the makers of the blade tools did from time to time come together in a common habitation, at least in the later stages of the Yabrudian (*sic*)" (*op cit*: 228) (Note 2).

The question of whether a distinct population was responsible for the Amudian and Pre-Aurignacian industries has frequently been linked to another fundamental problem. The similarity of the typological character of the Amudian artifacts to early industries of the classic Upper Palaeolithic of Western Europe (particularly the Châtelperronian) has been remarked upon from the time of their initial description. The possible relationship of the Amudian to those European industries has been treated from several perspectives. Although Garrod at one time saw the blade tools of Layer E as potentially antecedent to the Châtelperronian (Note 3), by the early 1950s she had considerably modified this view (Garrod 1953: 33). At issue in these and subsequent discussions (e.g. Bordes 1955, 1960; Garrod 1956) were two basic related questions. At the heart of the matter was the question of whether the typological resemblances of the Amudian and Pre-Aurignacian to the European industries were sufficient in and of themselves to demonstrate a generic link between the cultures of the two regions. In a more fundamental sense, this aspect was one of relative probabilities of single *versus* multiple origins of technological innovation. Related to the resolution of these arguments was the second question, which concerned the chronological relationships of the Near Eastern and European industries. This latter question seems now to have been resolved in a general sense; the Near Eastern industries appear to be significantly earlier than the earliest Upper Palaeolithic of Western Europe (see Jelinek 1982a). Radiocarbon dates from Unit I (Layer C) at Tabun show an age of about 50 000 years. The Unit XI (Layer Ea-Eb) deposits appear to be significantly earlier on the basis of the considerable depth of deposits and the complex series of events that separates them from Unit I. An age of 70-80 000 BP now seems reasonable for Unit XI. Thus the chronological evidence, as presently interpreted, suggests that it is extremely unlikely that any generic relationship existed between the Amudian of the Levant and the Upper Palaeolithic of Western Europe. Given this recent knowledge, our major concern with the Amudian might more appropriately relate to a better understanding of the appearance, context, and disappearance of this curious phenomenon, rather than to its improbable links to the industries associated with the first appearance of anatomically-modern *Homo sapiens sapiens* in Western Europe.

THE CONTEXT OF THE AMUDIAN INDUSTRIES AT TABUN

The excavations at the Tabun Cave from 1967 to 1972 (Jelinek *et al.* 1973; Jelinek 1975, 1981, 1982a, 1982b) recovered characteristic Amudian artifacts from levels corresponding to Garrod's Layers Ea and Eb. These later excavations emphasized a reconstruction of the detailed geological stratigraphy of the site and the segregation of artifacts by natural level.

A preliminary stratigraphic sequence, based on the profile visible at the beginning of excavation, served as a working system during the excavation and was used in some preliminary reports (e.g. Jelinek 1975). In this preliminary sequence, the Amudian elements were assigned to Bed 48. A complete revision of the stratigraphy was undertaken following the excavation. This work resulted in the separation of 14 major stratigraphic units which were separated from each other by obvious disconformities.

The numerous interruptions in the sequence undoubtedly resulted from intermittent periods of karstic collapse of sediments into chambers underlying the present outer and inner chambers of the cave. Each of the major stratigraphic units (I-XIV) contains a number of geological beds (equivalent to the French *couches*), which are designated in a continuous sequence beginning with Bed 1 at the top of Unit I.

Following an examination of back-plots (at 20 cm intervals in both cardinal directions) of the position and concentration of artifacts in the sequence and an examination of the excavation records and profiles (graphically and photographically documented at one-metre intervals), a number of levels of artifact deposition were isolated in each bed. In general, these levels were divided in more detail when the beds were thick and when artifacts were abundant. Major divisions of the beds into 'superior' (S) and 'inferior' (I) levels were carried across the entire excavated area, while more detailed subdivisions (designated by numbers and letters – e.g. "75S1A") were generally confined to the central portions of the section.

Within this revised stratigraphy, the characteristic Amudian artifacts are found in Bed 75 of Unit XI. Sixteen stratigraphic contexts with sufficient numbers of artifacts to warrant comparative analysis were isolated in Unit XI. These contexts are distributed through the five beds of Unit XI as follows:

Contextual isolates in Unit IX at Tabun

Bed 73	73S	73I
Bed 74	74S	74I
Bed 75	75S 75S1 75S2	75I1 75I2
	75X	
Bed 76	76S1A+B 76S1C 76S2	76I1 76I2+3
Bed 77	77B	

Within these 16 contextual isolates there is considerable variability in virtually all aspects of industrial composition. Unit XI is the only major geological unit that includes evidence of all three of the industries (or facies) that characterize what I have proposed as the "Mugharan Tradition". The demonstration of gradual changes in the relative frequencies of bifaces (Acheulian artifacts) and scrapers (Jabrudian artifacts) in the sequence from Unit XIII through Unit X has already been remarked upon as strong evidence for cultural continuity among the people that produced and used those artifacts at Tabun (Jelinek 1982a). Were this the only major evidence of cultural change in these levels it would not have been necessary to suggest a new term to replace the old designation of "Acheuleo-Jabrudian". However, the presence of a third industrial variant in the form of Amudian artifacts cannot be accommodated under this term. If these artifacts can be shown to be a part of the same tradition that produced the Acheulian and Jabrudian tools, then a more comprehensive term is warranted for this tradition (and "Mugharan" is seen as preferable to "Acheuleo-Jabrudo-Amudian"!).

This view is obviously in disagreement with that of Garrod for the

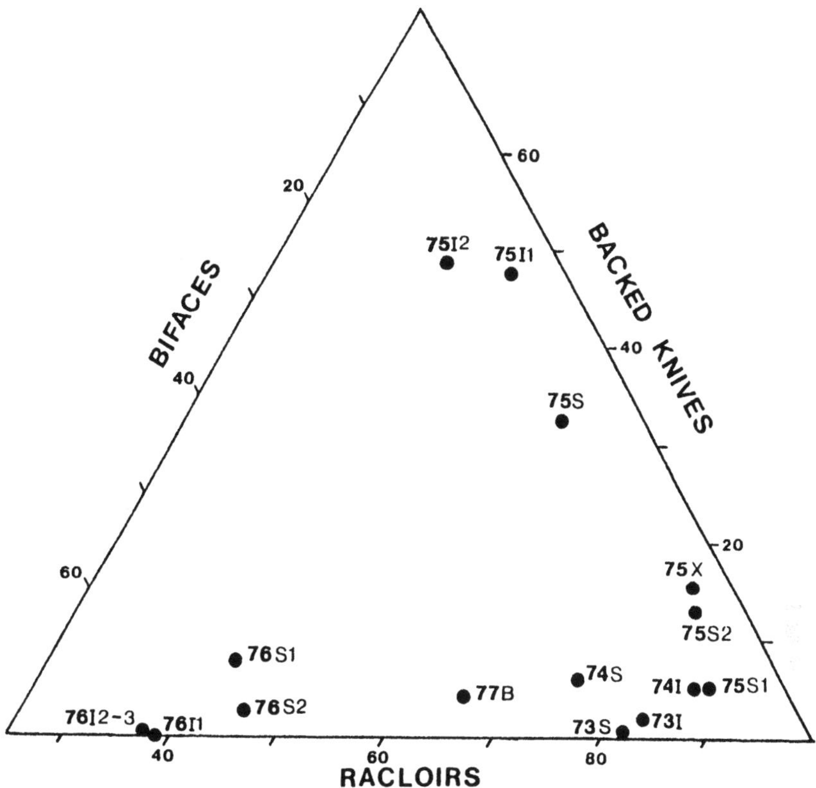

Figure 4.1. Triangular coordinate diagram illustrating the relative frequencies of backed knives, bifaces and racloirs in 15 stratigraphic contexts with over 50 of these artifacts in Unit XI at Tabun.

Amudian at Abri Zumoffen and of Rust for the Pre-Aurignacian, both of whom saw the producers of the blade industries as distinct human populations (see above). Evidence for the incorporation of the Amudian into the Mugharan tradition at Tabun is presented below.

THE AMUDIAN AS A FACIES WITHIN THE MUGHARAN TRADITION

An examination of the relative frequencies of 'Upper Palaeolithic' tools (Bordes types 30-37 and 40) in Unit XI indicates that tools other than backed knives vary more or less at random through the full sequence, although, as would be expected, they are present in somewhat higher frequencies during peak periods of manufacture of prismatic blades. Their presence may indeed reflect the fact that, given the presence of this flake type, there are few options for modification that do not produce 'Upper Palaeolithic' tools. Thus, the backed knives seem to be the only consistent typological characteristic of the Amudian. When their relative frequencies are viewed in a comparison with *racloirs* and bifaces, representing the Jabrudian and Acheulian, it is possible to show some segregation among

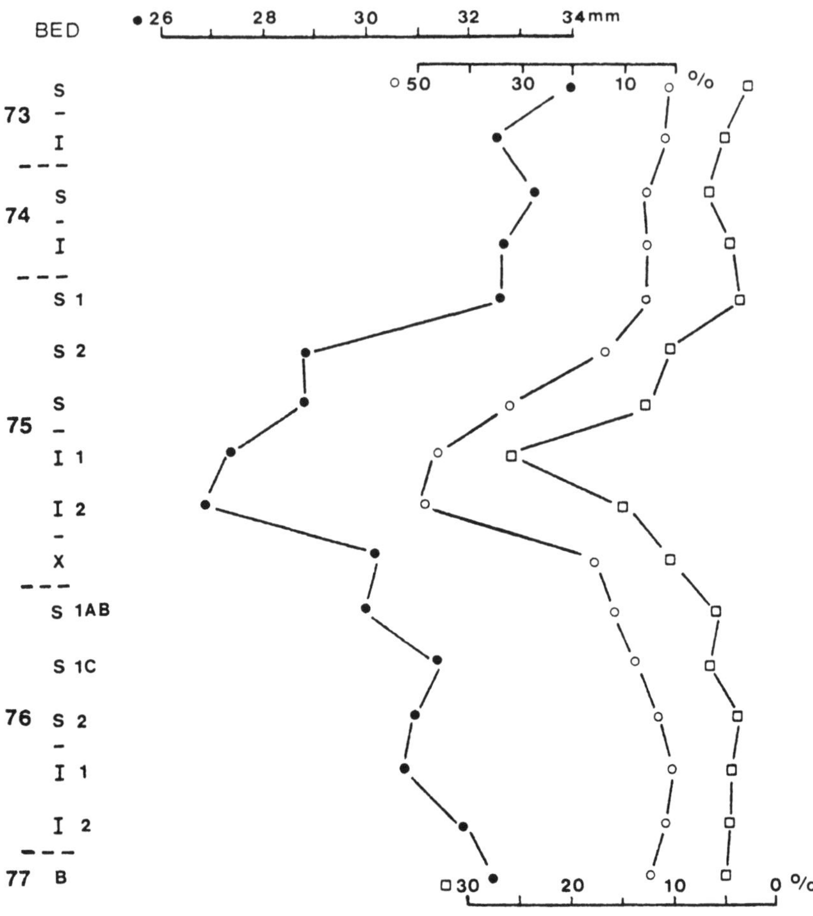

Figure 4.2. Frequencies of three variables related to the appearance and disappearance of the Amudian industries in 16 stratigraphic contexts in Unit XI at Tabun.

the facies of the Mugharan tradition (Figure 4.1). This triangular coordinate diagram illustrates the relative frequencies of the three kinds of artifacts for 15 samples with counts of over 50 of these artifacts (76S1A, B, and C are combined here). Beds 75I1, 75I2, and 75S clearly segregate as Amudian, with 75I1 and 75I2 most extreme. The Jabrudian facies is represented by all of beds 73 and 74, by 75S1 and 75S2 and by 75X, which underlies 75S as an isolate at the eastern end of the section (see note below). Beds 75S1 and 74I are the most extreme examples of the Jabrudian. The Acheulian facies consists of all of the Bed 76 samples, with the 76I samples most extreme. Bed 77B appears to be most similar to the

Jabrudian group. Thus, there are clear differences in the typological composition of the different kinds of industries present in Unit XI. However, it should be noted that all of the samples (except 76I1, which has no backed knives) have some representation of all of the three tool groups, and when a comparison is made that includes the relative frequencies of all retouched tool types the differences between the industries are much less extreme.

Other aspects of variability in the Tabun artifacts (e.g. basic inventory [retouched tools, complete flakes, broken flakes, cores, bifaces], heat alteration, platform, etc.) do not show variation conforming to the presence of the Amudian as indicated by backed knives. The only exception is the generally higher frequency of *couteaux à dos naturel* (type 38) in the Amudian – a not unexpected product of the manufacture of prismatic blades and long flakes. The relative thickness of the blades and their characteristic large plain platforms suggest production by simple direct percussion.

At no point in the stratigraphic sequence revealed by our excavations in Unit XI is there evidence of an isolated blade- tool industry, nor is there convincing evidence of the three major levels of concentration of these kinds of artifacts noted by Garrod (1956). Instead, there is a single gradual increase and decrease in the frequency of characteristic Amudian elements relative to other tool and flake types. Coincident with this single cycle of appearance and disappearance of Amudian elements is a decrease and increase in the median width of complete flakes. While this trend is most apparent when all flakes are combined in each sample, it is still evident when prismatic blades are removed from the samples. This suggests a that a basic technological shift toward and away from the manufacture of narrower flakes is an important factor in the appearance and disappearance of the Amudian.

Several coincident trends relating to the appearance and disappearance of the Amudian are shown in Figure 4.2. The similarity of the curves clearly reflects some significant level of redundancy in the frequencies of the variables shown. Median absolute width of flakes corresponds closely with relative frequency (among all flakes) of prismatic blades and burin spalls, and both of these vary with the frequencies of backed knives relative to bifaces and *racloirs* (the characteristic implements of the Acheulian and Jabrudian facies).

The most interesting information apparent in Figure 4.1 relates to the nature of the changes in these variables prior to the appearance of the Amudian. During the period of the deposition of all of Bed 76 there was a gradual decrease in flake width. Through the later part of this period, represented by the 76S beds, there was a gradual increase in the frequency of backed knives. The abrupt nature of the rise in both categories to the peaks of Bed 75I2 is probably more apparent than real and relates to time gaps in the sequence between Beds 75 and 76 (Note 4).

The general technological sequence suggested by Figure 4.1 is one of a gradual trend toward absolutely narrower flakes, with an increase in the manufacture of backed knives corresponding to the later period of this trend. It is interesting to note that the peak period of manufacture of prismatic blades (the 'hallmark' of the Upper Palaeolithic) corresponds only to the late peak of manufacture of narrow flakes (Bed 75I1 as opposed to the earlier Bed 75I2). This suggests that the emphasis on prismatic blade technology seen here may have been stimulated by an increasing typological emphasis on the production of narrow backed knives. Taken together, this evidence strongly supports an *in situ* evolution of technology that culminated in a brief period of relatively intensive manufacture of a tool type that is also characteristic of later cultural periods, the backed knife on a prismatic blade. This brief peak was, in turn, followed by a gradual decrease in emphasis on this pattern of tool production. There is no evidence here to support the presence of an intrusive human population or cultural pattern that was responsible for a distinctive Upper Palaeolithic technology and typology. The entire evolution took place in the context of a continuing pattern of technology and typology characteristic of the other two facies of the Mugharan tradition.

At present there is no environmental evidence to suggest that different conditions prevailed at Tabun during this interval (no pollen is present in the Bed 75I samples tested thus far and no animal bone is preserved in these levels). In the absence of confirming evidence, the full correlation of this sequence with the marine isotope record (Jelinek 1982a) must still be regarded as tentative. This correlation suggested that the Amudian corresponded to a peak cold period and that the following cold peak (for which some confirming evidence is present) also is characterized by a prolific production of blades (in Layer D). This very limited evidence suggests that resources utilized in the cave during cooler periods may have required more 'slicing activities' than was the case in other periods.

While the evidence strongly supports a gradual change on the part of local populations rather than exotic intrusion, the manufacture of these distinctive artifacts at Tabun does seem to represent a single restricted phenomenon without antecedents or descendants, and as such remains something of an enigma that requires an explanation. The evidence from Zumoffen and Jabrud I, where the manufacture of prismatic blades and blade tools seems to have taken place without as much manufacture of Jabrudian or Acheulian tools, introduces another dimension into the problem. The context of these materials, below and above Late Jabrudian and Acheulian levels, is similar to Tabun and suggests a widespread phenomenon approximating a synchronous 'stylistic' innovation whose presence can only be confirmed by a more detailed understanding of the technological and environmental sequences and the absolute chronological relationships between these sites and levels. In any event, one aspect of the problem of the Amudian that does seem to have been resolved at this point is that it had no generic relationship to the Upper Palaeolithic cultures of Europe or the Near East, nor does it bear any resemblance to the industries associated with the earliest examples of morphologically 'modern' *Homo sapiens sapiens* in the region in which it occurs.

NOTES

1. The initial reference to industrial designations is given in quotes in the orthography originally employed; subsequent use of the terms is in English orthography. The spelling of the site of Jabrud, and of the industry named from the site, here follows the spelling in the title of Rust's (1950) report.

2. This interpretation relating to the mixture of Amudian and Jabrudian at Abri Zumoffen is seen earlier in a paper presented by Garrod and Kirkbride in 1958 – e.g.: "This seems to suggest that the makers of the two industries continued to live side by side for a time" (Garrod and Kirkbride 1961b: 318).

3. "I suggest that these should be explained by contact between the Micoquian and a very early blade culture, possibly ancestral to the Châtelperron stage of Europe, whose centre of dispersion theoretically lies somewhere in Southern Central Asia" (Garrod 1937: 34).

4. Bed 75X may, in fact, correlate with Bed 76 or an even earlier period. It is a small pocket of material on the east side of the excavation, isolated from the deposits to the west by a limestone outcrop. It was assigned to Bed 75 in the absence of structural evidence of continuity with Bed 76. It is overlain by a layer of Bed 75 that is constricted as it passes over the outcrop. The lower portion of this layer (with very few artifacts) has been designated as 75I and the upper, which appears to correspond to the lowest portion of 75S2 or upper 75I1 to the west, as 75S. The apparently 'Jabrudian' assemblage in Bed 75X differs from the 'Acheulian' of Bed 76 and suggests that the 75X deposits may constitute a residue from several earlier periods.

REFERENCES

Bordes, F. 1960. Le Pré-Aurignacien de Yabroud (Syrie) et son incidence sur la chronologie du 1960. Le Pré-Aurignacien de Yabroud (Syrie) et son incidence sur la chronologie du Quaternaire en Moyen Orient. *Bulletin of the Research Council of Israel, Section G. Geosciences* 9G (2-3): 91-103.

Bordes, F. 1955. Le Paléolithique Inférieur et Moyen de Jabrud (Syrie) et la question du Pré-Aurignacien. *L'Anthropologie* 59: 486-507.

Garrod, D.A.E. 1956. Acheuléo-Jabroudien et "Pré-Aurignacien" de la Grotte du Taboun (Mont Carmel): étude stratigraphique et chronologique. *Quaternaria* 3: 39-59.

Garrod, D.A.E. 1953. The relations between South-west Asia and Europe in the Later Palaeolithic age, with special reference to the origin of the Upper Paleolithic blade cultures. *Journal of World History* 1: 13-38.

Garrod, D.A.E. 1937. The Near East as a gateway of prehistoric migration. In C.G. MacCurdy (ed.) *Early Man.* London: J.B. Lippincott: 33-40.

Garrod, D.A.E and Bate, D.M.A. 1937. *The Stone Age of Mount Carmel:* Vol. 1. Oxford: Oxford University Press.

Garrod, D.A.E. and Kirkbride, D. 1961a. Excavation of the Abri Zumoffen, a Paleolithic rock-shelter near Adlun, south Lebanon, 1958. *Bulletin du Musée de Beyrouth* 16: 7-46.

Garrod, D.A.E. 1961b. Excavation of a Palaeolithic rock shelter at Adlun, Lebanon 1958. In G. Besu and W. Dehu (eds) *Bericht Uber den 5 Internationalen Kongress für Vor- und Frühgeschichte, Hamburg vom 24 bis 30 August 1958*. Berlin: Mann: 313-320

Jelinek, A.J. 1982a. The Tabun Cave and Paleolithic Man in the Levant. *Science* 216: 1369-1375.

Jelinek, A.J. 1982b. The Middle Palaeolithic in the Southern Levant, with comments on the appearance of Modern *Homo sapiens*. In A. Ronen (ed.) *The Transition from Lower to Middle Palaeolithic and the Origin of Modern Man.* Oxford: British Archaeological Reports International Series 151: 57-108.

Jelinek, A.J. 1981. The Middle Paleolithic of the Levant: synthesis. In J. Cauvin and P. Sanlaville (eds) *Préhistoire du Levant.* Paris: Centre Nationale de la Recherche Scientifique: 299-302.

Jelinek, A.J. 1975. A preliminary report on some Lower and Middle Paleolithic industries from the Tabun Cave, Mount Carmel (Israel). In F. Wendorf and A.E. Marks (eds.) *Problems in Prehistory: North Africa and the Levant.* Dallas: Southern Methodist University Press: 297-315.

Jelinek, A.J., Farrand, W.R., Haas, G., Horowitz, A. and Goldberg, P. 1973. New excavations at the Tabun Cave, Mount Carmel, Israel: a preliminary report. *Paléorient* 1: 151-183.

Rust, A. 1950. *Die Höhlenfunde von Jabrud (Syrien)*. Neumunster: Karl Wachholtz.

Rust, A. 1933. Beitrag zur Erkenntnis der Abwicklung der vorgeschictlichen Kulturperioden in Syrien. *Praehistorische Zeitschrift* 24: 205-218.

Turville-Petre, F. 1927. *Researches in Prehistoric Galilee 1925-1926.* London: British School of Archaeology in Jerusalem.

5: A Technological Analysis of the Upper Palaeolithic Levels (XXV-VI) of Ksar Akil, Lebanon

K. OHNUMA AND C. A. BERGMAN

INTRODUCTION

The rock-shelter of Ksar Akil (Lebanon) has the longest known stratigraphic sequence of Upper Palaeolithic material in the Near East. The artifacts recovered during excavations in 1937-1938 (see below) come from some 25 geologically defined levels spanning 19 metres in depth. Most of the previous studies of this material have concentrated on the retouched tools and their relative percentages in each layer. It is clear from examining the collections, as well as from comparing them with the most recent excavations (Tixier 1970, 1974; Tixier and Inizan 1981) that some classes of retouched tools are underrepresented. Thus, we believe that typological analyses based on tool counts are of limited value, and the present study therefore concentrates on flaking technology, discussing the archaeological material in two major groups: the early Upper Palaeolithic levels (levels XXV-XVI) and the later Upper Palaeolithic levels (levels XIII-VI).

HISTORY OF RESEARCH AT KSAR AKIL

The Ksar Akil rock-shelter is located 10 kilometres northeast of Beirut at the base of a limestone cliff on the right bank of the Antelias River Valley, about 3 kilometres from the coastal plain (Figure 5. 1). Much of the history of research at the site has been discussed elsewhere (see Bergman and Copeland 1986; Bergman 1987: 1-6) so only a brief account follows below.

Excavations began in 1922 when 'treasure hunters' dug through the deposits to a depth of 15 metres. Work was continued by Day (1926a, 1926b) of the American University of Beirut, who recovered a small amount of flint and bone artifacts. On the advice of the Abbé Breuil, the rock-shelter was excavated again in 1937. The archaeological team came from Boston College, Massachusetts, and was directed by Doherty (Murphy 1938; 1939). During the first two seasons at the site, a datum point at 80. 9 m above sea level was established (Ewing 1947). This datum should not be confused with that used on Wright's section which is apparently 75 m above sea level (Figure 5. 2; see Copeland 1987: vii). Excavations began in 2 m x 2 m units (Figure 5. 3) and reached a depth

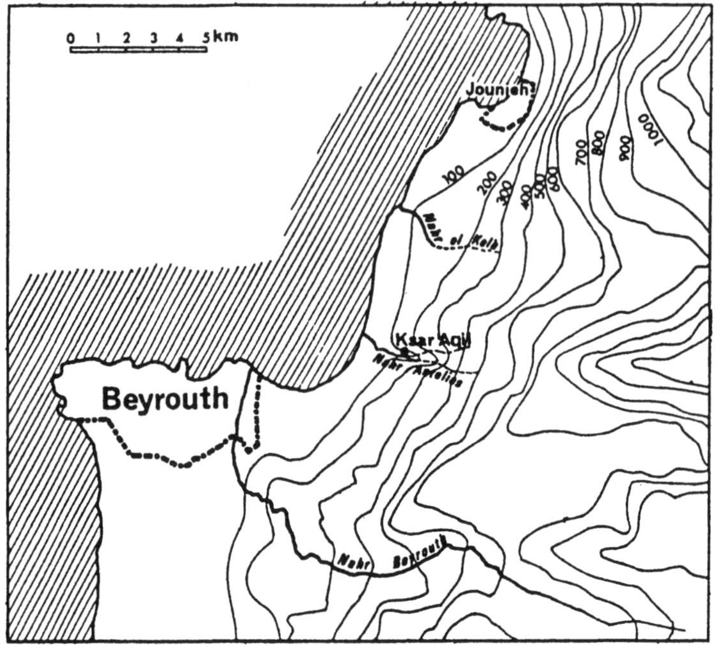

Figure 5.1. Map of the area surrounding Ksar Akil (after Tixier 1970).

of 19 m, which was about 4 m above bedrock (Newcomer 1972: 7).

There are two radiocarbon C14 dates related to the Boston College excavations (Vogel and Waterbolk 1963), the earliest of which is 43 750 ± 1500 BP (GrN-2579) obtained from a dark clay band presumably containing charcoal, in level XXVI (16 m below datum). The second date of 28 840 ± 380 BP (GrN-2195) was derived from shells collected at 6. 00-7. 50 m below datum. Although most authors assign this date to level VIII it actually covers parts of levels IX, VIII, and VII.

Ewing (1963) reported that a 'Neanderthaloid' maxilla fragment was recovered from level XXV/XXIV at approximately 15 m below datum, but this has never been described in detail. The stone tool kit from these levels consists mainly of end-scrapers, truncation burins and chamfered pieces (Azoury 1986: 107). An almost complete skeleton of *Homo sapiens sapiens* was recovered at 11. 46 m below datum along with some fragments of bone from a second individual (Ewing 1947). These remains occurred in archaeological level XVII which has a stone-tool assemblage made up of end-scrapers, backed blades and Ksar Akil points (Azoury 1986: 158).

In 1969 J. Tixier commenced an important new series of excavations at Ksar Akil. Using careful excavations techniques, such as the three dimensional recording of the artifacts, he established a finely divided stratigraphy. Tixier reached a depth of some 8. 75 metres corresponding roughly to the top of level X of the 1937-1938 excavations (Copeland 1987: Figure A; Tixier personal communication), before being forced to suspend work in 1975 due to the civil disturbances in Lebanon.

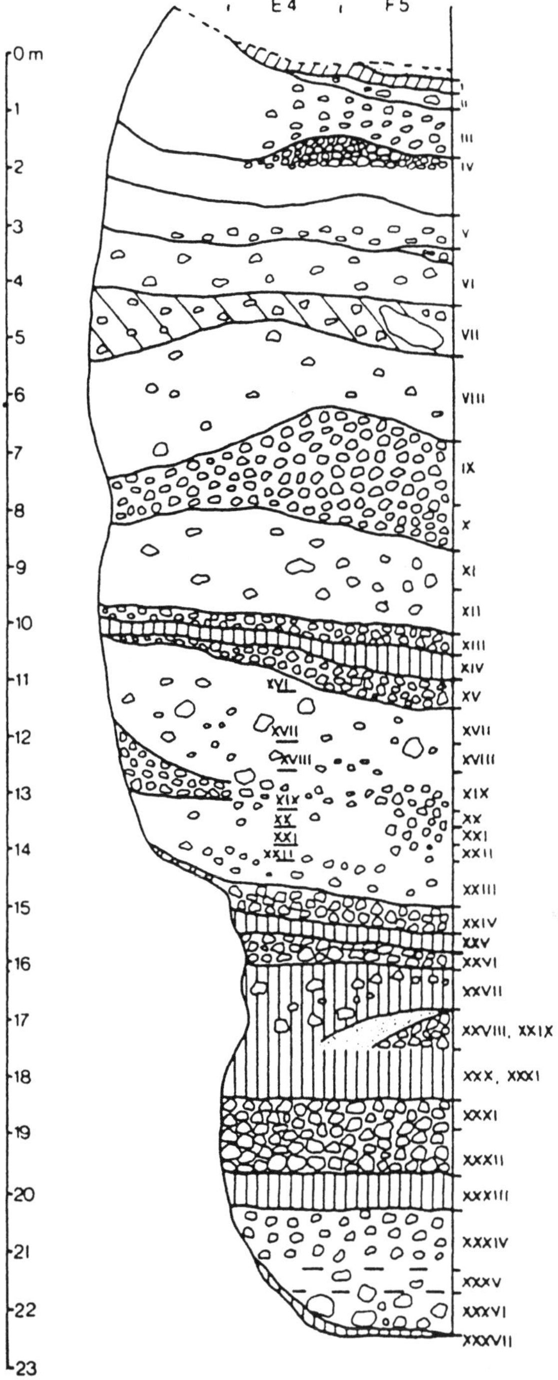

Figure 5.2. Stratigraphic section of Ksar Akil established in excavations by the Boston College team.

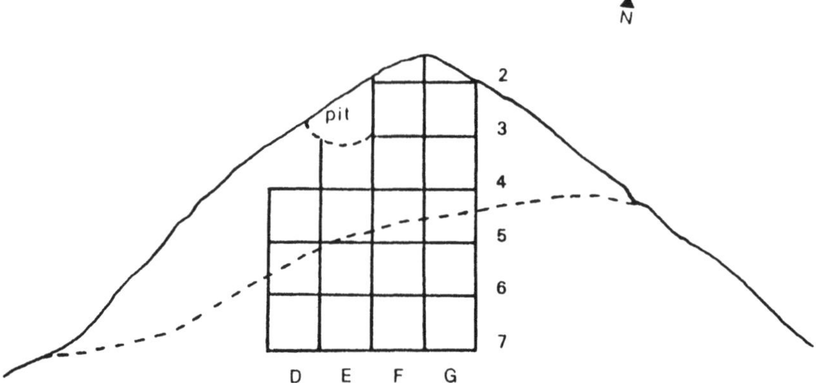

Figure 5.3. Grid system used in excavations at Ksar Akil by the Boston College team (excavation units are 2 metres square).

PROBLEMS IN STUDYING THE 1937-1938 EXCAVATIONS OF KSAR AKIL

The nature of the 1937-1938 excavations has placed certain limitations on this study. Several problems were encountered during the course of examining the collections:

1. Tixier's excavations revealed that the upper levels have a fine and complex stratigraphy. The thick geological levels recognised by the Boston College team caused some minor mixing of the archaeological layers and their artifacts. It is highly likely that this has led to some 'blurring' of breaks in the stratigraphic sequence. In one case, it can be proven conclusively that difficulty existed in separating two contiguous levels (IX and VIII), resulting in one of them appearing 'transitional' between two distinct phases (Bergman 1987: 101-2; see also Azoury 1986: 123). However it must be stated that, in the authors' opinion, *the overall degree* of mixing is relatively insignificant and does not affect the analysis of the stone tools in any serious way.

2. It would appear that artifact collection was somewhat selective. The general absence of 'chips' (retouch flakes; cf. Newcomer and Karlin 1986), burin spalls and microliths (Tixier and Inizan 1981) is probably in part related to the mesh size of the sieves used; a "medium grade" sieve was used on all sediments (Murphy 1938: 237; Bergman 1987: Plate 19).

3. The amount of material encountered caused the excavators to discard some of the artifacts (Murphy 1939: 212). It is impossible to be certain exactly what criteria were used when artifacts were thrown away (Azoury 1986: 88).

All of these factors should cause anyone dealing with these collections to exercise extreme caution. Most work on the site has focused on the tool typology and their proportional occurrences in the different levels. For the above reasons none of the artifact samples can be regarded as complete and studies based on tool or debitage frequencies alone may be misleading.

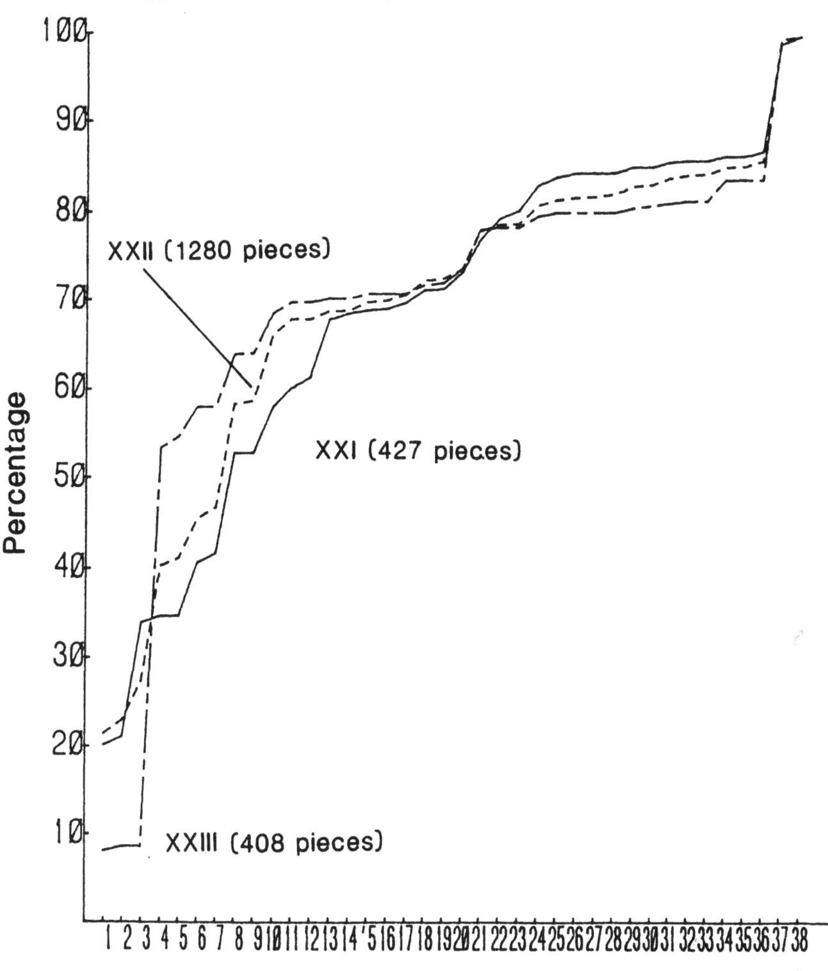

Figure 5.4. Cumulative graphs of retouched pieces from levels XXIII-XXI at Ksar Akil.

METHOD

The present study is derived from the Doctoral research on Ksar Akil conducted by both of the authors. Similar attribute lists were used in both studies, so that there is a good correlation between the kinds of technological features noted. The large amount of material collected in 1937-1938 necessitated the use of sampling. Two 2 m x 2 m units, E4 and F4 (Figure 5. 3), were selected for this purpose as artifacts from both squares occur in almost every level between XXV and VI; over 15 000 pieces were examined from these squares (Ohnuma 1986; Bergman 1987). The data on each artifact were recorded separately, and the entire sample was processed using the Statistical Package for the Social Sciences (SPSS: Nie *et*

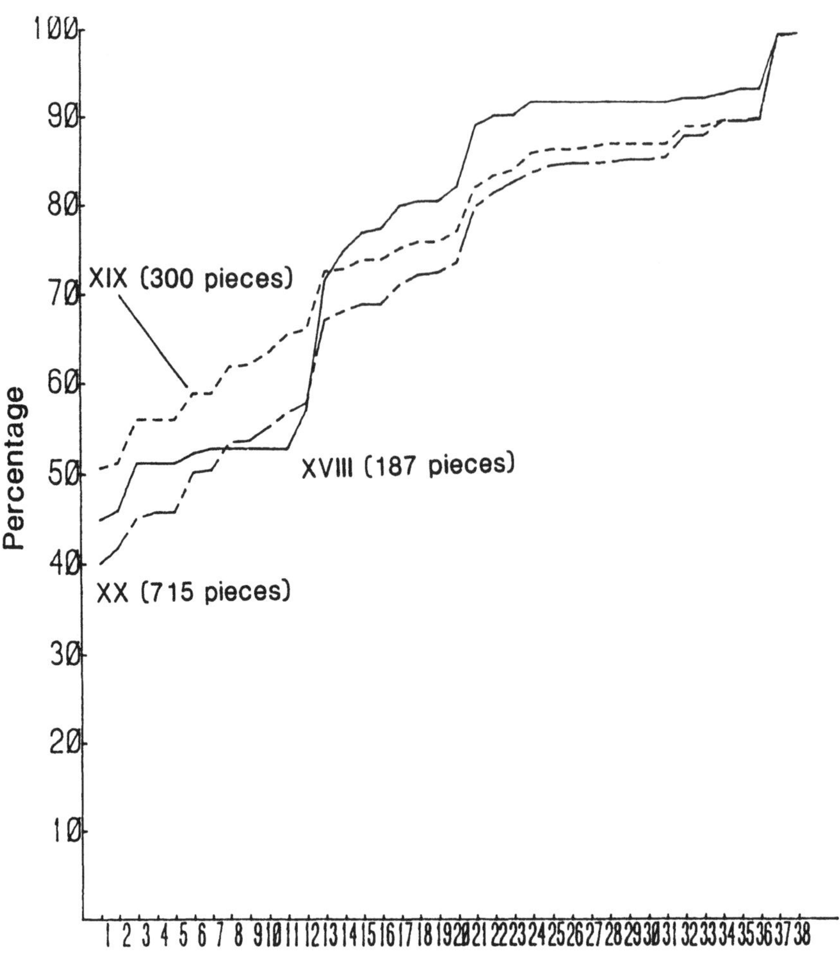

Figure 5.5. Cumulative graphs of retouched pieces from levels XX-XVIII at Ksar Akil.

al. 1975). In the technological studies below only complete blanks were analysed; broken pieces of debitage present incomplete information and were excluded from analysis.

LEVELS XXV-XIV

Two Doctoral theses written by Azoury (1971) and Newcomer (1972) previously discussed the material in these levels. Azoury (1971, 1986) established three sets of levels between XXV and XII based on technological changes in blank production and proportional fluctuations in the components of the tool kits.

Newcomer (1972) analysed the burins from all 25 Upper and Epi-Palaeolithic levels (XXV-I) and established three groups of levels on the bases of the technology of their blank production, typology and burin morphology. He concluded that the evolution of the Upper Palaeolithic sequence at Ksar Akil was continuous except between levels XIV and XIII, where a major stratigraphic break occurred.

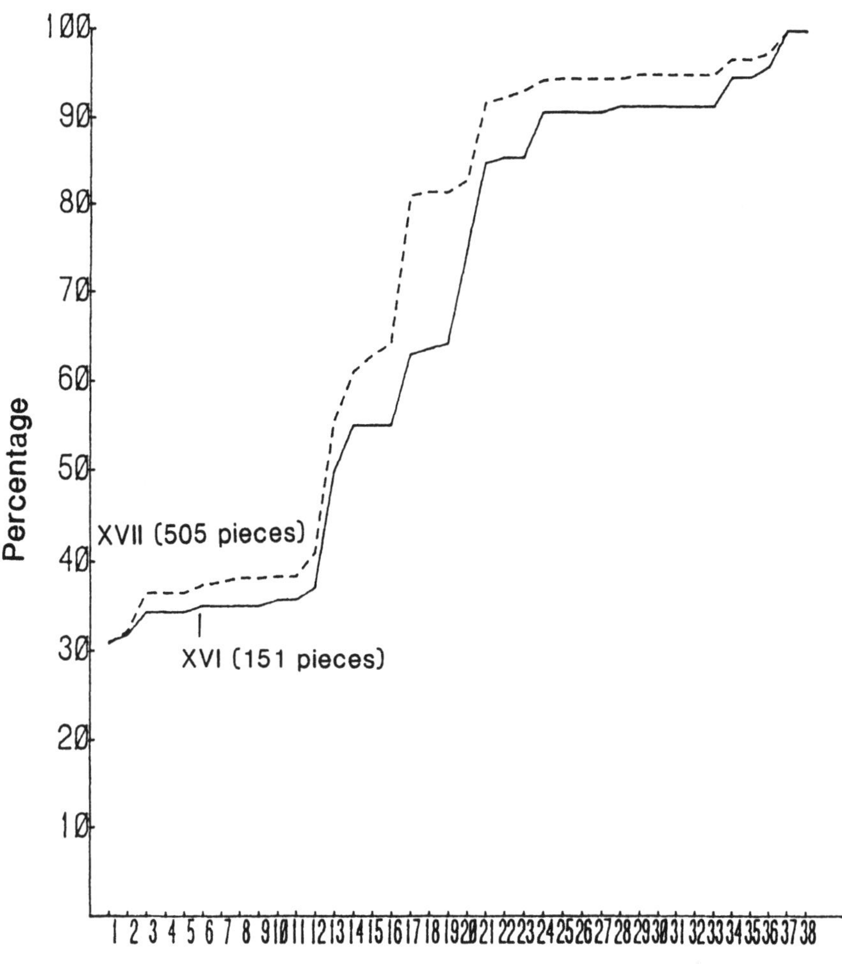

Figure 5.6. Cumulative graphs of retouched pieces from levels XVII and XVI at Ksar Akil.

In the present study the material was first classified into debitage, cores, retouched tools, 'chips' and hammerstones. Blanks were classified into cortical debitage, partially cortical pieces, non-cortical debitage and crested pieces. Non-cortical debitage was subdivided into Levallois flakes, Levallois points, elongated Levallois points, flake-blades (including Levallois blades), flakes, blades and bladelets. A flake-blade is defined as being non-cortical debitage with the length equal to or more than twice the width, but which does not bear parallel or sub-parallel ridges on the dorsal surface (Bordes 1961: 6). The definitions of blades and bladelets follow those provided by Bordes and Crabtree (1969:1) and Tixier (1963: 38-9).

Cores were classified morphologically into prismatic, pyramidal, bipyramidal, discoidal, globular and miscellaneous. If any showed clear traces of Levallois methods of flaking in their final form, they were regarded as Levallois and recorded according to whether they produced flakes, blades or points.

Fifty-four tool types were defined by Azoury in her type list (1986); due

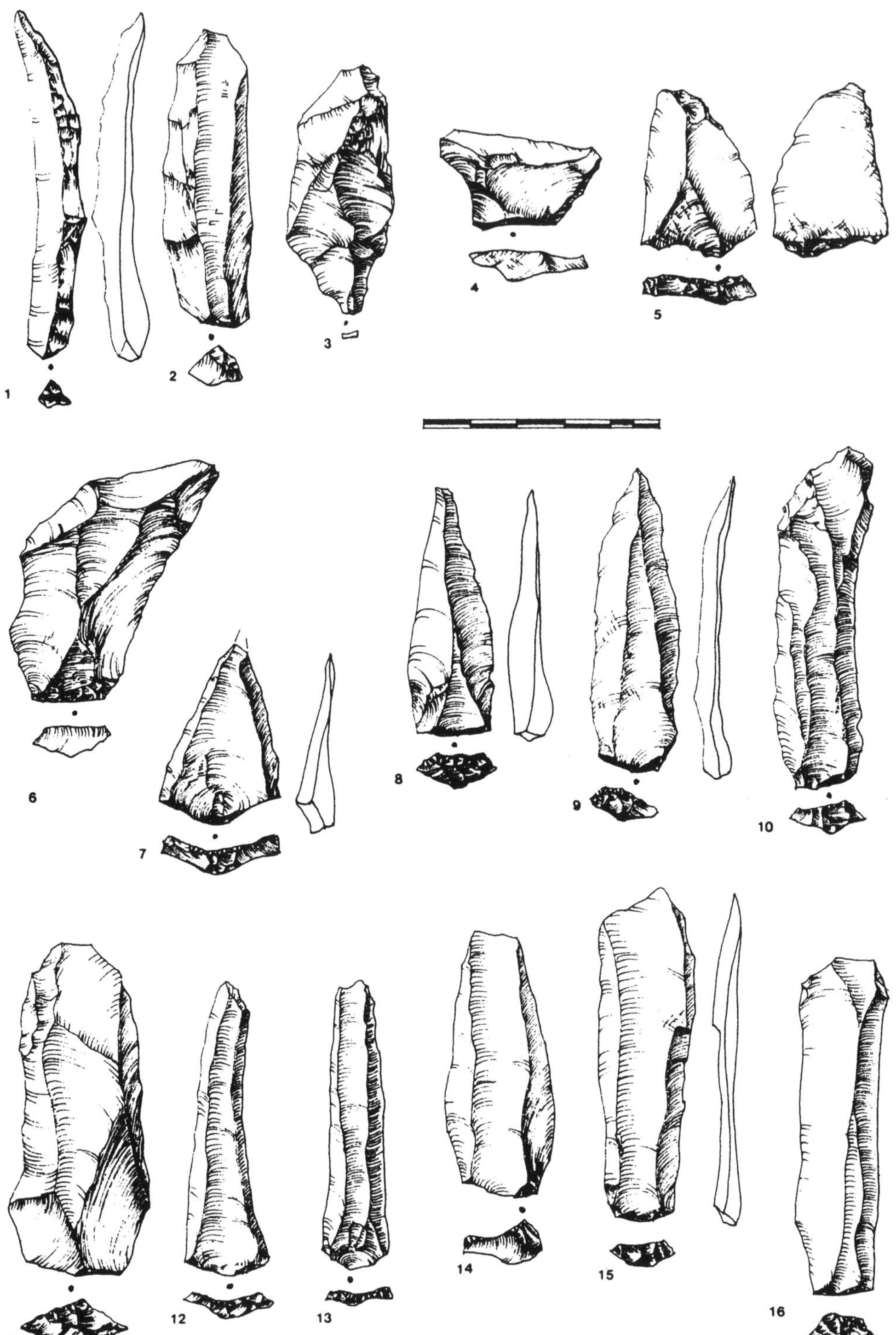

Figure 5.7. Debitage from levels XXIII and XXII at Ksar Akil.

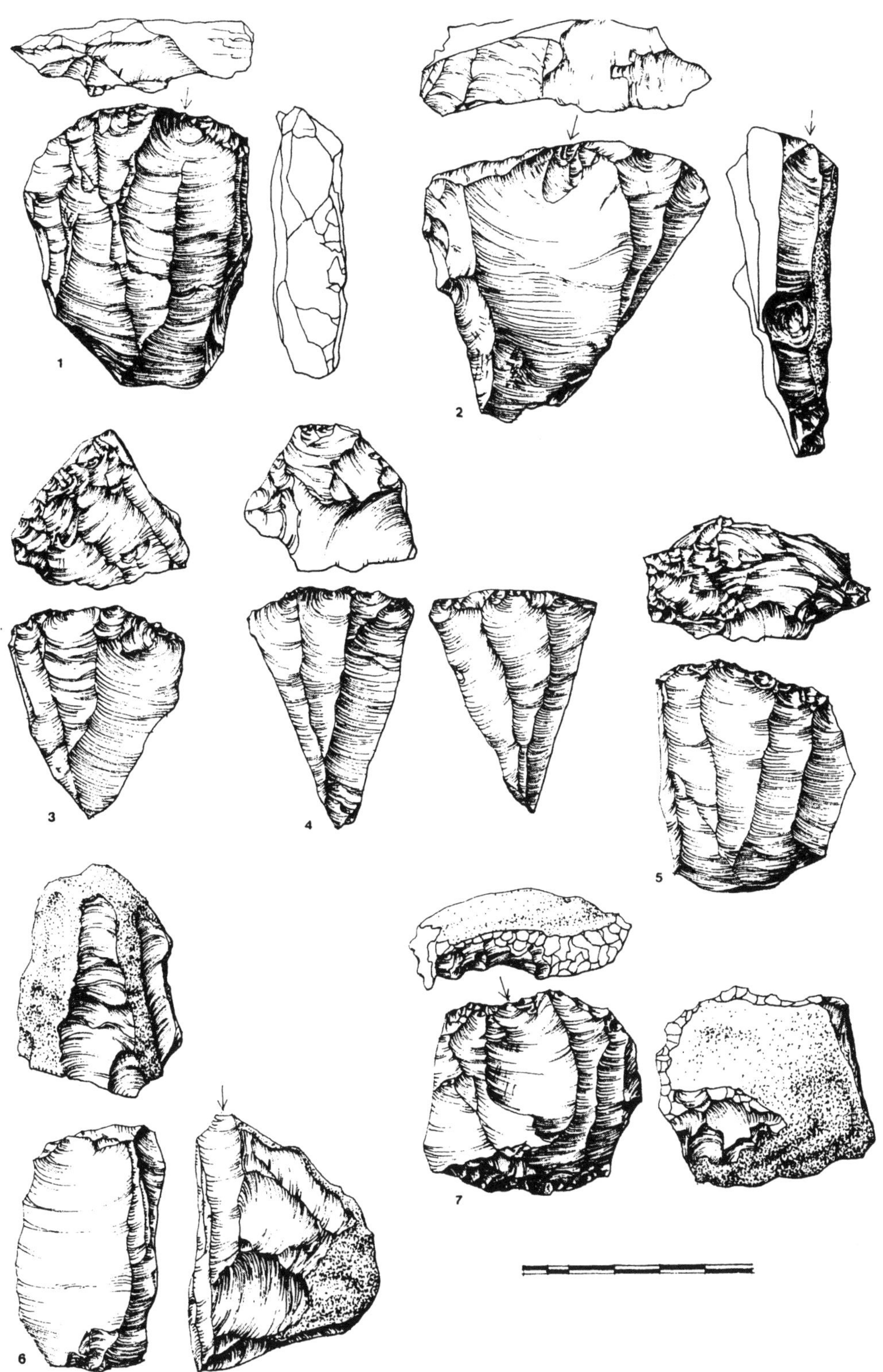

Figure 5.8. Cores from levels XXIII and XXII at Ksar Akil.

Table 5.1 Frequencies of debitage classes (totals and percentages).

Level	Total	Cortical	Partially Cortical	Non-Cortical flakes	Flake-Blades	Blades	Bladelets	Crested debitage
XIV	40	—	2	12	6	12	4	4
			(5.0)	(30.0)	(15.0)	(30.0)	(10.0)	(10.0)
XV	84	—	16	44	14	6	4	—
			(19.0)	(52.4)	(16.7)	(7.1)	(4.8)	—
XVI	394	2	72	105	54	55	69	37
		(0.5)	(18.3)	(26.6)	(13.7)	(14.0)	(17.5)	(9.4)
XVII	690	—	111	87	114	148	115	115
			(16.1)	(12.6)	(16.5)	(21.4)	(16.7)	(16.7)
XVIII	259	1	35	40	48	47	55	33
		(0.4)	(13.5)	(15.4)	(18.5)	(18.1)	(21.2)	(12.7)
XIX	360	6	60	74	76	53	56	35
		(1.7)	(16.7)	(20.6)	(21.4)	(14.7)	(15.6)	(9.7)
XX	556	3	62	51	206	127	66	41
		(0.5)	(11.2)	(9.2)	(37.1)	(22.8)	(11.9)	(7.4)
XXI	256	—	24	50	80	67	30	5
			(9.4)	(19.5)	(31.3)	(26.2)	(11.7)	(2.0)
XXII	871	2	52	177	344	205	75	16
		(0.2)	(6.0)	(20.3)	(39.5)	(23.5)	(8.6)	(1.8)
XXIII	296	1	21	81 138		31	22	2
		(0.3)	(7.1)	(27.4)	(46.6)	(10.5)	(7.4)	(0.7)
XXIV	41	1	3	12	25	—	—	—
		(2.4)	(7.3)	(29.3)	(61.0)			
XXV	29	—	6	13	7	—	1	2
			(20.7)	(44.8)	(24.1)		(3.4)	(6.9)

to the fact that the present study focuses on the technology of blank production, Azoury's list has been considerably reduced (see Azoury and Hodson 1973 on short type lists). Therefore, the retouched tools were classified using a type list made up of 38 categories. The definitions of the tool types are the same as those provided by Azoury (1986: 42-64) and de Sonneville-Bordes and Perrot (1954, 1955, 1956a, 1956b).

TECHNOLOGY AND TYPOLOGY OF LEVELS XXV-XIV

1. Debitage and cores

It can be seen that in the debitage classes flake-blades play a major role in the elongated debitage in levels XXV-XXII, while blades and bladelets gradually increase from level XXIII onwards and take the place of flake-blades in level XIX (Table 5. 1).

Prismatic shaped cores are always the most common, while Levallois cores occur only in levels XXV-XX (Table 5. 2). Among the prismatic cores, single platform cores are more common in levels XXV-XX, whereas those with opposed platforms increase in number from level XIX and are

Table 5.2 Frequencies of core classes (totals and percentages).

Level	Total	Prism-atic	Levallois			Discoidal	Pyramidal	Globular	Miscellan-eous
			Flake	Blade	Point				
XIV	4	1 (25.0)	—	—	—	1 (25.0)	—	—	2 (50.0)
XV	5	1 (20.0)	—	—	—	1 (20.0)	—	—	3 (60.0)
XVI	31	19 (61.3)	—	—	—	3 (9.2)	1 (3.2)	—	8 (25.8)
XVII	164	130 (79.3)	—	—	—	6 (3.7)	7 (4.3)	3 (1.8)	18 (11.0)
XVIII	38	26 (68.4)	—	—	—	1 (2.6)	2 (5.3)	1 (2.6)	8 (21.1)
XIX	123	73 (59.3)	—	—	—	8 (6.5)	3 (2.4)	3 (2.4)	36 (29.3)
XX	160	117 (73.1)	1 (0.6)	—	2 (1.3)	2 (1.3)	3 (1.9)	5 (3.1)	30 (18.8)
XXI	68	45 (66.2)	—	—	—	3 (4.4)	4 (5.9)	1 (1.5)	15 (22.1)
XXII	135	88 (65.2)	2 (1.5)	2 (1.5)	3 (2.2)	6 (4.4)	7 (5.2)	1 (0.7)	25 (18.5)
XXIII	105	74 (70.5)	1 (1.0)	2 (1.9)	1 (1.0)	5 (4.8)	2 (1.9)	—	20 (19.0)
XXIV	39	19 (48.7)	—	2 (5.1)	1 (2.6)	2 (5.1)	—	1 (2.6)	14 (35.9)
XXV	21	11 (52.4)	1 (4.8)	—	—	4 (19.0)	1 (4.8)	—	4 (19.0)

most numerous in levels XVIII-XVII. In levels XXV-XIV the most common striking platform is plain, with faceted platforms gradually decreasing in levels XXIV-XVII (Table 5. 3).

The butts on the blanks (Bordes 1961, 1967, 1972) show significant changes in the early part of the Ksar Akil sequence (Table 5. 4); those with convex dihedral facets and multiple (straight or convex) facets occur most often in levels XXV-XIX, while linear and punctiform butts are most common in XVIII-XVI.

In levels XXV-XVI the thickness of the blank and its butt are closely related (see Newcomer 1975: 97; Ohnuma and Bergman 1982). These measurements become smaller in levels XXV-XIX and are smallest in levels XVIII-XVI. The frequencies of hard and soft hammer mode (see Ohnuma and Bergman 1982) illustrate that although soft hammer-struck blanks always outnumber hard hammer-struck blanks, there is a clear break between levels XXI and XX; beginning in level XX the difference between the occurrence of the two flaking modes gets larger (Figure 5. 13).

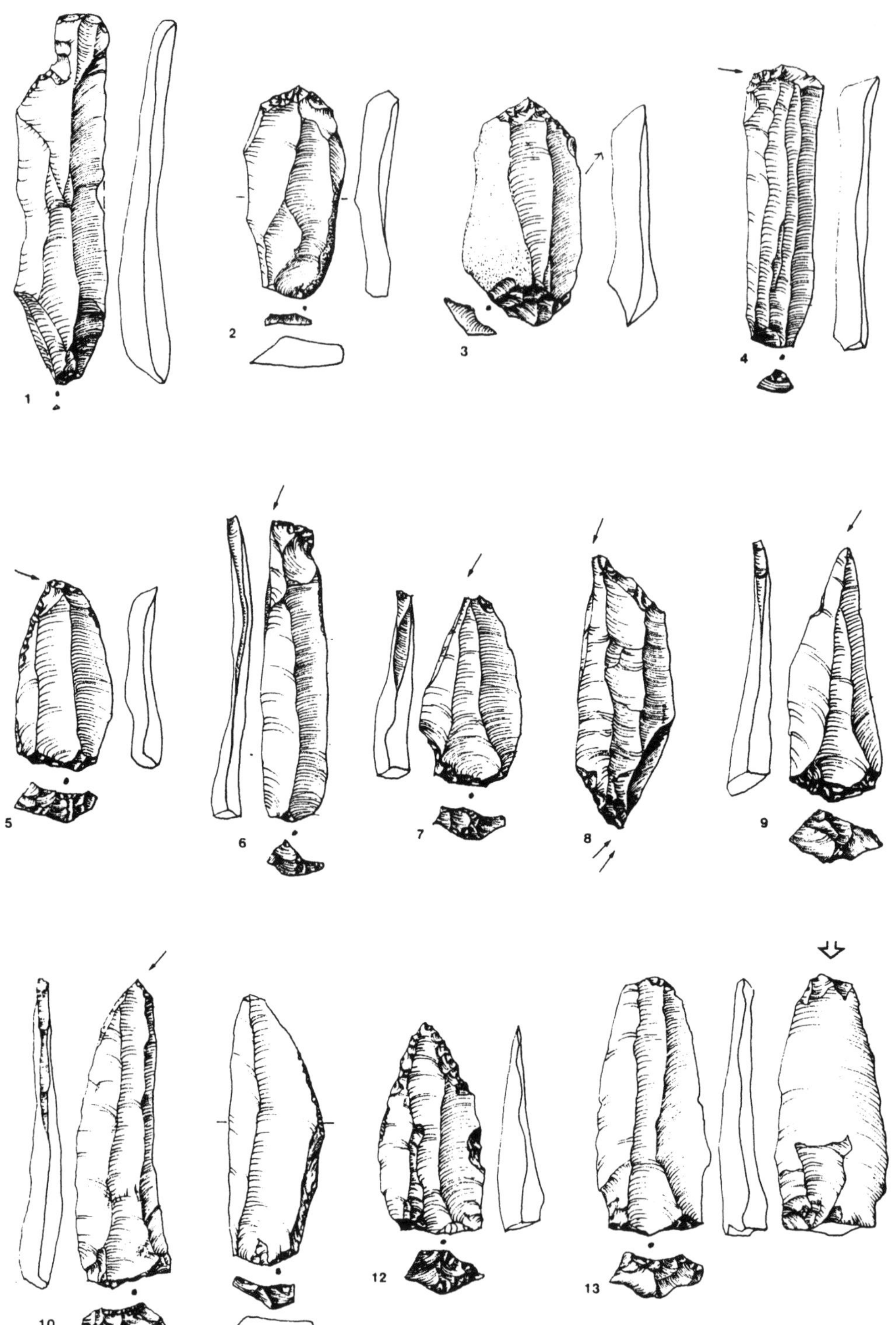

Figure 5.9. Tools from levels XXIII and XXII at Ksar Akil.

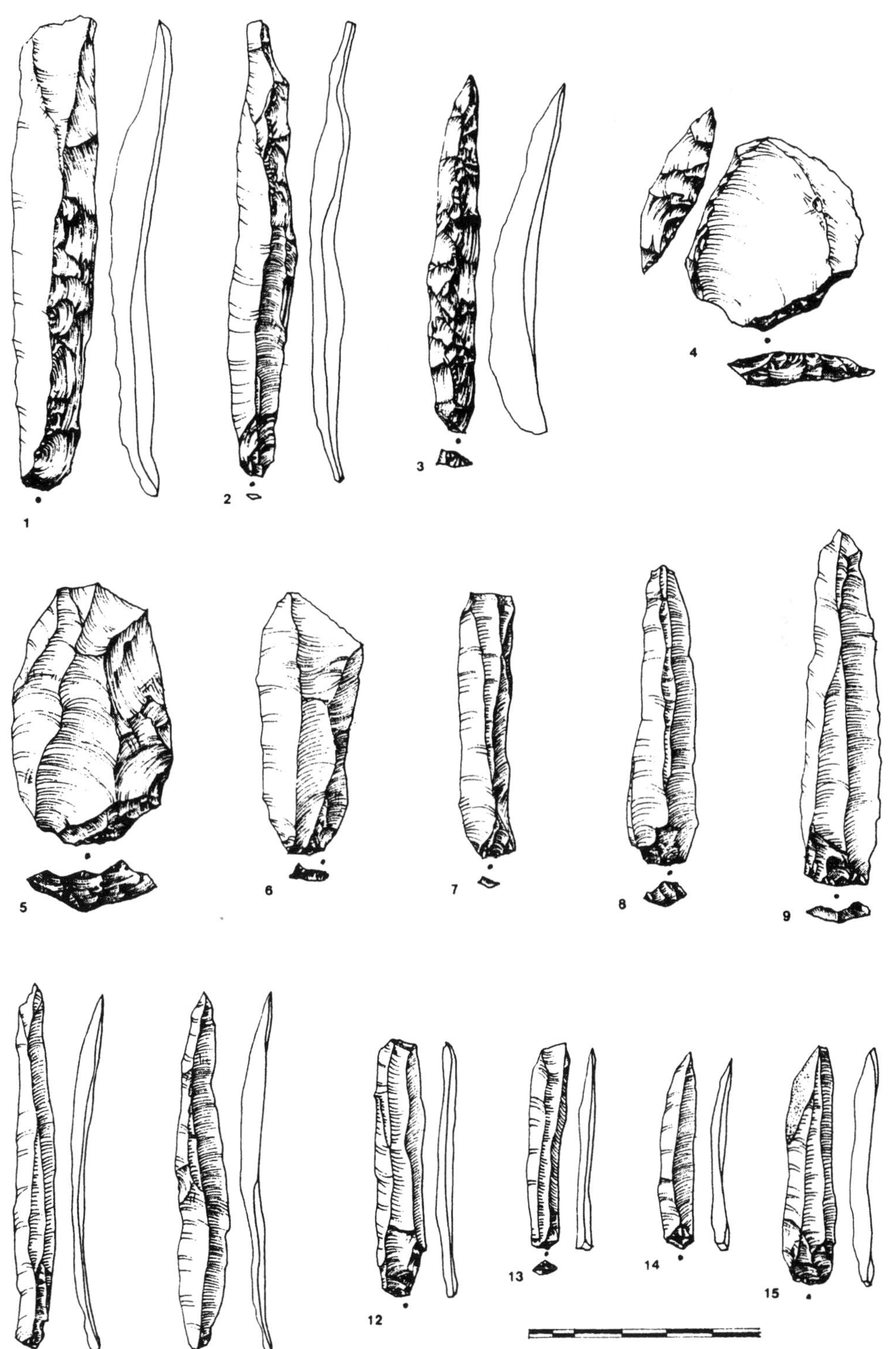

Figure 5.10. Debitage from levels XVII and XVI at Ksar Akil.

Table 5.3 Frequencies of striking platform types of cores (total and percentages).

Level	Total	Cortical	Plain	Convex dihedral facetted	Straight multiple facetted	Convex multiple facetted	Broken
XIV	7	2 (28.6)	3 (42.9)	2 (28.6)	—	—	—
XV	8	—	4 (50.0)	4 (50.0)	—	—	—
XVI	57	2 (3.5)	34 (59.6)	20 (35.1)	—	—	1 (1.8)
XVII	299	8 (2.7)	234 (78.3)	53 (17.7)	1 (0.3)	2 (0.7)	1 (0.3)
XVIII	69	2 (2.9)	53 (76.8)	10 (14.5)	—	3 (4.3)	1 (1.4)
XIX	197	13 (6.6)	132 (67.0)	44 (22.3)	—	7 (3.6)	1 (0.5)
XX	251	13 (5.2)	144 (57.4)	77 (30.7)	2 (0.8)	13 (5.2)	2 (0.8)
XXI	99	2 (2.0)	65 (65.7)	23 (23.2)	3 (3.0)	6 (6.1)	—
XXII	193	5 (2.6)	107 (55.4)	66 (34.2)	4 (2.1)	8 (4.1)	3 (1.6

There is a gradual but clear change through the sequence in the elongation of debitage. Blank length/width ratios increase over time, with peaks in levels XX and XVII. Thus, it can be stated that in levels XXIII-XXII both hard and soft hammers (which probably include soft *stone* hammers) were being used in the same style of non-marginal flaking and that in levels XVIII-XVI soft hammers (or soft punches for indirect percussion; Azoury 1971: 96; Newcomer 1972: 38) were being used to strike close to the platform edge (marginal flaking) in order to produce debitage which, on average, was longer and thinner.

Bi-directional opposed dorsal scars (Bordes and Crabtree 1969: 2-3) increase in levels XX-XVI at the expense of the unidirectional scar pattern common in levels XXIII-XXI. The scar patterns on cores roughly coincide with those on the debitage: in levels XXII-XX the unidirectional pattern is most common, whereas the opposed pattern becomes more frequent in levels XIX-XVI (Tables 5. 5 and 5. 6).

The plan-shapes of the debitage (cf. Marks 1976: 372) show blanks with parallel edges gradually increasing from level XXIV and becoming most numerous in level XVIII, while converging edges gradually decrease (Table 5. 7). An examination of core shapes indicates that they are mainly parallel in levels XXV, XXIV and XX-XVI, whereas in levels XXIII-XXII converging sided cores occur about as often as those with parallel sides (Table 5. 8). It is evident, therefore, that the overall shape of a core influences the plan-shape of the blanks it produces; debitage with parallel

Table 5.4. Frequencies of butt types of débitage (totals and percentages).

Level	Total	Cortical	Plain	Convex dihedral facetted	Straight multiple facetted	Convex multiple facetted	Partially facetted	'Chapeau de Gendarme'	Linear	Punctiform	Broken
XIV	40	—	10 (25.0)	15 (37.5)	1 (2.5)	4 (10.0)	2 (5.0)	—	5 (12.5)	2 (5.0)	1 (2.5)
XV	84	1 (1.2)	21 (25.0)	28 (33.3)	5 (6.0)	20 (23.8)	3 (3.6)	1 (1.2)	1 (1.2)	1 (1.2)	3 (3.6)
XVI	394	5 (1.3)	125 (31.7)	79 (20.1)	4 (1.0)	19 (4.8)	4 (1.0)	1 (0.3)	79 (20.1)	68 (17.3)	10 (2.5)
XVII	690	15 (2.2)	200 (30.1)	85 (12.3)	6 (0.9)	11 (1.6)	2 (0.3)	—	175 (25.4)	176 (25.5)	12 (1.7)
XVIII	259	3 (1.2)	71 (27.4)	33 (12.7)	9 (3.5)	9 (3.5)	3 (1.2)	—	56 (21.6)	62 (23.9)	13 (5.0)
XIX	360	7 (1.9)	121 (33.6)	85 (23.6)	15 (4.2)	40 (11.1)	7 (1.9)	—	44 (12.2)	32 (8.9)	9 (2.5)
XX	556	10 (1.8)	160 (28.8)	164 (29.5)	40 (7.2)	89 (16.0)	37 (6.7)	—	23 (4.1)	19 (3.4)	14 (2.5)
XXI	256	5 (2.0)	70 (27.3)	81 (31.6)	18 (7.0)	52 (20.3)	16 (6.3)	—	8 (3.1)	1 (0.4)	5 (2.0)
XXII	871	7 (0.8)	230 (26.4)	249 (28.6)	75 (8.6)	204 (23.4)	63 (7.2)	—	16 (1.8)	7 (0.8)	20 (2.3)
XXIII	296	5 (1.7)	80 (27.0)	102 (34.5)	11 (3.7)	44 (14.9)	21 (7.1)	—	17 (5.7)	3 (1.0)	13 (4.4)
XXIV	41	1 (2.4)	13 (31.7)	11 (26.8)	2 (4.9)	12 (29.3)	—	—	1 (2.4)	— —	1 (2.4)
XXV	29	—	10 (34.5)	9 (31.0)	4 (13.8)	5 (17.2)	—	—	—	—	1 (3.5)

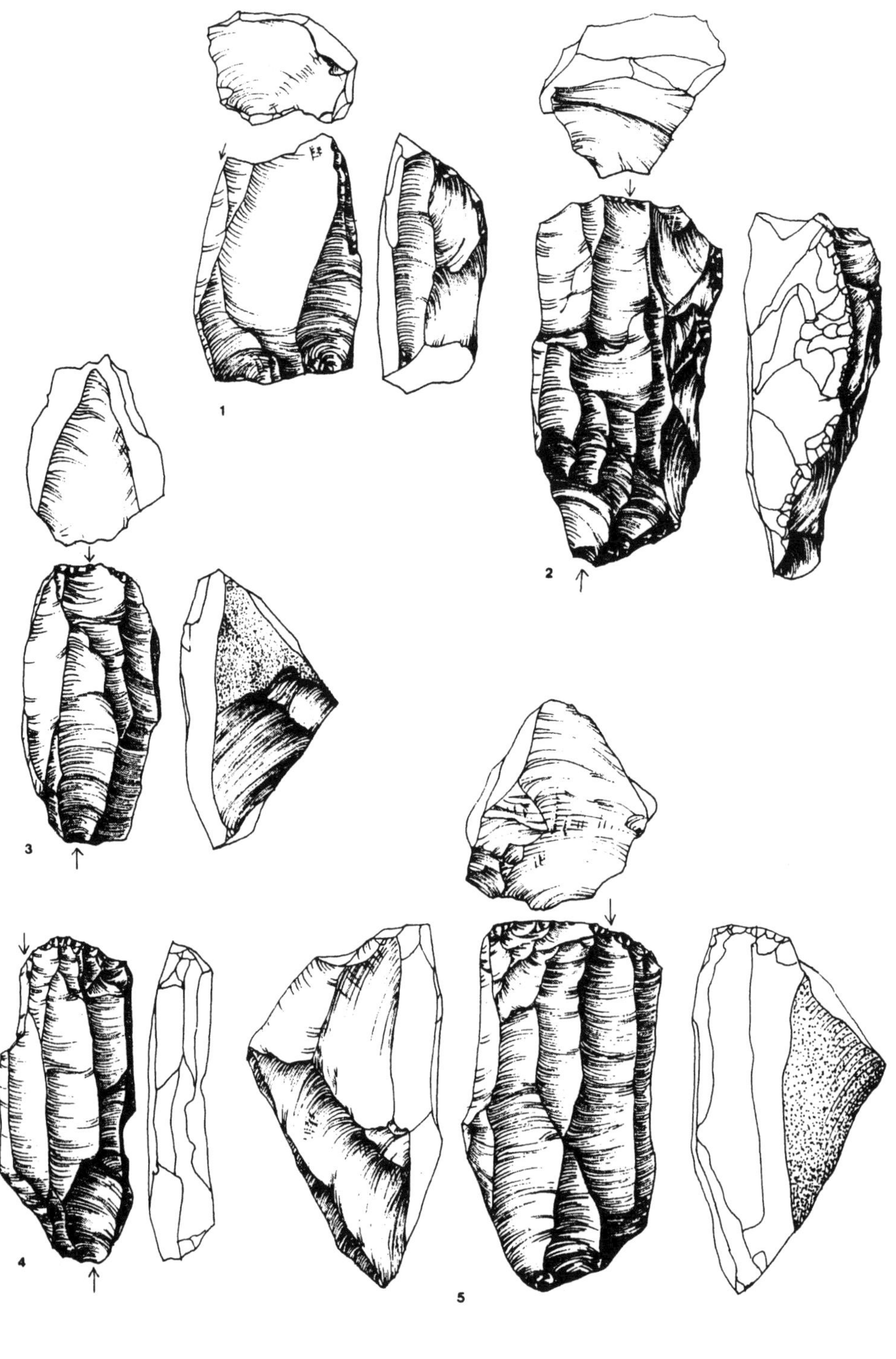

Figure 5.11. Cores from levels XVII and XVI at Ksar Akil.

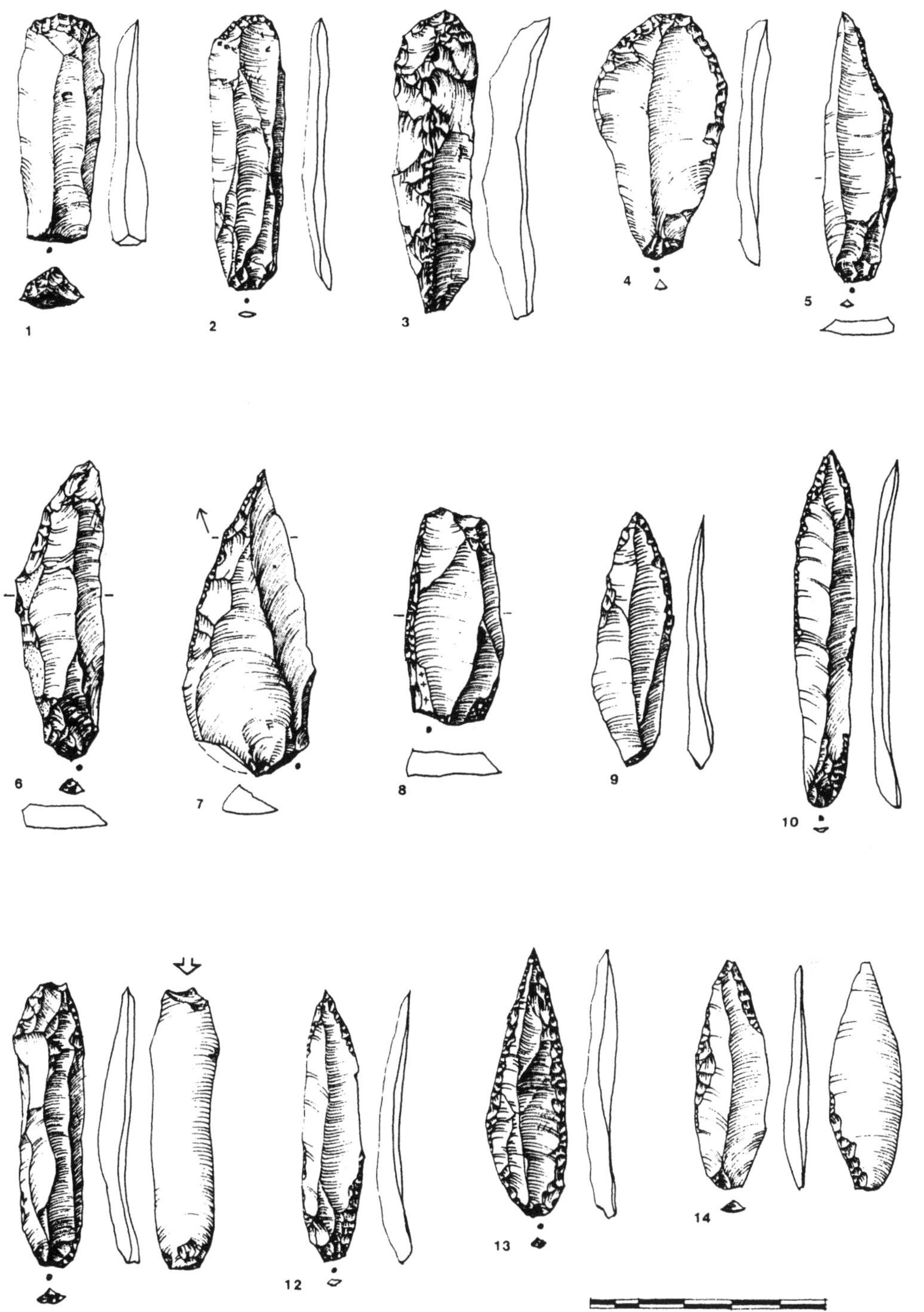

Figure 5.12. Tools from levels XVII and XVI at Ksar Akil.

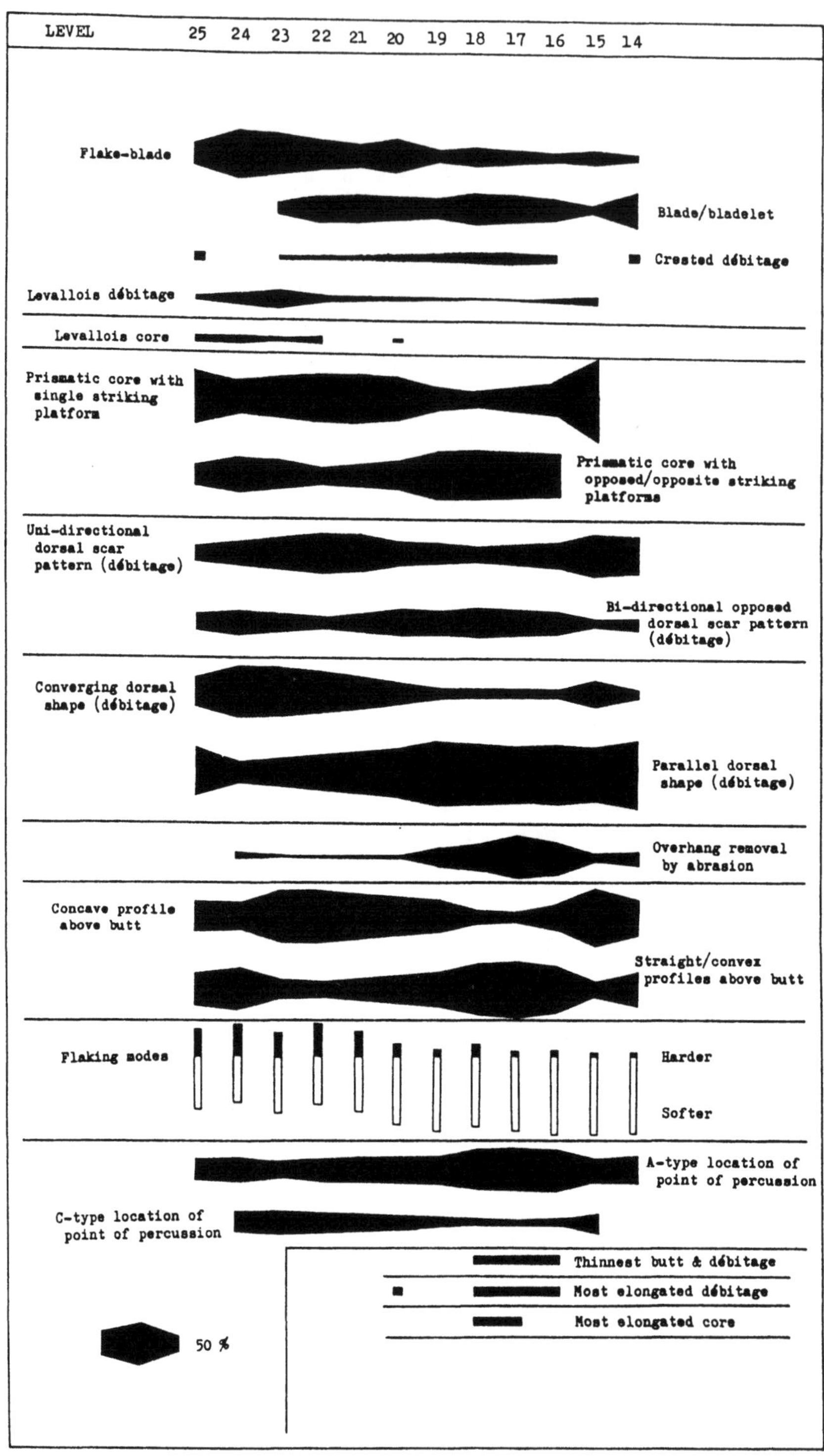

Figure 5.13. Frequencies of representative technological features in levels XXV-XIV at Ksar Akil.

Table 5.5 Frequencies of dorsal scar patterns of débitage (totals and percentages)

Level	Total	Unidirectional	Bi-directional opposed	Crossed	Centripetal	Trace of cresting	Miscellaneous
XIV	38	17 (44.7)	6 (15.8)	11 (29.0)	—	4 (10.5)	—
XV	80	40 (50.0)	9 (11.3)	30 (37.5)	—	1 (1.3)	—
XVI	384	121 (31.5)	112 (29.2)	107 (27.9)	3 (0.8)	40 (10.4)	1 (0.3)
XVII	685	161 (23.5)	250 (36.5)	156 (22.8)	—	118 (17.2)	—
XVIII	254	55 (21.7)	99 (39.0)	61 (24.0)	—	39 (15.4)	—
XIX	344	100 (29.1)	113 (32.9)	88 (25.6)	2 (0.6)	40 (11.6)	1 (0.3)
XX	513	168 (32.8)	176 (34.3)	104 (20.3)	2 (0.4)	63 (12.3)	—
XXI	240	118 (49.2	54 (22.5)	52 (21.7)	3 (1.3)	13 (5.4)	—
XXII	795	448 (56.4)	132 (16.6)	189 (23.8)	4 (0.5)	17 (2.1)	5 (0.6)
XXIII	270	110 (40.7)	70 (25.9)	89 (33.0)	—	1 (0.4)	—
XXIV	37	11 (29.7)	11 (29.7)	14 (37.8)	1 (2.7)	—	—
XXV	29	6 (20.7)	6 (20.7)	14 (48.3)	1 (3.5)	2 (6.9)	—

edges and blunt distal ends are closely related to cores with parallel sides, while pieces with converging edges and pointed ends are related to triangular shaped cores.

As mentioned earlier, Levallois cores in their final forms occur in levels XXV-XX only. The presence of Levallois debitage from level XIX upwards, where no Levallois cores are found, strongly suggests that the 'Levallois' debitage in these levels was produced during the reduction of non-Levallois cores.

2. *Retouched tools* (Figures 5. 4-12)

Judging from the cumulative graphs, there are at least three groups of levels with distinct sets of retouched pieces: levels XXIII-XXI, levels XX-XVIII, and levels XVII-XVI. The small samples from levels XXV-XXIV are not included here; both are characterised by numerous chamfered pieces.

i) Levels XXIII-XXI (Figure 5. 4)

Levels XXIII-XXI have tool kits composed of end-scrapers, chamfered

Table 5.6 Frequencies of scar patterns of cores (totals and percentages)

Level	Total	Uni-directional	Bi-directional	Crossed	Centri-petal	Single flake scar	Trace of cresting	Miscel-laneous
XIV	5	—	1 (20.0)	1 (20.0)	—	3 (60.0)	—	—
XV	6	3 (50.0)	—	— (50.0)	—	—	—	—
XVI	37	9 (24.3)	10 (27.0)	15 (40.5)	1 (2.7)	1 (2.7)	1 (2.7)	—
XVII	188	46 (24.5)	82 (43.6)	42 (22.3)	2 (1.1)	2 (1.1)	13 (6.9)	1 (0.5)
XVIII	43	11 (25.6)	18 (41.9)	7 (16.3)	—	1 (2.3)	2 (4.7)	4 (9.3)
XIX	140	37 (26.4)	51 (36.4)	42 (30.0)	3 (2.1)	3 (2.1)	1 (0.7)	3 (2.1)
XX	181	80 (44.2)	60 (33.1)	25 (13.8)	1 (0.6)	7 (3.9)	3 (1.7)	5 (2.8)
XXI	75	51 (68.0)	12 (16.0)	9 (12.0)	2 (2.7)	—	—	1 (1.3)
XXII	147	67 (45.6)	31 (21.1)	40 (27.2)	2 (1.4)	2 (1.4)	2 (1.4)	3 (2.0)
XXIII	110	35 (31.8)	36 (32.7)	38 (34.5)	—	—	—	1 (0.9)
XXIV	47	11 (23.4)	18 (38.3)	12 (25.5)	2 (4.3)	—	—	4 (8.5)
XXV	23	4 (17.4)	6 (26.1)	8 (34.8)	4 (17.4)	—	1 (4.3)	—

pieces and burins on lateral preparation.

ii) Levels XX-XVIII (Figure 5. 5)

Levels XX-XVIII have very high frequencies of end-scrapers (the highest is level XIX), end-scrapers on retouched blanks and backed pieces (partially backed pieces occur more often than those which are totally backed).

iii) Levels XVII-XVI (Figure 5. 6)

These levels are made up of a high percentage of end-scrapers, backed pieces (mainly partially backed) and 'Ksar Akil points'. The Ksar Akil points (generally known as 'el-Wad points' in the Southern Levant; see Bergman 1987: 13-4), which already appear in level XXII in small numbers, increase dramatically in level XVII.

In levels XV-XIV very few retouched tools were recovered but end-scrapers and notches seem to characterise both of them. It is highly likely that level XV is disturbed, while level XIV is virtually sterile and represents a major stratigraphic break (Azoury 1986: 172).

In spite of sharp changes in percentage of some of the tool types (such

Table 5.7. Frequences of dorsal shapes of débitage.

Level	Total	Parallel	Converging	Expanding
XIV	22	18: 81.8%	3: 13.6%	1: 4.5%
XV	28	19: 67.9%	8: 28.6%	1: 3.6%
XVI	185	139: 75.1%	16: 8.6%	30: 16.2%
XVII	377	283: 75.1%	34: 9.0%	60: 15.9%
XVIII	152	121: 79.6%	13: 8.6%	18: 11.8%
XIX	192	147: 76.6%	25: 13.0%	20: 10.4%
XX	402	264: 65.7%	114: 28.4%	24: 6.0%
XXI	188	104: 55.3%	76: 40.4%	8: 4.3%
XXII	651	278: 42.7%	352: 54.1%	21: 3.2%
XXIII	205	70: 34.1%	118: 57.6%	17: 8.3%
XXIV	26	8: 30.8%	15: 57.7%	3: 11.5%
XXV	8	5: 62.5%	3: 37.5%	—

as the sudden decrease in the number of chamfered pieces in level XXI and sudden increase of Ksar Akil points in level XVII), it seems that levels XXV-XVI represent a continuous developmental sequence in the "totals and proportions of type frequencies" (Bordes 1960: 109).

3. Blank selection

Table 5. 9 shows the types and frequencies of blanks selected to be retouched into tools. Non-cortical flakes as blanks gradually decrease within levels XXIV-XXII. Flake-blades also decrease through time, while blades slowly increase and are most numerous in level XVII; bladelets increase in number from level XIX and are most commonly used in XVI. Crested debitage is most often used in levels XVII-XVI.

DISCUSSION OF LEVELS XXIII-XIV

Levels XXIII-XXI exhibit a radically different approach to core reduction from that of levels XVIII-XVI. This difference seems to have resulted from a continuous and gradual process of technological evolution, with most of the technological features changing at different stages (Figure 5. 13).

From a technological viewpoint of core reduction, levels XXIII-XVI at Ksar Akil may be grouped into levels XXIII-XXI/XX and levels XX/XIX-XVI. These two sets of levels are characterised as follows:

Levels XXIII-XXI/XX

Single platform prismatic cores with converging sides are strongly related to unidirectional dorsal scars on the debitage, converging lateral edges, pointed distal ends, and debitage morphologically similar to Levallois points (see also Bergman 1981; Marks 1983). Other technological fea-

Table 5.8. Frequencies of overall shapes of cores (totals and percentages).

Level	Total	Parallel	Converging	Expanding	Miscellaneous
XIV	4	3 (75.0)	—	1 (25.0)	—
XV	5	3 (60.0)	2 (40.0)	—	—
XVI	31	18 (58.1)	8 (25.8)	—	5 (16.1)
XVII	164	99 (60.4)	33 (20.1)	3 (1.8)	29 (17.7)
XVIII	38	23 (60.5)	5 (13.2)	—	10 (26.3)
XIX	123	78 (63.4)	24 (19.5)	2 (1.6)	19 (15.4)
XX	160	70 (43.8)	48 (30.0)	—	42 (26.3)
XXI	68	29 (42.6)	29 (42.6)	—	10 (14.7)
XXII	135	65 (48.1)	61 (45.2)	2 (1.5)	7 (5.2)
XXIII	105	47 (44.8)	50 (47.6)	—	8 (7.6)
XXIV	39	26 (66.7)	12 (30.8)	—	1 (2.6)
XXV	21	12 (57.1)	4 (19.0)	4 (19.0)	1 4.8)

tures noted include convex dihedral faceted or convex multiple faceted butts, partially regularising facets (cf. Bordes 1947: 7) and a point of percussion between two ridges (type C on Figure 5. 13). There is no significant difference in butt thickness on blanks detached with hard and soft hammers. This seems to indicate that flaking tools of different hardness, such as hard or soft hammerstones and soft hammers like antler, were used in an identical manner. According to Azoury (1985: 34), phase 1 of Ksar Akil (levels XXV-XXII/XXI) is characterised by an Upper Palaeolithic typology plus "a specialised Levallois technique" which she regarded as similar to the blank production in levels B and C of Abou Sif, Jordan. The biggest difference between Azoury's work and this present study is the criteria used to define and identify Levallois cores and debitage, especially Levallois blades and their cores. Most of the cores and blanks classified by Azoury as Levallois have fallen into the categories of prismatic cores and non-Levallois flake-blades and blades. The Levallois blades and points in levels XXIII-XXI/XX are far more numerous than Levallois cores, probably because many of them were detached from prismatic cores during continuous production of blades (Bordes 1980; Bergman 1981; Marks 1983).

Table 5.9. Frequencies of blank classes for retouched pieces (totals and percentages).

Level	Total	Cortical debitage	Partially Cortical	Non Cortical	Flake-blades	Blades	Bladelets	Crested debitage	Cores	Older debitage	Natural Stones	Debris
XIV	18	—	—	3 (16.7)	8 (44.4)	4 (22.2)	1 (5.6)	2 (11.1)	—	—	—	—
XV	20	—	1 (5.0)	10 (50.0)	2 (10.0)	1 (5.0)	—	1 (5.0)	1 (5.0)	4 (20.0)	—	—
XVI	150	2 (1.3)	22 (14.7)	21 (14.0)	23 (15.3)	40 (26.7)	25 (16.7)	10 (6.7)	7 (4.7)	—	—	—
XVII	505	14 (2.8)	72 (14.3)	40 (7.9)	80 (15.8)	177 (35.0)	51 (10.1)	50 (9.9)	19 (3.8)	2 (0.4)	—	—
XVIII	187	5 (2.7)	39 (20.9)	28 (15.0)	35 (18.7)	44 (23.5)	16 (8.6)	8 (4.3)	2 (1.1)	9 (4.8)	—	1 (0.5)
XIX	300	9 (3.0)	68 (22.7)	44 (14.7)	61 (20.3)	66 (22.0)	25 (8.3)	13 (4.3)	4 (1.3)	6 (2.0)	4 (1.3)	—
XX	714	25 (3.5)	134 (18.8)	132 (18.5)	169 (23.7)	158 (22.1)	33 (4.6)	29 (4.1)	10 (1.4)	14 (2.0)	10 (1.4)	—
XXI	427	15 (3.5)	88 (20.6)	91 (21.3)	113 (26.5)	73 (17.1)	22 (5.2)	4 (0.9)	1 (0.2)	18 (4.2)	2 (0.5)	—
XXII	1,265	43 (3.4)	212 (16.7)	381 (30.1)	354 (28.0)	179 (14.2)	19 (1.5)	23 (1.8)	5 (0.4)	37 (2.9)	12 (1.0)	—
XXIII	396	9 (2.3)	59 (14.9)	167 (42.2)	106 (26.8)	26 (6.6)	6 (1.5)	11 (2.8)	3 (0.8)	5 (1.3)	1 (0.3)	3 (0.8)
XXIV	96	1 (1.0)	10 (10.4)	44 (45.8)	32 (33.3)	3 (3.1)	1 (1.0)	4 (4.2)	—	—	—	1 (1.0)
XXV	46	1	17 (2.2)	17 (37.0)	7 (37.0)	2 (15.2)	1 (4.3)	1 (2.2)	— (2.2)	—	—	—

Levels XX/XIX-XVI

Opposed platform prismatic cores with parallel sides are closely related to bi-directional opposed dorsal scars on blades and bladelets and blanks with parallel lateral edges and blunt distal ends. The techniques of debitage also include linear/punctiform butts, overhang removal by abrasion, or point of percussion located directly behind a clear ridge (type A on Figure 5. 13) and thinner blanks than before. Butts on blanks detached with soft hammers are thinner than those detached with hard hammers, suggesting that most of the debitage detached with soft hammers was flaked marginally by direct or indirect percussion.

Levels XXIII-XXI/XX of Ksar Akil, therefore, can be described as having a non-Levallois blade technology utilising single platform prismatic cores. Levels XX/XIX-XVI, on the other hand, exhibit an elaborate technology for producing blades and bladelets from parallel-sided prismatic cores with opposed striking platforms. The blade/bladelet production is far more refined than in the earlier levels in that cresting and re-cresting was used more frequently, as was overhang removal by abrasion to prepare the core's edge for a marginal blow.

The frequency trends of debitage classes are reflected in those blanks which were selected for tool manufacture (Table 5. 9) and it is evident that the changes in core reduction strategies are closely linked to changes in typology. The technology as a whole, however, seems to have changed more gradually, probably following the changes in typology, for the biggest discontinuity in the assemblages of retouched tools is between levels XXI and XX, whereas technological change is more marked from level XX or XIX onwards.

LEVELS XIII-VI

A major stratigraphic hiatus in level XIV separates the material of the lower levels from that of levels XIII-VI. Much of the published work on Ksar Akil has focused on the Mousterian, Transitional and early Upper Palaeolithic stone tool assemblages (e. g. Newcomer 1968-1969; Copeland 1975; Marks 1983; Azoury 1986; Marks and Volkman, in press) with surprisingly few publications concerned with the later Upper Palaeolithic levels XIII-VI. Indeed, three of these levels (XI, X and IX) have never been described in detail.

At a conference on the terminology of the Palaeolithic of the Levant held at the Institute of Archaeology, London, in 1969 (Bergman 1987: 7-9), Waechter presented a brief summary of levels XIII-VI. Following the suggestion of Hours and Bar-Yosef it was decided to designate all of the material from these levels as "Levantine Aurignacian" with three subdivisions: Phase A (levels XIII-XI), Phase B (levels X and IX) and Phase C (levels VIII-VI).

Although most of the material excavated by Tixier at Ksar Akil comes from the Epi-Palaeolithic levels, some Aurignacian artifacts were also recovered (Tixier and Inizan 1981: 360). From the description of the archaeological phases recognised by Tixier (Tixier and Inizan 1981), it is

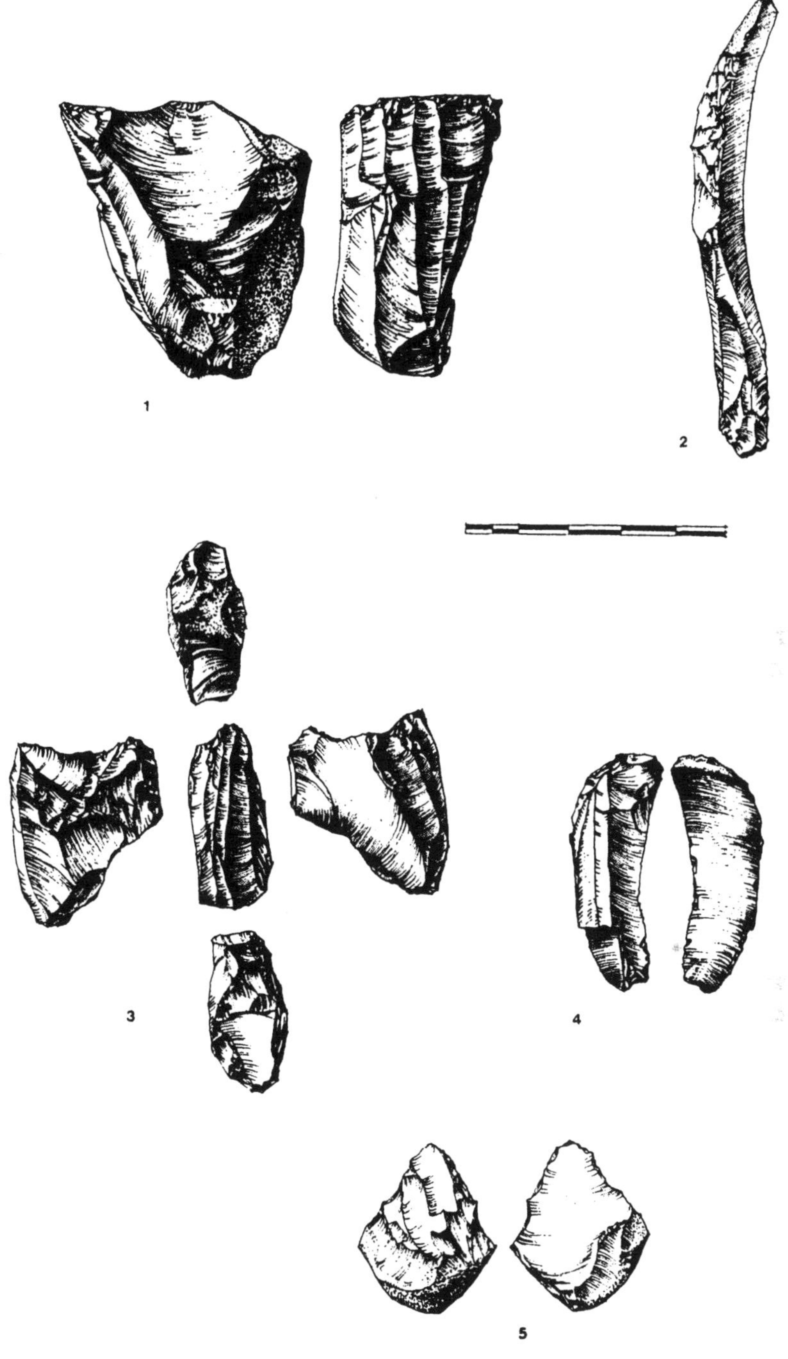

Figure 5.14. Cores and debitage from levels XIII-XI at Ksar Akil.

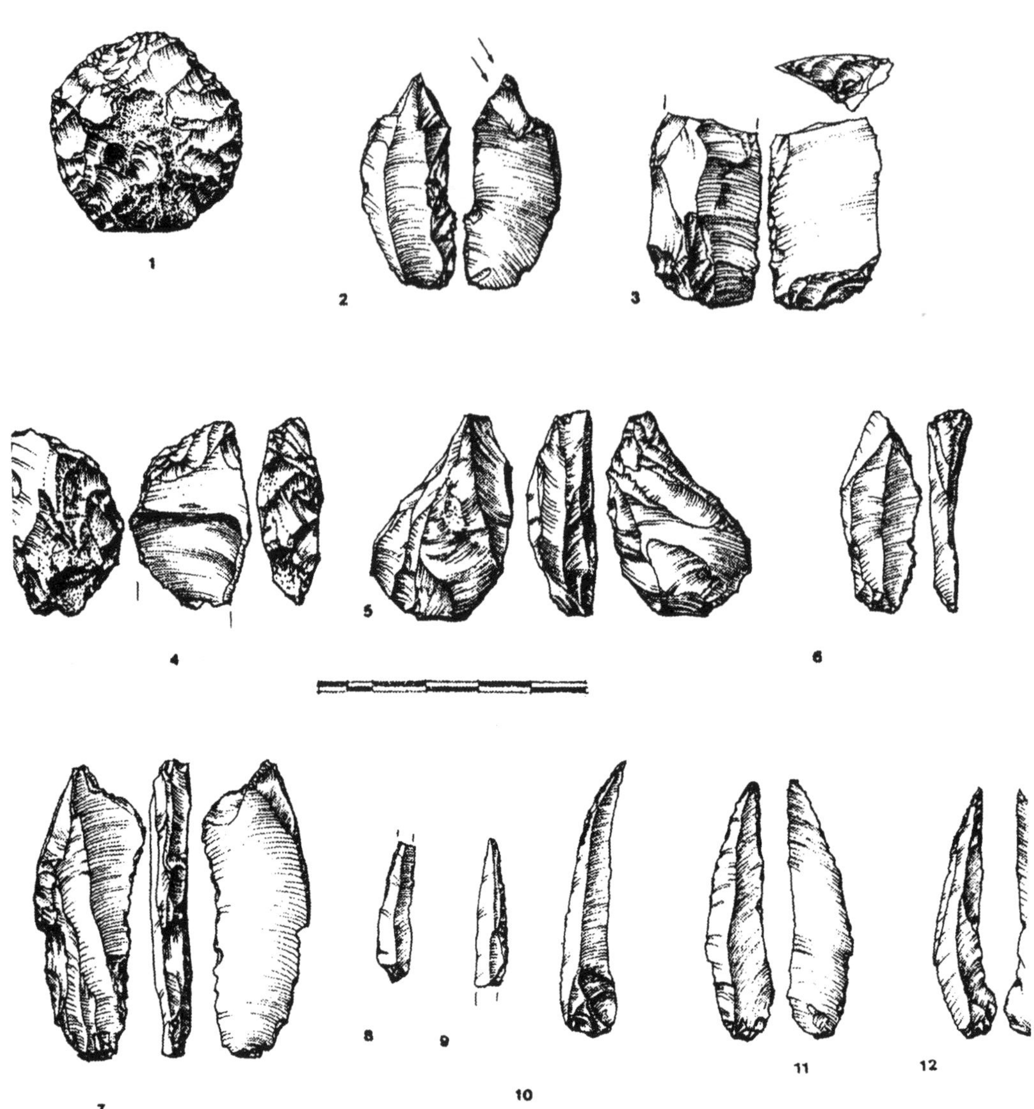

Figure 5.15. Tools from levels XIII-XI at Ksar Akil.

possible to provide the following *rough* comparison with the 1937-1938 excavations:

Boston College (1937-1938)	Tixier (1969-1975)	
level	*phase*	*level*
VI	IV	8AC-1OA
?	V	10B-10H(1)
VII-VIII	VI	10H2-11C
IX-(X)?	VII	12

TECHNOLOGY AND TYPOLOGY OF LEVELS XIII-VI

1. Debitage and cores

The most common kinds of blade and bladelet cores found in levels XIII-VI are prismatic and have single, plain platforms (from 64. 5% to 94. 1%). As in levels XVII-XVI, platform abrasion appears on most cores and was used to prepare them for a blow aimed close to the platform's edge (marginal flaking). Flake cores are more diverse in their morphology and may have one or more platforms.

Core preparation and maintenance incorporates many of the features seen in the earlier levels XVII-XVI, including the use of cresting in preparation and core tablets in maintenance. In addition, cresting appears to have been used to correct knapping accidents such as hinge fractures, as well as to maintain the curvature of the flaking face during reduction. One other feature noted on the bladelet cores is the occasional removal of a large flake from the side of the platform in order to narrow the platform and flaking face. It is essential to maintain a relatively narrow platform and flaking face during the manufacture of twisted bladelets. In levels XIII-XI blade debitage begins with the production of larger, wider blanks which gradually become smaller and narrower during reduction. While it is possible that these layers have separate reduction sequences for blades and bladelets, the evidence is less clear than in levels VIII-VI where the tiny bladelets used to make microliths may be detached from 'tools' like carinated scrapers or burins.

The kinds of blanks produced in each level vary in different parts of the sequence (Table 5. 10 and Figure 5. 22). In levels XIII-XI blades and bladelets outnumber flakes, and blades are more common than bladelets (cf. Tixier 1963). In levels X and IX blades and bladelets remain most numerous but the bladelets dominate. An important shift in the flaking technology occurs in level VIII. The assemblages in levels VIII-VI are all characterised by large numbers of unretouched flakes. Large numbers of tiny bladelets also occur in these levels but are certainly underrepresented in the 1937-1938 collections.

The butts on flakes, blades and bladelets are most often plain (Table 5. 13). In the case of flakes there is a greater variation in butt types; faceted and cortical butts are always more common on flakes. Flake butts are also larger than those on blades, which are usually quite small (Table 5. 12). This is due to the fact that the technique for detaching blades and bladelets relies on striking close to the platform edge, removing a small part of it.

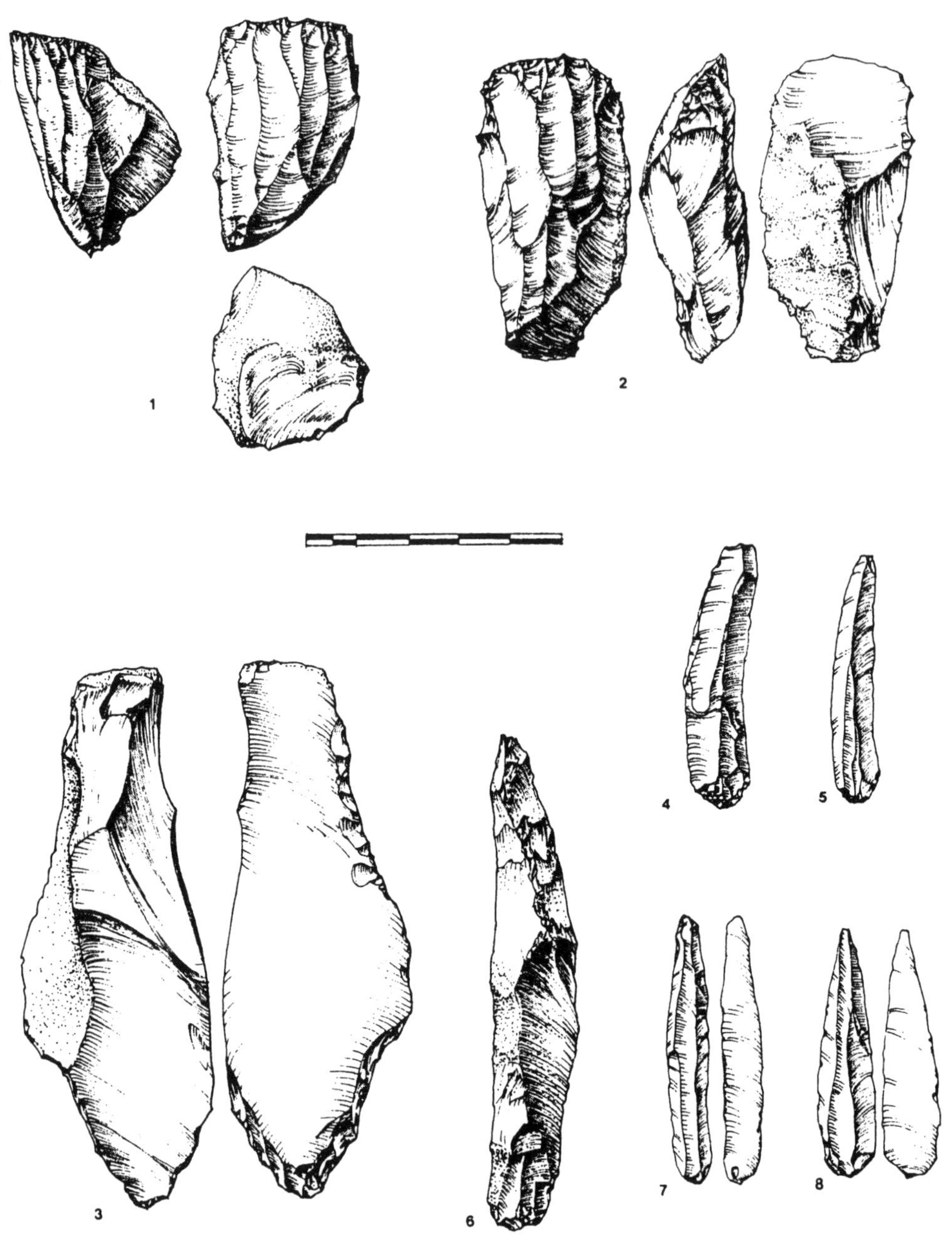

Figure 5.16. Cores and debitage from levels X and IX at Ksar Akil.

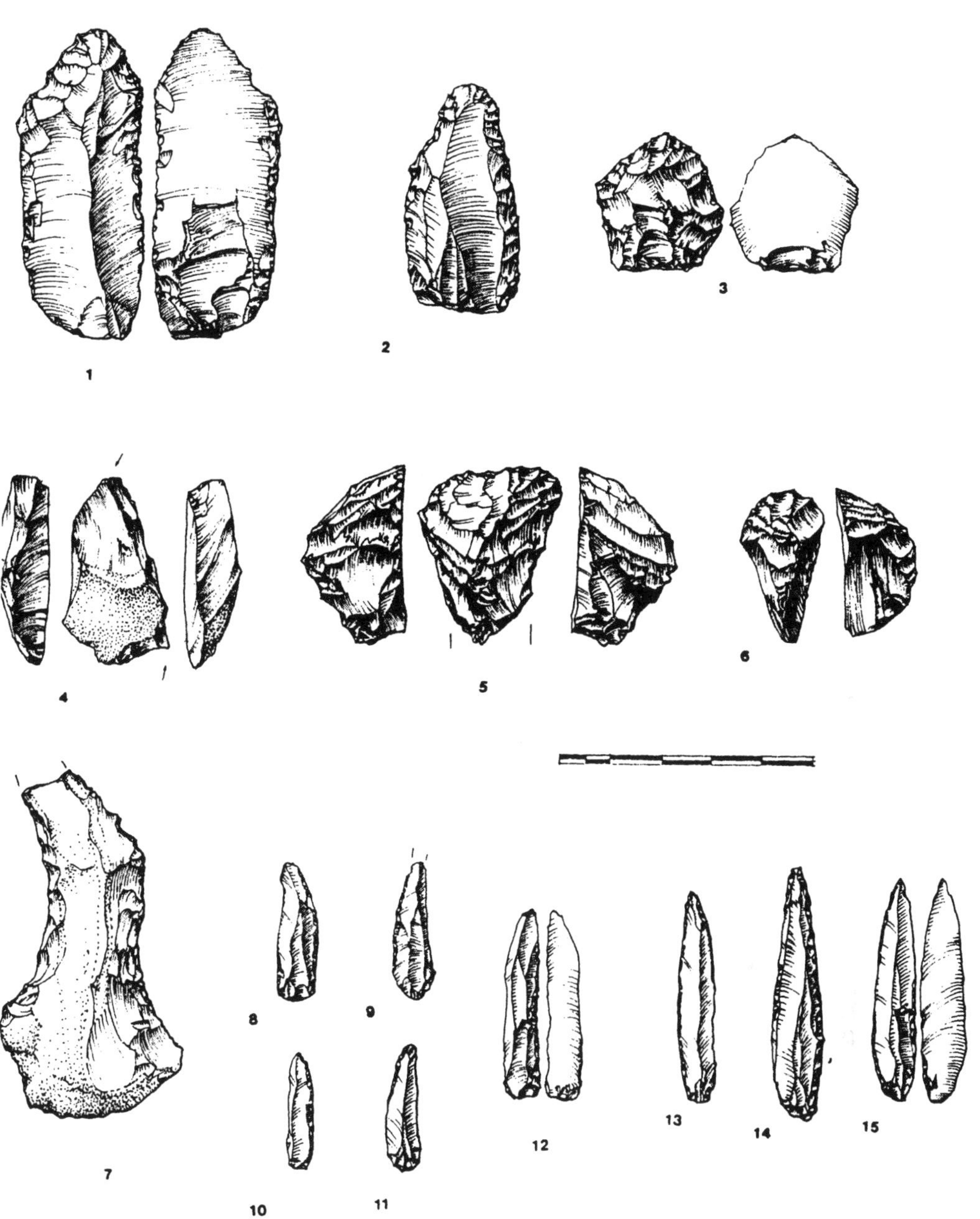

Figure 5.17. Tools from levels X and IX at Ksar Akil.

Table 5.10. Percentages of flakes, blades and bladelets in levels VI to XIII at Ksar Akil.

Level	Number of Pieces	Flakes %	Blades %	Bladelets %
VI	1588	50.63	14.30	35.08
VII	5115	90.24	8.35	1.41
VIII	6413	73.34	13.55	13.11
IX	31019	42.18	18.52	39.30
X	9926	26.99	33.77	39.24
XI	3842	48.54	36.34	15.12
XII	2953	30.44	63.53	6.03
XIII	358	42.46	41.06	16.48

Table 5.11. Thickness of flake and blade/bladelets blanks (means and standard deviations).

Level			Number of pieces	Flakes (mms)		Blades/Bladelets (mms)	
				TH	SD	TH	SD
IX	F4	7.25	467	6	3	3	2
IX	E4	7.65	199	6	3	3	2
IX	F4	7.75	303	9	4	4	2
X	E4	8.10	439	6	2	4	2
X	F4	8.40	454	6	3	3	1
X	E4	8.65	453	7	3	4	2
XI	F4	8.95	265	5	3	5	3
XI	F4	9.30	361	8	4	5	3
XII	F4	9.70	239	7	4	4	2
XII	E4	10.00	492	7	4	5	2
XIII	F4	9.90	124	9	4	5	3

The flaking hammers used to detach flakes, blades and bladelets throughout these levels were usually soft. The use of a hard hammer was more frequently observed on flakes and only rarely on blades or bladelets.

The flake attributes are generally less standardised than those noted on blades and bladelets, reflecting the fact that flakes can be produced in many different ways, while blades and bladelets are usually the result of careful and deliberate core preparation. Cortex occurs more often on flakes than on blades (Tables 5. 14 and 5. 15); as blades are detached at a later stage in the flaking sequence this is not surprising. Platform abrasion, which helps prevent crushing, is essential for blade production and is less common on flakes which are generally detached with a non-marginal blow.

Table 5.12. Butt dimensions (width and thickness) on flakes and blades/bladelets (means and standard deviations).

Level		Number of	Flakes (mms)				Blades/Bladelets (mms)			
		pieces	W	SD	TH	SD	W	SD	TH	SD
VIII	F3 6.24	190	14	9	6	3	3	1	1	—
IX	F4 7.25	467	12	8	5	3	3	1	1	—
IX	E4 7.65	199	12	7	4	3	3	2	1	—
IX	F4 7.75	303	14	7	5	3	2	1	1	—
X	E4 8.10	439	12	8	4	2	5	3	1	—
X	F4 8.40	454	11	6	4	3	3	2	1	—
X	E4 8.65	453	12	8	4	3	5	3	2	1
XI	F4 8.95	265	10	8	3	2	5	4	2	1
XI	F4 9.30	361	14	9	6	4	5	4	2	1
XII	F4 9.70	239	15	9	5	4	5	3	2	1
XII	E4 10.00	492	14	8	5	3	5	4	2	1
XIII	F4 9.90	124	13	8	5	4	6	4	3	2

Table 5.13. Form of butts on blades and bladelets.

Level			Number of pieces	Plain %	Facetted %	Cortical %	Crushed or broken %	Imitating a dihedral burin %	Unidentified %
VI	F4		107	99.07	0.94	—	—	—	—
IX	F4	7.25	193	90.67	0.52	1.04	7.77	—	—
IX	E4	7.65	82	90.24	—	1.22	8.54	—	—
IX	F4	7.75	238	85.29	0.42	1.68	10.08	—	2.52
X	E4	8.10	349	87.97	0.57	2.01	9.46	—	—
X	F4	8.40	272	80.88	5.15	4.41	8.82	—	0.74
X	E4	8.65	403	81.89	2.98	0.50	12.90	0.25	1.49
XI	F4	8.95	82	74.39	6.10	3.66	8.54	—	7.32
XI	F4	9.30	220	74.09	5.00	2.73	13.18	1.82	3.18
XII	F4	9.70	189	74.07	2.12	1.59	11.11	2.12	9.00
XII	E4	10.00	353	72.52	1.98	3.40	7.65	4.25	10.20
XIII	F4	9.90	62	72.58	6.45	8.07	8.07	4.84	—

On blades and bladelets parallel lateral edges are most frequent, although each level studied here has a significant percentage of pieces with converging edges (Table 5. 16). These usually have blunt distal ends, while pointed extremities occur more often than on flakes (Table 5. 17).

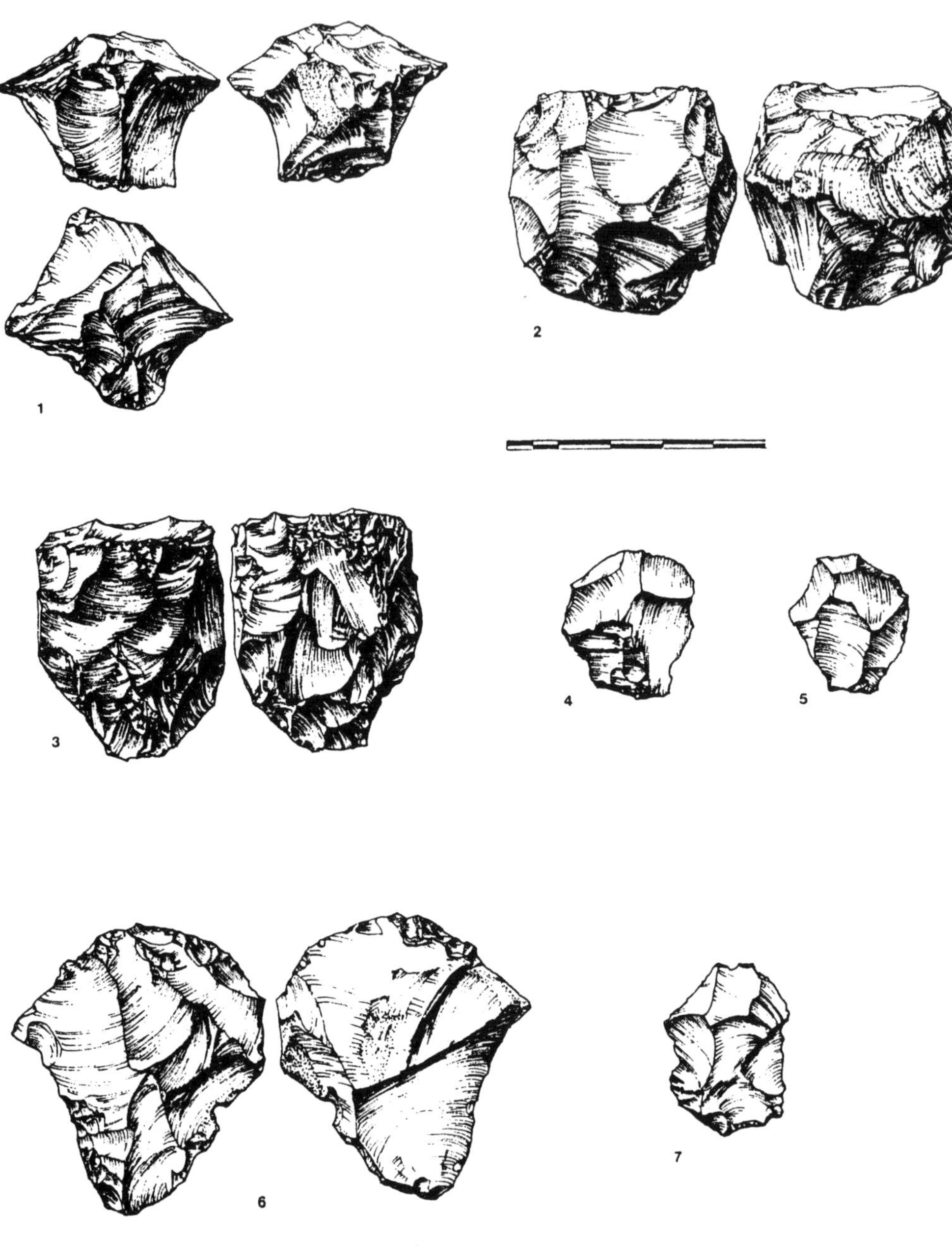

Figure 5.18. Cores and debitage from levels VIII and VII at Ksar Akil.

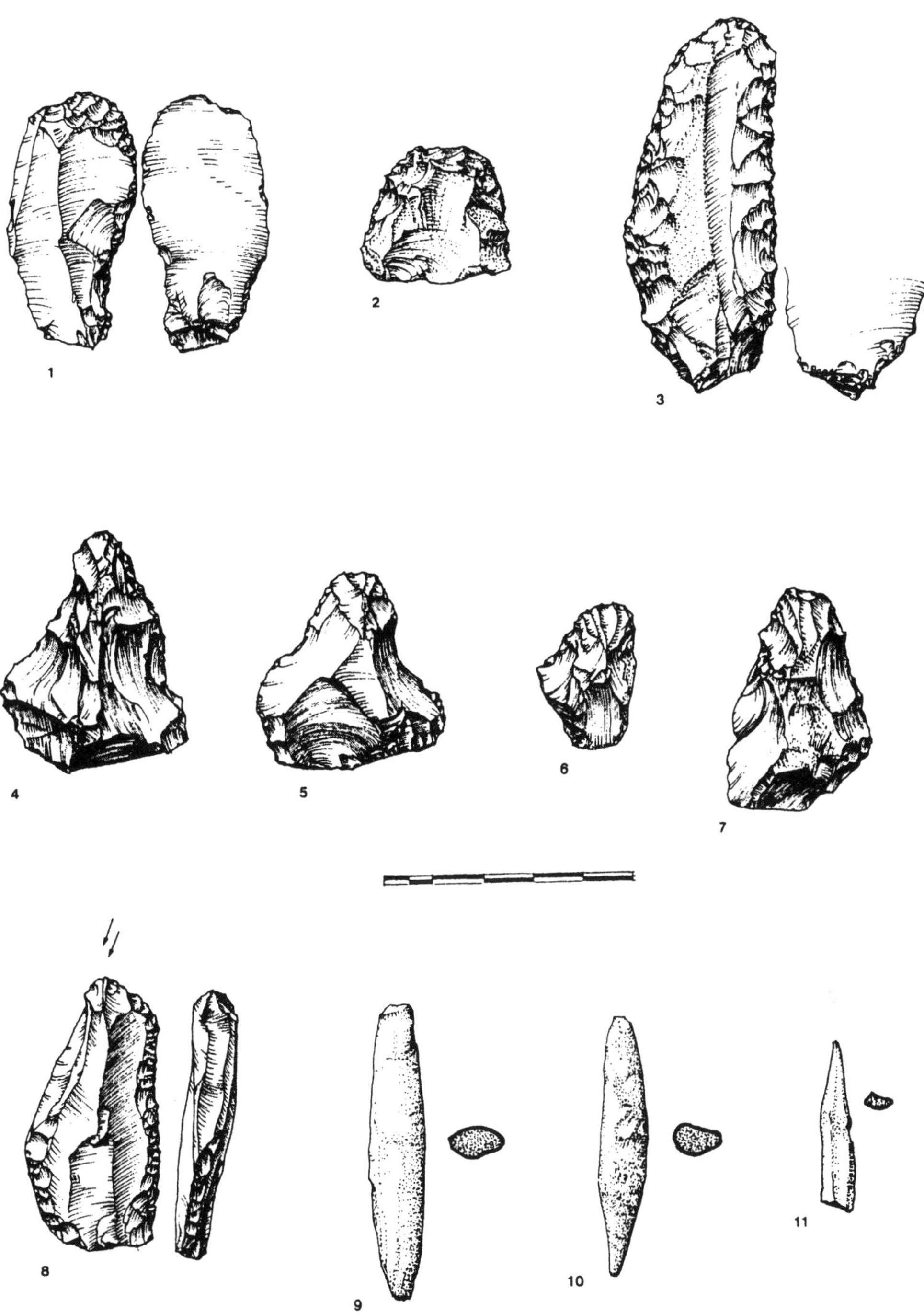

Figure 5.19. Tools from levels VIII and VII at Ksar Akil.

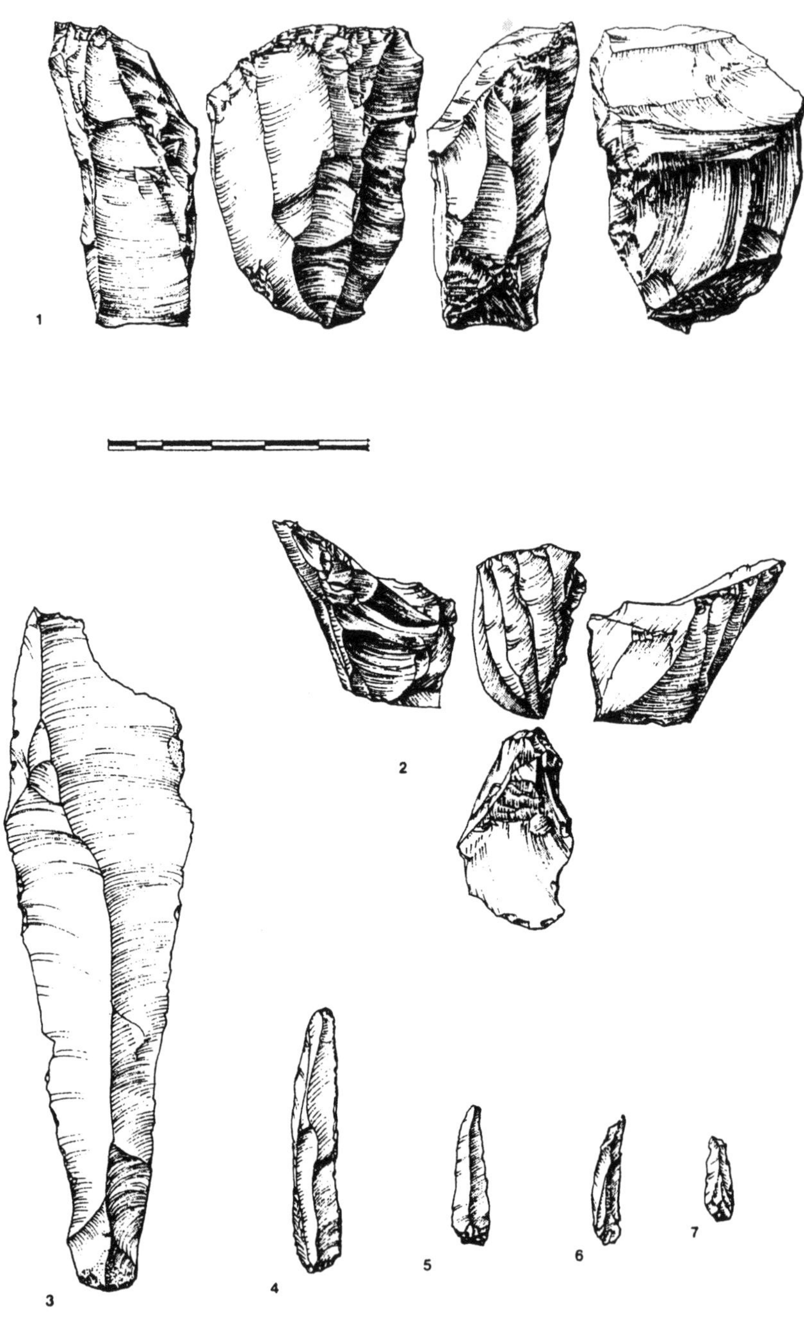

Figure 5.20. Cores and debitage from level VI at Ksar Akil.

Table 5.14. Frequencies of cortex and platform abrasion on flakes.

Level			Number of pieces	Cortex %	Platform abrasion %
VIII	F3	6.24	146	36.99	19.18
IX	F4	7.25	274	38.69	38.69
IX	E4	7.65	117	30.77	33.33
IX	F4	7.75	65	35.39	30.77
X	E4	8.10	90	26.67	54.44
X	F4	8.40	182	35.17	37.36
X	E4	8.65	50	38.00	40.00
XI	F4	8.95	183	38.80	30.06
XI	F4	9.30	141	51.77	34.04
XII	F4	9.70	50	40.00	32.00
XII	E4	10.00	139	39.57	33.09
XIII	F4	9.90	62	52.23	9.68

The dorsal scar pattern on blades and bladelets is overwhelmingly dominated by unidirectional removals (Table 5. 18). Crossed scars, when they do occur, are generally related to cresting, and the rest of the scars on the blank are unidirectional. This is consistent with the cores which usually have single platforms and are flaked unidirectionally.

A change noted in the profile shape of the blades and bladelets (Table 5. 19) is related to a shift in the techniques of flaking between levels XIII-XI and X-IX. In the first group of levels most of the blanks are twisted in profile; the amount of twisted debitage is never less than 55%. This is directly linked to the percentage of blanks with offset debitage which can reach over 54% in the samples (Figure 5. 23). In levels X and IX, however most of the blades and bladelets have straight or curved profiles, while offset debitage occurs on only 11% of the blanks.

2. *Retouched tools*

The retouched tools in levels XIII-VI separate into four distinct stratigraphic groups.

i) Levels XIII-XI (Figures 5. 15 and 5. 24)

These levels are characterized by a higher percentage of burins than end-scrapers (excluding carinated types; Bergman 1987: 144-5). The number of burins immediately separates levels XIII-XI from levels XX-XVI which have an exceptionally low burin index (see Newcomer 1972: 379 and Azoury 1986: 104). Carinated tools account for 15%-28% of the assemblages, while retouched bladelets and el-Wad points make up about 17%. The Aurignacian index ranges between 15. 44 (level XII) and 31. 11 (level XI).

Table 5.15. Frequencies of cortex and platform abrasion on blades and bladelets.

Level			Number of pieces	Cortex %	Platform abrasion %
VI	F4		107	4.67	91.59
IX	F4	7.25	193	18.65	92.75
IX	E4	7.65	82	8.54	93.90
IX	F4	7.75	238	15.55	94.54
X	E4	8.10	349	15.76	89.97
X	F4	8.40	272	19.12	91.18
X	E4	8.65	403	17.37	89.08
XI	F4	8.95	82	12.20	74.39
XI	F4	9.30	220	27.27	75.00
XII	F4	9.70	189	26.98	77.78
XII	E4	10.00	353	22.10	77.90
XIII	F4	9.90	62	25.81	83.87

ii) Levels X and IX (Figures 5. 17 and 5. 24)

Level IX is mixed and, therefore, it has been excluded from the present discussion (see Bergman 1987: 129 for data on level IX). Level X is made up of more scrapers than burins. Carinated tools decrease significantly in percentage from the previous level XI to around 11% (see also Newcomer 1972: 380-1). Retouched bladelets and el-Wad points make up 33% of the tool assemblage. The el-Wad variant with inverse retouch at the proximal end appears for the first and only time. The Aurignacian index for level X is 15. 63.

iii) Levels VIII and VII (Figures 5. 19 and 5. 24)

These two levels have tool kits composed of very high percentages of scrapers, especially nosed and shouldered types which account for up to 33% of all tools. Burins, carinated tools and el-Wad points occur in reduced numbers from the preceding levels; a small number of microliths, sometimes made on twisted debitage, were also recovered. The Aurignacian index in both levels is very high: 39. 25 in level VIII and 42. 45 in level VII. An important feature of both levels is the presence of numerous bone tools, consisting of points and awls; over 70% of the 131 osseous tools recovered in 1937-1938 come from here (Newcomer 1974; Bergman 1987: Appendix 3).

iv) Level VI (Figures 5. 21 and 5. 24)

This level marks a return of the dominance of burins (42. 50% of the tool kit) and contains a number of burins on 'Clactonian notches' (Newcomer 1971, 1972). Scrapers occur in reduced numbers, although the percentage of carinated tools increases. Also included in the collection from 1937-1938 are a number of non-geometric microliths, as well as

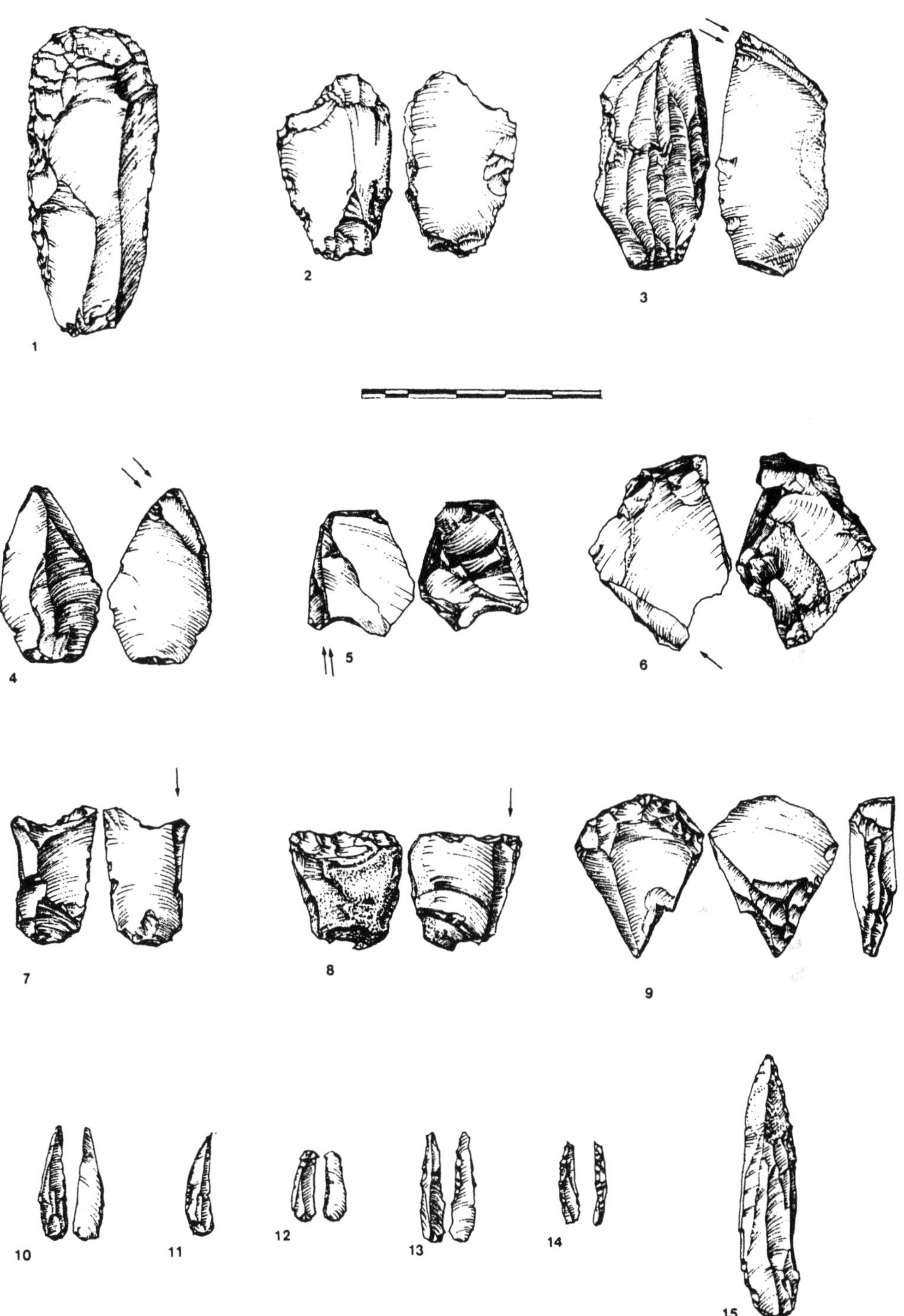

Figure 5.21. Tools from level VI at Ksar Akil.

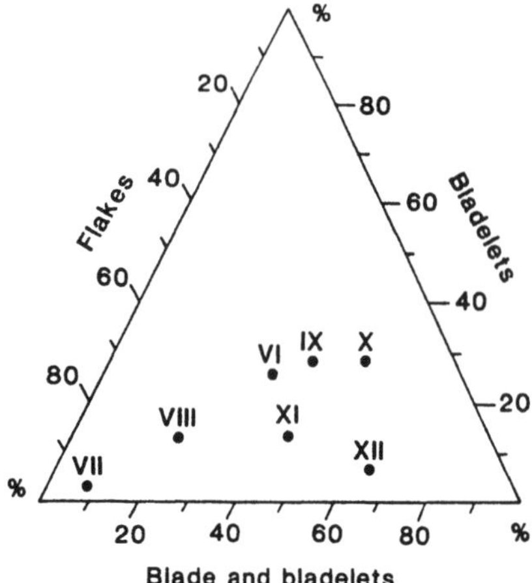

Figure 5.22. Three-pole graph of blank classes in levels XII-VI at Ksar Akil.

lamelles Dufour. The Aurignacian index is lower (12.15) than in the two previous levels. Sixteen bone and antler tools were recovered from level VI.

3. Blank selection

The pattern of blank selection throughout levels XIII-IX was remarkably consistent in that flakes, often with cortex, were most often used for end-scrapers, carinated tools and burins. El-Wad points were made on both blades and bladelets. However, in levels XIII-XI most el-Wad points were made on twisted blades. In levels X and IX the points were generally made on straight or curved blanks and there were greater numbers made on bladelets. In contrast to the scrapers, carinated tools and burins, cortex almost never occurs on el-Wad points.

Levels VIII and VII are characterised by flake production with a low bladelet index (< 30). It is hardly surprising, therefore, that most retouched tools are made on flakes. These levels have separate reduction sequences for blades and bladelets. Some tiny bladelets used as tool blanks probably come from pieces classified as tools.

Level VI is similar to levels XIII-IX, with most scrapers, carinated tools and burins being made on flakes. Blades and bladelets are used for el-Wad points, while the microliths and *lamelles Dufour* are made on tiny bladelets detached from cores, carinated scrapers or burins.

Crested pieces and core tablets are not usually selected for tool manufacture in any of the levels in this part of the sequence. When they are used as blanks it is always to make scrapers, carinated tools or burins.

DISCUSSION OF LEVELS XIII-VI

Levels XIII-XI

These levels are characterised by single platform bladelet cores with many of the technological features seen in levels XVII-XVI, including plain platforms, the extensive use of cresting, platform abrasion and the core

Table 5.16. Form of lateral edges on blades and bladelets.

Level			Number of pieces	Parallel %	Converging %	Expanding %	Unidentified %
VI	F4		107	65.42	32.71	0.94	0.94
IX	F4	7.25	193	53.89	37.31	7.77	1.04
IX	E4	7.65	82	68.29	31.71	—	—
IX	F4	7.75	238	68.91	28.15	2.94	—
X	E4	8.10	349	61.89	34.10	3.73	0.29
X	F4	8.40	272	61.03	30.15	8.09	0.74
X	E4	8.65	403	60.05	35.98	3.72	0.25
XI	F4	8.95	82	64.63	24.39	7.32	3.66
XI	F4	9.30	220	76.36	15.91	7.27	0.46
XII	F4	9.70	189	64.55	22.75	11.64	1.06
XII	E4	10.00	353	64.87	30.88	3.97	0.28
XIII	F4	9.90	62	62.90	24.19	9.68	3.23

tablet technique. The debitage is primarily composed of blades and bladelets, with blades being numerically dominant. The blades and bladelets are usually soft-hammer-struck with the blow landing behind or beside a ridge on the flaking face. A major difference between levels XIII-XI and the earlier levels is the high number of twisted blanks which never make up less than 50% of the bladelet sample.

The tool kits are composed of more burins than scrapers (excluding carinated types), while retouched bladelets and el-Wad points occur in smaller numbers. El-Wad points tend to be twisted in profile and the morphology of the blades and bladelets selected causes the pointed tips to be asymetrical and offset to the long axis of the blank (Bergman 1981: Figure 1-2). Aurignacian tool types occur in some numbers throughout these levels; level XII is characterized by carinated and flat-faced carinated burins (Newcomer 1972: 380), while level XI has large numbers of lateral carinated scrapers (Azoury 1986: Figure 197; Bergman 1987: 62).

Levels X and IX

Levels X and IX also have a blade and bladelet technology utilising single platform cores with some of the same technological features noted in the previous levels. Blades and bladelets dominate the unretouched pieces, while bladelets rather than blades are most common. Another significant difference occurs in the bladelet sample which has fewer pieces with twisted profiles and offset debitage.

As stated earlier, level X is made up of more scrapers than burins, while up to 33% of the tools are retouched bladelets and el-Wad points or

Table 5.17. Form of distal ends on blades and bladelets.

Level			Number of pieces	Hinge fracture %	Pointed %	Blunt or Cortical %	Unidentified %
VI	F4		107	1.87	62.62	34.58	0.94
IX	F4	7.25	193	6.22	32.12	58.03	3.63
IX	E4	7.65	82	6.10	26.83	67.07	—
IX	F4	7.75	238	6.72	26.05	66.81	0.42
X	E4	8.10	349	7.16	30.95	61.03	0.86
X	F4	8.40	272	8.82	33.82	51.10	6.25
X	E4	8.65	403	8.93	31.02	58.07	1.99
XI	F4	8.95	82	12.20	21.95	58.54	7.32
XI	F4	9.30	220	9.09	18.18	71.82	0.91
XII	F4	9.70	189	12.17	22.22	64.02	1.59
XII	E4	10.00	353	9.07	26.91	60.91	3.12
XIII	F4	9.90	62	14.52	14.52	70.97	—

variants. El-Wad variants with inverse retouch are absent in levels XIII-XI. The shift in the flaking technology is clearly seen in the points, which tend to be made on straight or curved blanks with the pointed tips being symmetrical in plan. Carinated tools occur less frequently than in the preceding level XI, giving this material a less pronounced Aurignacian appearance.

Levels VIII and VII

These two levels mark a dramatic shift in the flaking technology in the upper part of the Ksar Akil sequence (see also Dortch 1970: 183). For the first time flakes dominate the sample of debitage. The cores which produced these pieces are most often shapeless. They appear to have been flaked alternately, with one removal forming the platform for the next blow. Platform abrasion is rarely used as the blow lands well on to the platform (resulting in large butts) and, hence, the danger of crushing is negligible. There are separate reduction sequences for blades and bladelets. Tiny bladelets, used as blanks for tools, probably come from pieces categorised as tools.

The tools are primarily composed of scrapers (*c.* 60%) with nosed and shouldered scrapers making up around one-third of the assemblages. Burins and carinated tools occur much less frequently, as do tools like el-Wad points. Bone and antler tools are found in relatively large numbers in both levels; there are only six examples of these tools in the underlying levels (XIII-IX).

Level VI

Level VI probably represents the start of the terminal Upper Palaeolithic at Ksar Akil and is also dominated by flakes. The sample of debitage

Table 5.18. Scar patterns on blades and bladelets.

Level			Number of pieces	Uni-directional %	Opposed %	Crossed %	Multi-directional %	Unidenti-fied %
VI	F4		107	97.20	—	2.80	—	—
IX	F4	7.25	193	91.19	1.04	7.77	—	—
IX	E4	7.65	82	90.24	2.44	7.32	—	—
IX	F4	7.75	238	87.82	4.62	6.30	1.26	—
X	E4	8.10	349	85.10	2.87	8.88	—	3.15
X	F4	8.40	272	88.97	2.57	6.62	0.74	1.10
X	E4	8.65	403	86.10	1.49	10.92	1.24	0.25
XI	F4	8.95	82	67.07	2.44	13.42	—	17.07
XI	F4	9.30	220	62.72	4.55	20.00	4.55	8.18
XII	F4	9.70	189	73.02	1.59	15.34	1.59	8.47
XII	E4	10.00	353	76.20	3.97	13.03	1.42	5.38
XIII	F4	9.90	62	91.94	4.84	1.61	—	1.61

contains more bladelets than blades. As in the two previous levels there are separate core reduction sequences aimed at the production of blades and bladelets; the latter appear to result from 'true' bladelet core reduction, while some tiny examples are probably the by-products of 'tool' manufacture.

This level marks the return of a burin-dominated tool kit after their decline in levels X-VII. The 'burin on a notch' is virtually confined to level VI, which also contains other truncation burins made on small flakes. Nosed and shouldered scrapers appear in greatly reduced numbers, while carinated tools are more common than in the two previous levels. There are a number of non-geometric microliths and *lamelles Dufour* in level VI, although they occur in lower percentages than in Tixier's sample (Tixier and Inizan 1981).

There are a number of major differences in the percentages and morphology of the retouched tools between these four groups of levels. When combined with the data on the flaking technology it appears unlikely that the upper part of the Ksar Akil sequence represents a developmental evolution. It is suggested that each group of levels (e. g. XIII-XI, X-IX and VIII-VII) displays a relatively short-term evolution which is separated from the other groups by breaks in the archaeological sequence. However, it must be stated that Tixier (Tixier and Inizan 1981) has not identified any breaks in his sequence, which extends down to the top of level X of the 1937-1938 excavations, and regards his material as part of an Aurignacian evolving *in situ* at Ksar Akil.

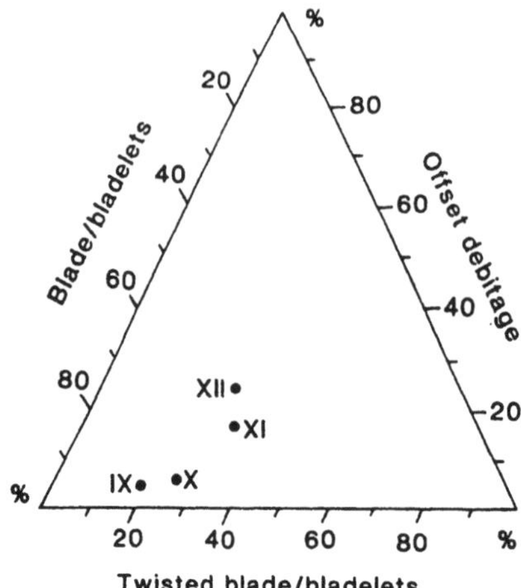

Figure 5.23. Three-pole graph of technological attributes of blades/bladelets in levels XII-IX at Ksar Akil.

THE LEVANTINE AURIGNACIAN

It seems worthwhile to consider here the relevance of the new interpretation of the term 'Aurignacian', proposed by prehistorians working in Israel, to the site of Ksar Akil. Gilead (1981a, 1981b: 339) states that the Aurignacian is "technologically dominated by the production of flakes over blades and typologically by the end-scrapers (steep or flat), the burins or both (> *c.* 50%) as well as low quantities of bladelet tools (< *c.* 20%)".

The use of European terminology to describe Stone Age cultures in the Levant has a long history. At Ksar Akil the material in levels XX-XV was originally described as Châtelperronian (Ewing 1947), because of the large numbers of backed and partially backed blades, as well as the presence of 'Châtelperron points'. Stone tool assemblages with significant numbers of carinated tools, 'Font Yves points' and Aurignacian blades were naturally called Aurignacian. Copeland (1986) has rightly pointed out that the use of a European term like 'Aurignacian' should imply a certain degree of similarity between assemblages in Europe and the Levant. On this point it is worth noting that François Bordes (Wenner Gren Symposium 1969; Bergman 1987: 8) saw sufficient similarities between level X at Ksar Akil and the Aurignacian of Font Yves (France) to warrant calling the Lebanese material Aurignacian. However, according to the definition of Aurignacian proposed in the Southern Levant, level X would be excluded from this designation because it is technologically a blade and bladelet-dominated industry. This raises the important question of whether it is appropriate to use a European cultural term in a manner which departs from its original meaning.

It may be that part of the problem concerning the definition of the term 'Levantine Aurignacian', at least in regard to flaking technology, is due to

Table 5.19. Blade/bladelet profiles.

Level			Number of pieces	Straight %	Curved %	Twisted %	Unidentified %
VI	F4		107	1.87	62.62	34.58	0.94
IX	F4	7.25	193	11.40	69.95	18.14	0.52
IX	E4	7.65	82	13.42	54.88	31.71	—
IX	F4	7.75	238	8.40	56.72	29.83	5.04
X	E4	8.10	349	9.46	61.89	28.65	—
X	F4	8.40	272	8.82	51.10	38.60	1.47
X	E4	8.65	403	8.19	46.40	37.97	7.44
XI	F4	8.95	82	10.98	23.17	59.76	6.10
XI	F4	9.30	220	5.91	26.82	66.82	0.46
XII	F4	9.70	189	4.76	19.58	68.78	6.88
XII	E4	10.00	353	5.10	27.76	60.06	7.08
XIII	F4	9.90	62	14.52	27.42	56.45	1.61

the apparent absence of material similar to Ksar Akil levels XIII-IX in Israel. These levels have varying percentages of Aurignacian tools associated with blade and bladelet producing technologies. It would seem that most of the 'Levantine Aurignacian' in the Southern Levant (e. g. Hayonim D, el-Wad D1 and D2, Kebara D1 and D2, el-Quseir C) is related to Ksar Akil levels VIII and VII. These Ksar Akil levels are technologically dominated by unretouched flakes with numerous scraper and burin types. If these assemblages are the only type recovered so far in Israel, it is hardly surprising that they are the only ones regarded as 'Aurignacian'.

To conclude, if the new interpretation is accepted by prehistorians in the Northern and Southern Levant, then the only assemblages studied here from Ksar Akil which can be called 'Aurignacian' are levels VIII-VII. Until some kind of consensus can be achieved by prehistorians working in the Levant it has been decided to label the assemblages in levels XIII-VI by reference to stages (Bergman 1987: 148) which follow those described by Azoury (1986) and Ohnuma (1986).

CONCLUSION

The Upper Palaeolithic sequence at Ksar Akil shows a tendency from the beginning towards the production of elongated debitage. The earliest material, dated to sometime after 43 000 BC, can be divided technologically into two subphases: (1) levels XXV-XXIV produced a small sample which has a number of opposed platform cores with parallel sides; (2) levels XXIII-XXI/XX utilise single platform cores which are flaked unidirectionally and have faceted platforms and converging sides. The debitage is characteristically thick with large faceted butts and common by-products of blade manufacture are pieces morphologically identical to

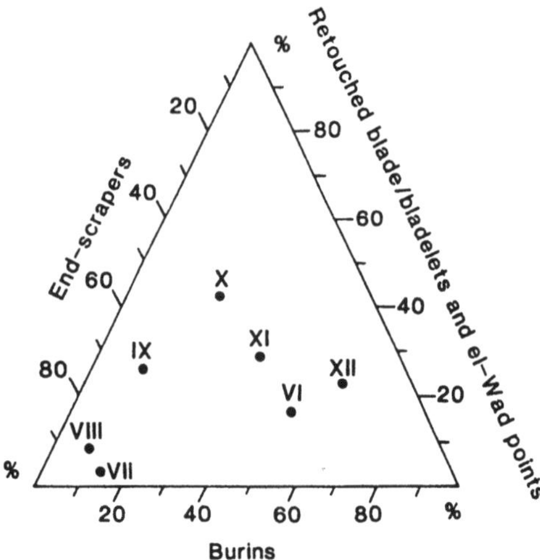

Figure 5.24. Three-pole graph of tool typology in levels XII-VI at Ksar Akil.

Levallois points.

We would agree with Marks (1983) that Ksar Akil levels XXIII-XXII are remarkably similar to level 4 at Boker Tachit in the Negev (dated to around 40 000 BC: Marks and Volkman, in press). It is the authors' opinion that the Ksar Akil material is part of the developmental sequence also seen at Boker Tachit. It is in this sense that levels XXV-XXI/XX represent specific stages in the technological transition into the Upper Palaeolithic. The tool kits in levels XXV-XXII of Ksar Akil, however, contain numerous chamfered pieces and differ substantially from those of Boker Tachit.

The 'Transitional' and early Upper Palaeolithic levels XXV-XVI are characterized by a developmental evolution of the stone industries (Newcomer 1972: 332-83; Besançon *et al.* 1977). The technological changes occur over a relatively long period of time, perhaps several thousand years, and seem to follow changes in the retouched tool typology. The evolution of the Transitional industries (levels XXV-XXI/XX) into the retouched point and backed bladelet assemblages (i. e. levels XVII-XVI) is clearly seen at Ksar Akil, which establishes a developmental link with assemblages called Ahmarian in the Southern Levant. At Ksar Akil skeletal remains of anatomically modern type are associated with a retouched point and backed blade tool assemblage in level XVII.

Levels XX/XIX-XVI are characterised by a gradual change in the blade technology. In these levels the blanks are detached from opposed platform cores which are parallel-sided. The striking platforms are plain and show extensive signs of platform abrasion. Cresting is more often used in these levels, and the core tablet technique, rather than faceting, is employed to correct faulty platforms.

Many of these flaking techniques continue after the stratigraphic hiatus in level XIV (e. g. blade technologies which utilise cresting, the core tablet

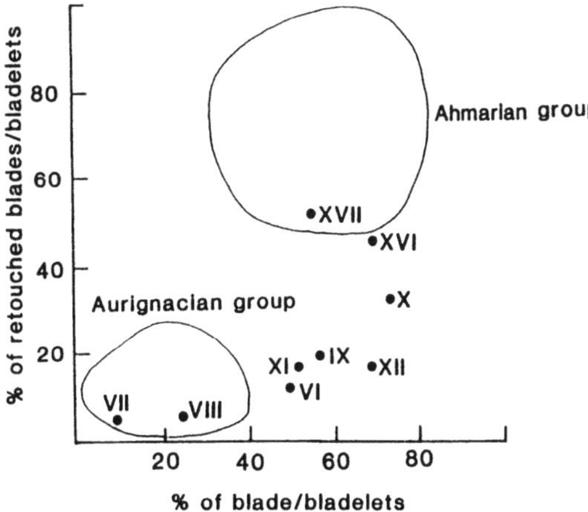

Figure 5.25. Graph taken from Gilead 1981a (Figure 7.1) comparing Ksar Akil levels XVII-VI with the Ahmarian and Aurignacian industries of the Southern Levant.

technique, plain platforms and platform abrasion). One important technological difference between levels XIII-XI and the earlier material is the greatly reduced number of opposed-platform cores and blanks with opposed dorsal scars. An important technological feature found for the first time in levels XIII-XI is a high percentage of blades and bladelets with twisted profiles and offset debitage. Levels X and IX also have a predominance of blades and bladelets but with a reduced percentage of twisted blanks; offset debitage is also relatively rare.

A dramatic change in the flaking technology occurs between levels IX and VIII: for the first time in the Upper Palaeolithic sequence at Ksar Akil there are assemblages (levels VIII and VII) dominated by unretouched flakes. Many of the cores in these levels are shapeless and appear to be flaked alternately. Level VI, which is also flake dominated, marks the beginning of the terminal Upper Palaeolithic at Ksar Akil.

It is the authors' opinion that the stratigraphic sequence of levels XIII-VI does not represent a developmental continuum, and should not be referred to as (Levantine Aurignacian) A, B and C. It is suggested that this sequence displays short-term evolution within several groups of levels (XIII-XI, X-IX and VIII-VII) but these are separated from each other by breaks in the archaeological sequence. Indeed, it is worth noting that, unlike levels XXV-XVI, changes in the technology of blank production occur simultaneously with changes in the tool kits.

In common with most other Near Eastern sites, Ksar Akil demonstrates a sequence where blade and bladelet industries are stratigraphically older than those with flakes (Gilead 1981a, 1931b; Marks 1981). Recent dating of Tixier's phase VI, roughly comparable to the flake-dominated levels VIII-VII of the 1937-1938 excavations, reveals that these industries may be as old as 32 000 BP (Mellars and Tixier 1989). Some differences do occur at Ksar Akil and probably reflect a local Lebanese evolution. These include the earliest levels which contain numerous chamfered pieces, a tool type that is virtually unknown outside of Lebanon. Another significant feature is the presence of several assemblages (levels XIII-IX) which are blade- and bladelet-based with varying percentages of Aurignacian tools. This facies is known at a few other sites in the Northern Levant and may suggest that the Levantine Aurignacian initially developed out of the local blade based cultures; however, none of the sites have a complete unbroken sequence where this evolution is visible. To conclude, the Ksar Akil sequence fits into the broad general framework proposed for the Upper Palaeolithic in the rest of the Levant, but at the same time has elements which are specifically Lebanese.

REFERENCES

Azoury, I. 1971. *A Technological and Typological Analysis of the Transitional and Early Upper Palaeolithic Levels of Ksar Akil and Abu Halka.* Unpublished PhD Thesis, University of London.

Azoury, I. 1986. *Ksar Akil, Lebanon: a Technological and Typological Analysis of the Transitional and Early Upper Palaeolithic Levels of Ksar Akil and Abu Halka.* Oxford: British Archaeological Reports International Series S289.

Azoury, I. and Hodson, F. R. 1973. Comparing Palaeolithic assemblages: Ksar Akil, a case study. *World Archaeology* 4: 292-306.

Bells, V. 1938. Removal of a side in the transformation of keeled scrapers at Mount Carmel, Lebanon. *Bulletin of the American School of Palaeolithic Research* 14: 55-61.

Bergman, C. A. 1981. Point types in the Upper Palaeolithic sequence at Ksar 'Akil, Lebanon. In J. Cauvin and P. Sanlaville (eds) *Préhistoire du Levant.* Paris: Centre National de la Recherche Scientifique: 319-330.

Bergman, C. A. 1987. *Ksar Akil, Lebanon: a Technological and Typological Analysis of the Later Upper Palaeolithic Levels.* Oxford: British Archaeological Reports International Series S329.

Bergman, C. A. (in press). Hafting and use of bone and antler tools from Ksar Akil, Lebanon. In D. Stordeur (ed.) *Manches et Emmanchements Préhistorique.* Lyon: Maison de l'Orient Méditerranéen. In Press.

Bergman, C. A. and Copeland, L. 1986. Preface. In I. Azoury *Ksar Akil, Lebanon: a Technological and Typological Analysis of the Transitional and Early Upper Palaeolithic Levels of Ksar Akil and Abu Halka.* Oxford: British Archaeological Reports International Series S289: i-viii.

Besançon, J. , Copeland, L. and Hours, F. 1977. Tableaux de préhistoire Libanaise. *Paléorient* 3: 5-45.

Bordes, F. 1947. Etude comparative des différentes techniques de taille du silex et roches dures. *L'Anthropologie* 51: 1-29.

Bordes, F. 1960. Evolution in the Palaeolithic cultures. In S. Tax (ed.) *The Evolution of Man: Man, Culture and Society.* Vol 2: *Evolution After Darwin.* Chicago: University of Chicago Press: 99-110.

Bordes, F. 1961. *Typologie du Paléolithique: Ancien et Moyen.* Bordeaux: Publications de l'Institut de Préhistoire de l'Université de Bordeaux, Mémoire I.

Bordes, F. 1967. Considérations sur la typologie et les techniques dans le Paléolithique. *Quatär* 18: 25-55.

Bordes, F. 1972. Du Paléolithique moyen au Paléolithique supérieur: continuit; ou discontinuité. In F. Bordes (ed.) *The Origins of Modern Man.* Paris: UNESCO.

Bordes, F. 1980. Levallois débitage et ses variantes. *Bulletin de la Société Préhistorique Française* 77: 47-49.

Bordes, F. and Crabtree, D. 1969. The Corbiac blade technique and other experiments. *Tebiwa* 12: 1-21.

Copeland, L. 1975. The Middle and Upper Paleolithic of Lebanon and Syria in the light of recent research. In F. Wendorf and A. E. Marks (eds) *Problems in Prehistory: North Africa and the Levant.* Dallas: Southern Methodist University Press: 317-350.

Copeland, L. 1986. Introduction to volume 1. In I. Azoury *Ksar Akil, Lebanon: a Technological and Typological Analysis of the Transitional and Early Upper Palaeolithic Levels of Ksar Akil and Abu Halka.* Oxford: British Archaeological Reports International Series S289: 1-24.

Copeland, L. 1987. Preface to volume 2. In C. A. Bergman *Ksar Akil, Lebanon: a Technological and Typological Analysis of the Later Upper Palaeolithic Levels.* Oxford: British Archaeological Reports International Series S329: iv-ix.

Day, A. E. 1926a. The rock shelter of Ksar 'Akil near the cave of Antilyas. *Palestine Exploration Fund*: 158-160.

Day, A. E. 1926b. The rock shelter of Ksar 'Akil. *Al-Kulliyyah* (AUB Issue) 12: 91-97.

Dortch, C. 1970. *The Late Aurignacian Industries of Levels 8-6 at Ksar 'Akil, Lebanon.* Unpublished MA Thesis, University of London.

Ewing, J. F. 1963. A probable Neanderthaloid from Ksar 'Akil, Lebanon. *American Journal of Physical Anthropology* 21: 101-104.

Gilead, I. 1981a. *The Upper Palaeolithic in the Sinai and the Negev.* Unpublished PhD Thesis, Hebrew University, Jerusalem.

Gilead, I. 1981b. Upper Palaeolithic tool assemblages from the Negev and Sinai. In J. Cauvin and P. Sanlaville (eds) *Préhistoire du Levant.* Paris: Centre National de la Recherche Scientifique: 331-352.

Gingell, C. and Harding, P. 1981. A method of analyzing the technology in Neolithic and Bronze Age assemblages. *Staringia* 6: 73-76.

Hours, F. 1974. Remarques sur l'utilization de Listes-Types pour l'étude du Paléolithique supérieur et de l'Epipaléolithique du Levant. *Paléorient* 2: 3-18.

Marks, A. E. 1976. *Prehistory and Paleoenvironments in the Central Negev, Israel:* Vol. 1*: The Avdat/Aqev Area (Part 1)*. Dallas: Southern Methodist University.

Marks, A. E. 1981. The Upper Palaeolithic of the Negev. In J. Cauvin and P. Sanlaville (eds) *Préhistoire du Levant.* Paris: Centre National de la Recherche Scientifique: 343-352.

Marks, A. E. 1983. The Middle to Upper Palaeolithic transition in the Levant. In F. Wendorf and A. Close (eds) *Advances in World Archaeology* Vol. 2. New York: Academic Press: 51-92.

Marks, A. E. and Volkman, P. (in press). The Mousterian of Ksar Akil: Levels XXVIA through XXVIIIB. *Paléorient.* In Press.

Mellars, P. A. and Tixier, J. 1989. Radiocarbon-accelerator dating of Ksar 'Aqil (Lebanon) and the chronology of the Upper Palaeolithic sequence in the Middle East. *Antiquity* 63: 761-768.

Murphy, J. W. 1938. The method of pre-historic excavations at Ksar 'Akil. *Anthropology Series, Boston College Graduate School 3:* 272-275.

Murphy, J. W. 1938. Ksar 'Akil, Boston College expedition. *Anthropology Series, Boston College Graduate School* 4: 211-217.

Newcomer, M. H. 1968-69. The chamfered pieces from Ksar Akil (Lebanon). *Bulletin of the Institute of Archaeology* 8-9: 177-191.

Newcomer, M. H. 1971. Un nouveau type de burin à Ksar Akil (Liban). *Bulletin de la Société Préhistorique Française* 68: 267-272.

Newcomer, M. H. 1972. *An Analysis of a Series of Burins from Ksar Akil (Lebanon)*. Unpublished PhD Thesis, University of London.

Newcomer, M. H. 1974. Study and replication of bone tools from Ksar Akil (Lebanon). *World Archaeology* 6: 138-153.

Newcomer, M. H. 1975. Punch technique and Upper Paleolithic blades. In E. Swanson (ed.) *Lithic Technology: Making and Using Stone Tools*. The Hague: Mouton: 97-102.

Newcomer, M. H. and C. Karlin 1986. Flint chips from Pincevent. In M. H. Newcomer and G. de G. Sieveking (eds) *The Human Uses of Chert*. Cambridge: Cambridge University Press: 33-36.

Ohnuma, K. 1986. *A Technological Study of the Upper Palaeolithic Material from Levels XXV-XIV from Ksar Akil*. Unpublished PhD Thesis, University of London.

Ohnuma, K. and Bergman, C. A. 1982. Experimental studies in the determination of flaking mode. *Bulletin of the Institute of Archaeology* 19: 161-170.

Passemard, E. 1927. Mission en Syrie et au Liban. *Bulletin de la Société Préhistorique Française* 24: 70-72.

Sonneville-Bordes D. de and Perrot, J. 1954. Lexique typologique du Paléolithique supérieur. *Bulletin de la Société Préhistorique Française* 51: 327-335.

Sonneville-Bordes, D. de and Perrot, J. 1955. Lexique typologique du Paléolithique supérieur. *Bulletin de la Société Préhistorique Française* 52: 75-79.

Sonneville-Bordes, D de. and Perrot, J. 1956a. Lexique typologique du Paléolithique supérieur. *Bulletin de la Société Préhistorique Française* 52: 408-412.

Sonneville-Bordes, D. de and Perrot, J. 1956b. Lexique typologique du Paléolithique supérieur. *Bulletin de la Société Préhistorique Française* 53: 547-559.

Sonneville-Bordes, D. de. 1960. *Le Paléolithique Supérieur en Périgord.* Bordeaux: Delmas.

Tixier, J. 1963. *Typologie de l'Epipaléolithique de Maghreb*. Paris: Centre de Recherches Anthropologiques, Préhistoriques et Ethnographiques.

Tixier, J. 1970. L'abris sous roche de Ksar 'Aqil: La campagne de fouilles 1969. *Bulletin du Musée de Beyrouth* 23: 173-191.

Tixier, J. 1974. Fouille à Ksar 'Aqil, Liban (1969-1974). *Paléorient* 2: 187-192.

Tixier, J. and Inizan, M. -L. 1981. Ksar 'Aqil, stratigraphie et ensembles lithiques dans le Paléolithique supérieur: fouilles 1971-1975. In J. Cauvin and P. Sanlaville (eds) *Préhistoire du Levant.* Paris: Centre National de la Recherche Scientifique: 353-367.

Vogel, J. C. and H. T. Waterbolk. 1963. Groningen radiocarbon dates IV. *Radiocarbon* 5: 163-202.

Wenner-Gren Symposium, London. 1969. *Unpublished Transcript of the Proceedings.*

6: Middle to Upper Palaeolithic Transition: the Evidence for the Nile Valley

PHILIP VAN PEER AND PIERRE M. VERMEERSCH

In the recent focus of prehistoric research on the Middle to Upper Palaeolithic transition, the Nile Valley has not formed a major area of interest. Indeed, Nile Valley Palaeolithic prehistory in general has not received too much attention in the literature, apparently because of its supposed paucity of data. But in recent years, substantial new data have been collected, which throw new light on this area. Some of this new evidence is relevant to the problem of the Middle to Upper Palaeolithic transition, and requires us to rethink some of the traditional ideas on this topic (Figure 6.1).

HISTORICAL PERSPECTIVE

Ever since the investigations in the Nile Valley by Sandford (1934) and Sandford and Arkell (1929, 1933, 1939), the main point of interest has been the establishment of a chronology based essentially on the Nile terrace deposits. Prehistoric industries were considered in relation to these terraces and were looked upon as chronological markers for the deposits. A similar approach was adopted by Caton-Thompson (1946). Her account of the Palaeolithic industries assumed that all artifacts recovered from the (presumed) same terraces could be regarded as a unity for the whole length of the Valley. Clearly there was little concern in these studies for the kind of problems which concern prehistorians at the present day.

Caton-Thompson was the first to present some general observations on the topic of the Middle to Upper Palaeolithic transition, when dealing with what she termed the "Epi-Levalloisian industries". These are defined (1946: 59) as "...those varied regional industries of Levalloisian technique and descent, which anachronistically occupy the Upper Palaeolithic period and are physiographically linked with the silt regime in the Nile Valley which succeeded the gravel regime". These industries were supposed to illustrate "...the capacity of a specific Stone Age industry to transform itself, without the least evidence of outside human interference or influence, into something totally different" (1946: 59).

From the early sixties on, the 'Combined Prehistoric Expedition' carried out an intensive survey of Nubia, within the general framework of the Nubian Monuments Salvage Campaign. Many new Palaeolithic sites were discovered and excavated (Wendorf 1965, 1968). Some of these

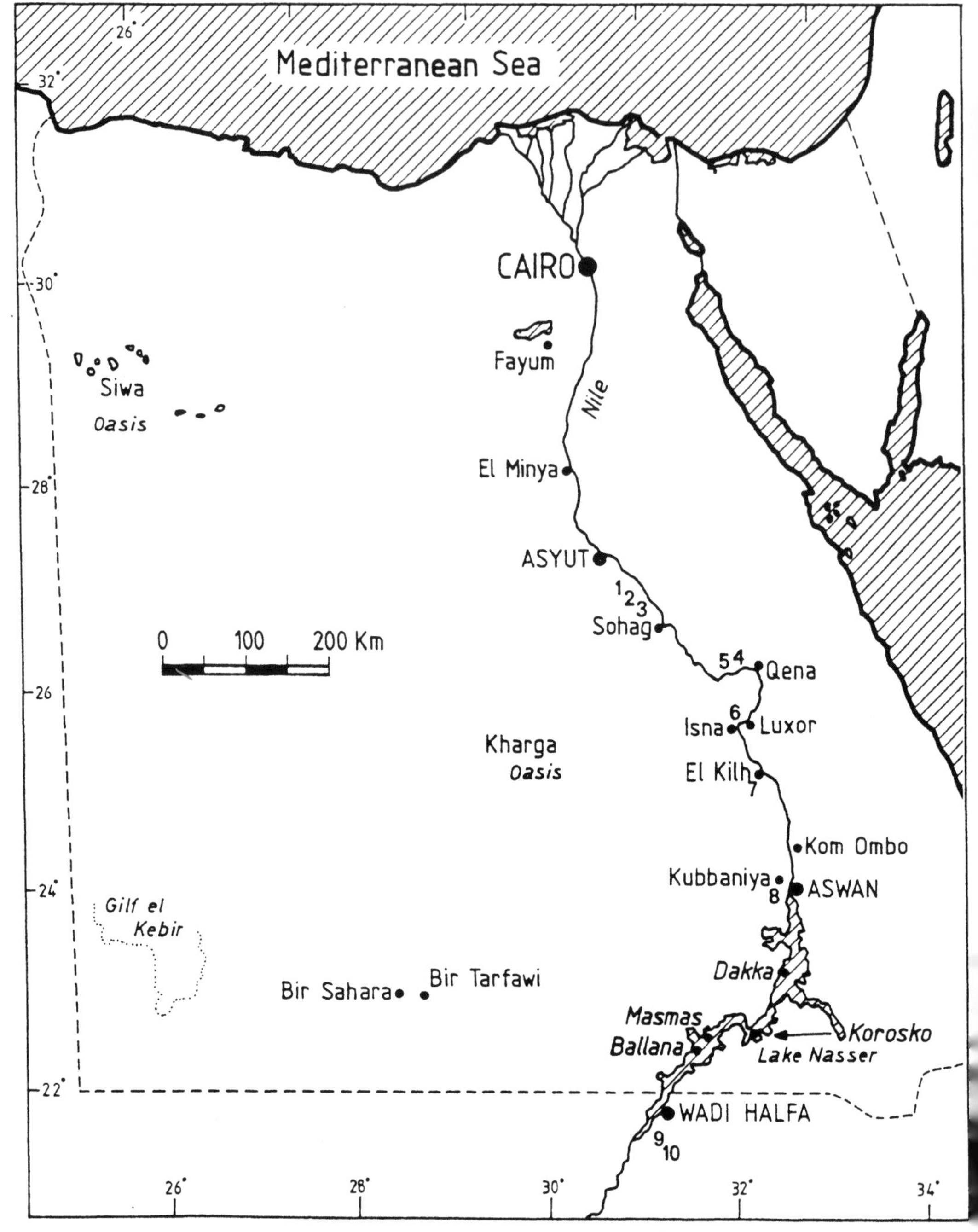

Figure 6.1. Map of Egypt, showing location of sites referred to in the text: 1. Nazlet Khater-2; 2. Nazlet Khater-3; 3. Nazlet Khater-4; 4. Makhadma-6; 5. Shuwikhat-1; 6. E71K9; 7. E71P1; 8. E82-5; 9. 1018; 10. 443.

provided assemblages with a mixture of Middle Palaeolithic technology (in particular Levallois technology) and Upper Palaeolithic typology, stratified in relation to Nilotic deposits. Some radiocarbon dates of up to 25 000 BP were obtained for either the archaeological samples or the associated deposits. Because of their presumed age and their typological content, such industries were considered 'Upper' or rather, 'Late' Palaeolithic. It was also believed that the modern sedimentation system of the Nile was a fairly recent phenomenon, not older than 25 000 BP (de Heinzelin 1968: 52). Viewed in these terms, one had access to an excellent tool of relative dating, which suggested that Levallois technology had persisted in the Nile Valley until well after 20 000 BP, into the Late Palaeolithic.

The situation changed however when some of the samples that had provided the radiocarbon dates were redated to a much older age (Wendorf and Schild 1976a: 239). At about the same time, the Belgian Middle Egypt Prehistoric Project (BMEPP) excavated a site in Egypt where Nile deposits, consisting of gravels and silts, contained fresh Middle Palaeolithic artifacts *in situ* (Paulissen and Vermeersch 1987: 55). Increasingly, it came to be realized that the modern sedimentation system of the Nile was of a far greater antiquity than was previously recognized (Wendorf and Schild 1976a: 239; Said 1981; Schild and Wendorf 1986: 30-1; Paulissen and Vermeersch 1987: 60). Its potential as a conclusive chronological framework has yet to be established. The chronological position of many assemblages related to Nilotic deposits, once believed to be firmly established, is now seen to be uncertain. New discoveries in Egypt by the BMEPP revealed further problems. In the course of this work two non-Levallois blade assemblages were found, dating to well before the Late Palaeolithic Levallois assemblages.

All this new evidence has certainly not improved our understanding of the course of the events in the Nile Valley (Paulissen and Vermeersch 1987; Close 1987) during the Late Pleistocene. Since the whole question, in terms of the lithic industries, seems to be related in one way or another to Levallois technology, the latter will form the main focus of this review.

THE MIDDLE PALAEOLITHIC

Nubia

As a result of the work of the Combined Prehistoric Expedition, the Middle Palaeolithic is documented best in Nubia. An important characteristic of the Middle Palaeolithic in Nubia is its use of locally available sandstone as raw material. A striking technological feature is the presence of two special variants of the Levallois technology (see below). In addition to these special variants, more conventional centripetal Levallois technology, comparable to that of Western Europe, is present. The latter will be referred in the remainder of the review as "classical Levallois".

The 'Nubian Levallois' techniques were defined initially by J. and G. Guichard (1965: 68-9). They described two types of so-called 'Nubian cores', of which the essential characteristics were as follows:

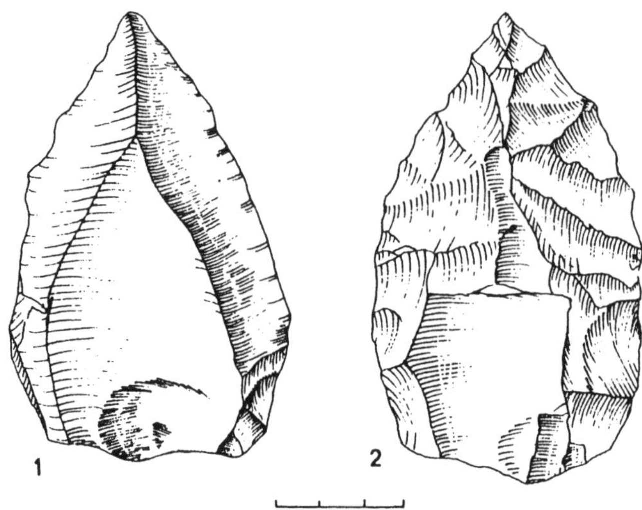

Figure 6.2. Characteristic core types of the Nubian Middle Palaeolithic; 1. Nubian core type I; 2. Nubian core type II (after J. and G. Guichard 1965).

1. Nubian core I: On a more or less triangular nodule, a central ridge is created by the removal of two intersecting elongated flakes from the distal end. Subsequently, a Levallois point is struck off. From our own experience with the lithic material, we found that central-ridge preparation is often restricted to the distal part, in order to produce pointed flakes, rather than real points (Figure 6.2, no. 1).

2. Nubian core II: A triangular core with protruding distal end is peripherally prepared in order to produce pointed flakes. In the most typical cases, it would seem that a central ridge is created by a transversal preparation. In many instances, however, it is hard to distinguish these cores from 'classical' Levallois cores, except for their triangular shape (Figure 6.2, no. 2).

J. and G. Guichard (1965, 1968) also presented a general classification for the Middle Palaeolithic, based on sites from the Eastern desert near Wadi Halfa. They differentiate between two main industries:

(a) Nubian Middle Palaeolithic: assemblages with bifacial foliate objects, side-scrapers and Nubian cores. Based on the "degree of perfection of tools" and the degree of eolisation, two sub-groups are distinguished:

(i) Early (or Phase I) Nubian Middle Palaeolithic

(ii) Upper (or Phase II) Nubian Middle Palaeolithic

(b) Non-Nubian Middle Palaeolithic: characterized principally by negative traits: absence of bifacial foliate objects, side-scrapers and Nubian cores. All forms of retouched tools are very rare.

Marks (1968a) studied a number of Middle Palaeolithic sites to the north of Wadi Halfa and close to the Nile, together with two further sites located in the Guichard area. He proposes the following classification (1968a: 280-93):

– Denticulate Mousterian: Few tool types are present, among which, denticulates and notches are dominant. Nubian cores are lacking. This industry is said to be comparable to the European Denticulate Mousterian.

– Nubian Mousterian: More tool types are present than in the Denticulate Mousterian among which side-scrapers and Upper Palaeolithic tool types are the best represented. Nubian cores are generally present.

Two facies can be distinguished within this group:

(i) Nubian Mousterian A (without bifaces)

(ii) Nubian Mousterian B (with bifaces)

The assemblages on which both the classifications have been based, seem hardly comparable. In general, the Guichard assemblages give a much more 'massive' impression than the Mousterian assemblages of Marks. A detailed comparison of the Nubian Middle Palaeolithic and the Nubian Mousterian (Marks 1968a: 296-300) revealed important differences. Most striking in this respect is the fact that in the Nubian Middle Palaeolithic industries Nubian type II cores are the most numerous, whereas type I cores are most common in the Nubian Mousterian.

It has been suggested that the presence of bifacial foliate objects in the Nubian Middle Palaeolithic might point to an affiliation with the Aterian (Guichard 1965: 111; 1968: 184; Wendorf and Schild 1976b: 22). This however is highly unlikely, since true pedunculated tools are absent in the Nubian sites and because of the numerous bifaces present in most of these assemblages. Although the Aterian is present in the Eastern Sahara (Wendorf and Schild 1976b: 21-2; 1979), there is no clear evidence of its presence in the Nubian Nile Valley. On the other hand, a sub-Saharan affiliation, as has also been suggested (Guichard and Guichard 1965: 111; 1968: 184), is possible, but would require a more detailed study. To us it seems likely that these assemblages are related to earlier Acheulian industries and that they are much older than the Mousterian assemblages.

In addition to the Mousterian industries, Marks (1968b: 315-91) has defined the 'Khormusan' industry (a flake industry) in which classical Levallois technology is abundant and Nubian technique is almost absent. The typology is dominated by burins on flakes. On the basis of its stratigraphic position within Nilotic deposits and available radiocarbon dates (see above), the Khormusan was originally considered to be Upper Palaeolithic (Wendorf *et al.* 1965: xxii-xxvii) and in that sense an industry of 'mixed' character. As noted earlier, the Khormusan samples have now been redated. As a consequence, the Khormusan was reattributed to the Middle Palaeolithic (Wendorf and Schild 1976a: 239; 1976b: 23).

Egypt

Since 1976, The BMEPP has carried out research on the Middle Palaeolithic of the Egyptian Nile Valley (Vermeersch *et al.* 1978, 1979, 1980, 1986). Work has been concentrated mainly on sites in firm stratigraphic context, together with systematic collections from some surface sites.

Tools are always rare in these assemblages, and the sites seem to represent primarily chert exploitation sites. From a technological point of view, the assemblages are quite similar to the 'Mousterian' industries of Nubia: differences that do exist are likely to be due mainly to differential use of raw material. Nubian technique is a feature of some sites, whereas in others it is almost entirely lacking.

A possible Aterian surface site from Wadi Kubbaniya has been reported by Singleton and Close (1980: 229-38). A few pedunculated tools, in addition to foliates do indeed occur, but unfortunately, the assemblage is very poor: few tools are present and no debitage has been collected. The site is important however since it represents the best (albeit scanty) evidence so far for the hypothesis of an Aterian presence in the Nile Valley. Recently, attention has been drawn to the fact that the Nubian technique, so common in the Mousterian of the Nile Valley, is present in some Aterian assemblages from the Sahara and the Maghreb (Van Peer 1986: 321-4). Some relationships with the Nile Valley thus seem to be suggested by the data. A full assessment of the question of Aterian presence in the Nile Valley will clearly have to await recovery of more detailed evidence.

A number of other assemblages have also been reported from Wadi Kubbaniya (Wendorf and Schild 1986: 33-9), such as E82-5 and E82-4, the latter being a surface site apparently related to the Khormusan.

Chronology

All except two of the Middle Palaeolithic and Mousterian sites so far investigated in Nubia are surface sites and therefor provide no firm evidence for chronology. They are all located beyond the reach of the Nile floodplain. The Khormusan sites on the other hand are correlated with Nilotic silt and sand deposits and some of them have been found stratified within these deposits. The available dates indicate that the Khormusan lies beyond the range of radiocarbon dating (for a discussion, see Wendorf *et al.* 1979: 220-1). The fact that the Khormusan is restricted to Nilotic deposits and has never been found outside these contexts may suggest that arid conditions prevailed at that time (Wendorf and Schild 1976b: 17-18).

Most of the Egyptian sites which are in stratigraphic context, are reworked in local wadi deposits. This indicates that they have been redeposited during a period of humid conditions. One of these sites, Nazlet Khater-2, provided a radiocarbon date of >35 700 BP (GrN-10578). On the other hand, some sites are almost *in situ*: At Makhadma-6, a largely *in situ* site was found on top of a gravel layer, which in turn contained numerous rolled Middle Palaeolithic artifacts, and was covered by only a few centimetres of slopewash deposits. Here therefore we have evidence for two chronologically separated units. Site E82-5 at Wadi Kubbaniya was *in situ* within dune sands and in the base of overlying aeolian sands (Wendorf and Schild 1986: 33). The latter provided several TL dates (Schild 1987: 16) which suggest that a major arid period had begun around or before 60 000 BP (Schild 1987: 16; Paulissen and Vermeersch 1987: 56).

Based on this rather scarce information, we would propose at least three chronological stages for the sites that have been mentioned above (from younger to older):

1. The Khormusan in Nubia, and at some sites in Egypt.
2. The 'Mousterian' sites of Nubia and most of the Egyptian sites, dating to before 60 000 BP.
3. Nubian Middle Palaeolithic. So far, this has only been documented in Nubia.

TRANSITIONAL INDUSTRIES

Ever since the work of Caton-Thompson (1946), the occurrence of assemblages of 'dualistic' character has been recognized in the Nile Valley. Technologically, these industries rely to a large extent on Levallois technology. In addition to Levallois technology however, there is evidence of systematic microblade production, represented by both single and opposed platform (mostly microblade) cores (see Note 1). Amongst the retouched tools, 'Upper Palaeolithic' forms such as backed elements and burins appear which are made both on Levallois flakes and microblades. We will describe here two such assemblages, one from Site 1018 in Nubia (Marks 1968c: 413-23) and one from Site E71P1 in Egypt (Wendorf and Schild 1976a: 27-41, 243-50). Through a technological analysis, we will try to demonstrate the dynamic character of this Levallois technology, eventually leading to the local development of blade and microblade production.

Site 1018

Site 1018 is situated on the east bank of the Nile, to the southwest of Wadi Halfa. Here, a surface concentration of artifacts was found. Excavations revealed an occupation layer on top of a dune.

In the lithic industry, chert was the dominant raw material and was derived from small Nile pebbles, which were abundant in the immediate vicinity. Most of the Levallois flakes and cores in the asssemblage were produced by what has been described as the 'Halfa' method. This is a specialised Levallois technique, involving very intense distal, lamellar preparation. It aims primarily at the production of secondary Levallois flakes – i.e. flakes presenting a large negative scar from an initial Levallois flake removal on their dorsal surface. In addition to this special technique, a few examples of classical Levallois technology were present. Nubian technique is absent. Levallois (Halfa and classical Levallois) cores account for 47 percent of the total cores, and Upper Palaeolithic type cores (single and opposed platform) for 42 percent. Of these, single-platform cores are the most numerous. In general these cores were not extensively flaked. Most of them produced small, relatively elongated flakes, rather than microblades.

Among the tools, backed elements and burins are well represented, while notches, denticulates and various scrapers also occur.

Site E71P1

This site is situated on the west bank of the Nile, in the El Kilh region of Upper Egypt. A large surface concentration occurred on the southeast slope of a small hill, with no obvious artifact clusters within this concentration. Trenches disclosed the presence of artifacts *in situ* in loose sandy silt and in the underlying slightly consolidated silt. Artifacts were vertically dispersed up to 50 centimetres, with the main concentration between 20 and 30 centimetres below surface. The site was divided into four arbitrary areas in order to collect the surface material, since a preliminary inspection of the tools suggested some differentiation within the site. Wendorf and Schild (1976a: 243) however state that no technological differences could be detected between the different areas. The authors (1976a: 243) suggest that such a large concentration is probably composed of several clusters, possibly reflecting several occupations. In view of the obvious problems inherent in large surface concentrations, we will limit ourselves here to the material from sector C of this site.

This assemblage, consisting of both surface and fresh excavated materials from trench 5, appears to be homogeneous from a technological point of view. Levallois technology plays a major part in the reduction strategy. This Levallois technology is of the same nature as that described for Site 1018. Nubian technique is lacking, whilst Upper Palaeolithic type cores (single and opposed platform) are well represented. Single platform cores are dominant. Only few of these cores have been intensively flaked and produced real blades or (most commonly) microblades.

As regards typology, there is a complication in the sense that the excavated collection comprises many 'Ouchtata' retouched microblades, whereas these forms are lacking in the surface collection. In other respects, the typologies of both surface and excavated material are very similar. Notches and denticulates are the most common form, while burins and scaled pieces are well represented. A few backed elements are also present. Some of the burins are made on large Levallois flakes.

Levallois technology

In recent studies, the variability of the Levallois concept has been demonstrated in different parts of the world. In the Nile Valley too, such variability is attested by the occurrence of Nubian and Halfan technique, in addition to the classical Levallois technique. We will try to investigate now if Halfa Levallois technology in particular fits an evolutionary technological pattern. Therefore we need to begin by looking at the Middle Palaeolithic Levallois technology.

We have stated above that the 'Mousterian' industries of both Nubia and Egypt are very similar from a technological point of view. Some differences are probably related to the use of different raw materials: in Egypt chert was used, whereas in Nubia sandstone was the main raw material. The existence of two Levallois technological groups in the Nile Valley, which we have already noted, can be demonstrated in a detailed way for the Egyptian assemblages, since these have been studied specifically from this point of view. One group shows an important use of the Nubian technique in addition to the classical Levallois technology (N-group), while the other group almost exclusively used the classical technique (K-group). Although the distinction between the two technological groups rests mainly on the absence/presence of the Nubian technique, it can be seen that there are also major differences in the character of the classical Levallois technology of both groups. As an example of the N-group, the Nazlet Khater-3 (NK3) assemblage will be described, and for the K-group, Nazlet Khater-2 (NK2). Both sites were excavated by the BMEPP. They are situated on the left bank of the Nile, some 12 kilometres west of Tahta. Both sites provided abundant lithic material. Tools however are rare at these sites, which is not surprising in view of their apparent function as raw material exploitation sites.

The characteristics of the classical Levallois technology of both NK3 and NK2 (Figure 6.3) have been summarized in Table 6.1. For the flakes, the patterns of preparation are the same: the lateral preparation (Crew 1975: 13) is over 30 percent, and the pattern in both cases is of the radial type. Frequencies of butt types are very similar too. Differences can be observed between the average numbers of preparatory scars, the length, and the width/thickness ratios of the flakes. The two former differences are statistically significant at the 5 percent level. For the cores, the difference between the means of the number of distal preparatory scars is also statistically significant (5% level).

In addition to the differences in Levallois technology, the blade index and the frequency of Upper Palaeolithic type cores is higher at NK2.

It would thus seem that these groups differ not only in their use of certain Levallois methods, but also in certain other features of both classical Levallois technology and of non-Levallois technology.

The characteristics of the Levallois technology of the 1018 and E71P1C assemblages are summarized in Table 6.1. For the flakes, the amount of lateral preparation is low as compared to the Middle Palaeolithic industries, especially at 1018. The mean number of preparatory scars is high at E71P1C; the lower value at 1018 is certainly due to the fact that only secondary Levallois flakes are present. At E71P1C, as in the Middle Palaeolithic industries, both primary and secondary flakes occur. As for the butt types, we notice an increase of the frequency of '*chapeau de gendarme*' butts. Also, the flakes are small. This may be due either to a differential selection of raw material or to the character of the particular raw materials available. This is certainly the case at 1018, where Levallois technology was applied to small chert pebbles, apparently the only sources available at the site. For the cores, the emphasis on the distal preparation is clear. Overall, the Levallois technology of 1018 and E71P1C seems to be closer to the NK2 Levallois technology than to that of NK3. Some examples of Levallois flakes and cores (Figure 6.4) illustrate a close resemblance with NK2, especially in the emphasis on the distal preparation. There is thus a clear technological link between the Middle Palaeolithic K-group and the 1018 and E71P1C assemblages. Within the latter

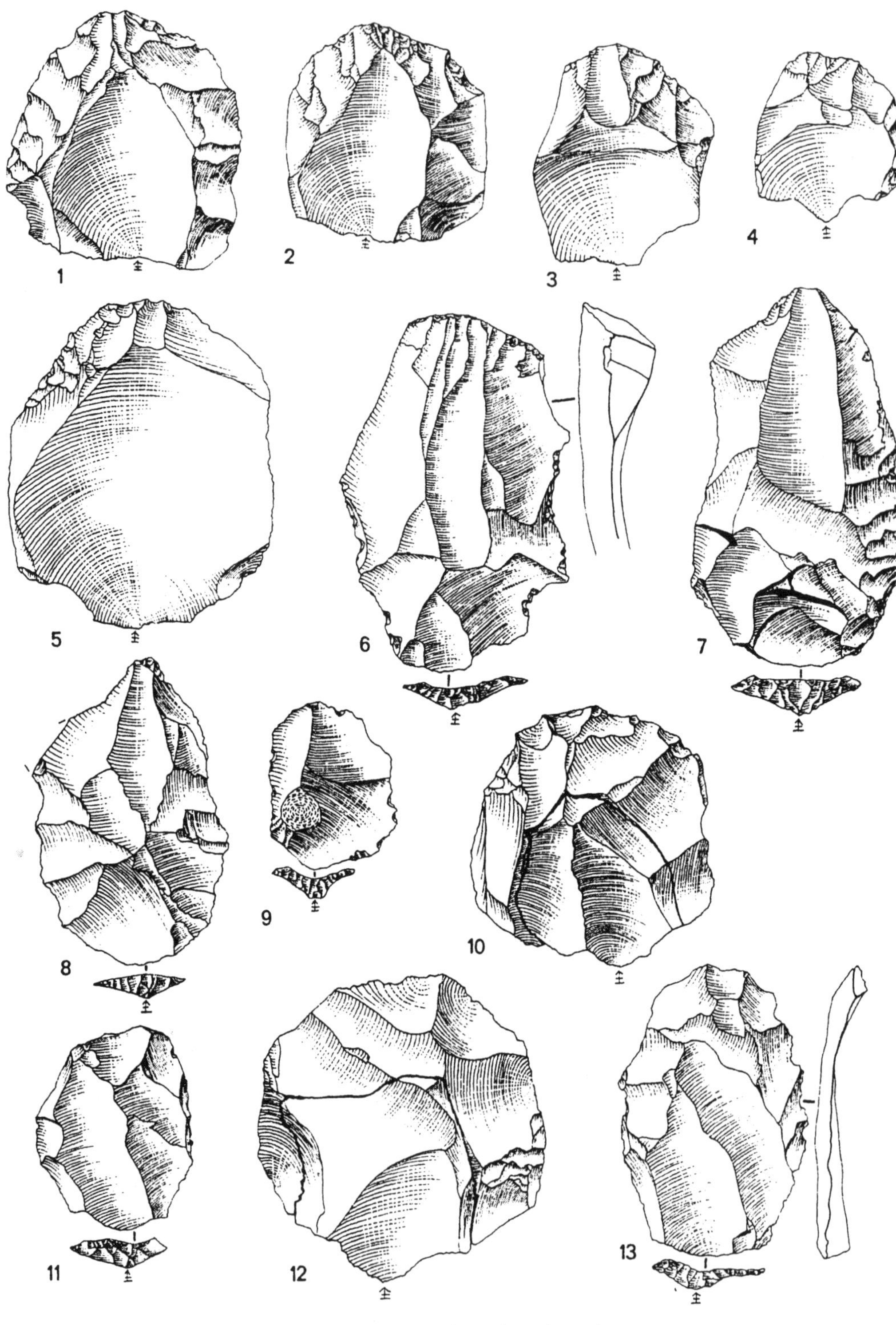

Figure 6.3. Levallois technology from Nazlet Khater-2: 1-5: Levallois cores; 6-9, 11, 13: Levallois flakes; 10, 12: Levallois cores with Levallois flakes refitted.

Table 6.1. Levallois technology of Middle Palaeolithic and transitional industries in the Nile Valley. Note that the individual percentages of butt types do not add up to 100%, since some types have been discarded.

Sites	NK-3 (N-group)	NK-2 (C-group)	1018	E71P1C
Levallois Flakes				
number	108	145	105	101
percentage of preparatory scars per sector (Crew 1975: 13)				
bidirectional	62	61	85	73
distal	25	27	45	30
proximal	37	34	40	43
lateral	38	39	15	27
butt types (percentages)				
plain	9	6	2	4
dihedral	10	8	3	12
en chapeau de gendarme	15	14	22	34
faceted convex	50	59	63	32
faceted straight	5	4	2	3
number of preparatory scars				
mean	6.92	7.94	7.30	7.83
standard deviation	2.12	2.43	1.92	2.78
length (mm)				
mean	51.97	45.72	23.52	37.36
standard deviation	14.28	10.40	4.19	9.72
width/thickness ratio				
mean	6.67	7.09	7.10	7.14
standard deviation	2.47	2.47	2.29	2.15
Levallois Cores				
number	21	59	51	69
number of distal preparatory scars				
mean	4.00	4.89	6.45	6.35
standard deviation	1.84	1.63	1.82	1.64

however there is a tendency towards more bi-directional preparation, greater importance of the *chapeau de gendarme* butt types, and smaller flakes.

At this stage, we have no idea of how the Khormusan, which is said to be more recent than the 'Mousterian' industries, is related to this evolutionary pattern. It seems clear however that this cannot be intermediate between the K-group and assemblages like that from Site 1018.

Levallois technology – blade production

The evolutionary development of one particular form of Levallois technology from the Middle Palaeolithic through the mixed industries can thus be demonstrated. In its final stage, this Levallois technology gave way to

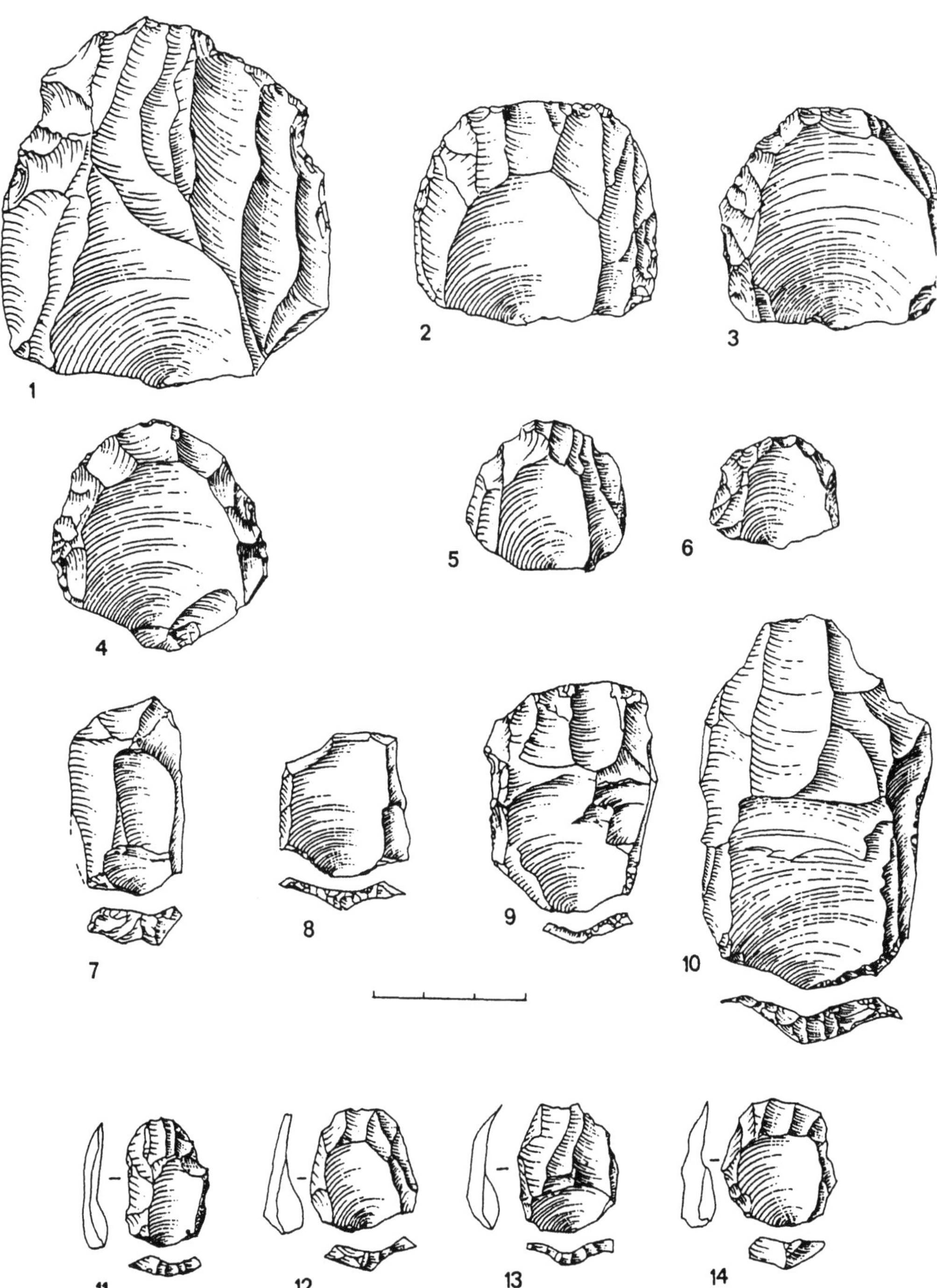

Figure 6.4. Levallois technology from sites E71P1C and 1018: 1-4: site E71P1C — Levallois cores; 7-10: site E71P1C — Levallois flakes (after Wendorf and Schild 1976a); 5-6: site 1018 — Levallois cores; 11-14: site 1018 — Levallois flakes (after Marks 1986c).

a special form of microblade or blade production, which seems to be different from Upper Palaeolithic blade production strategies in other regions. At sites 1018 and E71P1C, Levallois technology is accompanied by considerable numbers of cores of Upper Palaeolithic type. An analysis of the reduction patterns of both Levallois and Upper Palaeolithic type cores shows that it is essentially the same. The flaking is carried out on the flat surface of a nodule. Whereas among the Levallois cores, the microblade production from one striking platform on one end (the distal end) serves as preparation for a final flake removal from the proximal end, this microblade production is the final product of the flaking sequence on the Upper Palaeolithic type cores. Only the idea of producing one final flake had to be abandoned to arrive at this end. The technique of setting up a crest in order to start systematic blade production is unknown in these industries. Crested blades that do occur are of a special type: they are produced in the course of the flaking process, whenever the flaking reaches the lateral edge of the core. Their cross-section is in most cases asymmetric. Similar crests occur in Middle Palaeolithic industries.

The convergence of Levallois technique through microblade preparation with true microblade (blade) technique, is very clearly attested in these assemblages by the occurrence of the two techniques on the *same* cores. Indeed there are examples of Halfa Levallois cores on which the 'Halfa' surface, after having produced a Halfa flake, served as a striking platform for microblade production on the other face. On other cores we can observe the opposite pattern.

Thus it would seem that Halfa Levallois technology in this case provided the idea of regular microblade production (or blade production; the distinction between the two seems to be merely determined by the size of the raw material, rather than being the result of a different technological approach) (Marks 1968c: 459). Once this technological shift had occurred, it would seem a logical step to abandon the older technique. Industries like those of 1018 and E71P1C, showing direct evidence of this transformation, can be considered as 'transitional' industries.

UPPER PALAEOLITHIC INDUSTRIES

The particular form of blade production described above is the dominant reduction strategy in a number of assemblages in Nubia and Upper Egypt which on technological and typological grounds are Upper Palaeolithic. In some of these industries, Halfa Levallois technology is still present to a limited extent, thus demonstrating explicitly their relationship to the transitional industries. Some of the best documented of these assemblages deserve to be described in detail.

Site 443
(Marks 1968c: 429-46)

Site 443 is situated on the same dune formation as Site 1018, somewhat further to the east. The two sites are separated by a wadi which cuts to a depth of two metres below the dune crest. The lithic material again occurred in an occupation layer on the dune sands. The same raw material as at Site 1018 was employed, with an even greater emphasis here on chert pebbles. The site provided an assemblage in which 73 percent of all cores are of single or opposed platform type, with the characteristics described above. Single platform cores are the most numerous. Halfa technology is present, but to only a limited extent. The blade category consists mainly of poor quality microblades.

It would seem that here the shift towards microblade or blade production had been completed. This technological shift is also reflected in the tool typology. Seventy percent of the tools are manufactured on blades or (mostly) microblades, in which backed elements are overwhelmingly dominant (more than 80 percent). Burins, notches, denticulates and scrapers are much less frequent. Although a shift in the reduction strategy is apparent, the affinities with Site 1018 are clear. Several assemblages of this type employing both Levallois and microblade production as the basic reduction strategy, were grouped together under the name 'Halfan' by Marks (1968c), who has already pointed to its apparent technological and typological continuity.

Site Shuwikhat-1
(Paulissen *et al.* 1985: 7-14)

This site is situated on the east bank of the Nile, some 15 kilometres to the west of Qena. The lithic material occurred almost *in situ* in Nilotic silts, which were named 'Shuwikhat' silts. The raw material used was exclusively chert, which was locally present in the form of cobbles in the nearby wadis. Some of these cobbles were quite large – up to 20 centimetres. The reduction strategy, intended mainly for blade production (there is no systematic microblade production), fits the technological pattern described above, and is in several aspects is reminiscent of Levallois technology. In the first place, the dominant core type is a flat opposed-platform core with debitage confined to one side (Figure 6.5, nos 1, 3, 5). In some cases these cores produced 'accidental' Levallois flakes (Figure 6.5, no. 9), which in fact are simply miss-struck blades. If the core was abandoned at that stage, it looks very much like a Halfa Levallois core (Figure 6.5, nos 2, 4). Many blades have prepared butts, either convex facetted or '*en chapeau de gendarme*' (Figure 6.5, nos 6, 7). This special type of individual butt preparation is also a 'Levallois' tradition. Typologically, the assemblage is dominated by characteristic denticulated blades and burins, together with some end-scrapers and some fragments of backed blades.

Site E71K9
(Wendorf and Schild 1976a: 244-53)

This site is situated near Isna and consists of two concentrations. The first was a large, dense concentration on the surface of a dune. Two trenches were dug here, revealing artifacts that occurred to a depth of 40 centimetres in the dune sand. The other locality yielded only a small surface

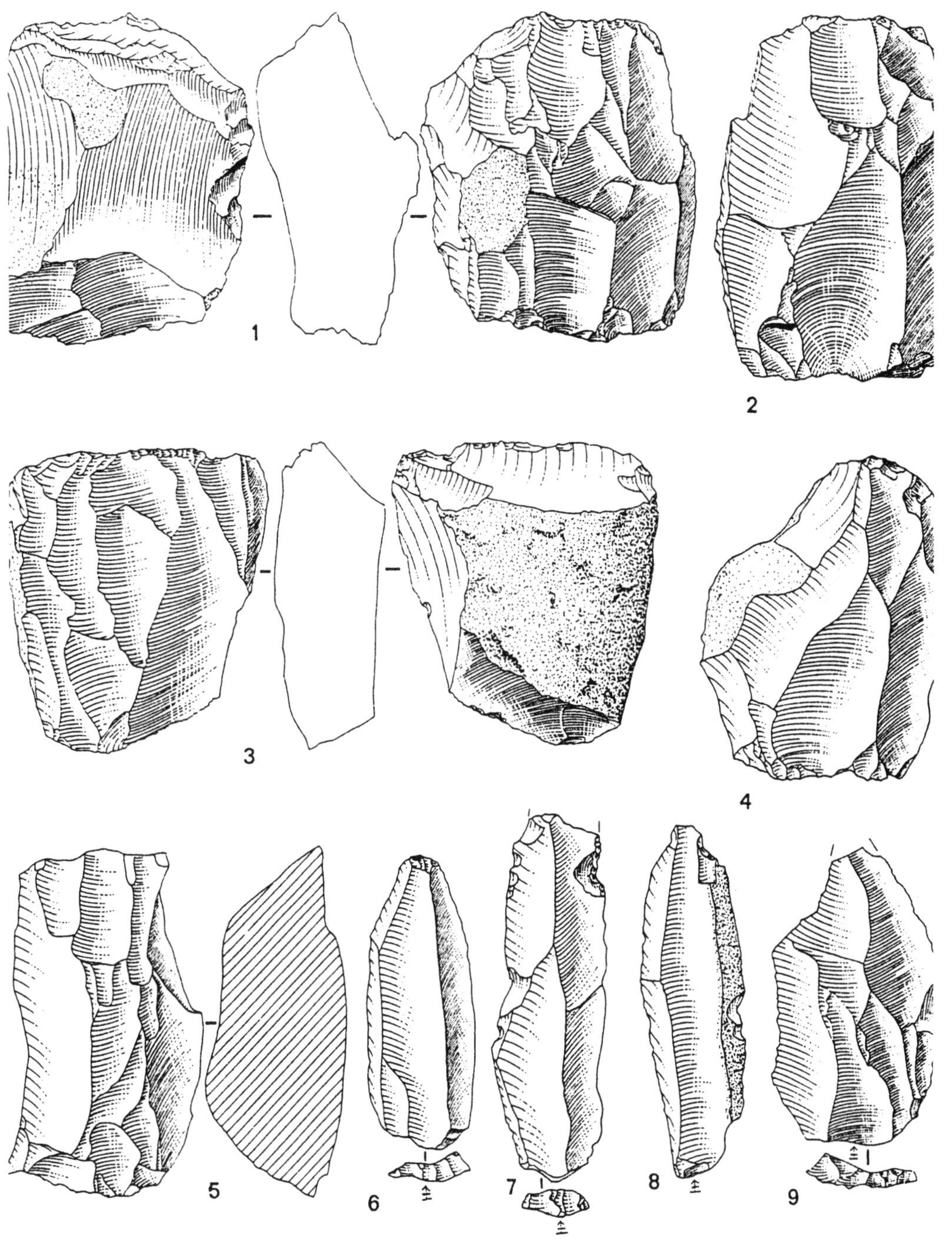

Figure 6.5. Artifacts from Shuwikhat site 1: 1, 3, 5: opposed platform cores; 2, 4: opposed platform cores with 'Levallois' removals; 6-8 blades; 9: accidental Levallois flake.

concentration which has only been briefly described. Opposed-platform blade and microblade cores are dominant, together with some single-platform cores. Some cores passed from blade to microblade cores at the end of the reduction sequence. From the illustrations, it would seem that the butts of the blades were sometimes prepared, either convex facetted or '*en chapeau de gendarme*' (Wendorf and Schild 1976a: 251). Typologically, the assemblage is characterized by large numbers of burins and blades with basal blunting. Denticulated blades similar to those from Shuwikhat-1 also occur. Backed elements are rare.

The Shuwikhat industry

It will be clear that the assemblages from Shuwikhat-1 and E71K9 are very similar in all aspects. The technological strategy seems to fit the pattern already foreshadowed by the transitional industry of E71P1C. Both E71P1C and E71K9 have been described as belonging to the 'Idfuan' industry by Wendorf and Schild (1976a: 243). We would propose however to distinguish between a 'transitional' and a true 'Upper Palaeolithic' industry. Therefore, we would group E71K9 and Shuwikhat-1 together under the term of the Shuwikhat industry.

Nazlet Khater-4
(Vermeersch *et al.* 1982, 1984)

The chert mining site of Nazlet Khater-4, situated in the immediate vicinity of the Middle Palaeolithic sites of NK2 and NK3, produced a blade industry. Blades were obtained mainly from single platform cores, worked on large cobbles. The reduction strategy here is quite different from that of the industries described above. The cores cannot be integrated into the Levallois technology-related pattern as described earlier. Often the flaking was carried out on the side of a cobble (either with or without a prior crest) and was then extended on one or two flat surfaces. In other instances, thick nodules were used, on which single platform flaking was carried out almost peripherally. The blades are quite irregular, and microblades are not numerous. The special function of the site may be partially responsible for the observed differences. These however seem too important for this factor to be the only one responsible. Tools are rare; denticulates are the most numerous type, together with some burins and end-scrapers. In addition to these types, a few characteristic bifacial axes have been found. A modern *Homo sapiens* skeleton was associated with the NK4 industry, buried at some distance from the site.

CHRONOLOGY

Chronology undoubtedly remains the major problem in interpreting the industrial sequence in the Nile Valley.

For the 'Halfan' industries, the following radiocarbon dates have been obtained (see Irwin *et al.* 1968: 110; Marks 1968c: 400; Wendorf *et al.* 1979: 221):

Site 2014	19 150 ± 375 BP	WSU-332	
Site 8859	18 600 ± 500 BP	WSU-318	
Site 443	16 500 ± 375 BP	WSU-201	
	17 620 ± 410 BP	SMU-576	(rerun —
	17 200 ± 330 BP	"	two counts)
Site 6G29	14 970 + 1420/-1730 BP	GXO-576	
Site 6B32	25 700 + 2500/-3700 BP	GXO-410	

Since there were thought to be some problems with the WSU dates, some of these samples were rerun. The revised and slightly older date for Site 443 was obtained after incomplete treatment of the sample. The dates for 2014 and 8859 belong to the same series of WSU dates, but no remeasurement has been performed for those samples. We suggest that all these dates should be used with caution. The date for Site 6G29 is even younger than the preceding dates. This date however was obtained on dispersed charcoal collected from different areas, the relationship between which is not clearly described. Site 6B32, where Halfa technology is abundant, is considerably older. This date was obtained on charcoal from a yellow-red sandy silt layer, in which some artifacts were also found (Irwin *et al.* 1968: 36). At first sight, the dates are consistent with the proposed pattern of technological evolution – that is to say, the younger dates are associated with industries with few or no Levallois cores, while the older date is associated with a Levallois-dominated assemblage. This could be seen as an argument to accept the dates as they stand. On the other hand, it could be that the whole of the Halfan chronology implied by the ^{14}C dates is too recent.

Site E71P1 has a consistent series of radiocarbon dates on shells from different trenches, indicating an age of around 17 000 BP (Wendorf and Schild 1976a: 32). This however is more recent than the TL dated Upper Palaeolithic industries. The Shuwikhat-1 date of 24 700 BP ± 2500 (OxTL-253) (Paulissen and Vermeersch 1987: 47) was on burned clay, of which the association with the lithics is beyond doubt. The E71K9 date of 21 588 BP ± 1518 (OxTL-161-C-1) (Wendorf and Schild 1975: 138) was also on burned clay. Paulissen and Vermeersch (1987: 51) have contested the value of the E71P1 dates for the lithic industry because of the unproven association of the dated material with the archaeological remains. The transitional character of the industry from sector C makes an older date far more likely, at least for this part of the site.

The Nazlet Khater-4 blade industry is dated by radiocarbon dates on charcoal samples to around 33 000 BP (Vermeersch *et al.* 1984: 342). This site is therefore isolated from the others both by its lithic industry and its early chronological position.

The isolated position of Nazlet Khater-4 may be an argument in favour of the hypothesis of the introduction of blade technology into this region from elsewhere, and in particular from areas to the north, around the Mediterranean coast. In Cyrenaica, a blade industry (the Early Dabban) occurs from about 40 000 BP in the Haua Fteah cave (McBurney 1967: 167). An interesting feature of this industry is the presence of some bifacial tools and 'adzes'. As noted above, bifacial technique also occurs at Nazlet Khater-4. On the other hand, backed elements are always present in the Early Dabban whereas these are completely absent at Nazlet Khater-4. Unfortunately, little information on the technology of the Early Dabban is available, so that we cannot compare it to Nazlet Khater-4 from that point of view. Early blade industries which are present in the Sinai, such as the Lagaman in Gebel Maghara (Bar-Yosef and Belfer 1977: 42-84) and the Ahmarian (Phillips 1987: 109-15) do not seem at all comparable to Nazlet Khater-4. In the Eastern Sahara, no such early blade industries have been found.

CONCLUSION

In the Nile Valley, there is evidence of a dynamic Levallois technology, which eventually led the way to a particular form of blade production and Upper Palaeolithic industries. The chronological framework to support this model however is poorly documented. At present, we can only show that Upper Palaeolithic industries of this type existed in Egypt by 25 000 BP. In Nubia, they may be more recent. In addition to this internal transformation, there is possibly also an introduction of blade technology from the Mediterranean coast at an earlier date. Cyrenaica is an area that has to be considered in that respect.

This hypothesis would imply two mechanisms for the appearance of blade technology in the Nile Valley, with geographical implications: an intrusion from the north into Middle Egypt, and an internal transformation occurring within Upper Egypt and Nubia.

It must be stressed however that a great deal of these interpretations are at present speculative and will require confirmation from new sites and new dates. More essential field work is needed in the coming years. Despite the lack of sufficient data, however, it is clear that the Nile Valley, in its own way, was integrated in the pattern of changes that are observed in other parts of the world between 45 000 and 30 000 BP.

ACKNOWLEDGEMENTS

We would like to thank Dr. F. Wendorf, A. Close and A.E. Marks, Department of Anthropology, Southern Methodist University, Dallas, for their kind permission to study their collections. Also, we owe them and the staff of the Department of Anthropology our gratitude for the help received during our visits. Discussions on this subject were greatly appreciated.

NOTE

1. Microblades have been defined by Marks (1968c: 394) as blades that are less than 3 centimetres in length.

REFERENCES

Bar-Yosef, O. and Belfer, A. 1977. The Lagaman industry. In O. Bar-Yosef and J.L. Phillips (eds) *Prehistoric investigations in Gebel Maghara, Northern Sinai.* Qedem, Monographs of the Institute of Archaeology, the Hebrew University of Jerusalem, 7.

Caton-Thompson, G. 1946. The Levalloisian industries of Egypt. *Proceedings of the Prehistoric Society* 12: 57-120.

Close, A.E. 1987 (ed.). *Prehistory of Arid North Africa: Essays in Honor of Fred Wendorf.* Dallas: Southern Methodist University Press.

Crew, H. 1975. *An Evaluation of the Variability of the Levalloisian Method: its Implications for the Internal and External Relationships of the Levantine Mousterian.* Unpublished PhD Dissertation, University of California, Davis.

de Heinzelin, J. 1968. Geological History of the Nile Valley in Nubia. In F. Wendorf (ed.) *The Prehistory of Nubia.* Dallas: Fort Burgwin Research Center and Southern Methodist University Press: 19-55.

Guichard, J. and Guichard, G. 1965. The Early and Middle Paleolithic of Nubia: A Preliminary Report. In F. Wendorf (ed.) *Contributions to the Prehistory of Nubia.* Dallas: Fort Burgwin Research Center and Southern Methodist University Press: 57-166.

Guichard, J. and Guichard, G. 1968. Contributions to the study of the Early and Middle Paleolithic of Nubia. In F. Wendorf (ed.) *The Prehistory of Nubia.* Dallas: Fort Burgwin Research Center and Southern Methodist University Press: 148-193.

Irwin, H.T., Wheat, J.B. and Irwin, L.F. 1968. *University of Colorado Investigations of Paleolithic and Epipaleolithic sites in the Sudan, Africa.* University of Utah Anthropological Papers 90.

Marks, A.E. 1968a. The Mousterian Industries of Nubia. In F. Wendorf (ed.) *The Prehistory of Nubia.* Dallas: Fort Burgwin Research Center and Southern Methodist University Press: 194-314.

Marks, A.E. 1968b. The Khormusan: an Upper Pleistocene industry in Sudanese Nubia. In F. Wendorf (ed.) *The Prehistory of Nubia.* Dallas: Fort Burgwin Research Center and Southern Methodist University Press: 315-391.

Marks, A.E. 1968c. The Halfan industry. In F. Wendorf (ed.) *The Prehistory of Nubia.* Dallas: Fort Burgwin Research Center and Southern Methodist University Press: 392-460.

McBurney, C.B.M. 1967. *The Haua Fteah (Cyrenaica) and the Stone Age of the South-East Mediterranean.* Cambridge: Cambridge University Press.

Paulissen, E. and Vermeersch, P.M. 1987. Earth, man, and climate in the Egyptian Nile Valley during the Pleistocene. In A.E. Close (ed.) *Prehistory of Arid North Africa: Essays in Honor of Fred Wendorf.* Dallas: Southern Methodist University Press: 29-68.

Paulissen, E., Vermeersch, P.M. and Van Neer, W. 1985. Progress report on the Late Palaeolithic Shuwikhat sites (Qena, Upper Egypt). *Nyame Akuma* 26: 7-14.

Phillips, J.L. 1987. Sinai during the Paleolithic: the early periods. In A.E. Close (ed.) *Prehistory of Arid North Africa: Essays in Honor of Fred Wendorf.* Dallas: Southern Methodist University Press: 105-122.

Said, R. 1981. *The Geological Evolution of the River Nile.* New York: Springer.

Sandford, K.S. 1934. *Paleolithic Man and the Nile Valley in Middle and Upper Egypt*. Chicago: Oriental Institute Publications 18.

Sandford, K.S. and Arkell, W.J. 1929. *Paleolithic Man and the Nile-Fayum Divide*. Chicago: Oriental Institute Publications 10.

Sandford, K.S. and Arkell, W.J. 1933. *Paleolithic Man and the Nile Valley in Nubia and Upper Egypt*. Chicago: University of Chicago Oriental Institute Publications 17.

Sandford, K.S. and Arkell, W.J. 1939. *Paleolithic Man and the Nile Valley in Lower Egypt*. Chicago: Oriental Institute Publications 46.

Schild, R. 1987. Unchanging contrast? The Late Pleistocene Nile and Eastern Sahara. In A.E. Close (ed) *Prehistory of Arid North Africa: Essays in Honor of Fred Wendorf*. Dallas: Southern Methodist University Press: 13-28.

Schild, R. and Wendorf, F. 1986. The geological setting. In A.E. Close (ed.) *The Prehistory of Wadi Kubbaniya*. Vol. 1: *The Wadi Kubbaniya skeleton: a Late Paleolithic burial from Southern Egypt*. Dallas: Southern Methodist University Press: 7-32.

Singleton, W.L. and Close, A.E. 1980. Report on site E-78-11. In A.E. Close (ed.) *Loaves and Fishes: the Prehistory of Wadi Kubbaniya*. Dallas: Southern Methodist University Press: 229-237.

Van Peer, P. 1986. Présence de la technique nubienne dans L'Atérien. *L'Anthropologie* 90: 321-324.

Vermeersch, P.M., Otte, M., Gilot, E., Paulissen, E., Gijselings, G., and Drappier, D. 1982. Blade Technology in the Egyptian Nile Valley: some New Evidence. *Science* 216: 626-628.

Vermeersch, P.M., E. Paulissen, G. Gijselings, J. Janssen. 1986. Middle Palaeolithic chert exploitation pits near Qena (Upper Egypt). *Paléorient* 12: 61-65.

Vermeersch, P.M., Paulissen, E., Gijselings, G., Otte, M., Thoma, A. and Charlier, C. 1984. Une minière de silex et un squelette du Paléolithique Supérieur ancien à Nazlet Khater, Haute Egypte. *L'Anthropologie* 88: 231-244.

Vermeersch, P.M., Paulissen, E., Gijselings, G., Otte, M., Thoma, A., Van Peer, P. and Lauwers, R. 1984. 33 000-yr old mining site and related *Homo* in the Egyptian Nile Valley. *Nature* 309: 342-344.

Vermeersch, P.M., Paulissen, E., Otte, M., Gijselings, G. and Drappier, D. 1978. Middle Palaeolithic in the Egyptian Nile Valley. *Paléorient* 4: 245-252.

Vermeersch, P.M., Paulissen, E., Otte, M., Gijselings, G. and Drappier, D. 1979. Prehistoric and geomorphologic research in Middle Egypt. *Palaeoecology of Africa* 11: 111-115.

Vermeersch, P.M., Paulissen, E., Otte, M. Gijselings, G., Drappier, D. and Van Peer, P. 1980. Excavations at Nazlet Khater (Middle Egypt). *Bulletin de l'Association Internationale pour l'Etude de la Préhistoire Egyptienne* 2: 73-76.

Wendorf, F. 1965 (ed.). *Contributions to the Prehistory of Nubia*. Dallas: Fort Burgwin Research Center and Southern Methodist University Press.

Wendorf, F. 1968 (ed.). *The Prehistory of Nubia*. Dallas: Fort Burgwin Research Center and Southern Methodist University Press.

Wendorf, F. and Schild, R. 1975. The Paleolithic of the Lower Nile Valley. In F. Wendorf and A.E. Marks (eds.) *Problems in Prehistory: North Africa and the Levant*. Dallas: Southern Methodist University Press: 127-169.

Wendorf, F. and Schild, R. 1976a. *Prehistory of the Nile Valley*. New York: Academic Press.

Wendorf, F. and Schild, R, 1976b. The Middle Paleolithic of Northeastern Africa: new data and concepts. In F. Wendorf (ed.) *Deuxième Colloque sur la Terminologie de la Préhistoire du Proche-Orient.* Proceedings of Colloquium 3 of the Ninth Congress of the International Union of Prehistoric and Protohistoric Sciences, Nice, 1976: 8-34.

Wendorf, F., Schild, R. 1986. The archaeological sites. In A.E. Close (ed.) *The Prehistory of Wadi Kubbaniya.* Vol. 1: *The Wadi Kubbaniya skeleton: a Late Paleolithic burial from Southern Egypt.* Dallas: Southern Methodist University Press: 33-48.

Wendorf, F., Schild, R. and Haas, H. 1979. A new radiocarbon chronology for prehistoric sites in Nubia. *Journal of Field Archaeology* 6: 219-223.

Wendorf, F., Shiner, J.L. and Marks, A.E. 1965. Summary of the 1963-1964 field season. In F. Wendorf (ed.) *Contributions to the Prehistory of Nubia.* Dallas: Fort Burgwin Research Center and Southern Methodist University Press: 57-166.

7: The Szeletian and the Stratigraphic Succession in Central Europe and Adjacent Areas: Main Trends, Recent Results, and Problems for Resolution

P. ALLSWORTH-JONES

SUMMARY

New evidence relating to the Szeletian and the transition from Middle to Upper Palaeolithic in Central Europe is presented and assessed. Significant new finds from Hungary and Moravia are summarised. They include a number of open-air sites in the vicinity of the Bükk mountains, some of which have produced industries which are of relevance when considering the antecedents of the Szeletian. Research in Hungary has also recently concentrated on the analysis of the variegated lithic raw materials in that country, including the dominant raw material at Szeleta itself. New excavations in Moravia include those at Vedrovice V, Bohunice, and Stránská Skála, and important surface collections have also been published. The concept of the 'Bohunician' as an entity distinct from the Szeletian is discussed. Some scepticism is expressed with regard to the geological age claimed for, and the nature of, the supposed early Aurignacian sites of Vedrovice II and Kupařovice I in southern Moravia, and also their derivation from the 'Krumlovian' in the same area. General questions concerning the interpretation to be given to cave and open-air sites are discussed, particularly the so-called 'cave bear hunters' hypothesis. The concept of the 'Olschewian' should be finally abandoned following the full publication of the finds from Potočka Zijalka. Consideration is then given to areas outside Central Europe which are closely linked to it in terms of their archaeological record for the period in question. So far as Northwestern Europe is concerned, some doubt is expressed as to whether there was really a continuous 'leafpoint tradition' stretching in effect all the way from the Micoquian to the Solutrian constituting a particular kind of 'transition model' from Middle to Upper Palaeolithic. Our knowledge of the situation in Southeastern Europe has been transformed by a large number of new ^{14}C dates, particularly in Romania, and new sites, particularly in Soviet Moldavia. The new dates have confirmed the antiquity of the Mousterian in the area, throwing into doubt the idea of a 'transitional' or 'pre-Szeletian' late phase of the local Middle Palaeolithic. Attention is drawn to the frequent co-occurrence of bifacial leafpoints, backed blades, and Aurignacian-style end-scrapers in the early Upper Palaeolithic of Moldavia. Soviet researchers have tended to split the finds into a number of different 'cultures', but it is considered that they probably represent broadly a single entity, parallel to but probably not identical with the Szeletian of Central Europe. The Kostienki-Streletskaya variant is considered to be probably still quite separate, in spite of attempts made to link it to this area. In the concluding section, the presently available dating evidence for Central and Southeastern Europe is summarised as a whole, with 46 ^{14}C dates for the Aurignacian and 13 for the Szeletian and its cognates. The model whereby the Szeletian can be seen to run in part in parallel with the Aurignacian has analogies in Western Europe, where the same can be observed in the case of the Châtelperronian and the Uluzzian. Some recently published information concerning the hominid record is summarised and considered in relation to its archaeological context. In particular, the contrast is noted between the situation at Máriaremete and Bacho Kiro; in the first case the Jankovichian is associated with Neanderthal man, whereas in the second an early Aurignacian is associated with anatomically modern man at >43 000 BP. Overall, it is maintained that the new evidence is consistent with a replacement/acculturation model for the Szeletian and its cognates, rather than a model of independent indigenous transition from Middle to Upper Palaeolithic and from *Homo sapiens neanderthalensis* to *Homo sapiens sapiens*.

INTRODUCTION: DEFINITION AND ROLE OF THE SZELETIAN

There can be no doubt that, if one is studying the transition from Middle to Upper Palaeolithic in Central Europe, the Szeletian is an integral part of it, just as the Châtelperronian is in Western Europe. The evidence concerning the Szeletian and related entities, in so far as it was available up to the end of 1983 and in so far as even rough justice could be done to it by one person, was recently summarised in book form (Allsworth-Jones 1986). It is not the intention here to go over again the matters considered in laborious detail in that book; rather the attempt is made to evaluate certain newly published data, and to draw attention to certain aspects which seem to have a broader significance. In line with this approach (and to save space) the bibliography for this article does not include any of the references listed in my book; it contains only works which either were published after 1983 or were for one reason or another not available to me at that time. Some of the major conclusions then reached were as follows.

In the years before the First World War, when leafpoints began to be found in Palaeolithic contexts in Central Europe, they were ascribed by analogy with Western Europe to the 'Solutrian' or the 'Proto-Solutrian'. Later, when it became clear that some of these contexts were Middle rather than Upper Palaeolithic, the term 'Prae-Solutréen' was coined for them instead, as advocated by Freund and Zotz. The term 'Szeletian' really came into vogue, thanks to the work of Prošek, Vértes, and Valoch, after the Second World War. The key role which the Szeletian might have played in the transition from Middle to Upper Palaeolithic is clear from Prošek's classic formulation that, "Le Szélétien s'est formé; directement du Moustérien du Bassin Karpathique sous l'influence de l'Aurignacien".

Such a formulation, implying that the Szeletian came into being as the product of an acculturation process at the junction of Middle and Upper Palaeolithic, is taken to be basically correct, and there is an obvious parallel to the Châtelperronian in Western Europe as understood by Leroi-Gourhan and Lynch.

There is in general a need for care in the definition of the Szeletian, because leafpoints similar to those with which it is associated are known to have occurred in a number of different and unrelated Palaeolithic contexts, due almost certainly to the effects of convergence alone. For that reason, the use of blanket phrases such as 'leafpoint cultures' and the like is considered undesirable, and emphasis is laid on assemblages which occur in closed stratigraphic situations. More particularly, there are problems of definition when it comes to determining the various component parts of the Szeletian and to elucidating the relationships which might have existed between the Szeletian and its predecessors. The second question especially has been bedevilled by the tendency to search for so-called 'typogenetic' connections between the entities concerned, whereas the intellectual validity of this process has usually gone both unexamined and unjustified.

Apart from Szeleta itself, there are seven other cave sites in the Bükk mountains of Hungary which are commonly regarded as Szeletian, and one (the Herman Ottó cave) which may be, although its status is still debated. In addition, there are a number of open-air sites in the vicinity (particularly Avas and Eger) which Vértes regarded as Mesolithic but which other authors have suggested might be at least in part related to the Szeletian. A similar suggestion has been made in respect of two sites west of the Bükk (Hont-Babat and Hont-Csitár). To the northeast, in what is now Slovakia, an open-air site was located at Veľký Šariš and another has now been found at Petrovany (Kaminská 1985).

The sites west of the Danube, mainly but not entirely in the Pilis and Gerecse mountains, were referred to by Vértes as the 'Trans-Danubian Szeletian'. Apart from Jankovich, there are seven other cave sites which are usually regarded as belonging to this group, as well as one open-air site at Lovas near lake Balaton. One of the cave sites, Dzeravá Skála (Pálffy), is separated from the others, on the western slopes of the Little Carpathians in present-day Slovakia. Nevertheless, it is not only for historical reasons that it is linked to the other sites in this area. Gábori-Csánk has suggested that this group deserves a separate name, the 'Jankovichian', to emphasise its distinctiveness *vis-à-vis* the Szeletian of the Bükk. The view taken by this author is that the Jankovichian sites do still form part of the Central European Szeletian in a wider sense.

Apart from Dzeravá Skála, there is one other cave site in western Slovakia which is conventionally classified as Szeletian (Čertova Pec), and six open-air sites, of which one (Trenčín IV) can only rather tentatively be regarded as belonging to the group. Most of these sites are situated in the valley of the river Váh. I have suggested that three caves in the Moravian Karst (Křížova, Pod Hradem, and Rytířská) are so similar to Dzeravá

Skála that they may be regarded as constituting a westward extension of the Jankovichian. Apart from these three, all other localities claimed as Szeletian in Moravia and southern Poland are open-air sites, and the great majority even now consist of surface finds only. In the few cases where the finds were stratified, as at Předmostí and Zwierzyniec, there are great difficulties with regard to their interpretation, and the likely associations of the leafpoints discovered at those two sites remain unsure. A case can be made for saying that they belonged together with the Aurignacian, since we know from sites elsewhere in the area (e.g. Gottwaldov-Louky) that leafpoints did at times form an integral part of Aurignacian inventories. The abundant material from Ondratice I and (to a lesser extent) Otaslavice evidently represents more than one phase of settlement, but it can now be sorted out essentially on the basis of typology, technology, and raw material alone. The best stratified site, originally published by Valoch as "an early stage of the Levallois facies of the Moravian Szeletian", was considered to be Bohunice. There are some further stratigraphic indications at Rozdrojovice, Neslovice, and Dzierzyslaw I. All told, 12 open-air sites conventionally classified as Szeletian (excluding Ondratice I and Otaslavice but including Bohunice) were compared by this author with 14 sites similarly classified as Aurignacian, in an attempt to analyse the statistical structure of these industries in Moravia and southern Poland (Allsworth-Jones 1986: 170-7).

Also in southern Poland are four cave sites, of which the most important is Nietoperzowa, regarded by Chmielewski as forming a separate entity, named by him the 'Jerzmanowician'. The view taken by this author, following Valoch, is that (as in the case of the Jankovichian) it is better to regard these sites as just one special complex within the Szeletian as a whole. The cave of Nad Kačákem in the Bohemian Karst provides a further possible parallel, as does the Ranis 2 industry. Chmielewskí's somewhat far-flung comparison to Kostienki-Tel'manskaya is not considered to carry conviction; more convincing is the possibility of a link to the Belgian and British Early Upper Palaeolithic, as represented at the sites of Spy, Paviland, and Kent's Cavern.

With regard to the antecedents of the Szeletian, there are a number of competing claims, not all of them equally convincing. One of the most important undoubtedly is the Micoquian, where leafpoints did have a significant role to play in some cases, most notably at Rörshain, Staroselʼye, and Muselievo. Some interesting new information in this respect is provided by Valoch (1981) in his study of the earlier layers at Kůlna. The Micoquian at this site may possibly have extended back as far as the last interglacial period. The likelihood that the material from the lower travertine at Ehringsdorf belongs to this same period has recently been reasserted on palaeontological grounds (Cook *et al.* 1982: 43), although there is a suggestion on the basis of uranium/thorium age determinations that it may go back to 225 ± 26 kyr BP. *Pace* Behm-Blancke, the archaeological assemblage cannot be classified as a 'Pre-Szeletian Mousterian', but I have suggested that it may be regarded as some kind of ultimate predecessor of the Micoquian.

The Altmühlian, defined principally on the basis of Mauern, and excluding both Kösten and Ranis, has been found at a total of ten cave sites in southern Germany, and is exclusively Middle Palaeolithic in character. It is legitimate to regard it as deriving ultimately from the Micoquian, but since it very likely overlapped chronologically with the Szeletian it is less of a predecessor than a parallel development to it. Hungarian authors, particularly Vértes and Gábori, have tended to favour a local origin for the Szeletian in the Bükk mountains, especially at Subalyuk, and at the two small sites of Kecskésgalya and Ballavölgy, which Vértes considered to be 'intermediate' between Szeleta and Subalyuk. That argument appears to rest upon a misunderstanding, and the conclusion reached by this author in regard to Subalyuk was that while it may well have formed part of the general background from which the Szeletian sprang, it was unlikely to have led directly to it. A more convincing contender is the site of Büdöspest, also in the immediate vicinity of Szeleta. The 'Preszeletian Mousterian' (no relation to Ehringsdorf) distinguished by Nicolaescu-Plopşor in the caves of the southern Carpathians rests upon a very insecure foundation; but leafpoints do undoubtedly occur in a Levallois-Mousterian context elsewhere in Southeastern Europe, e.g. at Samuilitsa 2, Kokkinopilos, and Mamaia, and this may have had some bearing on the ultimate origin of the Szeletian.

On stratigraphic grounds it is likely that the Szeletian as defined above (except for the sites attributed to the Jankovichian) extended from some time early in the Hengelo interstadial to a period corresponding approximately to the Denekamp interstadial. Up to 1983 there were to my knowledge eight ^{14}C dates from Szeleta, Nietoperzowa, Čertova Pec, and Bohunice, with a range extending from 43 000 to 32 620 BP and a median date of 40 786 BP. This is broadly consistent with the stratigraphic indications. The Jankovichian sites (not ^{14}C dated) are likely to be significantly earlier, mainly before the Tokod faunal phase in the Carpathian Basin, although they may also have extended into Hengelo. The Jankovichian at Dzeravá Skála and Máriaremete is associated with Neanderthal remains, and this was considered likely to be true of the Szeletian as a whole, as well as the Châtelperronian. By contrast, the Aurignacian, so far at least, can be shown to be exclusively associated with anatomically modern forms. This author therefore (following Vértes and Leroi-Gourhan) expressed the view that the end of the Szeletian and the Châtelperronian coincided with the demise of Neanderthal man, as one would expect on the assumption that an acculturation model was correct (Allsworth-Jones 1986).

How then do these conclusions stand up in the light of recent results, and what general issues of interpretation arise?

THE SZELETIAN AND ITS ANTECEDENTS IN HUNGARY

Following the preliminary reports, the results of the excavations at Máriaremete have now been published in full (Gábori-Csánk 1983). In practically every respect, this is a site which can be taken as typical of the Jankovichian, or as Vértes would have had it, the Trans-Danubian Szeletian. According to Gábori-Csánk's account, the Upper Remete cave consists of two chambers(front and rear) joined by a narrow corridor. The maximum thickness of the deposits varies between 1.2 and 2.0 metres. The Palaeolithic remains are concentrated in layer 4, described as yellowish loessy at the entrance to the cave but more clayey in the interior and reddish at the base; in the rear chamber the layer could be divided into two distinct units, the sparse Palaeolithic remains being confined to the upper horizon. Mainly on faunal grounds, the layer is regarded as having spanned the epoch from the end of the last interglacial period to the "maximum of Würm I", i.e. the period immediately preceding Hengelo. Twenty four animal species, excluding rodents, were recognised by Kretzoi. There is a little charcoal, identified as belonging exclusively to *Larix/Picea*. The Palaeolithic remains were concentrated in the front chamber, particularly at the boundary of squares 5 and 6. They consist of 13 stone artifacts, plus 1 fossil marine shell which must have been brought in by man. According to Gábori-Csánk, the stone artifacts include 5 bifacially worked tools, 3 side-scrapers, and 5 flakes. The bifacially worked tools are classified by her as "bifaces" or "racloirs-bifaces" rather than leafpoints, and the whole industry therefore (in common with the rest of the Jankovichian) as a "modified form of the Micoquian" (Gábori-Csánk 1983: 284). In my opinion, this assessment is mistaken, because the bifacially worked tools can perfectly well be regarded as leafpoints (Gábori-Csánk 1983: Figure 16: 1, 2, 3, 7) and the strongly marked Levallois characteristics of the industry suggest affinities elsewhere, e.g. in the Levallois-Mousterian of Southeastern Europe. The hominid remains ascribed to Neanderthal man were found in the front chamber in layer 4, at the boundary of squares 5 and 6, and consist of three right mandibular teeth (two incisors and a canine) much worn and attributed to a single individual.

The contrast between the industries from Jankovich and Tata has been re-emphasised by the publication of 431 artifacts from the late István Skoflek's bequest to the Hungarian National Museum (Dobosi 1983). Dobosi re-affirms the validity of Vértes's analysis of the material from Tata, including his categorisation of certain tools as miniature handaxes (Dobosi 1983: Figure 9, where 'szakóca' is incorrectly translated as 'celt'), though it is doubtful whether she would agree with this author that the assemblage could be classified as Micoquian. She also re-affirms the likelihood that the settlement at Tata dates to about the time of the Brørup interstadial (not much earlier, as claimed by Schwarcz) and quotes in that connection the interesting parallel offered by Pécsi's Basaharc Lower fossil soil; I would entirely agree with that assessment.

So far as the Bükk mountains are concerned, further details have been given relating to the excavations carried out at Diósgyőr-Tapolca in 1973, but apart from that no new cave investigations have taken place. The centre of interest has shifted to the open-air sites east of the Bükk, particularly Avas and Sajóbábony, and the creation of an entirely new taxonomic unit (the 'Bábonyian') has been proposed. In addition, further intensive research into the raw materials employed in prehistoric Hungary has thrown new light on the Szeletian and allied industries, and has to some extent corrected or modified descriptions earlier given.

As a result of the excavations carried out in the 1930s, two Pleistocene layers were recognised at Diósgyőr-Tapolca, the lower of which was divided into two cultural levels on archaeological grounds. The archaeological material was attributed both to the Gravettian and to the Szeletian, and Vértes suggested that a Mousterian level might also have been present at the site. The excavations conducted in 1973 are of particular importance for the light which they throw upon the early occupation of the site (Hellebrandt, Kordos and Tóth 1976). No undisturbed material was discovered inside the cave, but the brick wall closing the entrance was removed together with the cement steps in front of it, and two sections were established on either side of the steps, II to the north and I to the south (Hellebrandt *et al.* 1976: Figures 1, 3, 4). Section I had layers numbered 1-4, and section II layers numbered 1-5. Layers 1 and 2 in both cases were the same, a holocene humus horizon and a sterile deposit of clay with stones. Layers I/3 and II/3 and 4 consisted of large limestone blocks in a clay matrix, underlain in the case of layer I/4 by a deposit of clay with fine debris. In section II a stratigraphic break was identified beneath the limestone blocks, and the basal layer II/5 (another clay with fine debris) is held to be quite distinct from all the others on sedimentological grounds. Its characteristics are interpreted as indicating a cold damp climate with some traces of solifluction, whereas the layers above are said to exhibit a drier more continental type of environment. Layer II/5 alone is considered to represent a truly homogeneous deposit (Hellebrandt *et al.* 1976: 14, 32). The artifacts recovered from sections I and II layers 3 and 4, and section II layer 5, respectively, are as shown in Table 7.1 (see Hellebrandt *et al.* 1976: 14 and Tables I-VII).

Section II layer 5 is technologically distinct, in that 52% of the artifacts are said to have been made from porphyritic pebbles, a far higher proportion than in the case of the other layers, and so far as could be ascertained there was an abundant use of the so-called "Clactonian" technique (Hellebrandt *et al.* 1976: 10, 13, Figure 5). Typologically, it is characterised among other things by some rather ill-defined points and a few seemingly rather poor handaxes (Hellebrandt *et al.* 1976: Tables III.2 and VI.3). The industry as a whole is considered to be most comparable to the small assemblage excavated by Vértes in layer V of the Lambrecht Kálmán cave and described by him as 'Prämoustérien'. The tools from the upper layers seem to be broadly similar to those earlier found in the cave with, disappointingly enough, an apparent admixture of Gravettian and, presumably, Szeletian artifacts (Hellebrandt *et al.* 1976: 9-10, 32). The authors draw particular attention to the fact that a backed blade (possibly a fragment of a micro-Gravette) was found at the boundary of section II layers 3 and 4 (Hellebrandt *et al.* 1976: Table II.5) and that Mousterian-

Table 7.1. Artefact totals for Diósgyör-Tapolca cave (1973 excavations)

	Sections and Layers	
	I and II/3 and 4	II/5
Burins	—	7
Awls	3	11
Retouched blades	5	—
Backed blades	1	—
Points	1	13
Sidescrapers	6	14
Choppers	—	2
Handaxes	—	3
Total Tools	16	50
Cores	2	—
Blades and flakes	51	47

type side-scrapers (e.g. Hellebrandt *et al.* 1976: Tables I.3 and II.6) can be found at the same stratigraphic level. There are other pronouncedly Upper Palaeolithic elements such as retouched blades and an advanced type of blade core (Hellebrandt *et al.* 1976: Table 1, 2 and 4, and 5 respectively; the blade core is described as a "core scraper") and the backed blade is of course perfectly congruent with the fine Gravette point earlier found in the cave (Allsworth-Jones 1986: Figure 12.7).

The newly excavated layers have been dated palaeontologically (Hellebrandt *et al.* 1976: 30-1). Apart from fish, amphibians and reptiles, snails, and birds, 15 micromammalian species not previously recorded were recovered, and 22 large mammalian species. The distinctiveness of layer II/5 is again confirmed, and particularly because of the presence in it of *Lagurus lagurus* and *Asinus hydruntinus* it is assigned to the Varbó phase of the Hungarian faunal succession, corresponding approximately to the end of the last interglacial and the beginning of the last glacial period. These two species do not reappear in the upper layers, but nonetheless these layers are also considered to pre-date the Tokod phase or the so-called "cold peak of Würm I", largely because of the presence in them of *Megaloceros giganteus* and a large form of Pleistocene horse. As the authors themselves recognise, however, there is a manifest contradiction between a palaeontological placement of layers I and II/3 and 4 in a period corresponding to the lower layers at Subalyuk, and the archaeological data. There is no way in which the Gravettian or other markedly Upper Palaeolithic elements in the layers could be assigned to a period as early as this. The evidence regarding the dating of these layers therefore remains contradictory and unsatisfactory.

One of the most important open-air sites east of the Bükk is Avas, in the immediate vicinity of Miskolc. Our knowledge of the prehistoric occupation of this area has been transformed, thanks to the work of K. Simán, who has not only provided a thorough reassessment of the old finds but has also carried out new excavations of her own (Simán 1979a, 1979b, 1982, 1983, 1986a, 1986c). As a result, it is clear that there were several different occupations of Avas; these differences were masked by Vértes's attempt to treat the bulk of the material together as one unit, which he labelled "Mesolithic". The heterogeneity of the finds also does not allow us to attribute them as a whole to the Szeletian; while some Szeletian elements were certainly present, their exact role remains to be determined (cf. Allsworth-Jones 1986: 99-100). The catalogue of work done at Avas over the years is given by Simán, and she also provides a map indicating the different localities concerned (Simán 1979a: 12-13 and Figure 1).

The first find from Avas dates back to 1905, when a glassy quartz porphyry leafpoint from the Reformed Church cemetery on the northern slopes of the hill came into the possession of Ottó Herman. In the same year another leafpoint was found at No. 12 Petőfi Street within Miskolc itself, and it was as a direct result of these finds that Ottó Herman successfully urged that excavations be commenced within the caves of the Bükk. On his visit to Hungary in 1923, Breuil agreed that Ottó Herman's find from the cemetery was indeed comparable to the leafpoints at Szeleta. In 1909 and 1913 excavations were carried out in the cemetery, the latter being conducted by I. Gálffy, the Director of the Museum in Miskolc. Vértes himself agreed that it was reasonable to call these finds at least in part Szeletian, and more recently this view has been confirmed by Ringer (1983: 127).

The next excavations at Avas were of a very different character. They were carried out by Hillebrand in 1928-1935 on the summit of the hill on the northern side; he discovered traces of extensive hydroquartzite mines, which he attributed to the "Proto-Campignian". Vértes regarded them as probably Neolithic or later, and this view has been confirmed by Simán (1979b, 1986a) following her exhaustive examination of the old records. She also put down two test pits (Simán 1979b: Figures 3 and 4) which convinced her that Hillebrand's "ideal cross-section" was no more than an unrealistic abstraction. Instead, she has worked from the drawn sketches of individual sections, usually prepared by Hillebrand's subordinates, in order to reconstruct a convincing sequence of events at the site. Prehistoric man was interested in this area because, under a variable thickness of surface deposits, there were three layers of andesite or rhyolite tuff, separated from each other by beds of marly clay. The lower two tuff layers were penetrated by siliceous solution, forming the hydroquartzite which was used for making stone tools. Simán has ascertained that there were at least two stages of use of the mines. In the earlier stage, relatively simple pits were made in order to break through the upper tuff layer (Simán 1979b: Figure 6: 'A' and 'H'); in the later stage, shafts were sunk (Simán 1979b: Figure 7/2: 'E') and in one case at least an underground gallery, passing beneath the upper tuff layer, joined two such shafts ('G' and 'E';

Simán 1979b: Figure 1; 1986a: Figure 1). Summing up all the evidence, Simán concludes that mining probably commenced at some point during the Neolithic (certainly not as early as the Szeletian, *pace* J.K. Kozlowski 1972-3: 13) and continued intermittently until the 16th century A.D., when it is known that flintlocks were being exported from Miskolc.

In reaching this conclusion, Simán has quite correctly drawn a distinction between the artifacts which were indubitably found at the level of extraction, and those which were discovered in the deposits above the upper tuff layer and which cannot be proved to have had anything to do with the mining operations. The former include hammerstones of various kinds (Simán 1979b: Figures 9/1 and 9/2) and retouched flakes interpreted as mining tools (Simán 1979b: Figures 10/3 and 11/1). There are parallels for these categories of tools at Neolithic or Chalcolithic mining sites in Hungary such as Sümeg-Mogyorósdomb and Tata-Kálvári-adomb, which have ^{14}C dates ranging from 5960 ± 95 to 3810 ± 65 BP (cf. Bácskay 1986). Also found at the level of extraction were fragments of pottery, including, in the 'Z' pit, two pieces belonging to the base of a large vessel (Simán 1979b: Figure 12/2) and the remains of a human skeleton of anatomically modern type (Simán 1979b: 99). In contrast to this, a glassy quartz porphyry leafpoint which, as Simán says, is quite comparable to others found on the surface at Avas (Simán 1979b: Figure 11/2) was discovered in the 'B' pit in a brown fossil soil above the upper tuff layer; it cannot be connected with the mining operations because the tuff layer was not broken through and there was no mining in this area (Simán 1979b: Figure 6: 'B'). In the 'A' pit, in the same fossil soil, a polished stone axe was discovered (Simán 1979b: Figures 12/1 and 6: 'A'). Evidently these finds cannot originally have belonged together; hence, Simán argues, both are in a secondarily disturbed position within a soil which was itself displaced. While therefore it is reasonable to connect the leafpoint with the Szeletian, the mining operations as such belong to a much later period.

No further excavations were conducted on Avas until 1961, when Vértes and Korek put down soundings at various different places in the localities referred to as Avastető and Alsószentgyörgy. According to Vértes's account, the excavators found traces of an undisturbed culture layer, at an average depth of 40-70 cm, according to Ringer (1983: 128-9). Vértes's published inventory for excavated and surface material taken together amounted to 80 retouched tools, including 11 leafpoints and 16 bifaces, as well as several unretouched blades and a few cores. Then in 1975 Korek and Hellebrandt carried out rescue excavations at other places on Avastetö, which were continued by Simán in 1976-79 at Alsószentgyörgy, Felsőszentgyörgy, and Tűzköves, as well as Avastetö. Not much more can now be done at the sites, because Avas is being transformed increasingly into a built-up area. Of these excavations, the most important seems to have been that of 1979 at Alsószentgyörgy (Simán 1982, 1983: 44-6, Tables IV and V.1, 1986c). According to Simán, the area excavated amounted to only 7 x 2 metres, but she believes that the original extent of the site may have been about 40 x 20 metres. Pleistocene and Holocene deposits together were about 1-1.5 metres thick at this point, the archaeological finds being situated in the upper part of the Pleistocene layer. They occurred in a concentration some 10-20 cm thick. The finds were originally characterized by Simán as "Levalloisian", Middle Palaeolithic but with a pronounced lamellar technique (Simán 1983: Table IV: 1, 3, 4, 5). Subsequently, she seems to have played down the Levallois element, but she still agrees that the lower industry from Subalyuk provides the best parallel (Simán 1986c: 55). 2-300 tools were found, as well as cores, flakes, and manufacturing debris, indicating that the site served as a workshop. 95% of the raw material consisted of local hydroquartzite. Side-scrapers made up 35% of the tools, but the industry contained no traces of bifacial technique. Whatever it is called therefore it is clearly very different from the finds made by Vértes and Korek in the same area.

Apart from these excavations, surface collections have been made on Avas by various persons over the years ever since the first discoveries were made, in more recent times by A. Saád and L. Tóth (Simán 1979a). According to Ringer (1983: 127-8) some of these surface finds can be regarded as belonging to the 'Bábonyian', especially those from Tűzköves and the Tréki-Török vineyard at Felsöszentgyörgy. But it would surely be an error to attribute all bifacially worked artifacts found on Avas to the 'Bábonyian'. The glassy quartz porphyry leafpoint discovered by Hillebrand at the mining site can reasonably be regarded as Szeletian and, as already noted, it is perfectly comparable to others found on the surface at Avas. Tóth's collection contains a number of such pieces, which he was kind enough to show me prior to the handing over of his entire collection to the Miskolc Museum in 1975 (Simán 1979a). It is known also that much use was made of hydroquartzite from Avas at Szeletian cave sites in the Bükk, including Szeleta itself (e.g. Allsworth-Jones 1986: 89-90: Figure 16.6 and Plate 5.9). This material was probably not mined, but was collected in sufficient quantities on the surface (Simán 1979b: 100). There were evidently close links between Szeleta and Avas, and this fact should not be lost sight of in any reassessment of the relevant finds.

However that may be, Hungarian authors who have recently studied the evidence are unanimous in their opinion that Vértes's so-called "macrolithic Mesolithic", otherwise known as the "Eger culture", must be abandoned (Simán 1982; Ringer 1983: 132-4, 147-8; Biró 1986b). It is considered to be a "pseudo-unit", a mixture of different industries, which has led to many misunderstandings. This conclusion applies not only to Avas, but also to Eger-Köporostetö, the eponymous site, and to Korlát-Ravaszlyuktetö, in northeastern Hungary, some 50 kilometres from Miskolc (Simán 1983: Figure 1). For reasons which are not all together clear, Ringer regards the bifacial element at Eger-Köporostető as representing, not the 'Bábonyian', but a Mousterian of Acheulian Tradition (cf. Biró 1986b). The situation at Korlát, which has featured in the study of the Hungarian Palaeolithic ever since Roska's time, has been much clarified by Simán's recent surveys and excavations (Simán 1983, 1985,

1986a). It turns out that there was not one but several sites, mainly concerned with the exploitation of the local hydroquartzite (Simán 1986a: Figure 2). Apart from a number of Neolithic sites, there was one Middle Palaeolithic and one Upper Palaeolithic surface occurrence; between them, in 1983 and 1984, Simán excavated an area of 43 square metres which is described as an extraction site (cf. Biró 1986b). Two phases of use are indicated, the first major one Middle Palaeolithic, the second minor one Upper Palaeolithic. The Middle Palaeolithic finds included Micoquian-type handaxes and leaf-scrapers, as well as flakes, blades and cores. In view of the nature of these tools, as well as others reported earlier by Vértes, Ringer (1983: 129-130) believes that the industry from Korlat is related to the 'Bábonyian', more particularly since hydroquartzite from this area is found in the principal 'Bábonyian' sites on the eastern margins of the Bükk.

This brings us to the 'Bábonyian' itself, defined by Ringer (1983) as a Middle Palaeolithic "*Blattwerkzeugindustrie*". According to him, there are at least 20 sites which can be said to belong to this new taxonomic unit, but his published work relates essentially to the five most important of them only. Geographically, they fall into three groups. Sajóbábony-Méhésztetö and Sajóbábony-Kövesoldal are adjacent sites about 8 kilometres north of Miskolc, at the junction of the Sajó river and the Bábony stream, the name of which has been used to designate the new unit as a whole. Then there are two sites in the immediate vicinity of Miskolc, Kánástetö and Szabadkatetö. All these are exclusively surface sites, the largest being Miskolc-Kánástető which occupies an area of about 70 x 30/40 metres. The last site, Mályi-Öreghegy, about 10 kilometres southeast of Miskolc, is the only one to provide some indications of stratigraphy. According to Ringer's account, there are actually two different localities here. The first is again an exclusively surface occurrence, but at 100 metres to the north there is a second for which we have at least some kind of stratigraphic profile (Ringer 1983: 137). The profile, in the side of a brickyard, reveals two cultural layers (A and B) superimposed in loessy clay above a pebble layer at the base. The finds from layer A are said to be Upper Palaeolithic in character, whereas those from layer B are assigned to the 'Bábonyian'. Only one such find is actually illustrated by Ringer (1983: 38, Figure 3), a glassy quartz porphyry biface which is said not to be broken, although the illustration seems to belie that description. Unfortunately, it is not at all clear what relationship this profile bears to the one earlier described by Vértes, in the vicinity of which he discovered a leafpoint which he regarded as Szeletian (cf. Allsworth-Jones 1986: 99). Apart from this rather meagre stratigraphic information, Ringer provides an interesting re-evaluation of the geomorphological history of the Szinva valley. Ever since the investigations of K. Papp at the beginning of this century, the deposits in the valley have been regarded as Holocene in age, but Ringer provides good grounds for thinking that they may in fact have been Pleistocene in origin. In that case, the three bifacial artifacts found in 1891 at Bársony's house in Miskolc could have been genuinely *in situ*,

Table 7.2. Tool totals for the "Bábonyian" (5 major sites combined).

Endscrapers	16
Burins	8
Awls	10
Truncated flakes	3
Leafpoints/leafscrapers	45
Limaces	5
Sidescrapers	81
Denticulates	3
Miscellaneous	56
Biface knives	40
Handaxes	37
Total tools	304

and Ringer clearly regards them as being related to the 'Bábonyian'. In addition to the sites mentioned by Ringer, there is another at Sajószentpéter-Nagykorcsolás, north of Miskolc, where excavations were carried out by Simán in 1982 (Simán 1983: 46-8 and Table V, 2-11; cf. Biró 1986b). Simán reports that no cultural layer nor organic material was recovered, and that the finds themselves are not homogeneous. On typological and raw material grounds they are divided into two groups, one attributed to the 'Bábonyian' and one to the Gravettian. The 'Bábonyian' collection itself is quite impressive, consisting according to Simán of over 100 tools as well as cores, flakes, and manufacturing debris, but obviously we are none the wiser concerning its stratigraphic provenance. Considering the 'Bábonyian' as a whole, indeed, Ringer is obliged to agree (1983: 13-4) that for the moment it lacks an adequate stratigraphic basis and that its analysis and evaluation has proceeded essentially on the basis of technology and typology alone. This does suggest the need for caution in assessing his results.

Ringer illustrates 70 tools from his five major sites, out of a total which he calculates as 300 (Ringer 1983: 76-80). It seems from Ringer's own figures that the total should actually be 304, which may be classified as shown in Table 7.2. These figures are based on those given by Ringer, amalgamating many of his classes but otherwise with very few amendments. The rather large miscellaneous class includes 29 pieces in Bordes' categories 45-50, mainly *retouches abruptes épaisses* and *retouches bifaces*. Obviously, for the classification of the industry as Micoquian-related, the leafpoints/leaf-scrapers, biface knives, and handaxes are of particular importance. Levallois technology and typology is virtually or wholly absent. Ringer has coined the term 'Bábonyian' for this industry because he believes that it does not exactly correspond to any of the existing Micoquian variants, although he would agree that broadly speaking it does belong with them (Ringer 1983: 50-3, 120). Its most characteristic element is its high percentage of bifacial working (57% of all tools according to Ringer 1983: 87-8). Obviously there is a weakness in Ringer's diagnosis in that one is dealing with a combined list, which, unlike the illustrated figures, is not separated out according to provenance; a single stratified occurrence would have provided a far surer basis on which to proceed. Indeed, Ringer himself recognises (1983: 112-4) that the 'Bábonyian' sites may have occupied a long span of time, with significant differences between them, which he interprets developmentally. Significantly enough for a bifacial industry, 81% of the tools in the overall total are made of glassy quartz porphyry, whereas 15% are made of hydroquartzite from Avas and Korlát, and the remainder of other materials (Ringer 1983: 58-62).

Hungarian authors have greeted Ringer's finds with considerable interest and enthusiasm (Dobosi 1983: 13 and Note 27; Gábori-Csánk 1983: 285 and Note 67). Gábori-Csánk indeed remarks that in view of these finds it can be concluded that the Szeletian of the Bükk did not develop from the local Mousterian but from industries of the type found at Sajóbábony; further, in a remarkable *volte-face,* she goes so far as to say that this conclusion applies also to Subalyuk ("Le rapport génétique entre le Moustérien du Bükk (grotte Subalyuk) et le Szélétien que nos chercheurs se sont efforcés tant de fois de démontrer, ne nous paraît plus guère plausible aujourd'hui"). This newly sceptical appraisal of Subalyuk agrees largely with the views of this author (Allsworth-Jones 1986: 101-4), but it seems to me that considering the limited nature of the evidence which we have so far for the 'Bábonyian' it would be wise to exercise restraint in this respect also. The finds, while interesting in themselves, are essentially from the surface only and their appearance, with many broken pieces and some probably unfinished ones, has something in common with those from Kösten and Jezeřany, both workshop sites which have proved particularly difficult to classify. It would, in my opinion, be a pity to replace a tendency to call any bifacially worked artifact from the Bükk area 'Szeletian' with a tendency to call it 'Bábonyian' instead.

With regard to the raw material aspect, much progress has been and continues to be made. The Carpathian Basin and Hungary in particular are rich in raw materials of various kinds, the different properties and patterns of exploitation of which exercised a profound influence on the course of prehistory in the area. J.K. Kozlowski in a pioneering survey (1972-73) drew attention to some of the region's salient features, emphasising the changes through time, particularly the emergence of more or less specialised workshops in the Upper Palaeolithic, within a context of pronounced regional specialisation. In 1978 Dobosi published a further general review of raw materials employed in the Hungarian Palaeolithic, but the study of this aspect has been much improved and extended since 1980 thanks to a systematic survey carried out under the auspices of the Hungarian State Geological Institute, directed by J. Fülöp (Biró and Pálosi 1983; Biró 1986b). Partly as a result of this work, an international conference on prehistoric flint mining and lithic raw material identification in the Carpathian basin took place in Budapest and Sümeg in 1986, some of the results of which are of direct relevance to our theme (Biró 1986a; Dobosi 1986a; Simán 1986a, 1986b).

So far as the Szeletian in the Bükk is concerned, Vértes and Tóth performed an invaluable service when they demonstrated that the predominant raw material employed (particularly for leafpoint manufacture) was not "ash-grey chalcedony" (as assumed by Kadić and his successors) but what they termed glassy or vitreous quartz porphyry ("glasiger Quarzporphyr" (cf. Allsworth-Jones 1986: 88-9, 107). The properties of this material have now been further examined, and it seems that even this expression is not quite correct. According to Biró and Pálosi (1983: 411, 415) modern classificatory systems tend not to make use of the term 'quartz porphyry' at all; if it is still employed for the archaeologically significant material found in the Bükk, then the suggestion is that this should be referred to as 'felsitic quartz porphyry' (Simán 1986b) or, if we wish to emphasise the historical aspect, even 'Szeletian felsitic porphyry' (Biró and Pálosi 1983; Dobosi 1986a). Strictly speaking, however, in terms of modern usage this material should be referred to as rhyolite; and I think that this would be the better course in future. These developments do not mean, of course, that the work of Vértes and Tóth has been invalidated in any way; on the contrary, as Biró and Pálosi emphasise (1983: Figure 2, Table 3), their own analyses essentially confirm the earlier results; what we are dealing with is a question of nomenclature. The rhyolite used at Szeleta was derived from the area of the Kaán-Károly spring and was the product of volcanic activity occurring during the Ladinian stage of the Upper Middle Triassic. It reveals traces of later silicification and metamorphosis, as shown for example by the considerable amount of calcite which it contains. A number of thin sections were analysed on Dobosi's behalf by L. Ravasz-Baranyai at the Hungarian State Geological Institute, among them pieces which had been macroscopically identified as glassy quartz porphyry from Avas, Sajóbábony, Eger-Köporostetö, and Mályi-Öreghegy; they turned out to be in fact felsitic – or hyalitic – striped rhyolite (Dobosi 1978). Thus it may be appropriate to extend this more precise description to all the rhyolite emanating from the Kaán-Károly spring area (Biró and Pálosi 1983; Dobosi 1986a; Simán 1986b).

Another set of raw materials of particular interest to us is what Biró refers to as the 'hydroquartzite-limnoquartzite group', a heterogeneous but genetically-related series of rocks arising as a result of post-volcanic activity, which is especially important in the eastern part of Hungary (Biró and Pálosi 1983; Biró 1986a; cf. Dobosi 1978). Both hydroquartzite and limnoquartzite originated as hot solutions rich in silicon acid, but whereas hydroquartzite crystallised in a non-biogenic environment (either by impregnating the country rock or by filling clefts in the rock) limnoquartzite formed in a lacustrine environment and thus incorporated silicified microfossils and plant remains. Both terms have been used very loosely in the past, and as we have seen, the deposits on Avas and the artifacts derived from them have commonly been referred to as 'hydroquartzite', a term which Dobosi (1978) would like to retain. Strictly speaking, again, it seems that 'limnoquartzite' is the more exact appellation in this case (Simán 1986a; Biró 1986b). With regard to radiolarite, or radiolarian flint, which is so important at Jankovich and other Trans-Danubian sites, some attempts have been made to distinguish between the different types (e.g. Biró and Pálosi 1983: Plate III, 3 and 4, Figure 5) but the results so far are still inconclusive. Traditionally, the radiolarite used at Jankovich is said to have come from Dorog, but Biró (1986b) lists a number of other sites where extraction using distinctive mining tools may have been carried out for a very long time. Her model of a 'workshop complex' (Biró 1986b: Figure 10) applies, as she says, essentially to post-Palaeolithic times, but one should not perhaps underestimate the complexity of what went before.

Finally, in this section, mention should be made of new finds at Korolevo I. This site is not in Hungary, but it is not far from the Hungarian border, on the left bank of the Tisza in Transcarpathian Ukraine. It seems convenient to treat it at this point, and it is highly likely that the finds (when fully published) will be seen to have much in common with their Hungarian counterparts. The site was excavated in 1974-80, and a sequence of deposits 5-6 metres thick on the river's 120 metre terrace has been described by Adamenko and his colleagues; further information concerning the archaeological material is provided by Gladilin and Soldatenko (Adamenko *et al.* 1981; Gladilin 1983; Soldatenko 1983). As Adamenko says, the profile is remarkably complete, reflecting climatic oscillations over the greater part of the Quaternary, at least to the Brunhes/Matuyama boundary. He records a series of six fossil soils, some of them doubled, above weathered bedrock (Adamenko *et al.* 1981: Figure 1). Fossil soil II is held to correspond to the 'Riss-Würm' interglacial and fossil soil V to the 'Mindel-Riss' interglacial period. Cultural layers have been found throughout this sequence, numbering 14 in all. The lowest seven (V, Va-b-c, VI, VII, and VIII) have been attributed to the Acheulian, extending from fossil soil III to the base. The material includes Levallois cores and flakes, and among the retouched tools there is mention of choppers, cleaver-like implements, rare (usually atypical) handaxes, side-scrapers, denticulates and notches. Then there are six layers (I, II, IIa, III, IV, and IVa) attributed to the Mousterian. Layer I, above fossil soil I, is classified as a Denticulate Mousterian, made (as elsewhere in the sequence) almost exclusively of local andesite and dacite. Denticulates and notches account for over 60% of the tools, and for this reason Adamenko *et al.* compare the assemblage with those from Šipka and Čertova díra. Layer II, below fossil soil I, is described as a "bifacial Mousterian variant". Rather confusingly, it is compared at one and the same time with the lower layer at Subalyuk, Kůlna layer 7a, and Švedův Stůl. This suggests (although only the full publication of the material could confirm it) that it may be related to the Micoquian. Technologically, it is said to be a non-Levallois industry. Layers IIa-IVa are said to be "related to the Levalloisian-Mousterian industries and are genetically connected with the Acheulian complexes occurring below" (Gladilin 1983). Layer IIa is situated above fossil soil II, whereas the remaining layers III-IVa are within it. Chronologically therefore, these assemblages should belong to the very beginning of the last glacial and to the last interglacial period. This agrees with evidence from elsewhere, as does the possible superposition of a Micoquian-related industry above a Levallois-dominated one. The latter (excluding layer IIa) is referred to by Soldatenko (1983) as the "Korolevo culture". Fossil soil I is regarded as the equivalent of Brørup (Adamenko *et al.* 1981: 89) hence the Middle Palaeolithic in layer I should extend into post-Brørup times, a situation which again has many parallels elsewhere. The only layer which appears to be quite exceptional is Ia; this is said to occur (like layers II and IIa) beneath fossil soil I, hence it should be pre-Brørup, yet it is said to be early Upper Palaeolithic in character, employing a 'prismatic' technique and containing a large number of end-scrapers (Gladilin 1983). This conjunction appears to be well-nigh impossible; but even if the presumed age of fossil soil I were shifted to a more recent interstadial, the fact would still remain that an early Upper Palaeolithic horizon is sandwiched between two Middle Palaeolithic ones. That in itself is a highly unusual and significant circumstance which deserves to be further investigated. (Valoch 1986a, states that layerIa is "Aurignacoid" in character and that it has a ^{14}C date of >38 000 BP, but this information is not contained in the reference cited). Clearly, full publication of this site is essential so that it can be compared in detail with Hungarian and other Central European evidence.

THE SZELETIAN AND ITS CONTEMPORARIES IN MORAVIA

New excavations have been conducted by Valoch at Bohunice and Vedrovice V in 1981 and 1982 respectively (Valoch 1982, 1984), and two important collections broadly comparable to Bohunice have been published in full: the quartzite industry from Ondratice I and the surface collection of R. Ondráček from Podolí (Svoboda 1980; Oliva 1981; according to Svoboda however the last-named site should properly be referred to as 'Líšeň-Lepiny': Svoboda and Svoboda 1985: 506). In addition, significant new investigations have been carried out at Stránská Skála , long known for its Aurignacian surface collections and for its early Palaeolithic site with Upper Biharian fauna (Svoboda 1984b), and at the main site at Líšeň, scarcely 2 kilometres distant from it. Excavations were conducted by Valoch at Stránská Skála on behalf of the Moravian Museum in 1982, and by Svoboda on behalf of the Czechoslovak Academy of Sciences from 1982 onwards. The Moravian Museum finds have not yet been published, but preliminary accounts of Svoboda's work here and at Líšeň (an exclusively surface site) have appeared (Svoboda 1985, 1986b, 1986c; Svoboda and Svoboda 1985). Summary details of the tools and cores from these seven sites are presented in Table 7.3; those for Stránská Skála and Líšeň are incomplete, since on the basis of the published information it is not possible to estimate the true numerical

Table 7.3 Tool and core totals and calculated indices for 7 Moravian open-air sites ("Bohunician" and Szeletian)

	Vedrovice V 1982	Bohunice 1981	Ondratice I quartzite	Podolí	Stránská Skála III Layer 5	Stránská Skála IIIa Layer 4	Líšeň
Endscrapers	5	8	251	60	16	19	592
(including carinate/nosed	—	2	26	2	1	5	63
Endscraper – burins	—	—	2	1	—	—	—
Burins	3	3	77	10	—	3	77
(including carinate)	—	—	7	—	—	—	—
Awls	1	1	15	2	—	—	—
Truncated blades	—	—	5	3	—	—	—
Retouched blades	1	3	25	11	2	—	42
Leafpoints: bifacial	4	1	5	3	—	—	59
unifacial	—	—	—	3	1	1	48
Levallois: flakes	1	}42	146	}189			
blades	—		105				
points	—		34	21			
retouched points	—	—	10	—			
Mousterian points	1	—	70	1	2	—	14
Sidescrapers	10	7	341	36	7	10	219
Notched pieces	31	}15	63	22	}3	16	153
Denticulates	7		12	6			
Splintered pieces	—	—	10	4	—	—	—
Miscellaneous	1	—	58	46	9	9	187
Biface Knives	—	—	2	1	—	—	—
Flat handaxes	—	—	—	1	—	—	—
Total Tools	65	80	1231	420	40	58	1391
Initial cores	—	—	193	—			
Precores	—	—	385	—			
Cores	96	55	494	265	40	37	1291
(including Levallois)	—	—	229	56			
Cores as % of tools and cores	59.6	40.7	46.5	38.7			
Unretouched Levallois as % of tool total	1.5	52.5	23.2	50.0			
Major tool groups %s (totals minus Levallois):							
endscrapers	7.81	21.05	26.53	28.57	40.00	32.76	42.56
burins	4.69	7.89	8.14	4.76	—	5.17	5.54
leafpoints	6.25	2.63	0.53	2.86	2.50	1.72	7.69
sidescrapers	15.62	18.42	36.05	17.14	17.50	17.24	15.74

Table 7.4. Tool and core totals for Hostějov and Troubky

	Hostějov	Troubky
Endscrapers	5	10
(including carinate/nosed)	—	2
Endscraper – burins	—	1
Burins	5	7
(including carinate)	—	3
Awls	1	—
Truncated flake/blades	1	—
Retouched blades	2	4
Retouched pointed blades	1	—
Leafpoints: bifacial	2	—
unifacial	1	—
Sidescrapers	5	3
Notches and denticulates	2	1
Splintered pieces	1	—
Total Tools	26	26
Cores: complete	5	10
fragments	1	38

strength of the Levallois component (Svoboda and Svoboda 1985: 513, Table V). The raw figures correspond to those of the authors concerned, with very slight modifications, but the indices have been recalculated in order to ensure a uniform presentation here. Apart from this, Valoch has published details of two smaller surface collections from Central Moravia, Hostějov and Troubky, and has compared them to two other similar ones in the area, Stříbrnice and Zdislavice, as shown in Table 7.4 (Valoch 1985, 1986b); and, last but not least, he and his colleagues have now made available full information concerning Vedrovice II and Kupařovice 1, which are claimed to represent the early Aurignacian in southern Moravia (Valoch, Oliva, Havlíček, Karásek, Pelíšek and Smolíková 1985). Summary details of the collections from these two sites, based upon indications given by Valoch and his colleagues, are given in Table 7.5. Altogether, therefore, there is a lot of new information available from Moravia, and in particular stratified information. That is especially welcome since – at least so far as open-air sites are concerned – there has long been a startling contrast between the earlier and the later stages of the Upper Palaeolithic in this area: the latter represented by rich, even dramatic, sites like Předmostí, Dolní Věstonice, and Pavlov, the former (at least until the discovery of Bohunice) seemingly condemned to a monotone existence practically bereft of reliable context or of any other feature of interest beyond the stones themselves.

Valoch's excavations at Bohunice in 1981 embraced an area of only two square metres, but, as he says, they confirmed the previous stratigraphic observations, and the nature of the industry recovered is the same as that described earlier. The overwhelming majority of the 1625 artifacts are of Stránská Skála chert; only 19 (or 1.17%) are of other materials, including chert from the region of the Krumlovský les. As Valoch emphasises, this is an industry characterised by an advanced Levallois technique, the low Levallois technical index of 12.44 no doubt being accounted for by the fact that Levallois blades can only with difficulty be distinguished from non-Levallois ones (Valoch 1982: 40). Apart from 42 Levallois flakes, blades, and points, the stratified collection includes one bifacial leafpoint and two carinate or nosed end-scrapers (Valoch 1982: Figures 6:1, 7:5, 8:1). The excavations at Vedrovice V in 1982 were far more extensive, covering an area of 120 square metres, although the artifacts were concentrated only in parts of this area. They have revealed an industry classified as "unequivocally Szeletian" (Valoch 1984: 8) in the lower part of a fossil soil presumed by Valoch to be the same as that at Bohunice and to date to the same period, i.e. Hengelo. The accuracy of this supposition has been borne out by the two ^{14}C dates now available for the site: 37 650 ± 550 (GrN-12374) and 39 500 ± 1100 BP (GrN-12375) (Valoch 1986a). For the first time therefore we have really reliable stratified data from the region of the Krumlovský les, and it is to be hoped that further excavations will add to the available material. As it is, to my mind, a clear parallel to Bohunice already exists not only stratigraphically but also in terms of the artifacts represented. Of the 4946 pieces so far recovered, the great majority (3772) are tiny chips, and refitting of parts demonstrates the excellence of the stratigraphic context. The raw materials consist mainly of local chert and quartzite, but there is also some radiolarite and Nordic and/or south Polish flint. The tools recovered up to now include one Levallois flake and four bifacial leafpoints (Valoch 1984: Figures 3:1, 5: 1-3).

The finds from Ondratice I and Podolí are also compared by Svoboda and Oliva with those from Bohunice, although the two authors differ radically in their understanding of what the 'Bohunician' is. Ondratice of course forms only one part, albeit a very important one, of a wider pattern of Palaeolithic settlement on the Drahany Plateau (Allsworth-Jones 1986: 144-51), an area which is characterised among other things by the widespread occurrence of quartzite blocks and boulders (Štělcl 1986: Figure 1). The petrological characteristics of these materials have been re-examined by Štělcl, in the light of a new occurrence found in the vicinity of Ruprechtov (Štělcl 1986: Figure 2). Svoboda excavated nine trenches in 1977 in an endeavour to clarify the stratigraphy at Ondratice; he did not discover any artifacts *in situ*, but his observations confirm those of earlier workers, particularly Pelíšek. It appears that loess up to 3 metres thick is present above Miocene sands, but it thins out and disappears near to the summit of the locality, and it is surmounted by a Holocene para-brownearth up to 80 cm thick (Svoboda 1980: Figure 2:3). Since there are

Table 7.5. Artefact totals and calculated indices for Vedrovice II and Kupařovice I.

	Vedrovice II	Kupařovice I
Endscrapers	28	69
(including carinate/nosed)	9	11
(including core tools)	3	38
Endscraper – burins	—	2
Burins	66	102
(including carinate)	14	6
(including core tools)	4	10
Awls	3	—
Truncated blades	9	8
Retouched blades	8	3
Retouched pointed blades	7	1
Backed blades	1	1
Leafpoints: bifacial	—	4
Tayacian points	2	—
Sidescrapers	66	38
(including core tools)	8	4
Notched pieces	32	6
Denticulates	7	6
Splintered pieces	4	2
Choppers/chopping tools	25	12
Miscellaneous	2	12
Bifaces	2	4
Total Tools	262	270
Total minus core tools	247	218

no exact details concerning the provenance of the artifacts from the site, there is no objective way to determine whether the quartzite and non-quartzite components belonged together or not. Hypothetically, Svoboda suggests, one can suppose that there were older settlement layers mainly or wholly dependent on local quartzite, and younger layers where imported chert was preferred (Svoboda 1980: 11). This suggestion, of a progressive shift in raw material emphasis, is the same as that put forward by this author (Allsworth-Jones 1986: 148-9). Svoboda's work is based on the quartzite artifacts kept in the Moravian Museum Brno and in the Prostějov Museum, consisting of 10 784 and 427 pieces respectively. Svoboda's tool total of 980 pieces does not include Levallois flakes and blades, but it can be deduced from his account that these amounted to approximately 251 items, giving an overall total of 1231. That compares quite well with my count of 1273 tools, which was however based on two or three other collections in addition to those of Brno and Prostějov. Allowing for some individual differences, there is overall agreement on the typological balance of the industry as such, and Svoboda documents very well certain features of it. According to him, there were 26 carinate end-

Major tool groups % (totals minus core tools bracketed):	Vedrovice II	Kupařovic I
Endscrapers	10.69 (10.12)	25.56 (14.22)
Burins	25.19 (25.10)	37.78 (42.20)
Leafpoints	—	1.48 (1.83)
Sidescrapers	25.19 (23.48)	14.07 (15.60)
Cores:		
Precores	3	—
1 – platform flake/blade	162	202
2 – platform flake/blade	12	42
others	116	74
Total cores	293	318
Struck pieces and fragments	347	446
Total including the above	640	764
Cores as % of tools and cores	52.80	54.08
Ditto including struck pieces and fragments	70.95	73.89
Hammerstones	9	2
Lames à crête	57	120
Partially retouched or utilised flakes and blades and fragments of tools or cores	142	48
Flakes	923	4016
Blades	97	446

scrapers and 7 carinate burins, as well as 7 bifacial leafpoints (but two of the latter are in fact biface knives: Svoboda 1980: Figure 40:1 and 3). Svoboda was able to devote much more attention than myself to the technological aspect of the industry, and there is no doubt that his detailed account now constitutes the authoritative reference on that subject. He identified 385 pre-cores and 494 cores (compared with my totals of 65 and 249 respectively) as well as many core trimming flakes including 314 one- and two-sided *lames à crête*. Svoboda agrees that 229 (or 46.36%) of the cores could be classified as Levallois in the Bordean sense, and that "the Levallois character of the Ondratice quartzite industry is obvious at first glance" (Svoboda 1980: 55-6); hence (despite the different classification procedures adopted) his overall conclusion with regard to the technological characteristics of the industry is also the same as mine. What is interesting and new in Svoboda's analysis is that he has been able to demonstrate the dynamic of the process whereby an evolved Levallois technique could have led on to one characteristic of the early Upper Palaeolithic. In his view, not only is there a connection between certain types of pre-cores and their corresponding cores; there is also an "evolutionary progression" from flat Levallois cores to Levallois cores with 'upright' preparation, which by a process of narrowing and lengthening led on to cores with frontal crests; the latter were eminently suitable for blade production and formed the prototype of the advanced cores known from the later stages of the Upper Palaeolithic (Svoboda 1980: 58, 98; Figures 5.3 and 4, 16.1-4, 20.1-4; by cores with 'upright' preparation is meant cores with preparation of the back at right angles to the striking surface: Svoboda 1980: 37-9). Svoboda calculates that 117 (or 23.68%) of the cores could have produced blades in this way (Svoboda 1980: 56).

The industry from Podolí is completely unstratified, but, as Oliva says, it is very similar to Bohunice in its technological and typological characteristics. The principal raw material used was chert, particularly that from Stránská Skála, which is only 3 kilometres away; there was also a little quartzite, radiolarite, and quartz. There are 141 complete and 124 fragmentary cores. Fifty-six of the complete cores are classified as Levallois, 34 are described as "Upper Palaeolithic prismatic", and there are 51 others; but the boundary between Levallois and non-Levallois cores is often very unclear (Oliva 1981: 8). There are 210 Levallois flakes, blades, and points, making up half the tool total. Mention is made of 2 carinate or nosed end-scrapers (Oliva 1981: Figures 2:8, 9:1), and there are 6 leafpoints, 3 bifacial and 3 unifacial (Oliva 1981: Figures 6:3, 5.2); Oliva also mentions one flat handaxe and one biface knife, although on the basis of the illustrations I do not find these descriptions convincing (Oliva 1981: Figures 6:1 and 4).

The attention of Valoch and Svoboda was drawn to Stránská Skála when excavations on the summit of the limestone outcrop revealed traces of Palaeolithic occupation beneath an extensive workshop belonging to the Eneolithic Funnel Beaker culture (Svoboda 1985: Figures 1 and 2; cf. Svoboda 1986a). This locality has been named Stránská Skála III. Excellent stratigraphic profiles have been obtained here and at other localities named IIIa, IIa, and IV (Svoboda 1985: Figure 1; Svoboda and Svoboda 1985: Figures 3, 4). Archaeological material belonging to the 'Bohunician' has been found at Stránská Skála III layer 5, IIIa layer 4, and IIa layer 5, whereas an Aurignacian horizon has been discovered higher up the sequence at Stránská Skála IIIa layer 3. Svoboda (1986b: 12-13; 1986d: 240) makes a convincing case, backed by ^{14}C dates, for a three-fold early Upper Palaeolithic occupation: (1) before 40 000 BP, at the end of the 'First Würmian Pleniglacial'; (2) at around 38 000 BP during the 'First Interpleniglacial amelioration' (corresponding to Hengelo); and (3) at around 31 000 BP during the 'Second Interpleniglacial amelioration' (corresponding to Denekamp). The 'Bohunician' is associated with the first two of these phases and the Aurignacian with the third. The relevant ^{14}C dates are as follows (Svoboda 1985): 'Bohunician': 41 300$^{+3100}_{-2200}$ BP (IIIa layer 4, charcoal displaced by cryoturbation to the level of layer 3: GrN-12606), 38 500$^{+1400}_{-1200}$ (GrN-12298) and 38 200 ± 1100 BP (both III layer 5: GrN-12297); Aurignacian: 30 980 ± 360 BP (IIIa layer 3, material from an intact hearth: GrN-12605). As Svoboda comments, the earliest of the 'Bohunician' dates is comparable to those obtained at the type-site itself. Valoch always made it clear that, although the artifacts at Bohunice were found in the lower 10 cm of a fossil soil identified with Hengelo, they must already have been situated in the underlying loess, the upper part of which was subsequently transformed by pedogenesis. Their age therefore was always considered to correspond to the final stage of early Würm loess deposition (cf. Allsworth-Jones 1986: 141-4, 267-8). In addition, Svoboda has now ascertained that, at least in some parts of the site, there is a solifluction-disturbed horizon immediately below the fossil soil with which the artifacts may also have been associated; lithologically at least this would present a closer analogy to the earliest occupation at Stránská Skála (Svoboda 1986d: 239; Svoboda and Svoboda 1985: Figure 2, right). Stránská Skála also reveals quite clearly that the 'Bohunician' occupation did extend into Hengelo, whereas an Aurignacian horizon at a time corresponding to Denekamp is of course entirely normal (although, as Svoboda says, this is in fact the first ^{14}C date we have for an open-air Aurignacian site in Moravia: Svoboda 1986c: 39-40).

The main characteristics of the 'Bohunician' industry from Stránská Skála III layer 5 and IIIa layer 4 (unfortunately minus the Levallois component) are evident from the figures in Table 7.3. As would be expected at workshop sites, cores are relatively frequent in relation to retouched tools. In classifying them, Svoboda has followed the same system which he elaborated at Ondratice, according to which flat Levallois cores account for 21 of the 40 at Stránská Skála III (5) and 16 of the 37 at Stránská Skála IIIa (4)), whereas Upper Palaeolithic type cores account for 5 and 3 pieces respectively, and the remainder belong to other categories. Although they are not included in the type list, unretouched Levallois flakes, blades, and (in particular) points obviously played a prominent role at both localities (Svoboda 1985: Figure 3, 1-5, 10; Figure 4, 1-4). As Svoboda emphasises, the retouched tools particularly at Stránská Skála IIIa (4) do include a few carinate or nosed end-scrapers, as well as isolated examples of unifacial leafpoints. Again as expected, the raw material used was almost exclusively local chert obtained from the limestone outcrop itself, but there were other materials as well, including chert of Krumlovský les type. There are indications that these 'imported' raw materials were worked on the spot (Svoboda 1985: 267; 1986c: 37) and proportionately they appear to be rather more frequent among the retouched tool types (Svoboda 1985: 267; 1986b: 9; 1986c: 37). Only the barest details have so far been given of the collection from Líšeň, but with a total of 27,653 artifacts (Svoboda 1986c) it is obviously of considerable importance; of those, 1291 are cores and 1391 are retouched tools (Svoboda and Svoboda 1985). The largest category of cores is described once again as flat Levallois (546 out of 1291), with in addition Upper Palaeolithic and other types (230 and 515 respectively). So far as the retouched tools are concerned, both Aurignacian-type end-scrapers and leafpoints – bifacial and unifacial – are more prominent than at Stránská Skála, although Svoboda emphasises that the possibility of admixture does exist. Forty-two "*pointes à retouches dorsales*" have been included here with the retouched blades, and 14 others have been classified as Mousterian points. It is to be hoped that this collection will be published in more detail, since it is regarded as an integral part of the 'Bohunician', a secondary workshop-cum-habitation site as compared with the primary workshops on the summit of the limestone outcrop itself.

The small new surface collections reported by Valoch are not in the same league as this, but they are interesting in so far as they reveal once again the difficulties in dealing with material of this kind and because they come from a part of Moravia which obviously merits further study. Neither Hostějov nor Troubky are large enough for statistical analysis (the total number of artifacts in each case being 170 and 180 respectively) but the tool totals as given by Valoch do nonetheless have a certain indicative value (Valoch 1985, 1986b; summarised here in Table 7.4). Hostějov is situated west of Uherské Hradiště and Troubky is southwest of Kroměvříž; both therefore form part of the Palaeolithic settlement along the middle course of the river Morava, which is already well known for a concentration of sites around Napajedla (investigated by Klíma) and in the vicinity of Kroměvříž (investigated by Oliva). Troubky has the additional distinction of being situated right at the source of a most interesting raw material occurring in the Litenčice hills and referred to as 'menilitic slate' (Valoch 1986b). The material is described as of good quality, appearing in the form of slabs up to 10 cm thick. The interest of the sites for us in the present context however resides rather in their taxonomic assessment. Valoch refers to both of them as so-called 'indifferent' industries in which either 'Aurignacoid' or 'Szeletoid' elements may be present. The meaning of this statement will be clearer if we examine a little more closely the circumstances at the two sites and the comparison made between them and their neighbours, Stříbrnice and Zdislavice.

Obviously the small assemblage from Hostějov does contain 'Szeletoid' elements, in that it has one retouched pointed blade, one unifacial leafpoint, and two bifacial leafpoints (Valoch 1985: Figure 2:1-4). At Stříbrnice, another small surface site 6 kilometres away, there is a generally similar inventory – except that there are no leafpoints and there are 'Aurignacoid' elements in the shape of two carinate or nosed end-scrapers. Troubky also possesses 'Aurignacoid' elements, in that it has two carinate or nosed end-scrapers, three carinate burins, and one combined end-scraper-burin of Aurignacian type (Valoch 1986b: Figures 2:3, 6, 8; 3:5-7). But again its neighbouring site of Zdislavice has 'Szeletoid' characteristics in the shape of three bifacial leafpoints (Valoch 1986b: Figure 1:2, 3, 5). Because of the discovery of these artifacts, Valoch was obliged to change his original classification of Zdislavice as a 'pure' Aurignacian site – a sequence of events similar to that which occurred further south at Křepice, except that in that case Klíma felt no obligation to alter his classification (cf. Allsworth-Jones 1986: 162, 167). One might dismiss these ambiguities simply as a result of trying to attempt too much too quickly on the basis of small surface collections – and the difficulties certainly are typical of those which occur on a larger scale throughout Moravia – but nonetheless there are some signs that these so-called 'indifferent' or 'mixed' industries may to some extent reflect reality. They correspond essentially to what Klíma has called an "Aurignacian of Morava type", which he has defined not only on the basis of surface collections but also at the significant little stratified site of Gottwaldov-Louky, which, I have suggested, may provide a useful key to understanding not only the sites on the middle Morava but also a broader category of sites in Central and Southeastern Europe with an apparent conjunction of leafpoints and Aurignacian forms (cf. Allsworth-Jones 1986: 163-4, 168, 177, 244).

A more important ambiguity, and a more pressing need for definition, arises when we attempt to clarify what is meant by the 'Bohunician' in relation to the Szeletian, and to determine whether the use of the former term is actually justified. The seven inventories listed in Table 7.3 clearly add materially to our knowledge of the early Upper Palaeolithic of Moravia and all, in my opinion, are more or less comparable to the industry originally described by Valoch from Bohunice; but, in place of his first simple and logical formula, there are now two new conceptions of what constitutes the 'Bohunician'. The first of these has been put forward by Oliva (1981, 1984b, 1986a) and has been accepted by Valoch (1982, 1983, 1984, 1986a), whereas the second is due to Svoboda (1980, 1984a, 1985, 1986b, 1986c; Svoboda and Svoboda 1985).

In the course of a re-examination of the artifacts originally recovered by Valoch from Bohunice, Oliva claimed to have ascertained that all the leafpoints and some of the side-scrapers were made of materials other than the predominant Stránská Skála chert: a different kind of chert either from the region of the Krumlovský les or from Boskovice (Oliva 1981, 1984b: 210). Oliva therefore asserted that these items – which he agreed were found *in situ* with the remainder of the inventory (Oliva 1981: 12) – were acquired by exchange from the Szeletian. Oliva also claimed that the same phenomenon could be observed at Podolí and Líšeň-Čtvrtě, and that a process of exchange in the opposite direction accounted for the appearance of Levallois elements in the Moravian Szeletian and Aurignacian (Oliva 1981: Figure 3). Olivás method of reasoning and his conclusion both appear to be unacceptable (Allsworth-Jones 1986: 143-4). In the first place, as Svoboda points out (1984a: 365 and Note 1; 1986c: 37; Svoboda and Svoboda 1985: 513), the actual situation at Bohunice may not have been so simple, since according to him there are 'Szeletian types' made of Stránská Skála chert, and whole nests of debris indicating the working of non-local raw materials on the spot. There clearly was not a straightforward exchange therefore. The same argument applies *a fortiori* to Podolí and Líšeň. Oliva himself agrees (1981: 11-2) that the unifacial leafpoints at those localities were made of Stránská Skála chert, and that there was *in situ* production of artifacts from other raw materials. As noted above, the same situation has been observed in an excavated context at Stránská Skála itself. In the second place, more importantly, there are many instances in the Palaeolithic record where certain raw materials were preferred for particular purposes, and it is quite unnecessary to label such materials 'intrusive'. According to Svoboda (1984a: 365) this is precisely the case at Líšeň, where there is a definite link between specific kinds of raw materials and specific types of tools. As he remarks, it is a strange kind of logic that seeks to defend the homogeneity of surface collections, but at the same time artificially sunders stratified well documented assemblages!

Svobodás own conception of the 'Bohunician' is different, but it must be noted that it has undergone an evolution with time. It was first defined on the basis of his study of the material from Ondratice. He suggested (1980: 73-4) that there were three different phases of settlement at that site: (1) a settlement of 'Bohunice type', corresponding to the bulk of the quartzite industry (as detailed in Table 7.3); (2) a Szeletian settlement, corresponding to Valoch's numbered localities III-VII and to part of the quartzite and chert artifacts from Ondratice I; and (3) an Aurignacian settlement, as demonstrated at Ondratice II-Zadní Hony. (New localities numbered VIII-XII have now been discovered by J. Ječmínek, and they too are regarded by Valoch as probably Szeletian; in his article, Valoch confirms an observation made by this author that glassy quartz porphyry from the Bükk can be found occasionally at Ondratice, thus providing an intriguing hint of contact between the two areas: Valoch 1983: 8). Svoboda's idea of the 'Bohunician' as originally put forward, is that it constituted a "transitional stage" between Middle and Upper Palaeolithic, a "horizon" which could have formed the basis for the later development of different "Upper Palaeolithic cultures", such as the Aurignacian, the Szeletian, and the Jerzmanowician; i.e. it is near to the idea of the "synthetotype" as put forward by Laplace (Svoboda 1980: 87-9, 98-9). In fact there are good grounds for objecting to Laplace's formula (Allsworth-Jones 1986: 21-2), and it seems to me unfortunate that Svoboda should have chosen to embrace this particular model to explain the quartzite industry at Ondratice.

In his later writings, the idea of the "synthetotype" seems to have been quietly dropped, but, largely as a result of his excavations at Stránská Skála, Svoboda has complicated his original notion by suggesting that the "Bohunician, Levallois-leptolithic industries" can be divided into two contemporaneous (not successive) variants (Svoboda 1985: 267; 1986c: 37-8, 40-2; Svoboda and Svoboda 1985: 512-3). Variant I corresponds to the originally-analysed quartzite industry from Ondratice I, as well as Bohunice itself and Stránská Skála III layer 5. This variant contains very few Aurignacian elements, but Svoboda is now prepared to admit that they do exist, even at the type site itself (Svoboda 1986c: 41; cf. Allsworth-Jones 1986: 142). Variant II is said to include Stránská Skála IIIa layer 4, as well as Líšeň and Podstránská, and here there is a much increased representation of Aurignacian types as well as unifacial and bifacial leafpoints. Obviously in these circumstances the fixing of boundaries *vis-à-vis* the Aurignacian and the Szeletian becomes much more difficult, as Svoboda himself admits (Svoboda 1986c: 40; Svoboda and Svoboda 1985: 505). In defining the geographical coverage of the 'Bohunician', Svoboda suggests that it extended in a band about 50-60 kilometres long from Ondratice via Stránská Skála to the Bobrava valley (Svoboda and Svoboda 1985: Figure 1). As such it embraces sites like Želešice and Ořechov I and II (attributed by Valoch to the Szeletian) as well as Podstránská (attributed by the same author to the Aurignacian); hence Svoboda suggests an arbitrary boundary of 7.5% as a maximum proportional representation for leafpoints in the 'Bohunician'; but (apart from the fact that that is the actual percentage occurrence at Ořechov II) what is the rationale for such a figure? In these circumstances, Valoch for example (1985: 14) equally well can, and does, speak instead of a "more or less strong influence of the Bohunician" on basically Szeletian assemblages, and – when the argument is conducted in these terms – who is to say which one of them is right? It is for reasons such as this that Svoboda is obliged to play down the typological element, arguing instead that the 'Bohunician' is essentially a technological phenomenon, a "Levallois-leptolithic complex" based upon the exploitation of Stránská Skála chert and of quartzite of the type occurring at Ondratice (Svoboda 1985: 267; 1986c: 41; Svoboda and Svoboda 1985: 513). This limitation allows Valoch to conclude that the 'Bohunician' is "an utterly local and short-term phenomenon" (Valoch 1986a: 11). Svoboda himself speaks of the Szeletian '*sensu stricto*' and '*sensu lato*', and I would rather take his careful work as defining the place of the 'Bohunician' industries within the latter than as laying the foundations for an entirely independent entity.

The meaning to be attached to the assemblages from Vedrovice II and Kupařovice I, both claimed to be early Aurignacian, is a different problem again. Hitherto, in the absence of a full publication, I took a cautious attitude in this regard, having in mind particularly the possibly parallel case of Byčí Skála (Allsworth-Jones 1984: 14-5, 164-5, 192-3). On the basis of preliminary reports, some scepticism was also expressed by Svoboda (1984a: 362-3) and J.K. Kozlowski (1982: 163, 170). Svoboda in particular pointed out that the preliminary data did not appear to be sufficient to locate either site precisely within the bounds of the last glacial period, and he raised a doubt about the validity of the analysis carried out, since according to him the supposed end-scrapers and burins at these sites are in fact cores. This is the same kind of mistake as was made at Býčí Skála. It is fortunate then that we do now have the full publication concerning these sites and it is possible to come to an informed judgement about them (Valoch *et al.* 1985). The two critical points obviously concern, first, the stratigraphic/chronological position claimed for the industries and, second, their actual nature.

The site of Vedrovice II is situated approximately 100 metres northwest of the western wall of a former brickyard, which extends for about 180 metres in a north-south direction (Valoch *et al.* 1985: Figure 27, Plate I). Surface finds were made over an area of about 100 x 100 metres, and in 1976 Valoch excavated four strips of 20 x 1.5 metres each in the centre of this area. Five small artifact concentrations were found in loess beneath plough soil with an approximately 30 cm vertical distribution (Valoch *et al.* 1985: Plate II). One hundred and sixty-seven artifacts were found altogether, including 24 tools or other retouched pieces and 84 cores. These artifacts are not datable as such, hence the importance of the western wall of the brickyard which reveals a loess succession divided by a number of fossil soils or lenses all of which are inclined in a northwest-southeast direction. Beneath plough soil, a loessy substratum, and loess, there is an 80 cm thick fossil brown soil marked by solifluction at the base, the characteristics of which have been studied by Smolíková (Valoch *et al.* 1985: Figure 28, 188-91). This is underlain by a thick loess deposit, which is subdivided by three weaker brown soils or lenses, and then at the base by a thick double soil complex, the age of which according to Valoch was not established but which Pelíšek clearly regarded as belonging to the last interglacial period (Valoch *et al.* 1985: 184-5). At various times up to 1976 (Valoch *et al.* 1985: Plate III.1) a total of nine artifacts were found in the wall of the brickyard in loess beneath the upper fossil brown soil and the first of the three weaker brown soils, i.e. between the first and second of the latter (Valoch *et al.* 1985: Figure 27, x). The artifacts appear to consist of three struck chert pebbles and six flakes of the same material. As such they are not particularly diagnostic, but (having regard to the inclination of the fossil soils) they are regarded as providing the stratigraphic key to the main body of finds 100 metres away. If that connection is accepted, then, as Valoch says, obviously everything hinges on the date which can be attributed to the upper fossil brown soil studied by Smolíková. Smolíková (who also studied the fossil soil at Bohunice) is quite clear on this point. According to her, this fossil soil cannot be the equivalent of PK I (Stillfried B); rather it corresponds to the basal part of PK II (the upper member of Stillfried A). This would imply a date early in the last glacial period, and that creates a difficulty for Valoch, since, as he himself pointed out in relation to Klíma's interpretation of the stratigraphy at Dolní Vevstonice, it is extremely unlikely that the Upper Palaeolithic commenced by that time (cf. Allsworth-Jones 1986: 37-9). Valoch therefore rejects Smolíková's attribution of the soil to PK II, not on geological, but on archaeological grounds, as the Czech summary of the article makes clear (Valoch *et al.* 1985: 198: "z archeologického hlediska však nejde o PK II"). Hence he suggests instead that the fossil soil corresponds to an unspecified 'Middle Würm' interstadial (but Smolíková did not compare it to Bohunice!) and that the finds therefore belong to some point in the latter part of 'Altwürm' (Valoch *et al.* 1985: 167-9). Unfortunately we can see from this summary on what a slender basis the supposed dating of the artifacts from Vedrovice II really rests. Valoch's hypothesis assumes that a reliable connection can be made between the nine (undiagnostic) artifacts found in the western wall of the brickyard and those discovered at the main excavated (though largely surface) site 100 metres away. But that connection leads to insuperable contradictions if we accept the geological dating of the upper fossil soil to an early part of the last glacial period, since everyone agrees that the Upper Palaeolithic is unlikely to have commenced by then. In the circumstances it seems better not to press the case for a connection between the two sets of finds, which for the moment leaves the main collection still essentially undated.

Kupařovice I is at a distance of about 7.5 kilometres from Vedrovice II, on the west bank and about 750 metres from the present course of the river Jihlava (Valoch *et al.* 1985: Figure 32). Surface finds were made over an area of about 200 x 50 metres just at the edge of the 4 metre terrace above the river, and the geological circumstances of the site were investigated by means of a number of test pits dug for various purposes (profiles A-A[1], B-B[1]; Valoch *et al.* 1985: Figures 33-4). The original position of the artifacts was ascertained by excavations carried out in 1970 and 1974-5 and an illuminating geological study was conducted by Karásek (Valoch *et al.* 1985: Figure 29, Plate IV, 176-83). One hundred and eighty-nine artifacts were found altogether, including 22 tools or other retouched pieces and 50 cores. The artifacts constantly occur, at the level of the 4 metre terrace, beneath plough soil and loess, in the top part (up to 15 cm deep) of a coarse sandy and gravelly layer, above fine sands and river gravels with boulders. Many of the artifacts have edge damage, and it is suggested that they were subjected to at least some degree of water transport, perhaps due to a former tributary of the Jihlava, the course of which can be traced a little way to the northwest of the site. The basal river gravels with boulders constitute a former terrace of the Jihlava which is regarded as 'Altwürm' in age; since the finds were deposited on top of the terrace they are said to belong approximately to the 'Mittelwürm' period. Obviously this is a very generalised dating, and since the term 'Mittelwürm' can embrace any episode from Hengelo to Denekamp there is nothing to say that these finds are any earlier than any others attributed to the Aurignacian in Moravia. Neither in the case of Kupařovice I nor in that of Vedrovice II, therefore, are there convincing stratigraphic grounds for asserting that these sites stood at the beginning of the Upper Palaeolithic succession in the area.

The industries from the two sites have been fully described and well illustrated by Oliva (Valoch *et al.* 1985: 110-46). Details, following his indications, are given in Table 7.5. They require some commentary. In the first place, it is clear that at both sites (confirming what we saw already in regard to the excavated material) there are more cores than there are tools; if we include the struck pieces and fragments which cannot be categorised according to type, cores constitute 70.95 and 73.89% of the combined tool and core totals at Vedrovice II and Kupařovice I respectively. These percentage totals would be increased if we included among the cores those pieces categorised by Oliva as "core tools", of which there are according to him 15 at Vedrovice II and 52 at Kupařovice I. Oliva himself admits that, particularly at Kupařovice I, it is sometimes difficult to distinguish between cores and tools (Valoch *et al.* 1985: 140, 141, 144, 146), and he himself agrees that at Býčí Skála the numbers of 'core tools' were originally much exaggerated (Valoch *et al.* 1985: 147). But it seems to me, on the basis of the discussion and the illustrations, that Svoboda is correct and that a similar mistake has been made at these sites. That applies not only to the 'core tools' but to many others which are included in the type list, particularly the end-scrapers and the burins. So far as the burins are concerned, this is borne out by Oliva's careful morphometrical analysis; as he points out, the modal class for burin-blow width at both sites tends to be in the vicinity of 15 mm, "was kaum Analogien finden wird" (Valoch *et al.* 1985: Figure 26, Table 7, 144). If these objects are recognised as cores and the measurements are therefore held to relate to flake/blade removal width, it would not be so difficult to find analogies! Secondly, it is clear from all indications that these are 'advanced' rather than 'archaic' industries. The cores themselves are mostly of a good single-platform flake/blade variety, and there are large numbers of *lames à crête*, proportionately more, as Oliva says, than at any of the Central European Aurignacian sites described by Hahn (Valoch *et al.* 1985: 137). The characteristics of the striking platforms are in accordance with this, since according to Oliva the majority of blades at both sites have 'lipped' or punctiform platforms indicative of a soft (direct or indirect) percussion technique. Levallois technology was not employed. Thirdly, with regard to typology, there are some Aurignacian traits, but one would not say they were very pronounced. There are four bifacial leafpoints at Kupařovice I (Valoch *et al.* 1985: Figure 17:11) and a total of six artifacts are regarded as bifaces (Valoch *et al.* 1985: Figures 6:1-3, 17:2 and 3). To judge from the illustrations however these could well be unfinished pieces, such as occur most notably at Jezeřany, which is also in the vicinity (cf. Allsworth-Jones 1986: 27, 165-6). Svoboda has recently sounded a similar cautionary note in this regard (Svoboda 1984b: 177; 1986b).

All in all, therefore, it seems to me that these industries have to be classified very differently than in the scheme favoured by Valoch and Oliva. They are workshop sites, with an advanced blade technology, typologically not very distinctive, but with some Aurignacian traits as well as a few leafpoints. Geologically, at the moment, they cannot be precisely dated and they certainly cannot form the twin foundation pillars for the Aurignacian in Moravia which Valoch and Oliva would like to see. It is not surprising that Oliva has found it very difficult to discover an analogy for these industries (Valoch *et al.* 1985: 147). In particular, he admits that they are not similar to such stratigraphically-established early Aurignacian assemblages as Das Geissenklösterle, Istállóskö, and Bacho Kiro (Valoch *et al.* 1985: 149-50). It is, I believe, to these latter sites that we should look when we are searching for the origins of the Aurignacian in Central and Southeastern Europe. I agree with Valoch that there was such an early Aurignacian in the area and that it had a decisive role to play in the origin of the Szeletian (Valoch *et al.* 1985: 172-3), but on present evidence I do not think that Vedrovice II and Kupařovice I are convincing candidates for this role. Criticisms similar to the above have been succinctly expressed by Svoboda following the full publication of these sites (Svoboda 1986b).

The new finds from Hungary and Moravia prompt consideration of certain general issues relating to the interpretation of early Upper Palaeolithic cave and open-air sites in Central Europe, as follows.

CAVE SITES: NATURE OF THE OCCUPATION

These days there is in general among prehistorians a degree of wariness in dealing with cave sites, a recognition that they had several roles to play and that human occupation was only one factor among many. This holds true for both the European Middle and Upper Pleistocene, as recent studies by Cook *et al.* (1982) and Jenkinson (1984) for example demonstrate. New approaches such as these have profound implications for Central Europe, where Bayer and Vértes regarded the 'Olschewian' and the Szeletian respectively as confined exclusively to cave sites. The reality appears to be otherwise, and already-published data indicate both the extent to which cave inventories constitute highly skewed distributions and the degree to which cave occupations were influenced by other than human factors.

As an illustration of the first point, summary details of the artifacts found at eight Central European cave sites are presented in Table 7.6. The figures given for Mauern, Ranis, Nietoperzowa, and Křížova correspond, with slight modifications, to those of the authors of the site reports concerned. The stone tool totals for Jankovich and Szeleta are those established by this author, whereas the totals for flakes, blades, and debris are somewhat rough estimates based on the gap between these tool counts and the complete artifact totals provided by Vértes and Kadić respectively. The figures for Jankovich do not include artifacts attributed to the Gravettian, and it is likely that the bone points also do not belong together with the listed stone tools. It is probable that during Hillebrand's excavations not all flakes, blades, and debris were sorted out and kept. With regard to Szeleta, the figures derived from Kadić are likely to be exact, but they refer only to his own excavations. The figures for Istállóskö are those given by Vértes, and they relate solely to his 1950-1 excavations. The figures for Potočka Zijalka are derived from the recently-published monograph on the site (Brodar and Brodar 1983). They represent the totality of finds from the cave – layers 8, 7, 5, and 3 at the entrance, and layers 5 and 4 at the back.

As a first approximation certain contrasts can immediately be observed between these figures and those typical of open-air sites, as in Table 7.3. As a rule, except for Mauern G, there are proportionately fewer cores at the cave sites, and in two cases there are none at all. The numbers of unretouched flakes, blades, and debris, excepting Mauern G and Szeleta, are generally much reduced and it is not uncommon for tools and cores together to constitute at least 30-50% of the total, in one case as much as 95%. The figures provide quantitative support for the opinion voiced by this author and others that many of these sites served as no more than temporary refuges or stop-overs rather than permanent habitation centres. That impression would be considerably strengthened were smaller cave sites to be included, for these are only the larger ones in their respective areas. Further details given by the authors of the site reports support the idea of intermittent human occupation alternating with periods when animals alone were present.

Thus, at Mauern, Müller-Beck and his colleagues have emphasised that "only occasional occupation of the caves" is indicated for both layers G and F, and this is more markedly the case at Ranis. Hülle showed that the finds belonging to the Ranis 2 inventory were scattered over the whole cave system, and in view of the virtual absence of any signs of tool manufacture he concluded that the occupation was the work of hunting groups who only occasionally visited the site. Fifty-three of the 60 stone tools are leafpoints. The Ranis 3 inventory is more balanced, with indications (as at Mauern) of some artifact manufacture on the spot. Layer IX (between the two) contains a thick "bone plaster" now thought to indicate the presence of a hyaena den in the cave at that time. The three Jerzmanowician layers at Nietoperzowa contain progressively smaller stone-tool inventories, amounting to 62, 43, and 26 pieces respectively; 79 of the overall total are leafpoints, a percentage less than at Ranis 2 but still sufficient to reveal a remarkable imbalance in the industry. The few finds from Křížova are comparable in character and context to those from layers 15 and 16 at Pod Hradem (one or two leafpoints and a retouched blade) also attributed to the Szeletian; as Musil emphasises, Pod Hradem must have served basically as a cave bear den, and cannot in any sense have been a permanent human settlement.

Jankovich is by far the richest of the Trans-Danubian sites, yet the inventory can hardly be considered abundant, in view of the fact that the deposits in the side chamber reached a thickness of at least 5 metres. At Dzeravá Skála, Prošek ascertained that the Szeletian finds in layers 5-11 were scattered over the whole excavated surface, with no traces of hearths nor any cultural horizon as such; he stressed the paucity of waste material in the Szeletian collection, and considered that the site had served as no more than a temporary hunting shelter. The same is true of the new site at Máriaremete, as has been emphasised by Gábori-Csánk (1983: 265). She considers it probable that the rear chamber served as a lion or hyaena den during the deposition of the lower horizon of layer 4. As she says, the complete tool total from all the Jankovichian cave sites in the region west of the Danube amounts to hardly 180 pieces; but if these were only "transitory and occasional" sites, then the real habitation centres must have been elsewhere, in the open-air (Gábori-Csánk 1983: 285; all the stranger therefore that – contrary to the recently-restated position of Dobosi and Vörös – she wishes to exclude Lovas from the list of Jankovichian sites: Gábori-Csánk 1983: 276-7). The finds from Szeleta and the Bükk in general do not reveal quite the same imbalance as at Jankovich, but considering that the deposits at Szeleta reached a maximum depth of 12.5 metres they are scanty enough, as Hillebrand complained eloquently in his time. In this case, according to Kadić, the artifacts were concentrated around a few big hearths (two in layer 7 and four in layer 4) and a possible workshop area near the entrance to the cave, but elsewhere they were sporadic. The figures for tools and cores as a percentage of the total number of artifacts are very similar to those

Table 7.6. Artefact totals and calculated indices for 8 Central European cave sites

	Stone tools	Cores	Flakes, blades, debris	Total stone arte-facts	Bone and Ivory arte-facts	Cores as % of tools and cores	Tools and Cores as % of total stone artefacts.
Mauern G							
Mousterian	94	56	463	613	—	37.6	24.5
Mauern F							
Altmühlian	114	17	269	400	—	13.1	32.75
Ranis 2	60	—	3	63	—	—	95.0
Ranis 3	91	9	40	140	—	9.0	71.4
Nietoperzowa							
Layers 6, 5, 4	132	5	116	253	?1	3.7	54.15
Křižova							
Layers 7, 8, 9	10	1	19	30	—	9.1	36.7
Jankovich	102	—	14	116	21	—	87.9
Szeleta							
Layers 3 & 4	259	9	635	903	2	3.4	29.7
Layers 5–7	216	5	479	700	—	2.3	31.6
Istállóskő (1950-51)							
Layer 9 Aurignacian 1	17	2	27	46	114	10.5	41.3
Layers 7 & 8 Aurignacian 2/3	54	2	53	109	31	3.6	51.4
Potočka Zijalka	70	6	229	305	130	7.9	24.9

established for Mauern. From the accounts given by Kadić and Hillebrand, it is clear that the cave remained for long periods uninhabited by man, but rather by cave bear, several complete articulated skeletons of which were recovered in the deposits. The sparse finds from Istállóskö are also said to have been concentrated in and around hearths; the figures given for the Aurignacian 1 and 2/3 industries relate only to Vértes's excavations of 1950-51, because they are likely to be the most reliable, but those available for 1912-1948 tell essentially the same story.

So far as Potočka Zijalka is concerned, we are fortunate now to have a full and beautifully illustrated account of the site, which S. Brodar excavated as long ago as 1928-1935 (Brodar and Brodar 1983). The account throws much light on all aspects of the site, which (in some respects regrettably) provided the starting point for Bayer's theories concerning the 'Olschewian'. It is only tragic that the account can never now be complete, because some of the stone and bone tools disappeared from the Museum in Celje during the Second World War, whereas the entirety of the palaeontological collection was destroyed by bombing. Luckily, much of the species identification, including that of the microfauna, had already been carried out by O. Wettstein, and all the documentation of the excavation (apart from that for the first year, 1928) survived. Seventy-six stone tools and cores and 130 bone tools from the site have been described and illustrated by S. and M. Brodar, as shown in Table 7.7. Because of the loss of some material, these figures are obviously not complete, but they do give an excellent idea of the industry. Overall, as S. and M. Brodar emphasise, this is a clearly Aurignacian inventory. In my own opinion, there is quite a close comparison between it and the upper assemblage from Istállóskö as excavated from 1912 onwards (cf. Allsworth-Jones 1986: 92-3). There are proportionately rather fewer end-scrapers and burins at Istállóskö, but side-scrapers are relatively prominent at both sites, and in particular they are both dominated by retouched blades, including retouched pointed blades of a type which for many years were taken to be a virtual hallmark of Istállóskö (Brodar and Brodar 1983: 448, Plate 1). The Potočka Zijalka assemblage is clearly a blade industry, as S. and M. Brodar say, but they only illustrate and describe a few of the unretouched artifacts (e.g. S. and M. Brodar 1983: 88, 445, 433, 434, Plate 2) and the figure of 229 for flakes, blades and debris given here has been obtained by mere subtraction of the tools and cores from the grand total given by them (Brodar and Brodar 1983: 114).

The overall figures disguise the fact that the finds come, not only from different layers, but also from different parts of the 115 metres long cave: the back, excavated in 1928 and 1929, and the entrance, excavated between 1929 and 1935 (Brodar and Brodar 1983: Figure 9). The very careful planning tells us a lot about the kind of occupation which can be expected in caves of this kind (and we must remember that Potočka Zijalka is at a height of 1700 metres above sea level, above the present tree-line). At the entrance, on the western side, occupation began in layer 8, but the traces of human presence are very slight. Layer 7 represents the main cultural horizon, containing all the stone tools and cores and 34 bone points as well as one bone awl. It contains 16 hearths of various sizes, and it is noticeable that the stone tools are closely grouped around them (Brodar and Brodar 1983: Figure 48). A total of no less than two cubic metres of charcoal (belonging to two species, *Picea excelsa* and *Pinus cembra*) was removed from these hearths. Layer 6 (although it apparently represented a more favourable climatic interval) contained no direct evidence of human presence, but 9 musk ox canines are thought to have been brought in by man. Layer 5 constitutes a second important cultural horizon, but it differs from the first in that there are no stone tools (only a few unretouched flakes) to accompany 12 bone points, one of which is split-based. There are three hearths, but the bone points (with one exception) are not related to them, rather they are scattered over the entire excavated surface (Brodar and Brodar 1983: Figure 50; one could wish that this type of information were available for Jankovich and Dzeravá Skála!). Once again there are virtually no signs of human presence in layer 4, and layer 3 has indications which are sparse enough. At the rear of the cave, artifactual remains are confined to layers 5 and 4 (there is no direct relationship between the layers at front and rear) and there is only one

Table 7.7. Tool and core totals and calculated indices for Potočka Zijalka

	Totals	Major stone tool groups (%)
Endscrapers	19	27.14
(including carinate and nosed)	13	
Endscraper-burins	1	
Burins	11	15.71
Truncated flake/blades	1	
Retouched blades	15	} 28.57
Retouched pointed blades	5	
Mousterian points	1	
Sidescrapers	11	15.71
Notched pieces	4	
Miscellaneous	2	
Total Stone Tools	70	
Total Cores	6	
Non-split based bone points	126	
Split-based bone points	1	
Bone awls	2	
Bone needles	1	
Total bone tools	130	

hearth in layer 4. As S. Brodar emphasised from the beginning, there are no stone tools, only a few unretouched flakes. By contrast, there is a total of 80 non-split based bone points and one bone needle; as in layer 5 at the entrance, they are scattered over the entire surface and have no relationship to the single existing hearth (Brodar and Brodar 1983: Figures 52 and 53). In contra-distinction to the situation at the entrance, most of the bone points are undamaged (Brodar and Brodar 1983: Figure 59). As M. Brodar says, it is highly likely that the contrasts between front and rear reflect differential usage of the cave space, and he himself suggests that the rear may have been employed as a place to spend the night.

This was of course not the way in which the evidence was interpreted by J. Bayer, who published his account of the 'Olschewian' solely on the basis of the material excavated at Potočka Zijalka in 1928 and 1929. According to Bayer, the 'Olschewian' was distinguished by the following features: 1. There was a predominance of bone points over stone artifacts, which were few and uncharacteristic but certainly Upper Palaeolithic in character; 2. The bone points were exclusively non-split based, a fact which Bayer assumed to be of not only formal but also functional significance; 3. Many of the other animal bones were artificially pierced; 4. The inhabitants of the culture lived exclusively in caves and hunted mainly cave bear. Whereas they were contemporary with the Aurignacians, they lived in a degree of isolation and formed a separate race.

M. Brodar pointed out already in 1971, and he repeats in the monograph, that none of these arguments can be regarded as still valid, and that the 'Olschewian' in Bayer's sense did not exist. Bayer's statements, as the above account makes clear, were contradicted in many respects at Potočka Zijalka itself prior to the close of the excavations in 1935: notably with regard to the frequency and nature of the stone and bone tools. M. Brodar cites a number of other instances, including Istállóskö and Mokriška jama, where split-based and non-split based bone points are known to have occurred together in the company of characteristically Aurignacian stone artifacts, and Potočka Zijalka as such has now been justifiably published as "eine hochalpine Aurignacjägerstation". Moreover, bone points are not confined to Aurignacian cave sites, as the evidence from Willendorf for example (already available in Bayer's time) clearly shows. The so-called artificial piercing of animal bones, M. Brodar admits, is generally regarded with scepticism, and is in any case too general a trait to serve as a cultural determinant; although he himself believes, on the strength of the evidence from Potočka Zijalka and Mokriška jama, that some examples may be intentional (Brodar and Brodar 1983: Figures 57 and 58). But if Potočka Zijalka can no longer serve as the *locus classicus* for the 'Olschewian' culture, it does provide a beautifully documented example of the manner in which such sites were occupied and utilised by man during the early Upper Palaeolithic in Central Europe.

The eight cave sites mentioned above clearly give the impression of having constituted specialised, intermittent, and for the most part strictly temporary settlements. Usually this is interpreted to mean that they were hunting sites, and Vértes among others considered that the Szeletians were essentially cave bear hunters. The same idea has been put forward in relation to other Middle and Upper Palaeolithic communities, and formed part of Bayer's definition of the 'Olschewian'. Specialised hunting as such seems to be an increasingly important phenomenon from the Middle Palaeolithic onwards. Gábori (1979) has recently summarised a number of instances from Central and Eastern Europe where there was an evident concentration on animals such as reindeer, mammoth, ibex, bison, saiga antelope, and *Asinus hydruntinus*; he points out that there is no necessary correlation between type of industry and predominant fauna. There is also no doubt that cave bear could be hunted, and occasionally concentrated upon, as at Érd. Nor do we need to exclude altogether the idea of rituals connected with the cave bear, since ethnographic parallels indicate that special bear festivals and ceremonies may sometimes be performed by certain peoples (Severin 1973: 228-32: the Ainu; 291-5: the Lapps). The expressed opinion of this author however is that the idea of the 'cave bear hunt' has been much exaggerated, and that in respect of the Szeletian and the 'Olschewian' it may be considered largely a myth (Allsworth-Jones

1986: 109-11). The reasons for this are twofold. In the first place, there are numerous indications at individual caves that for much of the time man was not present: either (as at Pod Hradem) the total number of artifacts in relation to cave bear remains is infinitesimal, or (as at Szeleta) there are clear signs that human presence was restricted to certain areas and periods only. At Istállóskö, Vértes recorded that there were many undisturbed cave bear bones immediately above the lower cultural layer 9, and undisturbed use of the cave for long periods of time by cave bears is also evidenced by many fragments of rock incorporated in the deposits and rubbed smooth by continual passage of cave bears. In the second place, it has been demonstrated by Kurtén that certain patterns of mortality may be expected among cave bears which have nothing to do with human activities, but which earlier may have been wrongly interpreted. In particular, one can expect quite large numbers of juvenile as well as senile individuals who died of natural causes. At Nietoperzowa, Wójcik has shown that the cave bear population possessed great stability throughout, and that any major contribution by man to the extinction of this animal must be excluded, at least at that site.

Potočka Zijalka provides some interesting information in this regard. As elsewhere, cave bear was by far the dominant species represented, right through from layer 9 to layer 3 at the entrance and up to layer 2 in the rear of the cave. Because of the subsequent destruction of the fauna, no precise figures can be given, but already at the time of the excavation more than 1000 individuals were recognised on the basis of the numbers of canines present. The authors of the monograph recognise that the site was used extensively by cave bears for over-wintering, and some idea of the population structure can be gathered from the details given for layer 2 in the rear of the cave, where traces of human occupation were entirely absent. According to their account most of the individuals present were adult bears, with a few very old ones and some young ones, including complete skeletons of newly-born bear cubs (Brodar and Brodar 1983: 182). The authors very fairly admit that the cave provides no clear answer to the question as to whether man hunted the cave bear, and that the question of a possible cave bear cult is even more problematic (Brodar and Brodar 1983: 203). They suggest tentatively, although there is no proof, that cave bear bones were in fact used to make the bone points, and it is claimed that one fragmentary cave bear bone from layer 4 or 5 at the rear of the site served as a hafting device (Brodar and Brodar 1983: Figure 56). Nonetheless, it seems to me that one can deduce from the authors' account that on the whole the occupations of men and cave bears were mutually exclusive. None of the human occupations, except possibly for those in layer 7, were very intensive. Yet S. and M. Brodar record that, compared with other layers above and below, layers 5 and 7 had relatively fewer (though fragmentary) cave bear bones, and such as there were were not concentrated around the hearths (Brodar and Brodar 1983: 187, 201).

Some new evidence bearing on this problem is also available from Bacho Kiro and Istállóskö. The final Report on the excavations at Bacho

Kiro (Kozlowski 1982) contains much detail on the fauna in relation to the archaeological finds. Cave bear was present in practically all the Pleistocene layers. The average age of the population was 4.5 years, and according to T. Wiszniowska there were two noticeably higher periods of mortality: among young animals which had not yet reached sexual maturity, and among the old-aged (above 10.4 years old). This is what would be expected in a natural cave bear population. Meticulous excavation records have clarified the relationship between man and cave bear in several of the layers. So far as the Middle Palaeolithic is concerned, it was ascertained that the greatest concentrations of stone tools coincided with the largest numbers of bones of bovidae and cervidae, whereas in layers with sparse artifact assemblages the bones of cave bear were relatively more frequent; the conclusion was therefore that cave bear "was not among the animals hunted by the cave dwellers, but lived here in periods when man was not present" (Kozlowski 1982: 112). A similar situation was discovered in layer 11, the first Upper Palaeolithic layer containing a large 'Bachokirian' or early Aurignacian assemblage, since cave bear remains were said to be situated outside the principal hearths where the archaeological material and the remains of other animals were concentrated. An exception to the general rule is provided by layer 8, which contained very few archaeological remains at first classified as 'Olschewian' but now regarded as an integral part of the 'Bachokirian'. There are only 2 flakes, 4 sandstone slabs, and the tip of a bone point said to be non-split based (Kozlowski 1982: Figure 10.1). The bone point was found amid a cluster of cave bear bones, and is taken to directly reflect hunting activities (Kozlowski 1982: 161-2). In my submission therefore the new evidence from Bacho Kiro supports the author's thesis, that human and cave bear occupations of cave sites generally alternated, and that cave bear hunting (which could be carried out occasionally) was not the inhabitants' main preoccupation.

An entirely different conclusion has been reached by Vörös as a result of his reanalysis and reinterpretation of the bones from Istállóskö: he claims that at this site cave bear constituted the hunters' "main big game", the hunt for adults was "continuous", and there was a seasonal "slaughter" of foetuses and new-born cubs (Vörös 1984: 19). Attention has already been drawn to the intermittent nature of human occupation at Istállóskö and the signs that for long periods cave bear alone was present. Vörös's data-base therefore requires careful examination; undoubtedly he has added significantly to the information previously made available by Jánossy, Soltész and Tasnádi-Kubacska (summarised in Allsworth-Jones 1986: 109-11); but in my opinion the evidence does not sustain the interpretation offered. For his study, Vörös has used "directly" 6282 cave bear bones derived from the 1950-1 excavations of Vértes and "indirectly" (presumably on the basis of written records only) 2261 such bones from Mottl's excavations of 1938 and Vértes's of 1948 (Vörös 1984: 10, 15). The material excavated by Vértes in 1947 was not available; this included a large hearth in layer 7, most of the bones in which according to Vértes belonged to juvenile cave bears (Vörös 1984: 15, 18; layer 7 nonetheless continues to feature in Vörös's overall statistics which in the circumstances may be considered questionable). In 1950-1 a total of about 15 800 cave bear remains were found at Istállóskö, but only certain of them (consistently chosen) were brought for further study to Budapest. As Vörös says, the 40% representative value of the 6282 bones currently available from those excavations does decrease their employability for the kind of archaeozoological reconstruction which interests him (Vörös 1984: 15). Of the 4O mammal species represented at Istállóskö, 20 according to Vörös could be regarded as suitable for hunting (apart from the cave bear) and of these 14 were really important. On the basis of the available data from 1938, 1948, and 1950-51, he calculates the numbers of cave bears *versus* the other 20 mammals, in terms of actual bones and minimum numbers of individuals, as shown in Table 7.8.

Taking all the figures together, cave bear accounts for 88.07% of the total in terms of numbers of bones, and 73.75% of the total in terms of numbers of individuals represented. It is on the basis of the former figure that Vorós claims (1984: 13) "cave bear is absolutely dominant among hunted mammals". No one has ever doubted that cave bear is numerically predominant at Istállóskö, but whether the inference drawn about human behaviour is correct or not is another matter. The first point to note is that more than half of the cave bear totals (52 and 53%) are accounted for by material attributed to layer 9. If one considers the data for 1950-51 alone that figure would rise to 71%. But Vörös himself admits (1984: 11, 17-9) that the accumulation of cave bear bones in layer 9 is at least partly (or, I would suggest, almost wholly) due to natural factors; and it is therefore quite misleading to include all this material in order to arrive at the conclusion that cave bear accounted for the majority of the "eaten or potentially eatable meat" (Vörös 1984: 17 and Table 6). Secondly, one has to consider the number of juvenile individuals, which are "few except among cave bears" (Vörös 1984: 17) and which, contrary to Vértes's opinion, are likely to be the result of natural rather than man-made factors. Juveniles account for 184 of the 304 cave bears in layer 9 according to Vörös, or 60.53%, which is near to the total proportion of juveniles established for the 1950-51 excavated material as a whole by Soltész and Tasnádi-Kubacska, on the basis of the teeth studied by them. The proportion of juvenile individuals in layers 8, 7, and 3 drops, according to Vörös, to give an overall figure of 45.72%. But the figures for layer 7 do not include the remains from 1947 where according to Vértes 80% of the cave bears found were juveniles. One is not saying that there was no exploitation of cave bear at Istállóskö, but there as elsewhere the extent of it should not be exaggerated. It might be more profitable to examine more closely the role of the other major mammals, particularly chamois and reindeer (Vörös 1984: Figure 2). As Vörös says (1984: 18-9) these animals are likely to have followed annual migration routes along the Bán and Eger valleys in the immediate vicinity of Istállóskö, and their presence would have encouraged a seasonal occupation of the site.

Table 7.8. Faunal remains from Istállóskö (after Vörös 1984)

Layer	9	8	7	3	Totals
Bones:					
Cave bear	4460	2011	1702	370	8543
20 other mammals	361	585	161	50	1157
Individuals:					
Cave bear	304	183	73	13	573
20 other mammals	53	92	48	11	204

Neither Istállóskö nor Bacho Kiro are Szeletian sites (though both may have been not unrelated); their importance in this context lies in their reliable numerical and observational data which allows us to draw certain conclusions about the 'cave bear hunting' phenomenon as such; these I believe are applicable to the Szeletian cave-site occupations also, where for the most part we lack information of comparable precision. Though these occupations may not have been specifically linked to the 'cave bear hunt', I submit that the evidence presented is enough to demonstrate that they were indeed specialised, even rather peculiar, sites forming part of a larger network still mainly hidden from us.

OPEN-AIR SITES: NATURE OF THE VARIATION

The problems arising in the case of open-air sites are quite different. The bulk of such sites attributed to the Szeletian are in Moravia, and while some good stratigraphic indications do exist (as at Bohunice, Vedrovice V, and Stránská Skála) for the most part we are dealing with surface sites only. The same holds true for the Aurignacian in this area, which in any case is closely bound up with the Szeletian, both in terms of the nature of the material and of the models which have been held to be applicable to it. Recent interpretations offered for the Aurignacian in Moravia are therefore also relevant to the Szeletian, and they illustrate the epistemological considerations which apply to the study of surface sites in general.

Foremost among the interpreters of the Moravian surface sites has been Valoch, who has sought to arrange both Aurignacian and Szeletian occurrences in a seriated order based on technological and typological characteristics, reflecting according to him principally chronological and cultural-developmental factors. In doing this, he has made a number of assumptions about the industries concerned: particularly, that they contain so-called "archaic" and "progressive" elements, and that the former include what he refers to as "core tools". By extension of the same kind of argument, he has attempted to trace the so-called 'typogenetic' connections between both cultures and their Middle Palaeolithic predecessors:

the Micoquian in the case of the Szeletian and the 'Krumlovian' in the case of the Aurignacian. The 'Krumlovian' was originally referred to as a "Tayacian of Fontéchevade type" and again consists essentially of surface finds. Unfortunately, many of these assumptions are open to serious question, as can be illustrated by the case of Býčí Skála. The quartzite assemblage from this cave site was defined by Valoch as a "pre-Aurignacian" or "Aurignacian 0" industry standing right at the head of his Aurignacian series. But an examination of the original material in the Moravian Museum (Brno) confirms that the so-called "core tools" are in fact blade cores of quite an advanced type, so it is not appropriate to refer to them either as tools or as 'archaic' (Allsworth-Jones 1986: 14-5, 165; cf. discussion above in connection with Vedrovice II and Kupařovice I). Hahn has gone so far as to suggest that no direct link between this industry and the Aurignacian can be proved, and apparently it is now accepted also by Oliva (1984a: 617) that the assemblage is of "uncertain cultural appurtenance". This has not prevented Oliva from constructing a much more complicated seriation scheme for the Aurignacian of Moravia, and the subsequent dispute between him and Svoboda illustrates many of the issues at stake (Oliva 1982, 1984a, 1984b, 1986a, 1986b; Svoboda 1984a, 1984b, 1986c). Apart from the particularities of the dispute, it should be noted that there is a fundamental philosophical divergence between these two authors. As a constant theme in his consideration of the Czechoslovak Palaeolithic, Svoboda emphasises that "apart from the biological evolution of the toolmaker, other factors such as the raw material and changing environment were able to regulate the instrumental adaptation process" (Svoboda 1984b: 169). By contrast, Oliva adopts what seems to be an extreme idealist approach. He specifically states that he does not accept the principle of "least effort" and that he prefers "non-utilitarian" explanations (Oliva 1984b: 218). Hence he emphasises what he calls "psycho-social" motives in causing technological change: manufacturer's prestige, exchange of wealth, leisure and play activities and so forth (Oliva 1984b: 209, 219). The authors' views on individual topics naturally reflect their divergent philosophical positions.

According to Oliva's scheme (1982: Table 2) the Moravian Aurignacian sites can be divided into three stages, each consisting of two phases, and two facies. The scheme implies a constant chronological progression, as measured by a number of criteria, including: a relative decrease through time in the numbers of side-scrapers and "core tools"; a corresponding increase in the numbers of combined and multiple tools, and of end-scrapers and burins taken together; and rather complex changes in the mutual relations between specifically Aurignacian tools, such as nosed and carinate end-scrapers and carinate burins. The second stage is considered to be most similar to the French Aurignacian I. The earliest sites are said to be Vedrovice II and Kupařovice I, as we have already seen, and it is these sites which are said to form the link to the Middle Palaeolithic 'Krumlovian', which is found in the same area.

The points of controversy concern (1) the claims made for Vedrovice

II and Kupařovice I in relation to the 'Krumlovian'; (2) the validity of any seriation scheme such as Oliva's when it is based on surface collections alone; and (3) what alternative explanation may be offered for the observed variability in the early Upper Palaeolithic of Moravia.

1. The nature of the industries from Vedrovice II and Kupařovice I and their stratigraphic position have been fully considered above and attention has been drawn to weaknesses in the arguments presented by Valoch and Oliva. With regard to the 'Krumlovian' the situation has been very fairly summarised by Valoch (1984: 8-10) but it also is not such as to inspire much confidence. Of the four sites originally placed in this category (Vedrovice I and II, Maršovice I, and Kubšice I) only Maršovice I remains (though to it have been added others, including some in the area of Dolní Kounice, where more Szeletian surface sites have also been found). In the case of Vedrovice I and II, "the Aurignacian character of the industries became clear only later" (Valoch 1984: 16), whereas excavations at Kubšice I in 1970 and 1974 revealed that the stone tools there were from the fill of recent prehistoric pits and were associated with pottery (Valoch 1984: 10; cf. Svoboda 1986a, where further excavations of Neolithic pits at Vedrovice and Maršovice are described). There could hardly be a more eloquent demonstration of the dangers of coming to hasty conclusions on the basis of surface collections. To resolve these contradictions, Svoboda has proposed an entirely different solution linking Vedrovice, Maršovice, and Jezeřany in a single "lithic exploitation area" where Jezeřany functioned as the "central site"; "the character of these industries may well be explained by their functioning as workshops at the raw material sources" (Svoboda 1984a: 368; cf. Valoch 1984: 21, 16-19). Valoch (1986b) has however restated his opposition to this concept.

2. Svoboda (1984a: 363-4) draws attention to some of the weaknesses inherent in Oliva's supposedly chronological arrangement of the Moravian Aurignacian. How and why were his particular typological criteria chosen, and do they necessarily bear the interpretation put upon them? If there is little or no stratigraphic control, what kind of statements can be made about the material, and according to what principle can it legitimately be organised? As he says, in a situation where there are no independent checks, logically any solutions are possible, since they are neither demonstrable nor refutable. Svoboda therefore advocates putting the main emphasis on what stratified sites there are, but he also suggests a number of other factors which may help to account for the visible pattern in the data, particularly those concerning raw materials. That is not to say that there is no chronological axis at all. Oliva advances, I think, good reasons for regarding certain of the Moravian sites with predominant burins as being a late development, comparable to such stratified occurrences as Langmannersdorf and Bockstein-Törle (Oliva 1986b; cf. Allsworth-Jones 1986: 169-70, 173-4).

3. Both Oliva (1982, 1984a) and Svoboda (1984a) provide interesting information on the raw materials which were exploited in the early Upper Palaeolithic of Moravia. Oliva draws attention to clear differences of

emphasis between different parts of the area in terms of the use made of local and imported raw materials; but Svoboda goes much further when he points out that many of the quantitative distinctions between component parts of industries emphasised by Oliva and assumed by him to have a chronological significance may in fact owe their existence to such geographical factors. Thus, the further a site was from locally abundant raw material sources, the greater reliance one would expect to find placed on imported raw materials, and, because of the need to economise, the more one would be likely to encounter combined and multiple tools, as well as generally smaller sized artifacts. Conversely, one could expect to find many side-scrapers and cores at workshop sites where local raw materials were exploited, and there might be a tendency for such tools to be of larger size, as in the case of the Ondratice I quartzite industry. Svoboda does not suggest that these factors alone are sufficient to account for the variability observed in the early Upper Palaeolithic of Moravia, but they are certainly worth bearing in mind as an alternative explanation, both here and elsewhere.

There are two further problems raised by the dispute between Oliva and Svoboda, one technological and one social, which both require further study and are of general significance in any consideration of the Middle-Upper Palaeolithic transition in Central Europe. In the first place, there is the question of the role of the Levallois technique in the origin of the Upper Palaeolithic. As we have seen, Svoboda (1980) has indicated a way in which an evolved Levallois technique could have led on to one characteristic of the early Upper Palaeolithic at Ondratice I, although he does not claim that the cores with frontal crests at that site are anything more than prototypes of later Upper Palaeolithic cores. Oliva by contrast (1981: 19-24) is inclined to downgrade the role of the Levallois technique in this process, and does not consider that the technology characteristic of the Aurignacian owes anything to it. That may be an exaggeration, but the fact is that there is still a gap between the blade cores of Ondratice I and those recently so carefully studied from the Aurignacian at Das Geissenklösterle (Hahn and Owen 1985: Figure 2). Svoboda himself (1986c: 41) very fairly admits this point, stating that though he believes in a continuous transition from one to the other, a technological gap still exists between "evolved Bohunice-type industries" and the Aurignacian. As things stand, this constitutes the "main defect" of his model, which, he agrees, still requires further stratigraphic confirmation. The technological gap in question corresponds to the one so commonly detected at stratified sites between the Middle and the Upper Palaeolithic, e.g. at Bacho Kiro (Kozlowski 1982: 163, 170).

The second problem concerns the social aspect. Hahn, in his book on the Aurignacian, treated it implicitly as a strongly segmented society in which local differences could easily arise, and Otte (1985b) has recently suggested much the same thing for the Gravettian. In seeking an explanation for the way in which far-flung raw materials could have found their way to Moravia, Oliva (1984a: 622-5) has proffered a similar model. He suggests that if the Aurignacians had in fact a semi-settled rather than a nomadic way of life, such objects are more likely to have been obtained by exchange with neighbouring groups than by wide-ranging hunting expeditions (cf. J.K. Kozlowski 1972-73). This is no doubt somewhat speculative, but at a time when it is being suggested (e.g. Clark 1981; Smith 1985) that behavioural factors as much as anything else were responsible for the transition to the Upper Palaeolithic, clearly this matter does require further investigation; and it is hoped that the remarks made here about cave and open-air sites may be helpful in this regard.

NORTHWESTERN EUROPE

Commonly the Szeletian is regarded as being confined to the eastern part of Central Europe (Valoch 1986a), but in discussion it is frequently linked to the Altmühlian which is characteristic of the western part of the area (Allsworth-Jones 1986: 7-8, 66-73). Some of the sites classified as 'Altmühlian' or 'Jerzmanowician' have in turn been linked to the early Upper Palaeolithic in Belgium and Britain (Allsworth-Jones 1986: 19-20, 181-8). Northwestern Europe, therefore, is not irrelevant for our purposes. Recently some new finds have been made and a number of additional 14C dates have been obtained, but for the most part attention has been focused on the reinterpretation of the old finds. M. Otte has over the last few years done much to clarify our understanding of the position in Belgium. He, at first, identified Spy and Goyet as the only early Upper Palaeolithic sites with which leafpoints could be associated and he hesitated over whether to regard these artifacts as an integral part of the Aurignacian or as evidence of a preceding and separate occupation. He performed a very useful service by drawing a clear distinction between the unifacial leafpoints occurring at Spy and Goyet and the retouched pointed blades which are characteristic of the Upper Périgordian at Maisières (Allsworth-Jones 1986: 30-1, 181-2). J.B. Campbell, in his re-evaluation of the British Early Upper Palaeolithic, at first treated it as a single unit, but he subsequently suggested a three-fold succession which does accord better with the continental evidence: a 'Lincombian' occupation (characterised essentially by leafpoints comparable to those from Ranis and Nietoperzowa) succeeded by the Aurignacian and then by the "Maisièrian" (characterised principally by tanged points of Font-Robert type). Otte has now provided a new synthesis, which differs in some respects from his earlier account, and which seeks to demonstrate continuity between Middle and Upper Palaeolithic in the area (Otte 1983, 1984, 1985a).

Otte, like Campbell, now supposes that the leafpoints of the early Upper Palaeolithic can be definitely separated from and are antecedent to the Aurignacian, which he regards as appearing from outside in a relatively advanced form. As he explains (Otte 1985a: 7-8) he was particularly influenced in coming to this view by the posthumous publication of Hülle's finds from Ranis (a site which McBurney in his day also regarded as providing the "most reliable continental clue" to the British early Upper Palaeolithic) and also by the suggestion made by J.K. and S.K. Kozlowski that there was such an entity as the "Ranis-Mauern culture" (cf. Allsworth-Jones 1976). But Otte now seeks to carry the argument much further by establishing links both backwards and forwards in time from the early Upper Palaeolithic. He suggests that there was a definite tradition ("la tradition des pointes foliacées", "la tradition des outils appointés par retouches plates") which can be traced back to the Middle Palaeolithic, e.g. at the Grotte du Docteur and Couvin, and forward to the Upper Périgordian at Maisières (Otte 1984: 161). He suggests further (again echoing in a new form an idea previously put forward by McBurney) that the 'Maisièrian' may have had a role to play in the origin of the French Solutrian; but even he has to admit that, in view of the possibilities of convergence and the lack of connecting links, this is a very frail hypothesis (Otte 1985a: 19). We may note in passing that J.K. Kozlowski has also put forward a slightly different version of what he calls the "Jerzmanowice-Ranis-Lincombian complex", which, he believes, stretched across the entire North European plain and may have lain at the origin of a migration towards the east in post-Denekamp times (J.K. Kozlowski 1983: this article contains in my opinion various disputable statements concerning the description and chronological placement of the artifacts at Mamutowa, Koziarnia and Zwierzyniec).

We are however mainly concerned with Otte's transition model for Middle to Upper Palaeolithic in the area. Points of controversy concern (1) the supposed Middle Palaeolithic predecessors of the early Upper Palaeolithic; (2) the nature of the early Upper Palaeolithic itself; and (3) the possibility of a stable 'leafpoint tradition' continuing later into the Pleistocene, particularly at Maisières and analogous sites in Britain.

1. Two sites belonging to the Middle Palaeolithic in Belgium are singled out by Otte as being of particular importance, the Grotte du Docteur and Couvin. The industry from the Grotte du Docteur was originally classified by Mme. Ulrix-Closset as a "Moustérien à retouche bifaciale"; she described Couvin as "une industrie qui, chronologiquement, doit appartenir au début du Paléolithique supérieur, mais qui conserve des traditions du Paléolithique moyen", and she considered it to be generally comparable to the Altmühlian (Ulrix-Closset, personal communication). Otte states a general model of progression from one to the other and thus to the Upper Palaeolithic as such: "Dès le Paléolithique moyen de nos régions, on voit apparaître en effet la pratique d'aménagement par retouches plates bifaces... C'est peut-être de ce phylum que dérivent des industries techniquement plus évoluées et où les pièces à retouches bifaces sont en partie façonnées sur des supports laminaires" (Otte 1984: 160). It is my contention that this generalised language conceals the differences between the sites and attaches too much 'evolutionary' significance to a particular technological device (flat bifacial retouch) which may not warrant it.

So far as the Grotte du Docteur is concerned, it is clear that this is in fact a Micoquian site, as Ulrix-Closset herself recognised by comparing it to Klausennische and Schambach (cf. Allsworth-Jones 1986: 48, 57-8).

The situation at Couvin is more complex. Tracing the history of the site, Cattelain and Otte (1985) explain that the upper stage of the cave was excavated by Lohest and Braconnier in 1887-8, whereas parts of the terrace were excavated by Maillieux and de Loë on behalf of the Musées Royaux d'Art et d'Histoire in 1905. It appears that the material excavated by Lohest and Braconnier cannot be traced, but that from the excavations of 1905 is lodged at the Musée du Cinquantenaire in Brussels. Otte has described the old finds as technologically intermediate or at the boundary between Middle and Upper Palaeolithic (Cattelain and Otte 1985: 125; cf. Otte 1983: 311). New excavations on the terrace in 1984 succeeded in locating intact Pleistocene deposits (although they seem to have been secondarily transported, perhaps from the cave) and an industry similar to that discovered earlier (Cattelain and Otte 1985: Figure 4). The finds are said to include four leafpoints (although one of the illustrated pieces looks more like a biface knife) as well as retouched flakes. The new discoveries at Couvin are obviously very significant (particularly since they also include a human milk tooth) and a detailed publication of all the material is of course needed; but so far as the old finds are concerned (judging from those I was able to study personally at the Musée du Cinquantenaire and those illustrated by Ulrix-Closset) it seems to me that this is an overwhelmingly Middle Palaeolithic industry, with a marked predominance of side-scrapers over all other tool forms. This, together with the presence of flake-blades and the use of plano-convex retouch, makes a comparison between this site and Mauern entirely apposite; but there are not in my opinion any convincing grounds for regarding the industry as transitional to the Upper Palaeolithic. This assessment is consistent with the new dating evidence which we now have. Faunally, according to J.M. Cordy, the Pleistocene layer at the site corresponds to the Les Cottés interstadial. A previous ^{14}C date obtained on bone from the 1905 excavations (Lv-720: 25 800 ± 770 BP) was obviously at variance with this; but there is now a new date of 46 820 ± 3290 BP (Otte and Ulrix-Closset, personal communication) which is much more consonant with the rest of the evidence – although in fact too early for an interval which is regarded as the equivalent of the Hengelo interstadial.

The external affinities suggested for Couvin bring us on to a consideration of the Altmühlian itself, particularly Mauern. Can it be lumped together with Ranis in a single "Ranis-Mauern culture" as J. K. and S. K. Kozlowski suggested? I think not (Allsworth-Jones 1986: 66-73). There are considerable similarities, both technological and typological, between layer G (Mousterian) and layer F (Altmühlian) at Mauern, and both are overwhelmingly Middle Palaeolithic in nature. The Ranis 2 industry presents a contrast, in that of the 60 tools only four can be classified as side-scrapers, there are no handaxes or biface knives, and although there are no cores the industry itself creates a strongly lamellar impression. Of the 53 leafpoints, 25 are bifacial and 28 are unifacial, the bifacial examples being much less standardised than those at Mauern. Consequently I agree with the view expressed in Hülle's monograph that the affinities of this industry lie more with Nietoperzowa than with Mauern. I am at a loss to understand Otte's description of the Ranis I industry, which he suggests is equivalent to layer F at Mauern and from which he derives the Ranis 2 industry (Otte 1985a: 9-10), since the layer in question contains only seven tools and two cores and these may represent (according to Hülle) traces of two different industries attributed respectively to a Micoquian of Klausennische type and a Mousterian comparable to Königsaue B.

Apart from Couvin, there are some other new indications of late Middle Palaeolithic industries which may be comparable to the Altmühlian as represented at Mauern. In Germany, two open-air sites at Zeitlarn I and II, near Regensburg, have quite surprisingly been attributed to the Szeletian (Schönweiss and Werner 1986; cf. Valoch 1986a). Both unfortunately are devoid of stratigraphy and consist of surface finds only, although they are considered to have originated from a reddish-brown sandy loam found in the vicinity above sandstone bedrock. The major components of this industry are said to consist of bifacial leafpoints, side-scrapers, and end-scrapers. Judging by the illustrations, the leafpoints are very similar to those which occur in the Altmühlian. The industry is attributed to the Szeletian mainly on account of the end-scrapers, which as Schönweiss and Werner point out do not occur in layer F at Mauern. At least one of the illustrated end-scrapers however could belong to the category of '*grochakis*' or small round retouched flakes as known in the Micoquian at Schambach (Schönweiss and Werner 1986: Figure 3.8), and some of the leafpoints look like biface knives (Schönweiss and Werner 1986: Figure 2.7 and 13), which suggests quite different affinities for this collection. Schönweiss and Werner admit that in comparison with the Szeletian there are very few blade tools and that this does militate against the comparison they have made (Schönweiss and Werner 1986: 8). In the circumstances, a ranking of these sites with the Altmühlian rather than the Szeletian looks likely, although with surface finds one can never be sure.

In Britain, so far, only one stratified site seemed to have produced material comparable to the Altmühlian at Mauern, that of Soldier's Hole, excavated by Parry in the 1920s (Allsworth-Jones 1986: 185-6). There are now four new radiocarbon dates for Parry's layer 4, of which three are judged to be acceptable (Gowlett *et al.* 1986b): OxA-691: >34 500 (spit 12); OxA-692: 29 300 ± 1100 BP (spit 13); OxA-693: >35 000 (spit 14). The date of c. 29 300 BP is comparable to the majority of reliable dates from the British Early Upper Palaeolithic, but the fact that the other two are minimal estimations only suggests that the real age of this layer is older than that. Another site which may possibly turn out to be comparable is that of Picken's Hole, about 6 kilometres to the west (Apsimon 1986). Apsimon thinks that the most likely analogy for the industry in unit 3 at this site exists at the Wookey Hole Hyaena Den, where Boyd Dawkins' excavations in the 19th century produced a number of bifacially retouched 'ovoids' which may reasonably be regarded as belonging to a Mousterian of Acheulian Tradition. Alternatively, may not the one distinctive trimming flake from Picken's Hole rather relate to bifacial leafpoints such as we have at Soldier's Hole? The evidence is too scanty to be sure, although the ^{14}C dates now available from unit 3 seem to be too recent for the Mousterian: BM-2117: 27 540 ± 2600 and BM-654: 34 365^{+2600}_{-1900} BP (Apsimon 1986).

2. With regard to the early Upper Palaeolithic as such, the supposed second stage in the development of Otte's 'leafpoint tradition', it is difficult to reach a satisfactory conclusion because so many of the finds were made a long time ago and their stratigraphic circumstances can probably now never be fully determined. Otte himself admits that at Spy there is no objective way in which the leafpoints can be separated out from the other components in de Puydt and Lohest's "deuxième niveau ossifère" which contained both Aurignacian tools and some of Mousterian type, and the same uncertainty exists at Goyet. The Aurignacian itself is said by Otte to have commenced in the cold phase before the Denekamp (Arcy) interstadial and to have continued right up to the equivalent of the Tursac interstadial (Otte 1984: Figure 57); the late phase was already dated at the Trou du Renard at 24 530 ± 470 BP, and another late site has now apparently been found at Sprimont (La Grotte de la Troweye Rotche) (Otte 1983).

A disentangling of the various elements in the early Upper Palaeolithic of Britain is even more difficult (Allsworth-Jones 1986: 183-7). Aurignacian components are predominant, particularly at the two largest sites of Kent's Cavern and Paviland, and it is difficult to agree with Campbell's revised idea that the first site may be taken as typical of the 'Lincombian' and the second as typical of the Aurignacian. The apparent association of leafpoints and Aurignacian artifacts is repeated at smaller sites, including in particular Ffynnon Beuno (Allsworth-Jones 1986: Figure 50, 1 and 3). In these cases it is hard to escape the impression that Aurignacian elements and leafpoints really did belong together. For Paviland we have two satisfactory new ^{14}C dates, and two which are not so satisfactory. The two dates OxA-365: 29 600 ± 1900 and OxA-366: 28 000 ± 1700 BP are in agreement with the date previously obtained for the main occupation horizon of 27 600 ± 1300 BP. As Gowlett and his colleagues comment, the dates demonstrate "human presence in the cave at a time consistent with the presence of earlier Upper Palaeolithic artifacts, and contemporary with a phase of the Aurignacian in France" (Gowlett *et al.* 1986a). On the other hand, OxA-140: 38 800^{+8000}_{-4000} BP "shows that faunal remains have been deposited in the cave from mid-Devensian times, but does not prove human presence". There is no new date for the 'Red Lady' of Paviland to supplement the existing one of 18 460 ± 340 BP, which, as Stringer (1986) comments, is " difficult to relate ... to the early Upper Palaeolithic occupation of the site and more generally to patterns of human settlement in Western Europe". Sollas in 1912 found additional fossil hominid evidence recognised as Paviland 2, but this has produced an unexpectedly young date (OxA-681: 7190 ± 80 BP) which, as Stringer says, further complicates the picture, but adds nothing to our knowledge of the Palaeolithic occupation of the cave. Three hominid specimens found by Balch at

Badger Hole have likewise been shown not to belong to the Palaeolithic (Gowlett *et al.* 1986b; Stringer 1986).

The predominantly Aurignacian character of the British early Upper Palaeolithic, which seems so obvious in the southwest, never appeared however to extend to the sites at Creswell Crags (Nottinghamshire), which have now been exhaustively re-examined by Jenkinson (1984). As he says, the early Upper Palaeolithic occupation is confined to two sites, Pin Hole and Robin Hood's Cave, i.e. it is even more scanty than the Mousterian, one layer of which at Pin Hole has now been dated to 38 850 ± 2500 BP. Jenkinson (1984: Figure 38) has been able to clarify the situation at Pin Hole, thanks to his discovery of how Armstrong's recording system was supposed to work, and some rearrangement of the material in comparison with Campbell's inventory has resulted. Seventy-two artifacts are attributed to the early Upper Palaeolithic, 25 tools and 47 flakes and blades. The tools include a tanged point and two bifacial leafpoints, all of which were found in the upper part of the deposits, albeit at the front and rear of the cave respectively. The evidence is not conclusive, but one may suggest therefore that both formed part of a relatively late 'Maisièrian' ensemble. The remainder of the early Upper Palaeolithic assemblage is not particularly diagnostic. Practically all the artifacts from Robin Hood's cave come from the 19th century excavations of Mello and Dawkins and Laing, who between them virtually cleared out the interior of the cave system (Jenkinson 1984: Figure 14). In the circumstances, their attribution to any particular horizon must be somewhat problematical. Jenkinson considers that 41 artifacts can be confidently assigned to the early upper Palaeolithic, 11 tools and 30 flakes and blades, whereas Campbell gives a total of 35 tools. There are no tanged points and no diagnostic Aurignacian forms, but (according to Campbell's count) there are 8 unifacial and 2 bifacial leafpoints. The unifacial leafpoints are similar to those found elsewhere in the British early Upper Palaeolithic, but Campbell considers that at least one of the bifacial leafpoints may be attributed to the 'Maisièrian' rather than the 'Lincombian' (cf. Allsworth-Jones 1986: Figure 50.2). During Campbell's excavations, a date of 28 500^{+1600}_{-1300} BP was obtained at Robin Hood's Cave. Three further dates have now been obtained on Campbell's material, but they do not add substantially to our knowledge of the site, nor do they increase our confidence in the existing date (Gowlett *et al.* 1986a): OxA-380: 4250 ± 75; OxA-199: >36,000; OxA-362: 3100 ± 80 BP. As Gowlett and his colleagues comment, "the dates show that the deposits had been disturbed or penetrated by later charcoal... Dating of talus contexts demonstrates that they are particularly vulnerable to mixing". The importance of Robin Hood's Cave seemed to have been enhanced by the fact that hominid remains were found within it, first by Laing in 1888 and then by amateurs in 1974. It has since been discovered that the remains are in fact quite recent, with a direct ^{14}C date (OxA-736) of 2020 ± 80 BP. As Gowlett *et al.* (1986b) comment "in several cases direct dating has shown putatively Palaeolithic human remains to be intrusive from later periods".

3. Mention of Pin Hole and Robin Hood's Cave brings us to a consideration of the third aspect, the supposed continuation of the 'leafpoint tradition' into the later Pleistocene at Maisières and elsewhere. It seems to me that this constitutes the weakest part of Otte's hypothesis. He himself has convincingly emphasised the distinction between the unifacial leafpoints occurring at Spy and Goyet and the retouched pointed blades which are characteristic of Maisières. As he also admits, his model requires that the 'leafpoint tradition' should somehow carry on undiminished through the 'intrusive' Aurignacian occupation, only to emerge in another form later. To achieve this it would be necessary to bridge quite a considerable chronological gap (Otte 1984: Figure 57). The occupation horizon at Maisières is dated to 27 965 ± 260 BP, whereas other sites, including Spy, Trou Magrite, and L'Hermitage at Huccorgne are much later than this (Otte 1985b: Figure 4). As we have seen, it is very difficult to disentangle the British evidence, but the few characteristic tanged points at Kent's Cavern, Paviland, and Pin Hole (as well as at open-air sites such as Forty Acres Pit and Bramford Road Pit) do probably correspond to the 'Maisièrian' and therefore follow the Aurignacian-dominated early stage of the Upper Palaeolithic. The very interesting collection of artifacts from Pulborough remains to be fully published, but according to the selection of the material which I have seen, as well as the full corpus of drawings made by R.M. Jacobi, and following discussions with him, I think that this material too is most analogous to Maisières (and not, *pace* Otte, to the early stage of the Upper Palaeolithic). (I am grateful to Roger Jacobi for the opportunity to study and comment on this material). In considering the origin of assemblages similar to Maisières, Otte suggests (1985a: 18) that Ranis may once again have had a crucial role to play, and that the Ranis 3 industry, while leading on from Ranis 2, may also be considered as a forerunner of the Gravettian. This is quite surprising and, I think, unacceptable; in common with most other authors I take the view that the Ranis 3 industry – while certainly having some idiosyncrasies – is broadly comparable to the Aurignacian (Allsworth-Jones 1986: 68-70). I also think (unlike Otte 1985a: 19) that the occurrence of retouched pointed blades in the Central European Gravettian can be perfectly well explained by convergence and that it is quite unnecessary to invoke any 'cultural tradition' to account for it (Allsworth-Jones 1986: 31-2).

The same mechanism in my opinion accounts for many of the phenomena discussed by Otte under the rubric of a 'leafpoint tradition' which is supposed to extend all the way from the Micoquian to the Solutrean. The links proposed between the various entities are in many cases no more than the thinnest of threads, and the language employed has debatable implications, as Hole and Vencl pointed out some years ago in another context (cf. Allsworth-Jones 1986: 24). Now, as then, one may have an uncomfortable feeling that the argument as presented gives a "spurious life to stone tools", and one should remember that leafpoints in particular are after all implements, not the "materialisation of an abstract idea invested with an inherent will to develop". In my opinion, the evidence from Northwestern Europe is much more episodic and disconnected than Otte suggests. There are undoubted links to Central Europe, but it is just because of these links that I prefer to interpret the evidence for the transition from Middle to Upper Palaeolithic in the area not as a continuous locally unbroken chain but in terms of an acculturation model which implies an input from outside.

SOUTHEASTERN EUROPE

In this area of Europe, considerable advances have been made in the last few years; some previously excavated sites have been re-examined and their inventories published in full; but above all large numbers of new sites have been discovered and, for the first time, a significant body of ^{14}C dates is available which challenges some of the 'traditional' ways of ordering archaeological material in the area. The area is of relevance to us because, rightly or wrongly, several sites have in the past been linked to the Szeletian or regarded as possible Szeletian forerunners (Allsworth-Jones 1986: 18, 54-6, 73-82). The presence of leafpoints in various Middle and Upper Palaeolithic contexts is an undoubted fact; it is their interpretation which is still a subject of debate.

In Bulgaria, full published information is now available concerning the excavations at Bacho Kiro (Kozlowski 1982) and at Samuilitsa (Sirakov 1983). So far as Samuilitsa is concerned, I concluded (Allsworth-Jones 1986: 74) that the great bulk of the material consists of a "homogeneous Levallois-Mousterian which cannot be appropriately labelled Szeletian at all". Dzambazov did at first call it that, although it appears that even he has now abandoned this terminology and, for him, Starosel'ye provides the closest analogy for what he now prefers to call the "Muselievo culture" (Dzambazov 1977). Most of the material from Samuilitsa 2 has recently been published in detail by Sirakov, and his figures (slightly amended and with recalculated indices) provide the basis for those given in Table 7.9. His account is very valuable in several respects, and his conclusion (Sirakov 1983: 87) that the assemblages in this cave are generally speaking homogeneous, corresponding almost entirely to a Typical Mousterian of Levallois facies, is similar to that of this author. But there are one or two differences to which attention should be drawn.

The total assemblage studied by Sirakov comes to 2893 pieces (not including 9 leafpoints which are treated separately by him) which occur in ten stratigraphic levels bearing a given relationship (but not an entirely clear one) to six different layers in the cave. The ten levels are grouped into four units and two series, an upper series (levels 1 and 2, 3 and 4) and a lower series (levels 5 and 6, and 7-10). This compares tolerably well with Dzambazov's succession of layers numbered (M) to (A) which I also grouped into four separate units. A great virtue of Sirakov's study is that he has been able to incorporate full details of the unretouched Levallois flakes, blades, and points, which I was not able to do; but certain other categories have been left out, which means that his overall appraisal of the industry is less than complete. In particular, the Upper Palaeolithic elements which occur at the top of the succession have been deliberately excluded, and, for reasons which are not explained, the cores were also excluded from the study (Sirakov 1983: 10-12, 43-5). The latter omission is especially regrettable in a work which emphasises the technological aspect. Sirakov estimates that there may have been up to 100 cores; I counted 82, of which I classified 43 as disc cores and 17 as Levallois blade or point cores. Sirakov's total tool count comes to 857 pieces (to which we have to add 9 leafpoints) compared to mine of 640 pieces, which would be increased to 779 if the extra unretouched Levallois elements were added in, so the overall agreement is tolerably close. Of Sirakov's 9 leafpoints, 6 have decipherable markings, but he regards them all as having come from levels 5 and 6 at the top of the lower series. By contrast, I list 12 leafpoints, 2 of which appear not to be illustrated by Sirakov (Allsworth-Jones 1986: Figure 8, 2 and 3); of these, the majority with decipherable markings were located in Dzambazov's layers (H) and (G), at a depth of approximately 100-130 cm, which would agree with Sirakov's account. But there are indications that others were located higher up. Hence while I agree with Sirakov that the lower levels at the site did not contain leafpoints, I think that they continued to form part of the industry for longer than he allows. This would agree with Dzambazov's own account. The total number of side-scrapers listed by Sirakov is also much less than in my own inventory. But having pointed out these differences, one should emphasise again that there is no disagreement between us regarding the overall balance and significance of this industry. The site can still be regarded as an essential component of the Levallois-Mousterian in Southeastern Europe.

Also regarded as part of this province, but less certainly attached to it, are four open-air sites in northern Bosnia discovered and excavated by Basler: Londža, Kamen, Crkvina, and Visoko Brdo. All have both Middle and Upper Palaeolithic occupations, and at all except Londža a few leafpoints have been found, in inadequate stratigraphic contexts. Gábori suggested that the "late" phase of the Middle Palaeolithic in each case might be regarded as transitional to the Upper Palaeolithic (Allsworth-Jones 1986: 78-9, 177). It has been difficult to appraise these sites properly both because of the lack of direct dating evidence and because of the seemingly rather enigmatic nature of their deposits. In both respects we are now much the wiser thanks to the work of A. Montet-White and her colleagues over the last ten years (Montet-White *et al.* 1986). This work has been carried out above all at Kadar, on the banks of the Sava, but also at Zobište and Luščić higher up and in the vicinity of the other four sites mentioned. Middle Palaeolithic occupations were discovered at Kadar and Zobište, both held to be comparable to those at Londža and Kamen, whereas Luščvić produced an Upper Palaeolithic of Aurignacian type in a stratigraphic sequence similar to that at Visoko Brdo. Sedimentological studies were carried out at all the sites, but only Kadar had pollen, enough to construct a pollen diagram interpreted by reference to that from

Table 7.9 Artefact totals for Samuilitsa 2 (after Sirakov, 1983)

Levels	1 & 2	3 & 4	5 & 6	7-10	Totals	
Endscrapers	12	2	12	1	27	
Burins	8	4	3	3	18	
Awls	8	5	5	3	21	
Truncated flake/blades	—	2	1	1	4	
Tanged points	2	—	—	—	2	
Bifacial leafpoints	—	—	9	—	9	
Levallois:						
flakes	141	31	39	19	230	} 303
blades	20	9	20	8	57	
points	6	3	6	1	16	
retouched points	3	3	6	2	14	
Pseudo-levallois points	1	—	—	1	1	
Mousterian points	—	2	3	—	5	
Sidescrapers	71	22	63	12	168	
Notched pieces	9	4	3	3	19	
Denticulates	26	9	23	13	71	
"abrupt alternate retouch", etc.	63	8	26	13	110	
Miscellaneous	62	11	18	2	93	
Total Tools	432	115	237	82	866	
Unretouched Levallois as % of tool total	38.7	37.4	27.4	34.1	35.0	
Non-Levallois un-retouched flake/blades	1195	211	488	142	2036	

Tenaghi Philippon. As is general in northern Bosnia, there were practically no faunal remains. The sediments in most cases witnessed a change from clayey to more silty layers with occasional sandy horizons, indicating a shift through time from humid colluvial to drier aeolian conditions. There were numerous stratigraphic breaks. The climatic/chronological interpretation is backed up by a series of thermoluminescence dates on burnt cores from all three sites. Thus, the Mousterian in layer 3 at Kadar (separated by a marked erosional episode from the Epigravettian in layer 2) is equated with the Elevtheroupolis interstadial (= Odderade), and the same is held to be true of layer 3 at Zobište. There, however, the Mousterian spans layers 2, 3, and 5, of which 2 is equated with a rigorous episode succeeding the Elevtheroupolis and preceding the Heraklitsa interstadial (= Moershoofd) whereas 5 is equated with the Doxaton interstadial (= Amersfoort). There is again a sharp erosional contrast between layer 2 and layer 1, which contained a disturbed Epigravettian industry. The Mousterian in northern Bosnia on this evidence therefore extends from Amersfoort via Brørup and Odderade to pre-Moershoofd times. Sample 16/84Z1817 from layer 5 at Zobište produced TL dates of 97 500 ± 7000 and 85 500 ± 8500 BP. Layers 2 and 3 have yet to be dated, but Montet-White *et al.* comment that an age of *c.* 85 500 years for layer 5 agrees well with the other information available from the site. The Aurignacian at Luščić is situated at the base of layer III, which is equated with an episode of aeolian deposition succeeding the Krinides interstadial (= Denekamp). Sample 16/84L1894 produced TL dates of 29 900 ± 2800, 28 600 ± 1400 and 27 000 ± 3000 BP, of which that at *c.* 28 600 BP is regarded as the best estimate. These dates are of course perfectly comparable to those obtained for the Aurignacian elsewhere.

With regard to the nature of the industries, Montet-White *et al.* (1986) emphasise the importance of raw material factors, particularly the common reliance on radiolarite pebbles which helps to create an impression of homogeneity in the Palaeolithic succession of northern Bosnia. At Zobište it seems that the working of the pebbles in the Mousterian layers was deliberately organised so as to produce two types of blanks: naturally backed knives and Levallois flakes. Most Levallois flakes were not retouched, but there is a well-defined category of obliquely truncated Levallois flakes allied to Levallois points (Montet-White *et al.* 1986: Figure 27: 1, 2, 4, 5). The most common retouched tools are side-scrapers, whether made on naturally backed knives or not (Montet-White *et al.* 1986: Figure 27: 6-10). Montet-White and her colleagues emphasise that the northern Bosnian Middle Palaeolithic industries, with their important Levallois component, are however quite distinct from the side-scraper-dominated industries of Krapina, Veternica or Vindija, and constitute a true 'regional variant'. No leafpoints were found either at Zobište or at Luščvić. The total assemblage at Luščvić consists of 1406 pieces, but these include only 15 cores and 17 tools. The industry is classified as a 'final' phase of the Aurignacian largely because of the presence of a series of small carinate end-scrapers.

In all this we are a far cry from the picture hitherto presented for the Palaeolithic of northern Bosnia, and in particular from Gábori's suggestion that a "late" phase of the Middle Palaeolithic may have been transitional to the Upper Palaeolithic. The dates now available show that the two are far apart; the Mousterian cannot be regarded as in any way "late" and, as Montet-White *et al.* (1986: 87) comment, the early Upper Palaeolithic is in fact poorly represented in the region. According to their own assessment, the industries previously classified by Basler as Aurignacian at Londža and Visoko Brdo are Gravettian, and as we have seen the Aurignacian at Luščvić belongs not to an early but to a 'final' phase. The Upper Palaeolithic artifacts from Kamen are still regarded as Aurignacian in character, but as Montet-White *et al.* (1986: 34) point out, the site was disturbed during the Bronze Age and today its contents can only be ordered typologically. No mention is made of Crkvina. We do not know any more than before about the real provenance of the leafpoints, but it seems most likely that they in fact formed an integral part of the Middle

Palaeolithic, which we are still justified in regarding as broadly comparable to the more markedly Levallois-Mousterian industries of Southeastern Europe.

New developments as far as Romania is concerned relate above all to the emerging framework of ^{14}C dates, which imply or compel radical reconsideration of existing interpretative schemes, especially in so far as they concern the transition from Middle to Upper Palaeolithic (Honea 1984, 1986a, 1986b). In my book I already drew attention to some of the weaknesses and contradictions apparent in the 'traditional' Romanian approach to this subject, which essentially goes back to the teachings of the later Professor C.S. Nicolaescu-Plopşor. Thus I criticised his idea about a so-called 'Preszeletian Mousterian' in the caves of the southern Carpathians, and, in part following arguments advanced by Vértes, I suggested there was no need to suppose that the Middle Palaeolithic here was particularly 'late' nor did it possess any signs of transition to the Upper Palaeolithic (Allsworth-Jones 1986: 79-82). While praising the work done by M. Cârciumaru in producing pollen diagrams for many individual Romanian sites, I considered that his general interpretative scheme (with suggested parallels to named interstadial episodes in Western Europe) was unsatisfactory, particularly inasfar as it concerned Ripiceni-Izvor: it had "drastically 'rejuvenated' the finds and should be rejected" (Allsworth-Jones 1986: 42-3). In my opinion, developments since then have reinforced this point of view.

In 1980, Cârciumaru published a book summarising and generalising his views on the climatic, chronological, and archaeological succession in Romania from the last interglacial period to the end of the Pleistocene. This brought together the material previously contained in scattered articles, but it repeated unchanged the main lines of his argument. As before, a Nandru interstadial complex was held to continue all the way from the beginning of Amersfoort to the end of Hengelo, to be followed by an Ohaba interstadial complex embracing the equivalents of Arcy, Stillfried B, and Tursac. The Tursac equivalent corresponded to a unit named Herculane I, but whereas Laugerie-Lascaux previously had been held to correspond to Herculane II, an extra Româneşti oscillation was now added to account for the latter part of this period. As before, the major archaeological events, in comparison with the rest of Europe, were said to have happened very late. According to Cârciumaru (1980: 264, 267-8) it was during the Ohaba interstadial that there took place the true passage from Middle to Upper Palaeolithic, a transitional period lasting for some 4000 years, during which Middle and Upper Palaeolithic communities continued to coexist. Cârciumaru allowed however (1980: 263) that the first manifestations of the Upper Palaeolithic appeared already in the glacial stage preceding Ohaba A in the Ceahlau basin – as for example at Bistricioara-Lutarie and Ceahlau-Dîrţu (cf. Bitiri and Cârciumaru 1978: Table 2). He also stated (1980: 263, 267) that the cave of Nandru-Spurcata was occupied at this time, and mentioned that whereas the occupation had initially been attributed to the 'Szeletian culture' it was now regarded as a 'Mousterian facies'.

An equally important statement concerning the Palaeolithic succession in Romania was provided in 1983 by F. Mogoşanu. In it, he both restated some Romanian 'traditional' positions while also providing trenchant and perceptive criticisms of others. Like Cârciumaru, he accepted that the Mousterian in Romania continued to a late date and that it overlapped with the early, and even with the late, Upper Palaeolithic. This applied above all to the caves of the southern Carpathians, which now, interestingly enough, are said to have their open-air analogues in the extreme west of the country, near Arad, at Cladova, Conop and Zabrani, compared in their situation and inventory with Érd (Mogoşanu 1983: 36-7). By contrast, Mogoşanu adopted a very critical attitude towards two sets of occurences which have been held to represent important points of transition between Middle and Upper Palaeolithic, the Oaş territory in northwestern Romania and Mitoc-Valea Izvorului in the northeast of the country. If, methodologically, his remarks are valid, they have quite important implications for Romanian archaeology as a whole.

In the Oaş territory, Bitiri stated (1972) that Mousterian levels were stratified beneath Aurignacian deposits at Boineşti and at Remetea-Somoş I and II. I have already stated my reasons for regarding these occurrences as Micoquian (Allsworth-Jones 1986: 55), but Bitiri (1972: 131) classified them as "late Mousterian with leafpoints". She claimed (1972: 131, 133) that this late Mousterian was contemporary with the first phase of the Upper Palaeolithic and that it was in fact transitional to it. Mogoşanu however emphasised (1983: 40) that one should be cautious in interpreting these results because of the stratigraphic circumstances at the sites: all are on hill tops or hill slopes which have been subject to aeolian deflation, leaving the successive Middle and Upper Palaeolithic deposits directly superimposed above each other. Thus the 'stratigraphy' had to be established principally on the basis of typology and raw material with, in effect, a seriation moving from local to imported raw materials and from 'archaic' to 'advanced' tool types. Hence it would be risky to place too much reliance on any supposed transition at these three sites (cf. Honea 1986a).

With regard to Mitoc-Valea Izvorului, Mogoşanu commented (1983: 41) that here there were "des éléments d'incertitude, plus grâves même que dans les cas précédents". The commune of Mitoc is situated on the middle course of the river Pruth 20 kilometres north of Ripiceni-Izvor. It contains a number of Palaeolithic sites. Mitoc-Valea Izvorului was first excavated by Bitiri in 1963, and on the basis of her brief report it seemed reasonable to regard the finds as essentially Micoquian, although an 'Upper Palaeolithic' blade element was also said to be present (Allsworth-Jones 1986: 55). Further excavations were carried out in 1977, as a result of which we have a much fuller and rather different picture of the site (Bitiri and Cârciumaru 1978). The site is described as being on a dissected terrace and has produced a 3.2 metre thick profile, with pollen but without any faunal remains (Bitiri and Cârciumaru 1978: Figure 1). There is a stratigraphic discordance at a depth of about 1.2 metres, above which the deposits are attributed to the late glacial period. Below, on the basis of pollen analysis, there is evidence for an interstadial episode above basal deposits of a stadial character. The interstadial deposits are at a depth of about 1.2-2.4 metres and are attributed by Cârciumaru to the Ohaba interstadial. The archaeological material, treated together as a single occurence, begins at a depth of about 2 metres and continues up to a depth of about 8O cm, but the upper part (above the stratigraphic discordance) has been disturbed by krotovinas, hence it is agreed that no reliable upper delimitation can be made. On this basis, in terms of Cârciumaru's scheme, the archaeological occupation commenced sometime between Ohaba A and B and continued until about the Herculane II period (Bitiri and Cârciumaru 1978: Table 2). Hence, again in terms of his scheme, it would be contemporary both with the final Mousterian in the Carpathian caves and Ripiceni-Izvor and with the early Upper Palaeolithic in the Ceahlau basin. This chronological placement, as well as the nature of the industry recovered, no doubt encouraged Bitiri and Cârciumaru to suggest the creation of a special "facies of Mitoc type" whereby the Mousterian became transformed into "a specific industry of local incipient Upper Palaeolithic type" (Bitiri and Cârciumaru 1978: 178; this formulation explains why in their Figure 1 and Table 2 the archaeological occurrence is labelled simply "Upper Palaeolithic"). This claim has not unnaturally attracted attention in a wider European context (e.g. Valoch 1986a) and deserves closer examination.

According to Bitiri and Cârciumaru, the site produced a total of 6307 artifacts, the great majority of which were small splinters indicative of its role as a workshop exploiting the locally available flint. There were 1105 so-called 'typical' pieces, including 123 cores, 521 blades, and 461 flakes. The tool totals, compiled after indications provided by Bitiri and Cârciumaru, are given in Table 7.10, together with recalculated indices for the major tool groups. It is quite obvious from the figures that various distinct and contrasting elements are present at this site. Upper Palaeolithic type tools include burins, awls, and rather Aurignacian-like end-scrapers (Bitiri and Cârciumaru 1978: Figure 7: 1-5). The remainder of the tools are principally Middle Palaeolithic in character. They are dominated by denticulates and notched tools, but there are other components as well – notably unretouched Levallois flakes (Bitiri and Cârciumaru 1978: Figure 4: 1-3) and bifacial pieces of which unfortunately only two are illustrated, both broken (Bitiri and Cârciumaru 1978: Figure 8.1 and 8.2), so it is not possible to be sure whether they were leafpoints or biface knives. On the strength of the earlier published material from Mitoc-Valea Izvorului, both types are likely to have been present. In the light of the above, it is not surprising that the authors of the report sought analogies for this industry, not indeed in the Upper Palaeolithic, but in the final Mousterian at Ripiceni-Izvor and in the Denticulate Mousterian of Stinka I upper layer in Soviet Moldavia (cf. Amirkhanov *et al.* 1980: Figure 2).

The weakness of the interpretation advanced for this site, according to

Table 7.10. Tool totals and calculated indices for Mitoc-Valea Izvorului

	Totals	Major tool groups (%) (totals minus Levallois)
Endscrapers	31	12.70
(including carinate)	2	
Burins	11	4.51
Awls	7	
Truncated flake/blades	12	
Levallois flakes	58	
Points	5	
Sidescrapers	8	3.28
Notched pieces	35	} 65.57
Denticulates	125	
Bifaces	10	4.10
Total tools	302	
Unretouched Levallois as % of tool total	19.21	

Mogoşanu (1983: 41), lies in its insecure stratigraphic foundations. He points out that the apparent association of different types of tools (Upper and two variants of Middle Palaeolithic) is quite unusual. The excavation methods, he suggests, may have been quite correct, but did not take sufficient account of the geomorphological circumstances prevailing in the area; on the steep banks of the middle Pruth a great deal of erosion and redeposition took place throughout the Pleistocene,and this affected almost all the Palaeolithic sites in the area. In other words the excavators may have failed to take account of possible inclination of the deposits and mixing of the material due to natural causes. (Their difficulties will have been accentuated if in such circumstances they followed the normal practice of Nicolaescu-Plopşor's pupils of digging by arbitrary horizontal levels: cf. Honea 1986a). This suggests caution in interpreting the results for, as Mogoşanu (1983: 41) put it, "nous estimons que, pour un problème aussi important que le passage du Moustérien au Paléolithique supérieur en Roumanie, on ne saurait se prononcer en l'absence de preuves qui soient, sous tous les rapports, indubitables".

The new framework of ^{14}C dates for Romania challenges many other 'traditional' interpretations in that country, and in particular throws into doubt the supposed long continuation of the Mousterian and its overlapping with the early Upper Palaeolithic in so relatively recent a period as Arcy and Stillfried B. These results are due above all to the dating programme embarked on by K. Honea in collaboration with his Romanian colleagues (Honea 1984, 1986a, 1986b). The principal new age determinations of relevance to us are summarised in Table 7.11.

A number of points emerge clearly from the examination of these dates.

1. The antiquity of the Middle Palaeolithic at Ripiceni-Izvor is decisively confirmed. Mousterian level III previously had a ^{14}C date >36 950 BP (Bln-811) which was not unacceptable but can now be regarded as superseded by the two new finite dates. The archaeological material is comparable to the Levallois-Mousterian in the rest of Southeastern Europe, e.g. Samuilitsa 2 (Allsworth-Jones 1986: 73-4). Mousterian level IV also had a ^{14}C date of 28 780 ± 2000 BP (Bln-810) which I considered to be quite unacceptable (Allsworth-Jones 1986: 43); as Honea says (1984: 32) it should now be "rejected with finality". The average age determination for level IV now comes to 43 700 BP. I have already stated my reasons for regarding Mousterian levels IV and V at Ripiceni-Izvor as belonging to the Micoquian (Allsworth-Jones 1986: 54-5, 64).

2. The antiquity of the Carpathian Mousterian can also be regarded as having been solidly established. The six dates obtained for the cave Cioarei de la Boroşteni are, as Honea says (1986a, 1986b), the longest continuous series yet available for this variant of the Middle Palaeolithic, with the very top of the sequence put at 37 750 ± 950 BP. The two dates from Ohaba Ponor relate to Mousterian level IIIa, the lowest unit in that level, and Honea suggests (1986a) that taking the two together "an older range" seems likely. It may be "essentially contemporaneous" with Mousterian level IV at Ripiceni-Izvor (Honea 1984: 36). The layer numbering presumably follows that established by Nicolaescu-Plopşor; layer III is particularly interesting both because of the comparisons which have been made between it and Subalyuk and because of the presence in it of three human phalanges of Neanderthal type (cf. Allsworth-Jones 1986: 80, 207). The date for the cave of Gura Cheii Rîşnov is considerably younger than the others, but note must be taken of its exact provenance which has been carefully explained by Honea: the dated material consists of cave bear bone, not associated with any artifacts, at the interface of the Mousterian level and the sterile unit above (Honea 1986a, personal communication). Hence, "the assay can be interpreted as only a maximum one for the overlying sterile unit and a minimum one for the underlying Mousterian unit; in other words, that unit remains undated". In the circumstances it seems regrettable that Cârciumaru still seeks to use this date to prove the long continuation of the Mousterian in the Carpathians (Cârciumaru 1985: 243 and Figure 2; this figure repeats in a hardly changed form the synchronisation scheme in Bitiri and Cârciumaru 1978: Table 2).

3. The new dates for the Aurignacian at Mitoc Malu Galben and in the Ceahlau basin are generally satisfactory and comparable to those obtained elsewhere. With regard to Mitoc Malu Galben, W.G. Mook comments (in Honea 1986a) that the date reported at >24 000 BP would at one standard deviation be in the range 30 000^{+6500}_{-4500} BP "but considerably older is not excluded". Honea comments that still earlier dates can be expected from Aurignacian levels I and II at the site. The date obtained at Ceahlau Dîrţu is rather imprecise, but the three earlier dates from Bistricioara-Lutarie II are perfectly satisfactory. The fourth date of 23 560^{+1180}_{-980} BP (GX-8845-G) obviously presents problems, since it is no more than burnt bone extracted

Table 7.11. Radiocarbon dates for Romanian sites (after Honea 1984, 1986a, 1986b)

Ripiceni-Izvor
Mousterian level III
GrN-11230 $46,400^{+4700}_{-2900}$ BP
GrN-11571 $45,000^{+1400}_{-1200}$ BP

Mousterian level IV
GrN-9208 $44,800^{+1300}_{-1100}$ BP

GrN-9207 $43,800^{+1100}_{-1000}$ BP
GrN-9209 $42,500^{+1300}_{-1100}$ BP
Mousterian level IV/V
GrN-9210 $40,200^{+1100}_{-1000}$ BP

Peştera Cioarei de la Boroşteni
Basal deposit
GrN-13004 > 45,000 BP
Mousterian level I, base
GrN-13003 > 50,000 BP
Sterile unit MI/MII, base
GrN-13002 $49,500^{+3200}_{-1100}$ BP
Mousterian level II, base
GrN-13001 $43,000^{+1300}_{-1100}$ BP
GrN-13000 > 46,000 BP
Mousterian level II, top
GrN-13005 37,750 ± 950 BP
Gura Cheii Rîşnov
Interface Mousterian/
Sterile unit above
GrN-11619 $29,700^{+1700}_{-1400}$ BP

Ohaba Ponor
Mousterian level IIIa
GrN-11618 $39,200^{+4500}_{-2900}$ BP
GrN-11617 > 41,000 BP

Mitoc Malu Galben
Aurignacian level III
GrN-12637 31,850 ± 800 BP
GrN-13007 > 24,000 BP
Aurignacian level IV
GrN-12636 28,910 ± 480 BP

Bistricioara-Lutărie II
Aurignacian level
GrN-11586 28,010 ± 170 BP
GrN-10529 27,350 ± 1300 BP
GX-8844 $27,350^{+2100}_{-1500}$ BP
GX-8845-G $23,560^{+1180}_{-980}$ BP

Ceahlau-Dîrtu
Aurignacian level
GX-9415 $25,450^{+4450}_{-2850}$ BP

Mitoc Malu Galben
Early Gravettian
GX-9418 26,700 ± 1040 BP
GrN-12635 27,150 ± 750 BP

from the charcoal sample GX-8844 which produced a date of $27\,350^{+2100}_{-1500}$ BP (Honea 1984: 28, personal communication). Hence it is anomalous and, as Honea says (1986b), "not scientifically acceptable". It is again unfortunate that Cârciumaru seems not to understand the situation, and uses this date among others to buttress his supposed synchronisation scheme (Cârciumaru 1985: 243-4). The two dates quoted for the Gravettian at Mitoc Malu Galben are the earliest reliable ones available, prior to the sorting out of the exact context of the others, having regard to the inclined stratigraphy of the site (Honea 1986a, 1986b). They are among the earliest for the Gravettian in Central and Eastern Europe and are quoted merely to show the speed with which the Gravettian succeeded the

Aurignacian in the area.

If in Romania the new developments concern above all the chronological framework provided by ^{14}C dates, in Soviet Moldavia it is the sheer profusion of new sites which is astounding. According to Borziyak (1983: 36) more than 250 such sites have been discovered since 1972, both in caves and in the open air, many multi-layered and/or covering large areas, with frequent preservation of faunal remains (cf. David 1983). The principal authors who have reported on these developments are I.A. Borziyak, G.V. Grigorieva and N.A. Ketraru. It seems to be generally agreed that the finds from Romanian and Soviet Moldavia form a continuum, a whole "historical-cultural region" (Rogachev and Anikovich 1984: 174-5) which Borziyak describes as the "Dniester-Carpathian" area. Situated as it is at the junction of the Balkans, the Russian Plain, and Transcarpathian Ukraine, and with such a plethora of sites, it is perhaps not surprising that in terms of current analyses the Palaeolithic succession in the area appears to present a "complex mosaic picture" (Borziyak 1983: 59). Obviously nothing less than monographic treatment could really do justice to it. All that can be done here is to indicate some of the main points of interest in the context of the transition from Middle to Upper Palaeolithic. I already drew attention to the apparent combination of leafpoints and Aurignacian forms in the early Upper Palaeolithic at Ceahlau-Cetațica, Ripiceni-Izvor, and Molodova V layer 10 (Allsworth-Jones 1986: 177-8) but it seems that these three sites are only part of a much wider pattern observable in this area. Borziyak has attempted to synthesise the available information in terms of a so-called three-fold path of gradual transition from the Mousterian to the Upper Palaeolithic and in terms of a four-stage grouping of the Upper Palaeolithic in the area (Amirkhanov *et al.* 1980; Borziyak 1983). He himself admits that these schemes are only provisional, and without personal inspection of the material it is difficult to come to an independent judgement about them. At least they provide a convenient frame of reference, which can serve as a basis for discussion, and in that spirit they are summarised below.

The first part of the argument rests on the fact that the Middle Palaeolithic of the area is not uniform. We have already noted the contrast between levels I-III and IV-V at Ripiceni-Izvor, attributed respectively to the Levallois-Mousterian and to the Micoquian. Borziyak refers to these two variants as "ordinary Mousterian of Levallois facies" and "bifacial Mousterian of Levallois facies". The former is represented at Molodova I and V, but in Borziyak's opinion it did not have any role in the origin of the Upper Palaeolithic (cf. Rogachev and Anikovich 1984: 174-5). Apart from Ripiceni-Izvor, the "bifacial Mousterian" variant has also been found at the cave of Buteshty (Amirkhanov *et al.* 198O: Figure 4; cf. indices in Sirakov 1983: 96-7). Borziyak suggests that industries such as Mitoc-Valea Izvorului and Ceahlau-Cetațica could be derived from this variant, which in tha light of the information given above may not seem too convincing in the case of Mitoc. More pertinent is Borziyak's description of the third Middle Palaeolithic variant in the area, the Denticulate

Mousterian, which has also been referred to above in connection with Mitoc. It is found in the cave of Stinka I lower and upper layers; the latter is characterised both by bifacial leafpoints and by end-scrapers some of which do seem to have an Aurignacian appearance (Amirkhanov *et al.* 1980: Figure 2: bifaces, 1-4; end-scrapers, 7, 8, 12, 15, 16, 18). It is also found in the cave of Buzduzhany I, where seven layers of Denticulate Mousterian were excavated by Ketraru in 1971-73 (Ketraru 1976). According to him, the Levallois technique can be discerned in all layers (as at Stinka I), and there were some bifaces in layers 5 and 6. According to the scheme favoured by Borziyak, the Denticulate Mousterian either gave rise directly to the Upper Palaeolithic (as at Klimautsy I), or Denticulate and 'bifacial' variants combined to do so (as at Brynzeny I layer 3 and Bobuleshty VI). These are said to constitute therefore his second and third paths of gradual transition to the Upper Palaeolithic.

The three sites mentioned, together with the cave of Chuntu, form the essential components of Borziyak's first Upper Palaeolithic group or stage. Both the cave of Brynzeny I and the open-air site of Bobuleshty VI were excavated by Ketraru; he at first suggested that they were related to the Szeletian, but he later changed his mind and proposed the creation of a separate "Brynzeny culture" instead (Ketraru 1973: 63, 65, 73). The cave of Chuntu, excavated by Borziyak and Ketraru in 1975 (Borziyak and Ketraru 1976), is also regarded as belonging to this culture. Borziyak accepts the idea that the Szeletian should not be regarded as extending beyond the Carpathians; he therefore agrees with the concept of the Brynzeny culture, which he regards as having existed parallel to the Szeletian as well as to the early Upper Palaeolithic on the Dniester. The illustrated material from Brynzeny I layer 3 (Amirkhanov *et al.* 1980: Figure 5: 12-20) includes bifacial leafpoints, backed blades, and Aurignacian style end-scrapers. (The site has also produced an abundance of faunal remains, 43 species according to A.I. David, among which horse and bison are predominant – an interesting contrast with Central Europe in itself: David 1983; Borziyak 1983: 39). The illustrated material from Bobuleshty VI (Amirkhanov *et al.* 1980: Figure 5: 1-11) includes the same elements: bifacial leafpoints, backed blades, and Aurignacian style end-scrapers. This is aside from the presence of some Levallois flake/blades and denticulates or notches which allows Borziyak to connect both sites to the local Middle Palaeolithic. The total inventory from layer 3 of Chuntu cave is said to consist of about 300 pieces; the tools include end-scrapers, backed blades, and a fragment of a bifacial leafpoint (Borziyak and Ketraru 1976). Borziyak contrasts with the three sites of the 'Brynzeny culture' the early Upper Palaeolithic inventory from Klimautsy I, which he links directly with the Denticulate Mousterian. Of the 525 tools, 252 according to Borziyak are denticulates or notches. But as before the illustrated material includes bifacial leafpoints and Aurignacian-style end-scrapers (Amirkhanov *et al.* 1980: Figure 3; bifaces, 1-4; end-scrapers, 9-19) and Borziyak specifically mentions that there are 5 backed blades. One therefore wonders whether the distinction made between Klimautsy I and the other sites attributed to the 'Brynzeny culture' really amounts to much. What is far more striking is the constant occurrence together of bifacial leafpoints, backed blades, and end-scrapers of Aurignacian type. It was precisely this conjunction to which I drew attention at Molodova V layer 10 (Allsworth-Jones 1986: 178), and I suggest therefore that we are essentially dealing with a single early Upper Palaeolithic entity in this area. With regard to the proposed links to the Middle Palaeolithic, one would like to be reassured on two points: (1) that the kind of factors operating at Mitoc-Valea Izvorului have been taken into account, i.e. that we are not dealing with mixed assemblages; and (2) that the proposed "typogenetic" connections are solidly based. Notches and denticulates for example are hardly enough to support such links, and bifacial pieces taken by themselves are notoriously unreliable, as the story of Szeleta and Subalyuk, among many others, shows.

The principal representative of Borziyak's second upper Palaeolithic group or stage is Gordineshty I, only 14 kilometres east of Mitoc-Valea Izvorului, and excavated by him in 1974-6 (Borziyak 1984). It is said to be characterised by blade cores (Anikovich 1983: Figure 2:9) and by a tool inventory in which end-scrapers are prominent; contrary to Borziyak's statement, these do include carinates (Anikovich 1983: Figure 2:4), and there are also some backed blades (Anikovich 1983: Figure 2:2). As before, these elements are accompanied by leafpoints, of which Borziyak distinguishes three different types: (1) points with rounded base (Anikovich 1983: Figure 2:6); (2) subtriangular points with concave base (Anikovich 1983: Figure 2: 8); and (3) points with partial bifacial work, like Jerzmanowice points. According to Borziyak (1983: 46) leafpoints with rounded bases like those from Gordineshty have been found at over 40 sites altogether in the region, including 25 sites along the Dniester, but many of these are surface occurrences. Despite the recurrence of the same three diagnostic types as before, Borziyak believes that the industry from Gordineshty is sufficiently distinctive to warrant the creation of a new taxonomic unit which he calls the "Pruth culture"; to this he attaches not only Gordineshty itself, but also Ceahlau-Cetațica and the three Aurignacian levels at Ripiceni-Izvor. Largely because of the leafpoints, and in contradistinction to his expressed view of the "Brynzeny culture", Borziyak also considers (1983: 52) that the Gordineshty complex may represent "one of the links in the further development of one or other Szeletian site", although he admits that the lack of geographical intermediaries makes this a very tenuous hypothesis.

We are not so much concerned here with Borziyak's groups 3 and 4, which he divides into several "cultural variants", because they relate primarily to the later stages of the Upper Palaeolithic. But there are two sites which deserve mention, Korpach-mys and Korpach, both open-air sites on the right bank of the river Rakovets, some 200-300 metres from each other, just above its junction with the Pruth. Korpach-mys was excavated by Borziyak in 1975-6 (Borziyak 1976, 1977); according to him (Borziyak 1982: 52-3) the cultural layer was situated in a fossil soil at a depth of 1.9-2.2 metres, and contained as particularly characteristic elements a series of artifacts reminiscent of "many-facetted burins/high end-scrapers/micro-bladelet cores", although no micro-bladelets as such were found at the site. Also found were two bifacial leafpoints, and two bone points the dimensions of which are given as 12.2 x 2.7 x 1 and 11.7 x 2.3 x 1.3 cm. These are the only bone points known from the area, just as those from Baia de Fier are the only ones known from the early Upper Palaeolithic of Romania, although there is another such occurrence from the cave of Molochnyi Kameń in Transcarpathian Ukraine, as described by Gladilin and Pashkevich (Allsworth-Jones 1986: Map 3). Largely on the basis of the bone points, Borziyak compares the site to the upper Aurignacian layer 8 at Istállóskö, observing quite correctly that this also contained a fragmentary bifacial leafpoint. The comparison is made only tentatively, and one wonders whether there is a need to mention any specific analogue in this context, because, on the basis of the description given, the stone tools alone are sufficient to indicate that there was a strong Aurignacian component at Korpach-mys, as in the early Upper Palaeolithic of Soviet Moldavia generally.

Korpach was excavated by Grigorieva in 1975-76 (Grigorieva 1983a, 1983b, 1983c, n.d.), and, although it is so near, has produced an industry which is markedly distinct in some respects. The 5-metre thick profile contains six stratified units of which (3) and (5) correspond to an upper and lower fossil soil respectively. According to the pollen analysis carried out by G.M. Levkovskaya, both fossil soils are interstadial in character, the upper containing indications of more severe climatic conditions than the lower. There are four cultural layers, I and II in unit (2), III in unit (3), and IV in unit (5), towards the base of the lower fossil soil at a depth of 3.6-3.8 metres. Layer IV has a ^{14}C date of 25 250 ± 300 BP (GrN-9758), which is comparatively recent when compared with those for the early Upper Palaeolithic in Romania quoted above. Grigorieva emphasises (1983b) that all the cultural layers have some traits in common, in that all are workshops based on the exploitation of local flint in which cores and tools occupy a very small part of the total, most being taken up by small splinters, flakes and blades. Thus the total numbers of artifacts, cores, and tools respectively in layers II-IV according to her are as follows (layer I is mixed and is therefore not included): layer II – 12 391: 65: 65; layer III – 1226: 18: 27; layer IV – 14 397: 171; 2O3. In layers II and III retouched bladelets are the predominant tool type (44 in layer II, 7 in layer III), and these include backed bladelets, which, as Grigorieva says, underlines the similarity between them (Grigorieva 1983a: 219). She compares these industries with those from the later layers at Molodova V, and evidently it would be reasonable to regard them in a general sense as belonging to the Gravettian (cf. Otte 1985b).

The situation is somewhat different in layer IV, which industry is defined by Borziyak (1985: 54) as a "separate Upper Palaeolithic culture variant". As already stated, the great majority of artifacts from this layer consist of small splinters, flakes and blades, to which one must add a few pebbles used as hammerstones and a number of limestone or sandstone slabs. The 171 cores are said to be mainly one- or two-platform blade cores. The tool totals given in Table 7.12 (with the exclusion of 33 retouched flakes) follow the indications given by Grigorieva (1983b: 2O); the indices for the major tool groups have been recalculated. It can be seen at once that, all distinctions apart, there are similarities between this industry and those from layers II and III. In particular, retouched bladelets are, as before, the predominant tool type, and Grigorieva specifically says (1983c: 205) that these include 5 backed bladelets. But two elements are quite new; the 19 backed lunates and the 8 bifacial leafpoints, 7 of which are broken, and 6 of which are also regarded as unfinished roughouts (Grigorieva 1983b: Figure 1: backed lunates, 1-5; bifacial leafpoints, 6-9). It is because of these elements – and in particular the conjunction of backed lunates, bifacial leafpoints, and side-scrapers – that there is talk of a new "culture variant" at this site. Grigorieva considers (1983b: 23) that there are some similarities between Korpach layer IV and early Upper Palaeolithic sites in Soviet Moldavia such as Brynzeny I layer 3, Bobuleshty VI, and Gordineshty I; but the closest parallel appears to be offered by Zwierzyniec I in southern Poland (Grigorieva 1983a, n.d.; cf. Allsworth-Jones 1986: 154-9). Quite correctly in my opinion, Grigorieva queries whether a separate "Zwierzyniec culture" can be recognised at that site; and she suggests that backed lunates, bifacial leafpoints, and side-scrapers could well have existed together in a single complex. I suggested that the backed lunates at Zwierzyniec might perhaps have been associated with the Gravettian. It is of course highly unlikely that there was any direct connection between Zwierzyniec and Korpach; nonetheless it is certainly intriguing that an apparently parallel development occurred at about the same time in these different areas. Since layers II and III at Korpach can be classified as broadly speaking Gravettian, and there are clearly some points of comparison between them and layer IV, the backed lunates in this layer may possibly be regarded once more as forming part of that technocomplex. Aurignacian elements are specifically said to be absent (Grigorieva, n.d.: 6O), and the ^{14}C date puts the site well within the Gravettian sphere.

Any attempt to sum up the early Upper Palaeolithic in Soviet Moldavia as a whole can obviously be only very tentative. Soviet authors, in reaction against too crude a stadial division of the Upper Palaeolithic, have suggested the creation of numerous distinct taxonomic units (the "Molodova", "Brynzeny", "Pruth", and – by implication – "Korpach" cultures for example). Whether such distinctions are really justified remains to be seen. What is striking to this observer at least is the persistent recurrence in the early Upper Palaeolithic of the area of three conjoint elements: bifacial leafpoints, backed blades, and end-scrapers of Aurignacian type. Soviet authors themselves, particularly Ketraru and Grigorieva, have repeatedly drawn attention to the presence of Aurignacian elements, which are more pervasive than the single comparison between Istállóskö and Korpach-mys would suggest. Leafpoints occur with remarkable

Table 7.12. Tool totals and calculated indices for Korpach layer IV.

	Totals	Major tool groups (%)
Endscrapers	14	8.24
Burins	21	12.35
Retouched bladelets	45	26.47
Retouched pointed bladelets	8	
Backed bladelets	5	
Backed lunates	19	11.18
Bifacial leafpoints	8	4.71
Sidescrapers	16	9.41
Notches/Denticulates	21	
Miscellaneous	13	
Total tools	170	

profusion in the area and are associated with many different kinds of assemblage, from Middle Palaeolithic through Aurignacian to Gravettian. To that extent they conform to the majority Romanian view that such leafpoints occurred randomly at various times in different contexts, without a specific cultural connotation (Allsworth-Jones 1986: 82). Nonetheless their concentration in the early Upper Palaeolithic in specific association with certain other tool types does suggest a more than accidental phenomenon, and it is for this reason that Soviet authors have been tempted to compare such sites as Brynzeny and Gordineshty with the Szeletian. It looks to me indeed as though there are obvious analogies in terms of context and time range, and I suggest that we may have here what I would prefer to regard as a single entity parallel to, but probably not identical with, the Szeletian of Central Europe. What were previously scattered sites and slight indications have come together to form a body of evidence that cannot be ignored. As for the anthropological associations of this entity we as yet know nothing.

One more matter remains to be discussed, and that is the possible relationship between the early Upper Palaeolithic in this area and the Kostienki-Streletskaya variant on the Don. The view hitherto taken by this author, following G.P. Grigoriev, is that this variant should be regarded as entirely independent from the Szeletian or any other such group in Central Europe (Allsworth-Jones 1986: 18O-1). But M.V. Anikovich (1983) has recently raised the possibility that this may not be so, and a brief consideration should be given to this question. The earliest sites belonging to the variant (Kostienki VI-Streletskaya II, Kostienki XII-Volkovskaya layer 3) occur in the lower humic bed at Kostienki, which according to Anikovich (1982) is still tentatively correlated with the Hengelo interstadial. The later sites (Kostienki XII-Volkovskaya layer 1a, Kostienki XI-Anosovka II layer 5, Kostienki I-Polyakova layer 5) occur in the upper humic bed, which as before is tentatively correlated with Stillfried B (Anikovich 1982; the stratigraphic position assigned to Kostienki I layer 5 differs from that stated in Allsworth-Jones 1986: 180; cf. Amirkhanov *et al.* 1980: 8, Note 8). There is a ^{14}C date for Kostienki XII layer 1a of 32 700 ± 700 BP (GrN-7758), so the earliest sites are presumed to be older than this (Amirkhanov *et al.* 1980: 8). Two dates presently available for the lower humic bed are however in approximately the same time bracket: GrN-10512: 32 200^{+2000}_{-1600} and LE-1436: 32 780 ± 300 BP (Ivanova *et al.* 1983; Rogachev and Anikovich 1984; Valoch 1986a). Both relate to Kostienki XVII-Spitsynskaya layer 2, the type site for the Spitsynskaya variant, which together with the Kostienki-Streletskaya variant initiated the early Upper Palaeolithic on the Don (Amirkhanov *et al.* 1980: 10; Kostienki XII layer 2 is also considered to belong to this variant). The present contrast between the ^{14}C dates and the supposed geological date for the lower humic bed presumably accounts for Olga Soffer's view (1989) that the Upper Palaeolithic in this area began relatively late. There are two Groningen dates for Sungir' of 24 430 ± 400 and 25 500 ± 200 BP, as well as a number of others which Soffer quotes to illustrate the difficulties of the ^{14}C method as hitherto applied in the Soviet Union (Soffer 1985: 228, Table 4.4). However the chronological controversy is finally resolved, the existing ^{14}C dates are of course broadly consistent with those already mentioned for the early Upper Palaeolithic in Romania and Soviet Moldavia.

The nature of the Kostienki-Streletskaya variant can be clearly seen from Anikovich's excellent account of the industry from Kostienki XII-Volkovskaya layer 3 (Anikovich 1977), details of which are summarised in Table 7.13. It is instructive to compare this with Szeletian inventories as such. At Kostienki XII both layers 2 and 3 occurred in the lower humic bed and they were not distinguished separately by Rogachev in every one of his trenches. As a result, there is a slight uncertainty over which artifacts should be attributed to which layer in some cases. There is however a marked raw material distinction between the two layers, and on this basis alone it was usually possible to separate the pieces satisfactorily. The artifacts in layer 2 were mainly of Cretaceous flint, whereas those in layer 3 were mainly of different varying-coloured flint. The varying-coloured flint occurred as slabs (there is an interesting parallel to Szeleta here) which affected the form of the cores and the blanks. A division between layers was particularly difficult in the case of cores, of which there are 29 in all. The most common category are termed flat, one-sided cores, all made on varying-coloured flint slabs and all considered to have come from layer 3 (Anikovich 1977: Figure 2:1 and 4). They contrast with prismatic flake/blade cores which are regarded as having belonged primarily in layer 2 (Anikovich 1977: Figure 2: 5 and 6). The tool totals given in Table 7.13 follow the indications given by Anikovich (1977: Table 2) except that all those pieces whose provenance he considered doubtful have been removed. Indices have been recalculated for the 66 remaining pieces. The complete absence of Levallois technique is noticeable, and Anikovich draws attention to the frequent use of bifacial retouch (often plano-convex) on other tools besides the leafpoints. Anikovich divides the leafpoints into a number of different categories. It is noteworthy that concave-based leafpoints, which are commonly regarded as particularly characteristic of the Kostienki-Streletskaya variant, account for only four out of the total (Anikovich 1977: Figures 6:2 and 3, 7:1). The remainder have differing plan-forms but generally rounded bases (Anikovich 1977: Figures 4:3; 6:1, 4, 5). All told, the industry certainly does create what Anikovich calls an "archaic" impression, i.e. one little influenced by Upper Palaeolithic forms and techniques, although the importance of the raw material factor must be kept in mind. The impression of "archaism" is strengthened if one compares the assemblage from layer 3 with that from layer 2. There are only a few tools, 50 in all, but as Anikovich says they are of a uniformly contrasting technique and typology. According to him, there are 11 end-scrapers, 22 burins, 4 retouched pointed blades, 9 splintered pieces, and 4 miscellaneous items. On that basis, it is not difficult to understand why Soviet authors draw such a sharp distinction between the Kostienki-Streletskaya and the Spitsynskaya variants of the early Upper Palaeolithic.

It has commonly been suggested that the Kostienki-Streletskaya variant might be traced back to the Micoquian of European Russia, as for example, at Il'skaya and Starosel'ye (Allsworth-Jones 1986: 21, 56, 209-10). Anikovich (1983) suggests that this may not in fact be so, for two reasons, chronological and typological. Praslov's excavations of 1963 and 1967-9 and Anisyutkin's re-examination of the materials from excavations by Zamyatin and Gorodtsov in 1925-8 and 1936-7 at Il'skaya have confirmed that the site was early rather than late in the Middle Palaeolithic, and that it was divided into two layers (Anisyutkin 1968; Beregovaya 1984). A considerable time gap may therefore separate it from the Kostienki-Streletskaya variant. Moreover the tools which have commonly been quoted as evidence for a link between the two appear to belong to the lower rather than to the upper layer. The latter, according to Anisyutkin is characterised by a virtual absence of bifacially worked tools and the predominance of Levallois technique (Anisyutkin 1968: 121). In Anikovich's opinion, the triangular or sub-triangular bifacially worked side-scrapers from the lower layer at Il'skaya cannot be taken as convincing forerunners of the Kostienki-Streletskaya concave-based leafpoints (Anikovich 1983: Figure 1: 12, 14-16). Starosel'ye is admittedly later than Il'skaya, and it certainly does have bifacially worked tools; nevertheless, in Anikovich's views an overall comparison of the industries from Starosel'ye and from Kostienki XII layer 3 reveals that they are quite distinct. As an alternative, he suggests that both the Kostienki-Streletskaya variant and the industry from Gordineshty I (which as we have seen also contains leafpoints with concave bases) may be traced back to the Middle Palaeolithic at Trinka III, a cave in Soviet Moldavia excavated by Ketraru and Borziyak in 1973. The upper layer (layer 3) at this site contained two leafpoints with concave bases (one complete and one fragmentary: Anikovich 1983: Figure 2: 11 and 12) and these are considered to be perfectly comparable to the ones on the Don. Anikovich suggests therefore that the assemblage from Trinka III layer 3 may form part of a more extensive

Table 7.13. Tool totals and calculated indices for Kostienki XII – Volkovskaya layer 3

	Totals	Major tool groups %
Endscrapers	17	25.76
Truncated flake/blades	2	
Bifacial leafpoints	15	22.73
Mousterian points	4	
Quinson points	3	
Sidescrapers	19	28.79
Miscellaneous	6	
Total tools	66	

Middle Palaeolithic horizon leading on directly to the two, geographically widely separated, Upper Palaeolithic entities. The trouble with this hypothesis is that, so far at least, it is based upon a very slender material foundation: one site, one layer, and only two characteristic retouched tools. Therefore it may hardly yet be said to be convincing. The Kostienki-Streletskaya variant still looks likely to be an independent phenomenon. But Anikovich's criticism of the established view concerning the link between this variant and the Micoquian of European Russia, and his suggestion of a possible connection between the "Dneister-Carpathian" area and the Don, are both worth keeping in mind.

CONCLUSION: THE SZELETIAN IN A EUROPEAN CONTEXT

In the foregoing pages inevitably a certain amount of attention has been focused on strictly archaeological issues, particularly on the definition of entities, e.g. the status of the 'Bohunician' in relation to the Szeletian in a larger sense, the appropriate way in which the early Upper Palaeolithic in Moldavia may be classified, and the sometimes difficult-to-determine boundary which exists between the Szeletian and the Aurignacian. Some of the difficulties encountered undoubtedly arise because of the unreliable nature of leafpoints as a cultural indicator; so much so that Svoboda has suggested that they can be treated only as an auxiliary and not as a determining factor (Svoboda 1986c: 38). As he says, their importance has probably been exaggerated in the past: "elles ne sont pas les indicatrices d'une seule culture, le Szélétien, mais l'expression commune des populations habitant l'Europe Centrale et les Balkans au moment du passage du Moustérien au Paléolithique supérieur" (Svoboda 1984b: 183). In addition to this, attention has been drawn to the many different elements which may contribute to the perceived variability in the archaeological record, as between cave and open-air sites, and within open-air sites themselves. Some scepticism has been expressed about certain industries which are held to be 'transitional' between Middle and Upper Palaeolithic, for three principal reasons: (1) they may be due to mechanical admixture of different components, as in the Oaş territory and at Mitoc-Valea Izvorului;

(2) there may be some apparent misunderstanding of the nature of the industries concerned, as at Vedrovice II and Kupařovice I; and (3) there may be an over-emphasis on certain typological elements coupled with a generalised use of language leading to the creation of what I would refer to as 'pseudo-historical phenomena', e.g. the supposed 'leafpoint tradition' in Northwestern Europe. I consider that there is still a significant technological gap between even the most advanced Levallois-based industries like Ondratice I and those characteristic of the Upper Palaeolithic as such, like the Aurignacian at Das Geissenklösterle. The Aurignacian remains, as I have said, "the first indubitably Upper Palaeolithic entity to be recognised throughout the length and breadth of Europe", and, one might add, the Near East as well (Allsworth-Jones 1986: 190). These archaeological phenomena should not however be considered in isolation, as the whole theme of the Symposium of which this paper forms part suggests (cf. Trinkaus 1986). Svoboda has stated (1986d: 237) that during the transitional period from Middle to Upper Palaeolithic there is a "lack of any clear-cut division valid for both archaeology and anthropology", but it is my contention that, at least so far as Europe is concerned, this statement is incorrect and that the archaeological and anthropological evidence do in fact go well together. Apart from the actual association of specific hominid types and archaeological inventories, it is above all accurate dating which is vital in this context, both for believers in a replacement/acculturation hypothesis and for those who favour a polycentric independent process.

New dating evidence has confirmed the antiquity of the Mousterian in Central and Southeastern Europe, where apparently anomalous suggestions that it may have continued up to a time corresponding to Denekamp or even beyond have been shown to be ill-founded. The new Hungarian finds and those from Korolevo I unfortunately lack direct dating for the most part, but, as at Kůlna, there are indications that a Micoquian-related occupation may have succeeded a Levallois-dominated one. In my opinion (*pace* Svoboda 1986d: 239, Figure 1) there is no justification for the suggestion that Šipka and Švédův Stůl should be dated to a period corresponding to Hengelo (Allsworth-Jones 1986: 205-6). That does not mean of course that no Mousterian industries survived to that time, and such a dating has always been claimed for the Altmühlian in layer F at Mauern (Allsworth-Jones 1986: 66-8). For the first time, it seems, we have a direct date for the Altmühlian, if it is accepted that Couvin can be related to it, and with this may be compared, in a somewhat tentative way, the dates now available for layer 4 at Soldier's Hole and unit 3 at Picken's Hole in southwest England.

Drawing on the evidence available to me at that time, I presented in my book (Allsworth-Jones 1986: 188-92, Table 2, Chart 2) 35 ^{14}C dates for the Aurignacian in Western Europe, 34 ^{14}C dates for the Aurignacian in Central and Southeastern Europe, and 7 and 8 such dates for the Châtelperronian and Szeletian/Jerzmanowician respectively. The 35 Western European Aurignacian dates ranged from 34 250 to 24 530 BP, with a median date of 31 000 BP; whereas the Central and Southeastern European dates ranged from 44 300 to 22 660 BP, with a median date of 30 815 BP. By comparison the dates for the Szeletian and Jerzmanowician ranged from 43 000 to 32 620 BP, with a median date of 40 786 BP. Hence, it could be shown that, while the 'climax' of the Aurignacian occupation in Central and Southeastern Europe and Western Europe was approximately synchronous, the range of dates in the former area was much greater than in the west, and that they extended back well into and even beyond Hengelo. Hence, I argued that there was a tendency for the Aurignacian to have commenced significantly earlier in Central and Southeastern Europe, and the fact that the dates for the Szeletian and Jerzmanowician ran parallel with the lower range of Aurignacian dates was what would be expected on the hypothesis that the Szeletian came into being by a process of acculturation with it.

It is interesting to compare the figures given then with those presented by G. Dombek in his study of the ^{14}C dates for the Aurignacian and the Gravettian/Upper Périgordian in Western and Central and Southeastern Europe (Dombek 1983). In order to reduce the weight given to extreme values, he preferred to slice off 10% of the relevant dates at either end. Thus he noted that the bulk of Aurignacian dates occur between 34 000 and 27 000 BP, whereas the bulk of Gravettian/Upper Périgordian dates occur between 27 000 and 20 000 BP, with a transitional period stretching from approximately 28 000 to 25 000 BP (Dombek 1983: Figure 4; cf. Allsworth-Jones 1986: 188). Dombek's study very well expresses the 'central tendency' embodied in both sets of dates, but does not differentiate between the different parts of Europe, and necessarily under-plays the evidence regarding the beginning of the Aurignacian, which is of particular concern to us.

It seems to me that the new evidence published since my book and Dombek's study were completed is consistent with the picture I previously presented. Seven new dates for the Aurignacian in Central and Southeastern Europe have so far been mentioned, from Stránská Skála IIIa layer 3, Mitoc Malu Galben, Bistricioara-Lutarie II, and Ceahlau-Dîrtu. This excludes one minimal date from Mitoc Malu Galben (GrN-13007) of >24 000 BP and one questionable date from Bistricioara-Lutarie II (GX-8845-G) of 23 560^{+1180}_{-980} BP; also no account is taken of the three, perfectly acceptable, TL dates from Lušćić (best estimate 28 600 ± 1400 BP). In addition to this, important new information has been made available from Das Geissenklösterle and from Willendorf. At Das Geissenklösterle there were already two ^{14}C dates from layer 12. There are now 5 others as follows: layer 13: 31 070 ± 750, 31 870 ± 1000, 32 680 ± 470 BP; layer 15: 34 140 ± 1000 BP; layer 16: 36 540 ± 1570 BP (Laville and Hahn 1981). All these dates belong to the Aurignacian. At Willendorf a new profile has produced ^{14}C dates for both the Aurignacian and the Gravettian (Haesaerts and Otte 1987). The three for the Aurignacian are as follows: layer 3: 34 100^{+1200}_{-1000} BP (GrN-11192); layer 2: 39 500^{+1500}_{-1200} BP (GrN-11190) and 41 700^{+3700}_{-2500} BP (GrN-11195). It is also legitimate to add to the total six ^{14}C

dates already known from Pod Hradem layer 8 since this layer contained a few "Aurignacoid" artifacts compared by Valoch with those from Istállóskö. The net effect of these additions means that there are now 55 (rather than 34) dates attributable to the Aurignacian in Central and Southeastern Europe, with a median date of 30 950 BP. The range of dates hitherto available has not been affected, though the median has been pushed back somewhat, so the new information can be regarded as broadly confirmatory of the existing situation. Das Geissenklösterle clearly takes its place as one of the most significant early Aurignacian sites in the area. The date from layer 16 at Das Geissenklösterle is only exceeded by two from Willendorf layer 2, two from Istállóskő layer 9, and one from Bacho Kiro layer 11. The last date is conventionally reckoned at >43 000 BP, but, as W.G. Mook comments, this would be equivalent to about $50\,000^{+9000}_{-4000}$ BP if a single standard deviation criterion was employed (Kozlowski 1982: 168). The new evidence therefore reinforces the case for the antiquity of the Aurignacian in this part of the world, and for its contemporaneity with the Szeletian. Hahn (1982) indicates that layers 15 and 16 at Das Geissenklösterle correspond to an "Aurignacien ancien" which precedes the "typical" Aurignacian with split-based bone points in layers 12 and 13. Laville (in Laville and Hahn 1981) clearly favours an interpretation of the deposits at Das Geissenklösterle whereby climatic phases 1-3 (containing the Aurignacian layers) would extend from the latter part of Hengelo up to and beyond Arcy (cf. Leroyer and Leroi-Gourhan 1983: Table 1), and since the ^{14}C dates are consistent with this it would be logical for Hahn to abandon the 'short' chronology he previously favoured. On the basis of the new dates from Willendorf, Haesaerts and Otte (1987) estimate that the Aurignacian in layers 2-4 may have extended from about 40 000 to 32 000 BP, but the actual mean value of the oldest date from layer 2 is of course in excess of that.

Of the eight dates previously attributed to the Szeletian and Jerzmanowician, three came from Bohunice. Similarly now three dates from Stránská Skála (III layer 5 and IIIa layer 4) are regarded as belonging to the 'Bohunician' and two (from Vedrovice V) are attributed to the Szeletian. On the assumption that all these dates can be treated together, we have at the moment therefore 13 dates for the Szeletian and its cognates in Central Europe. Once more the new dates do not affect the already established range, but the median at 39 500 BP is somewhat younger than the previous one. The range would however be disturbed to a considerable extent if another site not hitherto mentioned were included, that of Trenčianske Bohuslavice, in Slovakia, and a brief mention should therefore be made of this site.

Trenčianske Bohuslavice is on the western bank of the river Váh opposite Trenčin and not far from Nové Mesto and Váhom; large area excavations as well as stratigraphic investigations have been carried out by Bárta since 1981 at the locality Pod Tureckom. According to the available preliminary reports (Bárta 1982, 1983, 1984, 1985) the main find horizon was situated "at the base of Würm 3", with two subsidiary horizons, one earlier in a "Würm 2-3" fossil soil, and one later in "mid Würm 3" loess. Several thousand artifacts as well as substantial faunal remains have been recovered, and Bárta regards this as the first large Gravettian site in western Slovakia which can truly be said to be comparable to such well known Moravian sites as Dolní Věstonice and Pavlov. The artifacts in all three horizons are said to belong to the same cultural unit. Those discovered in the main find horizon, at both principal excavated points A and B, include a few bifacial leafpoints as well as, apparently, retouched pointed blades and unifacial leafpoints (Bárta 1983: Figure 5:2; 1984: Figure 2:1). Bárta himself emphasises that the discovery of the leafpoints stratified in a Gravettian horizon is a reminder that such finds taken in isolation cannot serve as a reliable cultural indicator and the most he is prepared to concede is that they may show traces of "Szeletian influence". The same viewpoint is expressed by Oliva (1986a: 3) when he refers to Trenčianske Bohuslavice, but only in the context of "the survival of Szeletian elements in some Pavlovian (i.e. Gravettian) sites". Bárta's geological placement of the site has received confirmation from the ^{14}C dates now available for it. According to J.K. Kozlowski (personal communication) three such dates have been obtained for the Gravettian occupation horizon (he does not specify which) ranging from 23 400 to 20 300 BP. But in addition, according to him, a Szeletian layer has been dated to 23 700 ± 500 BP (Gd-2490). This ^{14}C date is 9000 years younger than the previous youngest date for the Szeletian, which comes from layer 7 at the top of the sequence at Szeleta itself, and the idea that the Szeletian continued for that length of time would be very surprising. Since from the preliminary reports it appears that the few leafpoints at the site are an integral part of the Gravettian, it seems rather that the situation is comparable to that at Korpach layer IV, and that both sites belong to a technocomplex which is quite far removed from the Szeletian.

The model of the Szeletian running in part parallel with the Aurignacian in Central and Southeastern Europe has its analogues in Western Europe. In this area, a very clear picture has now been presented of the interaction between the Aurignacian and the Châtelperronian over a period extending from the latter part of the Hengelo interstadial, through a phase of climatic instability, to the beginning of the Arcy interstadial (Leroyer and Leroi-Gourhan 1983). Neanderthal remains are associated with the Châtelperronian in layer 8 at Saint-Césaire at a date estimated to be around 35-34 000 BP (Leroi-Gourhan 1984). The synchronism of the Aurignacian and the Châtelperronian is demonstrated not only by a comparison of different sites but by interstratification at Le Piage and Roc de Combe, and Leroyer and Leroi-Gourhan (1983) present a very plausible picture of the Aurignacian's "geographical progression" over the period in question. This situation seems to be matched by what is now known concerning the relationship between the Protoaurignacian and the Uluzzian in Italy (Palma di Cesnola 1980). The Uluzzian succession at the Grotta del Cavallo again extends over three climatic phases ranging from Hengelo to Arcy. The Protoaurignacian with Dufour bladelets is regarded by Palma di Cesnola as clearly intrusive, and is stratified above the Uluzzian at the Grotta de la Fabbrica and the Grotta de Castelcivita. At the latter site it replaces the Uluzzian at a time regarded as equivalent to the second phase at Cavallo. Therefore, as Palma di Cesnola says, it can be shown stratigraphically that the two industries proceeded at least in part in parallel. At the same time it should be noted that Aurignacian-type tools, particularly carinate end-scrapers, were present in the Uluzzian from the beginning, and that Aurignacian-type tendencies became stronger towards the end, so much so that Palma di Cesnola regards the Italian Protoaurignacian non-Dufour facies as possibly representing a logical continuation of the Uluzzian. There are therefore interesting similarities between Western Europe and the scenario which I have outlined for Central and Southeastern Europe.

The archaeological evidence also continues to be paralleled (as the case of Saint-Césaire quoted above shows) by the physical anthropological data. In a recent review of the evidence, Smith (1985) pronounced himself possibly ready to accept a pattern of gradual spread of 'progressive' alleles from a single centre of origin, but he went on to argue that in none of the regions of the Old World is there unequivocal evidence for the temporal overlap of archaic and modern *Homo sapiens*. According to him, Neanderthals may have continued until about 38 000 years ago in Central Europe and 33-31 000 years ago in Western Europe, whereas the earliest modern *Homo sapiens* appeared first in Central Europe about 36-34 000 years ago and in Western Europe probably less than 30 000 years ago. Although some of these figures may require revision, there is an undoubted time-cline here; yet Smith argued (1985: 214) that this "does not constitute contemporaneity in the same region". That is a surprising claim, since Central and Western Europe are by no means mutually inaccessible, and I submit that recent dating evidence has strengthened the case for contemporaneity of the two hominid forms in these parts of the continent. It may be noted in passing that two details of his case require amendment. As I pointed out elsewhere, there are no convincing grounds for associating Neanderthal remains with the Aurignacian at Vindija; and it is not reasonable to bracket the 'classic' Neanderthal remains from Subalyuk together with those from Gánovce and Krapina as 'Early'. The remains are associated with the later of the two Mousterian occupations in the cave and are likely to post-date the Brørup interstadial.

Of course, for any thesis to be sustained, it is absolutely vital that correct dating evidence should be available, or that at least good 'guesstimates' should be made. The site of Mladeč is a case in point. In my consideration of the evidence from Mladeč, I suggested that the hominids and Aurignacian-style archaeological remains might best be regarded as equivalent to Denekamp (Allsworth-Jones 1986: 210-11). In so far as a comparison can be made to Pod Hradem, that comparison is with layer 8, which has ^{14}C dates extending from 28 100 to 33 300 BP (Allsworth-Jones 1986: 38-9). I am glad to note that this dating has been accepted as most likely by Svoboda (1986b: 12-13; 1986d: Table on page 239), although if it is true that the human remains and the artifacts were thrown into the cave via a chimney or chimneys, then reliable estimates become even more difficult. In this connection, I am bound to say that I find the age estimates given for Mladeč by Frayer in his latest consideration of the hominid specimens misleading (Frayer 1986: 243-6). Frayer states that Mladeč is one of the earliest Upper Palaeolithic sites yet known, in part because of a correlation to Bacho Kiro "where a 'Mladeč point' is dated at about 40 000 BP". This statement is not correct. Non-split-based bone points are found at Bacho Kiro in layers 6a/7, 7, and 7/6b for the most part, which layers are regarded as the equivalent of Denekamp, and have ^{14}C dates of 29 150 ± 950 and 32 700 ± 300 BP. As already mentioned, a fragment of one possibly non-split-based bone point (Kozlowski 1982: Figure 10:1) was found in layer 8. There was also one split-based bone point in layer 9. The main 'Bachokirian' occupation occurs in layer 11 and has no bone points. Its ^{14}C date is >43 000 BP, but there is no way of dating layer 8 beyond saying that it is likely to be older than 32 700 BP, and a convincing correlation to Bacho Kiro on the basis of non-split based bone points must surely be based primarily on layers 6a/7, 7, and 7/6b. This is not in any way to deny the importance of Mladeč, the promised monograph concerning which should certainly be of great interest, and to settle the controversy perhaps the authorities of the Moravian Museum (Brno) could be persuaded to allow a small piece of hominid bone to go for direct dating. Nor of course am I doubting the antiquity of the Aurignacian and of anatomically-modern man in Central and Southeastern Europe as such. Bacho Kiro itself provides crucial new evidence in this regard. Thus, according to E. Glen' and K. Kaczanowski (Kozlowski 1982: 75-9) hominid remains are associated with both the 'Bachokirian' and the Aurignacian in layers 11, 7/6b, 7, and 6a/7 at the site. They consist of a fragmentary left juvenile mandible with one molar (layer 11), a fragment of a right adult parietal bone (layer 7), a fragmentary right juvenile mandible with two molars (layer 6a/7), and five isolated teeth, two of which may have belonged to a single individual. The authors base their conclusions mainly on their study of the teeth which, they say, "do not differ morphologically from contemporary forms", although in respect of their "absolute dimensions" they are intermediate between present-day specimens and known examples of Neanderthal teeth. This no doubt is why Kozlowski originally referred to the remains as representing "un *Homo sapiens* assez primitif", but however that may be they are not Neanderthal (Kozlowski 1982: 170). They therefore agree with all the other hominid remains discovered with the Aurignacian in Central Europe. By contrast, we have to note the Neanderthal-type hominid remains found with the Jankovichian at Dzeravá Skála and Máriaremete, details of which have been given above.

All in all, therefore, it is considered that the new evidence here presented, on both the archaeological and the physical anthropological side, is consistent with the thesis already advanced (Allsworth-Jones 1986), that the Szeletian was the product of an acculturation process at the junction of Middle and Upper Palaeolithic, it was most likely the creation of Neanderthal man, and both were replaced – not necessarily in anything like a catastrophic manner – by the Aurignacian and by anatomically-modern man.

ACKNOWLEDGEMENTS

An earlier version of this paper was presented at the World Archaeological Congress in Southampton in September 1986. I am most grateful to all those who commented on that paper and who subsequently provided me with much useful information by correspondence, particularly Dr. K.T. Biró, Mr. M. Frelih, Dr. K. Honea, and Dr. J. Svoboda. I am indebted to Dr. R. Jacobi for a number of bibliographical references. I would also like to thank Prof. Z. Böszörmenyi for kindly translating a number of Hungarian language articles into English at my request, notably those by Biró and Pálosi, V.T. Dobosi, Hellebrandt *et al.*, and K. Simán; nor can I forget to mention the magnificent Hungarian cooking which Dr. E. Balogh provided on those occasions, which helped to give a truly Central European flavour to the enterprise.

At the Cambridge Symposium in March 1987 Dr C. Farizy was kind enough to invite me to give a paper at the colloquium she was then organizing, subsequently held at Nemours in May 1988 under the title "Paléolithique Moyen Récent et Paléolithique Supérieur Ancien en Europe". Those who would like a more succinct summary of the topics treated here, written from a slightly different angle, will find it in the published proceedings of the colloquium under the title "Les industries à pointes foliacées d'Europe Centrale: questions de définitions, et relations avec les autres techno-complexes".

REFERENCES

Adamenko, O.M., Adamenko, R.S., Gladilin, V.N., Gnibidenko, Z.N., Grodetskaya, G.D., Pashkevich, G.A., Pospelova, G.A. and Soldatenko, L.V. 1981. Geologiya paleoliticheskoi stoyanki Korolevo I v Zakarpat'ye. *Sovietskaya Geologiya* 12: 87-92.

Allsworth-Jones, P. 1976. Review of J.K. and S.K. Kozlowski *Pradzieje Europy od XL do IV tysiaclecia p.n.e. Antiquity* 50: 152-154.

Allsworth-Jones, P. 1986. *The Szeletian and the Transition from Middle to Upper Palaeolithic in Central Europe.* Oxford: Oxford University Press.

Amirkhanov, Kh.A., Anikovich, M.V., Borziyak, I.A. 1980. K probleme perekhoda ot Must'ye k verkhnemu paleolitu na territorii Russkoi ravniny i Kavkaza. *Sovietskaya Arkheologiya*: 5-21.

Anikovich, M.V. 1977. Kamennyi inventar' nizhnikh sloyev Volkovskoi stoyanki. In N.D. Praslov (ed.) *Problemy Paleolita vostochnoi i tsentral'noi Evropy.* Leningrad: Nauka: 94-112.

Anikovich, M.V. 1982. On the hunting armament of the Kostenkian-Streletskian Sites. Abstracts: Vol. 2. 11th INQUA Congress: Moscow: 11.

Anikovich, M.V. 1983. O vozmozhnykh yugo-zapadnykh kornyakh Kostienkovsko-Streletskoi kul'tury. In N.A. Ketraru (ed.) *Pervobytnye drevnosti Moldavii.* Kishinev: Shtiintsa: 193-202.

Anisyutkin, N.K. 1968. Dva kompleksa Il'skoi stoyanki. *Sovietskaya*

Arkheologiya: 118-125.

Apsimon, A.M. 1986. Picken's Hole, Compton Bishop, Somerset: Early Devensian bear and wolf den, and Middle Devensian hyaena den and Palaeolithic site. In S.N. Collcutt (ed.) *The British Palaeolithic: Recent Studies*. Sheffield: Department of Archaeology and Prehistory, University of Sheffield: 169.

Bácskay, E. 1986. Prehistoric flint mining in Hungary. In *The Social and Economic Contexts of Technological Change*. Precirculated papers of The World Archaeological Congress, Southampton, 1986: 1-35.

Bárta, J. 1982. Mladopaleolitické sídlisko v Trenčianskych Bohuslaviciach. *Archeologické Výskumy a Nálezy na Slovensku v roku* 1981: Part 1: 27-29; Part 2: 319.

Bárta, J. 1983. Druhý rok výskumy na mladopaleolitickom sídlisku v Trenčianskych Bohuslaviciach. *Archeologické Výskumy a Nálezy na Slovensku v roku 1982:* 30-31, 272.

Bárta, J. 1984. Tretí rok výskumu na mladopaleolitickom sídlisku v Trencvianskych Bohuslaviciach. *Archeologické Výskumy a Nálezy na Slovensku v roku 1983:* 27-29, 232.

Bárta, J. 1985. Štvrtý rok Vyśkumy na mladopaleolitickom sídlisku v Trencvianskych Bohuslaviciach. *Archeologické Výskumy a Nálezy na Slovensku v roku 1984:* 37-38.

Beregovaya, N.A. 1984. *Paleoliticheskie mestonakhozhdeniya SSSR (1958-1970 gg.)*. Leningrad: Nauka.

Biró, K.T. 1986a. The raw material stock for chipped stone artefacts in the northern Mid-mountains Tertiary in Hungary. Proceedings of the International Conference on Prehistoric Flint Mining and Lithic Raw Material Identification in the Carpathian Basin, Budapest-Sümeg, May 1986: 183-195.

Biró, K.T. 1986b. Prehistoric workshop sites in Hungary. In *The Social and Economic Contexts of Technologial Change.* Precirculated papers of The World Archaeological Congress, Southampton, 1986: 1-40.

Biró, K.T. and Pálosi, M. 1983. A pattintott koeszközök nyersanyagának forrásai Magyarországon: sources of lithic raw materials for chipped artefacts in Hungary. *A Magyar állami Földtani Intézet évi jelentése az 1983 évrol:* 407-435.

Bitiri, M. 1972. *Paleoliticul in Ţara Oaşului: le paléolithique du Ţara Oaşului.* Bucharest: Institutul de Arheologie.

Bitiri, M. and Cârciumaru, M. 1978. Atelierul de la Mitoc-Valea Izvorului şi locul lui în cronologia paleoliticului României: l'atelier de Mitoc-Valea Izvorului et sa place dans la chronologie du paléolithique en Roumanie. *Studii şi cercetari de istorie veche şi arheologie* 29 (4): 463-480.

Borziyak, I.A. 1976. Raskopki mnogosloinoi stoyanki Korpach-mys. *Arkheologicheskie Otkrytiya 1975 goda.* Moscow-Leningrad: Nauka. 468.

Borziyak, I.A. 1977. Raskopki na verkhnepaleoliticheskoi stoyanke Korpach-mys. *Arkheologicheskie Otkrytiya 1976 goda.* Moscow-Leningrad: Nauka: 452-453.

Borziyak, I.A. 1983. Pozdnii paleolit Dnestrovsko-Karpatskogo regiona (Opyt sistematizatsii). In N.A. Ketraru (ed.) *Pervobytnye drevnosti Moldavii.* Kishinev: Shtiintsa. 33-64.

Borziyak, I.A. 1984. *Verkhnepaleoliticheskaya stoyanka Gordineshty I v Poprut'ye.* Kishinev: Shtiintsa.

Borziyak, I.A. and Ketraru, N.A. 1976. Issledovaniya paleoliticheskoi stoyanki v grote Chuntu. *Arkheologicheskie Otkrytiya 1975 goda.* Moscow-Leningrad: Nauka: 468-469.

Brodar, S. and Brodar, M. 1983. *Potočka Zijalka*. Ljubljana: Slovenska Akademija Znanosti in Umetnosti.

Cârciumaru, M. 1980. *Mediul geografic în pleistocenul superior şi culturile paleolitice din România: le milieu géographique en pléistocène supérieur et les cultures du paléolithique en Roumanie*. Bucharest: Editura Academiei Republicii Socialiste România.

Cârciumaru, M. 1985. Les cultures lithiques du Paléolithique supérieur en Roumanie: chronologie et conditions du milieu. In M. Otte (ed.) *La Signification Culturelle des Industries Lithiques: Actes du Colloque de Liège du 3 au 7 octobre 1984*. Oxford: British Archaeological Reports International Series 239: 235-255.

Cattelain, P., Otte, M. 1985. Sondage 1984 au "Trou de l'Abîme" à Couvin: état des recherches. *Helinium* 25: 123-130.

Clark, J.D. 1981. 'New men, strange faces, other minds': an archaeologist's perspective on recent discoveries relating to the origin and spread of modern man. *Proceedings of the British Academy* 67: 163-192.

Cook, J., Stringer, C.B., Currant, A.P., Schwarcz, H.P. and Wintle, A.G. 1982. A review of the chronology of the European Middle Pleistocene hominid record. *Yearbook of Physical Anthropology* 25: 19-65.

David, A.I. 1983. Impact of Paleolithic man on the mammalian fauna in Moldavia. Abstracts: Vol. 3. 11th INQUA Congress, Moscow: 69

Dobosi, V.T. 1978. A pattintott kőeszközök nyersanyagáról: über das Rohmaterial der retuschierten Steingeräte. *Folia Archaeologica* 29: 7-19.

Dobosi, V.T. 1983. Data to the evaluation of the Middle Palaeolithic industry of Tata. *Folia Archaeologica* 34: 7-30.

Dobosi, V.T. 1986. Raw material investigations on the finds of some Paleolithic sites in Hungary. Proceedings of the International Conference on Prehistoric Flint Mining and Lithic Raw Material Identification in the Carpathian Basin, Budapest-Sümeg, May 1986: 249-255.

Dombek, G. 1983. Die Radiocarbondatierung des Aurignacien, Gravettien, und Périgordien. *Archäologisches Korrespondenzblatt* 13: 429-435.

Dzambazov, N.S. 1977. Muselievskaya kul'tura i yeye mesto v paleolite. In N.D. Praslov (ed.) *Problemy Paleolita vostochnoi i tsentral'noi Evropy*. Leningrad: Nauka: 35-39.

Frayer, D.W. 1986. Cranial variation at Mladeč and the relationship between Mousterian and Upper Palaeolithic hominids. *Anthropos* 23: 243-256.

Gábori, M. 1979. Type of industry and ecology. *Acta Archaeologica Academiae Scientiarum Hungaricae* 31: 239-248.

Gábori-Csánk, V. 1983. La Grotte Remete 'Felsö' (Supérieure) et le 'Szeletien de Transdanubie'. *Acta Archaeologica Academiae Scientiarum Hungaricae* 35: 249-285.

Gladilin, V.N. 1983. Stratigraphy of the Palaeolithic in the Carpathians. Abstracts: Vol. 3. 11th INQUA Congress, Moscow: 87.

Gowlett, J.A.J., Hall, E.T., Hedges, R.E.M. and Perry, C. 1986a. Radiocarbon dates from the Oxford AMS system: datelist 3. *Archaeometry* 28: 116-125.

Gowlett, J.A.J., Hedges, R.E.M., Law, I.A. and Perry, C. 1986b. Radiocarbon dates from the Oxford AMS system: datelist 4. *Archaeometry* 28: 206-221.

Grigorieva, G.V. 1983a. Korpatch, un gisement stratifié du

paléolithique supérieur en Moldavie. *L'Anthropologie* 87: 215-220.

Grigorieva, G.V. 1983b. Mnogosloinaya pozdnepaleoliticheskaya stoyanka Korpach v Moldavii. In *Izyskaniya po Mezolitu i Neolitu SSSR*. Leningrad: Nauka: 18-27.

Grigorieva, G.V. 1983c. Palinologicheskie i radiouglerodnye dannye o mnogosloinoi stoyanke Korpach. In N.A. Ketraru (ed.) *Pervobytnye Drevnosti Moldavii*. Kishinev: Shtiintsa: 2O2-206.

Grigorieva, G.V. n.d. Pozdnepaleoliticheskie pamyatniki s geometriches kimi mikrolitami na Russkoi ravnine.

Haesaerts, P. and Otte, M. 1987. Nouvelles Recherches au Gisement de Willendorf (Basse-Autriche). UISPP, XI Congrès, Mainz, Preprint.

Hahn, J. 1982. Demi-relief aurignacien en ivoire de la grotte Geissenklösterle, près d'Ulm (Allemagne Fédérale). *Bulletin de la Société Préhistorique Française* 79: 73-77.

Hahn, J. and Owen, L.R. 1985. Blade technology in the Aurignacian and Gravettian of Geissenklösterle Cave, S.W. Germany. *World Archaeology* 17: 61-75.

Hellebrandt, M., Kordos, L. and Tóth, L. 1976. A Diósgyőr-Tapolca barlang ásatásának eredményei: Ergebnisse der Ausgrabungen in der Diósgyőr-Tapolca Höhle. *A Herman Ottó Múzeum Évkönyve* 15: 7-36.

Honea, K. 1984. Chronometry of the Romanian Middle and Upper Palaeolithic: implications of current radiocarbon dating results. *Dacia* 27: 23-39.

Honea, K. 1986a. Dating and periodization strategies of the Romanian Middle and Upper Palaeolithic: a retrospective overview and assessment. In M. Day, R. Foley and Wu Rukang (eds) *The Pleistocene Perspective*: Vol. 1. Precirculated papers of The World Archaeological Congress, Southampton, 1986: 1-50.

Honea, K. 1986b. Rezultatele preliminare de datare cu carbon radioactiv privind paleoliticul mijlociu din Peştera Cioarei de la Boroşteni (jud. Gorj) şi paleoliticul superior timpuriu de la Mitoc-Malu Galben (jud. Botosani). *Studii şi cercetari de istorie veche şi arheologie* 37 (4): 326-332.

Ivanova, M.A., Praslov, N.D., Sinitsyn, A.A. 1983. Problems of chronology and stratigraphy of the Paleolithic in the Kostionki-Borshchevo area on the Don river. Abstracts: Vol. 3. 11th INQUA Congress, Moscow: 111.

Jenkinson, R.D.S. 1984. *Creswell Crags, Late Pleistocene Sites in the East Midlands*. Oxford: British Archaeological Reports, British Series 122.

Kaminská, L. 1985. Nový nález listovitého hrotu z východného Slovenska. *Archeologické Rozhledy* 37: 195-197.

Ketraru, N.A. 1973. Pamyatniki epokh paleolita i mezolita. *Arkheologicheskaya Karta Moldavskoi SSR (1)*. Kishinev: Shtiintsa.

Ketraru, N.A. 1976. Raskopki v grote Buzduzhany I. *Arkheologicheskie Otkrytiya 1975 goda*. Moscow-Leningrad: Nauka: 471.

Kozlowski, J.K. 1972-3. The origin of lithic raw materials used in the Palaeolithic of the Carpathian countries. *Acta Archaeologica Carpathica* 13: 5-19.

Kozlowski, J.K. 1983. Le paléolithique supérieur en Pologne. *L'Anthropologie* 87: 49-82.

Kozlowski, J.K. (ed.) 1982. *Excavation in the Bacho Kiro Cave (Bulgaria): Final Report*. Warsaw: Państwowe Wydawnictwo Naukowe.

Laville, H. and Hahn, J. 1981. Les dépôts de Geissenklösterle et l'évolution du climat en Jura souabe entre 36 000 et 23 000 BP.

Comptes-Rendus de l'Académie des Sciences de Paris (série 2) 292: 225-227.

Leroi-Gourhan, Arl. 1984. La place du Néandertalien de St-Césaire dans la chronologie würmienne. *Bulletin de la Société Préhistorique Française* 81: 196-198.

Leroyer, C. and Leroi-Gourhan, Arl. 1983. Problèmes de chronologie: le castelperronien et l'aurignacien. *Bulletin de la Société Préhistorique Française* 80: 41-44.

Mogoşanu, F. 1983. Paléolithique et epipaléolithique. In V. Dumitrescu, A. Bolomey and F. Mogoşanu *Esquisse d'une Préhistoire de la Roumanie*. Bucharest: Editura Ştiinţifica şi Enciclopedica: 29-55

Montet-White, A., Laville, H. and Lezine, A.M. 1986. Le paléolithique en Bosnie du Nord: Chronologie, environnement et préhistoire. *L'Anthropologie* 90: 29-88.

Oliva, M. 1981. Die Bohunicien-Station bei Podolí (Bez. Brno-Land) und Ihre Stellung im beginnenden Jungpaläolithikum. *Časopis Moravského Musea* 66: 7-45.

Oliva, M. 1982. Estetické projevy a typologické zvláštnosti kamenné industrie Moravského Aurignacienu. *Časopis Moravského Musea* 67: 17-30.

Oliva, M. 1984a. Technologie výroby a použité suroviny Štípané industrie Moravského Aurignacienu. *Archeologické Rozhledy* 36: 601-628.

Oliva, M. 1984b. Le Bohunicien, un nouveau groupe culturel en Moravie: quelques aspects psycho-technologiques du développement des industries paléolithiques. *L'Anthropologie* 88: 209-220.

Oliva, M. 1986a. From the Middle to the Upper Palaeolithic: a Moravian perspective. In M. Day, R. Foley and Wu Rukang (eds) *The Pleistocene Perspective*: Vol. 1. Precirculated papers of the World Archaeological Congress, Southampton, 1986: 1-12.

Oliva, M. 1986b. Finds from the pleniglacial B from the territory of Czechoslovakia and the question of the "Epiaurignacian" settlement. In M. Day, R. Foley and Wu Rukang (eds) *The Pleistocene Perspective:* Vol. 2. Precirculated papers of the World Archaeological Congress, Southampton, 1986: 1-14.

Otte, M. 1983. Le paléolithique de Belgique: essai de synthèse. *L'Anthropologie* 87: 291-321.

Otte, M. 1984. Paléolithique supérieur en Belgique. In D. Cahen and P. Haesaerts (eds) *Peuples Chasseurs de la Belgique Préhistorique Dans Leur Cadre Naturel.* Brussels: Institut Royal des Sciences Naturelles: 157-179.

Otte, M. 1985a. *Les Industries à Pointes Foliacées et à Pointes Pedonculées dans le Nord-Ouest Européen.* Liège: Editions du Centre d'Etudes et de Documentation Archéologiques.

Otte, M. 1985b. Le Gravettien en Europe. *L'Anthropologie* 89: 479-503.

Palma di Cesnola, A. 1980. L'Uluzzien et ses rapports avec le Protoaurignacien en Italie. In L. Banesz and J.K. Kozlowski (eds) *L'Aurignacien et le Gravettien (Périgordien) dans leur Cadre Ecologique.* Nitra: Archeologický Ustav Slovenskej Akadémie Vied: 197-212.

Ringer, A. 1983. *Bábonyien: eine mittelpaläolitische Blattwerkzeugindustrie in Nordostungarn.* Dissertationes Archaeologicae: Régészeti Dolgozatok, Series 2, No. 11. Budapest: Eötvös Loránd University.

Rogachev, A.N. and Anikovich, M.V. 1984. Pozdnii paleolit SSSR. In P.I. Boriskovskii (ed.) *Paleolit SSSR:* Vol. 1, Part 3. Moscow-Leningrad: Nauka

Schönweiss, W. and Werner, H.J. 1986. Ein Fundplatz des Szeletien in Zeitlarn bei Regensburg. *Archäologisches Korrespondenzblatt* 16: 7-12.

Severin, T. 1973. *Vanishing Primitive Man*. London: Thames and Hudson.

Simán, K. 1979a. Régészeti ásatások es gyűjtések az Avason 1905-1978. *A Miskolci Herman Ottó Müzeum Közleményei* 17: 12-15.

Simán, K. 1979b. Kovabánya az Avason: Silexgrube am Avasberg. *A Herman Ottó Múzeum Evkönyve* 17-18: 87-102.

Simán, K. 1982. New results in the palaeolithic study of north-east Hungary. Abstracts: Vol. 1. 11th INQUA Congress, Moscow: 290.

Simán, K. 1983. Köeszközleletek Borsod-Abaúj-Zemplén megyében (1978-1982). *A Miskolci Herman Ottó Müzeum Közleményei* 21: 37-49.

Simán, K. 1985. Korlát-Ravaszlyuktetö. *A Herman Ottü Müzeum Evkönyve* 22-23: 83.

Simán, K. 1986a. Limnic quartzite mines in north-east Hungary. Proceedings of the International Conference on Prehistoric Flint Mining and Lithic Raw Material Identification in the Carpathian Basin, Budapest-Sümeg, May 1986: 95-99.

Simán, K. 1986b. Felsitic quartz porphyry. Proceedings of the International Conference on Prehistoric Flint Mining and Lithic Raw Material Identification in the Carpathian Basin, Budapest-Sümeg, May 1986: 271-275.

Simán, K. 1986c. Mittelpaläolithisches Atelier am Avasberg bei Miskolc (Ungarn). In *Urzeitliche und frühhistorische Besiedlung der Ostslovakei in Bezug zu den Nachbargebieten*. Nitra: Archäologisches Institut der Slowakischen Akademie der Wissenschaften: 49-55.

Sirakov, N. 1983. Reconstruction of the Middle Palaeolithic flint assemblages from the cave Samuilitsa II (northern Bulgaria) and their taxonomical position seen against the Palaeolithic of South-eastern Europe. *Folia Quaternaria* 55.

Smith, F.H. 1985. Continuity and change in the origin of modern *Homo sapiens. Zeitschrift für Morphologie und Anthropologie* 75: 197-222.

Soffer, O. 1985. *The Upper Paleolithic of the Central Russian Plain*. Orlando: Academic Press.

Soffer, O. 1989. The Middle-Upper Palaeolithic transition on the Russian Plain. In P. Mellars and C. Stringer (eds) *The Human Revolution: Behavioural and Biological Perspectives on the Origins of Modern Humans*. Edinburgh: Edinburgh University Press: 714-742.

Soldatenko, L.V. 1983. Mousterian in the Transcarpathia. Abstracts: Vol. 3. 11th INQUA Congress, Moscow: 248.

Štělcl, J. 1986. Knollensteine des Drahany-Hochlandes, Rohstoff der paläolitischen Steinindustrie. Proceedings of the International Conference on Prehistoric Flint Mining and Lithic Raw Material Identification in the Carpathian Basin, Budapest-Sümeg, May 1986: 207-210.

Stringer C.B. 1986. Direct dates for the fossil hominid record. in J.A.J. Gowlett and R.E.M. Hedges (eds) *Archaeological Results from Accelerator Dating*: Oxford: Oxford University Committee for Archaeology: 45-50.

Svoboda, J. 1980. *Křemencová industrie z Ondratic: k problému počátků mladého paleolitu*. Studie Archeologického Ustavu Československé Akademie Ved v Brně 9 (1). Prague: Academia.

Svoboda, J. 1984a. K některým aspektům studia exploatačních oblastí kamenných surovin. *Archeologické Rozhledy* 36: 361-369.

Svoboda, J. 1984b. Cadre chronologique et tendances évolutives du paléolithique tchecoslovaque: essai de synthèse. *L'Anthropologie* 88: 169-192.

Svoboda, J. 1985. Neue Grabungsergebnisse von Stránská Skála, Mähren, Tschechoslowakei. *Archäologisches Korrespondenzblatt* 15: 261-268.

Svoboda, J. 1986a. Primary raw material working in neolithic/eneolithic Moravia. Proceedings of the International Conference on Prehistoric Flint Mining and Lithic Raw Material Identification in the Carpathian Basin, Budapest-Sümeg, May 1986: 277-285:

Svoboda, J. 1986b. Origins of the Upper Palaeolithic in Moravia. In M. Day, R. Foley and Wu Rukang (eds) *The Pleistocene Perspective*: Vol. 1. Precirculated papers of the World Archaeological Congress, Southampton, 1986: 1-23.

Svoboda, J. 1986c. K počátkům mladého Paleolitu v Brněnské kotlině,stratigrafie, ekologie, osídlení. *Archeologické Rozhledy* 38: 32-45.

Svoboda, J. 1986d. The *Homo sapiens neanderthalensis-Homo sapiens sapiens* transition in Moravia: chronological and archaeological background. *Anthropos* 23: 237-242.

Svoboda, J. and Svoboda, H. 1985. Les industries de type Bohunice dans leur cadre stratigraphique et écologique. *L'Anthropologie* 89: 505-514.

Trinkaus, E. 1986. Les Néandertaliens. *La Recherche* 17 (180): 1040-1047.

Valoch, K. 1981. Einige mittelpälaolithische Industrien aus der Kůlna-Höhle im Mährischen Karst. *Časopis Moravského Musea* 66: 47-67.

Valoch, K. 1982. Neue paläolithische Funde von Brno-Bohunice. *Časopis Moravského Musea* 67: 31-47.

Valoch, K. 1983. Příspěvek k paleolitickému osídlení Prostějovska. *Časopis Moravského Musea* 68: 5-19.

Valoch, K. 1984. Výzkum Paleolitu ve Vedrovicích V (okr. Znojmo). *Časopis Moravského Musea* 69: 5-22.

Valoch, K. 1985. Paleolitická Stanice v Hostějově (o. Uh. Hradiště). *Casopis Moravského Musea* 70: 5-16.

Valoch, K. 1986a. Stone industries of the Middle/Upper Palaeolithic transition. In M. Day, R. Foley and Wu Rukang (eds) *The Pleistocene Perspective*: Vol. 1. Precirculated papers of the World Archaeological Congress, Southampton, 1986: 1-19.

Valoch, K. 1986b. Příspěvek k poznání zdrojů surovin v mladem paleolitu na Moravě. *Časopis Moravského Musea* 71: 5-18.

Valoch, K., Oliva, M., Havlíček, P., Karášek, J., Pelisek, J. and Smolíkova, L. 1985. Das Frühaurignacien von Vedrovice II und Kupařovice I in Sudmähren. *Anthropozoikum* 16. Prague: Ustřední Ustav Geologický: 107-203.

Vörös, I. 1984. Hunted mammals from the Aurignacian cave bear hunters' site in the Istállóskö cave. *Folia Archaeologica* 35: 7-28.

8: Chronological Change in Perigord Lithic Assemblage Diversity

JAN F. SIMEK AND HEATHER A. PRICE

INTRODUCTION

For many years, palaeoanthropologists and archaeologists have been interested in a set of European Late Pleistocene biocultural changes collectively referred to as the 'Middle to Upper Palaeolithic transition'. Recently, debate concerning this time period has intensified (e.g. Binford 1982; Mellars 1973, 1986; Rigaud n.d.; Simek and Snyder 1988; Smith and Spencer 1984; Stringer *et al.* 1984; Trinkaus 1983, n.d.; White 1982). Contributing to the increased interest are various new data, including information from the archaeological record (e.g. Champagne and Espitalié 1967), human palaeontology (Lévêque and Vandermeersch 1981), and DNA studies (Cann *et al.* 1987). The effect of these new data has been to complicate our understanding of this critical time period, resulting in several divergent interpretations of Late Pleistocene events in Europe.

A central concern in analysing the transition is the relation between human biological change and observed, contemporary behavioural change. Guided by this concern, some researchers have chosen to compare between a generalized Middle Palaeolithic cultural unit (composed of Mousterian industries) and a generalized Upper Palaeolithic cultural unit (formed by grouping Châtelperronian through Azilian industries). This widely-used method tends to emphasize discontinuity between the two major categories and seeks to explain biological changes in terms of behavioural selection (e.g. Trinkaus 1983). Several dimensions of the archaeological record have been treated from this point of view – archaeofaunas (e.g. Binford 1985; Mellars 1973; White 1982), 'symbolic' materials such as art and jewellery (White 1982, this volume), and lithic assemblages (Harrold 1978, 1983; Mellars 1973). In nearly all comparative studies, discontinuity is cited as evidence for increasing complexity (sophistication, variability, etc.) among cultures borne by fully modern humans. And quite often, this notion of increased complexity is considered in terms of changing assemblage diversity (faunal or lithic) across the transition.

Recently, we have begun to present results of ongoing research into variability among and between Late Pleistocene materials from southwestern France, treating variability in spatial order within sites (Simek 1987) and changing patterns of faunal assemblage diversity (Simek and

Snyder 1988). In this paper, we will continue our considerations by examining patterns of diversity in lithic assemblages over time. Our analysis will be historical (i.e. we consider time as a relevant variable in examining behavioural change) and treats data from 235 assemblages derived from 104 Périgord Mousterian and Upper Palaeolithic sites. We hope to show that evolutionary patterns in assemblage diversity do not support vitalistic interpretations based on discontinuities observed using very simple chronological units. Moreover, temporal patterning in lithic assemblage diversity does not coincide with patterning observed in other archaeological dimensions either in the nature of change or its timing.

BACKGROUND

For many years, researchers have recognized that a series of important changes in human adaptations, both physical and cultural, occurred in the northern Old World somewhere between 100 000 and 30 000 years ago. There has always been some debate as to the precise timing of these changes (cf. Smith and Spencer 1984), and much debate over their nature (e.g. Mellars 1973, 1982, *versus* White 1982). Still, most scholars have devoted their attention to events at the end of this time range (i.e. *c.* 45-30 000 years BP). This emphasis seems to be due to several important shifts that become apparent at this time.

First, 'Archaic' or 'Neanderthal' types of *Homo sapiens* were replaced on the Pleistocene landscape by skeletally modern forms of the species. Second, Mousterian flaked stone tool assemblages, so long prevalent in European human occupation sites, disappear. In their place, varied blade technologies collectively referred to as Upper Palaeolithic occur over most of Western Europe. (In the earliest stages, these include Châtelperronian and Aurignacian industries in the Périgord region of southern France.)

One group of analysts, composed primarily of physical anthropologists (but including prehistorians), believes that Late Pleistocene changes in the human organism and in the behaviour of groups of humans constitute an important evolutionary Rubicon, a revolutionary 'transition'. This group sees these changes as directly related in causal fashion (e.g. Trinkaus n.d.). A second group of scholars would allow that biological change may be independent of contemporary behavioural shifts during the Late Pleistocene (e.g. Conkey 1985; Dennell 1986; Rigaud n.d.; Simek and Snyder 1988; Smith, *et al.* 1989). Indeed, this group might avoid using the term 'transition' at all, preferring to examine different dimensions of the fossil and archaeological records as separate axes of variability before attempting explanatory integration.

Those that view behavioural and biological shifts as representing fundamentally related changes in human adaptations have performed much of their analyses using a comparative technique. This method is designed to detect differences between the Middle and Upper Palaeolithic. As conventionally applied, comparisons involve selecting a set of variables that are believed to have 'cultural' import (i.e. can be used to measure the degree to which a culture has developed). Next, variables are measured for the Middle and for the Upper Palaeolithic. Measurement normally comprises either establishing the presence or absence of a trait based on certain identifying criteria, or the calculation of derived indices designed to express the proportional contribution of individual characteristics to the culture as a whole. Finally, the completed measurements are used to show how Upper Palaeolithic industries differed from the Mousterian in cultural (or behavioural) terms.

One criterion that has frequently been employed to distinguish the Middle and Upper Palaeolithic is assemblage *diversity*. It is argued that the Upper Palaeolithic must be classified using more stone tool types and therefore has a greater variety of retouched stone tools than the Mousterian. Thus, the Upper Palaeolithic represents a more diverse culture than its predecessor (e.g. Mellars 1973; White 1982, 1986). At the same time, faunal assemblages associated with Upper Palaeolithic technologies are said to be composed of fewer animal species with one or a few species dominant (White 1982); thus, Upper Palaeolithic faunas are less diverse than Mousterian ones. Both characteristics (high lithic diversity and low faunal diversity) are seen as representing the same thing: techno-economic specialization related to a marked increase in cultural complexity across the Middle-Upper Palaeolithic boundary.

Diversity as an analytical concept was derived by ecologists to characterize natural communities in terms of the number and relative proportions of species present (Pielou 1975). Thus, properly considered, diversity has two components: *taxonomic richness* (the number of classes present) and *assemblage evenness* (how the classes are filled by the sample). Up to the present, only richness has been invoked in most discussions of the Middle-Upper Palaeolithic transition. However, that measure is demonstrably dependent on sample size (e.g. Grayson 1984; Kintigh 1984) – something usually not considered in using the concept – and represents only part of an assemblage's diversity. In fact, it is the second component of diversity, i.e. evenness, that is less affected by sampling and more amenable to comparative study (Pielou 1975).

To test the conventional view that Upper Palaeolithic stone tool assemblages are more diverse than Mousterian assemblages and represent a radically different kind of techno-economic system, we will measure diversity for 235 lithic assemblages from the Périgord region of southwestern France. We consider both relevant dimensions, richness and evenness. To examine change over time, we will relate the assemblages to two available chronological frameworks: the classic archaeological traditions on the one hand, and more-detailed sedimentary sequences (after Laville 1975; Laville *et al.* 1980) on the other. Unfortunately, only 71 of the 235 sample assemblages can be related to the regional chronostrata, so two sets of analyses, one for the large sample and one for the smaller set, will be performed. Before turning to our measurement of diversity and time variables, and discussing the techniques we employ, a brief description of the data sample is in order.

Table 8.1. Archaeological traditions defined for analyses and relevant sample information for each tradition. *N* is the total number of artifacts for all assemblages within a tradition. *k* is the number of types observed among all assemblages for each tradition.

Tradition	Sites	Assemblages	*N*	*k*
Mousterian	17	48	19995	62
Châtelperronian	23	14	9524	75
Aurignacian	26	45	31393	82
Upper Perigordian	14	33	18946	85
Solutrean	7	16	10427	80
Magdalenian	39	62	38941	85
Azilian	5	8	2318	71

MATERIALS AND METHODS

The Sample Assemblages

Data employed in the analyses comprise typological counts of retouched stone artifacts recovered from 104 sites in the Périgord region of southwestern France. Table 8.1 gives general information concerning the assemblage sample. Assemblages are composed of artifacts recovered in a single geological layer or archaeologically defined level, depending on how the original excavators collected and curated materials. The 'cultural' referent of a given assemblage was assigned following either the excavator's ascription or Sonneville-Bordes' (1960) study of Upper Palaeolithic assemblage systematics in the region. For all assemblages, one of the two Palaeolithic typologies developed for local lithic materials (Bordes 1961; Sonneville-Bordes and Perrot 1954-56) was used to classify the data. Thus, 48 Mousterian assemblages from 17 different sites, all classified by the Bordesian Middle Palaeolithic type list, and 187 assemblages classified using the Upper Palaeolithic list are examined here.

The Variables

As discussed above, diversity must be considered in terms of its two components, taxonomic richness and evenness, and separating the two components is even more important here since we consider assemblages classified according to two schemes. Many widely used measures of diversity (e.g. Shannon's H' statistic [1949]) are particularly sensitive to the classification scheme employed (Pielou 1975; Simek and Snyder 1988). Evenness, however, is calculated by standardizing diversity by the size of the classification applied to a given assemblage (Pielou 1975):

$$\text{Evenness} = H' / H(max) \qquad (1)$$

where H' is Shannon's general diversity statistic calculated as usual and $H(max)$ is the highest possible diversity value attainable using a classification. Because it is standardized for classification size, evenness is more appropriate than richness for comparative work, but it has not been used until now in the context of Palaeolithic lithic assemblage variation.

Table 8.2. Observed mean values for Diversity Indices and standard errors for index by archaeological tradition.

Tradition	Richness	Error	Evenness	Error
Mousterian	27.00	1.705	0.603	0.012
Châtelperronian	28.87	2.464	0.587	0.018
Aurignacian	40.02	1.761	0.667	0.013
Upper Perigordian	41.24	2.057	0.667	0.015
Solutrean	41.56	2.954	0.581	0.021
Magdaltenian	38.23	1.501	0.606	0.011
Azilian	32.25	4.177	0.576	0.03

Taxonomic richness is calculated by counting the number of classes present in an assemblage. Table 8.2 shows richness and evenness characteristics of the seven assemblage traditions considered here.

Of course, in southwestern France as elsewhere in Europe, Middle and Upper Palaeolithic assemblages are typed according to schemes of different sizes. The Middle Palaeolithic typology contains 63 retouched stone tool types (including 4 debitage types) while the Upper Palaeolithic typology used here contains 92 classes. This in and of itself would not be a problem if the two classifications dealt in the same dimensions of variation, but it is clear that they do not. As far as is presently known, the Middle Palaeolithic list reflects variability within a single archaeological tradition, the Mousterian. On the other hand, the Upper Palaeolithic list was designed to distinguish among a variety of traditions and is therefore composed of temporally diagnostic artifact types in addition to a large suite of mundane, ubiquitous classes. To make diversity indices more comparable among assemblages, we figured them *within* the Upper Palaeolithic traditions (e.g. Aurignacian, Magdalenian). In this way, only those diagnostic types relevant to a single tradition's assemblage profile were included in calculations. Mundane types were also included, so the resulting set of classes reflects a given tradition's general assemblage profile. The effect of classification size should be most pronounced for richness counts, since evenness standardizes by the number of possible classes. But if there is any effect, classification size variation should elevate evenness values for the Upper Palaeolithic assemblages, thereby favouring support for the transition model.

One simple way to test the effects of classification on diversity measures is to compare indices for the same assemblages classified according to the two schemes. Because Harrold (1978) did such a series of descriptions for 18 Châtelperronian sites from the Périgord, this test can be performed. When the number of classes present are considered, assemblages classified by the Mousterian list have a mean richness of 20.33 and variance of 40.35. The same set classed by Upper Palaeolithic types have a mean richness of 27.11 and variance of 86.1. A t test between these values indicates a significant difference between the two schemes ($t = -2.557, p = 0.015$). Mean assemblage evenness is 0.5724 ($s^2 = 0.0099$) when the

Middle Palaeolithic classes are used ($k = 53$) and 0.6138 ($s^2 = 0.0077$) when the Upper Palaeolithic list is employed ($k = 69$). Here, $t = -1.325$ for comparing the mean values ($p = 0.194$); the two values are equivalent. Thus, typology size does seem to effect assemblage richness: cases classified by the Upper Palaeolithic system may have more classes *because* of the scheme itself. As we will see, this problem with richness can usually be identified when present. Evenness, on the other hand, effectively standardizes data for varying classifications and is amenable to comparative analysis without reservation.

There is another potential problem with diversity indices. As was discussed above, richness is very sensitive to overall sample size for an assemblage (e.g. Grayson 1973, 1984). The relation between evenness and sample size is less well understood, but probably should also be assessed for sampling effects prior to use in chronological comparisons. This assessment can be performed in several ways (Simek 1989), but we use Kintigh's Monte Carlo simulation technique for reasons outlined elsewhere (Simek and Snyder 1988). In brief, the method compares observed diversity measures at given sample sizes to those predicted by randomly sampling a distribution with the same characteristics; it has proven quite useful in a variety of archaeological applications (see papers in Leonard and Jones 1989). For a complete description of the technique, the reader is referred to the original presentation (Kintigh 1984).

Time will be the second variable of concern here. Unfortunately, there is variation in temporal resolution among the 235 sample assemblages. Relatively few have radiometric dates associated with the assemblages. Variability in the dating techniques employed makes use of these dates problematic in comparisons of the kind we perform. Lower resolution, but still some precision, can be obtained using Laville's regional chronostratigraphy based on sedimentary sequences from Périgord sites. Just over 30% of the sample (71 assemblages from 21 different sites) can be placed within the relative scheme, and these will be analysed separately to examine fine-scale temporal change.

The only chronological measure that is applicable to all 235 lithic assemblages is provided by their archaeology: tradition. Admittedly, tradition provides only a general, category-level measure of time (Laville *et al.* 1980). Yet, for the larger Périgord sample, it is the best we can do at present.

Methods of Analysis

As is usual in statistical analyses of any kind, the nature of the variables being considered dictates the analytic techniques employed. In our studies, measurement of archaeological assemblage diversity is identical for all assemblages, involving calculation of taxonomic richness and evenness. Both are interval scale measures and can support a variety of treatments.

It is the measurement of time, conceptually the 'independent' or predictor variable, that differs among the analytic units. Using tradition as

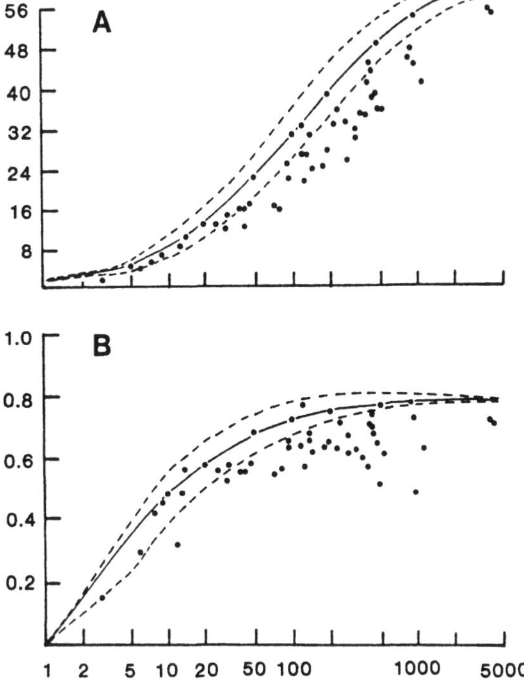

Figure 8.1. Assessment of sample size effects on Mousterian lithic assemblage diversity. A = assemblage richness; B = assemblage evenness. Sample size is the independent variable (X-axis) in both assessments.

the chronological measure, variables are interval scale on the one hand (richness and evenness) and nominal on the other (tradition). Thus, analysis of variance is the most appropriate technique for examining the effects of time on assemblage diversity for the large sample (e.g. Neter *et al.* 1985; Zar 1985).

For the 71 assemblages that can be related to Laville's local chronostratigraphic sequence, regression is an appropriate and powerful analytic tool (Neter *et al.* 1985). By attempting to fit a variety of linear models (curving polynomials or straight lines), the nature of change in lithic assemblage diversity over the span of the Later Pleistocene can be examined without reference to archaeological traditions. This, in turn, allows the timing of any observed trends to be assessed in relation to the 'Middle-Upper Palaeolithic transition' without emphasizing discontinuity *a priori*. However, because chronology is measured using an extended ordinal sequence, certain advantages of regression cannot be employed here; in particular, utilization of significant regression parameters to model rates of change is inappropriate although desirable.

Thus, two kinds of analysis will be performed to assess the effects of time on Palaeolithic lithic assemblage diversity, depending on which scale

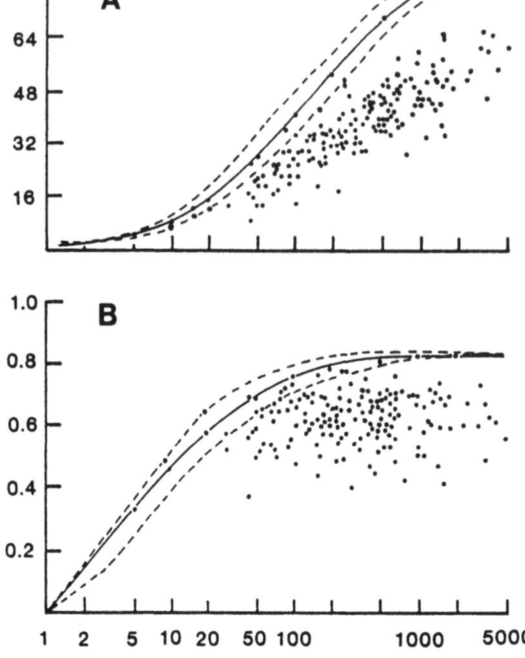

Figure 8.2. Assessment of sample size effects on Upper Palaeolithic lithic assemblage diversity. A = assemblage richness; B = assemblage evenness. Sample size is the independent variable (X-axis) in both assessments.

measures time. Of course, the transition model predicts that the Mousterian should differ significantly from the Upper Palaeolithic in ANOVA and that any significant regression trends should point to the Mousterian/Aurignacian boundary as important. The Châtelperronian – usually considered Upper Palaeolithic (e.g. Harrold 1978; White 1986) – should lie with the Aurignacian. Before turning to chronological assessments, the effects of sample size on diversity measurements will be considered.

ANALYSIS

Diversity Measures and Sample Size

The relation between sample size and assemblage diversity is examined by comparing values calculated for the data cases and values obtained by repeated random sampling of a model distribution of known form (Kintigh 1984). The model for the simulation process in this case is the data themselves, i.e. the total number of tools in each category within each cultural tradition. To determine if given assemblage richness and evenness values might simply be factors of sample size, a series of random sampling trials are carried out. In each trial, an assemblage the same size as that being tested is formed at random, and diversity values are calculated. In

Table 8.3. Results of regression analyses for lithic assemblage richness. Significant parameters are indicated by "**" ($p > 0.05$).

Variable	Coefficient	*T*-value	Probability(>*T*)
CUBIC SOLUTION			
Intercept	15.9163	1.93	0.0574
Sequence	2.1028	1.72	0.0909
Sequence^2	-0.0526	-1.01	0.3159
Sequence^3	0.0004	0.63	0.5314
QUADRATIC SOLUTION			
Intercept	19.7907	3.64	0.0005**
Sequence	1.3874	3.04	0.0033**
Sequence^2	-0.0202	-2.43	0.0159**
SIMPLE LINEAR SOLUTION			
Intercept	30.8168	9.53	0.0001**
Slope	0.2891	2.71	0.0086**

some cases, when sample sizes are small, as many as 200 trials may be run. Descriptive statistics (means, standard deviations, and 95% confidence intervals) are produced for each set of trials, and these are compared to observed values to assess the effect of sampling on the assemblage of interest. If a given value lies outside the 95% confidence range, the diversity values for that assemblage can be considered useful for further analyses.

Figure 8.1 shows a plot of random and observed diversity indices for the 48 assemblages classified by Middle Palaeolithic types. Note that, except when samples are very small, all Mousterian assemblages have richness values below the random plot (Fig. 8.1A). This means that Mousterian assemblages are almost uniformly less rich in artifact types than a random model would predict. Figure 8.1B shows that a similar pattern characterizes assemblage evenness. Here again, most assemblages lie below the random curve's 95% confidence interval – i.e. are significantly less even than expected – but there are also random assemblage values lying within the confidence band. Overall, these results indicate that Mousterian assemblage diversity is mostly independent of sample size and that most assemblages have low diversity. This characteristic is predicted by the traditional transition model.

Assessments of Upper Palaeolithic assemblage diversity show almost identical patterns (Fig. 8.2). Figure 2A shows that with very few exceptions, Upper Palaeolithic assemblages are less rich than predicted by random sampling processes. The low evenness pattern noted for the Mousterian is more pronounced for the Upper Palaeolithic data. The Upper Palaeolithic, according to these results, is characterized by lithic diversity every bit as low as the Mousterian. This result contradicts the traditional view of transition systematics. As for Mousterian assemblages,

Upper Palaeolithic richness and evenness values are not size dependent and can be used for further analyses. This established, we now turn to regression analysis of change in assemblage diversity over time.

Fine-scale Chronological Change in Lithic Diversity

For regression, 71 assemblages from 21 sites were given a relative sequence number depending on their position within Laville's regional chronology. The 71 assemblages were assigned to a total of 49 different sequence numbers. Obviously, some of Laville's chronostrata are represented by more than one assemblage; however, most phases contain a single assemblage, and the maximum number for a single phase is three. Thus, the chronostratigraphic sequence provides a relatively continuous measure of time. Sequence number was defined as the independent predictor variable for regression. Assemblage richness and assemblage evenness were defined as dependent, response variables.

Table 8.3 shows results of cubic, quadratic, and simple linear regression solutions for the effects of sequence position on assemblage richness. When a cubic function is fitted to the data, the overall model is significant, but none of the regression terms attains the stipulated level ($p < 0.05$). However, the quadratic regression solution is significant in all its terms, a result supported by homoscedastic residuals (Figure 8.3). Despite significance, the quadratic model fails to account for much of the variability in assemblage richness ($R^2 = 0.1008$). This indicates that time, while certainly an important factor generating some of the observed variation, is not the only or even the major source of variability in the number of artifact classes composing an assemblage. Other relevant factors might include site function, occupation duration, etc. Nevertheless, change through time does account for over 10% of the sample variance.

The quadratic function that describes the relation between chronological sequence number and artifact assemblage richness is:

$$\text{Richness} = 19.79 + 1.387\,(\text{Seq}) - 0.02\,(\text{Seq}^2) \quad (2)$$

Figure 8.4 shows the plotted function and the data points used to derive it.

As the transition model predicts, a quadratic polynomial characterizes change in richness over time. However, the change process clearly occurs later in time than the shift from Mousterian to Upper Palaeolithic industries. According to the regression, maximum assemblage richness occurs around sequence number 34, after a long and gradual increase. This corresponds to Laville's Recent Würm Phase 12 and contains some of the earliest Solutrian materials in the region. Before the Solutrian richness maximum, no radical shift among the earlier Upper Palaeolithic traditions or between Mousterian and Early Upper Palaeolithic is evident. This pattern holds even though our comparative study indicated that assemblages classified by the Middle Palaeolithic type list are generally less rich than those classed with Upper Palaeolithic types.

Table 8.4 shows the results of cubic, quadratic and simple linear

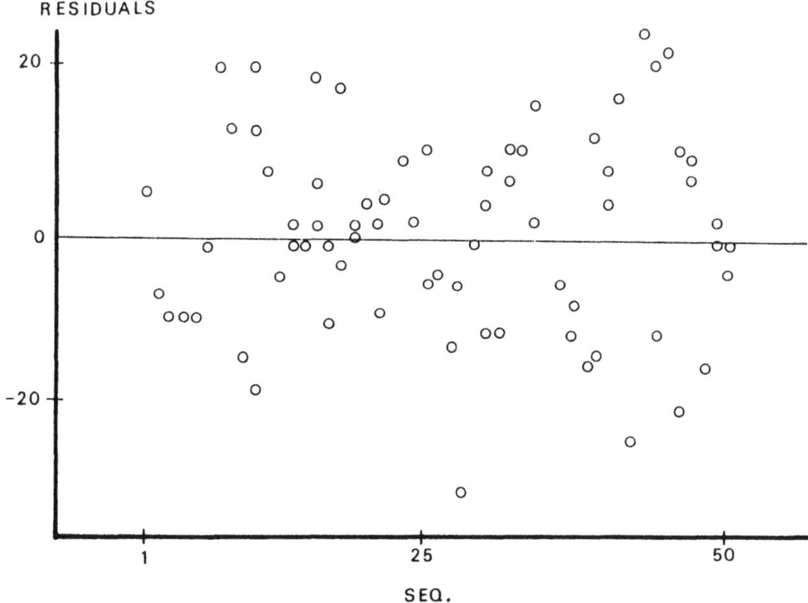

Figure 8.3. Residuals from quadratic regression defining the effects of chronostratigraphic sequence number on lithic assemblage richness (see text, Formula 2).

regressions testing the effect of time on assemblage evenness. As for richness, a quadratic function provides the best description of the analysed relationship; homoscedasticity in residuals confirms the model's success (Figure 8.5). This regression accounts for little more of the variance in assemblage evenness than did the solution for richness ($R^2 = 0.116$), but the result is statistically significant. Other factors that might have caused non-temporal variation in evenness have already been cited for richness.

The formal model derived by quadratic regression can be stated as follows:

$$\text{Evenness} = 0.602 + 0.005\ (\text{Seq}) - 0.0001\ (\text{Seq}^2) \qquad (3)$$

Figure 8.6 shows the regression function plotted with the 71 data points. Note the superficial similarity between this plot and that for richness; assemblage evenness increases gradually until it reaches its maximum predicted value and then decreases gradually until the end of the Palaeolithic.

While the forms of the derived curves are similar, change in assemblage evenness differs from richness in important ways. Evenness reaches its predicted maximum quite a bit earlier than richness, at sequence number 23. This period reflects the shift from Aurignacian to Upper Périgordian traditions in the region. There is no indication that a significant change occurs at the Mousterian/Châtelperronian or Châtelperronian/Aurignacian boundary.

Two important points should be noted here. First, analysis of change

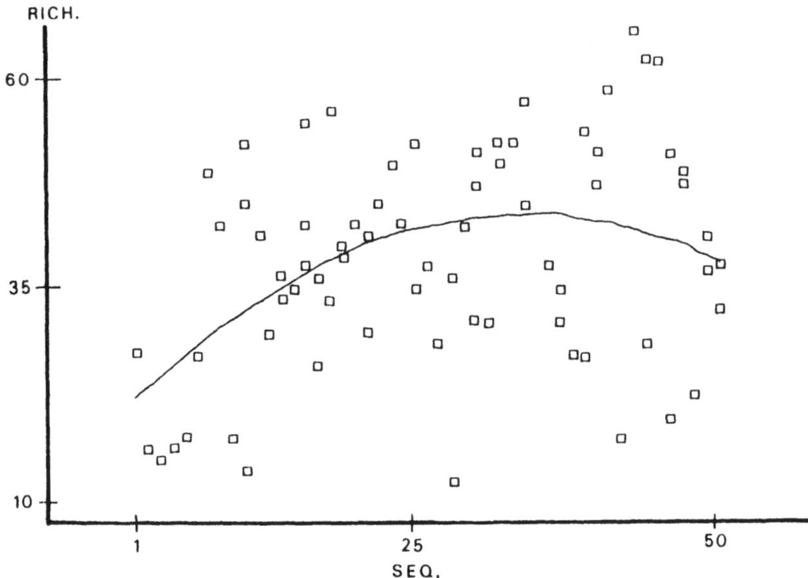

Figure 8.4. Plot of data points and regression derived by fitting quadratic function to relation between chronostratigraphic sequence number and assemblage richness (see text, Formula 2).

in lithic assemblage variation over time reveals a gradual increase in assemblage diversity that began during the Mousterian and ended during the later part of the Upper Palaeolithic. Second, the two components of diversity do not change at the same time or in the same way. We will return to both of these points shortly.

Change among Palaeolithic assemblage traditions

As was described earlier, the effects of time on assemblage diversity measures must be assessed in a general way in order to use the large Périgord Palaeolithic sample. Thus, analysis of variance is performed using archaeological tradition as the factor variable; richness and evenness comprise the response variables. Unbalanced ANOVA was employed in both tests, since the 235 sample assemblages were unequally distributed among the seven archaeological traditions.

Table 8.5 shows the results of ANOVA testing the effects of tradition categories on the variance in assemblage richness. The model is significant, and slightly more variability is accounted for here than by regression ($R^2 = 0.193$). Least-squares means analysis reveals some interesting patterns in the data. Two primary groups of traditions are defined. First, Mousterian, Châtelperronian and Azilian assemblages have very similar mean values for richness. Second, Aurignacian, Upper Périgordian, Solutrian and Magdalenian industries form another group with higher richness values. Here, the transition model for increased diversity with the Upper Palaeolithic might seem supported, since the early traditions form

Table 8.4 Results of regression analyses for lithic assemblage evenness. Significant parameters are indicated by "**" ($p > 0.05$).

Variable	Coefficient	*T*-value	Prob(>*T*)
CUBIC SOLUTION			
Intercept	0.5644	11.76	0.0001**
Sequence	0.0124	1.74	0.0873
Sequence^2	-0.0004	-1.42	0.1596
Sequence^3	0.0000	1.06	0.2942
QUADRATIC SOLUTION			
Intercept	0.6023	18.89	0.0001**
Sequence	0.0054	2.02	0.0475**
Sequence^2	-0.0001	-2.39	0.0195**
SIMPLE LINEAR SOLUTION			
Intercept	0.6648	35.17	0.0000**
Slope	-0.0008	-1.33	0.187

a group. However, the demonstrated effects of classification size on richness should again be noted in regard to these results.

Table 8.6 shows ANOVA results testing the effects of tradition classes on assemblage evenness. There is a very different pattern than that defined for richness. The model is significant and accounts for over 13% of the sample evenness variation ($R^2 = 0.139$). Two groups of traditions can again be defined by least-squares means analysis, but these are quite different than for richness. One group includes the Mousterian, Châtelperronian, Solutrian, Magdalenian, and Azilian traditions. The other group, with higher evenness values, comprises the Aurignacian and Upper Périgordian. As in regression, these two Upper Palaeolithic traditions are identified as critical points in the process of increasing assemblage diversity. Based on this second index, there is no evidence for significant, radical change in assemblage diversity with the coming of the Upper Palaeolithic.

DISCUSSIONS AND CONCLUSIONS

In concluding this paper, we first summarize results of quantitative analyses and then discuss some implications of our findings. All of the studies performed here represent assessments of the effects of time, variously measured, on lithic artifact assemblage diversity during the French Palaeolithic. In carrying out these tests, we examine the widely-held belief that increased cultural complexity characterized the coming of the Upper Palaeolithic in Western Europe, manifested by a shift from not very diverse, generalized techno-economic systems to more diverse, specialized ones. In the conventional view, this shift in technology had biological roots; fully modern peoples associated with the Upper Palaeolithic were physically better able to organize themselves because they were more 'advanced'. While this scenario has been widely disseminated, it has

Table 8.5. Results of analysis of variance and least-squares means analysis testing the effects of archaeological tradition on lithic assemblage richness. In LSMeans analysis, traditions are labelled as follows: (1) Mousterian; (2) Châtelperronian; (3) Aurignacian; (4) Upper Perigordian; (5) Solutrean; (6) Magdalenian; (7) Azilian. Significant parameters are marked by "**".

Variable	SS	Mean Square	*F*-value	Probability (>*F*)
ANOVA				
Model	7598.4597	1266.4099	9.07	0.0001**
Error	31827.9233	139.5962		
Total	39426.383			

LSMEANS
Probability > :T: HO: LSMEAN(I) = LSMEAN(J)

	1	2	3	4	5	6	7
1	/						
2	0.5333	/					
3	0.0001**	0.0003**	/				
4	0.0001**	0.0002**	0.6527	/			
5	0.0001**	0.0011**	0.6547	0.9292	/		
6	0.0001**	0.0014**	0.4383	0.2373	0.3149	/	
7	0.2458	0.4865	0.0878	0.0547	0.0700	0.1795	/

never been empirically tested in terms of assemblage diversity.

The first step in our analyses was the assessment of sampling effects on two standard measures of diversity (corresponding to the two theoretical components of the concept): assemblage richness and evenness. A Monte Carlo simulation technique was employed and little sample effect was detected. In general, neither Middle nor Upper Palaeolithic assemblages are very diverse.

The second step used assemblage diversity indices from assemblages placed within Laville's regional sedimentary sequence. Regression was used to test for the effects of time on diversity. Both richness and evenness are best characterized by quadratic equations, but in both cases less than 15% of the total diversity variation can be explained in terms of chronological change. While the curvilinear relations conform to the transition model, evenness reaches a maximum value as the Aurignacian changes to the Upper Périgordian, and richness increases gradually until the Upper Périgordian/Solutrian boundary. In other words, fine-scale measures of change in assemblage diversity show no breaks at the Middle/Upper Palaeolithic boundary. In addition, different components of diversity change at different times.

The third analytic step employed unbalanced analysis of variance to assess the effects of archaeological tradition on diversity indices for 235 Périgord Palaeolithic assemblages. Richness shows a break between the Châtelperronian and the Aurignacian, but evenness again points to the Aurignacian/Upper Périgordian boundary as pivotal in the change process. When the results of ANOVA are considered in light of regression

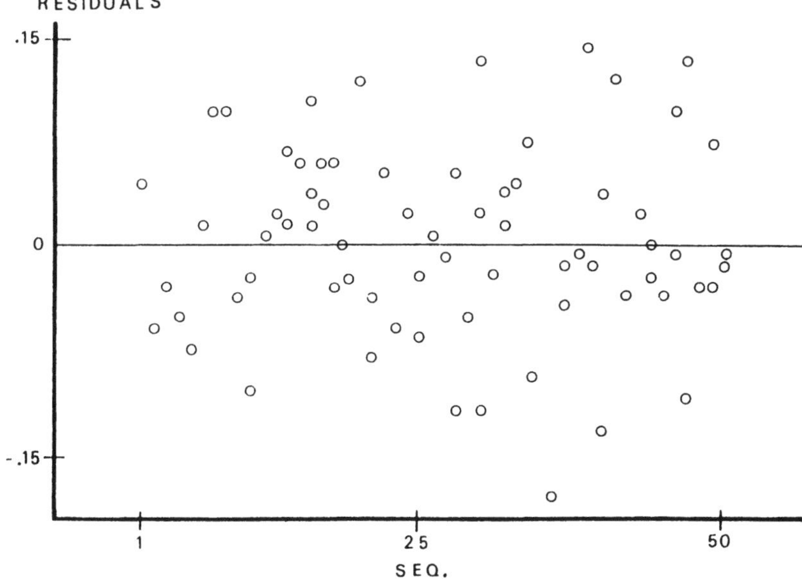

Figure 8.5. Residuals from quadratic regression defining the effects of chronostratigraphic sequence number on lithic assemblage evenness (see text, Formula 3).

results, the indicated change in richness can be seen as reflecting change *within* the Aurignacian rather than at tradition boundaries. Differences in classification size between the Middle and Upper Palaeolithic type lists might also have effected richness enough to influence the observed ANOVA solution. By contrast, both regression and ANOVA results accord well for assemblage evenness.

The implications of these findings for interpretations of Later Pleistocene biocultural changes are important. First, if a shift from generalized to specialized techno-economic strategies occurred during the later Palaeolithic in Western Europe, then the process was slow and gradual, began in the Middle Palaeolithic, and was complete (i.e. specialized) only during the Upper Périgordian. *No dramatic 'transition' in assemblage diversity is evident at the Middle/Upper Palaeolithic boundary*. Given the nature and timing of observed change patterns, it is highly unlikely that vitalistic explanations – those that seek to link the advent of modern humans to a more efficient economic system – are warranted for the shift in diversity. Second, there are indications that change was more subtle than previously believed. It has been proposed that the Middle/Upper Palaeolithic transition involved a basic shift from foraging to collecting land-use strategies (Binford 1982). This model implies greater *variability* in sites produced by collectors. The homoscedastic residual pattern observed for our data does not support such an implication, although there are other kinds of evidence that may indicate such a shift (most notably in site organization: Simek 1987).

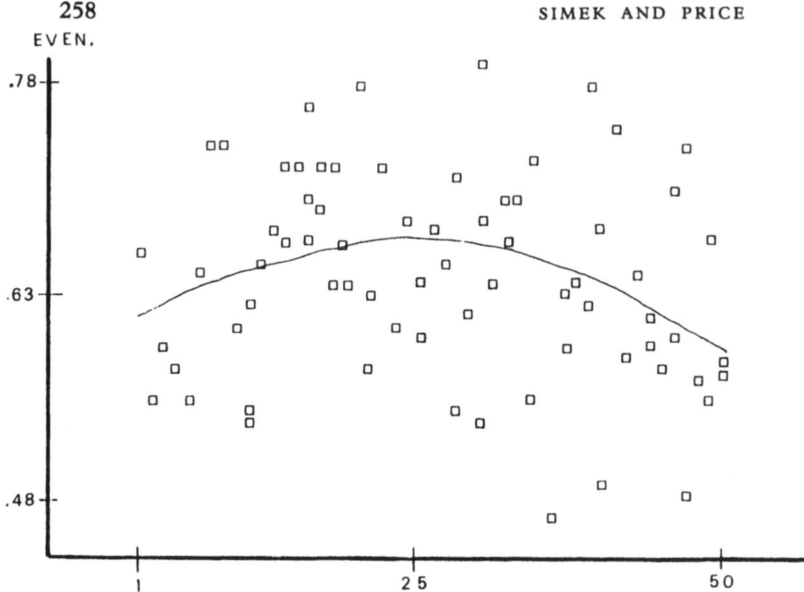

Figure 8.6. Plot of data points and regression derived by fitting quadratic function to relation between chronostratigraphic sequence number and assemblage evenness (see text, Formula 3).

What did cause the observed change? That is difficult to address at present. However, recent studies of changing faunal diversity from assemblages recovered in these same sites may provide a clue (Simek and Snyder 1988). A gradual decrease in faunal diversity occurred during the Recent Würm that mirrored a general deterioration in environmental conditions over the same period. Beginning with the Denekamp Interstadial and ending with the Cold Maximum of the Last Glacial, change in faunal diversity probably monitored global climates rather than any increased organization capability on the part of human predators. Thus, selective pressures from the environment may have resulted in increased specialization simply as the most direct response to environmental change. In any case, change in faunas seems to coincide with the curves defined here for lithic assemblages, i.e. major behavioural shifts occurred during the Upper Palaeolithic, when modern humans were already established in Western Europe. Economic specialization was probably not related in causal fashion to either the origins of modern people in the West or their relation to archaic *Homo sapiens*.

Another important implication of these results concerns the nature of change itself during the Late Pleistocene. In attempting to account for what are viewed as radically changing 'systems', we have often failed to recognize that relations among system components must be *demonstrated* before they can be utilized in modelling systemic change. There has been a strong tendency to assume that changes in biology and in behaviour that appear contemporaneous must be related. We have shown that changes

Table 8.6. Results of analysis of variance and least-squares means analysis testing the effects of archaeological tradition on lithic assemblage evenness. In LSMeans analysis, traditions are labelled as follows: (1) Mousterian; (2) Châtelperronian; (3) Aurignacian; (4) Upper Perigordian; (5) Solutrean; (6) Magdalenian; (7) Azilian. Significant parameters are marked by "**."

Variable	SS	Mean Square	*F*-value	Probability (>*F*)
ANOVA				
Model	0.2593795	0.0432299	6.11	0.0001**
Error	1.6120005	0.0070702		
Total	1.8713799			

LSMEANS
Probability > :T: HO: LSMEAN(I) = LSMEAN(J)

	1	2	3	4	5	6	7
1	/						
2	0.4705	/					
3	0.0003**	0.0003**	/				
4	0.0009**	0.0006**	0.9975	/			
5	0.3682	0.8132	0.0006*	0.0009**	/		
6	0.817	0.3517	0.0003**	0.001**	0.2781	/	
7	0.4037	0.7404	0.0054**	0.0067**	0.8914	0.3336	/

that appear contemporary may not be, and that changes appearing simple may be quite complex. Having recognized the complexity, we must begin to define and relate change processes in a variety of dimensions. We suggest that, as a first step, biological change and behavioural change be decoupled analytically in Europe, as it has been elsewhere, so that *empirical* relations between these dimensions can be examined. We do not wish to imply that no changes in behaviour marked the development of modern humans in Europe. We do insist that only by recognizing the complexity of Late Pleistocene change can we begin to unravel the problem.

ACKNOWLEDGEMENTS

We thank K. Kintigh for providing the software used to perform diversity calculations and assessments. F. Harrold generously provided data on some of the Châtelperronian assemblages and made cogent comments on an earlier draft of this paper. H. Bricker and T. Volman also improved the presentation with their comments. Of course, the final product is our responsibility alone. The support of the United States National Science Foundation (Award BNS-8606536) made collection of some of these data possible.

REFERENCES

Binford, L.R. 1982. Reply to White 'Rethinking the Middle/Upper Paleolithic transition'. *Current Anthropology* 23: 177-181.

Binford, L.R. 1985. Human ancestors: changing views of their behavior. *Journal of Anthropological Archaeology* 4: 292-327.

Bordes, F. 1961. *Typologie du Paléolithique Ancien et Moyen*. Bordeaux: Delmas.

Cann, R.L., Stoneking, M. and Wilson, A.C. 1987. Mitochondrial DNA and human evolution. *Nature* 325:31-36.

Champagne, F. and Espitalié, R. 1967. La stratigraphie du Piage: note préliminaire. *Bulletin de la Société Préhistorique Française* 64: 35-40.

Conkey, M.W. 1985. Ritual communication, social elaboration, and the variable trajectories of paleolithic material culture. In T.D. Price and J.A. Brown (eds) *Prehistoric Hunter-Gatherers: the Emergence of Cultural Complexity*. Orlando: Academic Press: 299-323.

Dennell, R. 1986. Needles and spear-throwers. *Natural History* 95: 70-78.

Grayson, D.K. 1979. On the quantification of vertebrate archaeofaunas. In M. Schiffer (ed.) *Advances in Archaeological Method and Theory*: Vol. 2. Orlando: Academic Press: 199-237.

Grayson, D.K. 1984. *Quantitative Zooarchaeology: Topics in the Analysis of Archaeological Faunas*. New York: Academic Press.

Harrold, F.B. 1978. *A Study of the Châtelperronian*. Unpublished Ph.D. Dissertation, University of Chicago.

Harrold, F.B. 1983. The Châtelperronian and the Middle/Upper Paleolithic transition. In E. Trinkaus (ed.) *The Mousterian Legacy: Human Biocultural Change in the Upper Pleistocene*. Oxford: British Archaeological Reports International Series S164: 123-140.

Leonard, R. and Jones, G. (eds) 1989. *Quantifying Diversity in Archaeology*. Cambridge: Cambridge University Press.

Kintigh, K. 1984. Measuring archaeological diversity by comparison with simulated assemblages. *American Antiquity* 49: 44-54.

Laville, H. 1975. *Climatologie et chronologie du Paléolithique en Périgord: Etude Sedimentologique de Depôts en Grottes et Sous Abris*. Provence: Université de Provence.

Laville, H., Rigaud, J.P. and Sackett, J. 1980. *Rock Shelters of the Périgord*. New York: Academic Press.

Lévêque, F. and Vandermeersch, B. 1981. Le néandertalien de Saint-Césaire. *La Recherche* 12: 242-244.

Mellars, P. 1973. The character of the Middle-Upper Paleolithic transition in southwestern France. In C. Renfrew (ed.) *The Explanation of Culture Change: Models in Prehistory*. London: Duckworth: 255-276.

Mellars, P. 1982. On the Middle/Upper Paleolithic transition: a reply to White. *Current Anthropology* 23: 238-240.

Mellars, P. 1986. A new chronology for the French Mousterian period. *Nature* 322: 410-411.

Neter, J., Wasserman, W. and Kutner, M.H. 1985. *Applied Linear Statistical Models: Regression, Analysis of Variance, and Experimental Designs*. Illinois: Richard D. Irwin. 2nd Edition.

Pielou, E.C. 1975. *Ecological Diversity*. New York: Wiley.

Rigaud, J.-P. 1988. The origins of modern man: the stone tools story. In E. Trinkaus (ed.) *Patterns and Processes in Later Pleistocene Human Emergence*. Cambridge: Cambridge University Press. In Press.

Simek, J.F. 1987. Spatial order and behavioural change in the French Paleolithic. *Antiquity* 61: 15-40.

Simek, J.F. 1989. Structure and diversity in intrasite spatial analysis. In R. Leonard and G. Jones (eds) *Quantifying Diversity in Archaeology*. Cambridge: Cambridge University Press.

Simek, J.F. and Snyder, L.M. 1988. Patterns of change in Upper Paleolithic archaeolfaunal diversity. In H. Dibble and A. Montet-White (eds) *Upper Pleistocene Prehistory of Western Eurasia*. Philadelphia: University Museum, University of Pennsylvania: 321-332.

Smith, F.H., Simek, J.F. and Harrill, M.S. 1989. Geographic variation in supraorbital torus reduction during the later Pleistocene (*c.* 80 000-15 000 BP). In P. Mellars and C. Stringer (eds) *The Human Revolution: Behavioural and Biological Perspectives on the Origins of Modern Humans*. Edinburgh: Edinburgh University Press: 172-193.

Smith, F.H. and Spencer, F. (eds) 1984. *The Origin of Modern Humans: a World Survey of the Fossil Evidence*. New York: Alan R. Liss.

Sonneville-Bordes, D. de. 1960. *Le Paléolithique Supérieur en Périgord*. Bordeaux: Delmas.

Sonneville-Bordes, D. de and Perrot, J. 1954-1956. Lexique typologique du Paléolithique Supérieur. *Bulletin de la Société Préhistorique Française* 50-53: 327-359.

Stringer, C.B., Hublin, J.-J. and Vandermeersch, B. 1984. The origin of anatomically modern humans in Western Europe. In F.H. Smith and F. Spencer (eds) *The Origins of Modern Humans: a World Survey of the Fossil Evidence*. New York: Alan R. Liss: 51-135.

Trinkaus, E. 1983. Neandertal postcrania and the adaptive shift to modern humans. In E. Trinkaus (ed.) *The Mousterian Legacy: Human Biocultural Change in the Upper Pleistocene*. Oxford: British Archaeological Reports International Series S164: 165-200

Trinkaus, E. 1988. The Upper Pleistocene transition: biocultural patterns and processes. In E. Trinkaus (ed.) *Patterns and Processes in Later Pleistocene Human Emergence*. Cambridge: Cambridge University Press. In Press.

White, R. 1982. Rethinking the Middle/Upper Paleolithic transition. *Current Anthropology* 23: 169-192.

White, R. 1986. *Dark Caves, Bright Visions*. New York: American Museum of Natural History.

White, R. 1988. Toward a contextual understanding of the earliest body ornaments. In E. Trinkaus (ed.) *Patterns and Processes in Later Pleistocene Human Emergence*. Cambridge: Cambridge University Press. In Press.

Zar, J.H. 1984. *Biostatistical Analysis*. New Jersey: Prentice-Hall. 2nd Edition.

9: The Middle-Upper Palaeolithic Transition in Southwestern France: Interpreting the Lithic Evidence

T. E. G. REYNOLDS

Understanding of the origins of modern man in southwestern France has derived, in archaeological terms, from the study of the Middle-Upper Palaeolithic transition. As such, this understanding is currently based on two archaeological phases separated out in the nineteenth century on the basis of a variety of contrasts in the material evidence, particularly the associated fauna, the stone tool traditions, bone working and the production of art works, as well as the assumed association of material evidence patterns with hominid type. These two phases have been described and explained independently and little attention has been specifically addressed to the nature of the interface between them. So it is that our best picture of the Middle-Upper Palaeolithic transition is based on a 'before and after' approach which takes the two phases and compares them in terms of their respective contents, complexity and rates of change (Mellars 1973; White 1982). Such an approach has been most successful in revealing what has changed, but is less so when an attempt is made to understand the nature of these changes, their scheduling and how they interact. The general approach uses a varied and complex set of data without granting sufficient attention to the problems involved in any single particular dataset, let alone providing adequate theoretical or methodological account as to how these different types of data may be effectively integrated. The different time-scales involved in deposition of individual datasets and their resolution cause blurring of historical sequence and mask the possibilities of identifying the actual scheduling of changes and the interaction of the behaviours involved, surely the very necessary components for explaining the Middle-Upper Palaeolithic transition. In short, explanation as to why the changes occurred is currently limited by the general way in which such changes are documented. It is in response to these issues that in this paper I shall focus solely on one aspect of the complex dataset, the lithic evidence, in an attempt to reveal some of the problems and issues involved in a single area of the data. This will highlight the need for further research to provide a more relevant and specifically orientated study of the lithics which may then reveal some of the scheduling of changes in behaviour in stone material use. Further such studies on other aspects of the data may then also be undertaken and integrated to more finely develop our understanding of the Middle-Upper Palaeolithic transition.

In southwestern France, the study of the Middle-Upper Palaeolithic transition has involved examination of two industrial groups, the Mousterian complex and the Châtelperronian. These two groups are both subsets of larger Palaeolithic phases, the Mousterian being a final Middle Palaeolithic whilst the Châtelperronian is an early Upper Palaeolithic industry. In more precise terms, the Mousterian may be regarded as a complex of assemblage types (or variants) of flake-tool dominated assemblages dating from between the onset of the last glacial (approximately 105 000 years BP) and the Würm II/III interstadial (at approximately 35 000 years BP) (Bordes 1984; Dennell 1983; Mellars 1986). The Châtelperronian is identified by possession of a type fossil, the Châtelperronian point, its many blade-based tools and its occurrence overlying Mousterian layers and is dated to either the Würm II/III interstadial or the early Würm III stadial 35-32 000 years BP.

These two industries are, however, classified using different typological systems reflecting their place and relevance in studies of their respective larger periods, the Lower and Middle Palaeolithic, and the Upper Palaeolithic. Both typological systems were devised to isolate and inform upon the time-space systematics of industries assigned to the Lower and Middle Palaeolithic or the Upper Palaeolithic, these typologies being that of Bordes (1953a, 1953b, 1953c, 1954, 1961, 1984) and de Sonneville-Bordes and Perrot (1953, 1954, 1955, 1956a, 1956b).

The Bordes system employs a type list of 62 tool forms, plus an open, miscellaneous category, with which to compare between assemblages. The frequencies of each tool type are counted and displayed in the form of a cumulative percentage frequency graph and coupled to the use of technological and typological indices to identify the particular assemblage form. The system is fully explained elsewhere (see especially Bordes 1984) and so will not be further described here. Use of the system demonstrated that four repeating assemblage forms could be identified; these forms have become known as the Mousterian variants, which are: the Typical, the Denticulate, the Charentian and the Mousterian of Acheulian tradition (MAT). These forms are also best described in Bordes (1984) and it must be noted that one variant form in particular, the MAT has been observed as being typologically similar to the first Upper Palaeolithic industry, the Châtelperronian, despite the description of the latter using a different typology (Bordes 1972b). The Châtelperronian is classified using the de Sonneville-Bordes and Perrot system which although based on that of Bordes and aimed at similar objectives, used a tool list of 91 types with an additional miscellaneous category. The similarity between these two industries (the MAT and the Châtelperronian) was such as to lead Bordes to suggest an *in situ* development of the Upper Palaeolithic in southwestern France (Bordes 1972b). Given that it is generally accepted that the Châtelperronian is the first Upper Palaeolithic industry of the region, it must also be accepted that the industry is, therefore a *product* of the transition and that the origins of the Upper Palaeolithic must lie in earlier industries. So, can any directional change be identified among the Mousterian variants that might be taken to represent internal change towards the Upper Palaeolithic on either a long term or short term basis?

In 1965, Mellars first noted chronological patterning of some of the Mousterian variants and has enlarged upon these observations at various times to date (Mellars 1965, 1967, 1969, 1986). The first observation to note on chronology is one of a short term similarity between the MAT and the Châtelperronian. Mellars observed that the MAT is a final Mousterian given its consistent occurrence at the top of Mousterian sequences. He further argues that it clearly post-dates the Charentian variant at a significant number of sites and as yet no reversals of this stratigraphy occur.

Additional strength for these observations may be derived from the similarity of the MAT to the succeeding Châtelperronian (which often overlies it) although to do so relies upon the acceptance of certain assumptions, such as typological similarity reflecting time-space relations. It is also worth stressing at this point that the recent application of thermoluminescence dates by Valladas *et al.* (1986) has dated MAT layers at the site of Le Moustier and that such layers are indeed relatively late in Mousterian terms (e.g. 42 500 ± 2000 BP for MAT type B and 50 300 ± 5500 BP for MAT type A).

The MAT itself further provides evidence for a directional change towards the Upper Palaeolithic typologically, for it comprises three 'types' – A, A/B and B which occur in stratigraphic succession within individual site sequences. The MAT type A possesses relatively high numbers of handaxes and lesser frequencies of backed knives whilst the MAT type B possesses relatively greater numbers of backed knives and fewer handaxes. In addition to the frequencies of these two tool types, the frequencies of the notches and denticulates, although variable, are often high and tools of Upper Palaeolithic type are also frequent. Bordes noted that there was a trend for the Upper Palaeolithic tools to move from atypical to typical forms as the industry moved from type A to B, and that a typologically intermediate form, type A/B could be identified (Bordes 1972a, 1972b). Technologically there does not appear to be any marked directional shift towards increased blade use. So the evidence for directional typological change within the MAT is strong and observations as to its date emphasise the similarities of this industry with the Châtelperronian. It may, therefore, be argued that a sequence of MAT A, MAT A/B, MAT B, Châtelperronian represents the process of lithic typological transition but that technologically the switch to an essentially blade-based technology occurs with the Châtelperronian itself. The process of transition may, from this evidence, be considered to begin in terms of the typological lithic base, within the Würm II stadial dated to sometime around 50 000 BP (Valladas *et al.* 1986). It is worth asking, however, whether further, longer term, directional change occurs within the Mousterian variants.

Once again Mellars' work on chronology is significant for he has observed patterning both between the Charentian and the MAT (noted earlier) and within the Charentian. The internal pattern within the

Charentian is based upon the use of Levallois technique, for two types or 'facies' may be identified; one, the Ferrassie, which uses Levallois technique whilst the other, the Quina, is a non-Levallois industry. Gradual change in technological indices do occur within the Ferrassie as the use of Levallois technique declines and Mellars additionally notes a chronological pattern of Quina Mousterian succeeding the Ferrassie based on site stratigraphy. One can add to this a study of the Quina variant itself by Le Tensorer which also identified directional patterns of techno-typological change (Le Tensorer 1978) and the suggestion by Bordes (1972a), that the Quina evolves out of the Ferrassie. The Charentian and the MAT are the most distinctive of the Mousterian variants and application of the Bordes typology has clearly demonstrated not only the existence of these variants but certain temporal relations as well. Thus, the typology must be considered successful in fulfilling its aims of identifying time-space systematics. What these patterns mean, however, remains to be seen.

The more general variant forms, the Typical and the Denticulate are not so easily understood in chronological terms although some of the following points may be of interest.

The Denticulate variant is defined principally on the basis of high frequencies of notched and denticulated forms. Such pieces may be created by natural processes such as cryoturbation and so caution should be exercised in attributing a cryoturbated layer to the Denticulate variant, particularly if overall tool numbers are low. Further, the Denticulate may be susceptible to archaeological sampling processes. It has been generally observed that denticulate tools occur in greater frequency in Mousterian layers towards the rear of cave and rock-shelter sites whilst side-scrapers occur in higher frequencies towards the front (Bordes 1972a). In multi-layer sites where rear wall recession is a feature a single sounding will be collecting samples reflecting this variance of distribution even if occupational debris is identical. Thus, higher levels will tend to contain more side-scrapers. However, the degree to which this factor operates is not yet isolable from publications and the nature of the rear of the site will vary the impact of this phenomenon. A site with a rear wall sloping inwards towards the site may fill up sufficiently rapidly to reverse the sampling problem, for successive occupations would then be placed relatively further forward resulting in upper levels yielding more denticulates. Such a phenomenon may be present in the Würm II sequence at Combe Grenal (Bordes 1972a). An example of rear wall recession without rapid deposition of cultural material affecting tool recovery would account for the presence of the Denticulate in the lower levels of Peyrony's excavations at La Ferrassie (see Bordes 1984).

This aspect of recovery does not only affect vertical point sampling but should also involve horizontal distribution. If a trench runs parallel to the site rather than sectioning it at right angles to the rear wall, then once again sampling may recover relatively more of a tool group. It is interesting to observe that at the site of Combe Grenal the upper layers, 1 to 35, were excavated in an area in which the long axis ran parallel to the rear wall of the site (Bordes 1972a). Material from these layers, then, has been sampled differently from the underlying deposits and it is these upper layers which show the greatest typological and chronological patterning. Such sampling issues have remained little discussed, probably as a result of use of the Bordes system which presents all material comparably regardless of the recovery method employed and its use of percentages which link the frequency of occurrence data between the different tool types. Larger scale excavations and presentation of horizontal distributions may go a long way towards overcoming such sampling problems.

The Typical variant is a generalised form which lacks distinctive artifact construction and has been well documented in Rissian contexts at the Abri Vaufrey (Geneste 1985). Its predominant occurrence at Combe Grenal in the Würm I may reflect these earlier roots and have chronological significance while further research into forms of the variant both rich in side-scrapers and alternatively poor in them may provide further relevant information. The separation of flake tool from biface dominated assemblages in the Middle Palaeolithic may be presenting a more complicated and confused picture of industrial succession than is useful. Additionally, it is worth asking whether the separation of the Mousterian from the other Middle Palaeolithic assemblages is useful to contemporary research. This separation currently stresses a historical artifact which is not necessarily representative of a prehistoric phenomenon. Questions of directional change, succession and rates of innovation should be as inclusive as possible if they are to be informative.

To summarise the discussion so far, it has been shown that chronological patterns of typological change occur both between and within the most distinct of the Mousterian variants. The patterning shown by the MAT in particular is important in revealing the actual scheduling of the typological lithic transition to the Upper Palaeolithic as beginning in the Würm II stadial. The significance of the differences between the Charentian and the MAT and the chronological separation of them is currently not clear. Other variant forms occur in typical form in the preceding glaciation and intermittently throughout the Würm I and II. Issues of sampling, chronology and recovery need careful consideration before these variants may be effectively evaluated.

It should be noted at this point that the chronological observations of Mellars have been disputed using a sediment-based inter-site composite stratigraphy (Laville 1973, 1975; Laville *et al.* 1980). This scheme was originally presented when the Mousterian period was considered to span only half the time now recognised and so to support it requires a stretching of the sequence (Dennell 1983). The possibility of omissions in the complex series of climatic and palaeoenvironmental records must be made greater by this lengthening of the period. Several of the sites used to construct the sequence show hiatuses and so potentially the scheme is flawed. Additionally, serious questions have been raised as to the methodology employed, particularly in terms of sampling and reconstruction (Reynolds 1985); these will not be repeated here, except to note that our understanding of cave taphonomy and weathering processes are still being developed and application of a technique employing such methods entails many assumptions as to the comparability of the data and so requires caution. If the scheme could incorporate a full account of the sampling employed and use statistics to show the relative strengths of correlations between site stratigraphies, then it might be a useful chronological tool. Until such a presentation takes place, the internally consistent observations of Mellars retain strength in their immediate simplicity and lack of derived assumptions, and also the fact that since being presented over twenty years ago, no single site has yet produced contradictory evidence.

The chronology issue cannot as yet be resolved by reference to absolute dating methods as these have not yet been adequately applied to the Charentian variant – a variant which plays an important role in Mellars arguments. So for the moment, the picture of industrial succession and chronology presented by Mellars remains the strongest view. It shows typological transition towards the Upper Palaeolithic in the final Mousterian, can any technological trends be isolated?

Generally, the Upper Palaeolithic is characterised by blade as opposed to flake technology. Explanation as to why this change should occur has focussed on two aspects, economy and cognitive ability (Bordes 1969; Demars 1983; Geneste 1985). The economic argument states that blades present a greater amount of usable edge than flakes and are thus more economic a use of flint, but many blade tools are focussed upon the ends of the blade and so this argument is weak. It is further weakened in discussing relative ideal forms rather than being based on assemblage analysis. Linking the economic and cognitive views is the area of raw material selection and preference but currently approaches to this remain limited. The motivations of the prehistoric knapper are beyond the reach of present methodology. The cognitive aspect is a further problem when trying to assess the technology of solely the Middle Palaeolithic where attention concentrates mainly upon the prepared core and flake techniques, especially Levallois technique.

This latter technology will be discussed as an example of the poverty of current approaches to lithic technology. It is hoped that this will reveal something of the need to develop specific, problem-oriented studies which operate outside the existing framework with its many built-in assumptions as to evolution in cognitive ability. Until this is achieved, interpreting the significance of changes such as the Middle-Upper Palaeolithic transition will be self-fulfilling.

A Levallois flake is a flake whose shape has been predetermined by a careful preparation of the core prior to the removal of that flake (Bordes 1961). The technology may not always be the same but the critical element is the idea of a predetermined form (Bordes 1980). From this it may be gathered that the Levallois technique is one aimed at fulfilling a mental template of the knapper. Often related to this is the idea of a correlation of morphology with function. So it was that the Binfords could interpret Bordes tool types patterning within the Mousterian variants to be that of toolkits (Binford and Binford 1966). If the knapper were successful in his preparation, a minimum of retouch would be required for the Levallois blank to be incorporated into its functional sphere. The correlation of function with blank/tool morphology is a problem that has become inherent in the application of Bordesian typology. Bordes himself noted that the Levallois technique was used for one of three different purposes; first, to obtain flakes that were twice as long as they were wide, second, to obtain blades and third, to obtain triangular or pointed blades or flakes (1980). Correlating with these differing purposes were specific core and/ or blanks. However, Bordes did also note that it is possible to obtain morphologically 'Levallois' points from cores other than the classic form he described (1980: 47). Significantly, he further noted that, as described above, Levallois technology has vast time and geographical distribution (1961: 26). As it is currently understood Levallois technique is witnessed in the Upper Palaeolithic blade cores of Europe (Bordes 1947; Newcomer 1975), and in the complex microblade technology of Japan (Kobayashi 1970), in addition to its more commonly recognised occurrence in the Lower and Middle Palaeolithic of Africa and Eurasia. The question must therefore be asked, what is the significance of Levallois technique and how useful is such a concept?

A Levallois blank is to be recognised by possession of truncated negative dorsal scars, in a classic case these are radially oriented on the flake, whilst they tend to be uni- or bi-directional on the points and blades. Platform facetting, although often associated with Levallois technique, is not a necessary element (Bordes 1961: 26), and so shall not be included in this discussion. It is reasonable to ask if such features are sufficiently distinctive. The majority of flakes removed from a core in a reduction sequence, once decortication has taken place, will show truncated dorsal scars and the significance of their orientation is difficult to discern without the ability to refit the flake to the core. Often it is not possible to tell from where specifically, and in what sequence, the flakes causing the negative dorsal scarring were removed. It may, perhaps, be possible to use the location of maximum thickness and dorsal convexity as significantly Levallois features but, in the author's experience of flakes from Mousterian contexts, this is not an infallible method either. Replication experiments aimed at reproducing classic Levallois flakes have suggested that an angle of intersection between the platform and the dorsal surface of between 60°-80° is significant (Boëda and Pèlegrin 1979), but given a large collection of flakes use of this method would be fallacious as these angles fall within the 'normal' range of such measurements – an angle of less than 90° being generally necessary for predictable flaking. So, it may be seen that in applying the Levallois concept to lithic assemblages, there is no clear means of identifying its products from a collection of blanks. This problem is accentuated by the inclusion of atypical Levallois flakes in the Bordes system. Points and blades of Levallois aspect may be produced from cores not classically Levallois and so cannot help shed any further light on this problem.

Turning to the cores, these are often characterised by the presence of a large, single, negative flake scar dominating one surface which truncates other 'preparatory' removals. However, there is no necessary singularity in the removals aspect of Levallois technique definition. The definition concentrates on the idea of preparation of the flake on the core rather than on core preparation itself (an important distinction which should be noted), and so recognition by core alone also presents problems. A single, major negative flake scar may be the product of a rejuvenation technique, removing surplus overburden on the core surface and aiding the maintenance of platform angles. Another problem is determining whether discarded core form reflects the desire for a given shape, or an aim to consistently reduce a given piece of stone. Patterning in the core form and its preparation may reflect a problem-solving approach to a specific raw material or materials. Most study involving Levallois technique is simply descriptive; intention has already been assumed. The problem with Levallois concepts is that there is an inherent assumption of intent and so the potential information on problem-solving is overlooked. The intention involved in reduction sequences is best studied through a combination of refitting and analysis of frequency of occurrence of relatively fixed morphology. Should a specific morphology recur then the likelihood may be that this form has been deliberately selected. However, should the purpose of the Levallois blank-tool not commonly occur at a site such a frequency approach would fail to recognise a Levallois presence. The identification of a Levallois aspect depends totally upon recognising intent and it is this which negates the utility of the Levallois concept to contemporary archaeology. By definition, archaeological recognition of Levallois technique is *post hoc* and therefore, intent is unlikely to be satisfactorily demonstrated. This point is significant particularly because an experimental replicative approach to the study of lithic technology can be applied to Levallois technique. This may produce a 'recipe' by which to achieve a blank of given form (Boëda and Pèlegrin 1979), but intent in the past cannot be correlated specifically with the final form of the blank removed. Indeed, given the occurrence of a number of hinged and step-terminated specimens from the Mousterian assemblages of Combe Grenal, one may question the degree to which, if a specific form were sought, the worker was capable of achieving it! It is a fact that the only true Levallois technique is that produced by replicative experimenters today.

All this is not to deny the existence of lithic technologies aimed at deriving specific objects, the blank patterning of Levallois flakes at Bakers Hole, England is a convincing example (based on the frequency of form approach), of the desire of prehistoric knappers to produce blanks of a given form. Further, few would argue that Upper Palaeolithic prismatic blade cores were not designed to produce blades or that the Japanese microblade technologies were not designed specifically for that purpose. The point is that all these technologies are identifiable at a more appropriate and informative level, within the temporal and spatial contexts of the sites in which they occur and the raw materials that were available. The all-pervading Levallois technique masks an interesting and potentially more informative level of analysis. The author would therefore argue for a replacement of Levallois concepts by more specific labelling and study. The concept of Levallois technique was framed at a time when a more general perspective was used, and it would now be more useful to study a somewhat more limited level of variability that is better suited to current interests. In this light the more specific work of Geneste examining reduction sequences in the southwestern French Middle Palaeolithic by a combination of refitting and experimental knapping is an excellent example of what can be achieved (Geneste 1985).

Turning to blade technology, it can be observed that this is often taken to indicate a cognitive advance over the preceding flake-based industries. This preconception is probably derived from the observation of blade frequency in the Upper Palaeolithic in industries generally associated with anatomically modern man. However, this picture has now changed following the discovery of Neanderthal remains associated with the Châtelperronian (Lévêque and Vandermeersch 1981). In an attempt to retain a higher mental image of modern man, it may be suggested that this phenomenon represents Neanderthals influenced from elsewhere by modern man. This may be the case but it should be asked why the association of blades with modern man is stressed, for earlier industries in the Middle Palaeolithic may comprise large numbers of blades (in certain Mousterian cases, 40% of an assemblage may be blades (Bordes 1984)). There is no reason to suggest this was an accident, blades occur in many assemblages dated to around the last interglacial in northern France (Bordes 1984). Why, then, is there a need to explain away the Neanderthal association with the blade-based Châtelperronian?

The real need is for study into the reduction sequences of these blade assemblages, their use of raw material and selection for tool manufacture in the light of raw material sources, which may be compared to similar data on flake-based assemblages. Comparison should also involve associated remains such as fauna, site context, industrial richness, etc. It may be that these assemblages comprise part of a complementary set of behavioural regimes which change over time to favour a blade-based orientation at the time of the Middle-Upper Palaeolithic transition. The typological and technological systems currently employed do not facilitate the study of a full behavioural system over a long time span.

This question of archaeological classification is also of critical importance in understanding the Middle-Upper Palaeolithic transition for, as was noted earlier, two different systems span the important industries. It is worth reviewing these to examine whether certain interpretations and assumptions as to the nature of the transition are based on a misunderstanding due to the use of different systems. In this review the main theme to be discussed is related to the techno-typology of the industries in relation to the typologies employed, that of the nature of the 'tool', its reflection of prehistoric mental imaging and interpretive use to examine rates of innovation and complexity. By investigating these issues, it is hoped that the comparability of the two typological systems will be revealed. Both systems derived the tools listed from a wide experience of Palaeolithic materials where frequently seen forms were recognised intuitively and incorporated into the list. So the final list comprised, in both cases, a set of intuitively recognised types which were included on the basis of frequency within the typologist's own previous experience. The concept of the 'tool' employed was based on the assumption that a successful form will become increasingly fixed through time and so tools will essentially reflect the mental templates of their manufacturer. Pieces that are atypical or intermediate were considered as 'noise' and regarded as insignificant, being probably due to the variable abilities of individuals in fulfilling their intentions (Bordes 1967, 1969, 1970). It is important to note that both systems stress frequency of occurrence as the reasoning behind tool selection for use in the type list and yet neither system provides a means of assessing such frequencies. The existence of a relatively fixed form of tool is assumed but this assumption is not yet adequately studied. It is possible that a continuous dataset is being arbitrarily divided making recent work like that of Dibble (1987) of particular note. What significance to attach to a tool in terms of interpretation has been subsumed within the typological systems but must be discussed. The type lists for the Mousterian comprise a set of variable types some of which have atypical forms and others have subtype forms. Certain types are based on frequency of occurrence at a single site and yet are rare elsewhere, the particular relevance of any type in interpretation of the list as a whole requires much greater consideration if effective discussion is to be undertaken as to the significance of the Mousterian variants. Thus, whilst the order of type occurrence along one axis is fixed for the cumulative percentage frequency graphs, it is possible that any individual artifact may be placed at a variety of different positions. Until the question of the validity of frequency-based types is tackled, the intermediate and atypical types are of influence, for the placing of these within the system may skew a graph form towards a typologist's preconceived impressions. For example, a transversely retouched piece from a Mousterian assemblage may be either a transverse side-scraper, a truncation or an end-scraper depending upon the 'feel' of an assemblage and the worker's guess as to original blank form. If other pieces are of Quina retouch the worker would probably place the piece with the side-scrapers; if, however, it derives from a high stratigraphic position, or an assemblage yielding handaxes, a placement amongst Upper Palaeolithic types may be preferred.

In addition to complicating the current picture of Mousterian variability, questions as to the reality of tool types have relevance to understanding the Middle-Upper Palaeolithic transition, for it has been argued that the Upper Palaeolithic represents a period of greater innovation and complexity in tool form when compared to the Mousterian (Mellars 1973). How far can this be attributed to the use of different typologies? The Mousterian tool is based generally upon blanks of irregular or semi-regular form; it is often difficult to determine what should constitute an 'edge' from an 'end'. Indeed, on irregular flakes, the only consistent reference point is the point of percussion and it is to this that all the other variables in tool description are referred in the Bordes system for the Lower and Middle Palaeolithic. In the Upper Palaeolithic the most common blank employed is the blade which, in addition to the point of percussion, has a greater regularity of form permitting more easy reference to laterals and an enhanced perception of 'end'. This regularity of form goes beyond the imposition of a flake axis for Mousterian tools and so additional factors are introduced into type description in the Upper Palaeolithic. Given a more regular form the terms proximal and distal can be successfully employed and oblique retouching identified. The major variables by which a tool is described, however, remain the same, these being edge form, retouch form and placing of retouch upon a piece, regardless of whether Mousterian or Upper Palaeolithic systems are being employed. As the number of reference points on a blank increases so do the numbers of tools that can be classified, even holding the major variables constant, so the use of a more regular blank form in the Upper Palaeolithic will naturally present the possibility of classifying a larger number of tool types than the Mousterian. Additionally, the Upper Palaeolithic typology uses size as a variable, differentiating between blade and bladelet tools. So, when one examines the tool type lists of both systems it becomes clear that both include subdivisions of what are more generally identifiable tool groups (e.g. side-scrapers, end-scrapers, burins, etc.) and that whilst the Upper Palaeolithic system would superficially appear to document more complex types and greater type variety, this is, in fact, an illusion. The retouching technology remains the same with one possible exception in the assumed pressure-flaking of Solutrian pieces (a later Upper Palaeolithic phenomenon) and it is blank selection in the Upper Palaeolithic that facilitates the recognition of more forms.

One further observation to be made is that the two systems operate different selection procedures when complex and compound tools are present. In the Mousterian these are simply placed into the least common type of the tool types present on the blank whilst in the Upper Palaeolithic many compound types are included into the list as a single compound entity. Reflection upon the type lists of both typological systems reveals the same broad tool categories with remarkably little innovation in real terms. It must be stressed that the use of typology and its creation of varying numbers of tool types is a reflection of our own ability to classify material and much less a measure of the mental imaging of earlier times. Comparing rates of innovation and the development of complexity require use of the same system of typology where such comparisons will at least be internally consistent.

It is here suggested that there are major difficulties in the interpretation of both Bordes and de Sonneville-Bordes and Perrot's typologies due to the emphasis on frequencies of intuitive types with no means of identifying such frequencies in 'real' terms. The Middle-Upper Palaeolithic transition as it is pictured through a 'before and after' approach highlights differences in the two typological systems in the treatment of compound forms and also the technological move towards increased use of regular blank forms with additional types being size-based. As such, this approach presents an image of increased typological complexity and innovation that is not, in fact, substantiable when the final Mousterian (MAT) and first Upper Palaeolithic industries are taken alone. This fact might suggest two things. Firstly, both typologies are demonstrably successful in their aims of documenting time-space systematics but what the behavioural information encoded means is unclear. There is a need for coherent problem-orientated studies of the lithic evidence using specifically-developed typologies.

Secondly, that the process of change that led to the Middle-Upper Palaeolithic transition in southwestern France probably began, in lithic terms, prior to, or during, the time at which the MAT was dominant in the region. The transition appears to represent a behavioural change in the manufacture of tools, possibly reflecting a change in activities undertaken. This being the typological shift during the MAT, followed by a later change towards habitual selection of blades for the manufacture of these tools. This move to blade technology may have been prompted by any number of things, amongst these possibly a specialisation of roles within society, a change in the social organisation or even a desire to exploit a standard blank form which would then provide a more readable 'field' for the input of social information.

It is interesting to speculate that as Neanderthals are associated with both earlier Mousterian industries and the Châtelperronian, they are likely to have been associated with the process of change itself and so, in southwestern France at least, the Middle-Upper Palaeolithic transition may be a Neanderthal phenomenon. Whether this represents a possible 'bow-wave' effect of ideas from elsewhere or was internally generated cannot be determined from present evidence.

This review has focussed on the lithic evidence, studies in more detail of this, and of other behavioural evidence such as bone working, settlement complexity, and art are needed to isolate the different aspects of the transition. It may appear that considerable differences in the scheduling of change in these various areas occur and thus the nature of the process of transition would become much clearer. This enhanced picture would then more easily show the interaction of behaviours in the process of physical and cultural evolution during the last glacial and the role played by Neanderthal populations in such changes

REFERENCES

Binford, L.R. and Binford, S.R. 1966. A preliminary analysis of functional variability in the Mousterian of Levallois facies. *American Anthropologist* 68: 238-295.

Binford, S.R. and Binford, L.R. 1969. Stone tools and human behavior. *Scientific American* 220: 70-82.

Boëda, E. and Pèlegrin, J. 1979. Approche technologique du nucléus Levallois à éclat. *Etudes Préhistoriques* 15: 41-48.

Bordes, F.H. 1947. Etude comparative des différentes techniques de taille du silex et des roches dures. *L'Anthropologie* 51: 1-29.

Bordes, F.H. 1953a. Notules de typologie paléolithique I: outils moustériens à fracture volontaire. *Bulletin de la Société Préhistorique Française* 50: 224-226.

Bordes, F.H. 1953b. Notules de typologie paléolithique II: pointes Levalloisiennes et pointes pseudo-Levalloisiennes. *Bulletin de la Société Préhistorique Française* 50: 311-313.

Bordes, F.H. 1953c. Essai de classification des industries 'moustériennes'. *Bulletin de la Société Préhistorique Française* 50: 457-466.

Bordes, F.H. 1954. Notules de typologie paléolithique III: pointes moustériennes, racloirs convergents et déjetés, limaces. *Bulletin de la Société Préhistorique Française* 51: 336-339.

Bordes, F.H. 1961. *Typologie du Paléolithique Ancien et Moyen.* Bordeaux: Delmas.

Bordes, F.H. 1967. Considerations sur la typologie et les techniques dans le Paléolithique. *Quartär* 18: 25-55.

Bordes, F.H. 1969. Reflections on typology and techniques in the Palaeolithic. *Arctic Anthropology* 6: 1-29.

Bordes, F.H. 1970. Réflexions sur l'outil au Paléolithique. *Bulletin de la Société Préhistorique Française* 67: 199-202.

Bordes, F.H. 1972a. *A Tale of Two Caves.* New York: Harper and Row.

Bordes, F.H. 1972b. Du Paléolithique moyen au Paléolithique supérieur: continuité ou discontinuité? In F.H. Bordes (ed.) *The Origins of Homo Sapiens.* Paris: UNESCO: 211-218.

Bordes, F.H. 1980. Le débitage Levallois et ses variantes. *Bulletin de la Société Préhistorique Française* 77: 45-9.

Bordes, F.H. 1984. *Leçons Sur Le Paléolithique.* Cahiers du Quaternaire 7. Paris: Centre Nationale de la Recherche Scientifique.

Bordes, F.H. and Sonneville-Bordes, D. de. 1970-1. What do Mousterian types represent? The significance of variability in Palaeolithic assemblages. *World Archaeology* 2: 61-73.

Demars, P.-Y. 1983. *L'Utilisation du Silex au Paléolithique Supérieur: Choix, Approvisionnement, Circulation. L'Example du Bassin de Brive.* Cahiers du Quaternaire 5. Paris: Centre Nationale de la Recherche Scientifique.

Dennell, R. 1983. A new chronology for the Mousterian. *Nature* 301: 199-200.

Dibble, H.L. 1987. The interpretation of Middle Palaeolithic scraper morphology. *American Antiquity* 52: 109-117.

Geneste, J.M. 1985. *Analyse Lithique d'Industries Moustériennes du Périgord: une Approche Technologique du Comportement des Groupes Humains au Paléolithique Moyen.* Unpublished PhD Thesis, University of Bordeaux I.

Kobayashi, T. 1970. Microblade industries in the Japanese archipelago. *Arctic Anthropology* 7: 38-58.

Laville, H. 1973. The relative position of Mousterian industries in the climatic chronology of the early Würm in the Perigord. *World Archaeology* 5: 323-329.

Laville, H. 1975. *Climatologie et Chronologie du Paléolithique en Périgord.* Marseilles: Laboratoire de Paléontologie Humaine et de Préhistoire, Université de Provence.

Laville, H., Rigaud, J.P. and Sackett, J. 1980. *Rock-Shelters of the Perigord.* London: Academic Press.

Lévêque, F, and Vandermeersch, B. 1981. Le Néandertalien de Saint-

Césaire. *La Recherche* 12: 242-244.

Mellars, P.A. 1965. Sequence and development of Mousterian traditions in southwestern France. *Nature* 205: 626-627.

Mellars, P.A. 1967. *The Mousterian Succession in South Western France.* Unpublished PhD Thesis, University of Cambridge.

Mellars, P.A. 1969. The chronology of Mousterian industries in the Perigord region of south-west France. *Proceedings of the Prehistoric Society* 35: 134-171.

Mellars, P.A. 1973. The character of the Middle/Upper Palaeolithic transition in south-west France. In A.C. Renfrew (ed.) *Explanation of Culture Change.* London: Duckworth: 255-276.

Mellars, P.A. 1986. A new chronology for the French Mousterian period. *Nature* 322: 410-411.

Newcomer, M.H. 1975. Punch technique and Upper Palaeolithic blades. In E.H. Swanson (ed.) *Lithic Technology: Making and Using Stone Tools.* The Hague: Mouton: 97-102.

Reynolds, T.E.G. 1985. Towards a Mousterian chronology. *Cave Science* 12: 129-31.

Sonneville-Bordes, D. de and Perrot, J. 1953. Essai d'adaptation des méthodes statistiques au Paléolithique supérieur: premiers résultats. *Bulletin de la Société Préhistorique Française* 50: 323-333.

Sonneville-Bordes, D. de and Perrot, J. 1954. Lexique typologique du Paléolithique supérieur. Outillage lithiques: I grattoirs; II outils solutréens. *Bulletin de la Société Préhistorique Française* 51: 327-335.

Sonneville-Bordes, D. de and Perrot, J. 1955. Lexique typologie du Paléolithique supérieur. Outillage lithique: III Outils composite – Perçoirs. *Bulletin de la Société Préhistorique Française* 52: 76-79.

Sonneville-Bordes, D. de and Perrot, J. 1956a. Lexique typologique du Paléolithique supérieur. Outillage lithique: IV burins. *Bulletin de la Société Préhistorique Française* 53: 408-412.

Sonneville-Bordes, D. de and Perrot, J. 1956b. Lexique typologique du Paléolithique supérieur. Outillage lithique (suite et fin): V outillage à bord abattu; VI pièces tronquées; VII lames retouchées; VIII pièces variées; IX outillage lamellaire. Pointe Azillienne. *Bulletin de la Société Préhistorique Française* 53: 547-559.

Le Tensorer, J.M. 1978. Le Moustérien type Quina et son évolution dans le sud de la France. *Bulletin de la Société Préhistorique Française* 75: 141-149.

Valladas, H., Geneste, J.-M., Joron, J.-L. and Chadelle, J.-P. 1986. Thermoluminescence dating of Le Moustier (Dordogne, France). *Nature* 322: 452-454.

White, R. 1982. Rethinking the Middle-Upper Paleolithic transition. *Current Anthropology* 23: 169-192.

10: The Early Upper Palaeolithic of Southwest Europe: Cro-Magnon Adaptations in the Iberian Peripheries, 40 000 – 20 000 BP

LAWRENCE GUY STRAUS

INTRODUCTION

As I recently pointed out (Straus 1983a), great contrasts exist between the Middle Palaeolithic (MP) and the *late* Upper Palaeolithic (LUP) in Southwest Europe in terms of technology, settlement pattern, subsistence, art and probably population density. The purpose of this paper is to explore the intervening period – the early Upper Palaeolithic (EUP) – and to discuss the nature of the transition which began with some technological changes and the anatomical change between *Homo sapiens neanderthalensis* and *Homo sapiens sapiens* and which ended with further technological developments and major organizational changes in the Solutrean and Magdalenian periods. The point is that the Middle-Upper Palaeolithic transition did not occur all at once at *c.* 35 000 BP. Aspects of the transition took at least 15-20 000 years to be achieved despite the appearance of anatomically modern humans sometime after that date in Southwest Europe.

This preliminary review of the EUP (= Châtelperronian, Aurignacian, Gravettian) evidence is focused on the Iberian Peninsula (Portugal and Spain) and on the adjacent regions of southernmost France (Pays Basque, Gascogne, Languedoc, Roussillon). Rather than an exhaustive survey of the subject, this contribution seeks to highlight some of the principal characteristics of human settlement, subsistence, technology and art during the period *c.* 40 000-20 000 BP in this important but relatively little synthesized region of prehistoric Europe. In addition, an attempt will be made to compare the Iberian EUP with the preceding MP (Mousterian) and with the succeeding LUP (Solutrean, Magdalenian/Epigravettian). A disproportionate amount of the available data comes from Vasco-Cantabrian Spain and the adjacent French Basque Country, areas with which I have some first-hand experience and for which there exist excellent summaries (e.g. Bernaldo de Quirós 1982; Arambourou 1976a, 1976b; Clottes 1976; Sacchi 1976a; Bahn 1984). EUP data are apparently scantier in Levantine Spain and in Portugal, and I have relied there mostly on up-to-date secondary sources (e.g. Fullola Pericot 1983; Cacho 1980, 1982; Zilhão 1985a, 1986).

Relatively few excavations of Aurignaco-Périgordian deposits have

Table 10.1. Radiocarbon dates for Middle and Upper Palaeolithic sites. Sources: Almagro & Fernández-Miranda 1978; Altuna 1972, 1984; Bahn 1984; Bofinger & Davidson 1977; Butzer 1981; Fortea & Jordá 1976; Freeman 1981; Fullola 1983; González Echegaray & Barandiarán 1981; Moure & García-Soto 1980; Sacchi 1976b; Straus & Clark 1986. For comments on recent radiocarbon-accelerator dating of Castillo and L'Arbreda, see text.

Site	Level	Date	Lab.No.	Cultural Attribution
Languedoc				
Canecaude	IV	24,510±400	Gif-2710	Aurignacian
Canecaude	III	22,980±330	Gif-2709	Aurignacian
St. Jean-de-Verges		24,200±600	Gif-2941	Aurignacian
St. Jean-de-Verges		21,500±400	Gif-2942	Gravettian
Catalonia				
Els Ermitons		36,430±1800	CSIC-197	Mousterian
L'Arbreda		25,830±400	Gif-6422	Early Aurignacian
L'Arbreda		22,590±290	Gif-6421	Evolved Aurignacian
L'Arbreda		20,130±220	Gif-6420	Late Gravettian
Roc de la Melca		20,900±400	?	Late Gravettian
Valencia				
Les Mallaetes	XII	29,690±560	KN-1/926	Aurignacian II
Les Mallaetes	VI	21,710±650	KN-1/920	Early Solutrean (?)
Parpalló		>40,000	BM-858	Pre-Solutrean
Parpalló	6.25-7,75m.	20,170+380 -370	Birm-520	Early Solutrean
Parpalló	6.50-7.00m.	20,490+900 -800	BM-859	Early Solutrean
Granada				
Carigüela		48-28,000	TL dates	Mousterian

been conducted in the last couple of decades, although there are important exceptions such as Cueva Morín in Santander/Cantabria (González Echegaray and Freeman 1971, 1973, 1978), Canecaude in Aude (Sacchi 1976a, 1976b), L'Arbreda in Gerona (Soler and Maroto 1987; Fullola Pericot *et al.* 1986). In addition, several sites excavated long ago have been the objects of recent confirmatory excavations of limited scale, including El Conde in Asturias (Freeman 1977) and Brassempouy in Les Landes (Delporte 1985). Other recent excavations are at present essentially unpublished (e.g. Gatzarría, Amalda [but see Laplace 1966; Altuna *et al.* 1989]). However, comprehensive publications of the hitherto basically unpublished earlier excavations at El Castillo and El Pendo (both in Santander) and Les Mallaetes (Valencia) have recently appeared (Cabrera 1984; González Echegaray 1980; Fortea and Jordá 1976). Excavations of remnant Aurignacian deposits at El Castillo are currently underway (V.Cabrera, personal communication). Nonetheless, the great majority of

Site	Level	Date	Lab.No.	Cultural
Gibraltar				
Devil's Tower	3	>30,000	GrN-2488	Mousterian
Gorham's Cave	G	49,200±3200	GrN-1556	Mousterian
Gorham's Cave	G	>47,000	GrN-1678	Mousterian
Gorham's Cave	G	47,700±1500	GrN-1473	Mousterian
Gorham's Cave	D	28,700±200	GrN-1455	Aurignacian (?)
Gorham's Cave	D	27,860±300	GrN-1363	Aurignacian (?)
Burgos				
Cueva Millán	1a	37,600±700	GrN-11021	Mousterian
Cantabria				
El Castillo	(flowstone)	31,450±1400	I-5149	Post-Mousterian
La Flecha	(flowstone)	31,640±890	SI-4460	Post-Mousterian
Cueva Morín	10	35,875±6777	SI-951	Chatelperronian
Cueva Morín	8A	27,655±556	SI-952	Early Aurignacian
Cueva Morín	8A	27,335±757	SI-952A	Early Aurignacian
Cueva Morín	8A	27,685±1324	SI-956	Early Aurignacian
Cueva Morín	7	28,655±865	SI-955	Typical Aurignacian
Cueva Morín	7	27,240±1535	SI-955A	Typical Aurignacian
Cueva Morín	7/6	31,570±901	SI-954	Typical Aurignacian
Cueva Morín	5sup.	20,110±350	SI-953	Gravettian
El Rascaño	7	27,240+950 -810	BM-1456	Aurignacian(?)
El Rascaño	9	>27,000	BM-1457	Aurignacian(?)
La Riera	1	20,860±410	BM-1739	Pre-Solutrean
La Riera	1	20,360±450	Ly-1783	Pre-Solutrean
La Riera	1	19,620±390	UCR-1270A	Pre-Solutrean
La Riera	4	20,970±620	GaK-6984	Solutrean
Ekain	IXb	>30,600	I-11506	Pre-Châtelperronian
Ekain	VIII	20,900±450	I-13005	?
Lezetxiki	IIIa	19,340±780	I-6144	Aurignaco-Perigordian

current excavations in Pyrenean France, in Vasco-Cantabrian and Levantine Spain, in Andorra and in Portugal concern late or terminal Upper Palaeolithic periods, so we will for some time be forced to rely heavily on a data base derived from early excavations, some dating to the early decades of the 20th century and even to the last decades of the 19th.

CHRONOLOGY

The Mousterian of Gascogne-Languedoc is undated and that of Iberia is virtually undated. All extant radiocarbon dates for the MP and EUP of these regions are listed in Table 10.1, with the exception of some which are patently erroneous (i.e. determinations for MP and EUP deposits in the range of 15-10 000 years or less). Most of the determinations in excess of 28 000 BP are either infinite dates, dates on materials from deposits of uncertain cultural attribution, or show that the Mousterian was older than about 47 000 BP in the case of Gorham's Cave, a fact which does not help

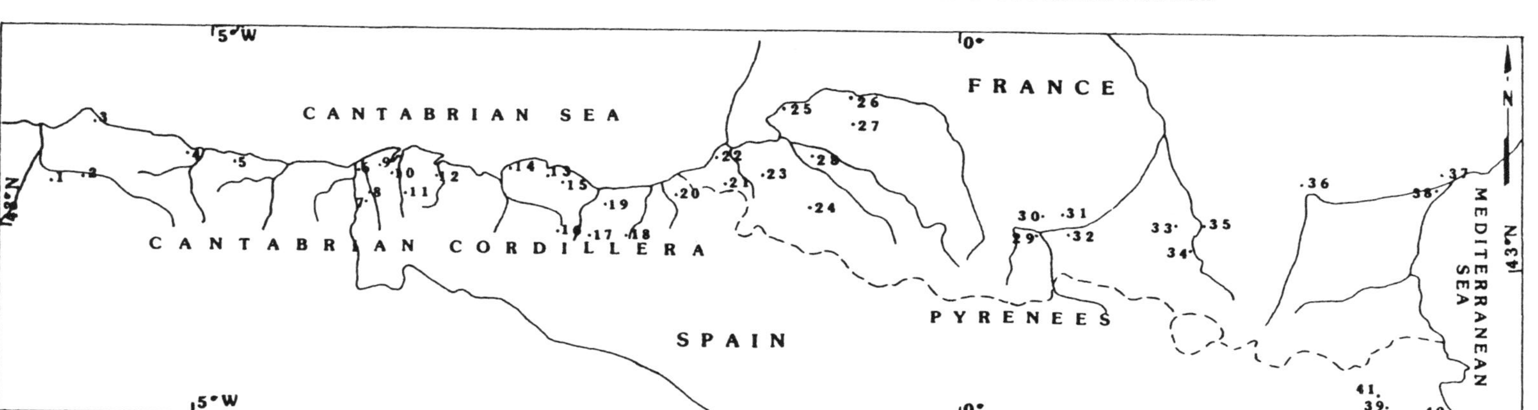

Figure 10.1. Early Upper Palaeolithic sites of north-central and northeast Spain and southern France. 1. El Conde; 2. La Viña; 3. Cueva Oscura de Perán; 4. El Cierro; 5. Cueto de la Mina, La Riera, Arnero; 6. El Cudón; 7. Hornos de la Peña; 8. El Castillo; 9. El Pendo, Camargo; 10. Cueva Morín; 11. El Salitre, El Rascaño; 12. El Otero; 13. Santimamiñe; 14. Kurtzía; 15. Lumentxa, Atxurra; 16. Bolinkoba; 17. Lezetxiki; 18. Usategi; 19. Ekain, Amalda; 20. Aitzbitarte; 21. Lezia; 22. Basté, Chabiague, Bidart, Mouligna, Villefranque; 23. Isturitz; 24. Gatzarria, Haréguy; 25. Tercis, Bénesse, St. Lon; 26. Montaut, Eyres; 27. Brassempouy, Pouillon, Gaujacq; 28. Saliès-de-Bearn, Labastide, Bidache, Sorde; 29. Gargas; 30. Montmaurin, Coupe Gorge, Gahuzère, Les Abeilles, Les Rideaux; 31. Aurignac I and II, Boussens, L'Hôpital; 32. Tambourets, Roquecourbère, Bouzin, Tarté, Marsoulas, Mas d'Azil, Téoulé, Rachat, Couteret; 33. Le Portel; 34. La Carane; 35. Tuto de Camalhot (Saint-Jean-de Vierge); 36. Canecaude, Les Cauneilles-basses; 37. Grande et Petite Grottes de Bize; 38. La Crouzade; 39. L'Arbreda, Reclau Viver; 40. Can Crispins, Calcoix; 41. Roc de la Melca.

give the age of the 'transition'. The only clear Mousterian date gives an age of 37-38 000 BP for Cueva Millán (Burgos), whereas the Châtelperronian of Cueva Morín has a date of about 36 000 BP, unfortunately with an error range of ± 6777 years (!). The 31 000 BP El Castillo and La Flecha (adjacent sites in Santander) travertine dates overlie Mousterian layers, and the Castillo one is apparently also associated with a 'typical Aurignacian' industry in Butzer's (1981) Unit 11a. Dates for 'Aurignacian' levels span a broad period, but the attributions of La Riera Level 1, Ekain Level VIII and Lezetxiki Level IIIa are problematical. The majority of the Aurignacian levels date between about 31 000-23 000 BP. There are fewer dates for the Gravettian (= Upper Périgordian), but they all cluster tightly around 21 000-20 000 BP, and overlap with the early Solutrean dates in both Cantabrian and Levantine Spain. In short, the early Upper Palaeolithic lasted from about 35 000 BP to about 20-21 000 BP, the latter being a rather arbitrary limit corresponding to the appearance of foliate Solutrean points. (The distinction is blurred by the presence of Gravettian elements such as Noailles burins in many Basque Solutrean assemblages and by the existence of shouldered points in Mediterranean Gravettian contexts.) This general time frame is similar to that of west-central France (Charentes, Périgord, Quercy, Auvergne).

Recently, however, J. Bischoff (Cabrera and Bischoff 1989; Bischoff *et al.* 1989) has obtained three radiocarbon-accelerator dates on charcoal from the basal Aurignacian at El Castillo ranging from 40 000 ± 2100 to 37 700 ± 1800 BP and four from the analogous level at L'Arbreda ranging from 39 900 ± 1300 to 37 700 ± 1000 BP. These series of dates are essentially stratigraphically coherent within the respective strata. In the case of L'Arbreda, the basal Aurignacian is directly underlain by a Mousterian deposit with two dates of 39 400 ± 1400 and 41 400 ± 1600 BP (plus one aberrant date of 34 100 ± 750 BP). The Aurignacian and Mousterian at El Castillo are separated by a sterile horizon. Bischoff (Bischoff *et al.* 1988) has also convincingly uranium-series dated travertines throughout the long Mousterian sequence at the Abric Romani, about 60 km south of L'Arbreda, between 60 000 and 40 000 years. The latter date represents the end of the Middle Palaeolithic at Romani, where it was overlain by an Aurignacian deposit (as yet undated). Bischoff's dates call into question:

1. The validity of all MP and EUP conventional ^{14}C dates;
2. The notion of an east-to-west migration migration of the makers of the European Aurignacian (or at least the vision of a relatively slow, 10 000 year migration); and/or
3. The idea of a relatively short period of chronological overlap between the Châtelperronian (preseumed Neanderthal) and Aurignacian (presumed Cro-Magnon) populations, and, by implication, an equally short period of potential acculturation and/or hybridization between these two populations.

GEOGRAPHICAL DISTRIBUTION

EUP sites have been definitely identified in the north-central Spanish provinces of Asturias (in its eastern half), Santander, Vizcaya and Guipuzcoa, but they are not known in the upper Ebro drainage or on the Meseta del Norte to the south of the Pyrenees and Cantabrian Cordillera respectively, although there are Acheulian and Mousterian sites on the Meseta. The distribution of EUP sites continues eastward along the 43rd parallel into the French Basque Country and Chalosse, where there are several major stratified cave sites (Isturitz, Gatzarría, Brassempouy) and a relatively large number of open air localities (Figure 10.1). This group of sites in extreme southwest France is separated from the abundant EUP sites of the Dordogne valley by the *terra incognita* of Les Landes de Gascogne, covered by sand dunes in Würm III and subsequently turned into swampland until drained in the 19th century (Arambourou 1979). So too are the sites of Haute-Garonne and Ariège (notably Gargas, Saint-Jean-de-Verges, and the type site of Aurignac) separated from the cluster of sites in Quercy and Agenais by a long empty stretch of the middle Garonne basin centred on Toulouse. The eastern French Pyrenean region *per se* is devoid of EUP sites, but slightly further north in the lower Aude drainage there are several rather isolated sites, notably La Crouzade and Canecaude.

However, Spanish Catalonia, particularly the area around Serinya (Gerona), is relatively rich in sites, notably L'Arbreda and Reclau Viver. Another cluster exists in coastal Valencia, the principal sites of which are Barranc Blanc, Parpalló and Les Mallaetes. Several EUP sites have also been found in Murcia and Almeria – notably Las Palomas, Las Perneras, Zajara, Ambrosio, Serrón and Los Morceguillos. There are references to at least a couple of sites in Málaga (El Chorro and El Higueron) (Cacho 1982), but the principal site in extreme southern Spain is Gorham's Cave (Waechter 1964). It is probably significant that, as in Cantabria, none of the Levantine sites (except Ambrosio) are more than about 50 km from the present coast, and most are much closer. Definite EUP sites are presently unknown in the interior of Spain.

The EUP of Portugal is represented by at most five sites – all in the region of Estremadura north of Lisbon (Zilhão 1985a, 1986). None of the materials come from modern excavations and all are currently being restudied, with the possibility that some may not be Aurignacian (J. Zilhão and A. Marks, personal communications). In addition, the only supposed 'Gravettian' level in the country is judged to be highly problematical and may be Solutrean (Zilhão 1985a: 103). As in Spain, all the sites are in lowlands near the present coast. This may *in part* be due to sampling problems (i.e. more intensive reconnaissance near the coasts where many of the major cities are located), but this cannot explain the absence of EUP sites even in the vicinity of major cities of the Iberian interior, where amateur and professional archaeologists have been active at least since the turn of the century or earlier. The high mesetas and sierras of the interior may have been relatively poor in food resources compared with the coastal zones, where many habitats are often to be found in close proximity to one another (even if EUP people did not make much or any use of marine foods). The more continental climate of the hinterlands may also have been a deterrent to significant human occupations during the Würm III stadial. (In that regard, it is worth speculating that Lower and Middle Palaeolithic occupations of the Iberian interior – so abundantly represented in the region of Madrid for example – may have taken place mainly under interglacial and early Würm interstadial conditions.) The distribution of LUP sites is fairly similar, but in northern Spain and in southern France, as well as to a lesser extent in Portugal, Solutrean and especially Magdalenian sites are also found in the mountains as human land-use patterns extended into the uplands with specialized (ibex-hunting) or seasonal occupations of the Cantabrian Cordillera, Pyrenees and outliers of the Serra da Estella. There is even some evidence for at least minor use of the Meseta in the form of rock art of probably LUP age (Mazouco in Tras-os-Montes, La Griega in Segovia, Penches, Atapuerca and Ojo Guareña in Burgos, Los Casares in Guadalajara and Maltravieso in Caceres: see Sauvet 1985). However the striking difference everywhere between the EUP and LUP is in the numbers of sites pertaining to each period.

SITE NUMBERS

In all of Vasco-Cantabrian Spain, a region so rich in Solutrean, Lower Magdalenian, Upper Magdalenian and Azilian sites, there is a total of only 20 Aurignaco-Périgordian sites plus a few others of problematical attribution. (Small, undated assemblages could in fact be of Solutrean age, albeit without foliate or shouldered points.) The number of sites is double for each of the succeeding periods, even though each of these only lasted about 3000 years as opposed to 15 000 for the aggregate EUP (Straus 1981). With the exceptions of clusters in eastern Asturias (around Posada de Llanes) and especially in central Santander, the EUP sites are isolated. Five of the EUP sites have both Aurignacian and Périgordian levels. As in the French Basque Country with the two sites of Isturitz and Gatzarría, these very few sites (Cueto de la Mina, El Pendo, El Castillo, Cueva Morín and Lezetxiki) contain most of the EUP materials for the region. However, at least 12 of the Vasco-Cantabrian sites are single-component and two (Bolinkoba and El Otero) have only multiple Gravettian and Aurignacian levels respectively. This situation is far different from that of the LUP, when multi-component sites (often major multi-purpose residential camps) and single-component sites abound throughout most parts of the region, frequently in major clusters (see Straus 1986). Indeed, discoveries of new Solutrean and especially Magdalenian sites are now common occurrences in Vasco-Cantabrian Spain, while EUP finds continue to be rare (a fact which may, admittedly, be due in part to the often deep, buried stratigraphic position of the latter in the caves of the region). Almost all the sites are in caves.

In the lowlands just north of the Pyrenees (up to the middle Adour in the west and the lower Aude in the east) latest available counts (Arambourou 1976a, 1976b; Clottes 1976; Sacchi 1976a) show about 40 EUP localities, about half of which are open-air sites – especially in the area around Biarritz, in Chalosse and near the Salat-Volp-Garonne confluences in lowland Ariège. Some of the cave sites are multi-component (e.g. Gargas, Les Rideaux at Lespugue, Tarté, La Tuto de Camalhot at St.-Jean-de-Verges, La Grotte du Pape at Brassempouy, Le Portel and Canecaude, besides Isturitz and Gatzarría). The 3000 year period of the Solutrean alone is represented by some 17 sites and find-spots in the same 400 x 100 km area. But it is "...with the Magdalenian that the distribution map fills up with a multitude of points, densest in the central Pyrenees, but in fact covering the whole chain" (Clottes 1976: 1221). As Clottes notes further,

> "This population density [is apparent] even despite the short duration of the Magdalenian [i.e. 17 000-11 000 BP]...It is only in the Magdalenian that people penetrated the deep valleys, above the moraines, and had access to the mountain sites...This vigorous expansion took place not only all along the chain, but also in depth. As has often been noted, Magdalenian people moved into the high Pyrenean valleys and even occupied upper altitude sites" (Clottes 1976: 1221).

This is true of the Solutrean and Magdalenian of Cantabrian Spain as well (e.g. Straus 1986, 1987a; see also Zilhão (1986) for central Portugal and Cacho (1986) for a high mountain Magdalenian site in Alicante).

In Catalonia there are apparently only six EUP sites and in Valencia only three, according to the recent summaries of Fullola Pericot (1979, 1983) and Cacho (1982). A few of these are open-air localities. Three of the cave sites (L'Arbreda, Reclau Viver and Les Mallaetes) have both Aurignacian and Gravettian components. In southeastern Spain there are 11 EUP sites reported by Cacho (1980, 1982). In contrast, there are 28 known Solutrean sites alone in all of eastern and southeastern Spain (Fullola Pericot 1979). Several of these sites have multiple Solutrean levels (e.g. L'Arbreda, Parpalló, Les Mallaetes, Barranc Blanc, Ambrosio). Magdalenian and/or Epigravettian sites are also relatively abundant (29) in Levantine Spain (Cacho 1982). The Solutrean and Magdalenian deposits are only rarely from the same caves and, of course, represent a much shorter time-span than the EUP. The four or five known Portuguese EUP sites are also outnumbered by at least 14 Solutrean sites. As of 1985 a half-dozen bonafide Magdalenian sites were known in Portugal, but more have subsequently been found and the number continues to grow (Zilhão 1985a and personal communication). Open-air sites are present for the Upper Palaeolithic of Portugal.

The EUP population of Iberia appears to have been very low and many areas seem to have been unpopulated or at least used with very low frequency – notably the vast area of central Spain and Galicia. Some

Table 10.2. Early Upper Palaeolithic ungulate and carnivore faunas from Vasco-Cantabrian Spain. 'A'=Aurignacian; 'C'=Châtelperronian; 'G'=Gravettian; 'p'=present; 'c'=common; 'a'=abundant. Numbers indicate MNI's per species.

Sites:	Riera	Conde			Rascaño			Pendo								
Levels:	1	4	3	1	9	8	7	VIIIb	VIII	VII	VI	Vb	Va	V	IV	III
Cultural attribution:	?	A	A	A	A	A	A	A	C	A	A	A	G	G	A	A
Equus caballus	5	1	1	1					1	3	1		1	1	1	3
Capra pyrenaica	4	1	2	3	1	3	2			1						
Rupicapra rupicapra	1					1										
Bos/Bison	4			1	1	1		1		1		1	1	1	1	3
Capreolus capreolus	1							1		1	1		1			
Cervus elaphus	7	1	2	1	2			2	1	5	7	1	4	5	8	12
Megaceros sp.											1					
Dama sp.																
Rangifer tarandus																
Sus scrofa															1	1
Dicerorhinus kirchbergensis																
Coelodonta antiquitatis																
Elephas primigenius																
Elephas antiquus																
Panthera pardus																
Felis spelaea										1				1		
Felis silvestris																
Lynx lynx																
Canis lupus			1							1		1				
Vulpes vulpes																1
Alopex lagopus																
Cuon alpinus																
Crocuta crocuta								1		1						
Ursus spelaeus												1			1	1
Ursus arctos																
Gulo gulo																

Sources: Altuna 1971, 1972, 1977, 1981, 1986, Altuna & Mariezkurrena 1984; Cabrera 1984; Fuentes 1980; Straus 1977.

Cueva Morín									Bolinkoba	Lezetxiki				Aitzbitarte	Ekain				Castillo				
10	9	8	S	7	6	5i	5s	4	E	4C	4A	3A	2	V	10a	9b	9a	8	18	16	14	13	12
C	A	A	A	A	A	A	G	G	G	A	A	?	G	A	C?	?	?	?	A	A	G	?	G
1	1	3	2	1	2	3	4	4	2	1	2	2	3	1					a	a	a	p	a
			1		1	2	3	3	11					2				2	p	p	p		p
			1			1	1	1	3	1	1	12	8	5	2	16	11	5	p	p	p	p	p
1	1	1	3	2	4	3	3	2	2					2	4	6	5	1	29	p	11		a
		1	1	1	4	6	5	4		1	1	1	2	1			3	1	p	p	p	1	1
1		1	7	5	4	10	9	10	1	4	2	8	3	7	5	4	3	5	216	c	10	p	
																						p	
										1		1									p		
		1	2				1					1						1	p		p		
											1	1							c	p	p		
												1											
								1															
																			p				
						1					1	1				1			p		p	p	
																			p				p
					1	1		1															1
								1															
						1		1			1	2	1	1		2	1	1		p	p	p	p
					1	1	1	1	3		1	1	2			2	1	2		p	p		p
							1				2					1					p		p
										4	3	7	2				11	6	>10	p	p		p
									2					1					p				
													1										

regions may have only been occupied sporadically. Even the relatively rich pre-Pyrenean distribution of sites seems spotty and geographically isolated from both the Périgord and Provence where there are major groups of EUP sites. Although there exist many gaps in the circum-Iberian distribution of EUP sites, obviously at least ephemeral sites must have existed in such regions as Tarragona, Castellón, Cadiz, Huelva, Algarve and Alentejo to provide continuity to the apparently coastal lowland settlement pattern, however low the density of population. However, EUP site numbers are still significantly greater in most regions than those for the MP, a period of at least 70 000 years, even though the distribution of MP and EUP sites seems essentially identical with the exception of the presence of some Mousterian sites on the Meseta.

SUBSISTENCE

There are very few substantial faunal studies from modern excavations of EUP deposits in southernmost France or Iberia (none at all really from Portugal, southeast Spain, Catalonia or the Pyrenees). The best data (mostly derived from the work of J.Altuna (1971, 1972, 1977, 1981, 1986; Altuna and Mariezkurrena 1984; Altuna *et al.* 1989), as well as by C.Fuentes (1980)) come from Vasco-Cantabrian Spain and are summarized in the form of minimum numbers of individuals (MNI) in Table 10.2. This Table also includes data from the 1910-1914 excavations of Obermaier at El Castillo, recently summarized by Cabrera (1984). The latter data show relative abundance in subjective fashion, although there are some supposed MNI figures for one Aurignacian and one Gravettian level. In assessing the phenomenal figures for bovines and red deer especially in El Castillo Level 18 it should be kept in mind that this stratum was about 70 cm thick and was excavated over a vast area of the cave vestibule (*c.* 130 m^2). Level 16 was slightly thinner on average (*c.* 55 cm), but was dug over an even larger area (*c.* 170 m^2), judging from plans, sections and descriptions in Cabrera (1984). Clearly the faunas of these levels are aggregates deposited over hundreds or even thousands of years during numerous occupations of the cave. The other levels listed in Table 10.2 are generally much thinner and were uncovered in much more modest-scale excavations, but are also undoubtedly palimpsests in their own right. Lezetxiki Level IIIa, in particular, was of substantial thickness (*c.* 70 cm).

Several observations can be made concerning the Vasco-Cantabrian faunas:

1. Carnivores of medium-to-large size are generally varied and abundant – more reminiscent of Mousterian cave faunas in the region than of the LUP ones (see Straus 1982a) and suggesting that these caves still frequently served as lairs/denning sites (or sources of carrion food) presumably between human occupations. Carnivores may also have been responsible for at least part of the ungulate faunal remains in the EUP deposits.

Such carnivore-rich faunas are absent in Solutrean and Magdalenian contexts in the region, suggesting greater intensity or permanence of human occupation of their chosen caves and more complete human responsibility for the ungulate faunas found in association with the artifacts. (Small carnivores [mustelids], as well as lagomorphs and rodents have been omitted from Tables 10.2 and 3, but they are generally quite rare.)

2. Total MNI's, and the numbers of identified specimens (NISP's) upon which they are based, are usually very low – compared with the typical MNI's for the LUP, often on the order of 15-20 animals or more just from the dominant species (red deer or ibex) in each thin, areally restricted excavation unit. With the exception of El Castillo Level 18 (discussed above), there is no evidence for mass hunting of *Cervus*, although some of the more recent levels (e.g. El Pendo IV and III, Morín 5 and 4) do have remains of 8-12 deer. Gross amounts of faunal remains are generally not impressive, given the volumes of sediment excavated in these sites.

3. The fast, wary, inaccessible ibex was very rarely hunted, except at Bolinkoba, where, however, Level E shows evidence of mixing with the overlying Solutrean sequence. Chamois is only moderately abundant in EUP levels of the sites of the hilly Basque Country. Both alpine species are abundant in certain specialized LUP occupations however.

4. Archaic megafauna (giant deer, mammoth, elephant, Merck's and woolly rhinoceros) are often represented in EUP assemblages, but usually only by a few scraps (e.g. teeth) not indicative of hunting or even scavenging for food.

5. Shellfish (except for very small numbers of items probably picked up as curiosities or for decoration: see below) and fish remains are very scarce or absent – in contrast with the situation in the regional Solutrean and especially Magdalenian, when there is substantial evidence for subsistence diversification including the exploitation of aquatic (marine and anadromous) resources at least as seasonal supplemental foods (Straus *et al.* 1980; Clark and Straus 1983). Bird remains are also rare, except at El Castillo (Cabrera 1984), where, lacking a study of cut marks, it is impossible to be sure that any but (or even) the aquatic species (unquantified) were procured by humans, as opposed to dying naturally in their natural cave roosts or being deposited there by raptors. All three types of birds (aquatic, cave-roosting and raptorial) are present in the EUP levels of El Castillo.

The quantified faunas (from old excavations) of the French pre-Pyrenean region are presented in Table 10.3, as summarized in MNI form by Bahn (1984). Bahn does not list the carnivores, but does indicate shellfish, lagomorphs and birds when present (which is infrequent, except in the case of birds in Isturitz, many of which are raptors and cave-roosting species). The minimum numbers of carnivores from the EUP levels excavated by R. and S. de Saint-Périer (1952) have been reconstructed here from their site monograph and show that both halls of Isturitz continued to be major dens as in the Mousterian (see Straus 1982a). A few

Table 10.3. Early Upper Palaeolithic faunas from Pyrenean France. 'A'=Aurignacian; 'G'=Gravettian; 'a'=abundant. Numbers indicate MNI's per species. *Sources:* Bahn 1984; R. & S.. de Saint-Périer 1952.

Sites:	Isturitz St Martin		Isturitz Grande Salle						Gargas		Aurignac	St. Jean de Verges	
Levels:	SIII	SII	V	IV	III	A	FIII	C					
Cultural attribution:	A	A	A	G	G	A	G	G	A	G	A	A	G
Equus caballus	2	3	a	4-5	4	10	6	3	7	1	12-15	4	2
Capra pyrenaica			1							1		3	1
Rupicapra rupicapra	1	2	2-3	6-7	5-6	2			3	7		2	1
Bos/Bison	2		1-2	12-14	2	1	27	4	8	8	12-15	2	3
Capreolus capreolus				2	6	1			1	1	3-4		
Cervus elaphus		1	2	4	1	1	7		3	8	1	3	1
Megaceros sp.	1		1	2					1	1	1		
Dama sp.												2	
Rangifer tarandus	2	3	2-3	20-22	4	1	15	5	2	17	10-12	25	1
Sus scrofa											1	1	
Coelodonta antiquitatis	1	1	1-2	2		1	1			1	1	1	
Elephas primigenius			1	3-4		1					1		
Panthera pardus			1										
Felis spelaea			1	1	1								
Felis silvestris		1		1	2-3								
Canis lupus	1	1	4	10	2								
Vulpes vulpes	3	1	5-6	25-28	15-20								
Alopex lagopus	2	4-5	8-9	4-5									
Cuon alpinus			1										
Crocuta crocuta	1	2	2-3	5-6	1								
Ursus spelaeus	1	1-2	5-6	6	4-5								
Ursus arctos			1	4-5	1								

Table 10.4. Early Upper Palaeolithic faunas from Les Mallaetes (MNI's). *Source:* Davidson 1976a.

	Aurignacian	Gravettian	Pre-Solutrean
Equus	1		
Capra	1	2	1
Bos		1	
Cervus	1	3	1
Felis sp.		1	
Oryctolagus	7	5	2

molluscs again seem to have occasionally been transported from the coasts for use as ornaments since some are perforated, but fish are absent from the EUP deposits of the Pyrenees – despite the richness in salmonids of the region's rivers, which were extensively exploited in the Magdalenian. As in north-central Spain, there are some deposits in the Pyrenean region which represent huge amounts of time: rich Level IV in the Grande Salle of Isturitz (a very big hall measuring about 600 m^2) was about 75 cm thick (R. and S. de Saint-Périer 1952: 80-1); Level A at Saint-Jean-de-Verges was 70 cm thick (J. and J. Vézian 1970); the Aurignac deposit listed by Bahn (1984) seems to be the entire lumped Aurignacian from the old excavations at that type site; the Gargas levels were similarly thick (50 cm for the Gravettian and 1.30 m for the Aurignacian) (Bahn 1984: 145). Thus it is not surprising that some of the MNI figures are rather high. In reality there is no strong evidence for the massive slaughter of herd animals (notably reindeer), so typical of the Magdalenian of the region. Ibex is virtually absent in the EUP, whereas in the Magdalenian and Azilian specialized ibex-hunting sites were established at higher altitudes in various parts of the Pyrenees (see Straus 1987a). Other data presented by Bahn (1984) in the form of NISP's (for Gatzarría and Lezía) or subjective estimates of relative species abundance (for Le Portel, Les Rideaux, Tarté, and Marsoulas) seem to confirm this picture of generalized hunting, different from the large-scale specialized hunting of the LUP. At the far eastern end of the region, the fauna from the old Aurignacian excavations at La Crouzade (Aude) included cave bear, lion and hyena along with reindeer, bovine and horse, and both EUP levels at Canecaude are said to be dominated by cave bear (Sacchi 1976a: 1176, 1178), which again suggests that these caves were lairs and that the ungulate faunal remains in them were not solely the product of human hunting.

Evidence for EUP subsistence in Levantine Spain and Portugal is all but non-existent. The early Aurignacian of L'Arbreda is said to contain remains of *Bos*, *Cervus elaphus*, and *Equus*, while *Oryctolagus cuniculus* is by far the dominant species (in terms of NISP?) in the upper Aurignacian level, followed by *Cervus* and then *Equus*. The Gravettian level is dominated by horse, but with abundant rabbit remains (Fullola Pericot *et al.* 1986). Cave lion is said to be present in the EUP deposit at Reclau Viver (Fullola Pericot 1979: 43). There do not seem to be faunal analyses for the pre-Solutrean deposits at Parpalló; the Solutrean and Epigravettian levels at that site were dominated by very large NISP's and MNI's of ibex and rabbit, together with significant numbers of red deer and some horse and aurochs (Davidson 1983). Davidson's (1976a) MNI counts for the EUP levels at Les Mallaetes are given in Table 10.4. These figures are particularly unimpressive when compared with the ibex and rabbit counts for the immediately overlying Solutrean levels.

The three levels of possible Upper Périgordian in the cave of Los Morceguillos in Almería (an old excavation) had small numbers of remains of red deer, boar, ibex and lagomorphs (Cacho 1978). The Aurignacian Level (D) of Gorham's Cave is overwhelmingly dominated by 447 rabbit bones, followed by 54 ibex and 26 red deer remains. The Level B (?Gravettian) assemblage is even smaller, with 171 rabbit bones and 49 bones of ibex (Davidson 1976b). Both levels have 0-3 remains each of horse, boar, and aurochs. (The Solutrean of Cueva Ambrosio is also noted for the abundance of rabbit remains: Ripoll 1961: 351.) The rabbit remains in EUP levels from eastern and southeastern Spain do not represent very much nutrition, particularly as rabbit meat is notoriously lean. Despite closeness to the sea and the presence of sometimes abundant shellfish in the Solutrean and Epigravettian/Magdalenian levels of at least Parpalló and nearby Volcán del Faro, there are virtually no molluscs in the EUP layers there or at Les Mallaetes (Davidson 1983: 84).

There are unquantified faunal lists from only two supposed EUP deposits in Portugal: the possible 'Aurignacian' or 'Moustero-Aurignacian' of Lapa da Rainha and the doubtful 'Gravettian' of Salemas (both caves). For the former, Roche (1971: 43) cites hyena, lynx, badger, wolf, fox, brown bear, Merck's rhinoceros, horse, boar, red deer, roe deer, aurochs, horse and rabbit, as well as birds and molluscs. (The shore is nearby and the birds are ones which can roost in caves.) Salemas Level 7 has hyena, wolf, fox, horse, boar, red deer, aurochs and rabbit, as well as a mollusc (Roche 1971: 45). The presence of so many carnivores again suggests that these caves were places of both hominid and carnivore habitation, and that not all the ungulate remains were necessarily the result of human hunting. In contrast, the Solutrean faunas from the on-going excavations at Gruta do Caldeirão include significant evidence for ibex hunting (Zilhão 1985b).

In several publications Freeman (1973, 1981) and I myself (Straus 1977, 1983a, 1983b, 1983c, 1986, 1987a, 1988; Straus *et al.* 1980; Straus and Clark 1986; Clark and Straus 1983) have underlined the subsistence intensification which characterized the LUP in Cantabrian Spain as well as in Pyrenean France. This included *both* the development of specialized techniques and weapons for the massive hunting of a few key herd ungulate species *and* the diversification of the subsistence base to include (where available) mammals which are difficult or dangerous to hunt, anadromous and marine fish, estuarine and littoral molluscs and birds – which, judging from cut marks on the bones, were butchered and presumably killed by humans (see Eastham 1984, 1985, 1986). In the

EUP of all the circum-Iberian regions there is no evidence for extensive herd-hunting of red deer, reindeer, ibex, bovines or horses, although some of the Isturitz material *might* suggest exceptions regarding *Equus*. (Obviously we need population structure data for all the assemblages.) Nor is there evidence for the exploitation of aquatic resources, despite the fact that the sites were certainly *no further* from rivers or the coast than they were in the Solutrean/early Magdalenian during the Last Glacial Maximum. On the other hand, subsistence in the MP – still little understood – may have combined some scavenging with limited, encounter hunting of individual, relatively facile big game prey, while totally ignoring aquatic food resources (except on Gibraltar, where *some* of the molluscs in Devil's Tower and Gorham's Cave *may* have been eaten, although without making much of a contribution to the diet: Freeman 1981). Neanderthal subsistence in these regions could probably be characterized as opportunistic foraging for both live and dead animal foods as well as possibly for vegetal ones where and when available. The Neanderthals probably competed with the wide variety of canids, felids, ursids and hyenids which abounded in Glacial Southwest Europe for prey, carcasses and habitation caves certainly to a much greater extent than did LUP or perhaps even EUP people (see Straus 1976, 1977, 1982a, 1983a). It is even unclear which (if any) of the recognized Mousterian lithic artifact types could have been used effectively as weapons. Bone or antler points are absent from the assemblages, so wooden spears, clubs and throwing stones may have been the only adjuncts to the Neanderthals physical strength and endurance in their possible hunting (see Trinkaus 1986). EUP subsistence probably lay somewhere in between the opportunistic, low-level strategies postulated for the MP and the highly organized, elaborately technological, diversified, efficient strategies fairly well documented for the LUP. LUP strategies often involved carefully planned logistical moves, as well as careful positioning at strategic points along game migration routes. These facts may be in part responsible for the proliferation of LUP sites as compared to the MP and EUP, but it is also likely that a combination of slow, cumulative, natural *in situ* growth and a gradual influx of people from Northern Europe – during the Last Glacial Maximum – led to a marked increase in regional population density (Jochim 1983; Straus 1977, 1981; Clark and Straus 1983).

TECHNOLOGY

EUP technology in Southwest Europe is dealt with in another paper in this symposium (Harrold 1989), however a few general observations are in order inasmuch as they are relevant to the subsistence evidence. Bone and antler artifacts (notably *sagaies*) are undoubtedly a novelty with the beginning of the EUP; such totally shaped osseous artifacts are absent in the Mousterian, despite arguments for 'flaked' bones at El Pendo and Cueva Morín by Freeman (1978, 1980). Although it is hard to imagine the broad, split-base bone 'points' of the 'typical Aurignacian' (which are present in several sites of the regions studied here) as actual weapon tips, there are some conceivably functional points among the *sagaies* of the EUP, and complex hafting systems seem to have existed.

The lithic technology does diversify fairly rapidly in the EUP and distinguishes itself from the regional Mousterian, although in areas dominated by quartzite bedrock (i.e. Asturias) true blades are always rare and Upper Palaeolithic assemblages maintain an 'archaic' appearance. (Indeed, because of the nature of available flint sources in many of the circum-Iberian regions except central Portugal, long blades so typical of the Dordogne Aurignacian are very rare.) Châtelperronian and early Aurignacian assemblages in circum-Iberia do indeed contain many Mousterian elements (see for example Bernaldo de Quirós 1982; González Echegaray and Freeman 1971; González Echegaray 1980; Freeman 1977) and there is no absolute technological *break* at 35 000 BP. Classic distinctions between Aurignacian and Gravettian (Upper Périgordian) assemblages (as measured by such things as the end-scraper, general burin, truncation burin and dihedral burin indices of D. de Sonneville-Bordes and J.Perrot and by the presence of particular fossil director types), as defined in the Périgord by Peyrony and de Sonneville-Bordes, often fail to hold up in Spain (see Bernaldo de Quirós 1982; Cacho 1980), and some degree of mental gymnastics is required to place particular Iberian assemblages into numbered Aurignacian or Périgordian 'phases' (e.g. González Echegaray and Freeman 1971, 1973; González Echegaray 1980). Basically there are industries with large but varying percentages of end-scrapers (made on blanks of differing thicknesses and, consequently, having differing edge angles), burins and backed pieces (some of which may have been knives, although Gravette points and some other types – if hafted – may well have been weapon tips). Perforators and various combinations and other specialized tools were also developed in the EUP.

By the end of the EUP, in Upper Périgordian times, some backed bladelets appear, marking the beginning of the development of microlithic technology which characterizes the LUP and Mesolithic. These replaceable elements in compound tools could, among other things, have served as tips and/or barbs for projected weapons. But this technology did not really take off until after the Solutrean experiment in foliate, shouldered and tanged lithic points, so well known in the circum-Iberian regions (and possibly associated with the invention of the bow and arrow). With the Magdalenian/Epigravettian both antler/bone points and microlithic armatures become extremely common. It is in this period that we have probable direct evidence for spear throwers.

Thus technological experiments in the development of various weapon types, as well as manufacturing, processing and maintenance tools, were clearly begun in the EUP and came to full fruit as they were increasingly needed in the efficient conduct of complex, carefully planned and timed activities in the LUP. It would seem that certain technologies, such as nets, traps and weirs, which leave no tangible traces, developed in the LUP – judging from the consistent presence of burrowing animals, fish and birds in the faunal assemblages beginning in the Solutrean. Such technological developments in post-EUP times also included the invention of the eyed needle, the harpoon, the leister and the fish gorge. The EUP represents an important technological step in the eventual cumulative development of the highly technological adaptive systems that were the LUP cultures of Southwest Europe.

ART

The Franco-Cantabrian region of Southwest Europe is famous for its cave art – including paintings, engravings, low-reliefs and, in the Pyrenees, clay sculptures. The European Upper Palaeolithic in general is also characterized by the appearance of both representational and non-figurative mobile art. This is true of Levantine Spain (notably Parpalló) as well as of Vasco-Cantabrian Spain and the Pyrenees. Whatever its significance and original meanings, the appearance of rupestral and portable art works is a novelty in the Upper Palaeolithic, which clearly sets this period apart from the Middle Palaeolithic. Whether the concept is emic or only etic, the social 'function' of the art has been argued in recent years (e.g. Conkey 1980; Gamble 1982; Jochim 1983; Straus 1982b, 1987b). If there is a relationship between art and territorialism, social identification, aggregation, communication systems etc., what can it tell us about the EUP in the circum-Iberian regions?

Mobile art is known from EUP contexts, but surprisingly most of the items come from only two sites: Isturitz and Parpalló. At the former site each of the Aurignaco-Périgordian levels (particularly in the Grande Salle) contained engraved stone plaquettes, often with schematic animal or even anthropomorphic representations. These are accompanied by engraved bone/antler 'points' (the famous *sagaies d'Isturitz*, whose transversal lines do not seem to have had an obvious practical function, unlike the 'anti-skid' striations on *sagaies* and linkshaft bevels), notched bones, perforated shells, stones, teeth and bones. These ornaments, decorated pieces and works of art seem far more common and more spectacular in the Gravettian levels than in the older Aurignacian ones (R. and S. de Saint-Périer 1952). Nonetheless, the 'typical Aurignacian' of the Salle de Saint-Martin did yield 73 *Littorina* shells (some perforated, some stained with ochre) found together with some possible bone beads, all of which had been burned in a small dug-out hearth at the base of which they were found (R. and S. de Saint-Périer 1952: 218-9). While rich for the period, the assemblages of art works and ornaments in the Isturitz EUP pale by comparison (both quantitatively and qualitatively) with the Magdalenian finds from that cave as well as from many other sites all along the length of the Pyrenees (e.g. Duruthy, Espalungue, Espélugues, Lorthet, Labastide, Gourdan, Mas d'Azil, Enlène, Bédeilhac, Belvis, etc.: see Clottes 1976; Bahn 1984; Arambourou 1976a, 1976b; Sacchi 1976a). Other EUP art is virtually non-existent in the Pyrenean region. There are, for example, a piece of engraved antler, an engraved plaquette and a couple of engraved bones from the Aurignacian of La Crouzade (Sacchi 1976a: 1175). At Gargas, Breuil (cited by Bahn 1984: 44) saw similarities between engravings on four items in the Gravettian deposit and some of the rupestral engravings in the same cave.

At Parpalló the first few engraved and painted plaquettes are from the late Gravettian, but at this site the same 'tradition' continued and flourished (with thousands of plaquettes) throughout the long Solutrean sequence and beyond, with stylistic continuity despite alterations in the canons for the representation of animals and the appearance of geometric designs in the uppermost levels (Fortea 1978). The Gravettian of nearby Les Mallaetes also yielded a few engraved plaquettes. However it was in the Solutrean, after c. 21 000 BP, that this Valencian 'school' of art exploded in productivity.

Mobile art and ornaments are very rare in the EUP of Cantabrian Spain (Barandiarán 1972; Straus 1987b). (The Châtelperronian levels – at Morín and El Pendo – have none.) There are a few perforated red deer canines in the Aurignacian of El Otero, El Pendo and Morín, as well as in the Gravettian of the latter sites. El Castillo, whose Magdalenian levels like those of El Pendo are so rich in art works and ornaments, yielded only a perforated shell and an engraved stone from the whole of its long EUP sequence (Cabrera 1984). From the entire region there are only four pieces with possible zoomorphic representations, most of which are of doubtful age, provenience or nature. (One of the credible cases is from the chronologically very late Gravettian of Cueva Morín: González Echegaray and Freeman 1971.) By contrast, even early Solutrean levels at La Riera are fairly rich in decorated bones (Straus and Clark 1986), and the Solutrean and Magdalenian of the whole region are famous for their engraved scapulae, *batons*, wands, etc. as well as for abundant perforated shells, teeth, bones, etc. In short, mobile art of such high quality and great quantity in the LUP of Vasco-Cantabrian and Levantine Spain and in Pyrenean France had its (often modest) beginnings in the regional EUP, and then mostly in its later phases and in only a few sites.

This brings us to the more difficult subject of cave art – difficult because of the problem of dating. Although since Breuil's day attempts have been made to assign some of the Pyrenean, Cantabrian and Levantine cave art to the EUP (as at Gargas, El Castillo, Les Trois Frères and La Pileta by Breuil himself), there are few *proofs* of such antiquity (Straus 1987b). There are three exceptions in the study area. The deeply engraved animal figures on the stalagmitic column in the Grande Salle of Isturitz were covered at the time of discovery by intact Middle Magdalenian archaeological deposits and thus date to the Solutrean or EUP. Recent discovered engravings (lines and animals) on the rock-shelter wall of La Viña in central Asturias were similarly covered over by Middle Magdalenian deposits and were thus made in the Solutrean or earlier (as yet undefined) occupations of the site. A series of deeply engraved lines in the nearby Cueva del Conde was covered by Aurignacian deposits. Aside from these three cases, all dating of cave art figures to the EUP is based on often rather general comparisons with mobile art figures (some of whose provenience information is problematical, as in the case of Hornos de la Peña:

Barandiarán 1972). On the other hand there are many convincing cases of close stylistic similarities between Solutrean and Magdalenian mobile art engravings and rupestral engravings (e.g. the famous Altamira and Castillo engravings on scapulae and cave walls, the Tito Bustillo plaquettes and rupestral engravings, etc.).

Finally, rather few decorated caves actually have EUP deposits. In Vasco-Cantabrian Spain there are at most eight such cases – and most of them are not major art sanctuaries. Most of even these sites also have LUP occupation horizons, and in fact at least 29 of the cave art sites of the region have Solutrean and/or Magdalenian deposits. The relationship between habitation and decoration would seem logical and quite probable, leaving EUP artists with only a minor role in the creation of the rich Vasco-Cantabrian corpus of cave art.

Similarly, in the Pyrenees few of the cave art sites have undoubted EUP deposits (Isturitz, Marsoulas, Le Portel and Gargas being exceptions), whereas many of the numerous sanctuaries have Magdalenian deposits (or are in caves located adjacent to others with Magdalenian habitation sites – such as Niaux which is surrounded by such sites, Le Tuc d'Audoubert and Les Trois Frères which are associated with the site of Enlène, etc.). Even *some* of the art in the Isturitz and Gargas karstic systems, and all the Marsoulas and Le Portel art, is probably best dated on convincing stylistic grounds to the LUP. In several cases the *only* cultural deposits in sanctuary caves are LUP in age. In such cases, to argue that some of the art is EUP in age (as was often done by Breuil, for example) would seem to be the least economical argument (see for example Sacchi 1984: 14 with regard to the case of Grotte Gazel in Aude). A recent study of the subject (Fortea 1978) concludes that the Palaeolithic rupestral art of Mediterranean Spain dates to the Solutrean and Magdalenian/Epigravettian, and indeed in the few sanctuaries with archaeological deposits, those levels date to the LUP. The Portuguese cave art site of Escoural also yielded a Solutrean point – the only diagnostic artifact found so far at that site (Zilhão 1986).

If Upper Palaeolithic art is indeed related to the development of territorialism (from the need to partition and be able to rely on specific resources) and more complex forms of social identification and organization, planning, information storage and transmission, then its predominantly late date is significant in light of the apparent growth of regional population densities and the development of more sophisticated settlement-subsistence systems in the Solutrean and Magdalenian. While such arguments are fraught with circularity, the expansion of artistic expression in the LUP may be informing us of the growth in organizational complexity associated with the adaptations of people in Southwest Europe between *c.* 20 000 and 10 000 BP. Again, as with hunting evidence and technology, cave and mobile art and personal ornamentation clearly had their origins in the EUP (namely in the early Aurignacian) in these regions, and there is some evidence for their growing importance and complexity by the end of the period as rather arbitrarily defined here (i.e. in the late Gravettian), leading to their 'explosion' in the LUP, when the 'need' for such expressions may have been significantly greater. Clearly these are ideas which require much further exploration, but the chronological convergence of the Last Glacial Maximum, regional population growth, subsistence intensification, greater complexity in weaponry, settlement expansion into the mountain zones, and the growth of art, is probably not a random coincidence.

STRUCTURES

As with the MP of these regions, there is scant evidence for EUP burial of the dead, and, in fact, skeletal remains are scarce – most probably precisely because of the lack of burials. (Apparently unburied Neanderthal remains – represented by isolated skeletal parts – have been found in the circum-Iberian regions at Lezetxiki, Le Portel, Malarnaud, Soulabé, Bañolas, Devil's Tower, Forbes Quarry, Carigüela, Cova Negra and Gruta Nova da Columbeira.) The only possible EUP burials are at Cueva Morín in the 'archaic Aurignacian', but there the trenches ('graves') are unaccompanied by skeletal remains (despite otherwise good faunal preservation. Only pseudomorphs or 'soil shadows' were found and preserved (Freeman and González Echegaray 1970). Notable but isolated EUP skeletal remains have also been found in Santander (and later lost) at Camargo (a skull) and El Castillo (a mandible) (Obermaier 1924). Isturitz also yielded several isolated remains, including mandibles (R. and S. de Saint-Périer 1952). Other sites have yielded similar fragmentary remains (e.g. La Crouzade: Sacchi 1976a: 1176). This evidence contrasts of course with often rich EUP burials in west-central France, northwestern Italy and Moravia, as well as with the Magdalenian (including the burials at Duruthy in the pre-Pyrenean area). Admittedly, however, LUP burials are also lacking in Cantabrian and Levantine Spain and in Portugal – possibly suggesting some long-lived differences in practices between these regions and Aquitaine.

Other archaeological features are also rare in the circum-Iberian EUP. A major exception again is the line of post-holes and the rectangular dug-out structure associated with the 'archaic Aurignacian' burials at Cueva Morín (Freeman and González Echegaray 1970). The presence of several dug-out hearths in the Gravettian of the Grande Salle at Isturitz and at least one in the 'typical Aurignacian' of the Salle de Saint-Martin was described and illustrated in section by R. and S. de Saint-Périer (1952: 80-1, 218-9). But such features are rare in the literature of these regions. In contrast, Solutrean and Magdalenian pits, hearths of various shapes and sizes, roasting structures, post-holes and cobblestone and sandstone slab pavements have been documented in Cantabrian Spain and Pyrenean France at sites including Las Caldas, Cueto de la Mina, La Riera, Tito Bustillo, El Juyo, Ekain, Erralla, Duruthy, Dufaure, La Vache, Enlène, among others. Fire-cracked rocks are abundant at most well-recorded LUP sites. The amount of construction activity in several of these sites, with significant artificial alteration of the living habitats, is striking and suggests continued technological developments in the uses of fire (for roasting, heat conservation, etc.) and possibly in methods for storage. Dry, stable surfaces were created with slabs and cobbles as a matter of course in both open-air and cave settings. These were regularly reconstructed as needed, particularly in the cases of Duruthy, Dufaure and probably Le Grand Pastou at the edge of the Pays Basque. Such rearrangements of natural locales, while not totally absent in the Mousterian or EUP (witness the elaborate hearths of the EUP rock-shelter sites of Abri Pataud and Le Facteur in Dordogne), are certainly rare and virtually absent in the circum-Iberian regions. Again the EUP/LUP distinction is one of degree. EUP use of fire, for instance, was certainly more common and probably more sophisticated than its use in the Mousterian, but the technology was carried further in the LUP as humans increased their control over their physical settings by modifying or constructing residential environments for themselves and their activities at regularly reused sites. Not surprisingly, the pavement structures of several open-air sites in the Isle Valley of Dordogne are dated to the LUP (Gaussen 1980), and such pavements are known at other Magdalenian sites in different European regions (e.g. Gönnersdorf, Germany). They represent major investments in sites, symptomatic of the level of organization achieved by LUP peoples in their elaborate adaptive strategies. Such constructions are not to be found in the circum-Iberian EUP, a fact which suggests either a less scheduled reuse of major residential sites or a lesser role or capacity for cultural mechanisms in dealing with food preparation and storage, shelter, warmth or physical comfort, than in the LUP.

CONCLUSIONS

This far-from-complete survey of the EUP of southern France and Iberia has touched on a variety of aspects of the adaptations of the first anatomically modern humans in these regions of Southwest Europe. The major conclusion suggested by the review of some of the data is that although biological evolution had run its course, hunter-gatherer cultural evolution was far from complete, perhaps because the need for more sophisticated tools and strategies had not yet fully arisen, as human populations in this part of the world had not yet grown beyond some threshold limit which would later, in the LUP, require that humans *use* their new bodies and brains in more complex ways and build on the technological advances of the EUP (e.g. blade production techniques, the concept of hafting, the systematic use of bone and antler, the frequent construction of hearths, etc.) to adapt to the environmental and demographic conditions of their world. LUP adaptations included the use of elaborate, highly planned, highly technological strategies for survival, with major social and cybernetic components (probably reflected by the explosion of mobile and rupestral art in this part of the world).

Anatomical and cultural evolution have already been shown, at Qafzeh and Skhūl (Israel) on the one hand and at Saint-Césaire (France) on the other, to have not gone strictly in synchrony. So too is it clear that the EUP represents a period of cultural transition during which new adaptations were being experimented with, using an anatomy and technologies new to Southwest Europe. The final stage in the development of the sort of highly sophisticated, intensive, efficient hunting and gathering economies and artistic traditions – so often taken to characterize the Upper Palaeolithic *as a whole* but in reality characteristic of the LUP *per se* – was built upon those 'advances' and was probably a coordinated response to the need to maintain a satisfactory adaptation, probably in the face of increased regional population densities. Culture is cumulative, but only progresses when innovations are relevant to the conditions of existence. While at a geological timescale the Middle-Upper Palaeolithic transition may look like a punctuation 'event', on a human timescale it was a gradual, uneven *process*.

ACKNOWLEDGEMENTS

Paul Mellars prodded me to venture out of my areas of expertise in the later Upper Palaeolithic of the 43rd parallel, to act as *agent provocateur* in the early Upper Palaeolithic of Southwest Europe in general. My apologies go out to the specialists in this field for whatever errors of fact or interpretation I may have committed in launching my overly simplistic opinions. I hope, however, that this will have served to stir up debate on the issues. My research over the last 19 years in the region has been mainly supported by the National Science Foundation, as well as by the Ford Foundation, the L.S.B. Leakey Foundation, the National Geographic Society and the University of New Mexico, to all of whom I am most grateful.

REFERENCES

Almagro Gorbea, M. and Fernández-Miranda, M. (eds) 1978. *C-14 y Prehistoria de la Península Ibérica.* Madrid: Fundación Juan March.

Altuna, J. 1971. Los mamíferos del yacimiento prehistórico de Morín. In J. González Echegaray and L. Freeman (eds) *Cueva Morín: Excavaciones 1966-1968.* Santander: Patronato de las Cuevas Prehistóricas: 367-398.

Altuna, J. 1972. Fauna de mamíferos de los yacimientos prehistóricos de Guipúzcoa. *Munibe* 24: 1-464.

Altuna, J. 1977. Fauna de la Cueva del Conde. *Boletín del Instituto de Estudios Asturianos* 90-91: 486-487.

Altuna, J. 1981. Restos óseos del yacimiento prehistórico del Rascaño. In J. González Echegaray and I. Barandiarán (eds) *El Paleolítico Superior de la Cueva del Rascaño.* Santander: Centro de Investigación y Museo de Altamira: 223-269.

Altuna, J. 1984. Dataciones absolutas. In J. Altuna and J. Merino (eds) *El Yacimiento Prehistórico de la Cueva de Ekain.* San Sebastián: Sociedad de Estudios Vascos: 43-44.

Altuna, J. 1986. The mammalian faunas from the prehistoric site of La Riera. In L. Straus and G. Clark (eds) *La Riera Cave.* Tempe: Anthropological Research Papers: 237-274, 421-480.

Altuna, J. and Mariezkurrena, K. 1984. Bases de subsistencia de origen animal en el yacimiento. In J. Altuna and J. Merino (eds) *El Yacimiento Prehistórico de la Cueva de Ekain.* San Sebastián: Sociedad de Estudios Vascos: 211-280.

Altuna, J., Baldeón, A. and Mariez Kurrena, K. 1989. The excavation of Amalda Cave. *Old World Archaeology Newsletter* 7: 22-25.

Arambourou, R. 1976a. Les civilisations du Paléolithique supérieur dans le Sud-Ouest (Landes). In H. de Lumley (ed.) *La Préhistoire Française*. Paris: Centre National de la Recherche Scientifique: 1243-1251.

Arambourou, R. 1976b. Les civilisations du Paléolithique supérieur dans le Sud-Ouest (Pyrénées-Atlantiques). In H. de Lumley (ed.) *La Préhistoire Française*. Paris: Centre National de la Recherche Scientifique: 1237-1242.

Arambourou, R. 1979. Préhistoire des Landes. *Bulletin de la Société de Borda* 104: 1-26.

Bahn, P. 1984. *Pyrenean Prehistory*. Warminster: Aris and Phillips.

Barandiarán, I. 1972. *Arte Mueble del Paleolítico Cantábrico*. Zaragoza: Seminario de Prehistoria y Protohistoria.

Bernaldo de Quirós, F. 1982. *Los Inicios del Paleolítico Superior Cantábrico*. Madrid: Centro de Investigación y Museo de Altamira.

Bischoff, J. Julia, R. and Mora, R. 1988. Uranium-series dating of the Mousterian occupation at Abric Romani, Spain. Nature 332: 68-70.

Bischoff, J., Soler, N., Maroto, J. and Julia, R. 1989. Abrupt Mousterian/Aurignacian boundary at *ca.* 40 ka BP: accelerator radiocarbon dates from L'Arbreda Cave. *Journal of Archaeological Science* 16: 563-576

Bofinger, E. and Davidson, I. 1977. Radiocarbon age and depth: a statistical treatment of two sequences of dates from Spain. *Journal of Archaeological Science* 4: 231-243.

Butzer, K. 1981. Cave sediments, Upper Pleistocene stratigraphy and Mousterian facies in Cantabrian Spain. *Journal of Archaeological Science* 8: 133-183.

Cabrera, V. 1984. *El Yacimiento de la Cueva de 'El Castillo'*. Madrid: Bibliotheca Praehistorica Hispana.

Cabrera, V. and Bischoff, J. 1989. Accelerator ^{14}C dates for early Upper Palaeolithic at El Castillo Cave. *Journal of Archaeological Science* 16: 577-584.

Cacho, C. 1978. La Cueva de los Morceguillos. *Trabajos de Prehistoria* 35: 81-98.

Cacho, C. 1980. Secuencia cultural del Paleolítico superior en el sureste español. *Trabajos de Prehistoria* 37: 65-108.

Cacho, C. 1982. El Paleolítico superior del Levante español en su contexto del Mediterráneo occidental. *Italica* 16: 7-32.

Cacho, C. 1986. Tossal de la Roca. In *Arqueología de Alicante, 1976-1986*. Cacho, C. Alicante: Instituto de Estudios J. Gil Albert: 19-21.

Clark, G. and Straus, L. 1983. Late Pleistocene hunter-gatherer adaptations in Cantabrian Spain. In G. Bailey (ed.) *Hunter-Gatherer Economy in Prehistory: a European Perspective*. Cambridge: Cambridge University Press: 131-148.

Clottes, J. 1976. Les civilisations du Paléolithique supérieur dans les Pyrénées. In H. de Lumley (ed.) *La Préhistoire Française*. Paris: Centre National de la Recherche Scientifique: 1214-1231.

Conkey, M. 1980. The identification of prehistoric hunter- gatherer aggregation sites: the case of Altamira. *Current Anthropology* 21: 609-620.

Davidson, I. 1976a. Les Mallaetes and Monduver: the economy of a human group in prehistoric Spain. In G. de G. Sieveking, I. Longworth and K. Wilson (eds) *Problems in Economic and Social Archaeology*. London: Duckworth: 483-499.

Davidson, I. 1976b. Seasonality in Spain. *Zephyrus* 26-27: 167-173.

Davidson, I. 1983. Site variability and prehistoric economy in Levante. In G. Bailey (ed.) *Hunter-Gatherer Economy in Prehistory*. Cambridge: Cambridge University Press: 79-95.

Delporte, H. 1985. Fouilles de Brassempouy en 1982, 1983 et 1984. *Bulletin de la Société de Borda* 110: 475-489.

Eastham, A. 1984. The avifauna of the Cave of Ekain. In J. Altuna and J. Merino (eds) *El Yacimiento Prehistórico de la Cueva de Ekain*. San Sebastián: Sociedad de Estudios Vascos: 331-344.

Eastham, A. 1985. The avifauna at Erralla Cave. *Munibe* 37: 59-80.

Eastham, A. 1986. The La Riera avifaunas. In L. Straus and G. Clark (eds) *La Riera Cave*. Tempe: Anthropological Research Papers: 275-284.

Fortea, J. 1978. Arte paleolítico del Mediterráneo español. *Trabajos de Prehistoria* 35: 99-149.

Fortea, J. and Jorda, F. 1976. La Cueva de Les Mallaetes y los problemas del Paleolítico superior del Mediterráneo español. *Zephyrus* 26-27: 129-166.

Freeman, L. 1973. The significance of mammalian faunas from Paleolithic occupations in Cantabrian Spain. *American Antiquity* 38: 3-44.

Freeman, L. 1977. Contribución al estudio de los niveles paleolíticos en la Cueva del Conde. *Boletín del Instituto de Estudios Asturianos* 90-91: 447-488.

Freeman, L. 1978. Mousterian worked bone from Cueva Morín. In L.Freeman (ed.) *Views of the Past*. The Hague: Mouton: 29-51.

Freeman, L. 1980. Hueso trabajado de El Pendo. In J. González Echegaray (ed.) *El Yacimiento de la Cueva de 'El Pendo'*. Madrid: Bibliotheca Praehistorica Hispana: 65-67.

Freeman, L. 1981. The fat of the land: notes on Paleolithic diet in Iberia. In R. Harding and G. Telecki (eds) *Ominvorous Primates*. New York: Columbia University Press: 104-165.

Freeman, L. and González Echegaray, J. 1970. Aurignacian structural features and burials at Cueva Morín. *Nature* 226: 722-726.

Fuentes, C. 1980. Estudio de la fauna de El Pendo. In J. González Echegaray (ed.) *El Yacimiento de la Cueva de 'El Pendo'*. Madrid: Bibliotheca Praehistorica Hispana: 217-237.

Fullola Pericot, J. 1979. *Las Industrias Líticas del Paleolítico Superior Ibérico*. Valencia: Servicio de Investigación Prehistórica.

Fullola Pericot, J. 1983. Le Paléolithique supérieur dans la zone méditerranéene ibérique. *L'Anthropologie* 87: 339-352.

Fullola Pericot, J., Soler, N. and Garcia-Argüelles, P. 1986. Nouvelles apportations et perspectives du Paléolithique supérieur en Catalogne. In A. ApSimon (ed.) *The Pleistocene Perspective*. Precirculated papers of the World Archaeological Congress, Southampton, 1986.

Gamble, C. 1982. Interaction and alliance in Palaeolithic society. *Man* 17: 92-107.

Gaussen, J. 1980. *Le Paléolithique Supérieur de Plein Air en Périgord*. Paris: Centre National de la Recherche Scientifique.

González Echegaray, J. 1980. *El Yacimiento de la Cueva de 'El Pendo'*. Madrid: Bibliotheca Praehistorica Hispana.

González Echegaray, J. and Barandiarán, I. 1981. *El Paleolítico Superior de la Cueva del Rascaño*. Santander: Centro de Investigación y Museo de Altamira.

González Echegaray, J. and Freeman, L. 1971. *Cueva Morín: Excavaciones 1966-1968*. Santander: Patronato de las Cuevas Prehistóricas.

González Echegaray, J. and Freeman, L. 1973. *Cueva Morín: Excavaciones 1969*. Santander: Patronato de las Cuevas Prehistóricas.

González Echegaray, J. and Freeman, L. 1978. *Vida y Muerte en Cueva Morín*. Santander: Institucion Cultural de Cantabria.

Jochim, M. 1983. Paleolithic cave art in ecological perspective. In G. Bailey (ed.) *Hunter-Gatherer Economy in Prehistory: a European Perspective*. Cambridge: Cambridge University Press: 212-219.

Laplace, G. 1966. Les niveaux castelperroniens, protoaurignaciens et aurignaciens de la grotte Gatzarría. *Quatär* 17: 117-140.

Moure, J. and Garcia-Soto, E. 1983. Radiocarbon dating of the Mousterian in Cueva Millán. *Current Anthropology* 24: 232-233.

Obermaier, H. 1924. *Fossil Man in Spain*. New Haven: Yale University Press.

Ripoll, E. 1961. Excavaciones en Cueva de Ambrosio. *Ampurias* 22-23: 31-48.

Roche, J. 1971. Le climat et les faunes du Paléolithique moyen et supérieur de la Province d'Estremadura. *Actas do II Congresso Nacional de Arqueologia*. Coimbra: 39-51.

Sacchi, D. 1976a. Les civilisations du Paléolithique supérieur en Languedoc occidental et en Roussillon. In H. de Lumley (ed.) *La Préhistoire Française*. Paris: Centre National de la Recherche Scientifique: 1174-1188.

Sacchi, D. 1976b. Les industries datées du Paléolithique supérieur a l'Epipaléolithique dans le bassin de l'Aude. *XXème Congrès Préhistorique de France*. Paris: 551-559.

Sacchi, D. 1984. *L'Art Paléolithique de la France Méditerranéenne*. Carcassonne: Musée des Beaux Arts.

Saint-Périer, R. and Saint-Périer, S. 1952. *La Grotte d'Isturitz: Les Solutréens, les Aurignaciens et les Moustériens*. Paris: Archives de l'Institut de Paléontologie Humaine.

Sauvet, G. 1985. Les gravures paléolithiques de la grotte de la Griega. *Bulletin de la Société Préhistorique de l'Ariège* 40: 141-168.

Soler, N. and Maroto, J. 1987. Els nivells d'occupatió del Paleolític Superior a la cova de l'Arbreda. *Cypsela* 6: 221-228.

Straus, L. 1976. Analisis arqueológico de fauna paleolítica del Norte de la Peninsula Ibérica. *Munibe* 28: 277-285.

Straus, L. 1977. Of deerslayers and mountain men: Paleolithic faunal exploitation in Cantabrian Spain. In L. Binford (ed.) *For Theory Building in Archaeology*. New York: Academic Press: 41-76.

Straus, L. 1981. On maritime hunter-gatherers: a view from Cantabrian Spain. *Munibe* 33: 171-173.

Straus, L. 1982a. Carnivores and cave sites in Cantabrian Spain. *Journal of Anthropological Research* 38: 75-96.

Straus, L. 1982b. Observations on Upper Paleolithic art: old problems and new directions. *Zephyrus* 34-35: 71-80.

Straus, L. 1983a. From Mousterian to Magdalenian: cultural evolution viewed from Cantabrian Spain and Pyrenean France. In E. Trinkaus (ed.) *The Mousterian Legacy: Human Biocultural Change in the Upper Pleistocene*. Oxford: British Archaeological Reports International Series S164: 73-111.

Straus, L. 1983b. Terminal Pleistocene faunal exploitation in Cantabria and Gascony. In J. Clutton-Brock and C. Grigson (eds) *Animals and Archaeology*. Vol. 1: *Hunters and their Prey*. Oxford: British Archaeological Reports International Series S163: 209-225.

Straus, L. 1983c. Paleolithic adaptations in Cantabria and Gascony: a preliminary comparison. In *Homenaje al Profesor Martin Almagro Basch*. Madrid: Ministerio de Cultura: 187-201.

Straus, L. 1986. Late Würm adaptive systems in Cantabrian Spain: the case of eastern Asturias. *Journal of Anthropological Archaeology* 5: 330-368.

Straus, L. 1987a. Upper Paleolithic ibex hunting in SW Europe. *Journal of Archaeological Science* 14: 163-178.

Straus, L. 1987b. The Paleolithic cave art of Vasco-Cantabrian Spain. *Oxford Journal of Archaeology* 6: 149-163.

Straus, L. 1988. Hunting in late Upper Paleolithic Western Europe. In M. Nitecki and D. Nitecki (eds) *The Evolution of Human Hunting*. New York: Plenum Press: 147-176.

Straus, L. and Clark, G. 1986. *La Riera Cave: Stone Age Hunter-Gatherer* Adaptations in Northern Spain. Tempe: Anthropological Research Papers.

Straus, L., Clark, G., Altuna, J. and Ortea, J. 1980. Ice Age subsistence in northern Spain. *Scientific American* 242: 142-152.

Stuckenrath, R. 1978. Dataciones de Carbono-14. In J. González Echegaray and L. Freeman (eds) *Vida y Muerte en Cueva Morín*. Santander: Institucion Cultural de Cantabria: 215

Vézian, J. and Vézian, J. 1970. Les gisements de la Grotte de Saint-Jean-de-Verges. *Bulletin de la Société Préhistorique de l'Ariège* 25: 29-77.

Trinkaus, E. 1986. The Neandertals and modern human origins. *Annual Review of Anthropology* 15: 193-218.

Waechter, J. 1964. The excavation of Gorham's Cave. *Bulletin of the Institute of Archaeology* 4: 189-222.

Zilhão, J. 1985a. Données nouvelles sur le Paléolithique supérieur du Portugal. *Actas, I Reunião do Quaternario Iberico (Lisbon)* Vol. II: 101-112.

Zilhão, J. 1985b. Néolithique ancien et Paléolithique supérieur de la Gruta do Caldeirão. *Actas, I Reuniao do Quaternario Iberico (Lisbon)* Vol. II: 135-146.

Zilhão, J. 1986. The Portuguese Estremadura at 18 000 BP. In A. ApSimon (ed.) *The Pleistocene Perspective*. Precirculated papers of the World Archaeological Congress, Southampton, 1986.

11: The Transition from Middle to Upper Palaeolithic at Arcy-sur-Cure (Yonne, France): Technological, Economic and Social Aspects

CATHERINE FARIZY

INTRODUCTION

Sites where Middle and Upper Palaeolithic levels occur together are not frequent; those which contain levels from the transition period are even fewer. Among them is Arcy-sur-Cure, a key locality for this period.

Situated in northern Burgundy, 200 km southeast of Paris, the caves of Arcy occur in a limestone massif on the left bank of the Cure river (Figure 11.1). Only the caves excavated from 1947 to 1963 by Professor André Leroi-Gourhan will be discussed here: the Grotte du Loup, Grotte du Bison, Grotte de l'Hyène and Grotte du Renne (Leroi-Gourhan and Leroi-Gourhan 1964). Mousterian assemblages have been found at all these sites. The Upper Palaeolithic is well-represented at the Grotte du Renne (Renne-Porche), while only sporadic Upper Palaeolithic artifacts exist at the nearby Bison and Loup caves. Despite the spatial proximity of the various sites at Arcy-sur-Cure the relationship between the different sites are relatively complex, due to their size and perhaps their function; as I will show, each habitat may differ from the next. Sedimentology and palynology have recently established correlations between the different horizons in the caves and evidence from the industries and faunal remains can be used to suggest hypotheses about their contemporaneity.

Mousterian groups occupied these caves at different times during Würm I and until late Würm II, as shown by radiocarbon dates for the Grotte du Renne of 37 500 ± 1600 BP (layer XII) and 33 700 ± 1400 BP (layer XI) (Ly-2164, Ly-2165). The top of the last Mousterian level is correlated with a temperate episode, after which the climate became progressively colder. The Châtelperronian groups came to Arcy at the end of the Hengelo-Les Cottés Interstadial and stayed until the first dry, cold episode of Würm III (Leroi-Gourhan and Leroyer 1983).

SITE TOPOGRAPHY AND OCCUPATION DENSITIES

The sites are located on a line along the Cure river (see Figure 11.2). They will be described separately, as follows.

Grotte du Loup is a cave *c.* 3 m x 4 m, opening to the south; in the northwest corner a low passage leads into the deepest part of the cave. The inhabitants of the main Mousterian level had available an area with a 3 m

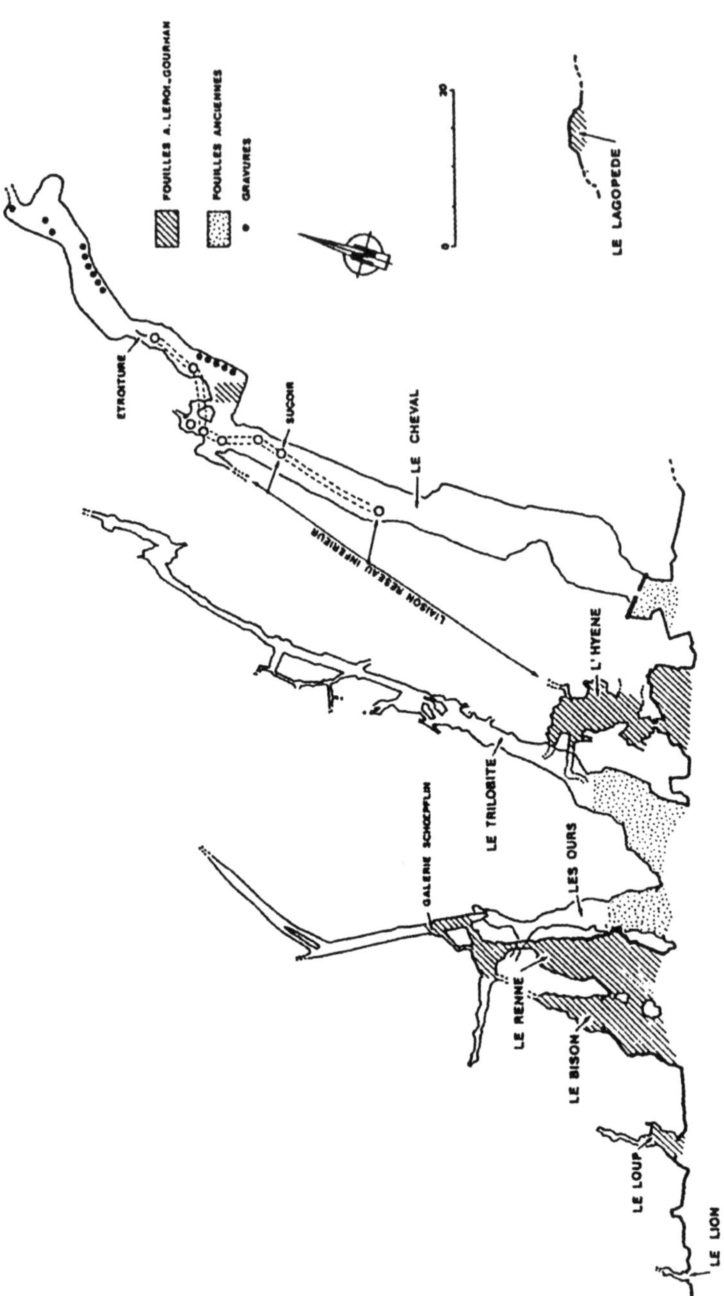

Figure 11.1. Map of the caves of Arcy-sur-Cure (Yonne, France).

clearance; by Upper Palaeolithic (Châtelperronian) times the ceiling was only 1.0 to 1.5 m high, and the area was habitable only if a hut was backed up against the rock wall. The Upper Palaeolithic material is preserved only in places protected from erosion, and therefore comprises only a limited industry.

Grotte du Bison consists of a narrow, low and long passage, only partially excavated, plus an exterior zone in front of the porch. The sequence of deposition begins with six Mousterian levels, followed by a horizon of fallen rocks containing Châtelperronian artifacts (this is level D, which has been correlated stratigraphically with Renne level VIII). The only Mousterian hearth found at Arcy was found in this cave: it consisted of an ashey-grey oval area (20 x 30 cm in diameter) containing bone fragments and burnt stones enclosed quite regularly by a circle of blackened and heated blocks. This fireplace seems to have functioned for only a short time. The relative poverty of the industry could be due to the small size of the excavated area.

Grotte de l'Hyène is a much larger cave, measuring 10 x 15 metres. At its base an ancient horizon contains a crude industry and rolled elements. An alluvial sand deposit (last interglacial?) containing blades, blade cores and a few retouched tools overlies this level, followed by the Würm series which extends from level IVb6 (with human remains) to IVa, with an interruption at level IVb2 (sands). Two types of occupation follow each other: in the early Würm levels (with relatively sparse occupation – 200-500 lithic remains) there is a dominance of retouched tools, some rare flaking debris and many cores. The presence of cores indicates that knapping took place at the site, although the raw material was probably not abundant since the smallest flake had been retouched and used. In contrast, at the end of the Mousterian sequence the site was intensely inhabited. The hard packed ground of IVb1 was covered by an enormous quantity of bone and lithic refuse. The habitation must have been crowded, and the density of the material is considerable; retouched tools are abundant, but unretouched flakes dominate the assemblage.

Throughout the entire Mousterian period, human occupation of the cave alternated with occupation by animals, as shown by coprolites and fragments of wolf and hyena bone found evenly stratified. The area of occupation varied little during the Mousterian due to the position of the entrance and the cave's configuration. But, at the beginning of occupation (level IVb6) the axis of circulation was from the entrance to right-rear and occupation debris was limited to the centre. Later (in level IVb1), the movement was to the left, and artifacts and bones are found mostly near the entrance. These two patterns should be compared with those showing differences of density and nature between the lower levels and the upper one. Thus it is possible to obtain information from spatial distributions, even if habitation structures and activity areas seem to be absent. The analysis indicates that archaeological horizons are well preserved and stratigraphically isolated; it also shows that if one asks the right questions one can extract information from apparently structureless data.

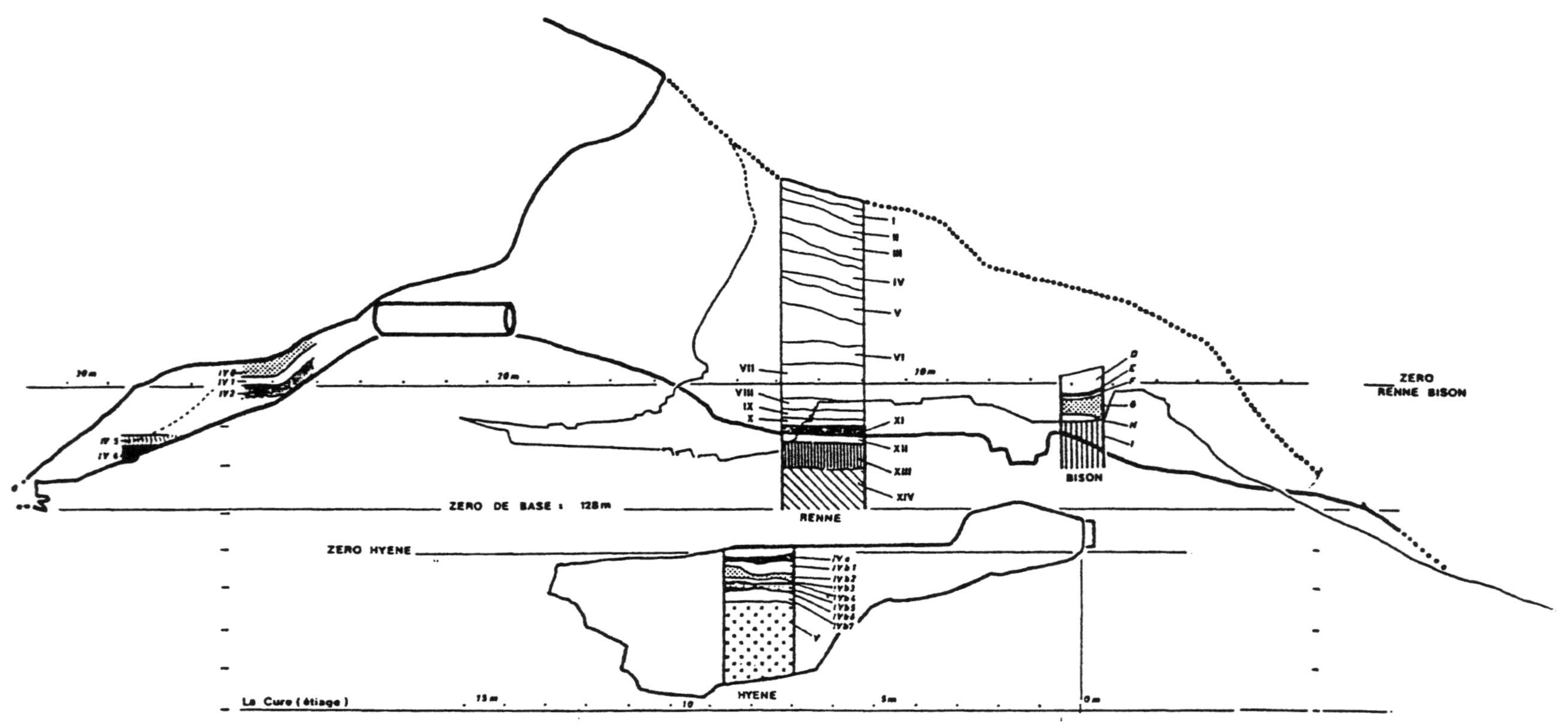

Figure 11.2. General stratigraphy of the caves of Arcy-sur-Cure: Bison, Renne-Porche, Renne-Galerie, Hyène.

Grotte du Renne. Mousterian material was uncovered in several zones of this cave complex: Renne-Rotonde (layer IV2), Renne-Galerie, and Renne-Porche where the most informative sequence was discovered. The latter area is largely open to the exterior, dominating the Cure valley, and has a good southern exposure.

Eight levels of Mousterian were uncovered in the deep gallery (Renne-Galerie, RGS IV2 to IV10) of which the richest was located on the ground surface at the time of discovery (level IV2). This floor seems to be the only known example of a Mousterian occupation surface which has remained intact and exposed to the air. The ceiling is fissured and all the walls are covered with calcite. To the north, the very narrow gallery is filled by stalagmitic deposits; to the south, five metres away, the ground is littered with limestone plaques, large fragments of bone, and flakes of chert and flint. The living floor was no more than 10 m^2 in area, and seems to correspond to the only dry zone in the gallery: beyond this area, gutters and stalagmites abound, but there are no further archaeological remains. The horizontal boundaries of the occupied area are so abrupt that one wonders whether a perishable wall may have stood there. The ceiling above the occupied area is less than 1.5 metres high, yet the analysis of the material shows unquestionably that Neanderthals stayed there for a while. This subaerially exposed assemblage is identical to the buried one uncovered at the Grotte de l'Hyène during ten years of meticulous excavation. This living floor offers two important pieces of information: (1) living space was restricted to the dry zone of the cave; and (2) refuse accumulation was not disturbed by systematic human intervention to clear away the most cumbersome elements. Thus humans seem to have lived in the midst of their garbage.

The upper Mousterian levels from Renne-Porche (levels XII and XI) are quite rich, each containing more than 2500 artifacts. Faunal remains are also abundant, though the excavated area is less than thirty square metres.

The Châtelperronian and Mousterian industries occur in clearly distinct stratigraphic levels at the Grotte du Renne (XI is yellowish, X is red) which cannot have been mixed during the excavation. The Châtelperronian sequence (levels Xc, Xb, Xa, IX, and VIII) is followed by an Aurignacian (level VII) and a Gravettian (level V) occupation.

The Châtelperronians occupied a living floor measuring 60 m^2 at Renne's terraced entrance. The living area is neither a true cave nor an open air site but a roofless shelter whose rear wall faces due south, thus warming up at the slightest ray of sunlight. Levels IX and X form a sequence very rich in remains (with 35 000 lithic pieces for level X alone) and might represent a time-span of hundreds of years. Because of continuous occupation, the Châtelperronian living floors at Arcy form a complex structure whose interpretation is made more difficult by gravity sorting and differential chemical action. The analysis is at present concerned mainly with the spatial organisation that appeared during excavation, and as yet only the obvious habitation structures have been studied;

spatial analyses of single artifact classes have not yet been carried out. Here, despite successive readjustments in the living areas, some intact structures have been studied, particularly those of the first Châtelperronian living floors (site northwest X-Y 12-14, northeast A-C 10-13) whose plan was circular, surrounded by limestone plaques which formed a bench. A. Leroi-Gourhan suggested that mammoth tusks and a wooden frame held the ensemble together, (marked by a series of post-holes), and that the huts covered a surface of about 12 square metres.

The formation of Level VIII was rapid, due to a collapse of limestone slabs from the roof. During this time the cave was visited by bears and hyenas. Limited human occupation has left no organized trace on the living floor, but the few tools are certainly related to the Châtelperronian.

THE MIDDLE PALAEOLITHIC INDUSTRIES: RAW MATERIAL, DEBITAGE AND TOOL MANUFACTURE

Before reviewing briefly the characteristics of the Mousterian industries from the different sites, it is important to remember that their composition could have been affected by a variety of external factors whose importance is at present difficult to estimate: type of site (e.g. porch, shelter or cavern), length of stay (continuous or intermittent), size of excavated area ($10m^2$ or $1m^2$) etc. One must keep all these constraints in mind, not only for evaluating the results and interpretations of the analyses, but to better understand the evidence despite the unknown factors. Are the observable differences due to differences in the nature of the occupation or to differences in cultural traditions?

The Mousterian assemblages are basically homogeneous. Taken separately they are small, and the industry is badly made and monotonous. But what makes Arcy particularly interesting is that there are several occupation sites, and that within each site there is a succession of different assemblages. What is the meaning of these assemblages and occupations that succeed one another? What affinities exist between the different horizons and different sites? How can their internal similarities be interpreted chronologically, techno-economically or culturally?

Chert is predominant in all Mousterian levels at Arcy-sur-Cure, along with flint nodules, metamorphic rocks and granite cobbles (the latter are found in all levels). The local chert was easy to procure (close to the massif of Saint-Moré, where erosion had exposed banks of chert in the caves). Flint is not found nearby and its origin is not yet exactly known.

The use of two very different raw materials allows some interesting behavioural interpretations. Generally speaking, tools with continuous abrupt retouch are more frequent in flint than in chert; the same is true for side-scrapers, but it would seem that there is no distinctive tool choice for flint. No tool well represented numerically was *exclusively* made of flint; chert and flint were used for the same tool types. Clearly, Mousterians valued flint though it was in short supply; those from the earlier levels fabricated a maximum of tools in flint and those from the more recent levels retouched their flint tools until exhausted.

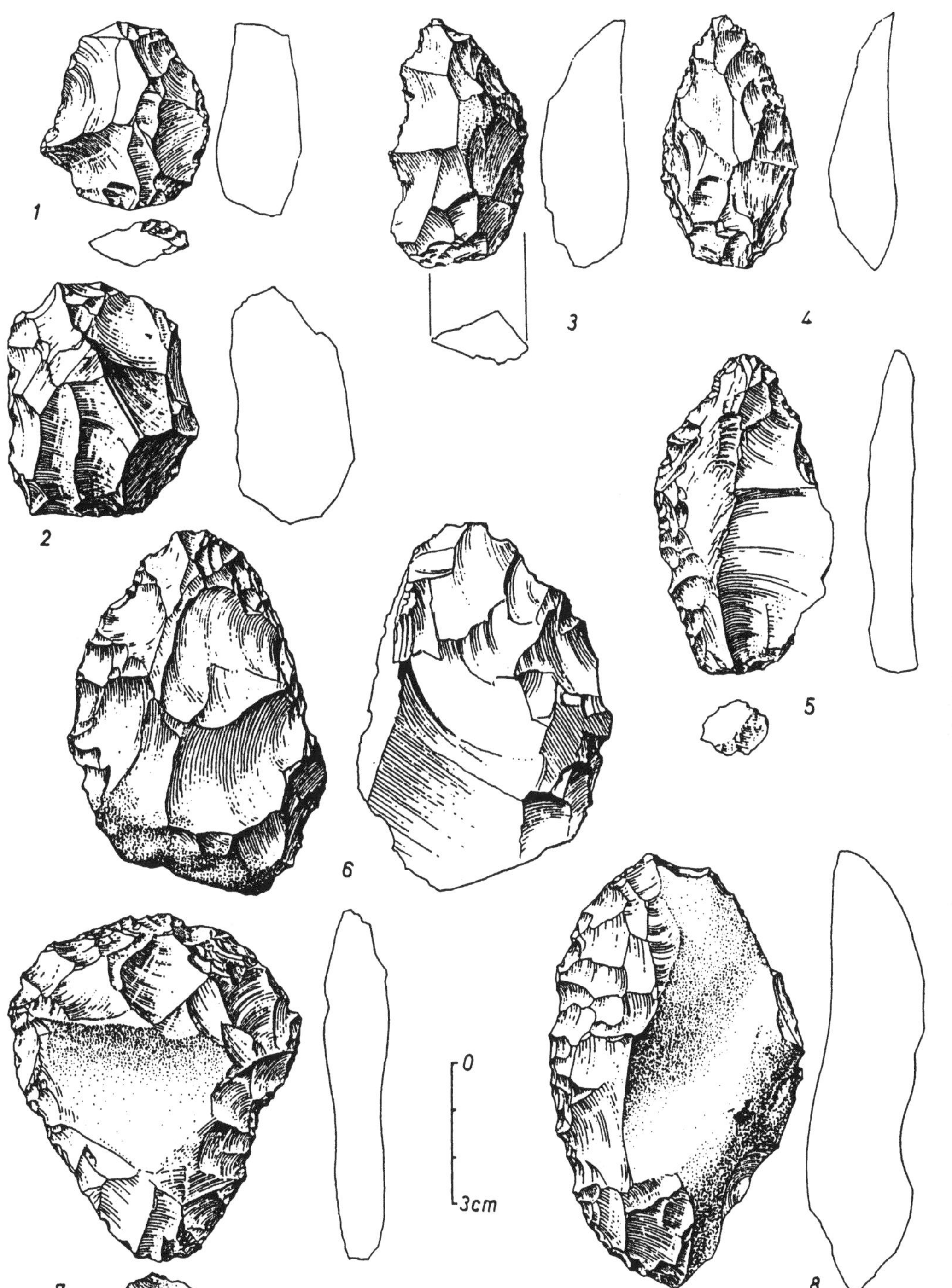

Figure 11.3. Grotte du Renne. Mousterian industry from level XIV.
1. scraper; 2. core; 3. denticulate; 4, 5, 6, 7, 8. side-scrapers; 6. bifacial tool.

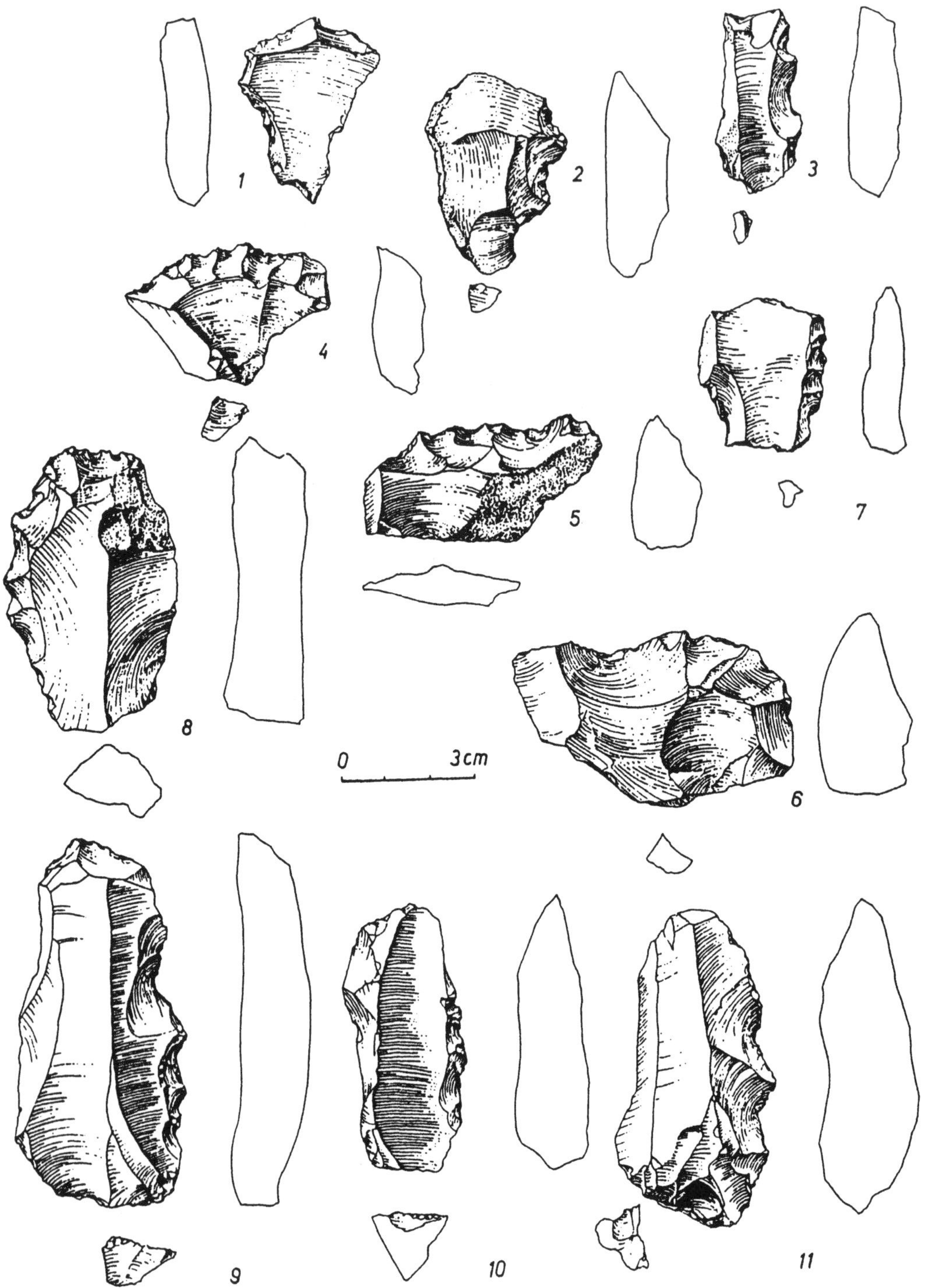

Figure 11.4. Grotte du Renne. Mousterian industry from level XI. Denticulated tools.

Mousterian cores are very numerous, often difficult to recognise (in chert) or reduced to a very small size by intensive flaking (in flint). However it has been possible to classify more than half of them into broad categories which can be found in identical proportions in all levels. It is the overall proportion of the cores themselves which varies from one horizon to another (from around 3% to 26%). Similar differences exist in the proportions of complete to broken pieces between levels (between 38% and 60%). The number of whole blades or flakes is always high and often comprises more than 50% of the objects recovered in particular assemblages.

Study of the debitage shows that Mousterian knapping products were of small dimensions; few levels have produced flakes larger than 7 cm (note that flakes are generally end-struck). Smooth, wide striking platforms with pronounced bulbs are always in the majority (50-80%) and the flaking angle is often quite large. Thus, when there is more than 15% of multiple facetted striking platforms in one level, we can say that the number is unusually high.

Flakes with areas of dorsal cortex also vary numerically, from 10% to more than 30%. Naturally backed pieces are also abundant (up to 30% of the debitage) and are associated with numerous 'pseudo-levallois points' in all levels.

Levallois debitage would seem negligible (rarely exceeding 8%) but this factor carries a certain weight in multivariate analysis because it is almost totally absent in certain levels, and has a strong correlation with certain industrial facies at Arcy.

The blade index ('I.Lam.' of F.Bordes) is high in all the Mousterian levels at Arcy and, as such, is not important for intrasite comparisons. The percentage of retouched pieces varies from one level to another while the proportion of partially retouched artifacts is very high in all Mousterian facies. The proportion of manufactured tools is not large (23% to 13.6%). Marginal retouch, often partially abrupt and narrow, is one of the characteristics that is constant in Arcy Mousterian, the only exception being Bison level J and Renne level XIV which yielded well-made Mousterian industries with real Charentian side-scrapers. The tool kit comprises side-scrapers, limaces and Mousterian points on the one hand, denticulates and notches on the other – to which can be added scrapers, borers, flat-faced burins, *raclettes* and truncated pieces.

TYPOLOGICAL DATA OF THE PRINCIPAL MOUSTERIAN ASSEMBLAGES

In the *Grotte de l'Hyène*, two principal assemblages appear in stratigraphic sequence.

Side-scrapers, denticulates and notched tools, though few in number, play a determining role typologically. A typical Mousterian in the lower level is followed by a characteristic Denticulate Mousterian in the upper level. A progressive diminution of the percentage of side-scrapers is correlated with an increase in the simpler and less well-made tool forms;

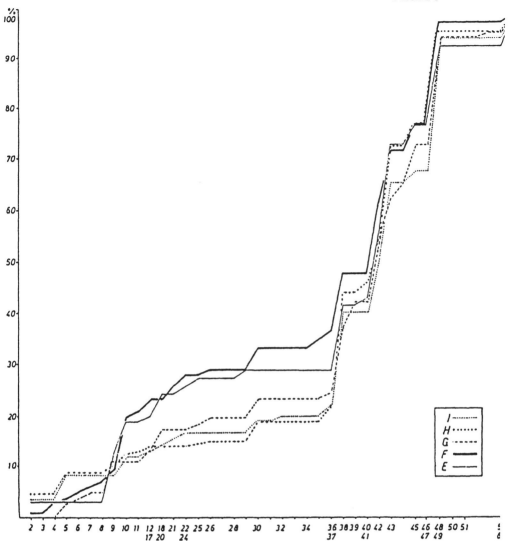

Figure 11.5. Grotte du Bison. Cumulative diagrams of industries from the Mousterian levels.

side-scrapers with two working edges disappear in favour of those with only one working edge.

The four distinct Mousterian levels at *Renne-Porche* are very unequally represented from the point of view of lithics: of the 5800 pieces studied, 5550 come from the two upper levels. Nonetheless, it is possible to compare specific variables in the four levels, and to show changes through time (Figure 11.3). The general trends are comparable to those at Grotte de l'Hyène and *Renne-Galerie*. Flint was the preferred raw material for tools, though its use decreased considerably from level XIV to XI. Cores

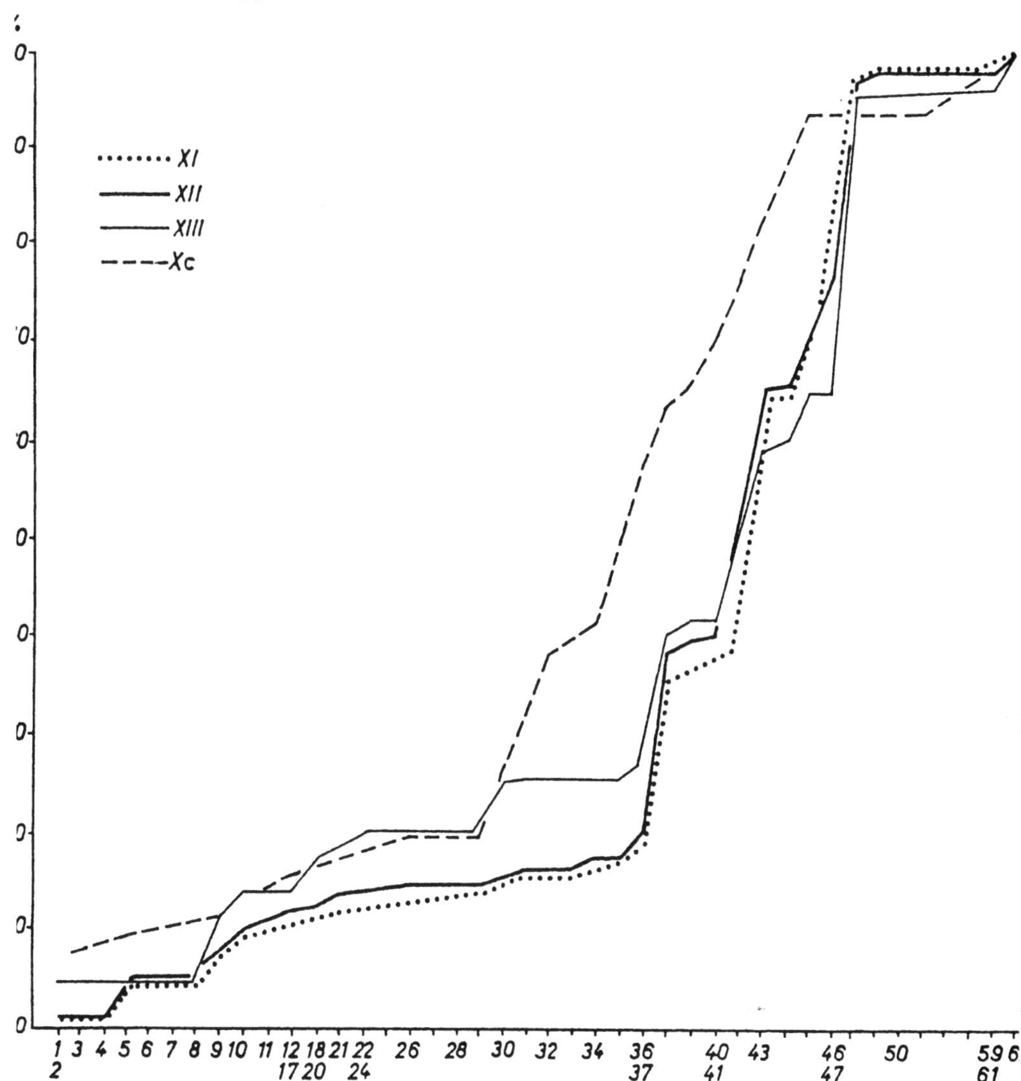

Figure 11.6. Grotte du Renne. Cumulative diagrams of industries from the Mousterian (XI, XII, XIII) and Châtelperronian (Xc) levels.

are relatively abundant in other caves, and particularly in the lower levels at La Grotte de l'Hyène and Renne-Galerie (15-30%) but are rare in *Renne-Porche* and very rare in the upper levels. There is a marked increase in the number of unbroken pieces as the number of cores decreases through time. As at Hyène, the proportion of retouched tools and the variety of tool types decrease simultaneously.

While the percentage (and the quality) of the side-scrapers diminishes from 45% in level XIV to 18.7% in level XI, side-scrapers (of the single and double variety) increase perceptibly in the upper horizons, as at Bison.

Upper Palaeolithic tools are few and consist mostly of end-scrapers. As in the case of Hyène, denticulates are quite abundant in all levels, but their percentage is higher in the upper horizons. This increase in denticulates is associated with an increase in notched pieces.

Thus least two different lithic complexes are superimposed at Renne: the first is formed by level XIV, representing a typical Mousterian; the second, consisting of levels XII and XI, is definitely a Denticulate Mousterian (Figure 11.4); level XIII is a very poor assemblage which could be either transitional between the two forms or a Denticulate Mousterian.

Even though the Mousterian industries of *La Grotte du Bison* are not abundant, a diagnosis of each series has been possible. The site shows a different succession of Mousterian facies from that recorded at the other sites, but this may be related to the configuration of the cave and its double character. The sequence may be summarized as follows:

(i) The industry of level J, composed essentially of side-scrapers, is typologically close to that of Renne XIV.

(ii) Horizons I and H seem typologically similar (though the first contains only half as many tools as the second) and have been classified as Denticulate Mousterian. Side-scrapers form 10% of the tool kit, which is dominated by denticulates; Upper Palaeolithic type tools are numerous. It is interesting to observe that, technically speaking, these two horizons are different, though the mean dimensions of the debitage are similar in the two levels, as are the proportions of facetted butts (dihedrals). There is a marked elongation of the flakes in level H and noticeable Levallois debitage in I. Only the assemblage from level H resembles the classical Denticulate Mousterian. Level I (which is poor) contains more cores and more broken pieces than H which has, above all, entire pieces.

(iii) The three upper horizons, G, F, and E, have assemblages with few homogeneous, distinctive techniques; each one seems to vary independently of the others, perhaps because they are too poor or are mixtures of many activity areas. Real facetted butts are rare in G, but abundant in F and E. Cortical flakes are abundant in E, while backed pieces are more numerous in levels F and G than in E. However these three levels have in common a small non-Levallois debitage, and the typological data seem coherent. Beginning in level G and afterwards, there is a progressive increase in the frequencies of side-scrapers over those of denticulates, accompanied by a net increase of double and convergent side-scrapers. With this change, the quality of the side-scrapers also improves.

Although it includes a large proportion of denticulates, this Typical Mousterian has typological characteristics which are more classical than those of the assemblages found at Hyène or Renne: side-scrapers are more numerous, and among them double and convergent ones are well represented. I do not know yet the exact chronological position of these industries, but it is clear that (1) they may be located chronologically between the Denticulate Mousterian and the Châtelperronian, and (2) that they show technical characteristics (e.g. a non-Levallois debitage,

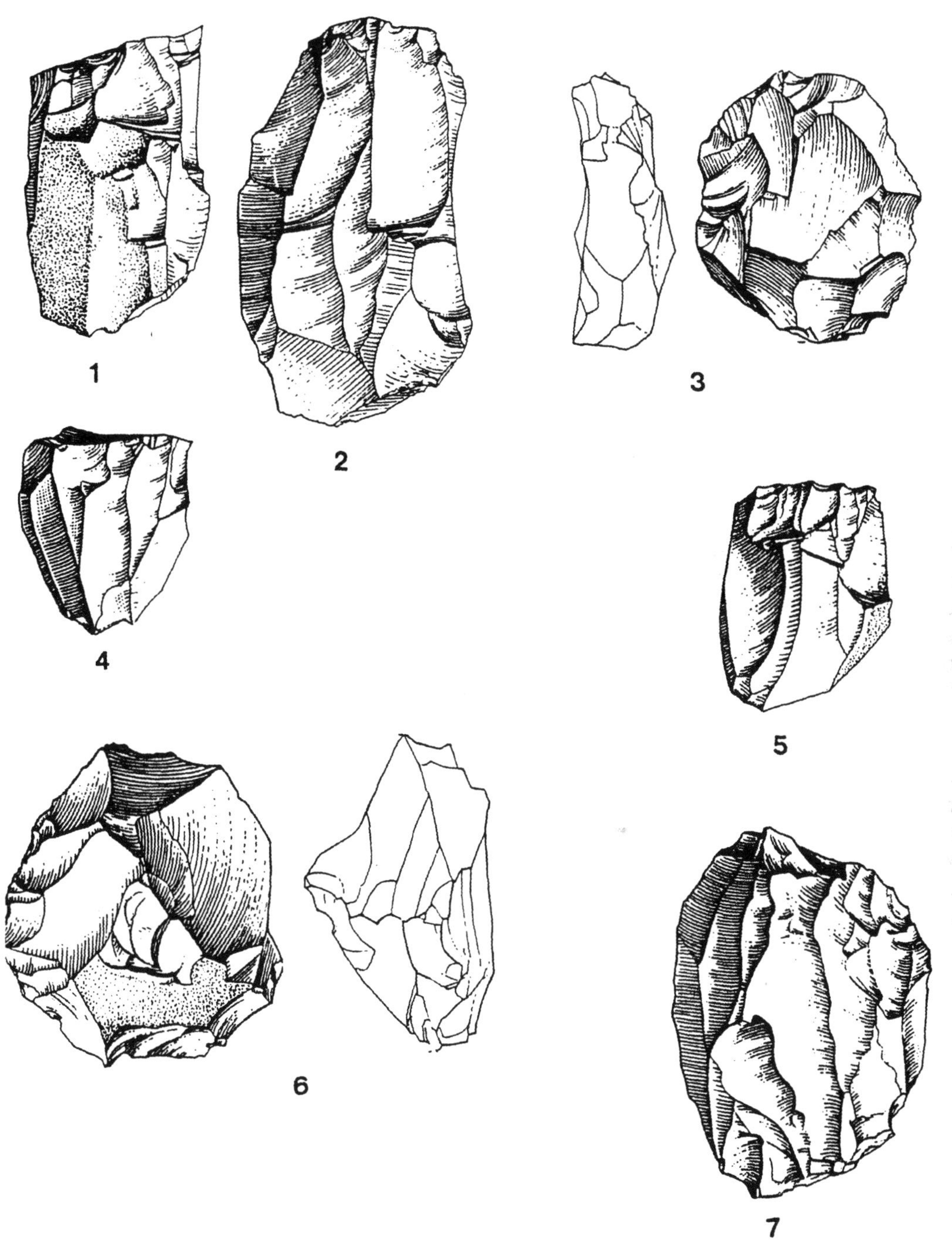

Figure 11.7. Grotte du Renne. Flint cores from level X (Châtelperronian).

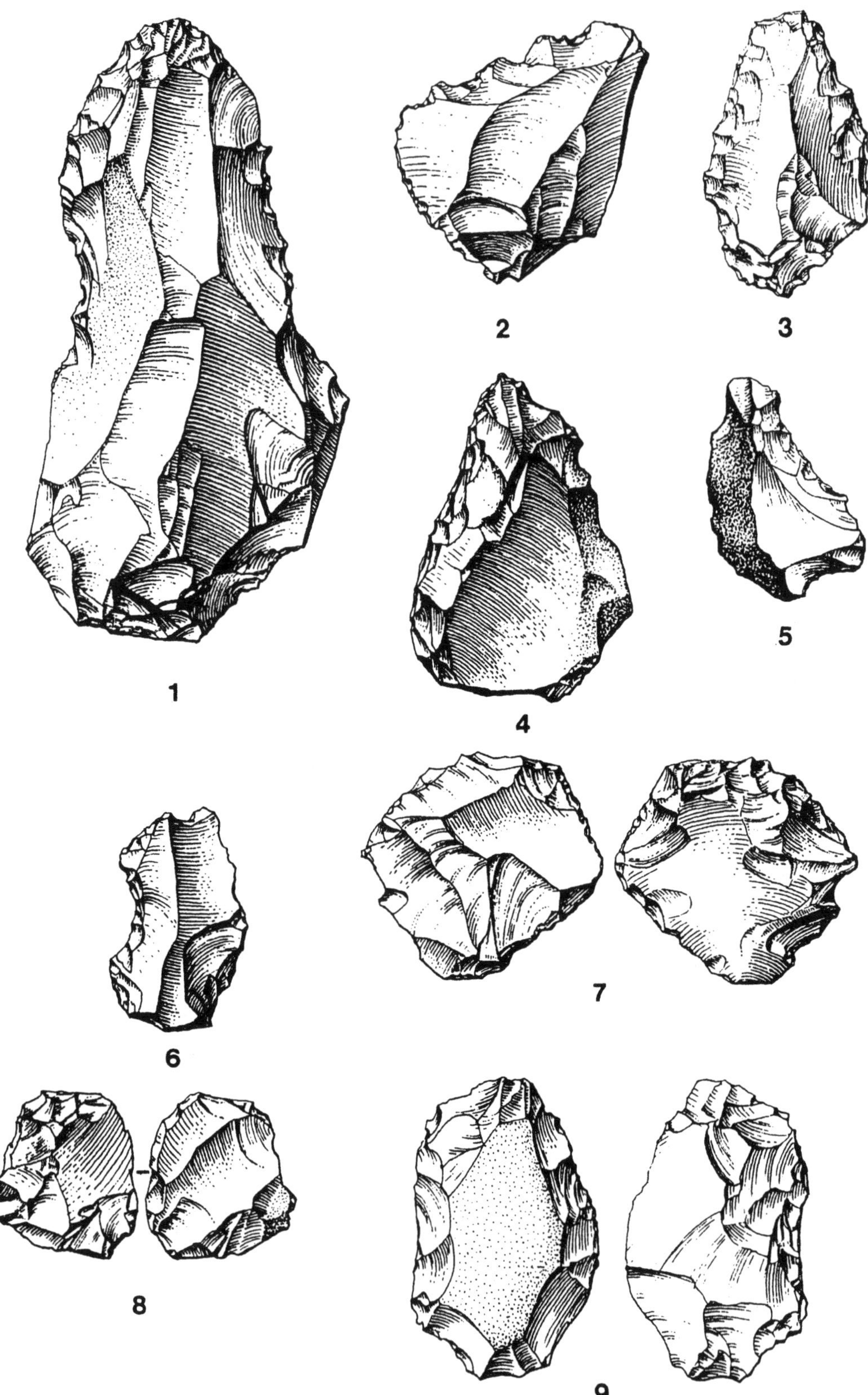

Figure 11.8. Grotte du Renne. Tools from the Châtelperronian level, level X: 1. scraper; 2. denticulate; 3, 4, 5. side-scrapers; 6. notch; 7, 8, 9. scaled pieces (*pièces esquillées*).

small size of flakes etc.) which in the other caves are associated with the uppermost levels of the Denticulate Mousterian (Figure 11.5).

In conclusion, if we use *techno-economic* variables, the Mousterian industries can be placed into two groups. The first one includes the lower levels of Hyène (IVb6 to IVb4), Bison I-J, the entire series of Renne-Galerie, and Renne XIV. A second group includes all the recent Mousterian – that is the Denticulate Mousterian and the rather mixed or transitional levels IVb3 at Grotte de l'Hyène and XIII at Grotte du Renne, in addition to the Typical Mousterian from the Bison levels E, F, and G. In typological terms, the Mousterian industries from Arcy are divided into two slightly different groups:

1. Denticulate Mousterian: Hyène IVa, IVb1, IVb3, Bison H and Renne XI and XII;
2. Other facies: Renne Galerie, Hyène IVb4 to IVb6, Bison E, F, G, I, J, Renne XIII, XIV, and Loup. Technically and typologically, the best affinities are to be found between Hyène IVb1 and IVa and Renne XII and XI (Figure 11.6).

EARLY UPPER PALAEOLITHIC LITHIC INDUSTRIES

One characteristic of the Châtelperronian at *La Grotte du Renne* is its richness. Level X alone has 35 000 pieces, of which nearly 4000 are tools. The raw material is similar to that of the Mousterian, but the proportions are different. Flint is more common than chert and is used more selectively. Blades, burins, scales pieces (*'pièces esquillées'*) are made of flint, while side-scrapers, notches and denticulates are made of chert. In other words, typical Upper Palaeolithic tools are in flint, while Mousterian ones are in chert. Curiously, Châtelperronian points may be made of either raw material; this is probably important but its significance remains to be explained.

With the Châtelperronian, a new technology appears, even though it is still not exclusively used. As noted above, blade-like flakes were numerous in all of the Mousterian levels at Arcy, but real blades were few. In the Châtelperronian, true blade technology is well established; Upper Palaeolithic blades are used almost exclusively to make Upper Palaeolithic tools and Châtelperronian points. Numerous prismatic cores appear, especially in flint; these were not present in the Mousterian (Figure 11.7).

The Châtelperronian tool kit contains many new tools (Farizy and Schmider 1985): 13-18% Châtelperronian points, 10% *pièces esquillées*, 13-16% burins (dihedral on truncation), 10-15% backed blades, 6% truncations, 10% scrapers on blades. To these may be added Arcy's well-known group of traditional Mousterian tools – 11-20% side-scrapers, 8% denticulates, and many flakes with thin abrupt retouch. Taken in isolation this group of Mousterian tools is surprisingly similar to that found in the earlier Denticulate Mousterian, although Châtelperronian side-scrapers are more abundant and more elaborate than in the Denticulate Mousterian ones. Aside from these traditional Mousterian elements, the Arcy Châtelperronian typology, which shows the same basic features through

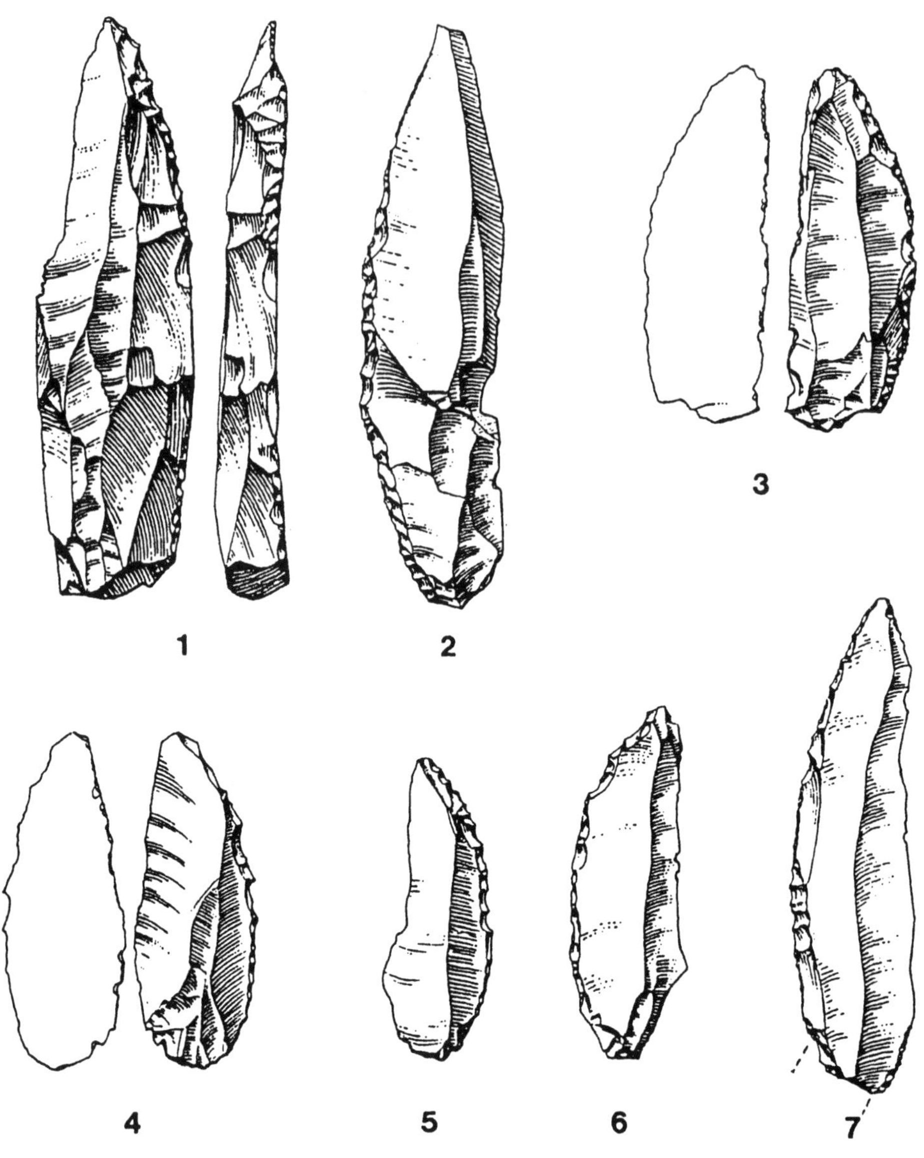

Figure 11.9. Grotte du Renne. Châtelperronian points from level X.

at least levels Xc, Xb, and Xa, is a true Upper Palaeolithic one. In fact some of its features are regarded as 'evolved' in other Châtelperronian assemblages which are later in time and do not include Mousterian-like tools (e.g. le Piage, Roc de Combe, Loup). Among these characteristics, three must be emphasised: the preponderance of burins over scrapers, the flatness and blade-likeness of the Châtelperronian points (Figure 11.9), and the importance of the 'Périgordian' group (up to 30%). Note that *pièces esquillées* were unknown in the Mousterian levels.

FAUNAL REMAINS

Differences between the Mousterian and Châtelperronian fauna can be summarized as follows:

(1) Faunal remains are, in general, much more abundant in the Mousterian than in the early Upper Palaeolithic levels. Some levels from *Grotte de l'Hyène* (IVb6, IVb1) are extremely rich (Girard 1978). During the Mousterian, horses predominated over reindeer, with the exception of the upper horizons at *Grotte du Bison*. Bone remains are rarely heavily fragmented; elements are large, complete long bones (particularly metapodials) are numerous, and all skeletal parts are represented.

There is no indication of use of fur during the Mousterian (in particular there seem to be no cut marks on phalanges). A. Leroi-Gourhan noted that wolves were the only carnivores whose remains showed cut marks and evidence of butchering. In general, carnivores lived at these sites alternately with men.

(2) The study of the Châtelperronian fauna is currently in progress. Based on personal communications provided by A. Leroi-Gourhan's and recent work by F. David one can present a few general preliminary observations. The three sublevels of layer X are unequally rich in remains, and the quantity of the fauna is proportional to the quantity of stone artifacts – i.e. rather poor in level Xc, abundant in Xb and in Xa. On the well preserved bones, tool-inflicted marks, marrow fracturing and post-depositional breaks are clearly visible. Bones from level X are coloured grey-ochre, which distinguishes them from the yellow-white bones of the lower levels. Tooth-inflicted marks do not occur on the bones although there are a few gnawing marks of small rodents. In the Châtelperronian levels, there is generally a biased representation of faunal elements, so that the entire skeleton is not present. There are few skull fragments and few teeth. No long bone is whole (except for one mammoth femur) and each element is broken into several pieces (F. David, study in progress). In short, there are no cumbersome bones within the immediate perimeters of the habitat. Among the herbivores, reindeer predominate (*c.* 60%), with mandible remains indicating a minimum of a dozen individuals. In order of decreasing frequency, horse, buffalo and mammoth also occur. A few bones of other cervids (including *Megaceros*) and rhinocerotids have also been identified. Carnivores are rarer in the Châtelperronian than in the Mousterian levels. They are represented by hyena, bear, lion, and fox; wolves are relatively abundant. One can attribute their fractured condition

to human activity. Clearly these animals were subjected to the same treatment as the herbivores. A bear whose long bones had been broken into small fragments had been cut up and eaten. The fur of these animals had evidently been used. The proportion of carnivore phalanges is larger than that predicted from the number of individuals based on other criteria. This over-representation of phalanges is correlated with cut marks on the hindquarters resulting from the dismembering of animals (wolf, hyena and even lion). The use of fur is certain. Hyena bones show the same breaks and the same biased representation of anatomical parts (few skulls, few elements of the axial skeleton) as herbivores. Hence there is no doubt that both herbivores and carnivores had been hunted and eaten. Because the minimum number of individuals obtained from carnivore phalanges is much higher than that obtained from other bones, this over-representation can be reasonably interpreted in economic terms. The number of mammoth tusks (used for erecting huts) is also greater than would be expected from the numbers of individuals calculated on other criteria.

MOUSTERIAN AND CHÂTELPERRONIAN BONE TOOLS

From the Mousterian levels of the *Grotte de l'Hyène* and *Grotte du Loup*, six horse premaxillary symphyses, broken behind the incisors, have a cutting edge which could have served as grinders or scrapers. Seventy-two fragments of bone with traces of intentional use come from Mousterian levels at Renne and Hyène, and are currently being studied by M. Julien and D. Baffier. These consist of fragments of long bones broken by percussion with areas of use-scars on their external faces, which have been interpreted as *retouchoirs*. One polished point was found at Renne-porche level XI.

The bone tools from the Châtelperronian levels have been published by Leroi-Gourhan and Leroi-Gourhan (1964). They consist of a bone awl with carefully shaped head, tubular bird bones severed at one end and broken at the other,fragments of long bones with the traces of a ring removed, rib fragments used as a pick, and more than twenty awls and various slender fragments of thin *baguettes*. *Retouchoirs* are numerous, as in the Mousterian.

PERSONAL ORNAMENTS

Most of the pendants from the Châtelperronian have been published briefly Leroi-Gourhan and Leroi-Gourhan (1964), but a new study is in progress by Y. Taborin. There are subcircular ivory fragments, and pendants of *Rhynchonella* shells and belemnites. Other pendants consist of grooved incisors of marmot and bovinds, perforated and grooved fox canines, a perforated deer tooth, a grooved wolf canine, notched reindeer incisors, a perforated reindeer phalanx, a stalactite fragment, and a crinoid fragment. Ochre fragments occur in abundance in the Châtelperronian levels.

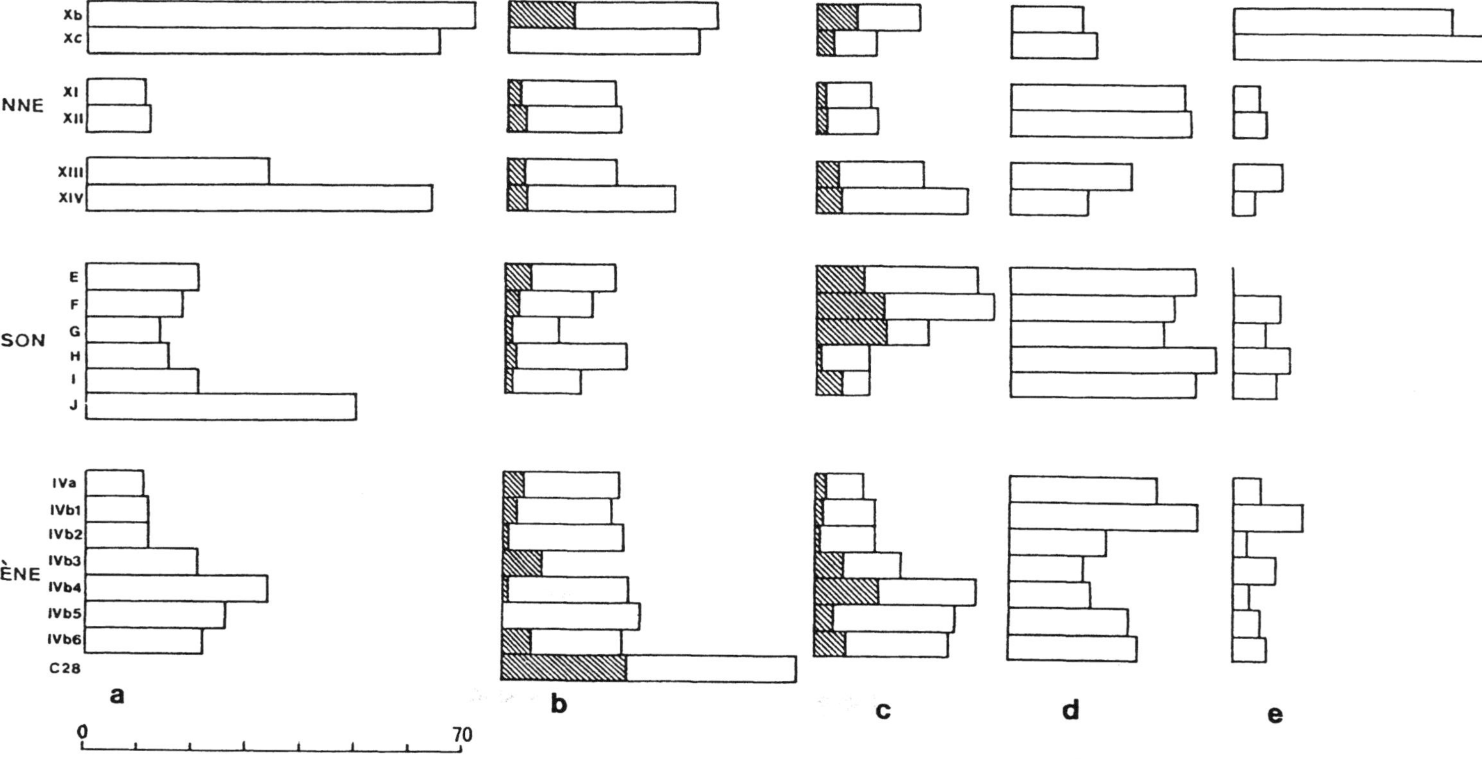

Figure 11.10. Graphs illustrating techno-economic differences between the Mousterian levels (Grottes de l'Hyène, Bison and Renne layers XI-XIV) and the Châtelperronian levels (Renne layers Xb, Xc) at Arcy-sur-Cure. From left to right: (a) = raw materials (flint %); (b) = laminary flakes and blades (shaded); (c) = side-scrapers (double-edged forms shaded); (d) = notched and denticulated tools; (e) = Upper Palaeolithic tool types.

DEPOSITIONAL CONTEXT OF THE HUMAN REMAINS

In the yellow clay at the top of level IV in *Grotte du Loup* (or at the base of level III) a series of stone blocks surrounded a hollow with irregular contours, 1.5 m wide and 0.35 m deep (A. Leroi-Gourhan 1950). Since the infill was similar in colour to the sediment in which the pit was dug, it was not possible to determine the exact contours of the feature. This hollow contained a human molar, two human incisors and two eroded cranial fragments. At present it is not clear whether these sparse remains represent all that is left of a complete burial, or whether this feature represents the deposition of a single head.

From level IVb6 at *Grotte de l'Hyène* a certain number of human remains are found intermixed with broken animal bones in a rounded ridge in the hall's periphery. According to A. Leroi-Gourhan these bones could represent traces of cannibalistic behaviour, though this remains to be confirmed.

In the Châtelperronian of level X at Grotte du Renne, many isolated teeth have been discovered. These show a series of 'archaic' features which A. Leroi-Gourhan (1958) has described as apparently Neanderthal in morphology.

CONCLUSION

If the discussion in the present paper has shown a primary bias in favour of the Mousterian data (and especially the lithic industries) this reflects largely the current status of research at Arcy. My own research on the Mousterian lithic assemblages has already been completed and published (Girard 1976a, 1976b, 1978, 1980, 1982) while analysis of the Châtelperronian stone artifacts by Beatrice Schmider and myself is still in progress. I believe that it is important to stress the characteristics of the Mousterian stone artifact assemblages because the Middle/Upper Palaeolithic transition at Arcy involved not only economic and social changes but also changes in technology.

Arcy-sur-Cure is an ideal case study because it provides a large and reliable body of data. The evidence is such that it is possible to review in detail all forms of cultural and behavioural changes, and to measure and understand their real meaning from the perspective of the transition from the Middle to the Upper Palaeolithic. It is clear that to fully understand the transition we must complete the analyses. The stone and bone artifacts, the fauna, the personal ornaments, the ochre, the habitation structures, the spatial distributions and the stratigraphy are still under study. Nevertheless, the preliminary observations summarized here allow us to draw some general conclusions.

Four caves yielded Mousterian assemblages in five different depositional contexts. Only one cave (the Grotte du Renne) contained rich Middle and Upper Palaeolithic levels in a single sequence. In spite of their proximity, the depositional sequences in the different caves cannot be accurately correlated. The caves are not located at the same level and there is no stratigraphic continuity between them. Thus some caves may have

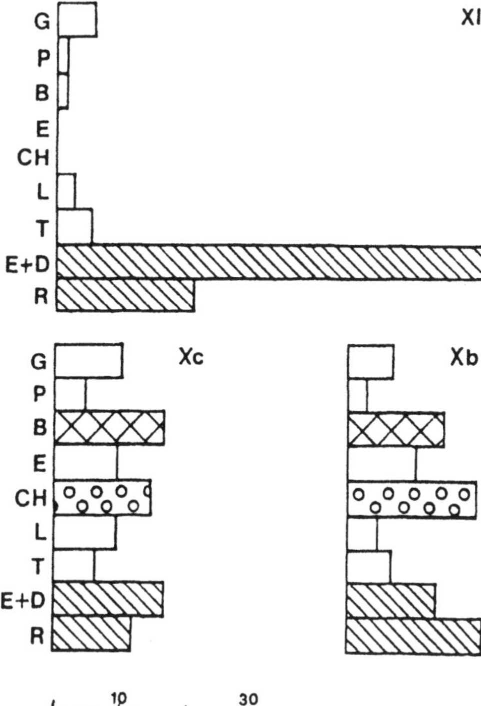

Figure 11.11. Grotte du Renne. Comparison of typological tool frequencies in the Mousterian and Châtelperronian (levels Xc, Xb) levels. G = scrapers; P = borers; B = burins; E = *pièces esquillées*'; CH = Châtelperronian retouched blades; T = truncated tools; E + D = notches + denticulates; R = side-scrapers.

been inhabited at particular periods while others were not. Moreover, we cannot overlook the possibility that in some caves layers may have been removed by erosion.

Considering the Mousterian from Arcy as a whole, we note some major characteristics which give a striking similarity to the lithic assemblages from all strata. This similarity does not exist, for instance, in a site like Combe Grenal but it is known at sites in the Mediterranean region. Nevertheless, despite this similarity, two major facies can be defined. One is close to a Typical Mousterian. The other is essentially a Denticulate Mousterian.

The Typical Mousterian is found in poor levels, towards the base of the depositional sequence, while the Denticulate Mousterian is found at the top of depositional sequences associated with rich occupation horizons. This density of material suggests long-lasting occupations with intense activities. Some levels seem to show transitional stages from Typical to Denticulate Mousterian, but due to the scarcity of data such an interpretation remains hypothetical. One exception to this sequence from Typical to Denticulate is reflected in the uppermost level of the Grotte du Bison which provides an assemblage with intermediate characteristics.

If I have stressed the difficulties in defining cultural changes at Arcy, it is because I feel it is far from easy to fully understand cultural processes through archeological materials. Much can be said about 'cultural breaks', 'evolutions' and 'contacts' but hypotheses are difficult to test (Figure 11.10). The picture is made more complex by unexpected features of the assemblages. For instance, relatively elongated, blade-like blanks and naturally backed blades are found in all Mousterian levels. The general reduction in the size of the debitage is pronounced in both the Denticulate Mousterian and Châtelperronian, and there is an abundance of notches and denticulates in both industries. Side-scrapers reappear in some numbers in the latest Mousterian levels at Bison and in a lesser measure at Renne, and continue in the Châtelperronian.

If the Châtelperronian seems to derive locally from some of the Mousterian facies, then we are entitled to consider that it may have originated in the Denticulate Mousterian as well as in the Mousterian of Acheulian Tradition. Arcy may not be an exception in this regard. At both La Quina and at Quinçay, the Denticulate Mousterian is located immediately below the Châtelperronian, and in several sites, the Denticulate Mousterian has been dated on geological grounds to the end of Würm II and sometimes to the beginning of Würm III. Moreover, the Mousterian assemblages recovered from Arcy are by no means identical to those of the 'classic' regions such as southwest France.

In any event, the Denticulate Mousterian does not really contain any definite elements announcing the Châtelperronian. And the Châtelperronian certainly cannot be regarded simply as a different or a new kind of Mousterian facies, even if similar techniques and tools exist in part of the assemblage. It exhibits, from the earliest stages, a series of relatively 'evolved' traits, which appear to reflect a new world conception (see Figure 11.11). It may well be that these features reflect prolonged and perhaps dramatic contacts with external human groups.

With respect to spatial distributions, we can observe some interesting similarities between the various Mousterian occupations at Arcy, regardless of the kind of sites involved (i.e. small caves, open air areas at the foot of a cliff, large caves with some light, and small cavities with no light at all). In all these different sites, we find the same seemingly random distribution of artifacts and bones; there is no evidence that the living surfaces were ever rearranged. This may have clear implications for understanding Neanderthal behaviour. The cleanliness of their living space seems not to have bothered the Neandertals.Only very large bones were thrown to the perimeter of the principal activity zone. On the other hand,the late Mousterian levels have yielded several curiously formed fossils; some nodular blocks of iron pyrites were found on a living floor, mixed in up with stone artifacts and bones. The behavioural significance of these remains has already been discussed by A. Leroi-Gourhan (1964: 69).

It must be concluded that the behavioural changes as seen through the Châtelperronian and Mousterian levels cannot be related directly to a different way of life – since the animals hunted in the different levels were the same, and most of the technologies employed in the Châtelperronian were already present in the Mousterian. The changes seem to relate rather to a different quality of life, in which the immediate surroundings of the human groups – the habitation zone – was perceived in a totally different way. Analogous changes can be seen in the character of the lithic industries, in which the forms of the tools appear to have a new, exclusive significance.

Clearly, problems of the origins and evolution of the Châtelperronian will not be explained exclusively by the data from Arcy-sur-Cure, although the richness of this site may provide answers to some important questions regarding the role of acculturation in behavioural and technological change. It is possible that as a result of external pressures, Neanderthal groups were able to adopt or develop many features of distinctively Upper Palaeolithic culture. It is certainly not necessary to reduce the Châtelperronian to the status of a Middle Palaeolithic industry to explain why anatomically Neanderthal hominids were associated with this early Upper Palaeolithic context.

ACKNOWLEDGEMENTS

I am grateful to Ethel de Croisset who provided the translation, to Francine David and Paola Villa who reviewed the manuscript, and to R. Humbert who drew the tools.

REFERENCES

Allsworth-Jones, P. 1986. *The Szeletian and the Transition from Middle to Upper Palaeolithic in Central Europe*. Oxford: Clarendon Press.

Bailloud, G. 1953. Note préliminaire sur des niveaux supérieurs de la grotte du Renne à Arcy-sur-Cure (Yonne). *Bulletin de la Société Préhistorique Française* 50: 338-345.

Bordes, F. 1961. *Typologie du Paléolithique Ancien et Moyen*. Bordeaux: Delmas.

Bordes, F. 1968. La question périgordienne. In D. de Sonneville-Bordes (ed.) *La Préhistoire: Problèmes et Tendances*. Paris: Centre National de la Recherche Scientifique: 59-70.

Bordes, F. 1971. Du Paléolithique moyen au Paléolithique supérieur, continuité ou discontinuité? In F. Bordes (ed.) *Origine de l'Homme Moderne*. Paris: UNESCO: 211-218.

Champagne F. and Espitalié R. 1967. La stratigraphie du Piage. *Bulletin de la Société Préhistorique Française* 64: 29-39.

Delporte, H. 1951. La grotte des Fees à Châtelperron (Allier). In *Congrès Préhistorique de France, Poitiers-Angoulême*. Paris: Société Préhistorique Française: 455-477.

Farizy, C. and Schmider, B. 1985. Contribution à l'identification culturelle du Châtelperronien: les données et l'industrie lithique de la couche X de la grotte du Renne à Arcy-sur-Cure. In M. Otte (ed.) *La Signification Culturelle des Industries Lithiques*. Oxford: British Archaeological Reports International Series S239: 149-169.

Girard, C. 1978. *Les industries moustériennes de la grotte de l'Hyène à Arcy-sur-Cure (Yonne)*. 11th Supplement to Gallia Préhistoire. Paris: Centre National de la Recherche Scientifique.

Girard, C. 1980. Les industries moustériennes de la grotte du Renne à Arcy-sur-Cure (Yonne). *Gallia Préhistoire* 23: 1-36.

Girard, C. 1982. Les industries moustériennes de la grotte du Bison à Arcy-sur-Cure (Yonne). *Gallia Préhistoire* 25: 107-129.

Harrold, F.B. 1983. The Châtelperronian and the Middle-Upper Paleolithic transition. In E. Trinkaus (ed.) *The Mousterian Legacy: Human Biocultural Change in the Upper Pleistocene.* Oxford: British Archaeological Reports International Series S164: 123-140.

Hours, F. 1959. Trous de poteaux dans un habitat châtelperronien à Arcy-sur-Cure (Yonne). In *Congrès Préhistorique de France, Monaco.* Paris: Société Préhistorique Française: 638-641.

Leroi-Gourhan, A. 1958. Etude des vestiges humains fossiles provenant des grottes d'Arcy-sur-Cure. *Annales de Paléontologie* 44: 87-148.

Leroi-Gourhan, A. 1961. Les fouilles d'Arcy-sur-Cure (Yonne). *Gallia Préhistoire* 4: 3-16.

Leroi-Gourhan, A. 1963. Châtelperronien et Aurignacien dans le Nord-Est de la France (d'après la stratigraphie d'Arcy-sur-Cure (Yonne). In Aurignac et l'Aurignacien: Centenaire des Fouilles d'E. Lartet. *Bulletin de la Société Meridionale de Spéologie et de Préhistoire* 6-9: 75-84.

Leroi-Gourhan, A. 1964. *Les Religions de la Préhistoire.* Paris: Presses Universitaires de France.

Leroi-Gourhan, A. 1965. Le Châtelperronien: problèmes ethnologiques. In E. Ripoll Perelló (ed.) *Miscela'nea en homenaje al Abate Henri Breuil*: Vol. 2. Barcelona: Instituto de Prehistoria y Arqueologia: 75-81.

Leroi-Gourhan, A. 1968. Le petit racloir châtelperronien. In D. de Sonneville-Bordes (ed.) *La Préhistoire: Problèmes et Tendances.* Paris: Centre National de la Recherche Scientifique: 275-282.

Leroi-Gourhan, A. Brezillon, M. and Schmider, B. 1976. Les civilisations du Paléolithique supérieur dans le centre et le sud-est du bassin parisien. In H. de Lumley (ed.) *La Préhistoire Française* 2. Paris: Centre National de la Recherche Scientifique: 1231-1338.

Leroi-Gourhan, Arl. and Leroi-Gourhan, A. 1964. Chronologie des grottes d'Arcy-sur-Cure (Yonne). *Gallia Préhistoire* 7: 1-64.

Leroi-Gourhan, Arl. and Leroyer, C. 1983. Problémes de chronologie: le Castelperronien et l'Aurignacien. *Bulletin de la Société Préhistorique Française* 80: 41-44.

Lévêque, F. and Miskovsky, J.C. 1983. Le Castelperronien dans son environnement géologique: essai de synthèse à partir de l'étude lithostratigraphique du remplissage de la grotte de la Grande Roche de la Plématerie (Quinçay, Vienne) et d'autres dépôts actuellement mis au jour. *L'Anthropologie* 87: 369-391.

Lévêque, F. and Vandermeersch, B. 1981. Le Néandertalien de Saint-Césaire. *La Recherche* 119: 242-244.

Mellars, P. 1973. The character of the Middle-Upper Palaeolithic transition in south-west France. In C. Renfrew (ed.) *The Explanation of Culture Change.* London: Duckworth: 255-276.

Pèlegrin, J. 1986. *Technologie Lithique: une Méthode Appliqué à l'Etude de Deux Séries du Périgordien Ancien (Roc de Combe, c. 8 La Côte III)*. Unpublished PhD Thesis, University of Paris X.

Sonneville-Bordes, D. de. 1960. *Le Paléolithique Supérieur en Périgord.* Bordeaux: Delmas.

Tavoso, A. 1986. L'outillage du gisement de San Francesco à San Remo (Ligurie, Italie). In M. Otte (ed.) *L' Homme de Néandertal.* Vol. 8: *La Mutation.* Etudes et Recherches Archéologiques de l'Université de Liège (ERAUL 35): 193-210.

12: Peopling Australasia: the 'Coastal Colonization' Hypothesis Re-examined

SANDRA BOWDLER

INTRODUCTION: THE 'COASTAL COLONIZATION' MODEL

Homo sapiens sapiens, fully anatomically humans, colonized Australia some 50 to 40 000 years ago. This assumption is supported by the archaeological and fossil evidence currently available. It has significant implications for our understanding of the evolution of modern humans. As has been often observed, this colonization, unlike that of the other New World, must have depended on some form of watercraft. It must also have depended on an economic basis with sufficient flexibility to allow its possessors to adapt to a completely unknown ecosystem. An understanding of the economic and ecological underpinnings of the colonization of Australasia is therefore important to our understanding of the nature of early *Homo sapiens sapiens*.

Ten years ago, I reviewed the evidence pertaining to the human colonization of Australia (Bowdler 1977). Pleistocene occupation of Australia had been conclusively demonstrated some thirteen years previously, many archaeological sites of late Pleistocene age had been discovered and various arguments about the nature of the initial occupation of the southern continent and subsequent adaptations had appeared in the literature. The significant discoveries of early occupation sites and human skeletal remains at Lake Mungo and Kow Swamp had been announced in 1970 and 1971 respectively (Bowler *et al.* 1970; Thorne 1971). My own work on Hunter Island had contributed to this body of evidence in ways which influenced my ideas at this time (Bowdler 1974) (see Figure 12.1).

In 1977, my basic hypothesis was that 'Australia was colonised by people adapted to a coastal way of life; that initial colonising routes were around the coast and thence up the major river systems; and that non-aquatic adaptations, such as desert and montane economies, were relatively late in the sequence' (Bowdler 1977: 204; there was a further argument about megafauna, which I will not pursue here). This argument was intended to challenge particularly the rather earlier work of Birdsell (1957), whose models of penetration of the Australian continent had waves of migration advancing from the north across Australia on the one hand, and, on the other, two lines of entry from the north, converging somewhere near Mount Isa, and pressing on in a more or less straight line

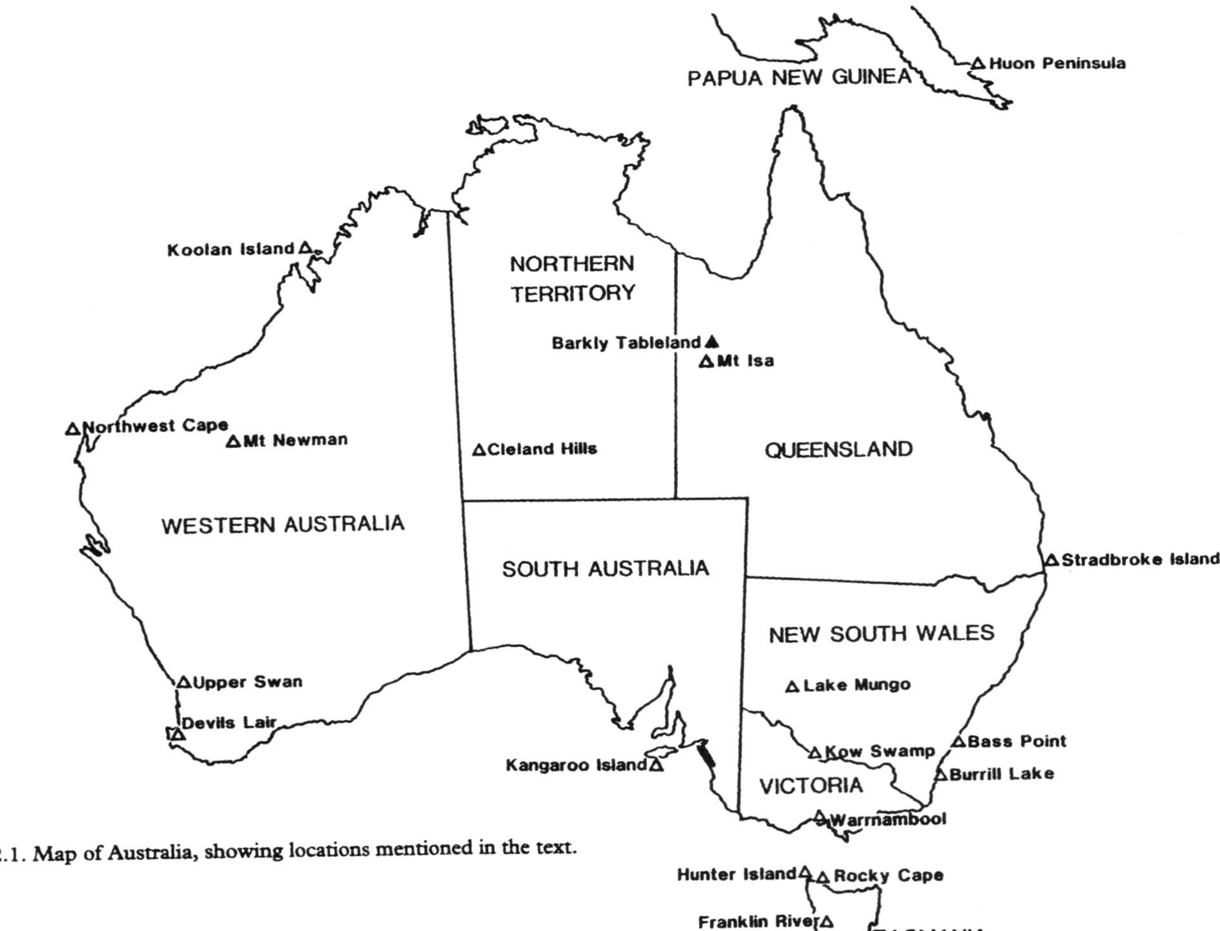

Figure 12.1. Map of Australia, showing locations mentioned in the text.

to Tasmania. At that time, and subsequently, Birdsell (1968, 1977) argued for 'maximum saturation' of the continent within a very short period', specifically '2204 years of total elapsed time'. In 1977, it appeared that Australian researchers supported Birdsell's predictions of rapid adaptation and saturation of Pleistocene Australia (Bowdler 1977: 206). My own interpretation of the evidence then available was that this was far from the case.

My argument was based firstly on *a priori* grounds, arising from the logic of the situation, and secondly on the archaeological evidence which had then accumulated.

In the first instance, it was, and is, generally recognized that Australia must have been colonized by sea. The submarine trough between Sundaland and Sahulland is too deep to have been exposed by any sea-level change since the Cretaceous; the dominance of marsupials on the Sahul side supports this argument. It therefore seemed unlikely that interior forest dwellers of Southeast Asia should find themselves adrift on a log or a canoe *en route* to Greater Australia, or that, on arrival, they would plunge into the unfamiliar desert heart. It seemed more likely that the voyage, whether purposeful or accidental, would have been made by coastal dwellers, probably with a litorally-oriented economy, who would have stayed close to those aquatic resources with which they were familiar.

This appeared to be supported by the available evidence. In Tasmania, many sites had evidence of human occupation dated to within the last 6000 to 8000 years, that is, to the time since the sea had reached its present level; few had earlier evidence of occupation. One of the few was Cave Bay Cave on Hunter Island, just off northwest Tasmania in Bass Strait. Here, humans appeared to have camped intermittently between *c.* 23 000 and 21 000 years ago, even more fleetingly between 21 000 and 19 000 years ago, and the site was then essentially abandoned until the sea reached its present level. It was then reoccupied by people with a highly developed coastal economy similar to that reflected in other sites such as Rocky Cape. The earlier occupation took place at a time when, due to lowered sea levels, Hunter Island was joined to the Tasmanian mainland, and I interpreted that evidence as representing transitory visits, by people on hunting trips away from their coastal base. This visitation was suggested to have ceased when the sea retreated to its maximum extent (Bowdler 1977: 215-20; 1984). A similar interpretation could be applied to other regions. For instance, on the eastern coastal plains of mainland Australia, many sites had been excavated and dated, yet only two pre-dated 6000 BP. Burrill Lake and Bass Point on the south coast of New South Wales were first occupied *c.* 22 000 and *c.* 17 000 years ago respectively; both are located virtually on the modern coast. At the maximum of the last sea-level retreat, due to the steep offshore profile of this particular stretch of coastline, the coast would then have been no more than 24-32 km away (Bowdler 1977: 220-1; 1976; Lampert 1971).

I will mention only briefly the further evidence adduced. In particular, it was suggested that the oldest sites then known for human occupation in

Australia – a series of sites in the Lake Willandra region which included Lake Mungo – showed a dietary reliance on fish and shellfish which I suggested represented a 'transliterated coastal economy'. I argued that the best access to these areas would have been up the major river systems from the south, requiring little modification of economic strategy or technology on the part of coastal migrants (Bowdler 1977: 223-5) Other sites and other regions showed similar patterns. Furthermore, no occupation of desert or montane areas could be convincingly shown to be earlier than terminal Pleistocene times, *c.* 12 000 years ago.

SOME SUBSEQUENT DEBATE

A considerable amount of criticism has been directed at the 'coastal colonization' model since 1977, some of which has been positive. There have however been objections raised at the level of both *a priori* grounds and logical reasoning as well as on the basis of a different reading of the evidence; and new evidence has come forward, which has been variously interpreted as supporting or refuting (or being irrelevant to) the coastal colonization model. I have myself suggested modifications and refinements (Bowdler 1981, 1983) without really addressing the overall structure. That is, to some extent anyway, what I now intend to do here. I will review what I have taken to be the major criticisms; look at some of the more recent relevant evidence, as well as my own 'refinements'; and see how well it seems to me that the original model has withstood the ravages of the last decade.

An early critique (Jones 1979) addressed the model more or less in passing. On the one hand, 'the evidence for man's presence before 32 000 BP seems to be sparse and is so far restricted to large lakesides or rivers close to the then presumed coast, giving some support to Bowdler's model that the economic adaptation of the very first men in Australia may have been restricted to the coastline or analogous situations' (Jones 1979: 453). On the other, 'Bowdler's view that man was not able to exploit the Australian landscape away from the coast or major rivers until terminal Pleistocene times is not supported by the site distributional evidence, though she is right in stressing that the highest human populations were related to the richest lacustrine and coastal resources as indeed they have been throughout Australia's prehistory until modern times' (Jones 1979: 455).

On the one hand, Jones recognized the *a priori* logic of the model; on the other, he disagreed with my interpretation of archaeological evidence, and it should be noted that at this time, it was essentially the same evidence. Indeed it has often been said that the logic of the model presents no problems, but that its duration in time is arguable (e.g. White and O'Connell 1982: 51). There might be said to be wide support for the concept of the initial colonization of Australia being coastal and for penetration of the interior to have been via river systems, but that adaptation to other environments was much more rapid than I supposed. My original argument was that a site plus a date did not indicate an adaptation, even if it were proof of simple human presence; 'site distributional evidence' needed to be carefully scrutinised, not simply accepted at its face value. This sort of difference in interpretation arose from different considerations of the same evidence. Other opinions have drawn from a more selective set of data.

One particular attempt to provide an alternative model suggested that the first humans in Australia were not particularly coastally adapted, but brought with them an 'all purpose economy' which 'could be applied in any area basically by just changing the staple food' (Horton 1981: 23). While I think it could be shown rather easily that no ethnographically recorded hunter-gatherer societies could possibly function in this way (despite Horton's use of the Anbarra as exemplars), this kind of suggestion is archaeologically untestable: any conceivable manifestation can be attributed to such an 'all purpose economy'. I would argue that the problem with this 'all purpose economy' conceptually is its apparent homogeneity; it is important to consider the structure of Australian prehistoric subsistence economies (see below). The author meanwhile was unable to refute the lack of arid zone sites older than 12 000 BP, and having created an all-purpose economy which should take its owners anywhere at all, had to create a reason for this lack. He therefore suggested that early Australians were incapable of digging wells or making water containers prior to this time. He suggests a maximum inland penetration by man (*sic*) between 50 000 and 30 000 years ago, of the well-watered woodlands, based, not on archaeological evidence but on the recorded presence of (largely undated) extinct animal remains, for which there is still no good evidence that they were regularly or even often eaten by human beings (Horton 1981: 24).

A somewhat more reasoned critique is contained in the detailed synthesis of White and O'Connell (1982), who argue that the speed of adaptation to non-coastal and non-riverine environments may have been more rapid than I envisaged. I certainly accept this possibility and will explore it further below. Once again, however, differences arise in interpretation of evidence. White and O'Connell cite the evidence from Devil's Lair in southwestern Australia as evidence of a well-established land economy; yet it fits my model just as well. Here was evidence of sparse occupation of a cave site between *c.* 30 000 and 12 000 (or possibly 6000) years ago (there are some problems with this site's dating), in a situation never further than 25 km from the sea. Furthermore, the faunal remains included not only small to medium-sized mammals but also fragments of marine shell, providing evidence of some sort of coastal contact (Dortch 1979).

Another account similarly argues for earlier non-coastal adaptations, using some newer evidence which I shall discuss below (including Upper Swan Bridge) but also using a collection of rather dubious evidence. Hallam (1983) like Horton sees an early occupation of a Pleistocene 'savannah belt' a well-watered non-coastal zone of more or less continuous 'savannah woodland conditions with a sufficiency of water resources, tuberous plants, and big game' running from the Kimberley across to the

Barkly Tableland down to the Darling/Lachlan/Murray drainage, across the southern Nullarbor to the present mallee belt of Western Australia, the upper/central Murchison drainage, and the Pilbara (Hallam 1983: 13). Did such an homogeneous and attractive zone ever exist? Was it somehow separate from the coastal plains? This seems to me an over-simplification of the environmental evidence summoned in support. Hallam (1983: 12) describes the first people in Australia as 'tropical Asians, with a technology which included seafaring and wood-working (including boat construction); and a subsistence base which emphasised plant and littoral resources'. Yet she claims that the 'coastal colonization' hypothesis (that people clung to the seashores) must be turned on its head. Pleistocene site distributions do not show people shifting from predominantly coastal to partially inland but from predominantly inland to partly coastal' (Hallam 1983: 14) What site distributions? 'A series of enigmatic sites whose precise dates are unknown' (Hallam 1983: 13).

SOME REFINEMENTS

My own revisions/refinements of the original 1977 model arose from two different interests. One was the nature of Aboriginal adaptations to life in the Australian highlands (such as they are), the other was an interest in Aboriginal use of rainforests. Both highlands and rainforests are restricted to eastern Australia, but the latter interest in particular can be seen to have led to some rather wider speculations.

I suggested in 1977 that 'non-aquatic adaptions, such as desert and montane economies, came relatively late in the sequence', when there was (and still is) little evidence at all for the antiquity of desert occupation, but there was rather more relevant to 'montane' areas. It seemed therefore appropriate to look at the latter in more detail. I therefore assembled a synthesis of ethnohistoric and archaeological evidence of Aboriginal exploitation of the Australian highlands and its antiquity (Bowdler 1981).

My initial starting point for this paper was that Aboriginal economies appeared to have relied upon specific staple foods, which provided a dependable resource whose importance was not necessarily in terms of calorific contribution alone; in coastal situations, shellfish was often such a staple (cf. Meehan 1977: 523-4, 527). These 'low key' staples appeared to provide an underpinning for an economy which encompassed a range of other resources, and it is unlikely they could be switched at will, requiring as they do detailed specific knowledge for their exploitation. This argument fits reasonably well with my original assumptions: it is easiest to exploit resources with which one is familiar. I therefore developed my earlier arguments to suggest that 'adaptations away from coastal/aquatic environments require one or more staples in the diet, a low key resource, probably vegetable, to take the place of shellfish' (Bowdler 1981: 100). The body of that paper was taken up with an attempt to identify what such staple foods might have been in the uplands context, along with a consideration of patterns of prehistoric occupation from the archaeological evidence.

Again, I do not wish to repeat that work in detail. The evidence was not uniformly tractable, particularly with respect to information as to traditional economies in the ethnographic present, but also with respect to archaeological research. I was able to put forward the following conclusions, with varying degrees of certainty.

A careful consideration of the archaeological evidence suggested there was little occupation of highlands areas before *c.* 12 000 years ago, and certainly nothing to suggest developed economic adaptations to these areas. Several sites were first occupied, albeit sparsely, in the period between *c.* 12 000 and 4000 BP. Widespread occupation of highland sites suggesting a much greater intensity of occupation occurred about 4000 years ago, concomitant with the appearance of the Australian Small Tool Tradition. I will simply note here that this phenomenon is seen to occur elsewhere in Australia, at this time, and is often labelled as 'intensification' (e.g. Lourandos 1985). There appeared also to be in some areas a connection between successful colonization of these areas and the successful exploitation of toxic macrozamia nuts, although these might not necessarily constitute a staple. In some areas, other vegetables appeared to fill this role, notably the daisy yam (*Microseris scapigera*). In general, and despite the patchiness of the material, I felt the case could be supported for montane economies coming late in the prehistoric sequence, certainly not before *c.* 12 000 years ago, and more intensively after *c.* 4000 years ago. I also felt that the conceptualization of economic systems as requiring the dependable underpinning of a low key staple, as suggested by Meehan's (1977) ethnographic work, was shown to be a useful analytical approach. These economies should be seen as embodying dual strategies or multiple strategies, often connected with a division of labour based on sex, and sometimes age.

Subsequently, I became involved with the New South Wales Forestry Commission, which made me think some more about trees and plants, and particularly rainforest (Bowdler 1983). I returned to an early work of Golson (1971) on Australian Aboriginal foodplants, and found there a passage which was both supportive and provocative.

> 'Should the plant exploitation of Australia's earliest settlers have been appreciably patterned by traditions established in Malaysia, expansion of early settlement may have been into areas typified by the greatest Malaysian element in their flora. These are less the interior regions, whose differences from the tropical north are so clearly registered for one large segment of those territories, than the coast where expansion may have been additionally recommended by the availability of seafoods. Furthermore, whereas at the northwest corner of the continent the coastal distribution of vegetation of the specified character ends within the tropics with a barren stretch of coast that must have been as inhospitable to man in the late Pleistocene as it is today, along the east coast it passes well beyond the tropics into the northern part of New South Wales' (Golson 1971: 209).

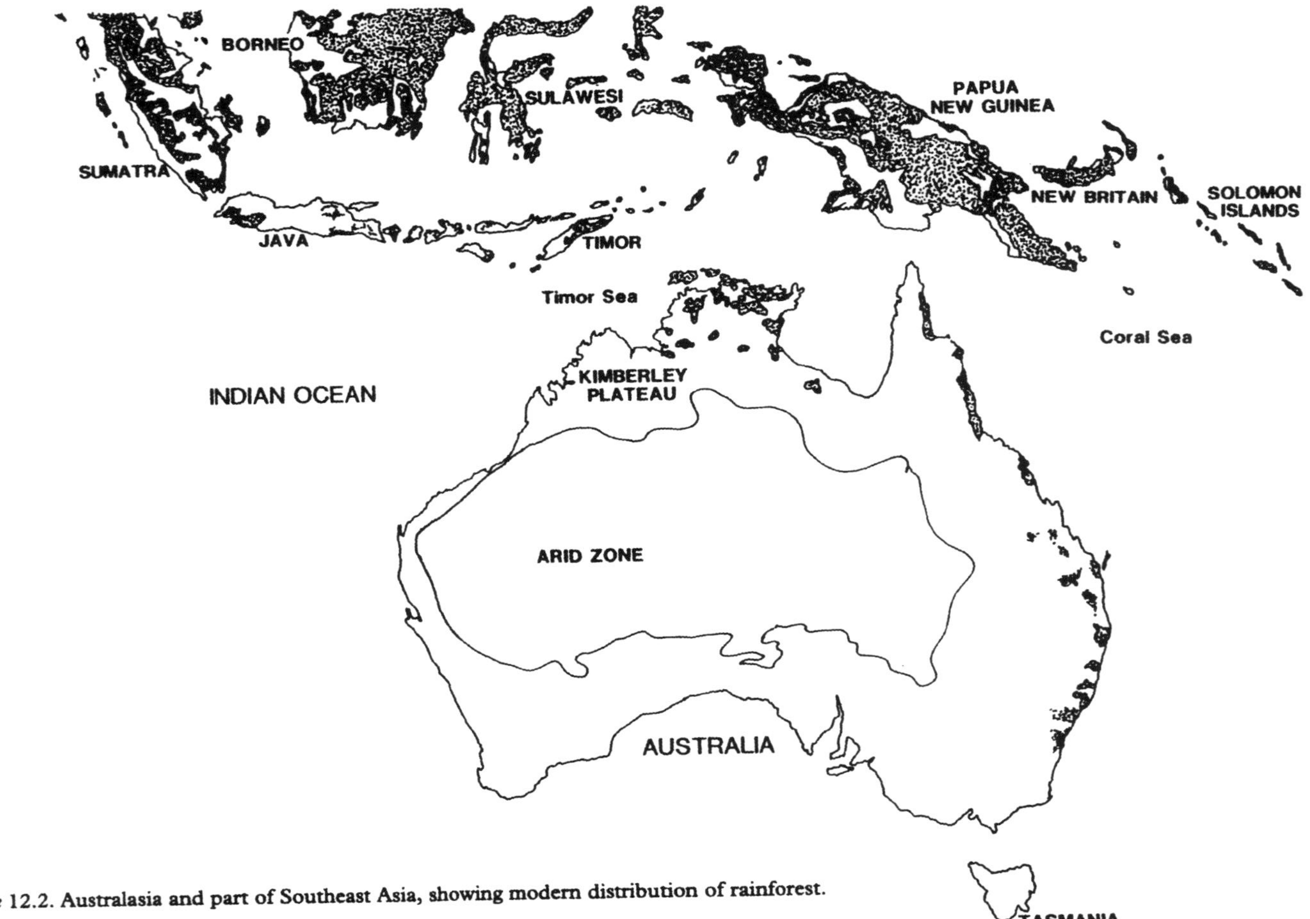

Figure 12.2. Australasia and part of Southeast Asia, showing modern distribution of rainforest.

To what extent might a coastal colonization have been linked with familiar rainforest plant resources (see Figure 12.2)? Palynological research has been carried out in north Queensland which documents the Pleistocene history of rainforest, which was locally dominant in the Atherton Tableland between some 80 000 and 38 000 years ago, was replaced by sclerophyll vegetation between 38 000 and 28 000 years ago, and had begun to re-invade *c.* 10 000 years ago (Kershaw 1974, 1975). To what geographical extent was rainforest present during the initial human colonization of Australia? I proposed a programme of research designed to investigate more closely Aboriginal usage of rainforests in recent times, as well as attempting to document more ancient relationships; this programme was not pursued by me personally due to my relocation to Western Australia.

'COASTAL COLONIZATION' TEN YEARS LATER

A certain amount of relevant new evidence has accumulated over the last ten years, particularly new 'old' dates in certain areas. I would also like to widen the focus somewhat at this point to incorporate Southeast Asia in the picture as slightly more than the assumed genetic and cultural reservoir from which Australians were drawn.

My initial starting point here is the observation that the last ten years has seen no convincing increase in overall antiquity for the arrival of human beings in Australia. In 1977, Lake Mungo and less certainly Keilor were held to demonstrate a human antiquity of the order of 40 000 years. In 1987, we may add Upper Swan Bridge in southwestern Western Australia which convincingly confirms an antiquity of 38 000 BP (Pearce and Barbetti 1981). Claims are made for greater antiquity, but there is no firm evidence. Two putative midden sites at Warrnambool are suggested to be of the order of 60-80 000 years old (John Sherwood, Warrnambool Institute of Advanced Education: *in litt*). There are however grave doubts as to whether these sites are in any way archaeological; there are no associated artifacts, and their presence can be otherwise explained (Denise Gaughwin, La Trobe University, personal communication). Other old favourites for a considerably greater than 40 000 year old antiquity remain as inconclusive as ever. The Kartan implements of Kangaroo Island remain undated, and Lampert has considerably modified his earlier thoughts (Lampert 1981, cf. 1977). The indirect pollen evidence from Lake George was always problematical, in that there were no associated archaeological manifestations to correlate with an inferred change in the fire regime about 120 000 years ago (Singh *et al.* 1981). Wright (1986) has now suggested an upward revision of that date, to *c.* 60 000 BP. Detailed investigation of the 'Murchison cement' has failed to demonstrate anything other than a human presence probably older than *c.* 17 000 BP (Bordes *et al.* 1983). It is true that the evidence may lie unrecognized around us, and will be elucidated by refinements in dating techniques. It can be observed however that unlike Europe or South Africa, we lack any firmly dated deposits, in caves or open sites, which overlie convincing occupational evidence which must be older than *c.* 40 000 BP. At the Willandra Lakes, for instance, the reverse is true: the lowest GolGol unit is quite bereft of archaeological content (Bowler 1976). At this time, it seems reasonable to put the initial colonization of Australia confidently within the 40-45 000 year time period, with a possible maximum of 50-60 000 years.

On this assumption, it would seem appropriate to turn to Southeast Asia. What archaeological evidence do we have for a putative parent culture and, more relevantly, its economic adaptations? Can we perceive in the archaeological record of Southeast Asia coastally adapted societies, possibly exploiting also rainforest species, at or before 50-60 000 years ago?

The Southeast Asian archaeological record prior to 40 000 BP is a veritable blank. There are of course the *Homo erectus* fossils from Java, most of which are considered to date from between *c.* 1.7 m.y. BP and 500 000 BP (Bellwood 1985: 43). There are the more problematical Ngandong fossil hominids, considered by some to be an early *Homo sapiens*, but perhaps more widely accepted as late *Homo erectus*, whose dating is uncertain in the extreme, with estimates ranging from 300 000 to 80 000 BP, with perhaps opinion favouring the earlier end of the range (Bellwood 1985: 50-2). None of these skeletal remains has any cultural context or associations whatsoever: no stone artifacts, no hearths, no evidence of economic or other activity. Some widely distributed surface assemblages of 'poorly differentiated pebble and flake industries' including an industry designated 'Pacitanian', are attributed to *Homo erectus*, but no proof of such an association exists (Bellwood 1985: 56-67). All Southeast Asian archaeological manifestations in the sense of cultural evidence are restricted to the last 40 000 years. On the mainland and Sumatra are found many Hoabinhian sites of terminal Pleistocene age including some coastal shell middens, and in island Southeast Asia a rather different flake tool technology has been identified at several sites, the oldest of which is of the order of 40 000 years (Bellwood 1985: 162-75; see also Jones 1989).

If it were not for the *Homo erectus* fossil material, on archaeological grounds alone it would be possible to argue that Southeast Asia was unoccupied until *c.* 40 000 years ago. It is indeed possible to suggest that the early fossil material represents an early population which did not reproduce itself, in that area, beyond say 300 000 years ago. Bellwood (1985: 67) comments that it could be argued that no humans crossed Huxley's line before 40 000 BP. If we set aside the *Homo erectus* material as representing a terminating population, we might go further and suggest that Southeast Asia and Australia were colonized simultaneously, possibly from China. The question then arises as to whether we should be discussing not the coastal colonization of Australia but the coastal colonization of Southeast Asia and Australia. This is really beyond the scope of this paper, but it may be noted that Bellwood's overview would lend support to an affirmative answer (Bellwood 1985: 161; see also Bowdler, in press).

Apart from referring to the Javanese fossil material, I have avoided discussion of the relevant physical remains of the early Australians. This is a difficult area, which has been discussed by others (e.g. Habgood 1989; Wolpoff 1989). There are, as is by now well known, perceived to be two different morphologies of early Australian. The gracile morphology is represented at Lake Mungo *c.* 28 000 years ago. The robust morphology is represented at Kow Swamp, and other places, but is nowhere reliably dated older than *c.* 13 000 BP. Thorne has argued that these morphologies represent the presence of two distinct populations in Australia during the Pleistocene. He has further argued that the robust populations derive from an early movement from Java of close *Homo erectus* descendants, and that the gracile populations represent a later movement from China (Thorne 1977, 1980a). Clearly, this demands an earlier movement to Australia than I am proposing, and negates the argument above about the non-continuation of Javanese *Homo erectus*. Although I find Thorne's arguments persuasive in some ways, it can be noted that there are other interpretations; and as an archaeologist, I await specifically archaeological confirmation of early robust populations in Australia. For the present I think it can be argued that at about 50 000 years ago there was a major colonizing push across Southeast Asia and Greater Australia by *Homo sapiens sapiens* with watercraft and a coastal adaptation. This of course pushes the initial question one step further back: whence came these people? Presumably from the north, which at least fits half of Thorne's argument and leaves the robust morphology to be accounted for perhaps as an indigenous variation.

Let us return to Greater Australia. Evidence of 'old' sites discovered since 1977 tends to support a modified version of the coastal colonization model, as follows. Australia and areas to the north were colonized by people adapted to a coastal way of life, and initial colonizing routes were around the coasts and thence up the major river systems, and desert and montane adaptions came relatively late in the sequence. There was however probably not one, or two, or any given number of colonizations, but, as Thorne (1980b) has suggested, 'one long continuous migration, all maritime, but with consequent adaptive cultural variations'. I am willing to accept that certain non-aquatic adaptations were made at a relatively early stage.

I have referred to the Warrnambool sites above; if indeed they are midden sites they clearly support an early coastal orientation, pre-dating any other known sites. As has been often stated however, one of the major problems is that most early coastal sites will now lie beneath the sea. It still remains for a concerted effort to attempt to locate such sites.

In southwest Tasmania, prehistorically occupied cave sites on the Franklin River constituted a discovery of major significance, being located in dense cool temperate rainforest country thought to have been not frequented by Aborigines in the ethnographic present. Analysis and dating of these sites has demonstrated that their occupation pre-dates the Holocene establishment of resource-poor rainforest here and that these sites provided bivouacs for hunters of wallabies between 20 and 15 000 years ago. While the sequence is not dissimilar to that of Cave Bay Cave, we see here a rather longer trek from the coast (6O km over mountains or 100 km up the river), through the period of maximum cold but also of mostly open vegetation (Kiernan *et al.* 1983).

To the north, in southern Queensland, an open site on modern Stradbroke Island near Brisbane has produced a long sequence of occupation beginning *c.* 20 000 years ago. The cultural evidence consists largely of stone artifacts particularly in the lower levels, but the location of the site was never far from the coast (Neal and Stock 1986). Further north again, Colless Creek rock-shelter on the Barkly Tableland in the Gulf country was first occupied over 18 000 years ago. While at the time it was some 200 km from the coast, it is located on a large drainage system (Flood 1983: 82-3).

New evidence has also come from New Guinea, where archaeological sites have been located on the Huon Peninsula dated to *c.* 36 000 BP. Cultural evidence found here includes a distinctive artifact form, a 'waisted axe', also known from undated contexts in Australia, from Kangaroo Island and Mackay in north Queensland (Groube *et al.* 1986). In a draft version of the published paper, Groube suggested that these artifacts were used to exploit 'forest edge' situations. Here we have an early coastal location, combined with a suggestion of rainforest species exploitation.

Finally, I will consider Western Australia. One of the most significant discoveries since 1977 has been the Upper Swan Bridge site, on the edge of metropolitan Perth. Here stone artifacts were associated with charcoal in an alluvial terrace of the Swan river. The charcoal was dated to *c.* 38 000 BP (Pearce and Barbetti 1981). A related manifestation in the Helena River valley confirmed this, being dated to *c.* 29 000 BP (Schwede 1983). These are obvious coastal/river situations, although devoid of economically informative evidence. It might be noted that the continental shelf is relatively narrow in the region of modern Perth.

In the region of Northwest Cape, the continental shelf is at its narrowest. Research is being carried out here by postgraduate student Kate Morse (1988). Excavations at Mandu Mandu Creek rock-shelter have revealed a discontinuous sequence of occupation beginning at about 33 000 years ago. It would appear that the site was abandoned for a considerable period of time shortly after about 20 000 years ago, and occupation was not resumed until over 2400 years ago. Nearby, open shell middens have been dated to *c.* 5000 BP. This overall sequence is very similar to that of Cave Bay Cave, with an early occupation between 33 000 and 20 000 years ago, followed by a hiatus coinciding with maximum sea level retreat, and with occupation of the region recommencing as the sea reached its present level. In all occupation levels of the rock-shelter, marine shellfish remains were found.

This situation until recently contrasted markedly with evidence from the Kimberley coast, where we might look for evidence of early occupation, given its geographical relationship to Southeast Asia. Investigations in several large shelters had failed to reveal any deposits suggestive of great antiquity (Blundell 1975). Detailed research by another postgraduate student Sue O'Connor (University of Western Australia) has been carried out in the Buccaneer Archipelago and adjacent mainland. Her excavations have demonstrated occupation on offshore Koolan Island in excess of 24 000 years (O'Connor 1989).

All this so far fits well into a not particularly modified version of the coastal colonization model. The area which does not conform is the Pilbara, an area with some curious archaeological and ethnographic features. A certain amount of research has been carried out here related to development projects, particularly mining, not a situation conducive to well co-ordinated research projects; the information to hand is therefore rather piecemeal and not well-digested. A number of rock-shelters have been investigated in the Pilbara interior. Two near Mount Newman incorporate sparse evidence for Pleistocene occupation. One is dated to *c.* 20 740 BP, the other to *c.* 26 300 BP (Maynard 1980; Troilett 1982; Houghton 1984; Williams 1986). The evidence is sparse throughout, and while there does not appear to be an obvious occupational hiatus in the region, it is also not possible to demonstrate a continuous occupation. On the Pilbara coast, a series of midden sites demonstrate occupation from about 7000 years ago (Veth 1984; Veth *et al.* 1984; Bevacqua 1974; Lorblanchet and Jones 1980; Lorblanchet 1978).

Mount Newman now is nearly 500 km from the coast, and at 20 000 BP, was at least another 15O km away. It can be argued that access could have been via the Fortescue River, but the fact remains that the area around Mount Newman is now semi-arid, situated as it is on the edge of the Gibson desert to the east. This is also the edge of the Hamersley Plateau, which contains within it gorges which form havens of moisture and coolness. Do we see here an early and successful non-coastal adaptation? It is interesting to note, on a 'tribal boundaries' map, the very small group territories recorded for the region – much smaller than would be expected given the usual equations made linking group size/density and rainfall/resources. Further systematic investigation of prehistoric land use in the Pilbara would be timely.

This evidence might however simply suggest opportunistic exploitation of a temporarily ameliorating arid zone fringe, rather than a permanent adaptation. Such an interpretation is also possible of Smith's (1987) evidence of 22 000 year old occupation in the Cleland Hills area of the Northern Territory. At none of these sites do we see clear evidence for continuing occupation from before 20 000 BP (the onset of maximum aridity) until the end of the Pleistocene. What is evident is intermittent early occupation, with a more obvious developed adaptation to these arid regions after 6000 BP (see also Veth 1987).

CONCLUDING REMARKS

In Western Australia, as elsewhere, I still see the coastal colonization model as serviceable, if requiring some modification here and there. I also see it as possibly being capable of extension to island Southeast Asia for Upper Pleistocene times. The main areas of modification are those required by new evidence from Tasmania (the Franklin River sites) and the Pilbara, that some non-aquatic adaptations may have occurred earlier than originally postulated.

The suggestion of the importance of certain Malaysian plant species, particularly those found in sub-tropical rainforests, originally put forward by Golson (1971) with a coastal adjunct, fits well with the new evidence from New Guinea. It hardly equates, however, with such early sites as Upper Swan Bridge or Lake Mungo, where such species are unlikely ever to have flourished. Are we, perhaps, looking at two kinds of colonization within the context of an envisaged continual movement of people into Greater Australia? Perhaps we can envisage on the one hand a movement round New Guinea and down the Australian east coast of coastally-oriented rainforest exploiters, gradually adapting to different plant species but staying in better-watered regions. On the other hand, perhaps we can see a more originally tightly coastal group who nevertheless learn to cope more quickly with more arid environments; getting to Upper Swan Bridge (without direct landfall by sea from Southeast Asia) must always have involved skirting the Great Sandy Desert. Perhaps we might see such adaptability arising not from an ungeneralised 'all purpose economy', but from an economy structured into different 'strategies', which are adaptable in due course to different environments. When we investigate the emergence of *Homo sapiens sapiens*, it may be that we need to consider such variables as economic adaptability as crucial to any definition of 'human'.

ACKNOWLEDGEMENTS

I thank Paul Mellars for soliciting this paper and my presence at the Conference. I am grateful to Peter Veth, Sue O'Connor and Kate Morse for discussion and for sharing their data. Thanks also to Cynthia Munday for typing and word-processing.

REFERENCES

Bellwood, P. 1985. *Prehistory of the Indo-Malayan Archipelago*. Sydney: Academic Press.

Bevacqua, R. 1974. The Skew Valley midden site: an Aboriginal shell mound on Dampier Island, Western Australia. Unpublished report to the Department of Aboriginal Sites, Western Australia Museum, Perth.

Birdsell, J.B. 1957. Some population problems involving Pleistocene man. *Cold Spring Symposia on Quantitative Biology* 22: 47-69.

Birdsell, J.B. 1968. Some predictions for the Pleistocene based on equilibrium systems among recent hunter-gatherers. In R.B. Lee and I. DeVore (eds) *Man the Hunter*. Chicago: Aldine: 229-240.

Birdsell, J.B. 1977. The recalibration of a paradigm for the first peopling of Greater Australia. In J. Allen, J. Golson and R. Jones (eds) *Sunda and Sahul: Prehistoric Studies in Southeast Asia, Melanesia and Australia*. London: Academic Press: 113-167.

Blundell, V.J. 1975. *Aboriginal Adaptation in Northwest Australia*. Unpublished Ph.D. Thesis, Wisconsin University.

Bordes, F., Dortch, C., Thibault, C., Paynal, J.-P. and Bindon, P. 1983. Walga Rock and Billibilong Spring. *Australian Archaeology* 17: 1-26.

Bowdler, S. 1974. Pleistocene date for Man in Tasmania. *Nature* 252: 697-698.

Bowdler, S. 1976. Hook, line and dillybag: an interpretation of an Australian coastal shell midden. *Mankind* 10: 248-258.

Bowdler, S. 1977. The coastal colonization of Australia. In J. Allen, J. Golson and R. Jones (eds) *Sunda and Sahul: Prehistoric Studies in Southeast Asia, Melanesia and Australia*. London: Academic Press: 203-246.

Bowdler, S. 1981. Hunters in the highlands: Aboriginal adaptations in the eastern Australian uplands. *Archaeology in Oceania* 16: 99-111.

Bowdler, S. 1983. Rainforest: colonised or coloniser? *Australian Archaeology* 17: 59-66.

Bowdler, S. 1984. *Hunter Hill, Hunter Island*. Terra Australis 8. Canberra: Department of Prehistory, Research School of Pacific Studies, Australian National University.

Bowdler, S. (in press). Early Southeast Asian prehistory: a view from down under. In M. Santoni (ed.) *Proceedings of the South East Asian Archaeological Conference*. Paris: Musée Guimet.

Bowler, J.M. 1976. Recent developments in reconstructing late Quaternary environments in Australia. In R.L. Kirk and A.G. Thorne (eds) *The Origin of the Australians*. Canberra: Australian Institute of Aboriginal Studies: 55-77.

Bowler, J.M., Jones, R., Allen, H. and Thorne, A.C. 1970. Pleistocene human remains from Australia: a living site and cremation from Lake Mungo, western New South Wales. *World Archaeology* 2: 39-60.

Dortch, C.E. 1979. Devil's Lair, an example of prolonged cave use in southwestern Australia. *World Archaeology* 10: 258-279

Flood, 1983. *Archaeology of the Dreamtime*. Sydney: Collins.

Golson, J. 1971. Australian Aboriginal food plants: some ecological and culture-historical implications. In D.J. Mulvaney and J. Golson (eds) *Aboriginal Man and Environment in Australia*. Canberra: Australian National University Press: 196-238.

Groube, L., Chappell, J., Muke, J. and Price, D. 1986. A 40,000 year-old human occupation site at Huon peninsula, Papua New Guinea. *Nature* 324: 453-455.

Habgood, P.J. 1989. The origin of anatomically-modern humans in Australia. In P. Mellars and C. Stringer (eds) *The Human Revolution: Behavioural and Biological Perspectives on the Origins of Modern Humans*. Edinburgh: Edinburgh University Press: 245-273.

Hallam, S.J. 1983. The peopling of the Australian continent. *Indian Ocean Newsletter* 4: 11-15

Horton, D.R. 1981. Water and woodland: the peopling of Australia. *Australian Institute of Aboriginal Studies Newsletter* 16: 21-27.

Houghton, G. 1984. An archaeological survey of temporary reserves 4883H and 4884H in the Rhodes River Ridge area of Western Australia. Unpublished report to the Department of Aboriginal

Sites, Western Australia Museum, Perth.
Jones, R. 1979. The fifth continent: problems concerning the human colonisation of Australia. *Annual Review of Anthropology* 8: 445-466.
Jones, R. 1989. East of Wallace's Line: issues and problems in the colonization of the Australian continent. In P. Mellars and C. Stringer (eds) *The Human Revolution: Behavioural and Biological Perspectives on the Origins of Modern Humans*. Edinburgh: Edinburgh University Press: 743-782.
Kershaw, A.P. 1974. A long continuous pollen sequence from northeastern Australia. *Nature* 251: 222-223.
Kershaw, A.P. 1975. stratigraphy and pollen analysis of Bromfield Swamp, northeastern Queensland, Australia. *New Phytologist* 75: 173-191.
Kiernan, K., Jones, R. and Ranson, D. 1983. New evidence from Fraser Cave for glacial age man in southwest Tasmania. *Nature* 301: 28-32,
Lampert, R.J. 1971. *Burrill Lake and Currarong*. Terra Australis 1. Canberra: Department of Prehistory, Research School of Pacific Studies, Australian National University.
Lampert, R.J. 1977. Kangaroo Island and the antiquity of Australians. In R.V.S. Wright (ed.) *Stone Tools as Cultural Markers: Change, Evolution and Complexity*. Canberra: Australian Institute of Aboriginal Studies: 213-218.
Lampert, R.J. 1981. *The Great Kartan Mystery*. Terra Australis 5. Canberra, Department of Prehistory, Research School of Pacific Studies, Australian National University.
Lorblanchet, M. 1978. Skew Valley (Dampier, Western Australia): shell middens and rock engravings. Unpublished report to the Australian Institute of Aboriginal Studies, Canberra.
Lorblanchet, M. and Jones, R. 1980. Les premières fouilles à Dampier (Australie de l'ouest et leur place dans l'ensemble Australien). *Bulletin de la Société Préhistorique Française* 76: 463-487.
Lourandos, H. 1985. Intensification and Australian prehistory. In T.D. Price and J. Brown (eds) *Prehistoric Hunter-Gatherers: the Emergence of Cultural Complexity*. Orlando: Academic Press: 385-423.
Maynard, L. 1980. A Pleistocene date from an occupation deposit in the Pilbara region, Western Australia. *Australian Archaeology* 10: 3-8.
Meehan, B. 1977. Man does not live by calories alone: the role of shellfish in a coastal cuisine. In J. Allen, J. Golson and R. Jones (eds) *Sunda and Sahul: Prehistoric Studies in Southeast Asia, Melanesia and Australia*. London: Academic Press: 493-531.
Morse, K. 1988. Mandu Mandu Creek rockshelter: Pleistocene human occupation of North West Cape, Western Australia. *Archaeology in Oceania* 23: 81-88.
Neal, R. and Stock, E. 1986. Pleistocene occupation in the south-east Queensland coastal region. *Nature* 323: 618-621
O'Conner, S. 1989. New radiocarbon dates from Koolan Island, west Kimberley, WA. *Australian Archaeology* 28: 92-104.
Pearce, R.H. and Barbetti, M. 1981. A 38,000 year old archaeological site at Upper Swan, Western Australia. *Archaeology in Oceania* 16: 173-178.
Schwede, M.L. 1983. Supertrench – phase two: a report on excavation results. In M. Smith (ed.) *Archaeology at ANZAAS 1983*. Perth: Western Australian Museum: 53-62.

Singh, G., Kershaw, A.P. and Clark, R. 1981. Quaternary vegetation and fire history in Australia. In A.M. Gill, R.H. Groves and I.R. Noble (eds) *Fire and the Australian Biota*. Canberra: Australian Academy of Science: 23-54.

Smith, M.A. 1987. Pleistocene occupation in arid central Australia, *Nature* 328: 710-711.

Thorne, A.G. 1971. Mungo and Kow Swamp: morphological variation in Pleistocene Australians. *Mankind* 8: 85-89.

Thorne, A.G. 1977. Separation or reconciliation? Biological clues to the development of Australian society. In J. Allen, J. Golson and R. Jones (eds) *Sunda and Sahul: Prehistoric Studies in Southeast Asia, Melanesia and Australia*. London: Academic Press: 187-204.

Thorne, A.G. 1980a. The arrival of man in Australia. A. Sherratt (ed.) *The Cambridge Encyclopaedia of Archaeology*. Cambridge: Cambridge University Press: 96-100

Thorne, A.G. 1980b. The longest link: human evolution in Southeast Asia and the settlement of Australia. In J.J. Fox, R.G. Garnant, P.T. McCawley and J.A.C. Mackie (eds) *Indonesia: Australian Perspectives*. Canberra: Research School of Pacific Studies, Australian National University: 35-42.

Troilett, G. 1982. Report on Ethel Gorge salvage project. Unpublished report to the Department of Aboriginal Sites, Western Australia Museum, Perth.

Veth, P.M. 1984. Report of an archaeological salvage programme at site P5431, Pope's Nose Creek, Pilbara, north-west Australia. Unpublished report to the Mains Roads Department, Perth.

Veth, P. 1987. Martutjarra prehistory: variation in arid zone adaptations. *Australian Archaeology* 25: 102-111.

Veth, P, Quartermaine, G. and O'Brien, B. 1984. Report of the archaeological salvage research programme at site P4665, Cape Lambert, north-west Australia. Unpublished report to Cliffs Robe River Iron Associates, Perth.

White, J.P. and O'Connell, J.F. 1982. *A Prehistory of Australia, New Guinea and Sahul*. Sydney: Academic Press.

Williams, S. 1986. *Analysis of the Lithic Component of the Ethel Gorge Collection, Newman, Western Australia*. Unpublished BA (Hons) Thesis, Centre for Prehistory, University of Western Australia, Perth.

Wolpoff, M.H. 1989. Multiregional evolution: the fossil alternatives to Eden. In P. Mellars and C. Stringer (eds) *The Human Revolution: Behavioural and Biological Perspectives on the Origins of Modern Humans*. Edinburgh: Edinburgh University Press: 62-108.

Wright, R.V.S. 1986. How old is zone F at Lake George? *Archaeology in Oceania* 21: 138-139.

B
General Studies

13: Middle Palaeolithic Socio-Economic Formations in Western Eurasia: an Exploratory Survey

NICOLAS ROLLAND

'Many ... archaeologists concentrate on problems that available data cannot answer. One can find few kinship relationships in a pile of flint flakes' (Patterson 1980: 497-8).

'The social life of early man is one central to prehistoric studies but which prehistorians ... refer to only by the way. It is easy to see why they should have been shy of this basic theme...they avert their eyes from the deeper shadows.... Attention is focussed first and foremost on the history of material culture .... an account of the daily life of baboons...tells us more about ourselves and our prehistory than we could ever hope to learn from a study of flaking-angles...' (Clark 1963: 437).

INTRODUCTION

This paper reviews current knowledge concerning socio-economic life during the Middle Palaeolithic. 'Socio-economic formation' refers to social units related to specific forms of home ranges, settlement systems and task groups, inasmuch as these are influenced by subsistence, demography and technology.

Resolution remains severely limited, given the patchy and uneven quality of the Middle Palaeolithic record, in addition to several theoretical uncertainties, for investigating such a topic.

Coverage will necessarily remain exploratory and confined to a review of models, accompanied by illustrative evidence from Western Eurasia, with special emphasis on documents from Western Europe. The discussion will also address one of the major themes of this symposium, namely, reconstructing the behavioural antecedents of modern hominids. A key issue emerging from this is to consider the proposition whether the Middle Palaeolithic represents a distinct epoch or substage within the Old Stone Age succession. Available documentation and current palaeoanthropological theory allow admittedly little more than provisional and open-ended answers. Toolmaking repertoires provide more familiar guidelines for subdividing the pre-Neolithic record but cannot constitute sufficient criteria for a logical periodization, unless they are demonstrably correlated with other aspects and trends in the development of ancient hominid lifeways.

The paper's structure is intended to convey such a focus on the problem. It begins by considering to what extent the Middle Palaeolithic can be isolated as a distinct horizon or developmental stage, on the basis of behavioural manifestations not confined to lithic repertoires. This is followed by surveying, to a limited extent, certain themes relating to subsistence and social organization and, next, by examining how current debates and alternative interpretations of Middle Palaeolithic inter-assemblage variability have contributed to reconstructing lifeways associated with this horizon. After assessing all of this information in terms of behavioural repertoires, we conclude by addressing another fundamental question, also bearing on the significance of the Middle Palaeolithic as a distinct substage: were the behavioural parameters of hominid populations contemporaneous with the Middle Palaeolithic quantitatively or qualitatively different from those of modern humans, and what mechanisms could account for such differences?

OPERATIONAL LIMITS

Prospects for reaching valid conclusions concerning Middle Palaeolithic lifeways face acute practical difficulties because they are contingent, firstly, on reconstructions from stringently limited direct evidence and, secondly, on a poorly developed theoretical framework. These limits appear at the level of preservation of the record, the current state of data-gathering methodology, and on available models for interpretation.

The record

Most research on Palaeolithic or recent hunting-gathering prehistory runs into familiar empirical problems of sampling bias and a poorly, differentially preserved record. Available evidence, particularly sites, could be regarded as the outcome of rare events, a situation with parallels in vertebrate palaeontology (Kurtén 1971: 15-6). Ethnographic data furthermore show, for instance, that the annual kill ratio among traditional caribou hunters in northern Canada's Barrenland would approach 19 500 animals for 650 people (Gordon 1981: 1). Only a fraction of this would preserve and find its way into Palaeolithic caves as bone midden from accumulated food refuse.

The cumulative effect of geomorphological change, mechanical or chemical erosion, in destroying, making inaccessible or invisible concrete evidence, increases with time. Current estimates for the Middle Palaeolithic time span approaching 200 000 convey the scope for loss of documentation.

The presumably greater mobility, lower densities and simpler portable material equipment of ancient hominid populations at that time would also have contributed to a sparse record.

Research methods

The opening citations illustrate two diametrically opposed approaches to the study of the past which bear on the present topic. The first constitutes an empiricist bias, with a plea for restricting research to problems for which only obvious concrete evidence exists. The other, holistic and more multidisciplinary in outlook (Rouse 1972: 6-11), gives priority to investigating phenomena because of their intrinsic interest, without imposing constraints due to temporary practical difficulties, and coincides more closely with this writer's approach. Fieldwork designs are largely determined by the questions being asked. Middle Palaeolithic research has focussed on chronostratigraphy and lithics. Interest in subsistence (but see Lindner 1950; Chase 1986), land-use patterns (see Freeman 1976), social life and ideology, has been mostly incidental and not methodical.

A bias in favour of sites as archaeological units has prevailed until recently, to the exclusion of data gathering on off-site evidence (Foley 1981). Another has pertained to enclosed (caves, rock-shelters) sites and home bases, especially in Europe. The need to find stratified sequences and preserved fossil fauna, along with awareness of differential preservation and the effect of geomorphological setting, have contributed to this. New researches in the Levant, however, have corrected these design imbalances and documented a representative range of site types and past activity relicts (Marks 1975: 351-2; Marks and Freidel 1977: 141; Larson 1979).

Retrieval of materials has concentrated on lithics and fossil mammals, neglecting macrobotanical remains (often preserved in Mediterranean cave sites). Off-site observations, while more difficult in the case of earlier Palaeolithic horizons in Western Eurasia, remain scarce. Comparative data on densities of remains or concentration indices are also uncommon (see Bailey *et al.* 1983, for Epirus and Ziegler 1973, for the Levant). Work on prehistoric blood residues on stone tools (Loy 1983) is only beginning. Reconstructions of natural habitats, enabling predictions of land-use strategies (see Ambrose and Lorenz, this volume) become difficult when dealing with Upper Pleistocene palaeoenvironments, but useful pioneering studies for the Late Glacial in Central Europe (Weniger 1982) indicate their potential for the Middle Palaeolithic.

Models

Middle Palaeolithic repertoires and activity remains present simpler, archaic-looking patterns with obscure information content, without real ethnographic parallels – even the most culturally impoverished recent hunter-gatherers possess minimal symbolic productions – making uniformitarian assumptions questionable (Bailey 1983: 174). Analogical reasoning, while still legitimate (Wylie 1982) risks being confined to the less satisfactory general comparative approach (Willey 1977: 86).

IS THERE A MIDDLE PALAEOLITHIC SUBSTAGE?

The Middle Palaeolithic could represent either a time segment within a continuous succession, or a discrete event possessing cumulative properties and discernible interfaces.

Most authors define the Middle Palaeolithic in Western Eurasia and the Mediterranean on the basis of lithic assemblage characteristics. It is described as an industrial complex, with standardized flake tool kits resulting mainly from the introduction of specialized (prepared-core) or 'mode 3' (Clark 1969: 31) primary flaking techniques (Bordes 1953a; McBurney 1960: 132; Leroi-Gourhan 1966; Skinner 1965). The concept of a 'Mousterian Complex' in Western Europe has been reserved for Early Würm occurrences. The term 'Mousterian' becomes altogether redundant since Middle Palaeolithic occurrences dating to pre-Würmian times are now more common.

The concept of technocomplex (Clarke 1968) if applied to the Middle Palaeolithic, would imply the existence of widespread adaptive development, at the level of toolmaking repertoires, cross-cutting a variety of biomes and covering continent-sized areas. This would also raise the issue of identifying and evaluating other changes or repercussions, associated with technological developments:

1. Tool-making repertoires changes may operate as self-contained events at the level of artifact manufacture methods and of style design;

2. They may be associated with changes in subsistence systems, social life or ideology, as a set of interacting changes;

3. They could function as an independent variable, triggering repercussions at other levels of ancient hominid life. In the last two instances, one can perceive lithic repertoires as either a key component in a closed system, which would require a series of readjustments throughout that system, following an initial change, or as an open subsystem, yielding diagnostic information about other aspects, because of their interrelationship (Binford 1962).

The introduction of new tool designs and of new technological procedures would indeed reflect important developments or changes in subsistence strategies (McGhee 1980: 40). Basic innovations of this kind do appear during the Upper Palaeolithic. Changes in the Middle Palaeolithic repertoires, however, only seem to have occurred at the level of flaking methods, resulting in greater attribute cohesion of artifacts, with few exceptions, suggesting a narrower range of transformation, making lithic evidence a less satisfactory criterion, if the Middle Palaeolithic is to be regarded as a separate horizon.

V.G. Childe (1956a, 1956b) has criticized the classic Stone Age subdivisions, particularly that equating the Palaeolithic and the Pleistocene epoch, thus relying on non-archaeological criteria for periodization. He has indicated his preference for subdivisions coinciding with more than chronological units and reflecting developmental events such as broad technological stages, correlated with significant economic and social transformations – exemplified, for instance, by the Bronze Age in Western Eurasia (Childe 1944). He suggested grouping together the Lower and Middle Palaeolithic on the one hand, and the Upper Palaeolithic and Mesolithic on the other, renamed 'Archaeolithic' and 'Miolithic' (or 'Leptolithic'), respectively to emphasize basic technological differences between them.

Recent findings and conclusions on the Palaeolithic and Mesolithic offer new guidelines in searching for major turning points in the developmental trajectory of pre-agricultural techno-economic prehistory (Clark 1980; Price and Brown 1985). Childe's suggested stages periodization (see Feustel 1968 for another attempt) could be adapted, as a heuristic device to the problem of immediate concern here – namely, defining the place of the Middle Palaeolithic in a periodization system, as follows:

I. Protolithic (or Basal Palaeolithic), comprising the Oldowan Complex (Toth 1985) of Subsaharan Africa.

II. Archaeolithic (or Early Palaeolithic), comprising the Lower and Middle Palaeolithic complexes of Africa and Eurasia.

III. Miolithic (or Advanced Palaeolithic), with the Upper Palaeolithic and Mesolithic of Africa, Eurasia and the New Worlds.

Further subdivisions could be introduced, if needed, enabling us to fit the Middle Palaeolithic with the final Early Palaeolithic or Archaeolithic episode (Combier 1962), which could coincide with significant techno-economic or socio-economic transformations.

The conventional procedure of relying on lithic industries as periodization markers, while still convenient, contains several drawbacks, however:

1. The Bordes classificatory system employed throughout Western Europe and much of the Mediterranean does not anticipate some artifact and industrial patterns encountered in Central and Eastern Europe;

2. The system also leaves out macrolithic industries of the African Middle Stone Age;

3. While lithic tool-making repertoires remain broadly applicable throughout Western Eurasia, the Mediterranean, Africa and India, this does not answer the question of whether developments in other areas of technology, subsistence or social life did or did not take place at the same time in areas such as Southeast Asia, where lithic industries are poorly developed (see Hayden 1977);

4. There are no *a priori* reasons for thinking that Middle Palaeolithic industries necessarily overlap or coincide with other distinct aspects of behaviour such as faunal exploitation or settlement patterns.

It is evidently premature and beyond the present paper's scope to attempt to demonstrate that the Middle Palaeolithic actually does represent a distinct substage with a set of correlated changes. The main objective will be to examine the extent to which various aspects of subsistence, social life and other activities are patterned to qualify as distinctive of the Middle Palaeolithic. Lithic repertoires, in this perspective, serve mainly as heuristic starting points for identifying trends (see Otte 1981, for a similar attempt relating to the Gravettian technocomplex of Central Europe).

REVIEW OF SOCIO-ECONOMIC VARIABLES

This survey covers evidence or conclusions pertaining to variables falling under the following headings: population, subsistence and home ranges, technology, settlements.

Population

Higher fertility ranks, as indicated by a steady geographical expansion from an African cradle into Eurasia and the New Worlds during the Pleistocene, appear among the traits distinguishing the hominid species from other primates (Bartholomew and Birdsell 1953; Lovejoy 1981).

Population remains a key variable of the socio-economic equation. Greater densities tend to covary with larger, more sedentary local groups.

Conventional demographic models of hunting-gathering populations have favoured stable equilibria (e.g. Birdsell 1968). Various cultural factors such as longer lactation periods, birth spacing, infanticide and mechanical devices do contribute more in managing population fluctuations than mortality (Dumond 1975). They do not neutralize completely, however, the effect of stochastic processes prevailing among small-sized reproductive groups (Ammerman 1975).

Middle Palaeolithic populations have occupied Western Europe continuously since late mid-Pleistocene times, apparently coping more successfully than their predecessors with palaeoclimatic cycles (as was also the case in other areas of temperate Eurasia). The Shanidar cave occupants seem to display social norms relaxing natural selective pressures (brought about by injuries or handicaps) thus improving average life expectancy (Trinkaus 1983). The high density of sites in southern Britain, from late Lower Palaeolithic through early Middle Palaeolithic, could also indicate more intensive regional occupation (Mellars 1974). Middle Palaeolithic groups initiated or completed the settlement of the Russian Plain, the steppes and highlands of central and inner Asia, perhaps as early as the penultimate glacial times (Bader 1965; Ranov 1972; Okladnikov 1972; Chard 1969; Gabori 1976), through migratory drift.

Population size and density estimates for the entire Middle Palaeolithic *oikoumene* of 1.5 million people and 0.030 people per km^2 respectively (Hassan 1981: Table 12.3) represent nothing more than educated guesses. They could indicate, however, only minimal density increases over the Lower Palaeolithic (while increasing the hominid realm) when compared with the Upper Palaeolithic (0.100/km^2). This hypothesis may receive support from the fact that site density in areas like the Périgord or Cantabrian Spain also remains lower than subsequently (Mellars 1973: Table 3; Straus 1977).

Environmental and demographic stochasticity (Ammerman 1975) must have operated in Western Europe throughout Middle Palaeolithic times, given the factor of numerous palaeoclimatic fluctuations recorded for that area. Site densities coinciding with the Early Glacial (Würm I) period in southern France seem much lower than for the Lower and Inter-Pleniglacial (Würm II and Würm II-III) (Rolland 1975: Table III-2a), especially when plotted against recently proposed revisions for the duration of the Upper Pleistocene (Dennell 1983). The time-space distribution of certain Mousterian Complex industry types, within restricted areas of southwestern France (such as the Würm II clustering of Quina sites in the Charente, and the high density of open-air MTA sites (Sireix and

Table 13.1 Examples of specialised or intensive exploitation of single game species in the middle Palaeolithic of Western Eurasia

Site	Unit/Layer	Location	Dominant Species	Per Cent	Reference
La Micoque	4	Périgord	horse	ca. 100	Patte 1971, Riguad and Laville, pers. comm. 1987
Combe-Grenal	late Riss	Périgord	reindeer	92-97	Bordes and Prat 1965
Combe-Grenal	50-52	Périgord	red deer	70-80	Bordes and Prat 1965
Roc-de-Marsal	IV	Périgord	reindeer	87	Bouchud and van Campo 1962
Le Roc		Périgord	bison	ca. 100	Rigaud 1980
Les Ourteix		Périgord	bison	80	Rigaud 1980
Sandougne		Périgord	bison	ca. 100?	Geneste, pers. comm. 1987
La Quina	C2-C3	Charente	bison		Bouchud 1966
Mauran		S. France	bison	ca. 100	Girard-Farizy and Leclerc 1981
Vergisson IV		E. France	reindeer	95	Combier 1959
Cotte St-Brélade	late Riss	Jersey	mammoth		Scott 1980
Ioton		S. France	horse	ca. 100?	Meignen 1976
Ehringsdorf/Taubach	Eem	Thuringia	elephant/rhino		Behm-Blancke 1960
Grotta del Poggio	Early Würm	C. Italy	red deer		Barker 1975
Molodova I	1-5	Ukraine	mammoth		Klein 1969
Staroselje	Early Würm	Crimea	steppe ass		Klein 1969, Vereschagin 1967
Shanidar	D	Kurdistan	wild goat	75.1	Perkins 1964

Bordes 1972: 324) – especially if the latter are penecontemporaneous with the same caves and shelters occurrences during Würm II in the Périgord – could signify population flux.

Subsistence and home ranges

Subsistence. The extent to which Middle Palaeolithic exploitation of fauna was specialized is currently the topic of considerable debate. The very notion of hunting, prior to the Upper Palaeolithic, has been challenged, with Binford (1982, 1984) arguing that no conclusive evidence for it (such as mass slaughter and some food storage) existed before then. Scavenging instead remained the common strategy. We make the following comments:

1. Hunting can be practised, admittedly on a less intensive scale, without using the above methods;
2. The same criteria employed for identifying scavenging has failed to confirm its role to any significant degree in the Middle Palaeolithic subsistence of southwestern France (Chase 1988);
3. Table 13.1 illustrates some instances throughout Western Eurasia for the specialized or intensive exploitation of a single species; concentration on only two species is also evidenced in the Levant – e.g. the exploitation of fallow deer and gazelle, at Kebara Cave (Davis 1977); gazelle and caprines (initially) then camelids and caprine (later) at Douara Cave (Payne 1983);
4. While authors have acknowledged long ago (Lindner 1950) the likelihood that individual kills may have accumulated at a site over an extended period, a number of Middle Palaeolithic sites exemplify instances when killing by driving game (a very simple technique) in unspecified numbers over cliff edges probably took place, such as La Cotte de St-Brélade (Scott 1980), Staroselje (Vereschagin 1967) and especially at La Quina (Jelinek *et al.* 1988). Furthermore, the prevalence of immature fossil elephants and rhinoceroses at La Cotte de St-Brélade, at Taubach-Ehringsdorf (Behm-Blancke 1960), at Tata and Krapina (Ga'bori 1976), could imply the use of artificial pitfalls, by taking advantage of herd structure and behaviour with juvenile individuals preceding the mother (Soergel 1922: 98-103; Lindner 1950: 152-60). The deliberate killing of very large mammals is evidenced since the Lower Palaeolithic in Europe as at Torralba, Ambrona and Lehringen. The dominance of juvenile deer remains at Cova Negra (Spain) also indicates specialized exploitation and knowledge of herd structures (Perez Ripoll 1977);
5. Food storage becomes necessary mainly when residence conditions compel semi-nomadic and especially semi-sedentary groups to live, during specific seasons, at locations some distance from game herd concentrations.

Comparative data for animal food refuse for both the Middle and Upper Palaeolithic cave and rock-shelter sites in southwestern France show clear evidence for concentration on one or two species. This specialization, however, intensifies significantly over time (notably for reindeer) but only from the Solutrian onward (Mellars 1973: Tables 1 and 2). This trend – recently confirmed statistically (Simek and Snyder 1988) – shows that full specialized exploitation takes place long after the Middle/Upper Palaeolithic interface. In Cantabrian Spain, specialization itself does not appear until the Solutrian and the Magdalenian (Freeman 1973). Most of the above evidence in Western Europe, however, comes from enclosed sites. Recent excavations of open-air sites, by contrast, frequently show an almost exclusive concentration on a single species, as at Mauran, Le Roc, and Maine d'Euche (Rigaud 1980).

Reconstructed Middle Palaeolithic subsistence patterns throughout Western Eurasia suggest successful exploitations of temperate and cold habitats by employing strategies ranging from broad-spectrum hunting (e.g. in the lowland open-air sites of Central Italy: Barker 1975: 120), opportunistic culling (e.g. in Cantabria: Freeman 1973), to more specialized methods.

Evidence for plant collecting is lacking, but the latter must have played a variable but at times significant dietary role in the North Mediterranean and the Levant (Payne 1983: 102) – areas lying beyond glacial permafrost belts – although strontium/calcium ratios in hominid remains indicate that meat consumption in the latter area remained important throughout (Schoeninger 1980: 215). Systematic retrieval of macrobotanical remains has been neglected (but see evidence for edible plants from Abri Agut, Spain: Freeman 1979).

Home ranges. Home ranges imply restricted areas where groups live and which they exploit throughout the year (Jewell 1966).

Spatial distribution patterns of Middle Palaeolithic sites vary between regions, reflecting different habitat characteristics, and contrast with those of the Upper Palaeolithic. In central Italy and Greece, Middle Palaeolithic sites cluster in the lowlands and the coast (the latter being more extensive during marine regression phases) in contrast to Upper Palaeolithic site distributions which included settlements in the highland and the interior (Higgs and Webley 1971; Barker 1975). Both Middle and Upper Palaeolithic occupations on the other hand, frequently overlap in the same areas and the same sites in southwestern France and Cantabrian Spain.

The more restricted Middle Palaeolithic home ranges of the North Mediterranean areas could be related to the local persistence of non-migratory large game, such as hippos, rhinoceroses and elephants, until the Last Inter-Pleniglacial, contributing to land-use systems closer to the 'restricted wandering' model (Meggers 1955) confined to lowlands (Barker 1975: 122). The generally cool, moist palaeoclimates of earlier glacial phases could, furthermore, reduce summer droughts enough to provide year-round fodder for most game herds (Rolland 1985). In the case of southwestern France, the generally favourable biomass conditions, favoured by a palaeoclimate combining Atlantic influences with longer insolation hours than in more boreal areas (Mellars 1985), would not have required extensive home ranges for the Middle Palaeolithic occupants. During the Upper Palaeolithic in the same area, on the other hand, greater site densities and episodic locational shifts (White 1985) could indicate semi-sedentary, rather than restricted wandering or semi-nomadic land use systems, without susbstantial changes in home ranges. A reverse trend, however, seems to have prevailed in the Southern Levant (Larson 1979).

Further clues about exploitation territories come from studies of the sources and spatial distributions of lithic raw materials. The opaline jasper at Fontmaure (western France) was quarried at the site. Tools and debitage made of this preferred material are common over hilltops and bluffs within a 5 km radius of the site, but become scarce beyond. Social territories inferred from this distribution also seem to coincide with a natural boundary formed by the left bank of the Vienne river (Pradel 1970). Quarrying locations tend to overlap with those for other activities (Binford 1979). Flint sources in the Périgord were abundant and easily accessible, making it less easy to delimitate prehistoric territories, in contrast to regions such as Thessaly or Elis, in Greece, where it has been shown that material was obtained further inland and transported back to occupation sites (Rolland 1988a).

The diversity of latitudes and habitat zones covered by Middle Palaeolithic occurrences in Western Eurasia could favour either residential or logistic mobility (Binford 1980) by different communities. The so-called 'Alpine' Mousterian (Jéquier 1975) apparently illustrates forays by small hunting groups, separately from the main community, during warmer seasons (Gábori-Csánk 1976).

Technology

'Technology' refers to activities designed to satisfy human needs which produce alterations in the material world and which possess a social dimension (Childe 1954: 38). It includes, beyond tools and constructions, technical behaviour contingent on human somatic characteristics such as running, manipulation, climbing and so on, or 'body techniques' (Mauss 1947).

Middle Palaeolithic technology *sensu lato* remains an inadequately explored topic because we are left with only non-perishable materials and because systematic inventories are generally lacking. Better information would improve prospects for measuring more objectively technological efficiency during that time by implementing criteria such as the amount of energy harnessed per year per capita (White 1949, 1959).

Notions about the 'expedient' nature of the Middle Palaeolithic toolkits (Binford 1973; Dennell 1983), along with observations about Neanderthal postcranial skeletal robustness, have led some authors to conclude that these populations placed greater reliance on physical strength and stamina than on innovative skills, implying less efficient foraging activities (Trinkaus 1983a; Zvelebil 1984).

We have just seen that faunal exploitation techniques, while undoubtedly opportunistic, hardly suggest haphazard strategies. They indicate, instead – even in regions with little specialization such as Cantabria – more intensive land-use than previously (Freeman 1973).

While technology provides one of the best discriminants for identifying prehistoric hominid behaviour and cumulative change (Wissler 1935: 520; Childe 1954; Leroi-Gourhan 1964a; Cazeneuve 1959), this should not lead us to overlook and underestimate the physical versatility and resilience of recent and fossil *Homo Sapiens* (Laughlin 1968: 311-3). Fossil *sapiens* postcranial skeletons also show robusticity (Wolpoff 1989).

A number of innovations appear by the end of the Lower Palaeolithic: (a) intensified use of fire-making including the deliberate firing of caves to eliminate insects and other cave dwellers (Perlès 1975); a wide range of hearth types (such as simple, built over slabs, tailed hearths etc.: Bordes 1971; Perlès 1977; Bonifay 1981); the use of heated pebble floors (e.g. at La Baume-Bonne, L'Aldène: Lumley and Bottet 1962; Lumley 1976); or the spread of ashbands, for insulation. One can only speculate about the possible use of simple techniques such as smoke-signal communication between separate groups while scanning their home ranges, as employed in the Western Australian desert (Gould 1981), or for deliberate bush-burning (Sauer 1952; Lewis 1982; Mellars 1976) to improve game fodder; (b) evidence for flint mining in Hungary (Gabori-Csank 1988); (c) the use of artificial screening devices by using infrastructural elements, such as timber, animal bones (e.g. the mammoth remains at Molodova or Ripiceni Izvor (Ivanova and Chernysh 1965; Paunescu 1986; Soffer 1989), or interlocking reindeer antlers at Raj cave (Kowalski *et al.* 1972: 138-9) or Roc-en-Pail (Bouchud 1966: 124)); (d) improved butchering techniques and, perhaps, the exploitation of fur-bearing animals (Leroi-Gourhan 1964a: 150).

Bone, antler and ivory technology remained, nevertheless, essentially expedient and limited in scope. Woodworking could have been more important, as inferred from some stone artifacts (notched, serrated tools, *raclettes*) but limited to portable equipment (Leroi-Gourhan 1964a: 146-147). Antlers were sawed deliberately and used for mining or prying off lithic materials from seams. Some bone artifacts replicate common lithic tools, such as *racloirs* or Tayac points (Bonifay 1981: 40).

Important new elements appear within lithic repertoires but consist largely of more efficient primary flaking techniques at the beginning of the Middle Palaeolithic. Few other innovations occur until the Upper Palaeolithic. The prepared-core technique itself clearly does represent a major breakthrough, whose consequences lasted until the advent of metallurgy (Leroi-Gourhan 1964a: 143-6). Its application to already existing tool forms, along with certain finishing techniques, increased the standardization of tool design even though tool repertoires retain an appearance of conservatism, and can be reduced to a very limited range of forms (Dibble 1987). The appearance of new designs could indicate a broadened range of task-specificity.

One possibly important innovation is the manufacture of Mousterian points (as distinct from pointed convergent *racloirs*), which could amplify the number of killing devices already documented from earlier periods, such as wooden spears, stone balls and clubs. Mousterian points possess hard tips, often broken from impact (Bordes 1961a). Their stress resistance indicate technical mastery of litho-mechanical properties (Kopper 1981). Early instances are known from La Cotte de St-Brélade (Callow 1986), more suitable for piercing elephant or rhinoceros hides. Raw material selection shows that points were made of hard yellow flint at Fontmaure in preference to the locally prevalent and common opaline jasper used for cutting tools (Pradel 1970: 488).

The otherwise virtually nil rate of lithic design innovations during the Middle Palaeolithic does not necessarily imply stagnation. The pace of change may appear either very slow or directionally random, but some indications for accelerating tempo, along with some trends and stochastic variations, become more evident closer to the Middle/Upper Palaeolithic interface (Rolland 1988b):

1. A growing tendency to use wood or bone working tools (e.g. denticulates, notches, Mousterian *raclettes*, to be superseded eventually by Aurignacian end-scrapers, burins and strangulated blades (Leroi-Gourhan 1963; Lumley 1965);

2. Besides increasing laminar flaking, tool uses seem at this time to broaden to include incipient Upper Palaeolithic designs, such as crude burins, *grattoirs*, carinated pieces, along with Upper Palaeolithic tool indices – as in the Mousterian of Acheulian tradition Type B, even in areas without the preceding Type A industries such as the Rhone Valley (Veyrier *et al.* 1951; Combier 1967).

Settlements

The systematic investigation of settlement structures from living-floor analyses, including the use of quantitative methods, has only recently begun (Whallon 1973, 1974; Freeman 1979; Simek 1984). Documentation of these features for the Middle Palaeolithic is very inadequate, given the number of known excavations where this information could have been exploited. For Middle Palaeolithic sites, the task of discriminating intentional structures from fortuitous associations continues to present difficulties, especially in caves and rock-shelter sites (Bordes 1975).

The regular use of enclosed sites as residences during the Middle Palaeolithic is probably related to permanent occupation of temperate zones during glacial cycles, to more intensive land use patterns and to improvements in fire-making techniques. Caves and shelters, in many instances, probably remained a marginal component or a last resort of settlement systems, as for example in the Cure Valley (France), prior to the late Mousterian horizons (Girard 1976). The Southern Levant, by contrast, provides evidence for the optimal use of caves (e.g. in the Mount Carmel area: Ronen 1979: 28), a pattern which seems to coincide with trends toward semi-sedentary settlement in both enclosed and open-air sites (Marks and Freidel 1977: 137-42; Larson 1979: 80-3).

Occupation intensity, inferred from artifact densities and contents (e.g. in West-Central Europe: Bosinski 1976) varied, ranging from constructed huts, open-air or cave screening structures (e.g. Molodova, Ripiceni Izvor,

Raj, Roc-en-Pail – to protect against weather or animals) to transitory uses such as Grotta Guattari at Monte Circeo (Piperno 1976-77). We note the following patterns, however:

1. The recurrence of simple artificial constructions (i.e. huts) in the open or inside caves or shelters, by the end of the Lower Palaeolithic – as at the Mas-des-Caves, Lunel-Viel (Bonifay 1981) or L'Aldène. The Mas-des-Caves cave contains artificial basins and vegetal fragments and impressions indicating the use of beddings. Hut traces probably also existed in the open at Fontmaure (Pradel 1970) and at Rheindahlen (Bosinski 1976). Remnants of stone walls have been reported from Pech de l'Azé I and Cueva Morín and traces of postholes at Combe-Grenal (Bordes 1972; Freeman 1979). Gullies protected against water at Mas-des-Caves. Floor-debris clustering suggests the use of hide screens at the Galerie Schoepflin, Arcy-sur-Cure (Girard 1976), and probably also at Combe-Grenal (Bordes 1961b: 87). A game observation outpost may also have been present at Fontmaure;

2. Sizes of living areas within Middle Palaeolithic caves and shelters in southwestern France rarely exceeds 30 m^2, in contrast to the Upper Palaeolithic (Mellars 1973). Living floors tend to display a concentration of lithic and faunal pieces pushed to the periphery, leaving a central area relatively uncluttered (Leroi-Gourhan 1964b; Mellars 1973) – as at Abri Lartet, Montgaudier (Duport and Vandermeersch 1972), Hauteroche (Debénath 1973) and Arcy-sur-Cure (Girard 1976). Furthermore, quantitative analyses of artifact clusters and spatial distributions do show non-randomness (Freeman 1979).

All these observations indicate, overall, less intensive site use and, perhaps, smaller residential groups than for the Upper Palaeolithic, albeit with significant regional variations (e.g. the Levant). In Europe, Upper Palaeolithic sites show more elaborate maintenance and periodic clearance of debris and refuse, illustrating what has been described as the beginning of the 'domestication of living space' (Leroi-Gourhan 1964b: 139-42). More careful maintenance of living areas seems to coincide with more intensive residence, according to ethnographic observations (O'Connell 1977; Murray 1980). Certain late Middle Palaeolithic occurrences, such as the '*dernier habitat*' at La Quina, or at Arcy-sur-Cure, contain traces of more concentrated occupations, with indications of artificial levelling of living surfaces (Henri-Martin 1964; Girard 1976).

Local group-size estimates from settlement sizes and structures, while useful, remain inconclusive. Certain site complexes such as Arcy-sur-Cure (Leroi-Gourhan 1961: Figure 1), La Chaise (Blackwell *et al.* 1983), Pech de l'Azé (Bordes 1954-55), and Vergisson (Combier 1959), contain evidence for penecontemporaneous layers in different loci, suggesting residence by larger groups who could have formed a troglodyte system occupying one or more valleys, by analogy with recent Kurdish cave dwellers (Solecki 1979: 328).

INTERASSEMBLAGE VARIABILITY AND SOCIO-ECONOMIC FORMATIONS

We now examine in this section how discussions and alternative models for interassemblage variability have contributed indirectly to reconstructing Middle Palaeolithic lifeways.

The Bordes model

François Bordes' manifold contributions to Middle Palaeolithic research include excavations of several key sites, the development of a classificatory system which is still widely used, and analyses of numerous assemblages and chronostratigraphic studies. His arrangement and interpretation of Early Würm Mousterian Complex industries in France provide the basic evidence and terms of reference from which current debates stem. The division of the industries into three groups, subdivided further into six industry types, rests on the following observations:

1. The discovery of a polymodal (rather than continuous) graphic distribution of *racloir* indices. The division of the industries into Quina, Ferrassie, and Mousterian of Acheulian tradition Type A and B (or MTA), Typical and Denticulate Mousterian (or MT and MD) follows from this polymodality – with some additional distinctive toolkit elements in the form of handaxes, specialized *racloir* forms, denticulates, and notches etc.

2. These industries represent long-lasting, coexisting repertoires throughout Early Würm times, although most of them hypothetically originated before the Last Glacial (e.g. the Quina industry in the layer 3 ('Tayacian') industry of La Micoque, the Ferrassie in layer 4 ('Premousterian') or at Ehringsdorf, and the Typical at Le Rigabe's layer X (Bordes 1968)). The MTA continues the Acheulian handaxe tradition. Evidence supporting penecontemporaneity of the different Mousterian Complex industries comes from their complex interstratification within individual site successions. The MTA occurs at the top at Combe-Capelle and Combe-Grenal but at the base of Le Moustier upper shelter and between two separate MT episodes at the lower shelter. MT and MD recur more or less randomly at several sites.

Bordes' behavioural interpretations (1961b, 1970), the object of the present discussion, came as a bi-product of these researches and findings, as an attempt to account for such puzzling patterns. The resulting model, despite many criticisms, retains several valid points. According to Bordes, the six Mousterian Complex industries are the stylistic expression of separate coexisting ethnic societies. Bordes considered other possible interpretations, such as the evolution of toolmaking repertoires, the effect of seasonal variations on assemblage composition, the influence of habitats or environment and the existence of specialized, task-specific toolkits, before preferring this conclusion (1970). Specifically, Bordes argued:

1. The notion of early researchers of an evolution within the Middle

Palaeolithic, beginning with the MTA, continuing with the MT and evolving finally into the Quina industry, became invalidated by subsequent research. According to Bordes, the MTA remains the only entity to undergo change from Type A to B and eventually, into the Châtelperronian (or first Upper Palaeolithic industry). It is also conceivable that the MT derived from the MTA but without replacing it.

2. The separate industrial types cannot represent seasonal variations since their number exceeds the number of seasons. Even thick archaeological layers fail to show evidence for variations indicative of tool-kit adjustments to seasonal tasks. (One could add to Bordes' observations that ethnographic data from northern Australia (Thomson 1959) shows that seasonally-variable repertoire manufacture normally coincides with shifts to different localities, instead of being found within a single site sequence, as would be the case in southwestern France.)

3. Differing environments such as the Maghreb and the Périgord nevertheless contain assemblages containing similar typological characteristics, indicating independence from direct environmental control. All Mousterian Complex industries, furthermore, were manufactured throughout Early Würm times, without displaying any evidence for the effect of the numerous palaeoclimatic fluctuations which took place during this time range. The only possible effect of ecological variations visible in assemblage composition is limited to degrees of secondary modification (i.e. retouching) of Levallois pieces, brought about by proximity or distance of access to flint sources, exemplified by the old dichotomy between 'Levalloisian' and 'Mousterian' occurrences, the former found in the loess quarry sites, the other in caves and rock-shelter sites (Bordes 1953b).

4. The identification of specialized activities, such as killing and butchering, quarrying or other discrete tasks, presents practical difficulties because these activities are rarely represented in the archaeological record, or because they form a palimpsest of remains in caves or shelters. Differing settlement contexts where toolkit differences would be expected – such as open-air and enclosed sites – maintain typological homogeneity (compare, for example, the Quina assemblages at Combe-Grenal and Chinchon, and the MTA assemblages at Pech de l'Azé and Toutifaut).

In summary, Bordes' model for Middle Palaeolithic socio-economic formations suggests the existence of low density, small-sized, territorially bounded semi- or fully sedentary, identity-conscious groups. Although without any conceivable modern ethnographic equivalents, they could conform to nucleated social entities (Yellen and Harpending 1972). Some similarities could also exist with the pastoralists from southwestern Iran (Barth 1959), where different ethnic groups monitor seasonal schedules throughout a common home range, avoiding direct contacts. The difference would be that Middle Palaeolithic communities lived under more favourable conditions in southwestern France, which made possible a settlement system that could be characterized as 'residential stability' as opposed to 'residential or logistic mobility' (Binford 1980). This kind of stability, combined with low population densities, would minimize group interaction and foster repertoire conservatism. Eventual territorial shifts would occur into previously unoccupied areas.

In addition to the criticisms made by others concerning this model (Binford 1966; Mellars 1967), we could mention the difficulties involved in reliably identifying, ethnic self-perception among foraging groups. The delimitation of cultural boundaries from the archaeological record, even among more complex later prehistoric societies such as those of the European Bronze Age (Hodder 1978; Shennan 1978) poses similar problems, reminiscent of those facing application of 'ecotone' concepts in biology (Rhoades 1985).

The Binford model

This model (Binford and Binford 1966, 1969) represents a permanent theoretical contribution by introducing a functional approach to the interpretation of Palaeolithic documents, and a general model for semi-nomadic hunting-gathering land-use systems. The Middle Palaeolithic case study remains only of historical interest, given various criticisms concerning flaws in evidence and statistical treatment (Bordes 1970; Mellars 1967, 1970; Jelinek 1988). Further analyses involving a wider range of data also proved to be inconclusive (Binford 1983). The issues it raised, however, have had protracted methodological repercussions, leading to seminal ethnoarchaeological investigations by the same author.

Binford's model draws from contributions in North American ethnology and prehistory, including the concept of structural poses (Gearing 1962), and also from human geography (Wagner 1960). It implements, above all, principles from the author's important methodological paper on archaeological research designs (Binford 1964), bringing up the issues of sampling and of functional variability. It states that mobile Palaeolithic groups must have abandoned, in discrete fashion, remains reflecting either extractive or maintenance activities, identifiable from work camps (killing, butchering, collecting, quarrying) and base camps (residential loci, where several activities were carried out), respectively. Middle Palaeolithic communities relied on an expedient technology (tools used and abandoned on the spot) rather than on a curated technology (tools, weapons transported and maintained), with the implication that assemblages in relatively undisturbed contexts provide representative samples of specific activities.

Implications of Mellars chronological findings

This research (Mellars 1969, 1986, 1988), while not addressing socio-economic issues as such, contains a conclusion having an impact on the choice of suitable models. It has established a cultural and palaeoenvironmental datum line for the Quina industry during Early Würm II. If the Ferrassie evolved into Quina, and both are stratigraphically earlier than the MTA in southwestern France, then (a) the notion of stable, coexisting, isolated ethnic groups manufacturing the different Mousterian industries becomes redundant; and (b) prospects for considering Middle Palaeolithic variability in functional terms, conceivable with penecontemporaneous industry types, run into serious difficulties.

RECENT APPROACHES

New analyses (Rolland 1977, 1981, 1986; Dibble 1984, 1985, 1987; Jelinek 1988) have examined methodically and in breadth the part played by the differential transformation or reduction intensity of stone tools (sometimes known as the 'Frison effect': Jelinek 1977) as a factor contributing significantly to Middle Palaeolithic variability. This aspect of assemblage variability was noted long ago, relating it to differences in settlement contexts (e.g. Commont 1914). We mentioned earlier Bordes' (1953b) explanation of the 'Levalloisian/Mousterian' dichotomy along similar lines, with respect to the Levallois tool component. Leroi-Gourhan, in an elegant essay (1966), regarded this effect as a widespread phenomenon.

Lithic reduction sequences comprise a number of factors contributing to variations in assemblage composition: primary and secondary flaking techniques, formal tool designs, kinetic or functional (tool use) aspects, degrees of tool wear and rejuvenation.

Middle Palaeolithic variability across Western Eurasia and the Mediterranean occurs along two major dimensions: a utilitarian one, comprising function as well as variations in differential transformation or reduction intensity; and a chorological (or spatially distributed) one (Childe 1956b: 15), involving the geographical clumping of idiosyncratic tool designs, such as Aterian stemmed pieces in North Africa, flake cleavers of the Pyreneo-Cantabrian region, Mousterian handaxes (cordiform, triangular) of Atlantic and Northwestern Europe, and leaf points of Central Europe and the Balkans. Other less conspicuous items such as plano-convex flaking, *déjetés racloirs*, and technical variations within Levallois technique, also show some spatial clumping.

My own findings relate to the first dimension of variability. They are summarized as follows (see Rolland 1977, 1981):

1. Unretouched non-Levallois flakes and blades must be included when computing artifact frequencies in assemblages, since many of these were tools, and because their frequencies covary unexpectedly with individual Mousterian Complex industries – i.e. these quantitative variations become a diagnostic element;

2. The outcome of calculating *racloir* frequency variations by including the above category results in a continuous, L-shaped grapho-statistical pattern (Rolland 1981: Figure 4);

3. Different Mousterian Complex industries occur non-randomly, along a continuous spectrum of variation representing implement frequencies (or all regularly retouched tool types from the Bordes type-list). Increases in *racloir* manufacture, however, account for this pattern (Rolland 1977: Figure 2). This means that variations in *racloir* manufacture, being continuous rather than polymodal, cannot be regarded as markers for partitioning Middle Palaeolithic assemblages into discrete groups and industry types. Most of the variability expressed by these named entities

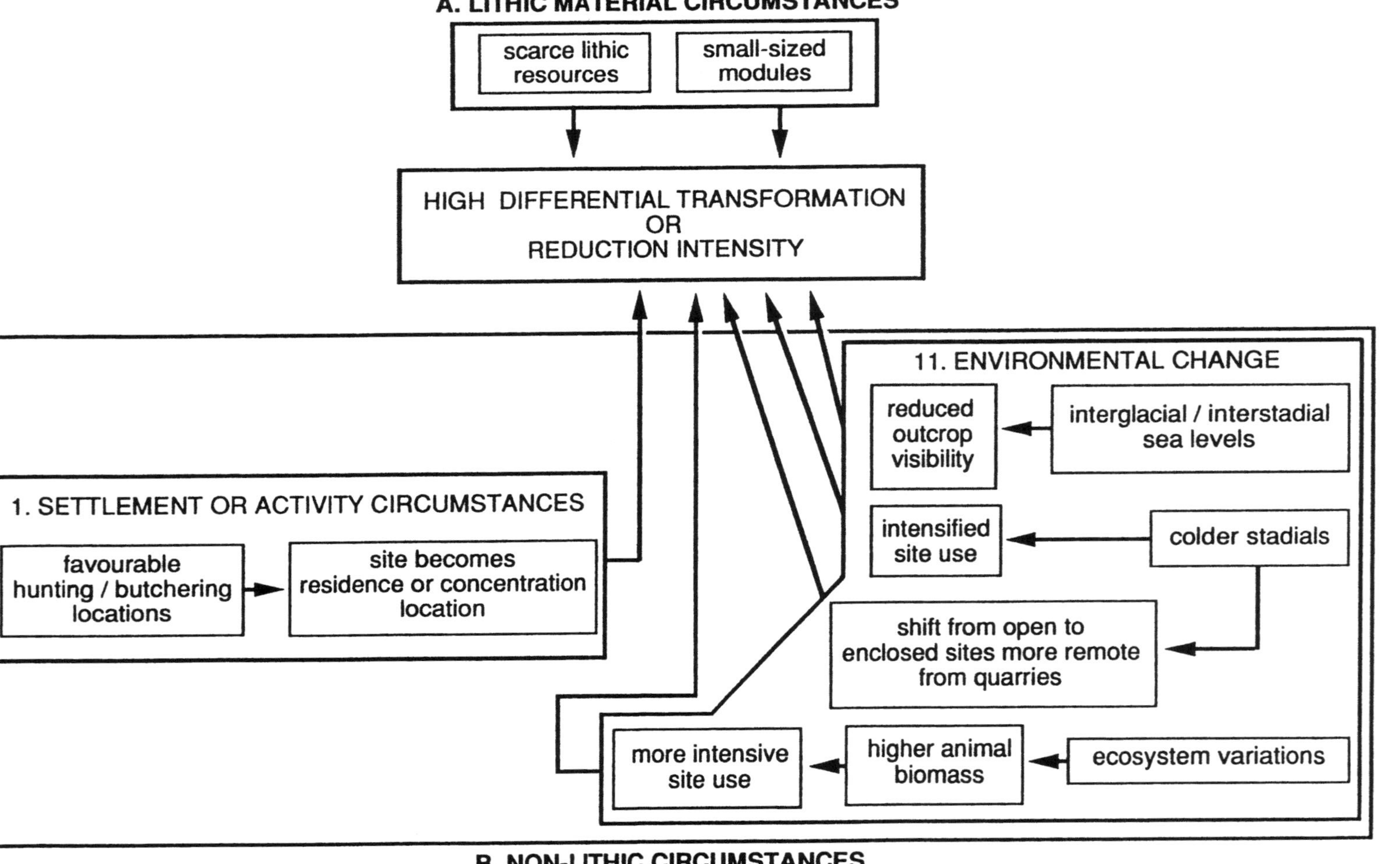

Figure 13.1. Input model for factors inducing parsimony in use of lithic raw materials.

consists, above all, in continuous variations in terms of differential intensity of transformation or reduction of the tools.

The basic, dynamic element underlying Middle Palaeolithic interassemblage variability, at least in Western Europe, is therefore made up of assemblages containing cutting instruments at various stages of utilization, wear and rejuvenation, ranging from unretouched Levallois and non-Levallois blanks or tools, through moderately retouched to intensively retouched, resharpened pieces such as limaces and transverse or convergent, bifacial *racloirs*. The other major element is represented by notched and serrated tools (see also Dibble 1988). These may form part of assemblages more or less dominated by *racloirs* but, when numerous, they tend to be accompanied by large quantities of unretouched or merely utilized cutting tools such as in the MD industry.

The present writer (and others before him) has argued that much of Middle Palaeolithic variability expresses more or less directly situations inducing either a profligate or a parsimonious use of available lithic raw materials, which influences assemblage structure – primarily the degree of secondary transformation of naturally cutting tools. Some of these circumstances may also coincide with the need to make greater use of notched and serrated tools.

Assemblages dominated either by *racloirs* or by denticulates and notches occur in a wide variety of contexts, habitats and palaeoclimates, however, suggesting the effects of heterogeneous circumstances. Assemblages labelled as 'Charentian' or as MD, could therefore be the outcome of converging factors, a possibility alluded to by Bordes (1962, 1969). This makes it unlikely that a single recurrent cause can account for the interassemblage variability described above.

What follows is a trial interpretation of interassemblage variability patterns, assessed in terms of both the lithic industries themselves, and associated patterns in the associated environmental and socio-economic variables.

Factors inducing lithic material parsimony and more intensive reduction fall into two main groups (see Figure 13.1):

1. Those related directly to lithic materials. These include variations in the access to raw materials (e.g. lessened outcrop visibility, such as at La Cotte de St-Brélade, during higher sea-level episodes: Callow 1986); distance travelled to reach sources (e.g. the Middle Palaeolithic, near coastal Elis in the Peloponnese: Chavaillon *et al.* 1969); local scarcity of suitable material (e.g. at La Chapelle-aux-Saints cave (Bouyssonie *et al.* 1913) or at the open-air site of Ioton (Meignen 1976)); and size variations affecting the choice of primary flaking techniques. The large-sized black flint nodules at La Quina were processed simply by detaching massive flake preforms, whereas at sites where nodules were smaller, disc-core or facetting techniques prevailed (Breuil and Lantier 1959: 154). It has been noted that the largest artifacts, even at Combe-Grenal, frequently coincide with layers where Levallois technique gives way to the La Quina-type primary flaking (Rolland 1988b: Figure 5);

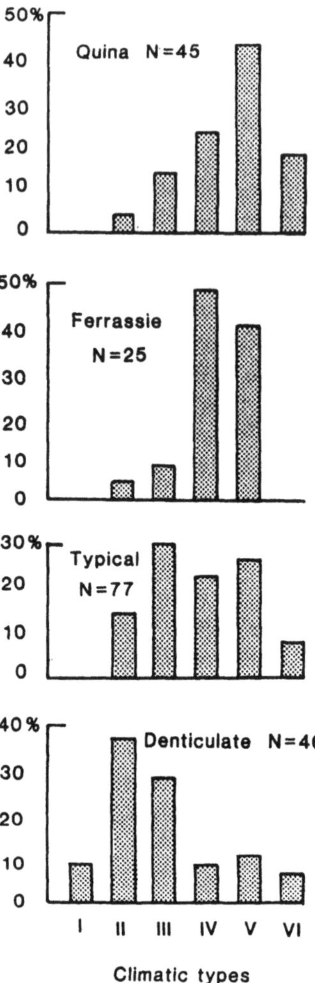

Figure 13.2. Palaeoclimatic phases associated with different industry types in the Mousterian complex. Climatic phases are as follows: I = temperate, very wet; II = temperate, wet; III = mild, wet; IV = cold, wet; V = cold, dry; VI = very cold, very dry. Modal tendencies among the different industry types range from temperate, wet (Denticulate); mild, wet (Typical); cold, wet (Ferrassie); to cold, dry (Quina). Palaeoclimates associated with the Denticulate industries cover the entire range. From Rolland 1981.

2. Those reflecting certain settlement or activity circumstances – some of which could overlap with the above. They include settlement types (in particular open-air *versus* enclosed sites, where occupation may be more or less transient or prolonged: Leroi-Gourhan 1966); settlement shifts, such as occupation of sheltered sites further away from quarries, with the onset of cold stadials (e.g. in the Maas river region, coinciding with Quina occurrences, in Belgium: Ulrix-Closset 1973); intensified use of expedi-

ent tools, such as the use of cutting tools in transit or killing and butchering sites. (Examples for this may exist in open-air sites in the south Russian plain (Il'skaya, Sukhaya Mechetka) or at Ehringsdorf); intensified site occupation resulting from local factors such as suitable topographic settings favouring, at the same time, mass killing and residence (e.g. La Quina: Jelinek *et al.* 1988) or, perhaps, periodically richer animal biomass (e.g. the Micromousterian horizon of Asprochaliko, Epirus: Rolland 1988a); or from more generalized events such as the effect on settlement intensity of more severe palaeoclimatic conditions (e.g. in southwestern France: Rolland 1981), or drier ones in the Levant, coinciding with the Yabrudian (Jelinek 1982: 72).

CONTEXTUAL EVIDENCE

Testing the effect of any of the above factors would require detailed investigations of individual site situations. Until such conditions are met, a preliminary step would be to search for one or more lines of collateral evidence compatible with consequences of palaeoclimatic changes on settlement intensity, reflected by raw material parsimony. Palaeoclimatic changes could lead to vegetation shifts and animal biomass fluctuations, all of these affecting hominid land-use strategies. The purpose here is to examine what kind of socio-economic units may correspond with such series of interrelated changes.

Evidence remains severely limited and will need future confirmation. Three sources can be examined in terms of their degree of covariation with separate Mousterian Complex industries (leaving out the MTA for which insufficient data exists for comparative purposes): palaeoclimatic diagnoses from site occurrences or layers; data on seasonal occupation, from reindeer teeth eruption patterns; and intensity of faunal exploitation, expressed by proportions of faunal to lithic remains.

1. Figure 13.2 shows that most or all of the entire spectrum of possible palaeoclimatic phases types can be identified with occurrences belonging to any of the Mousterian Complex industry types, but that each seems to coincide more often with particular segments of this palaeoclimatic continuum. MD occurrences correspond most frequently (over 70% of instances) with milder or temperate conditions, whereas the Charentian industries tend to coincide with severe conditions.

2. Tables 13.1 and 13.2 give data on seasonal occupation for southwestern France. The small number of observations, together with certain methodological flaws (Binford 1973; Guillien and Henri-Martin 1968) – superseded by the study of dental growth rings (Klein *et al.* 1983) – suggest that these observations are, at best, indicative and should be complemented by data from other animal remains, such as other cervids, equids and bovids. A bias appears in favour of warmer seasons. This could result from errors inherent in the method, or may actually tell us something about seasonal preferences, with a certain margin of error. The aim here is to detect possible differences between occurrences identified with specific industry types. We note (Tables 13.2 and 13.3) that Quina occur-

Table 13.2. Determination of seasonality of occupation for Early Würm sites in Southwestern France, based on analysis of reindeer remains (after Bouchud 1961, 1966; van Campo and Bouchud 1962)

			Seasonality of Occupation												
		Industry	Warm						Cold						Nos. of
Site	Layer	Type	M	J	J	A	S	O	N	D	J	F	M	A	Months
Roc-de-Marsal	XI	Quina						X	X		X	X			4
	IX	Quina	X	X							X	X			4
	VIII	Quina	X	X			X	X							4
	VII	Typical	X	X			X	X							4
	VI	Typical	X	X											2
	IV	Typical	X	X	X	X	X								5
	III	Denticulate	X	X	X	X	X								5
Hauteroche	1	Quina	X	X	X	X	X	X			X	X	X	X	10
	4	Denticulate	X	X	X	X	X	X					X	X	8
Roc-en-Pail	*5	Quina						X	X	X					3
	6	Denticulate	X	X	X	X	X	X	X	X	X				9
Les Cottés	2	Quina						X	X	X	X				4
La Quina	C2	Quina	X	X	X	X	X	X	X	X	X	X	X	X	12
Le Portel	*5	Quina	X	X	X	X	X	X	X	X					9
Chapelle-aux-Saints	—	Quina	X	X	X	X					X		X	X	7

* Number of observations is very small.

Table 13.3. Summary of seasonality patterns for different industry types indicated in Table 13.2.

Duration of Yearly Occupation (in months)

Industry Types	Nos. of Months 2	3	4	5	7	8	9	10	12	N (archaeological layers
Quina	—	1	4	—	1	—	1	1	1	9
Denticulate	—	—	—	1	—	1	—	1	—	3
Typical	1	1	1	—	—	—	—	—	—	3

Comparison of Preferred Season of Occupation (in months)

Industry Types	Seasons Warm	Cold	N (nos. of months)
Quina	54.3%	45.5%	57
Denticulate	77.2%	22.7%	22
Typical	100.0%	—	11
			90

rences are often spread over a larger portion of the year, along with a proportionately higher number of cold-season occupations than for the MD.

3. The frequencies of faunal remains over lithics remains (all tools and primary flakes and blades) may provide clues about the relative importance of animal exploitation residues between different Mousterian Complex industries (Figure 13.3). Faunal remains counts are underrepresented, since only bones identifiable according to species were retrieved or preserved in excavations from which data are available – in this case, southwestern France. This bias simply underestimates the amount of animal refuse but does not distort comparisons between industry types. Frequency ranges overlap with all the industry types represented here, but measures of central tendencies again show clinal variations from Charentian to MD, the latter possessing relatively lower proportions of animal remains, when compared with lithics.

Other potentially useful information, such as densities of remains per cubic metre or other units of deposits from stratified enclosed sites (see Bailey *et al.* 1983), are unfortunately not available.

The covariations observed above between Mousterian Complex industry types and palaeoclimates and fauna suggest finely modulated, continuous techno-economic patterns, rather than specialized activity clusters. They hardly indicate, however, the signatures of distinct stylistic or ethnic manifestations. Most of the tool-making repertoires and accompanying variations fit continuous patterns better. It seems that the cutting-tool component represents a morphologically elastic category of unretouched, utilized, trimmed and resharpened pieces whose modalities, in many instances, covary with environmental aspects. Charentian and MD illus-

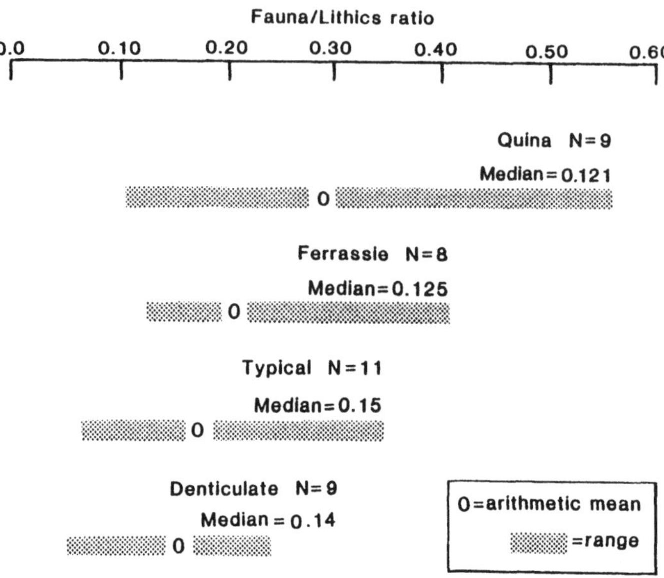

Figure 13.3. Proportions of faunal remains to lithic remains in different industrial types of the Mousterian complex in southwest France (higher values indicate higher frequencies of faunal remains). Data derive from the following sources: Denticulate Mousterian = Combe Grenal layers 11-15, 20, 38, 41; Pech de l'Azé II layer 4B; Typical Mousterian = Combe Grenal layers 7, 10, 36, 37, 40, 42-3, 50A, 50, 52; Pech de l'Azé II layer 4C; Pech de l'Azé IV layer X; Ferrassie Mousterian = Combe Grenal layers 27-33, 35 (layers 28-31 = 'attenuated Ferrassie'); Quina Mousterian = Combe Grenal

trate more conspicuous instances of these correlations.

An earlier version of the model (Rolland 1981) suggested that more protracted or intensive occupation episodes corresponded, on the average, with the more severe Early Würm phases, exemplified by the Charentian and some MT occurrences, whereas MD occurrences coincided frequently with milder phases and increased residential mobility in Western Europe. Raw material economizing, in areas rich in easily-accessible sources such as southwestern France, simply meant that quarrying trips during longer cold seasons became less frequent, carrying back larger nodules and preforms to residential sites (see Airvaux and Chollet 1975; Fish 1979; Turq 1985).

An additional repercussion of palaeoclimatic fluctuations on Middle Palaeolithic lifeways involved rearrangements of vegetational mosaics (Drury 1975; Paquereau 1974-75), influencing biomass densities of gregarious herbivore species: colder phases would favour the expansion of more open landscapes favourable to herd game increases (see also van Andel and Shackleton 1982, for Late Glacial Greece); milder episodes or phases would lead to higher proportions of arboreal vegetation, and lower herbivore biomass.

Drawing together these models and the empirical evidence discussed above, limited though it is, we can envision the following situations for Western Europe:

A. During colder phases, coinciding often with Charentian and some MT occurrences, settlements, while still semi-nomadic, become more intensive. Game herds become relatively more abundant. This would induce increased economy in the use of raw material or increased intensity of reduction, owing to the need to minimize quarrying journeys or movements in general, during the more extended cold seasons. Hunting also becomes relatively more important, as suggested by higher densities of animal bones;

B. Milder phases could lead to increased exploitation of plant resources as food (see Abri Agut, for an MD occurrence accompanied by macrobotanical remains of plant foods: Freeman 1979) and raw material for woodworking, hence the importance of assemblages conspicuously dominated by denticulates and notches. Semi-sedentary settlements involved greater mobility, partly correlated with longer warm seasons and evidence for more frequent quarrying trips, with more profligate use of lithic material.

These generalizations must be qualified by evidence for overlap showing some MD assemblages associated with colder palaeoclimates and higher densities in faunal remains. They refer thus to central tendencies along a continuum, not to discrete situations. They also remain tentative until they can be tested against a richer data base.

The situation in the Levant remains less conclusive. Researchers, as we have seen, conclude that the Levantine Mousterian (or Levalloiso-Mousterian) settlements were semi-sedentary, logistically organized, with more intensive cave occupation and widespread exploitation of different biomes (Marks and Friedel 1977; Larson 1979: 80-84). This appears to coincide with cooler, wetter conditions during the Last Glacial and with greater dependence on hunting. The subsequently drier, cool palaeoclimates accompanying the Upper Palaeolithic induced deteriorating subsistence conditions and a shift to nomadic land-use strategies.

Tool inventories, however, do not show intensive lithic reduction, with high frequencies of retouched or resharpened tools, as would be expected with more intensive occupation.

The preceding archaic Middle Palaeolithic horizon in the Levant, the 'Yabrudian', mostly known from cave sites, contains large quantities of intensively retouched tools, especially *racloirs*, roughly reminiscent of the Quina industry, suggesting lithic material economizing. It is more or less contemporaneous with the drier, warmer Last Interglacial (Jelinek 1982: Figure 5). Only future research will tell whether these characteristics were related to more intensive occupation, perhaps under 'oasis' conditions. The term 'semi-nomadic', with cyclic episodes of settlement flux, might be more suitable, however, for both Levantine horizons, than 'semi-sedentary' (by comparison, for example, with the use of this term in the early Postglacial Natufian).

The above situations are not incompatible with generalizations concerning a broad congruity between, on the one hand, latitudinal gradients from tropical, through subtropical, mild, temperate, boreal and arctic zones, and, on the other, a shift from emphasis on plant collecting to one on hunting, among recent foraging groups (Clarke 1976; Testart 1977; Binford 1980; Hayden 1981). This correlation, however, may be biased because of the disappearance in prehistoric times of most foraging societies from temperate and mild zones, leaving us with large and widespread gaps in the ethnoarchaeologically documented record. Hunting may also have been the main subsistence strategy in tropical savannas prior to the introduction of pastoralism (see Foley 1982). Palaeoclimatic fluctuations, furthermore, could produce the equivalent of latitudinal shifts, with concomitant variations in the relative importance of plant and animal protein foods, at least during the Middle Palaeolithic.

It would be premature to suppose that milder phases and an increased importance of collecting necessarily resulted in residential mobility (e.g. with the MD) as opposed to colder phases and greater reliance on hunting, with logistic mobility. The absence, so far, of fossil hominid burials in unquestionable association with MD occurrences (Vandermeersch 1965), could give an additional indication of shorter duration of occupation. Dying or deceased individuals would be abandoned or buried in graves too shallow for protection against disturbance by carnivores, as is often the case with some recent mobile hunter-gatherers (J. Woodburn, personal communication). We should expect more strategic flexibility and diversity under Mediterranean and temperate Atlantic latitudes. Hunting specialization, prior to later Upper Palaeolithic times, may nevertheless have favoured residential mobility, most of the time. Specialization in terms of the range of species hunted varies somewhat between open-air and enclosed sites, as noted previously for some MD occurrences. One interpretation of the spatial distribution of sites in Mediterranean France (Lumley 1972) perceived home bases, represented mainly by Charentian occurrences, surrounded by satellite sites (MD or MT occurrences).

We conclude – stressing again the tentative character of these interpretations and the unsatisfactory state of available evidence and some regional divergences – that Middle Palaeolithic socio-economic units throughout Western Eurasia consisted of predominantly semi-nomadic, anucleated groups (see Runnels 1983 for identification of such patterns in the Argolid). Semi-nomadism could also encompass varying degrees of residential intensity or flux (see Lee 1972), according to latitudes and environmental zones, but without ever reaching the level of true 'aggregation sites' (Conkey 1980). Enclosed sites were apparently occupied more frequently or intensively in southwestern Europe and the Mediterranean than in Central and Eastern Europe (Gabori-Csank 1976; Gamble 1983). The covariations just discussed should not be construed as the only possible interpretation of Middle Palaeolithic variability and, through this, a reconstruction of socio-economic life. Home-range exploitation and activity patterns by these communities involved adaptation to external constraints, but this was not necessarily a 'reactive' type of response.

Internal dynamics, search for stability, balancing between desirable goals, also played a part in decision-making, even if the sum of technological, social-organizational and cognitive repertoires did not supply as ample a 'superorganic' margin for mediating human-environment relationships as became the case with recent foraging communities. Fine-grained studies of cultural and environmental changes in prehistoric eastern North America (Marquardt 1985) suggest substantial time-lag periods in adjusting subsistence strategies to changing ecological conditions.

SUMMARY

The balance sheet of evidence for assessing the Middle Palaeolithic's developmental position, with special reference to technology and socioeconomic life, indicates that the different variables examined contribute unevenly to resolution and produce a mosaic effect. Some traits betray conservatism (i.e. retention of Lower Palaeolithic patterns), while others seem more innovative or evolve during the Middle Palaeolithic time-span. Some contrasts between the Middle and Upper Palaeolithic, however, become evident only during the latter's final episodes.

Hominid population increased mainly because of geographical expansion into temperate and colder habitats but without attaining significantly higher densities.

Subsistence became more specialized or intensive, but extreme specialization did not appear until the Solutrian, in the case of southwestern France.

Middle Palaeolithic settlements show more variability than those of the Lower Palaeolithic and include regular utilization of caves and shelters, together (in some cases) with the construction of simple artificial dwellings, hardly visible previously (Farizy 1985: 23). Organization of living space remains rudimentary or opportunistic until Early Würm times, although even then, not as elaborate as during the Upper Palaeolithic (Leroi-Gourhan 1961; Mellars 1973; Rigaud 1986). It should be observed, however, that Upper Palaeolithic settlements were not homogeneous in this regard: Aurignacian settlements appear less dense or semi-sedentary than those of the Gravettian in Europe (Movius 1966; Montet-White 1985).

Middle Palaeolithic technology reveals only few innovations or improvements, mainly with respect to fire-making, butchering or skinning techniques and (especially) lithic technology. Design repertoires show few changes from those of the later Lower Palaeolithic, but several tool forms will persist into the earliest Upper Palaeolithic horizons. Patterned interassemblage and time-related stochastic variability increase, however, during Early Würm, when compared with pre-Würmian times. The most decisive technological breakthrough, though, remains the appearance of Levallois technique.

Other new manifestations include burying the dead (documented since Early Würm times) together with some crypto-aesthetic activities such as certain shaped objects, bone incisions, the use of colouring matter, and collecting fossils and iron pyrites.

Social units may provide a reflection of restricted nomadism, or semi-nomadism, depending on circumstances, with local groups undergoing episodes of periodic flux.

The Middle Palaeolithic presents, therefore, a majority of traits reminiscent of, or persisting from, the Lower Palaeolithic but with several new elements, some of them appearing only with the Early Würm, thus justifying it as a final 'Archaeolithic' or perhaps, 'Mousterian Technocomplex' substage. Some manifestations continue well into the Upper Palaeolithic (see Dennell 1986) however, and give this Middle Palaeolithic a transitional character. Lower to Middle Palaeolithic boundaries are more blurred and protracted than those between the Middle and Upper Palaeolithic.

CONCLUSION: CURRENT PERSPECTIVES ON PALAEOANTHROPIC BEHAVIOUR

Determining the behavioural parameters of the hominid populations who manufactured Middle Palaeolithic repertoires adds another dimension to the problem of identifying techno-economic and socio-economic stages and substages within preagricultural prehistory.

Perspectives concerning Stone Age developments have undergone radical modifications over recent years. Characteristics regarded as consequences of food-production (Smith 1976), such as sedentary life, technological growth, elaborate social organization and structured group interactions, appear, in fact, during the later phases of the Upper Palaeolithic and, especially, the Mesolithic (Testart 1982; Price and Brown 1985). Several researchers see the major transformation taking place not with the advent of animal/plant domestication but one going back to the appearance of fully modern fossil hominids (Clark 1980; Zvelebil 1984). This latter event is matched by accelerated techno-economic change, and the emergence of various social and ideological manifestations connected with symbolic behaviour.

This raises, therefore, the fundamental issue of whether biological evolution, rather than culture-historical processes, made these developments possible. It raises, in turn, the question of possibly *qualitative* differences separating the behavioural capabilities of 'fully sapiens' and 'archaic sapiens' hominids contemporaneous with the Middle Palaeolithic. Some palaeoanthropologists, however, prefer to lump all of these together as '*Homo Sapiens*' (Tobias 1971).

We shall refer to the latter populations by the term 'palaeoanthropic', commonly used by French palaeontologists (Arambourg 1950; Piveteau 1957), to avoid ascribing taxonomic status or phylogenetic labels to an evolutionary group comprising fossil hominids known as 'progressive', or 'classic' Neanderthals, as well as 'early' Neanderthals. Palaeoanthropic populations include late mid-Pleistocene finds such as La Chaise, Biache St-Vaast, Fontéchevade, Saccopastore, Ehringsdorf, Krapina and all the Neanderthal variants found in Europe, Western Asia and North Africa, although their distribution could extend beyond these boundaries.

Biological linkage between palaeoanthropians and Middle Palaeolithic remains becomes less evident with the Middle/Upper Palaeolithic interface and the earliest Upper Palaeolithic horizons in Europe and the Levant – as demonstrated by the finds from Saint-Césaire, Djebel Qafzeh and Kebara.

Since we are dealing with significant biological evolutionary events and cultural transformations, with the passage from Middle to Upper Palaeolithic, it becomes legitimate to confront the issue about behavioural parameters mentioned above, since prior to the advent of fully modern hominids, slower rates of culture change were penecontemporaneous with ongoing hominization processes.

It is therefore more meaningful to consider the possible interplay between biological and cultural factors as a heuristic problem and approach it from a unified palaeoanthropological perspective *sensu lato* (Howell 1964).

Investigations using the concept of 'palaeoculture' cited earlier, could lead to either of the following conclusions:

1. The classic distinction (Boas 1940) between 'race' and culture as independent variables, established for all living and past representatives of the *Homo sapiens sapiens* species cannot be assumed for earlier hominid grades or palaeospecies. This could account for the expedient, 'stagnating' appearance of repertoires older than the Upper Palaeolithic, as well as the lack of explicit expression;

2. These archaic repertoires represent earlier developmental stages but show linear continuity with subsequent ones, in spite of their intrinsic simplicity, which falls outside the range of late prehistoric or recent foraging technologies. They can best be understood by reference to other cultural processes, more limited in scope, to ecological constraints, and to historical and social antecedents, without invoking biologically induced factors.

Resolving these issues is beyond both the scope of this paper and the capabilities of individual researchers. It will require input from various specialists, and sustained communication between appropriate disciplines. The body of concepts and evidence bearing on them comes from prehistoric archaeology, anthropological theory and several branches of human biology and cognate fields.

The discussion concludes by examining briefly certain relevant points. Although they fall on the side of a culturally-oriented approach, regarded as sufficient for understanding the Middle Palaeolithic record and its differences with later Palaeolithic, this should not be regarded as conclusive. They remain intrinsically open-ended and too incomplete in coverage to accomplish that.

Approach. By adhering to guidelines stressing a progression from 'the known to the unknown' – by analogy with the direct-historical approach employed in North American prehistory (e.g. Strong 1940) – it seems that the burden of proof falls on biological determinists, given the fact that they introduce variables, regarded as redundant when these involve modern hominid cultural behaviour. Their hypothesis calls for detailed and independent biological data, to establish that the material expression of palaeoanthropic and neoanthropic behaviour are qualitatively different, i.e. phylogenetically determined.

Anthropological theory. Several models could account for the simplicity of Middle Palaeolithic repertoires. Some have argued (e.g. Kroeber 1939) that a 'fit' appears to exist between the richness of environmental potential and cultural specificity or levels of complexity. A critical threshold may exist between resources, on the one hand, and population densities, available technologies, elaborate socio-economic systems and ritual behaviour on the other (Radin 1960; Price and Brown 1985; Johnson 1982). Middle Palaeolithic 'stagnation' could be a function of the time-binding properties proposed by technology and culture in general (Kroeber 1948). Upper Palaeolithic and Mesolithic societies became increasingly involved in exchange networks and interaction spheres (Bender 1978; Gamble 1980, 1982; Lourandos 1985). Proximity and contacts between clusters of communities increase the likelihood for innovation, by pooling and recombining techniques and ideas (Levi-Strauss 1958), conditions not yet attained during the Middle Palaeolithic.

Symbolic behaviour. The absence of explicit evidence for symbolic expression in the Middle Palaeolithic seems to present a serious objection to an exclusively cultural explanation for differences between the Middle and Upper Palaeolithic. Possible instances such as burials and burial objects, the use of colourants, the collection or shaping of non-utilitarian objects, 'skull cults' or 'cave bear cults' are either redundant, ambiguous or simply aesthetic manifestations, in the view of many researchers. The use of red ochre or manganese by palaeoanthropians is not sufficient to demonstrate that they endowed the colour red with 'blood-life' symbolism, as evidenced by ethnography (Turner 1966) and, apparently, by the Upper Palaeolithic burials at Grimaldi (Leroi-Gourhan 1964c), or as an ideological basis or taboo, for division of labour by sex among hunter-gatherers (Testart 1986; De Beaune 1986), already suggested for the Upper Palaeolithic (e.g. at Mal'ta). Living floor analysis does not permit the identification of a binary symbolic ordering of animals into 'living game' and refuse discard, suggested for La Cotte de St-Brélade (Leach 1977).

We are left with very few possible concrete indications for pre-neoanthropic symbolling expression: (a) possible scalping incisions on the face of the Bodo *Homo erectus* fossil, in East Africa (Clark 1984); (b) the Grotta Guattari Neanderthal calvarium, surrounded by stones (Blanc 1958); (c) the Shanidar burial flower pollens (Solecki 1971).

This virtual absence could mean either that the palaeoanthropic populations' hominization level had not yet evolved into such a capability, or that it was latent and remained unexpressed materially because Middle Palaeolithic repertoires were still too rudimentary and utilitarian to contain sociotechnic or ideotechnic manifestations, as opposed to merely technomic ones (Binford 1962).

Speech and skill. A certain indication of symbolic behaviour would be evidence for speech. This could come from direct fossil evidence such as brain endocasts, and more recently, flexion of skull bases, correlated with downward migration of the larynx and pharynx, making the hominid sound-range possible for articulated language. Skull flexion already appears with *Homo erectus* (Laitman 1986). Other lines of evidence tend to be controversial.

Leroi-Gourhan (1964a) has argued, from neurological evidence, for a close association between speech capability, gesticulation (and eventually, writing) and technical behaviour. Both occupy a substantial area of the brain cortex among living primates and are contiguous, although the speech function characterizes part of the human face. Speech and skill are contiguous on the brain and form a basic component of hominid social life (1964a: 163). Rudimentary forms of speech could have become possible as soon as toolmaking formed part of regular activities among Australopithecines and, especially, *Homo habilis*, in this perspective. Several prehistorians (Isaac 1969; Bordes 1971), as well as Leroi-Gourhan, regard the development of skill, expressed by the processes involved in Levallois technique, as an operational sequence sufficiently complex to imply symbolic behaviour and speech, on a level more or less comparable with that of modern humans.

We conclude with the following hypothesis:

1. Both palaeoanthropic and neoanthropic grades or evolutionary strains of *Homo sapiens* have more in common than not (in spite of some unquestionable differences), when compared with earlier hominid evolutionary groups. This indicates that Neanderthals and pre-Neanderthals had already attained an advanced level along the hominid biological hominization trajectory, including probably capabilities for speech and symbolic behaviour (Tobias 1971);

2. Historical and social factors remain exclusively responsible for culture's capacity for exponential growth rates and elaboration. The Middle Palaeolithic, despite a number of innovations, retains sufficient archaic-looking traits to qualify as the Early Palaeolithic's optimal point of expression, along with some transitional features. Palaeolithic industries contain elements of continuity throughout their development (Bordes 1960) but with a trend towards accelerated change, becoming gradually more pronounced during Upper Pleistocene times, in Africa and Western Eurasia (see Isaac 1972: Figure 5). The Middle Palaeolithic begins to show incipient tendencies for such an acceleration but without attaining scope for complexity displayed by some Upper Palaeolithic horizons;

3. The above features suggest a gap between the somatic *capabilities* for culture attained by palaeoanthropic populations and the Middle Palaeolithic's *actual* developmental level (see Holloway 1985). This gap corresponds perhaps more with limits inherent in the cultural repertoires themselves for cumulative growth, than with innate limitations of the populations who produced them.

REFERENCES

Airvaux, J. and Chollet, A. 1975. Le site moustérien de la Fontaine à Scorbé-Clairvaux (Vienne). *Bulletin de la Société Préhistorique Française* 72: 209-217.

Ammerman, A.J. 1975. Late Pleistocene population dynamics: an alternative view. *Human Ecology* 3: 219-233.

Arambourg, C. 1950. *La Genèse de L'Humanité*. Paris: Presses Universitaires de France.

Bader, O.N. 1965. The Palaeolithic of the Urals and the peopling of the north. *Arctic Anthropology* 3: 77-90.

Bailey, G.N. 1983. Concepts of time in Quaternary prehistory. *Annual Review of Anthropology* 12: 165-192.

Bailey, G.N., Carter, P.L., Gamble, C.S. and Higgs, H.P. 1983. Asprochaliko and Kastritsa: further investigations of Palaeolithic settlement and economy in Epirus. *Proceedings of the Prehistoric Society* 49: 15-42.

Barker, G.W. 1975. Prehistoric territories and economies in central Italy. In E.S. Higgs (ed.) *Palaeoeconomy*: 111-175. Cambridge: Cambridge University Press.

Barth, F. 1959. *The Land Use Pattern of Migratory Tribes in South Persia*. Indianapolis: Bobbs-Merrill Social Sciences Reprint.

Bartholomew, G.A. and Birdsell, J.B. 1959. Ecology and the protohominids. In M.H. Fried (ed.) *Readings in Anthropology*. New York: Crowell: 46-62.

Behm-Blancke, G. 1960. *Altsteinzeitliche Rastplätze im Travertingebiet von Taubach, Weimar, Ehringsdorf*. Weimar: Hermann Bühlaus.

Bender, B. 1978. Gatherer-hunter to farmer: a social perspective. *World Archaeology* 10: 204-222.

Binford, L.R. 1962. Archaeology as anthropology. *American Antiquity* 28: 217-225.

Binford, L.R. 1964. A consideration of archaeological research designs. *American Antiquity* 29: 425-441.

Binford, L.R. 1973. Interassemblage variability - the Mousterian and the 'functional' argument. In C. Renfrew (ed.) *The Explanation of Culture Change*. London: Duckworth: 227-254.

Binford, L.R. 1979. Organization and formation processes: looking at curated technologies. *Journal of Anthropological Research* 35: 255-273.

Binford, L.R. 1980. Willow smoke and dog's tails: hunter-gatherer settlement systems and archaeological site formation. *American Antiquity* 45: 4-20.

Binford, L.R. 1982. Comments on White: 'Rethinking the Middle/Upper Palaeolithic transition'. *Current Anthropology* 23: 177-181.

Binford, L.R. 1983. *In Pursuit of the Past*. London: Thames and Hudson.

Binford, L.R. 1984. *Faunal Remains from Klasies River Mouth*. Orlando: Academic Press.

Binford, L.R. and Binford, S.R. 1966. A preliminary analysis of functional variability in the Mousterian of levallois facies. *American Anthropologist* 68: 238-295.

Binford, L.R. and Binford, S.R. 1969. Stone tools and human behaviour. *Scientific American* 220: 70-84.

Birdsell, J.B. 1968. Some predictions for the Pleistocene based on equilibrium systems among recent hunter-gatherers. In R.B. Lee and I. de Vore (eds) *Man the Hunter*. Chicago: Aldine: 229-240.

Blackwell, B., Schwarz, H.P. and Debénath, A. 1983. Absolute dating of hominids and Palaeolithic artifacts of the cave of La Chaise-de-Vouthon (Charente) France. *Journal of Archaeological Science* 10: 493-513.

Blanc, A.C. 1958. Torre in Pietra, Saccopastore, Monte Circeo. On the position of the Mousterian culture in the Pleistocene sequence of the Rome area. In G.H.R. von Koenigswald (ed.) *Hundert Jahre Neanderthaler*. Utrech: Kemink en Zoon: 167-174.

Boas, F. 1940. *Race, Language and Culture*. Macmillan, New York.

Bonifay, E. 1981. Les plus anciens habitats sous grotte découverts à Lunel-Viel (Hérault). *Archaeologia* 150: 30-42.

Bordes, F. 1953a. Essai de classification des industries 'moustériennes'. *Bulletin de la Société Préhistorique Française* 50: 457-466.

Bordes, F. 1953b. Levalloisien et Moustérien. *Bulletin de la Société Préhistorique Française* 50: 226-235.

Bordes, F. 1954-1955. Les gisements du Pech de l'Azé (Dordogne). 1: Le Moustérien de tradition acheuléenne. *L'Anthropologie* 58: 401-432; 59: 1-32.

Bordes, F. 1960. Evolution in the Palaeolithic cultures. In S. Tax (ed.) *Darwin Centennial Celebration: the Evolution of Man*. Chicago: University of Chicago Press: 99-110.

Bordes, F. 1961a. *Typologie du Paléolithique Ancien et Moyen*. Bordeaux: Publications de l'Institut de Préhistoire de l'Université de Bordeaux, Mémoire I.

Bordes, F. 1961b. Mousterian cultures in France. *Science* 134: 803-810.

Bordes, F. 1962. Le Moustérien à denticulés. *Arheoloski Vestnik* 143: 43-49.

Bordes, F. 1968. *The Old Stone Age*. New York: McGraw-Hill.

Bordes, F. 1969. Comments on Collins: 'Culture traditions and environment of early man'. *Current Anthropology* 10: 301.

Bordes, F. 1971. Physical evolution and technological evolution in man: a parallelism. *World Archaeology* 3: 1-5.

Bordes, F. 1972. *A Tale of Two Caves*. New York: Harper and Row.

Bordes, F. 1975. Sur la notion de sol d'habitat en préhistoire paléolithique. *Bulletin de la Société Préhistorique Française* 72: 139-144.

Bordes, F. and Sonneville-Bordes, D. de. 1970. The significance of variability in Palaeolithic assemblages. *World Archaeology* 2: 61-73.

Bordes, F. and Prat, F. 1965. Observations sur les faunes du Riss et du Würm I. *L'Anthropologie* 69: 31-46.

Bosinski, G. 1976. Middle Palaeolithic structural remains from Western Central Europe. In L.G. Freeman (ed.) *Les Structures d'Habitat au Paléolithique Moyen*. Proceedings of Colloquium 9 of the UISPP, Nice, 1976: 84-77.

Bouchud, J. 1966. *Essai sur le Renne et la Climatologie du Paléolithique Moyen et Supérieur*. Périgueux: Magne.

Bouyssonie, A., Bardon, J. and Bardon, L. 1913. La station moustérienne de la 'Bouffia' Bonneval à la Chapelle-aux-Saints. *L'Anthropologie* 24: 609-634.

Breuil, H. and Lantier, R. 1959. *Les Hommes de la Pierre Ancienne*. Paris: Payot.

Callow, P. 1986. The Saalian industries of La Cotte de St. Brélade, Jersey. In A. Tuffreau (ed.) *Chronostratigraphie et Faciès Culturels du Paléolithique Inférieur et Moyen dans l'Europe du Nord-Ouest*. Paris: Supplement to the Bulletin de l'Association Française pour l'Etude du Quaternaire: 129-140.

Cazeneuve, J. 1959. Technical methods in the prehistoric age. *Diogènes* 27: 102-124.

Chard, C.S. 1969. Archaeology in the Soviet Union. *Science* 163: 774-779.

Chase, P.G. 1986. *The Hunters of Combe Grenal.* Oxford: British Archaeological Reports International Series S286.

Chase, P.G. 1988. Scavenging and hunting in the Middle Palaeolithic: the evidence from Europe. In H. Dibble and Montet-White (eds) *Upper Pleistocene Prehistory of Western Eurasia.* Philadephia: University Museum, University of Pennsylvania: 225-232.

Chavaillon, J., Chavaillon, N. and Hours, F. 1969. Industries paléolithiques de l'Elide II – région de Kastron. *Bulletin de Correspondence Hellénique* 93: 97-151.

Childe, V.G. 1944. Archaeological ages as technological stages. *Journal of the Royal Anthropological Institute* 74: 7-24.

Childe, V.G. 1954. Early forms of society. In E.J. Holmgard *et al.* (eds.) *A History of Technology*: Vol. 1. Oxford: Oxford University Press: 38-57.

Childe, V.G. 1956. *A Short Introduction to Archaeology*. New York: Collier.

Childe, V.G. 1956. *Piecing Together the Past.* London: Routledge and Kegan Paul.

Clark, J.D. 1984. The Way We Were. *AnthroQuest* 30: 1, 18-19.

Clark, J.G.D. 1963. Review of S.L. Washburn (ed.): *Social Life of Early Man. Proceedings of the Prehistoric Society* 29: 437.

Clark, J.G.D. 1969. *World Prehistory – a New Outline.* Cambridge: Cambridge University Press.

Clark, J.G.D. 1980. *The Mesolithic Prelude.* Edinburgh: Edinburgh University Press.

Clarke, D.L. 1968. *Analytical Archaeology*. London: Methuen.

Clarke, D.L. 1976. *Mesolithic Europe: the Economic Basis.* In G. de G. Sieveking, I.H. Longworth and K.E. Wilson (eds) *Problems in Economic and Social Archaeology.* London: Duckworth: 449-481.

Combier, J. 1959. Circonscription de Lyon. *Gallia-Préhistoire* 2: 118-120.

Combier, J. 1962. Chronologie et systématique du Moustérien occidental: données et conceptions nouvelles. *Atti del VI Congresso Internazionale della Scienze Preistoriche e Protoistoriche (Roma)* 1: 77-96.

Combier, J. 1967. *Le Paléolithique de l'Ardèche dans son cadre Paléoclimatique.* Bordeaux: Delmas

Commont, V. 1914. Les hommes contemporains du renne dans la vallée de la Somme. *Mémoires de la Société des Antiquaires de Picardie* 37: 207-646.

Conkey, M.W. 1980. The identification of prehistoric hunter-gatherer aggregation sites: the case of Altamira. *Current Anthropology* 21: 6099-6630.

Davis, S. 1977. The ungulate remains from Kebara cave. In B. Arensburg and O. Bar-Yosef (eds) *Moshe Stekelis Memorial Volume.* Jerusalem: Israel Exploration Society: 150-163.

De Beaune, S. 1986. A propos de la division sexuelle du travail chez les chasseurs-cueilleurs. *Bulletin de la Société Préhistorique Française* 83: 230-232.

Debénath, A. 1973. Un foyer aménagé dans le Moustérien de Hauteroche à Chateauneuf-sur-Charente (Charente). *L'Anthropologie* 77: 3299-3338.

Dennell, R. 1983. A new chronology for the Mousterian. *Nature* 301:

199-200.

Dennell, R. 1986. Needles and spear-throwers. *Natural History* 10: 70-78.

Dibble, H.L. 1984. The Mousterian industry from Bisitun Cave (Iran). *Paléorient* 10: 23-34.

Dibble, H.L. 1985. Raw material variability in Levallois flake manufacture. *Current Anthropology* 26: 391-393.

Dibble, H.L. 1987. The interpretation of Middle Palaeolithic scraper morphology. *American Antiquity* 52: 109-117.

Drury, W.H. 1975. The ecology of the human occupation at the Abri Pataud. In H.L. Movius (ed.) *Excavations at the Abri Pataud, Les Eyzies (Dordogne)*: Vol. 1. Cambridge (Mass): Peabody Museum: 187-193.

Dibble, H.L. 1988. Typological aspects of reduction and intensity of utilization of lithic resources in the French Mousterian. In H. Dibble and Montet-White (eds) *Upper Pleistocene Prehistory of Western Eurasia*. Philadephia: University Museum, University of Pennsylvania: 181-197.

Dumond, D.E. 1975. The limitation of human population: a natural history. *Science* 187: 713-721.

Duport, L. and Vandermeersch, B. 1972. La grotte de Montgaudier *La Recherche* 21: 280-281.

Farizy, C. 1985. Les cultures du Paléolithique moyen. In *Grand Atlas de l'Archéologie*: Vol. 24. Paris: Encyclopedia Universalis.

Feustel, R. 1968. Die Ur- und Frühgeschichte und das Problem der historischen Periodisierung. *Ethnographisch-Archäologisches Zeitschrift* 9: 120-147.

Fish, P.R. 1979. *The Interpretative Potential of Mousterian Debitage*. Tempe: Arizona State University Anthropological Papers 16.

Foley, R. 1981. *Off-Site Archaeology*. Oxford: British Archaeological Reports International Series S97.

Foley, R. 1982. A reconsideration of the role of predation on large mammals in tropical hunter-gatherer adaptation. *Man* 17: 393-402.

Freeman, L.G. 1973. The significance of mammalian faunas from Paleolithic occupations in Cantabrian Spain. *American Antiquity* 38: 3-44.

Freeman, L.G. 1979. The analysis of some occupation floor distributions from Earlier and Middle Paleolithic sites in Spain. In L.G. Freeman (ed.) *Views of the Past*. The Hague: Mouton: 57-116.

Freeman, L.G. 1976. *Les Structures d'Habitat au Paléolithique Moyen*. Colloquium 9 of the UISPP, Nice, 1976.

Gabori, M. 1976. *Les Civilisations du Paléolithique Moyen entre les Alpes et l'Oural*. Budapest: Akdemiai Kiadó.

Gabori-Csank, V. 1976. Le mode de vie et l'habitat au paléolithique moyen en Europe Centrale. In L.G. Freeman (ed.) *Les Structures d'Habitat au Paléolithique Moyen*. Proceedings of Colloquium 11 of the UISPP, Nice, 1976: 78-104.

Gabori-Csank, V. 1988. Une mine de silex paléolithique à Budapest, Hongrie. In H. Dibble and Montet-White (eds) *Upper Pleistocene Prehistory of Western Eurasia*. Philadephia: University Museum, University of Pennsylvania: 141-143.

Gamble, C. 1980. Information exchange in the Palaeolithic. *Nature* 283: 522-523.

Gamble, C. 1982. Interaction and alliance in Palaeolithic Society. *Man* 17: 92-107.

Gamble, C. 1983. Caves and faunas from last glacial Europe. In J.

Clutton-Brock and C. Grigson (eds) *Animals and Archaeology*. Vol. 1: *Hunters and their Prey*. Oxford: British Archaeological Reports International Series S163: 163-172.

Gearing, F. 1962. *Priests and Warriors*. Memoir 93. Menasha: American Anthropological Association.

Girard, C. 1976. L'habitat et le mode de vie au Paléolithique moyen à Arcy-sur-Cure (Yonne). In L.G. Freeman (ed.) *Les Structures d'Habitat au Paléolithique Moyen*. Proceedings of Colloquium 11 of the UISPP, Nice, 1976: 49-63.

Girard, C. and Leclerc, J. 1981. Les grandes chasses de Mauran. *La Recherche* 127: 1294-1295.

Gordon, B.C. 1981. Man-environment relationships in Barrenland prehistory. *Musk-Ox* 28: 1-19.

Gould, R.A. 1981. Comparative ecology of food-sharing in Australia and northwest California. In R.S.O. Harding and G. Teleki (eds) *Omnivorous Primates*. New York: University of Columbia Press.

Guillien, Y. and Henri-Martin, G. 1974. Croissance du renne et saison de chasse: le Moustérien à denticulés et le Moustérien de tradition acheuléenne de la Quina. *Inter-Nord* 13-14: 119-127.

Hassan, F. 1981. *Paleodemography*. New York: Academic Press.

Hayden, B. 1977. Sticks and stones: the Southeast Asian Upper Palaeolithic? In J. Allen, J. Golson and R. Jones (eds) *Sunda and Sahul*. New York: Academic Press.

Hayden, B. 1981. Subsistence and ecological adaptations of modern hunter-gatherers. In R.S.O. Harding and G. Teleki (eds) *Omnivorous Primates*. New York: University of Columbia Press.

Henri-Martin, G. 1964. La dernière occupation moustérienne de la Quina (Charente): datation par le radiocarbone. *Comptes-Rendus de l'Académie des Sciences de Paris* 258: 3533-3535.

Higgs, E.S. and Webley, D.P. 1971. Further information concerning the environment of Palaeolithic Man in Epirus. *Proceedings of the Prehistoric Society* 37: 367-380.

Hodder, I. 1978. Simple correlations between material culture and society: a review. In I. Hodder (ed.) *The Spatial Organisation of Culture*. London: Duckworth: 3-24.

Holloway, R.L. 1985. The poor brain of *Homo sapiens neanderthalensis*: see what you please.... In E. Delson (ed.) *Ancestors: the Hard Evidence*, 319-324. New York: Alan R. Liss: 319-324.

Howell, F.C. 1964. The hominization process. In S. Tax (ed.) *Horizons of Anthropology*. Chicago: Aldine: 49-59.

Isaac, G.L. 1969. Studies of early culture in East Africa. *World Archaeology* 1: 1-28.

Isaac, G.L. 1972. Chronology and the tempo of cultural change during the Pleistocene. In W.B. Bishop and J.A. Miller (eds) *Calibration of Hominid Evolution*. Edinburgh: Scottish Academic Press: 381-430.

Ivanova, I.K. and Chernysh, A.P. 1965. The Palaeolithic site of Molodova V on the Middle Dniestr (USSR). *Quaternaria* 7: 197-217.

Jelinek, A.J. 1977. The Lower Palaeolithic: current evidence and interpretations. *Annual Review of Anthropology* 6: 11-32.

Jelinek, A.J. 1982. The Middle Palaeolithic in the Southern Levant, with comments on the appearance of modern *Homo Sapiens*. In A. Ronen (ed.) *The Transition from Lower to Middle Palaeolithic and the Origin of Modern Man*. Oxford: British Archaeological Reports International Series 151: 57-104.

Jelinek, A.J. 1988. Technology, typology, and culture in the Middle

Paleolithic. In H. Dibble and Montet-White (eds) *Upper Pleistocene Prehistory of Western Eurasia*. Philadephia: University Museum, University of Pennsylvania: 199-212.

Jelinek, A.J., Debénath, A. and Dibble, H.L. 1988. A preliminary report on evidence related to the interpretation of economic and social activities of Neandertals at the site of La Quina (Charente) France. In M. Otte (ed.) *L'Homme de Néandertal*. Vol. 6: *La Subsistence*. Liège: Etudes et Recherches Archéologiques de l'Université de Liège (ERAUL 33): 99-106.

Jéquier, J.-P. 1975. *Le Moustérien Alpin*. Yverdon: Institut d'Archéologie Yverdonnoise.

Jewell, P.A. 1966. The concept of home range in mammals. In P.A. Jewell and C. Loizos (eds) *Play, Exploration and Territory in Mammals*. London: Academic Press: 85-109.

Johnson, G.A. 1982. Organizational structure and scalar stress. In C. Renfrew, M. Rowlands and B. Abbot-Seagraves (eds) *Theory and Explanation in Archaeology*. New York: Academic Press: 389-421.

Klein, R.G., Allwarden, K. and Wolf, C. 1983. The calculation and interpretation of ungulate age profiles from dental crown heights. In G.N. Bailey (ed.) *Hunter-Gatherer Economy in Prehistory: a European Perspective*. Cambridge: Cambridge University Press: 45-57.

Kopper, J.S. 1981. Palaeolithic tools: some design considerations. *Expedition* 24: 4-9.

Kowalski, K. and Kozlowski, J.K. 1972. Notes on chronology and palaeoecology. In Z. Rubinowski and T. Wróblewski (eds) *Studies on Raj Cave near Kielce (Poland) and its Deposits*. Kraków: Pan'stwowe Wydawnictwo Naukowe.

Kroeber, A.L. 1939. *Cultural and Natural Areas of Native North America*. Berkeley: University of California Press.

Kroeber, A.L. 1948. *Anthropology*. New York: Harcourt, Brace and World.

Kurtén, B. 1971. *The Age of Mammals*. London: Weidenfeld and Nicolson.

Laitman, J.T. 1986. L'origine du language artificiel. *La Recherche* 17: 1164-1173.

Larson, P.A. 1979. *Simulation Studies of Middle and Upper Palaeolithic Settlement Patterns for the Southern Levant*. Ph.D. Dissertation, Southern Methodist University. Ann Arbor: University Microfilms International.

Laughlin, W.S. 1968. Hunting: an integrating biobehaviour system and its evolutionary importance. In R.B. Lee and I. de Vore (eds) *Man the Hunter*. Chicago: Aldine: 304-320.

Leach, E. 1977. A view from the bridge. In M. Spriggs (ed.) *Archaeology and Anthropology*. Oxford: British Archaeological Reports Supplementary Series 19: 161-176.

Lee, R.B. 1972. !Kung spatial organization: an ecological and historical perspective. *Human Ecology* 1: 125-147.

Leroi-Gourhan, A. 1961. Les fouilles d'Arcy-sur-Cure (Yonne). *Gallia-Préhistoire* 4: 3-16.

Leroi-Gourhan, A. 1963. Châtelperronien et Aurignacien dans le nord-est de la France. *Bulletin de la Société Méridionale de Spéléologie et de Préhistoire* 6-9: 75-84.

Leroi-Gourhan, A. 1964a. *Le Geste et la Parole: Technique et Langage*. Paris: Albin Michel.

Leroi-Gourhan, A. 1964b. *Le Geste et la Parole: la Mémoire et les Rythmes*. Paris: Albin Michel.

Leroi-Gourhan, A. 1964c. *Les Religions de la Préhistoire*. Paris: Presses Universitaires de France.
Leroi-Gourhan, A. 1966. *La Préhistoire*. Paris: Presses Universitaires de France.
Levi-Strauss, C. 1958. Race and History. In UNESCO (ed.) *The Race Question in Modern Science*. Paris: UNESCO.
Lewis, H.T. 1982. Fire technology and resource management in aboriginal North America and Australia. In N.M. Williams and E.S. Hunn (eds) *Resource Managers: North American and Australian Hunter-Gatherers*. Boulder (Col): Westview Press: 45-67.
Lindner, K. 1950. *La Chasse Préhistorique*. Paris: Payot.
Lourandos, H. 1985. Intensification and Australian prehistory. In T.D. Price and J.A. Brown (eds) *Prehistoric Hunter-Gatherers: the Emergence of Cultural Complexity*. New York: Academic Press: 385-423.
Lovejoy, O. 1981. The origin of man. *Science* 211: 341-350.
Loy, T.H. 1983. Prehistoric blood residues: detection on tool surfaces and identification of species of origin. *Science* 220: 1269-1271.
Lumley, H. de. 1965. La grande révolution raciale et culturelle de l'Inter-Würmien II-III. *Cahiers Ligures de Préhistoire et d'Archéologie* 14: 133-135.
Lumley, H. de. (ed.). 1972. *La Grotte de l'Hortus*. Marseilles: Centre National de la Recherche Scientifique.
Lumley, H. de. 1976. Les structures d'habitat au Paléolithique inférieur. In H. de Lumley (ed.) *La Préhistoire Française*. Paris: Centre National de la Recherche Scientifique.
Lumley, H. de and Bottet, B. 1962. 'Sol empierré' dans le prémoustérien de la Baume-Bonne. *Cahiers Ligures de Préhistoire et Archéologie* 2: 3-9.
Marks, A.E. 1975. An outline of prehistoric occurrences and chronology in the Central Negev, Israel. In F. Wendorf and A. Marks (eds) *Problems in Prehistory: North Africa and the Levant*. Dallas: Southern Methodist University Press: 351-362.
Marks, A.E. and Freidel, D.A. 1977. Prehistoric settlement patterns in the Avdat/Aqev area. In A.E. Marks (ed.) *Prehistory and Paleoenvironments in the Central Negev, Israel*. Dallas: Southern Methodist University: 131-158.
Marquardt, W.H. 1985. Complexity and scale in the study of fisher-gatherer-hunters: an example from the eastern United States. In T.D. Price and J.A. Brown (eds) *Prehistoric Hunter-Gatherers: the Emergence of Cultural Complexity*. New York: Academic Press: 59-98.
Mauss, M. 1947. *Manuel d'Ethnographie*. Paris.
McBurney, C.B. 1960. *The Stone Age of Northern Africa*. Harmondsworth: Pelican.
McGhee, R. 1980. Technological change in the prehistoric Eskimo cultural tradition. *Canadian Journal of Archaeology* 4: 39-52.
Meggers, B. (ed.) 1955. *Functional and Evolutionary Implications of Community Patterning*. Seminars in Archaeology 1955. Washington (DC): Society for American Archaeology.
Meignen, L. 1976. Le site moustérien charentien de Ioton (Beaucaire-Gard): Etude sédimentologique et archéologique. *Bulletin de l'Association Française pour l'Etude du Quaternaire* 1: 3-17.
Mellars, P.A. 1967. *The Mousterian Succession in South-West France*. Unpublished Ph.D. Dissertation, University of Cambridge.
Mellars, P.A. 1969. The chronology of Mousterian industries in the Périgord region of south-west France. *Proceedings of the Prehistoric*

Society 35: 134-171.

Mellars, P.A. 1970. Some comments on the notion of 'functional variability' in stone-tool assemblages. *World Archaeology* 2: 74-89.

Mellars, P.A. 1973. The character of the Middle-Upper Palaeolithic transition in south-west France. In C. Renfrew (eds) *The Explanation of Culture Change*. London: Duckworth: 255-276.

Mellars, P.A. 1974. The Palaeolithic and Mesolithic. In C. Renfrew (ed.) *British Prehistory*. London: Duckworth: 41-99.

Mellars, P.A. 1976. Fire ecology, animal population and man. *Proceedings of the Prehistoric Society* 42: 15-46.

Mellars, P.A. 1985. The ecological basis of social complexity in the Upper Palaeolithic of southwestern France. In T.D. Price and J.A. Brown (eds) *Prehistoric Hunter-Gatherers: the Emergence of Cultural Complexity*. New York: Academic Press: 271-297.

Mellars, P.A. 1986. A new chronology for the French Mousterian Period. *Nature* 322: 410-411.

Montet-White, A. 1985. Review of H. Movius, H. Bricker and N. David (eds) *Excavations at the Abri Pataud. Quarterly Review of Archaeology* 6: 5-6.

Movius, H.L. 1966. The hearths of the Upper Périgordian and Aurignacian horizons at the Abri Pataud, Les Eyzies (Dordogne), and their possible significance. In J.D. Clark and F.C. Howell (eds) *Recent Studies in Paleoanthropology. American Anthropologist* 68 No. 2, Part 2: 296-325.

Murray, P. 1980. Discard location: the ethnographic data. *American Antiquity* 45: 490-502.

O'Connell, J.F. 1977. Room to move: contemporary Alyawara settlement patterns and their implications for aboriginal housing policy. *Mankind* 11: 119-131.

Okladnikov, A.P. 1972. Découverte du Paléolithique inférieur en Sibérie et en Mongolie. *Inter-Nord* 12: 191-206.

Otte, M. 1981. *Le Gravettien en Europe Centrale*. De Tempel: Brugge.

Paquereau, M.M. 1974-1975. Le Würm ancien en Périgord: étude Palynologique. *Quaternaria* 18: 67-159.

Patte, E. 1971. L'industrie de la Micoque. *L'Anthropologie* 75: 369-396.

Patterson, L.W. 1980. Comments on 'Archaeology as archaeology'. *Journal of Field Archaeology* 7: 497-498.

Paunescu, A. 1986. Structures d'habitat moustériennes mises au jour dans l'établissement de Ripiceni-Izvor (Roumanie) et quelques considérations concernant le type d'habitat paléolithique moyen de l'est des Carpathes. In Colloque International L'Homme de Néanderthal, Université de Liège: 183. Edition anticipée. Domaine Provincial de Wegimont.

Payne, S. 1983. The animal bones from the 1974 excavations at Douara Cave. In K. Hanihara and T. Akazawa (eds) *Paleolithic Site of the Douara Cave and Paleogeography of Palmyra basin in Syria*. Tokyo: Tokyo University Press: 1- 108.

Perez Ripoll, M. 1977. *Los Mamíferos del Yacimiento Musteriense de Cova Negra (Jativa, Valencia)*. Servicio de Investigacion Prehistorica, Valencia.

Perkins, D. 1964. Prehistoric fauna from Shanidar, Iraq. *Science* 144: 1565-1566.

Perlès, C. 1965. L'Homme préhistorique et le feu. *La Recherche* 60: 829-839.

Perlès, C. 1977. *Préhistoire du Feu*. Paris: Masson.

Piperno, M. 1976-1977. Analyse du sol Moustérien de la grotte

Guattari au Mont Circé. *Quaternaria* 19: 71-92.

Piveteau, J. 1957. *Traité de Paléontologie*. Vol. 7: *Primates, Paléontologie Humaine*. Paris: Masson.

Pradel, L. and Pradel, J.H. 1970. La station paléolithique de Fontmaure, commune de Vellèches (Vienne). *L'Anthropologie* 74: 481-526.

Price, T.D. and Brown, J.A. (eds) 1985. *Prehistoric Hunter-Gatherers: the Emergence of Cultural Complexity*. New York: Academic Press.

Radin, P. 1960. *The World of Primitive Man*. New York: Grove Press.

Ranov, V.A. 1972. Le peuplement préhistorique de la Haute-Asie. *L'Anthropologie* 76: 5-20.

Rhoades, R.E. 1978. Archaeological use and abuse of ecological concepts and studies: the ecotone example. *American Antiquity* 43: 608-614.

Rigaud, J.-P. 1980. Circonscription d'Aquitaine. *Gallia-Préhistoire* 23: 391-393, 395-396.

Rigaud, J.-P. 1986. Comportement de l' homme de Néanderthal. In Colloque International L' Homme de Néanderthal, Université de Liège: 103-106. Edition anticipée. Domaine Provincial de Wegimont.

Rolland, N. 1975. *The Antecedents and Emergence of the Middle Palaeolithic Industrial Complex in Western Europe*. Unpublished Ph.D. Thesis, University of Cambridge.

Rolland, N. 1977. New aspects of Middle Palaeolithic variability in Western Europe. *Nature* 255: 251-252.

Rolland, N. 1981. The interpretation of Middle Palaeolithic variability. *Man* 16: 15-42.

Rolland, N. 1985. Exploitation du milieu et subsistance au cours de la préhistoire ancienne de la Grèce. *Culture* 5: 43-61.

Rolland, N. 1986. Recent findings from La Micoque and other sites in south-western and Mediterranean France: their bearing on the 'Tayacian' problem and Middle Palaeolithic emergence. In G.N. Bailey and P. Callow (eds) *Stone Age Prehistory: Studies in Memory of Charles McBurney*. Cambridge: Cambridge University Press: 121-151.

Rolland, N. 1988a. Palaeolithic Greece: subsistence and socio-economic formations. In B.V. Kennedy and G. LeMoine (eds) *Diet and Subsistence: Current Archaeological Perspectives*. Calgary: Chacmool: 43-53.

Rolland, N. 1988b. Observations on some Middle Paleolithic time series in southern France. In H. Dibble and Montet-White (eds) *Upper Pleistocene Prehistory of Western Eurasia*. Philadephia: University Museum, University of Pennsylvania: 161-180.

Ronen, A. 1979. *Guide to the Prehistoric Caves of Mount Carmel*. Hof Hacarmel, Regional Council.

Rouse, I. 1972. *Introduction to Prehistory: a Systematic Approach*. New York: McGraw-Hill.

Runnels, C. 1983. The Stanford University archaeological and environmental survey of the southern Argolid, Greece. In D.R. Keller and D.W. Rupp (eds) *Archaeological Survey in the Mediterranean Area*. Oxford: British Archaeological Reports International Series S155: 261-264.

Sauer, C.O. 1952. *Agricultural Origins and Dispersals*. New York: American Geographical Society.

Schoeninger, M.J. 1980. *Changes in Human Subsistence Activities from the Middle Paleolithic to the Neolithic period in the Middle East*. Ph.D. Dissertation, University of Michigan. Ann Arbor: University

Microfilms International.

Scott, K. 1980. Two hunting episodes of Middle Palaeolithic age at La Cotte de Saint-Brélade, Jersey (Channel Islands). *World Archaeology* 12: 137-152.

Shennan, S.J. 1978. Archaeological 'cultures'; an empirical investigation. In I. Hodder (ed.) *The Spatial Organisation of Culture.* London: Duckworth: 113-139.

Simek, J. 1984. *A K-means Approach to the Analysis of Spatial Structure in Upper Palaeolithic Habitation Sites: Le Flageolet I and Pincevent 38.* Oxford: British Archaeological Reports International Series S205.

Simek, J.F. and Snyder, L.M. 1988. Changing assemblage diversity in Périgord archaeofaunas. In H. Dibble and Montet-White (eds) *Upper Pleistocene Prehistory of Western Eurasia.* Philadephia: University Museum, University of Pennsylvania: 321-332.

Sireix, M. and Bordes, F. 1972. Le Moustérien de Chinchon (Gironde). *Bulletin de la Société Préhistorique Française* 69: 324-336.

Skinner, J.H. 1965. *The Flake Industries of Southwest Asia: a Typological Study.* Ph.D. Dissertation, Columbia University. Ann Arbor: University Microfilms International.

Smith, P.E.L. 1976. *Food Production and its Consequences.* Menlo Park: Cummings.

Soergel, W. 1922. *Die Jagd der Vorzeit.* Jena: Gustav Fischer.

Solecki, R.S. 1971. *Shanidar: the First Flower People.* New York: Knopf.

Solecki, R.S. 1979. Contemporary Kurdish winter-time inhabitants of Shanidar Cave, Iraq. *World Archaeology* 10: 318-330.

Straus, L.G. 1977. Of deerslayers and mountain men: Palaeolithic faunal exploitation in Cantabrian Spain. In L.R. Binford (ed.) *For Theory Building in Archaeology.* New York: Academic Press: 41-76.

Strong, W.D. 1940. From history to prehistory in the northern Great Plains. In *Essays in Historical Anthropology of North America in Honour of J.R. Swanton.* Miscellaneous Collections 100. Washington (DC): Smithsonian Institution: 353-393.

Testart, A. 1977. Les chasseurs-cueilleurs dans la perspective écologique. *Information des Sciences Sociales* 16: 189-218.

Testart, A. 1982. *Les Chasseurs-Cueilleurs ou l'Origine des Inégalités.* Paris: Société d'Ethnographie.

Testart, A. 1986. La femme et la chasse. *La Recherche* 17: 1194-1201.

Thomson, D.F. 1939. The seasonal factor in human culture. *Proceedings of the Prehistoric Society* 10: 209-221.

Tobias, P.V. 1971. *The Brain in Hominid Evolution.* New York: Columbia University Press.

Toth, N. 1985. The Oldowan reassessed: a close look at early stone artifacts. *Journal of Archaeological Science* 12: 101-120.

Trinkaus, E. 1983. *The Shanidar Neanderthals.* New York: Academic Press.

Trinkaus, E. 1983. Neanderthal postcrania and the adaptive shift to modern humans. In E. Trinkaus (ed.) *The Mousterian Legacy: Human Biocultural Change in the Upper Pleistocene.* Oxford: British Archaeological Reports International Series S164: 165-200.

Turner, V.W. 1966. Colour classification in Ndembu ritual. In M. Banton (ed.) *Anthropological Approaches to the Study of Religion.* London: Tavistock: 47-84.

Turq, A. 1985. Le Moustérien de type Quina du Roc de Marsal (Dordogne). *Bulletin de la Société Préhistorique Française* 82: 46-51.

Ulrix-Closset, M. 1973. Le Paléolithique moyen dans le bassin mosan. *Bulletin de la Société Royale Belge d'Anthropologie*

Préhistorique 84: 71-96.

Van Andel, T.H. and Shackleton, J.C. 1982. Late Palaeolithic and Mesolithic coastlines of Greece and the Aegean. *Journal of Field Archaeology* 9: 445-454.

Van Campo, R. and Bouchud, J. 1962. Flore accompagnant le squelette d'enfant moustérien découvert au Roc de Marsal, commune du Bugue (Dordogne) et première étude de la faune du gisement. *Comptes-Rendus de l'Académie des Sciences de Paris* 254: 897-899.

Vandermeersch, B. 1965. Position stratigraphique et chronologie relative des restes humains du Paléolithique moyen du Sud-ouest de la France. *Annales de Paléontologie* 51: 59-126.

Vereschagin, N.K. 1967. Primitive hunters and Pleistocene extinction in the Soviet Union. In P.S. Martin and H.E. Wright (eds) *Pleistocene Extinctions*. New Haven (Conn): Yale University Press: 365-398.

Veyrier, M. E. Beaux and J. Combier 1951. Grotte de Néron, à Soyons (Ardèche): les fouilles de 1950 – leurs enseignements. *Bulletin de la Société Préhistorique Française* 48: 70-78.

Wagner, P. 1960. *The Human Use of the Earth*. Toronto: Collier-Macmillan.

Weniger, G.-C. 1982. *Wildbeuter und ihre Umwelt*. Universität Tübingen: Verlag Archaeologia Veratoria.

Whallon, R. 1973. Spatial analysis of occupation floors 1: application of dimensional analysis of variance. *American Antiquity* 38: 266-278.

Whallon, R. 1974. Spatial analysis of occupation floors 2: the application of nearest neighbour analysis. *American Antiquity* 39: 16-34.

White, L.A. 1949. *The Science of Culture*. London: Evergreen Books.

White, L.A. 1959. *The Evolution of Culture*. New York: McGraw-Hill.

White, R. 1985. *Upper Palaeolithic Land Use in the Périgord: a Topographic Approach to Subsistence and Settlement*. Oxford: British Archaeological Reports International Series S253.

Willey, G.R. 1977. A consideration of archaeology. *Daedalus* 106: 81-95.

Wissler, C. 1935. Material culture. In C. Muchison (ed.) *A Handbook of Social Psychology*. Worcester (Mass): Clark University Press: 520-564.

Wylie, M.A. 1982. *Positivism and the New Archaeology*. Ph.D. Dissertation, State University of New York at Binghamton. Ann Arbor: University Microfilms International.

Yellen, J. and Harpending, H. 1972. Hunter-gatherer populations and archaeological inference. *World Archaeology* 4: 244-253.

Ziegler, A.C. 1973. *Inference from Prehistoric Faunal Remains*. Reading (Mass): Addison-Wesley.

Zvelebil, M. 1984. Clues to recent human evolution from specialized technologies? *Nature* 307: 314-315.

14: Aspects of Behaviour in the Middle Palaeolithic: Functional Analysis of Stone Tools from Southwest France

PATRICIA ANDERSON-GERFAUD

INTRODUCTION

Some of the explanations offered for the transition from the Middle to the Upper Palaeolithic periods, with their characteristic stone tool assemblages, include, for example: changes in the nature of subsistence (Binford 1973, 1986) and of production and maintenance activities; the presence or absence of curation and planning (ibid); and prehension and hafting as it relates to differences in functional anatomy between the Neanderthals and *Homo sapiens sapiens* (Trinkaus, in press). Research into the function of Palaeolithic stone tools using experimentation with tool replicas combined with microwear analysis, has provided insights into some of these issues. In view of this, I will discuss certain recent functional observations of tools from the French Middle Palaeolithic, and make some comparisons with analogous studies of Upper Palaeolithic assemblages.

The first significant research into the function of Middle Palaeolithic tools using experimental methods and refined analysis of use-traces was applied to sites in the European USSR (see Note 6) (Shchelinskii 1974, and in Plisson 1988), carried out with the high-power microscopy techniques developed by Semenov (1964). My own research was the first to apply experimental methods, combined with a full range of microscopy and chemical analysis, to the functional analysis of the Middle Palaeolithic tools in Western Europe (Anderson 1979; Anderson-Gerfaud 1981; Anderson-Gerfaud and Helmer 1987). Keeley's (1980) innovative methodological work on the functions of British Lower Palaeolithic tools served as a guideline for this work. I have examined primarily sequences of Mousterian of Acheulian tradition levels from two rock-shelter sites in the Périgord region (Pech de l'Azé I and IV) both containing Type A and Type B variants of MAT (Bordes 1972, 1975); and from Corbiac, an open-air site with a large tool assemblage of the MAT-type, overlain by Upper Périgordian (Bordes 1968, 1973). These MAT levels are considered to date from the Würm II on the basis of sedimentological analysis (see Bordes 1972, with references) and there is a ^{14}C date of 40 000 BP for the hearth level at the base of the MAT sequence at Pech de l'Azé I. In these studies, approximately 1000 retouched tools were analysed for microwear traces, and of these 300 were sufficiently well preserved and showed sufficient traces to allow an interpretation in terms of specific functions, based upon comparison with experimentally-used tools.

Subsequently, Beyries (1987, 1988), using a study of microtraces, has analysed several hundred stone tools selected from small site assemblages of the other variants of the French Mousterian: Typical Mousterian, at Corbehem (northern France), Abri Vaufrey (Dordogne), Pié Lombard (southeast France) and Arcy-sur-Cure (Yonne); Denticulate Mousterian from Combe-Grenal, (Dordogne); Charentian Mousterian at Marillac (Charente); and the Middle Palaeolithic of Biache-St.-Vaast (Pas-de-Calais).

As yet, functional studies of stone tools dating from the crucial transitional periods of the Early Upper Palaeolithic are lacking, particularly for the Lower Périgordian (Châtelperronian). A large sample of tools from the Upper Périgordian levels at Le Flageolet I (Dordogne) has recently been analysed for microwear traces by Kimball (1989), and a preliminary microwear study has been carried out for tools from the Solutrian levels at Combe-Saunière (Dordogne) (Geneste and Plisson 1986) (Note 1). Functional studies have also been applied to a small number of tanged tools from the Gravettian of Belgium (Otte and Caspar 1987). By far the majority of microwear analyses have dealt with the later stages of the French Late Upper Palaeolithic, particularly the Magdalenian and the Azilian (e.g. Vaughan 1985, 1987; Moss 1983 and in press; Moss and Newcomer 1982; Keeley 1987; Audouze *et al.* 1981; Plisson 1985, 1987). A few of these studies of Upper Palaeolithic tools will be cited here to assess how far their use seems to differ from that of Middle Palaeolithic tools.

RESEARCH METHODS

Microscopic wear traces on prehistoric tools may be interpreted as due to tool use, tool hafting, transport, tool manufacture and retouch or, alternatively, to the effects of post-depositional modifications occurring either in the soil or after excavation. The diagnosis of their cause is usually made by observing attributes of the traces such as their appearance, orientation, location and distribution over the tool surface, followed by a comparison with similar traces produced on extensively-used (or experimentally trampled, water-rolled, etc.) copies of the particular prehistoric tool type in question (Note 2). Extensive replicative experiments continue, as the methodology is fine-tuned in response to new research questions stemming from different site economies, tool types, and time periods.

To infer the activity represented by microwear traces on stone tools requires determining the use-action, the general kind of material processed (wood, bone, skin, plants, shell, etc.) and its condition (hard, soft, dry, wet, with inclusion of abrasives, etc.). The kinds of traces which have been observed in studies of the French Palaeolithic – on tools in fresh-enough condition for microscopic analysis – include both edge chipping and rounding (usually seen with the stereoscope at magnifications of up to 100X), and various kinds of striations and 'polish' (normally identified using the metallographic microscope at magnifications of 100-400X). These traces are essentially due to the wearing-down and removal of material by various mechanisms (see below), with any residual component contributing very little, if at all, to the surface appearance of the tool as viewed at 100 or 200X. Experimental tools are usually treated with dilute acids and caustics, in soap and water, with alcohol or in an ultrasonic tank, in order to remove large residues covering the tool microsurface. Prehistoric tools are usually treated using some or all of these procedures in order to remove grease and other post-depositional contaminants, particularly if secondary deposits from the archaeological levels appear to mask the tool surface (Note 3). Finally, it is important to emphasize that our experimental data demonstrate that a combination of techniques of microscopic analysis, applied over a wide range of magnification levels, are needed in order to observe a sufficient variety of traces to diagnose reliably the use of prehistoric tools (see Anderson-Gerfaud *et al.* 1987 and Juel-Jensen 1988 for further references).

Finally, the scanning electron microscope (SEM) and energy-dispersive analyser (EDA) have also been used to study the above-mentioned categories of traces. These have proved particularly valuable in the higher magnification ranges (500-2000X) for revealing microscopic, essentially inorganic, residues detached from the actual material worked with the tool and preserved in minute depressions of the tool working edge. The functional residues we have identified using this method do not seem to affect the overall appearance of microwear polishes seen with the optical microscope, and tend to resist removal by dilute chemical baths used to clean contaminants from the tools.

EXAMPLES OF METHODOLOGICAL PROCEDURES AND APPLICATIONS

Residues associated with tool use have been particularly useful for narrowing down the most likely functions of certain stone and bone tools which microwear traces show were used to harvest or process plants (Anderson 1980; Anderson-Gerfaud 1986, 1988; Stordeur and Anderson-Gerfaud 1985; Briuer 1976; Shafer and Holloway 1979). For example, we analysed a small sample of experimentally-used tools and Mousterian tools using the SEM, and EDA (energy-dispersive analysis) of chemical components, which showed that in some cases identifiable residues were present, but only in areas of the tool where other wear traces (polish, striations, etc.) had been observed. Residues tended to be located on both the prehistoric tools and the experimentally-used tools in similar areas (Anderson-Gerfaud 1981, 1986). We were able to identify some of the residues found on the tools by comparing their micro-morphology with that of various living materials similar to those which are known to have been present during the Middle Palaeolithic in the Périgord.

We have found that different analytical methods (using light as well as electron microscopy, and both morphological and chemical techniques to identify traces) can give complementary information concerning tool

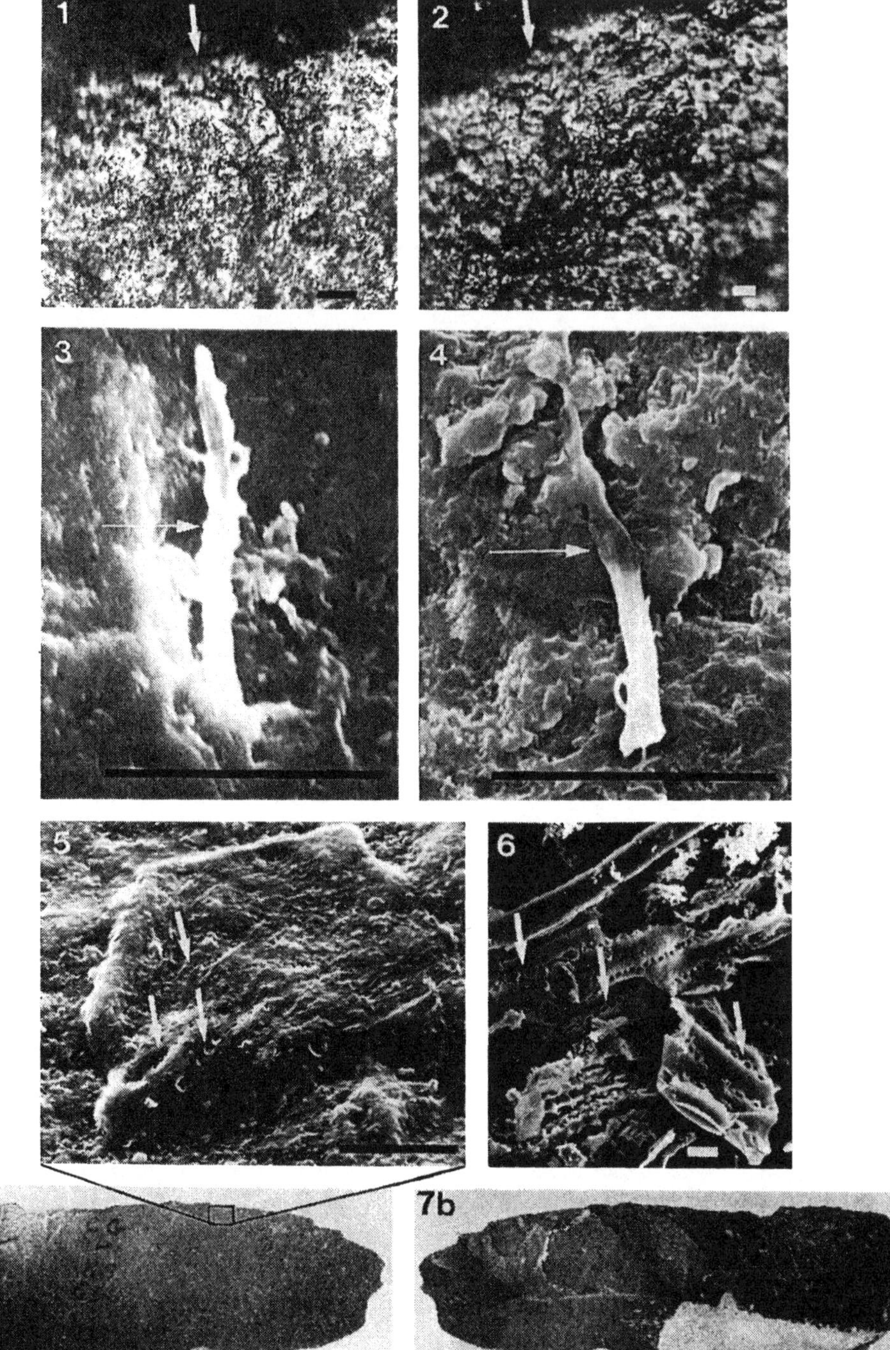

1
2
3
4
5
6
7a
7b

function. Each kind of data serves as a cross-check of the others. For example, microwear polish found on the bulbar face, along the edge of a Mousterian atypical end-scraper (Figure 14.1) is similar in appearance to the microwear polish on the edge of an experimental end-scraper used to scrape wood, as seen at 250X magnification (Figure 14.2). (Unused areas of the tools, not illustrated, are generally darker and more matt in appearance, and do not exhibit patterning of microtraces comparable to that produced by tool use). The use-motion of both tools can be deduced from the orientation and distribution of the microwear polish and striations in relation to the working edge, showing that both tools were used with a transverse (i.e. scraping) motion (see plate captions for further explanations).

Figures 14.1-7:

14.1. Microscopic view of edge surface, bulbar face of a Mousterian end-scraper from Corbiac, showing microwear polish (bright areas) whose appearance and orientation are characteristic of whittling wood. Original magnification = 590X, metallographic (optical) microscope.

14.2. View of microwear polish (bright areas) similar to that shown in Figure 14.1, on the edge (arrow) of the bulbar face of an experimental end-scraper used to whittle fresh poplar for 40 minutes. The arrow shows the orientation of the traces and the direction of the use-motion. Original magnification = 350X, metallographic microscope.

14.3. Residue of a plant micro-fibre (arrow), probably from wood worked using the tool edge, seen in an SEM photomicrograph taken in a microscopic area of the surface of the edge (bulbar face) of a Mousterian end-scraper from Corbiac. Original magnification = 1500X.

14.4. SEM view of a residue found on the edge surface, bulbar face of an experimental tool used to scrape birch. Arrow shows a residue of a micro-fibre from the wood scraped using the tool edge. Original magnification = 1500X. Microwear polish (not shown in Figures) on tools in Figures 14.3 and 14.4 was like that shown in Figures 14.1 and 14.2.

14.5. High-magnification SEM view of the edge micro-surface of the convex scraper from Combe-Grenal shown in Figre 14.7a (area in box not to scale). The residue comprises parallel long cells with rows of pits (arrows) where the cells join. These are characteristic of plant epidermis from certain siliceous plants (see Figure 14.6). The residue fragment is found where microwear polish characteristic of plant-cutting was seen, and is adhering in a slight depression of the tool micro-surface, approximately 100 micrometres from the edge limit. These data show this is a plant-harvesting tool. Original magnification = 2000X.

Figure 14.6. Reference specimen of fragments of siliceous epidermal long cells (phytoliths) extracted by chemical baths and centrifugation from a rush (Juncus). Although the epidermal tissue was torn by centrifugation, it can be seen that this structure (arrows) of parallel 'pits' between articulated long cells, is like that of the residue shown in Figure 14.5. As such cells are found only in grasses and sedges as well as rushes, the plant material harvested with this tool was most likely one or more of these. (Note: the size of the cells and pits can vary within one species of plant). SEM micrograph. Original magnification = 500X.

14.7a, b. Bulbar and dorsal-face views of a Mousterian convex side-scraper from a Würm I level of Combe-Grenal (Dordogne) with bright microwear polish on the edge. A residue located on its edge within the area of the box (Figure 14.7a) is shown in Figure 14.5. Comparison with botanical reference specimens and experimentally-used tools shows that it was used to harvest a grass, rush or sedge. Actual length of tool = 8.5 cms.

Horizontal bars in each photograph represent 10 micrometres.

If areas of the tool edge with microwear polish and striations are viewed using a scanning electron microscope at 500-2000X, sometimes residues of the material worked with the tool may be detected, as in Figure 14.3, showing a residue on the edge surface of an atypical end-scraper from Corbiac. The residue on the edge of an experimental tool used to scrape wood is shown in Figure 14.4. The elongated residues in these two figures (marked by arrows) are plant micro-fibres as seen at 2000X magnification, preserved in depressions of the microscopic surface of the working edge of the two tools. The residues are inferred to derive from the original use of the tools because they appear to be adhering to the surface of the tools, an effect characteristic of residues which were deposited on the surface of the edge during its use, rather than simple contaminations of the tools acquired at other points in time (see below and Note 3). In addition, these residues are found only in the area where other microwear traces characteristic of scraping wood are also found. Thus, the SEM study generally confirmed the initial interpretation of the use of the Mousterian tools derived from optical microscopy, in terms of the working of (woody) plant material – as it showed residues of plant micro-fibres in some of the same areas as the microwear polish and striations.

In other cases, SEM analysis was able to make a more precise diagnosis of the material worked with Middle Palaeolithic tools. For example, we observed bright microwear polish on a side-scraper from Combe-Grenal (Figure 14.7b) which resembled traces produced in two kinds of our experiments at the time: whittling fresh wood, and harvesting certain non-woody plants. Then, when the bulbar face of tool was studied using the SEM, we detected at some 500 micrometres from the working edge (Figure 14.7, box) and in a slight depression in the flint surface, a broken fragment of plant epidermis (Figure 14.5; Anderson 1980). Comparisons were made of this residue's micro-structure with that of leaf and stem epidermis from living plants analogous to those documented by pollen analysis to have been available in the area of Combe-Grenal during the Würm I (Bordes *et al.* 1966). We found that these residues are characteristic not of wood cells, but rather of silica phytoliths from epidermal long cells of grasses, sedges and rushes – as shown in Figure 14.6, illustrating a reference specimen of modern silica 'phytoliths'. Therefore, the bright polish visible along the edge of the Combe-Grenal tool corresponds most probably to cutting stems – perhaps the harvesting of silica-rich grasses, rushes or sedges (see below). This was confirmed in subsequent experiments in harvesting siliceous grasses, rushes and sedges which produced similar microwear polish and residues to those observed on the Mousterian tool (Anderson-Gerfaud 1983, 1986, 1988).

How can such minute residues be preserved? First, the epidermal cells mentioned earlier contain silica – as do wood micro-fibres, although in lesser amounts. This mineral is more resistant to decay than organic material, as recent finds of plant silica bodies (phytoliths) in Middle Palaeolithic (Matsutani 1979) and other archaeological sites can attest (see Rovner 1983 and Piperno 1988). Secondly, the residues are apparently 'protected' from mechanical removal during or after the use of the tool by having been caught in depressions in the tool micro-surface. Third, the residues are 'stuck' to the working edge of the flint tool as a result of hydrolysis (i.e. dissolution of small areas of the flint micro-surface from friction in the presence of water and abrasives: see Kamminga 1979 and Anderson 1980) which occurs on a scale of micrometres as the edge is intensively used to work various materials. We observed micro-morphological surface patterns on the utilized edge of flint tools, patterns which had been found experimentally by Le Ribault (1971) to be characteristic of dissolution (hydrolysis, amorphisation) (Anderson 1980; Anderson-Gerfaud 1981, 1986). In addition it was confirmed that this was taking place during tool *use* in particular by Andersen and Whitlow (1985) who found (using ion beam analysis to obtain hydrogen profiles of surfaces of both experimentally-used tools and tools from prehistoric sites) that more water was present in the surface of tool edges used to work materials containing liquids, than elsewhere on the tool surfaces.

Thus detectable microscopic alterations including abrasion and hydrolysis of the working edge surface can occur when a stone tool is used. This helps to explain why distinct 'microwear polishes' occur, whose appearance varies according to the particular hardness, humidity, and abrasiveness of the worked material.

Because of the general concordance of observations using physical methods of analysis with analysis of traces seen using optical microscopes, we think that for routine analysis of Mousterian tools, observations of microwear polish, striations, edge rounding and edge breakage by means of optical microscopes up to 400X are adequate, in association with experimental replication of these traces, to infer the general use (material worked and use-action) of Middle Palaeolithic tools which were not too soil-damaged for microscopic analysis. Subsequent SEM and EDA analyses used for a small proportion of the tools allowed us to check or to refine this inference (Note 4). Of course, the more kinds of microwear traces which can be found occurring together on a given tool (requiring a minimum of two microscopes for an adequate range of magnification), the more reliable the functional diagnosis.

DISCUSSION OF RESULTS

Can microwear analysis show whether differences in activity patterns, resource availability, or technological needs played a role in changes in material culture or explain internal variability during the Middle Palaeolithic in France (cf. Binford and Binford 1966; Binford 1971; Bordes and Sonneville-Bordes 1969; Mellars 1969, 1970; Rolland 1981; White 1982)? The sample of tools analysed at present, although too limited to necessarily be representative of function for the entire Mousterian, points to tendencies in regard to activities practiced, the relationship between tool form and function, and hafting and curation, with some patterns of tool use suggesting foresight.

Table 14.1. Functions of tools determined by use-wear analysis for Mousterian sites in southwest France. The samples derive from the following sites: A = Corbiac (Mousterian of Acheulian tradition); B = Pech de l'Azé IV, level F4 (Mousterian of Acheulian tradition type A); C = Pech de l'Azé IV, levels F3-F1 (Mousterian of

	Use-Motion	*Side-Scrapers*									
		Straight			*Convex*				*Concave*		
		simple	*inverse*	*double canted*	*simple*	*inverse*	*trans-verse*	*canted*	*conver gent*	*simple*	*inverse*
Wood	Scrape whittle or plane	A=1	B=1 D=2	D=1	B=4 C=1 D=2	D=1	A=8 B=9	A=6	A=7	D=2	
	chop			D=1							
	saw										
	complex undetermined use	B=1			B=1		A=1				
Skin; Hide	scrape				B=2 E=1		A=2 F=1	A=2			
Plants	cut								A=1		
	harvest				F=1						
	complex									B=1	
Unidentified Material	whittle scrape plane chop	A=1			B=2 C=2	D=2	A=1 B=1		A=1		D=1
	complex unknown					D=2		A=1	A=2		
	saw or cut										
	STONE										
	TOTALS	3	3	2	16	5	23	9	11	3	1
Hafting Observations	Clear traces of hafting tool used for:						A=1 (wood)		A=6 (wood) A=1 (plant)		
	Breaks thinning or edge damage (hafting? / tool used for:			A=1 (wood)	F=1 (plant)		B=1	A=1 (skin)			

Acheulian tradition types A/B and B); D = Pech de l'Azé I, level 4 (Mousterian of Acheulian tradition type B); E = Pech de l'Azé I, levels 5-7 (Mousterian of Acheulian tradition type B); F = Combe Grenal (Typical Mousterian).

Denticulates									*Hand axes*	*End scrapers*	*Flakes (debit-age)*	*Totals*
Straight			*Convex*			*Concave*						
simple	*trans-verse*	*conver-gent*	*simple*	*trans-verse*	*conver-gent*	*simple*	*trans-verse*	*conver-gent*				
A=1 B=4 C=3 D=2 E=2	B=2	B=1 C=1 C=1	A=1 B=15 C=1 D=5 E=4	B=6 C=1	B=1 E=2	B=2	B=1	B=1	A=1 D=3	A=16 B=9 E=2		132
										A=1		2
A=1 B=1 C=1	B=2					E=1				A=1		7
C=2		B=1	B=1 C=2 E=2			B=2 D=1			D=5	A=2 B=1		22
									A=2 B=7 C=2 D=1 E=1			21
											B=3	4
												1
						C=1					B=3	5
A=1 B=1 C=1 D=1			B=2 C=1 D=2			B=1				A=1 B=2		24
	B=1								D=3	A=1 (edge)		10
			A=1			C=1			D=1	(edge) A=1		4
D=1										B=2	B=1	4
22	4	4	37	7	1	11	1	1	13	52	7	236
											8	
	B=1 (wood)		B=1 (wood)	B=2 (wood)					A=1	A=2 A=1 B=2 B=2 E=1	(wood) (skin) (skin) (wood) (skin)	17

EVIDENCE OF SUBSISTENCE AND PROCUREMENT ACTIVITIES

Hunting and butchery tools

At present there are two conflicting interpretations of the subsistence strategies of Middle Palaeolithic groups. Binford (1986) has proposed that scavenging rather than hunting was the major technique of procurement of animal foodstuffs during the Lower and Middle Palaeolithic, and that hunting emerged as a major activity only during the the Upper Palaeolithic. Chase (this symposium) on the other hand has argued from a study of faunal assemblages from Middle Palaeolithic levels in southwest France that animals were not merely scavenged, but were competently and purposefully hunted. The Table accompanying this paper (Table 14.1) shows that no evidence for the use of projectiles and virtually none of butchery, was found in the samples of the major retouched tool types studied from Mousterian of Acheulian tradition levels in three sites (Anderson-Gerfaud 1981). Similarly, Beyries (1987, 1988), studying microwear polishes on samples of both debitage flakes and retouched tools, found no evidence for the use of projectiles, despite the fact that her sample included various kinds of 'points'. Butchery was apparently carried out using various retouched tools and debitage from her samples, although this was not one of their most frequent uses. Shchelinskii (in Plisson 1988) analysed tool samples (including Mousterian points) from three Mousterian sites in the European USSR and found none with microtraces characteristic of use as projectiles or weapons.

Of course, it is possible that hunting and butchery were carried out using artifacts in perishable material, or alternately that stone tools having served these purposes were absent from the sites or were not represented in the samples analysed. In addition, the 'level of detectability' of traces from contact with meat on tools from high-carbonate environments, such as those of the sites we studied, would be much less than that of, say, traces of woodworking. Indeed, experiments with different chemical treatments intended to roughly mimic the weathering of flint tools used to work various plant and animal materials, show that microwear traces seem to be of different compositions (i.e. differing according to the material worked with the tool edge) because they resist visually-detectable alteration from chemical treatment to different extents (Plisson and Mauger 1988). Therefore, the scarcity of tools diagnosed as butchery tools could be partially due to preservation conditions. As microwear traces of butchery have been found on tools from the Lower Palaeolithic of Africa (Keeley and Toth 1981) it would appear that preservation of such traces is not so much a question of age *per se*, as of particular conditions of soil and weathering.

A few other observations may indirectly concern projectile use in European Middle Palaeolithic sites. By far the majority of the tools studied from each typological category in our sample from the MAT have traces showing their use for various woodworking applications (see Table 14.1), and a certain number of the denticulates (nearly all of which were used to work wood) had individual notches used separately, apparently to shape narrow cylindrical objects (e.g. branches). Similarly, Beyries (1987) found that nearly all the Clactonian notches she analysed were used to work wood in this manner. Both of our results suggest that wooden spears or shafts (or stakes?) are objects which could have been manufactured using the above denticulates and notches. This is also a typical use of denticulates and scrapers by recent Aborigines in the Western Desert of Australia (Hayden 1977, 1979a). Indeed, the numerous other tools used to scrape, plane and whittle wood could also have been used to shape weapons, although here evidence is lacking as to the surface curvature of the objects made. Pointed wooden sticks (e.g. spears) have of course been found in a number of Palaeolithic sites (see Perlès 1977).

At present, however, the oldest stone tools from southwest France with microwear traces demonstrating their use as projectile armatures (in the absence of any microwear analysis for Aurignacian and Châtelperronian stone tools) are certain Gravette points and backed bladelets from the Upper Périgordian at Le Flageolet studied by Kimball (1989) using a full range of microscopy techniques. Some tools of these types however were used for other purposes as well. Microwear traces on twenty Font-Robert tanged points from the Gravettian of Belgium showed they had various functions (as knives, scrapers, burins, or piercers) but did not reveal any evidence for their use as projectiles (Otte and Caspar 1987).

Extensive experimentation and detailed microwear analyses have shown that Solutrian shouldered points from Combe-Saunière (Dordogne) were used as projectiles, and backed bladelets as projectile armatures or elements of composite butchery knives (Geneste and Plisson 1985). Backed bladelets from the Magdalenian at Pincevent (Moss 1983; Moss and Newcomer 1982; Plisson 1985, 1987) and Verberie (Keeley 1987; Audouze *et al.* 1981) were shown to have been used as projectile points, barbs and as elements of composite butchery knives, with most of the tools at these open-air sites in the Paris basin being used for hunting and butchery. This is further supported by the discovery at Pincevent of a projectile haft with laterally-hafted bladelets (Leroi-Gourhan 1983). Such tools can be located in particular site areas: Moss (1983) found that all the tools in the area around one of the hearths at Pincevent had microwear traces showing that they were used for butchery or as projectiles. Keeley (1987) showed that projectiles at Verberie were apparently retooled but not used there. Finally, other examples of European late Upper Palaeolithic stone tools with traces of use as projectiles are abundant (see Juel-Jensen 1988); harpoons and spear-throwers of bone also attest to Magdalenian projectile use (e.g. Stordeur 1981).

Tools used to harvest plant materials

What is the evidence for the gathering of plant material as food during the Middle Palaeolithic, as opposed to use in various artisanal or technological roles? Indeed, this question could be asked for the Upper Palaeolithic as well, as in sites of both periods only a small number of unretouched flakes and blades or of retouched tools in samples studied have traces of use from harvesting or processing plant material; no pattern emerges for these activities (see Table 14.1). In our sample, several tools from Pech de l'Azé IV (a denticulate, as well as several small unretouched flakes and a concave scraper found within three metre squares at the site), were used to cut or scrape soft plant materials, as well as a hafted convergent scraper from Corbiac (see below) – although the precise task in each case is difficult to glean. Beyries (1987) also found a small number of Mousterian tools which were used to work non-woody plant materials in an unspecified manner.

However, we were able to identify at least one plant-harvesting tool from the Middle Palaeolithic – a convex scraper on a blade from a Würm I level (Typical Mousterian) at Combe-Grenal, described earlier. This particular tool was significant in that it was clearly used with a curved, 'harvesting' motion, and edge damage on the edge opposite the one used suggests that it may have been used in a haft (Anderson-Gerfaud 1981; Anderson-Gerfaud and Helmer 1987). We then examined the tool with the scanning electron microscope to search for any minute fragments of residue material which might clarify its use (Anderson 1980). A residue located near the working edge, in a slight depression in the tool surface was found by comparison with microscopic cellular fragments (e.g. siliceous phytoliths) we extracted and studied from living plants) to be from a grass, or possibly a sedge (*Cyperaceae*) or a rush (*Juncus*).

It is not clear whether this tool was used to procure edible seeds, although certainly seeds of grass and probably sedge (Hillman, in Moore 1975) are edible foodstuffs. It is more likely that this tool, like other Palaeolithic tools of its nature, was used to gather or process plant materials for various artisanal or maintenance purposes (e.g. construction, basketry, fuel, etc.). Indeed, obtaining the plant stems seems to be the primary goal of harvesting plants with a stone tool, at least until cereals with domestic characteristics of ripening evenly and holding their grain at maturity are documented. This is probably because ripe seeds of wild grasses (and cereals, if we consider the Near Eastern Epi-Palaeolithic period) are efficiently gathered by hand-picking, or by stripping or rubbing of the plant over a basket, for example. Cutting with a sickle tends to cause ripe seeds to spill to the ground, but it is effective for harvesting stands whose grain is mostly still in the 'green' state (Anderson-Gerfaud 1983, 1987, 1988). Palaeolithic remains of edible seeds are as yet virtually unknown from French sites, nor are there clear examples of grinding stones used to process plant foods. We do not yet know whether this kind of harvesting tool was frequent in the Mousterian, as the tool discussed above was part of a sample of only a few tools we studied from Combe-Grenal. Future microwear research should provide further insights into these questions, as may careful flotation, soil sampling and separation methods applied to Palaeolithic sites to attempt to extract seeds and microscopic plant remains such as silica phytoliths (as done, for example, by Matsutani (1973) for Douara Cave in Syria; see Rovner 1983 and Piperno 1988). In summary, present evidence from microwear analysis of stone tools and archaeobotany does not provide unequivocal evidence of gathering and preparation of plant material as a foodstuff during the Palaeolithic in southwest France. At present, therefore, exploitation of plant foods cannot be shown to be a significant factor in explaining the Middle-Upper Palaeolithic transition. Indeed, few tools with traces of harvesting or processing of soft plant material have been reported to date for either the Middle or the Upper Palaeolithic. For example, Keeley (1987) finds plant-cutting tools are only a minor component of used tools in his sample from Verberie, and that these tools were usually made from local raw materials, were used at or near the site, and do not show traces of use with a haft.

Procurement of non-woody, soft plant material with stone tools for purposes of manufacture of other artifacts (or technological uses such as fuel), is most likely the activity represented by the occurrence of traces of plant-harvesting; few tools with such traces have been found to date for either the Middle or the Upper Palaeolithic in Europe. Other procurement of plant materials may be indicated by certain cases of traces on stone tools characteristic of wood-adzing, for example, although it is uncertain whether such tools were used primarily to procure wood, or to fashion it into various objects. The issue of woodworking will be discussed further below.

EVIDENCE OF MAINTENANCE AND MANUFACTURING ACTIVITIES

Plant-processing tools

A few of the tools with traces of the working of plant material would seem (to judge by the orientation, location and appearance of the use traces on the working edge) to correspond not to harvesting of plants, but to processing of plant stems and leaves (splitting, scraping, cutting). Again, it is unclear whether these traces correspond to production of artisanal, culinary, or medicinal products (Anderson-Gerfaud 1983).

Woodworking tools

By far the majority of the tools I have studied from each typological category of the Mousterian of Acheulian tradition have traces showing their use for various woodworking applications (see Figures 14.8, 15-17 and the breakdown of tool types shown in Table 14.1) and most of these were clearly used to shape wooden objects. (Some of these objects can be inferred to have had concave or convex surfaces). Some of the small number of tools with microwear traces characteristic of adzing, stripping or scraping bark could correspond to the procurement of these materials rather than to the shaping of various objects, but we think that virtually none of the retouched tools in our sample were used to intensively chop or saw trees, and that wood procurement tools comprise at most a small minority of our sample, as compared to woodworking. Some scrapers show clear traces of having been hafted (see, for example, Figures 14.8,

15-17, 21; see below).

In her analysis of tools from the other variants of the Mousterian, Beyries (1987) found that some tools were used to chop wood, and (as in our MAT sample) that woodworking was the use of most of the tools, although some of the sites Beyries studied are in northern France, where trees were even sparser than in the Périgord region during the Middle Palaeolithic. Similarly, in the Eastern European USSR, Shchelinskii (1974) found that some end-scrapers and planes at the Mousterian site of Erevan were used for woodworking, as were various side-scrapers and 'raclettes' from the Late Mousterian site of La Gouba. Woodworking was the use predicted by Hayden (1979a) for Middle Palaeolithic tools of southwest France based upon ethnographic observation: he found that woodworking was the overwhelming use of stone tools among the Western Desert Aborigines which are similar in form to Mousterian tools, and despite the fact that trees are no more frequent in this environment than in the tundra environments of last-glacial Europe.

As we have mentioned, nearly all of the denticulates, including the various sub-types with different edge morphologies (Anderson-Gerfaud 1981, and unpublished morphometric study) were used to work wood. The orientation of the use-traces show they often were used either to 'whittle', (i.e. a motion combining longitudinal and transverse motions

---

Figures 14.8-14:

14.8. Convergent convex scraper (dorsal and ventral views) from the Mousterian of Corbiac (Dordogne). Diagonal lines indicate the area of the tool with micro-traces caused by hafting; the box shows the area in which the micro-photo in Figure 14.10 was taken. Hafting traces, replicated by experiments, include bright linear features crossing the bulbar face, on the proximal half of the tool only (horizontal lines), and microwear polish in notches (circles). Dots show area of edge with microwear traces showing it was used to plane or adze wood. Actual maximum height of tool as oriented = 8.7 cms.

14.9. Experimental reconstruction of stone tool hafted in a split wooden haft made from a branch fork. The tool was effective when used for adzing, and microwear traces shown in Figure 14.11 were produced by the haft on the bulbar face (proximal half). (Reconstruction by G. Deraprahamian.)

14.10. Micrograph of area in box (not to scale), on the bulbar face of the scraper shown in Figure 14.8, taken using a metallographic (optical) microscope. The photomicrograph shows bright, horizontal linear wear features (a mixture of striations, abrasion traces and microwear polish) extending between opposing notches (shown by circles in Figure 14.8) and running across the lower half of the bulbar face of the tool. These traces were replicated in experiments working wood with tools used in split wooden hafts (Figures 14.9, 12, 14) and on the face of tools tied to hafts (Figure 14.20). Original magnification = 580X. Horizontal bar = 10 micrometres.

14.11. These bright linear traces resemble those in Figure 14.10 in location and micromorphology, and are found in areas of the stone tool surface in contact with the haft shown in Figure 14.9. Micrograph taken with a metallographic microscope. Original magnification = 300X. Horizontal bar = 10 micrometres.

14.12. Reconstruction of possible hafting system of a tool like that in Figure 14.8, in a notched wooden haft, with a tool secured by a hide strap.

14.13. Schematic drawing of reconstruction in Figure 14.12.

14.14. Reconstruction of possible hafting arrangement for tool in Figure 14.8, with the tool held firmly between the two bound hollowed-out halves of the wooden haft. The tool was used to scrape wood. Microscopic traces in the hafted area of the stone tool were similar (on the ventral face) to traces shown in Figures 14.10 and 14.11. (Tool reconstruction by D. Helmer.)

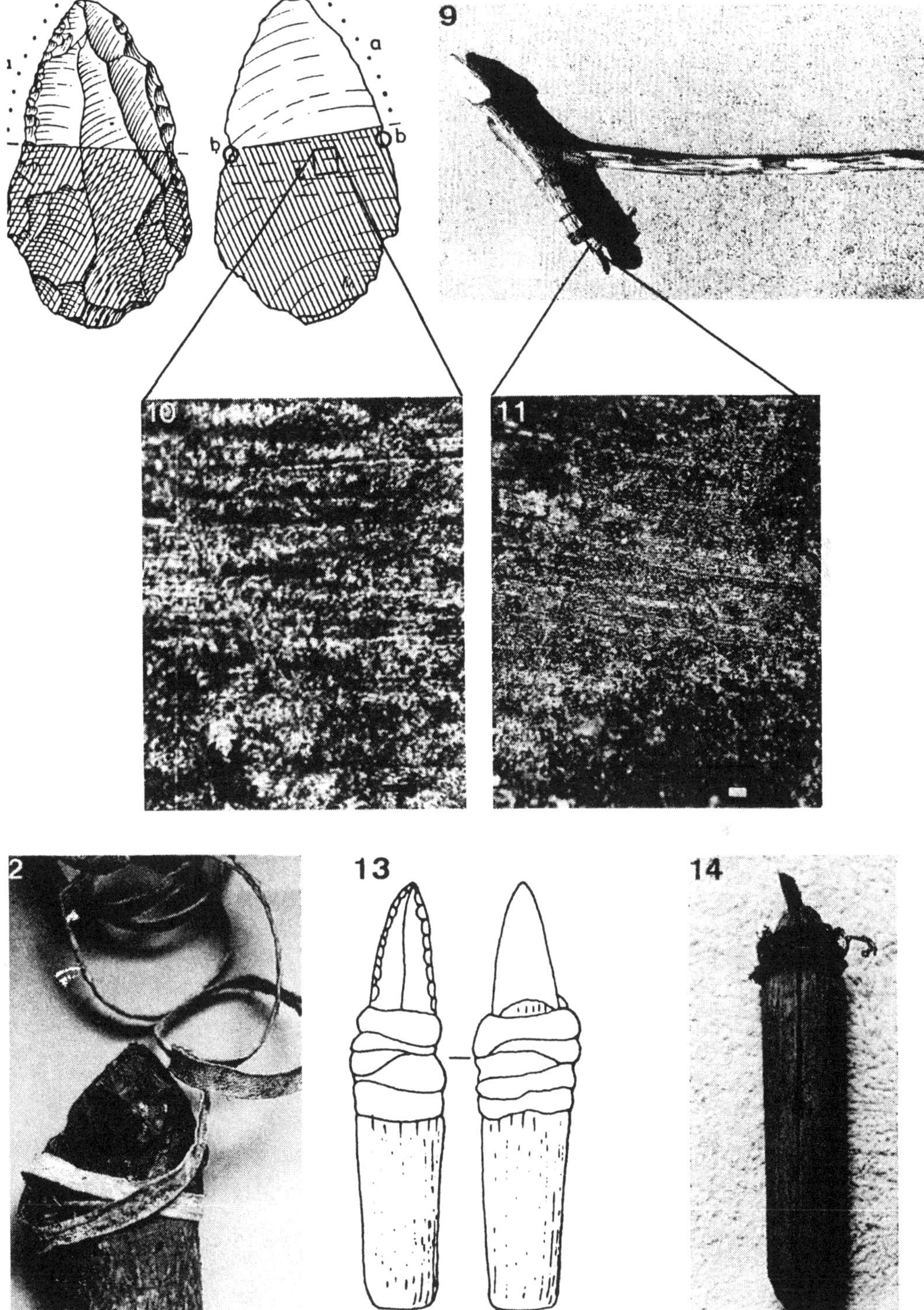

a
b
b
9
10
11
2
13
14

with the unhafted tool held at a low angle to the working surface – as Hayden (1977) shows for denticulates used by the Australian Aborigines) or to scrape or plane, both transverse motions. We have found no clear traces of hafting for the denticulates, and although we have found that some systems of hafting might not leave recognizable traces (Anderson-Gerfaud and Helmer 1987), we suppose they were more often hand-held than the scrapers (see below and Table 14.1).

Working of bone

It appears at present that shaped Middle Palaeolithic bone artifacts tend to have had only their working areas intentionally modified and not their general shape, in contrast to Upper Palaeolithic artifacts (Stordeur, personal communication; Vincent 1986). For example, in a recent microscopic and experimental study, Vincent (1987) finds clear evidence for working and use of bone artifacts in the Mousterian at Bois-Roche (Charente), with use of flaking, perforating, longitudinal splitting and hollowing-out. These artifacts appear to have been intentional tools, and she argues against any natural, carnivore or alimentary origin to these traces. Beyries (1988) finds that various kinds of Mousterian tools were used to work bone or antler, without specifying the use-motion. The fact that bone and antler artifacts were not found in the sites she studied may possibly be due to problems of preservation or of recognition. The samples we studied from the MAT, although not adequate to demonstrate the complete absence of bone-working tools, did not reveal any clear examples of tools used for this purpose, nor did those analysed by Shchelinskii (in Plisson 1988) for the European USSR. However, it is intriguing to note the technical sophistication of certain shaped personal ornaments recently found in collections of Early Aurignacian artifacts from southwest France by White (1989). It will be interesting to see if microwear analyses of stone tools from this period document a marked increase in bone-working as compared with the Middle Palaeolithic. Not surprisingly, examples abound of stone tools with microwear traces showing they were used to scrape, drill and groove bone and antler during the later Upper Palaeolithic. These include not only burins (see references in Anderson-Gerfaud *et al.* 1987), but also other tools such as some Font-Robert points. At least some of the latter tools (Otte and Caspar 1987), as well as bone-working tools from Verberie (Keeley 1987) are thought to have been hafted.

Working of stone

On a few tools from Pech de l'Azé IV we have found traces indicating that their working edges were in contact with stone and another material (soft plant material?), but it is unclear whether the stone was being worked with a longitudinal motion, or whether the tool edge was simply grinding against it in association with working of another material. Shchelinskii (see Plisson 1988) found that stone-scraping was the probable use of some obsidian tools he studied from the Erevan Cave in Armenia.

Hide and skin preparation

Approximately 10 percent of the retouched tool types we studied from the Mousterian of Acheulian tradition were used for preparing skin and hide. Hideworking tools, although found in different typological groups, are all convex-edged, representing either end-scrapers (e.g. Figures 14.22, 14.23) or side-scrapers. The major dichotomy of use-motions used for working hide in our sample – 'pushing' with the tool held at a low angle to the working surface, often for wetter or fresher hides, and 'pulling' the tool towards the user as it is held nearly perpendicular to the working surface (with the working angle as in Figure 14.20), and used to work hide in a dry state —has also been observed for stone hide-working tools of the Nunamiut Eskimo, as opposed to the Plains Indians (Hayden 1979). Shchelinskii (*op. cit.*) found that frequent use of the Middle Palaeolithic tools he studied was the scraping and piercing of hide, but that only the *earlier* phases of hide preparation were carried out using stone tools in the Middle Palaeolithic, whereas in the Upper Palaeolithic the later phases of hideworking are commonly indicated by microwear traces (see also Vaughan 1985, and Anderson-Gerfaud, unpublished observations for end-scrapers from La Madeleine; for further references in Juel-Jensen 1988). Working of hide is almost invariably the use documented for end-scrapers from the Upper Palaeolithic in France (ibid).

MOUSTERIAN CHRONOLOGY, VARIABILITY AND EVOLUTION

The results of my own analyses summarized in Table 14.1 indicate that in the French Mousterian of Acheulian tradition individual typological categories do not correspond in any simple way to distinct functions (with the exceptions noted above), at least at the level of precision at which we can infer their use (Anderson-Gerfaud 1981)). Similarly, for the Typical, Denticulate and Charentian industries studied by Beyries (1987) no correlation was found between use and typological category (with the exception of some scrapers from Biache-St.-Vaast: Beyries 1988), nor between overall activities demonstrated by use of tools and the particular kind of Mousterian 'tool kit' represented (Binford 1973; Binford and Binford 1966). Beyries does not mention whether her data show any detectable chronological patterns but her sample was deliberately varied in space, and she does not deal with any sequences of levels from one site or region.

Our data for the Mousterian of Acheulian tradition do reveal some possible changes through time. For example, in our sequence of MAT levels from Pech de l'Azé I, and particularly Pech de l'Azé IV, the variety in the form of woodworking tools decreases with time. This corresponds basically to the fact that side-scrapers (and end-scrapers, although most we analysed are atypical) are almost all woodworking tools, and that all varieties of side-scrapers become rarer as we approach the Upper Palaeolithic. However, variety in morphological sub-types of denticulates also decreases with time (Anderson-Gerfaud, unpublished study) although the overall proportion of denticulates among the retouched tools in fact increases. These tools tend to have higher edge angles and to be smaller in size in the later (Type B) than in the earlier (Type A) stages of the Mousterian. This could be due either to increasing 'economy of raw material' as regards woodworking, or to the fact the function of these tools in the MAT Type B was somewhat different than that during the MAT Type A (or of course to simple coincidence of changes in tool form and frequency independent of their function).

Woodworking tools are generally much less frequent in French later Upper Palaeolithic sites for which microwear analyses have been carried out, and butchery and hunting, or hideworking, are the uses of the majority of the tools (see references above). An exception to this generalization is the site of Cassegros, a Magdalenian cave/rock-shelter site in southwest France, where Vaughan (1985) found woodworking to be a frequent activity. Nonetheless various hide treatment activities were found to be the use of the majority of the tools at Cassegros, presumably a more wooded environment during the Magdalenian than at most of the sites studied from further north.

We found that although the majority of the Mousterian end-scrapers in our sample (nearly all of which are atypical) were used to work wood (Anderson-Gerfaud and Helmer 1987; Anderson-Gerfaud 1981), there are some specimens of the more typically 'Upper Palaeolithic' form in the MAT B sample, which were used for working hide and skin, rather than wood (e.g. Figure 14.22). Indeed, the MTA B levels of our sequence seem (despite very small samples) to show a dichotomy between 'degenerate' scrapers and denticulates which were used for woodworking, and end-scrapers used to work hide. Microwear studies of French later Upper Palaeolithic sites also show hide-working to be a consistent use for end-scrapers (Anderson-Gerfaud *et al.* 1987; Juel-Jensen 1988) and indicate similar functions for other types of Upper Palaeolithic tools.

Therefore these data suggest that changes in manufacturing and maintenance of non-stone artifacts and tools, due to a technological shift involving exploitation of hide, bone and antler, as opposed to wood, may explain some of the differences in stone tools found in Middle and Upper Palaeolithic assemblages. Furthermore, the data show that shifts in the choice of raw materials to be fashioned into artifacts using stone tools from the Middle to the Upper Palaeolithic are not related to the mere availability and density of these resources at different times and in different locations.

HAFTING OF TOOLS IN THE MIDDLE PALAEOLITHIC

One of the more intriguing results to come of our research on the use of tools in the French MAT, was the discovery of clear traces left by the rubbing of a haft for nearly all the convergent scrapers at Corbiac, of which six (e.g. Figures 14.8, 14.15, 14.17) were used for woodworking and one for working of other plant materials (Anderson-Gerfaud 1981). Similar traces were observed on a transverse scraper from this site (Figure 14.16) which was also used to work wood (see Table 14.1). Numerous hafting experiments and studies of prehistoric artifacts have shown that clear hafting traces are likely to be found only rarely on tools. For example, residues of adhesive have not been found on tools from the Palaeolithic of France, and it seems that adhesives would not have been preserved on a macroscopic level in most French Middle Palaeolithic contexts. Moreover, hafting with adhesives rarely allows the tool to move sufficiently in the haft as to leave traces of wear. In addition, it would seem that many uses of hafted tools do not involve sufficient force to repeatedly 'jam' the tool into the haft during use, thereby creating traces on the stone insert where it has rubbed against the haft. However, both the convergent scrapers and some of the transverse scrapers from Corbiac discussed above show a striking pattern of traces visible at 100-400X. On the bulbar face, linear abrasion and polish are limited to the proximal and medial portions of the tool, and occur mainly on surface convexities. Often wear can be seen in opposed lateral notches at each end of the hafting traces (see Figures 14.10 and 14.11 for examples of such traces). In some cases, markedly worn and 'polished' ridges occur on the dorsal face, but only in the proximal half of the tool, opposite the traces observed on the bulbar face. It is the combination of such traces and their limited location, effectively 'outlining' the hafted area of the tool, which distinguishes them from any traces caused by soil action, prehension, or tool manufacture. The traces described for the Corbiac scrapers were best replicated in our experiments by hafting copies of the tools into split hafts bound with various ligature materials (e.g. Figures 14.9, 14.12, 14.14: see Anderson-Gerfaud and Helmer 1987) and using them to scrape, plane and adze wood. When the hafting traces described above were confined to (or far more marked on) the bulbar face, the traces were best replicated by tying the tools to hafts using bindings of vegetable string (Figure 14.20) or animal materials, with the bulbar face against the wood haft, and using them to work wood (e.g. Figure 14.20). Moss (1987) has obtained similar kinds of hafting traces in similar experiments. Our experiments showed that both hafting methods are effective for woodworking with force, and particularly precision, and allow for rapid retooling of the implements.

Although we have seen that tools of similar morphology can show similar kinds of hafting traces, we do not think that there is a requisite morphology for hafting, nor that tool morphology is a clear guide to whether the tool was hafted or not. For example, some other tools from the Middle Palaeolithic sites we studied exhibit thinning in the bulbar area, together with lateral breaks (e.g. Figure 14.23) and other morphological features suggesting hafting, but they did not have clear microscopic traces indicating they were indeed hafted – possibly because they were hafted by means of adhesives and did not move against the haft during use. The convergent scrapers showing microwear traces of hafting discussed above did not exhibit marked thinning of the bulb or other kinds of macroscopic features mentioned above. However, it must be emphasized that removal of the bulb or other thinning or adaptation of the tool

Figures 14.15-24:
14.15. Convergent scraper from the Mousterian of Corbiac, showing traces due to hafting as shown (for example) in Figures 14.9, 12 or 14. Ventral face had microscopic traces like those shown in Figure 14.11. (c) = abraded ridges on the dorsal face of the tool; (b) = edge with microwear traces characteristic of use to whittle wood. Actual maximum height of tool as oriented = 8.5 cms.
14.16. Transverse scraper from Corbiac, with edge (a) used for planing or adzing wood, according to microwear traces. The tool shows abrasion of dorsal ridges (c) and microscopic bright linear traces on ventral face (as in Figures 14.10, 11) and bulbar area, suggesting hafting as in Figures 14.9, 12 or 18. Actual maximum height of tool as oriented = 7.5 cms.
14.17. Convergent scraper from Corbiac, with working edge used to whittle wood (a). Distinctive traces on edges (b) and ventral face, suggest hafting as in Figures 14.12 or 20. Actual maximum height of tool as oriented = 6.5 cms.
14.18. Schema of hafting technique used for Eskimo "women's knives", with the stone tool hafted in a split haft bound with ligatures (after Miles 1963).
14.19. Schematic drawing of right-angle haft, as used by North American Indian tribes, with stone tool tied or juxtaposed to the haft with lashings. This fixation technique may have been used for certain Mousterian side-scrapers and end-scrapers, with the overall shape of the haft as shown here or as in Figures 14.18 or 20, for example (after Brink 1978).
14.20. Reconstruction of stone tool lashed to a straight wooden haft which was thinned where the tool is tied. The tool was used to scrape wood. The microwear traces produced on stone tools by rubbing against such hafts during tool use resemble micro-traces found on certain Mousterian convergent scrapers and end-scrapers which were used to work wood in this way. Note that a right angle is formed by the haft and the hand and arm of the user. (Tool made by D. Helmer.)
14.21. Broken end-scraper from Corbiac. Edge (a) used to plane or adze wood. The tool shows hafting traces in area of (b) suggesting techniques shown in Figures 14.12 or 20. Actual height of tool fragment as oriented = 7.5 cms.
14.22. End-scraper from Pech de l'Az; I (Mousterian of Acheulian tradition type B level). The edge (a) was used to scrape hide. The lack of hafting traces observed on the ventral and dorsal faces suggests that the tool was used unhafted, or was glued to a haft, perhaps as in Figure 14.24. Actual maximum height of tool as oriented = 7.6 cms. (Drawing of tool after Bordes 1972: 82.)
14.23. End-scraper from Corbiac whose retouched edge shows microwear traces indicating that it was used to scrape skin. The bulbar area is thinned, but the tool does not show microwear traces from hafting like those obtained in our experiments with tied and split hafts. The tool may have been hafted as in Figure 14.24, with an adhesive, preventing the formation of microwear traces due to rubbing against the haft during use. Original maximum height of tools as oriented = 6.2 cms.
14.24. Experimental reconstruction of a woodworking tool, glued with resin to a thinned wood haft then wrapped with skin strips. (Reconstructed by D. Helmer.)
(Note: areas of tools marked with parallel diagonal lines have microwear traces indicating that these areas were covered by a haft, based upon comparisons with observations of hafting traces on experimental tools.)

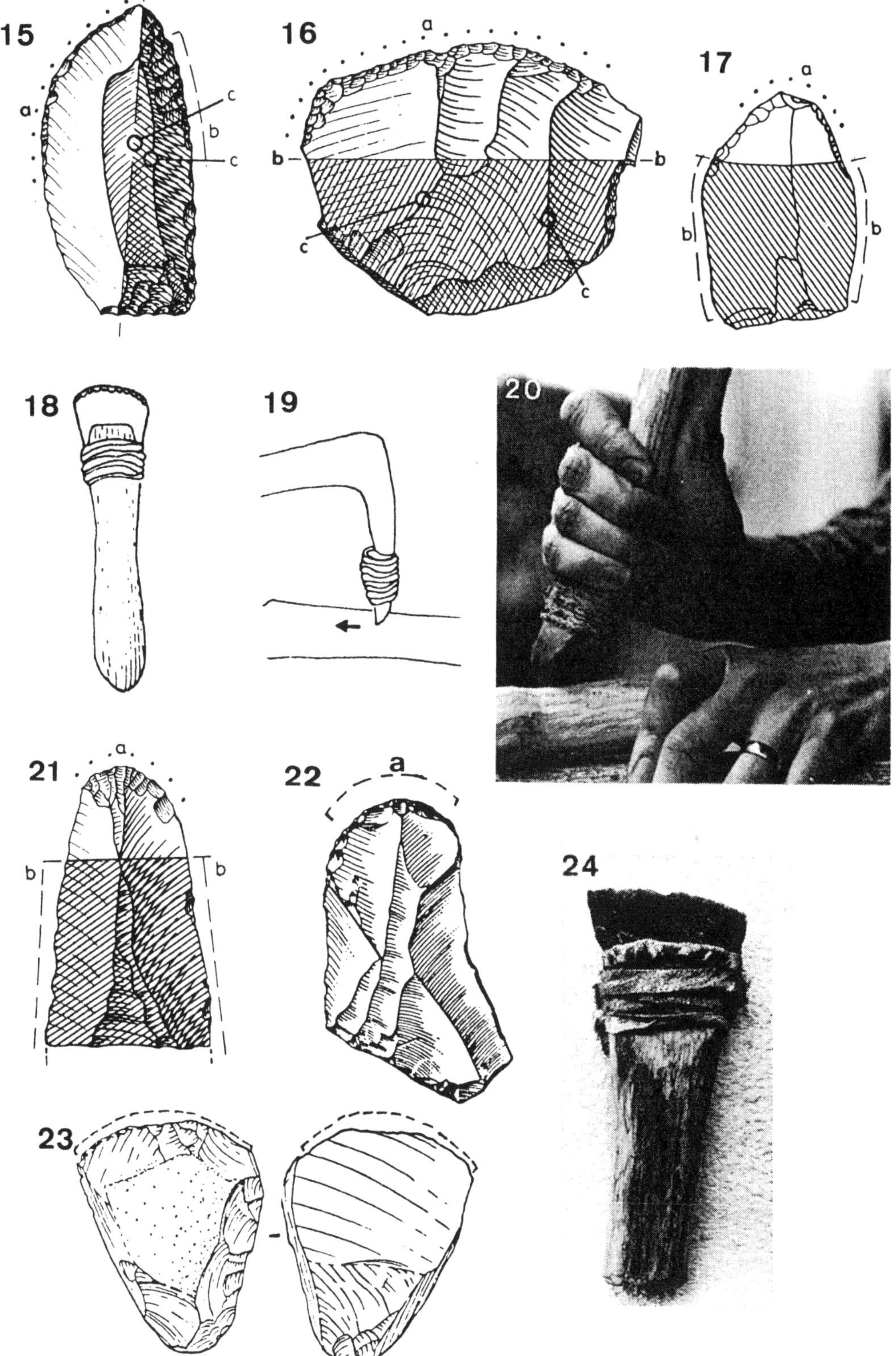
15
a
b
c
c
16
a
b
b
c
c
17
a
b
b
18
19
20
21
a
b
b
22
a
23
24

morphology is by no means a necessity for hafting. Even Neolithic stone tools from Swiss lake sites, inserted in slits in beautifully-preserved, carefully-fashioned wooden hafts, and glued using birch resin, did not often have their bulbs removed. The hafts were made to accommodate tools of a similar shape and size, and microwear analysis showed some of the tools had even been turned in the haft and reused, without the presence of the bulb's interfering with the good 'fit' of the stone tool in the haft (Anderson-Gerfaud and Plisson 1986). When systems of hafting by lashing or a split or hollowed haft are used, as described here, experiments and ethnographic examples show that the handle may be thinned to accommodate the tool's bulbar surface in the area where the tool is lashed to the haft – as shown in tool reconstructions illustrated in Figures 14.20 and 14.24. Similarly, in the case of split hafts, each half of the haft is thinned where the tool will be located, (Figures 14.9, 14.14) or the part of the haft in which the tool is placed is hollowed-out to accommodate its general form (Figures 14.12, 14.13).

The tool may be held firmly in place either by binding it directly to the haft, (Figures 14.12, 14.19, 14.20), or by gluing it with adhesives (Figure 14.24), or by pressure exerted on the haft by bindings (Figures 14.9, 14.13, 14.14, 14.18). These hafting arrangements have been documented ethnographically for tools from Patagonia (Mansur-Franchomme 1983) and for various knives and scrapers of the Nunamiut Eskimos (Miles 1963, and see Figure 14.18), and are thought to have been used for late Upper Palaeolithic tools (Keeley 1982; Plisson 1985). Moss (1987) has replicated hafting traces she found on Azilian tools at Pont d'Ambon (Dordogne) by using experimental tools hafted between two ribs which were bound together, and Vaughan cites a few tools at Cassegros which had traces apparently indicating either that they were tied to hafts or that their proximal ends were wrapped with bindings. It would seem that these kinds of hafting arrangements are used in the Middle as well as the Upper Palaeolithic.

CONCLUSIONS

It is interesting that according to our experimental observations (Anderson-Gerfaud and Helmer 1987), the kind and patterning of traces described above for the convergent and some of the transverse scrapers and end-scrapers (Anderson-Gerfaud 1981), correspond clearly to traces left by split or tied hafts, and it is in each case the slight motion of the tool as it 'jams' in the haft during use which causes clear traces to form. Evidence of these kinds of hafting traces and mechanisms implies that an actual handle was attached to the tool which altered its prehension significantly, and not merely a covering. Although of course we have no proof of the overall shape of the handle (which could have been straight, as shown in our experimental reconstructions, or bent, as in Figures 14.9 and 14.19), a right-angle or lever effect would be created by a straight simple haft and the arm and hand of the user, as in Figure 14.20, or could be incorporated into the haft form as in Figure 14.9 and 14.19. In summary, whether or not future studies prove such hafting was truly common in the Middle Palaeolithic (Beyries (1988) also cites hafting traces for three tools from Mousterian sites, as well as several convergent scrapers from Biache-St.-Vaast), we feel that its presence in this Mousterian of Acheulian tradition sequence has interesting implications for the question of Middle Palaeolithic modes of prehension, and provides an example of deliberate use of the fulcrum principle in the Mousterian of southwest France.

However, it cannot be shown that these examples from the Middle Palaeolithic correspond to anything but the hafting of a single stone tool at the end of a haft. The first clear examples to date of lateral hafting (usually inferred for backed bladelets hafted in a series to be then used as butchery knives or projectile elements) is found in the Upper Périgordian and continues through the later Upper Palaeolithic (see references above). This may be related to increased butchery and hunting carried out using stone tools from the Middle and the Upper Palaeolithic – although butchery is occasionally attested to by microwear analysis of stone tools in Middle Palaeolithic sites, and hunting (shown to have occurred in southwest France from study of faunal remains) could have been carried out using perishable weapons, or traces of stone tools used for hunting not detected or poorly preserved on the tool samples studied to date.

Finally, although the working of hide with stone tools was apparently not the main use of tools in southwest French Mousterian assemblages, in our sample it was always carried out with convex-edged tools and shows a dichotomy of angle and direction of use which also exists in the ethnographic record. Skin preparation is the principal use of tools in some of the sites studied by Shchelinskii from the European USSR, with woodworking also attested, as well as butchery and working of bone and stone. It appears that the latter stages of hide preparation were more often carried out using stone tools (often end-scrapers) in Upper Palaeolithic sites, than in Middle Palaeolithic sites.

I would suggest that a change in manufacturing and maintenance activities – especially the manufacture of artifacts in bone and perhaps hide, as opposed to wood – may be a part of the explanation for change in tool form from the Middle to the Upper Palaeolithic. This remains to be substantiated by microwear analysis of early Upper Palaeolithic tools. In the relatively small samples studied to date, the harvesting and processing of soft plant material is more anecdotal than a consistent function for either Middle or Upper Palaeolithic stone tools (Keeley even finds a small number of plant-harvesting tools in the Early Pleistocene at Koobi Fora: Keeley and Toth 1981). However, more careful attempts to recover microscopic plant remains (i.e. silica phytoliths) from the soil in Palaeolithic sites may provide further data on plant collecting and use.

In terms of the explanation of the significance of variability in tool form during the Middle Palaeolithic, it is unclear at present whether Middle Palaeolithic convergent scrapers show similar hafting traces because they were shaped and retouched in the course of their use with the haft (and therefore correspond to a phase in a reduction sequence, as proposed by Dibble 1987) or rather, whether they were made deliberately in a 'standardized' way with a specific view to use with a haft (Note 7). The latter situation in particular would imply technological strategy and a certain depth of planning, and also explain why hafting traces are almost exclusively found on convergent scrapers at present. Tools so similar may have been attached successively to given hafts, suggesting the hafts themselves were curated items.

Woodworking has been shown to be the most common use of retouched and unretouched tools studied in samples from French Middle Palaeolithic sites. It must be pointed out that relative rarity of certain other uses, in particular butchery, may be due to problems of preservation of the microwear traces; woodworking traces are relatively well-developed along the edge of tools used to work wood with transverse motions, and would tend to better-preserved and more resistant to alterations from the burial environment than certain traces of working animal material (see Plisson and Mauger 1988). Our experiments have shown that intensive woodworking can require frequent re-sharpening of the stone tool. This might be seen to support Dibble's (1987) suggestion that many Mousterian tools were in fact part of a successive reduction sequence, although our data can neither support nor refute the hypothesis of multiple use and resharpening of individual tools. The high frequency of woodworking tools as opposed to tools used for other purposes in Middle Palaeolithic sites, may of course be partially explained by the rapidity with which this particular activity produces wear on stone tools, or alternatively by differential preservation, as suggested earlier, and should not be taken to indicate, in any absolute sense, the frequency of this activity as opposed to others in French Middle Palaeolithic sites.

It is nonetheless significant that a positive identification of woodworking traces was made for the majority of the tools studied from French Middle Palaeolithic sites, particularly for most of the scraper types, and virtually all of the denticulates. The latter in our MAT sample were consistently used as wood-whittling tools, probably unhafted. Woodworking would thus seem to have been a common activity in both the sparsely-wooded Périgord region and northern France. Shchelinskii (in Plisson 1988) also cites woodworking at two sites in the European USSR. Therefore this apparent technological 'specialisation' in woodworking (most tools seem to have been used to shape objects rather than procure wood and bark) can hardly be attributed simply to relative availability of this resource in the surrounding environment.

Thus, these data from functional analyses of stone tools show woodworking specialisation, and reveal an 'invisible' wooden artifact technology of potentially highly varied function. Similar hafting patterns of scrapers which are morphologically very similar suggest that these may be stone elements possibly used successively for given hafts, and the hafts themselves may have been curated objects. The use of long hafts would serve as levers, increase precision and control of the tool, and is a practical but sophisticated way to adapt certain stone tools to particular woodworking functions. In conclusion, we feel our data provide examples of human behaviour in the Middle Palaeolithic which indicate planning and foresight, exceeding a mere accommodation of behaviour to immediate needs and conditions.

NOTES

1. Early work was done using a stereoscope by Bordes (1973)on the location of wear-traces on end-scrapers from the Upper Périgordian site of Corbiac.

2. For example, this experimental approach has been applied to the microwear analyses of Palaeolithic tools by Keeley (1980), Anderson-Gerfaud (1981, 1983; Anderson-Gerfaud and Helmer 1987), Vaughan (1985), Moss and Newcomer (1982), Moss (1983, 1987), Plisson (1985, 1987) and Schelienski (1974).

3. Standard procedures of chemically treating artifacts with dilute acids (HCl) and caustics (H_2O_2) and non-abrasive soap and water or alcohol, tend to eliminate problems of contamination of the tool surface by poorly-adhering soil from the archaeological level, or grease from handling, dust, etc. However, organic residue analysis of plant and animal oils, collagen or haemoglobin, for example, (Loy 1983; Briuer 1976; Fullager *et al.*, in press) may be impaired by any such chemical cleaning of tools prior to analysis.

4. In these particular cases, chemical analysis using EDA did not further clarify the situation, as plant material (probably siliceous) and silica comprising the flint were not able to be differentiated by the probe, which detected both as 'silicon'. However, the EDA microprobe was useful in other cases in narrowing down the identification of the worked material, where the residue material was made up of different minerals from those of the tool itself. For example, plant and bone-working residues composed of calcium (Anderson-Gerfaud 1981, 1986) and stone-working residues composed of magnesium (Del Bene 1980) were detected on flint tool edges. Plant silica was detected using the EDA and the SEM in notches of Neolithic bone tools shown to have most probably been used to strip seed heads from primitive cereals (Stordeur and Anderson-Gerfaud 1985). Finally, further avenues of research may be opened by applying methods used recently by Loy (1983) for identifying blood and other organic residues on prehistoric stone tools by use of chemical reagents (Briuer 1976; Shafer and Holloway 1979; Fullager *et al.*, in press) and applying techniques used to identify certain species of animal from fats extracted from the soil in Palaeolithic sites (Rottlander 1983).

5. Extensive experimentation has accompanied low-magnification study of Aurignacian stone and bone tools from the Middle East, showing that some were used as projectiles on the basis of certain special breakage patterns (Bergman and Newcomer 1983; Bergman 1987); detailed microwear analysis may demonstrate European Aurignacian bone points also were used as projectiles.

6. Shchelinskii carried out microwear analysis of obsidian tools from the cave site Erevan in Armenia, and flint tools from La Gouba (Monaseskaya) in the Prekuban (a final or late Mousterian) and Nosovo I (an Eastern European MAT in the PreAzov) (see Plisson 1988).

7. The clarity of the hafting traces on convergent scrapers in our MAT sample and Beyries' sample from Biache could conceivably be explained by the tools having been used in a haft for longer periods than other tool types, if they had been re-fashioned and re-used throughout a reduction sequence.

ACKNOWLEDGEMENTS

Research cited here would not have been possible without the generous help and advice given me from 1973-1981 by the late François Bordes, Director of the Institut du Quaternaire, University of Bordeaux, Talence, where I was a researcher. I thank my experimental collaborators Daniel Helmer and Gérard Deraprahamian (CNRS, Institut de Préhistoire Orientale, Jales, France) and I am grateful to Larry Kimball (Northwestern University) and Harold Dibble (University of Pennsylvania) for their critical discussion of various versions of this manuscript.

REFERENCES

Andersen, H.H. and Whitlow, H.J. 1983. Wear traces and patination on Danish flint artefacts. *Nuclear Instruments and Methods in Physics Research* 218: 468-474.

Anderson, P.C. 1979. A microwear analysis of selected flint artefacts from the Mousterian of southwest France. *Lithic Technology* 9: 32.

Anderson, P.C. 1980. A testimony of prehistoric tasks: diagnostic residues on stone tool working edges. *World Archaeology* 12: 181-194.

Anderson-Gerfaud, P. 1981. *Contribution Methodologique à l'Analyse des Microtraces d'Utilisation sur les Outils Préhistoriques*. Doctoral Thesis, University of Bordeaux I.

Anderson-Gerfaud, P. 1983. A consideration of the uses of certain backed and 'lustred' stone tools from Late Mesolithic and Natufian levels of Abu Hureyra and Mureybet (Syria). In M.-C. Cauvin (ed.) *Traces d'Utilisation sur les Outils Néolithiques du Proche-Orient.* Lyon: Travaux de la Maison de l'Orient 5: 77-105.

Anderson-Gerfaud, P. 1986. A few comments concerning residue analysis of stone plant-processing tools. In L. Owen and G. Unrath (eds) *Technical Aspects of Microwear Studies on Stone Tools.* Tubingen: *Early Man News* 9/10/11: 69-81.

Anderson-Gerfaud, P. and Plisson, H. 1986. Traces d'utilisation sur des outils emmanchés des lacs suisses. Paper presented to Symposium on *Typologie, Technologie et Tracéologie*, Centre de Recherches Archéologiques, Valbonne, France, 1986.

Anderson-Gerfaud, P. 1988. Using prehistoric stone tools to harvest cultivated wild cereals: preliminary observations of traces and impact. In S. Beyries (ed.) *Industrie Lithique, Tracéologie et Technologie.* Oxford: British Archaeological Reports International Series S411: 175-195.

Anderson-Gerfaud, P. and Helmer, D. 1987. L'emmanchement au moustérien. In D. Stordeur (ed.) *La Main et l'Outil: Manches et*

Emmanchements Préhistoriques. Lyon: Travaux de la Maison de l'Orient 15: 37-54.

Anderson-Gerfaud, P., Moss, E. and Plisson, H. 1987. A quoi ont-ils servi? L'apport de l'analyse fonctionnelle. *Bulletin de la Société Préhistorique Française* 84: 226-237.

Audouze, F., Cahen, D., Keeley, L.H. and Schmider, B. 1981. Le site magdalenien du Buisson Campin à Verberie (Oise). *Gallia Préhistoire* 24: 99-143.

Bergman, C.A. 1987. Hafting and use of bone and antler points from Ksar Akil, Lebanon. In D. Stordeur (ed.) *La Main et l'Outil: Manches et Emmanchements Préhistoriques*. Lyon: Travaux de la Maison de l'Orient 15: 117-126.

Bergman, C.A. and Newcomer, M.H. 1983. Flint arrowhead breakage: examples from Ksar Akil, Lebanon. *Journal of Field Archaeology* 10: 238-243.

Beyries, S. 1987. *Variabilité de l'Industrie Lithique au Moustérien: Approche Fonctionnelle sur Quelques Gisements Français*. Oxford: British Archaeological Reports International Series 328.

Beyries, S. 1988. Etude tracéologique des racloirs du niveau IIA. In Tuffreau and J. Sommé (eds) *Le Gisement Paléolithique Moyen de Biache-St.-Vaast (Pas de Calais): Stratigraphie, Environnement et Etude Archéologique*: Vol. 1. Paris: Mémoires de la Société Préhistorique Française 21: 215-230.

Binford L.R. 1973. Interassemblage variability: the Mousterian and the 'functional' argument. In C. Renfrew (ed.) *The Explanation of Culture Change*. London: Duckworth: 227-254.

Binford, L.R. 1986. Isolating the transition to cultural adaptations: an organizational approach. Paper prepared for the Advanced Seminar on The Origins of Modern Human Adaptations. School of American Research, Santa Fe.

Binford, L.R. and Binford, S.R. 1966. A preliminary analysis of functional variability in the Mousterian of Levallois facies. *American Anthropologist* 68: 238-295.

Bordes, F. 1968. Emplacement de tentes du Périgordien supérieur évolu; à Corbiac (près Bergerac, Dordogne). *Quartär* 19: 251-262.

Bordes, F. 1972. *A Tale of Two Caves*. New York: Harper and Row.

Bordes, F. 1973. Position des traces d'usure sur les grattoirs simples du Périgordien supérieur évolué de Corbiac (Dordogne). *In Estudios Dedicatos al Profesor Dr. L. Pericot*. Barcelona: University of Barcelona, Institute of Archeology and Prehistory: 55-60.

Bordes, F. 1975. Typological variability in the Mousterian layers at Pech de l'Azé I, II and IV. *Journal of Anthropological Research* 34: 181-193.

Bordes F. and de Sonneville-Bordes, D. 1969. The significance of variability in Palaeolithic assemblages. *World Archaeology* 2: 61-73.

Bordes, F., Laville, H. and Paquereau, M.-M. 1966. Observations sur le Pleistocène supérieur de Combe-Grenal. *Actes de la Société Linnéenne de Bordeaux* 103 B: 3-19.

Briuer, F.L. 1976. New clues to stone tool function: plant and animal residues. *American Antiquity* 41: 478-484.

Chase, P.G. 1989. How different was Middle Paleolithic subsistence? A zooarchaeological perspective on the Middle to Upper Paleolithic transition. In P. Mellars and C. Stringer (eds) *The Human Revolution: Behavioural and Biological Perspectives on the Origins of Modern Humans*. Edinburgh: Edinburgh University Press: 321-337.

Del Bene, T. 1979. An X-ray analysis of depositional polish. *Lithic Technology* 9: 36-37.

Dibble, H.L. 1987. Reduction sequences in the manufacture of Mousterian implements of France. In O. Soffer (ed.) *The Pleistocene Old World: Regional Perspectives*. New York: Plenum Press: 33-44.

Fullager, R., Meehan, B. and Jones, R. (in press). Residue analysis of ethnographic plant-working and other tools from northern Australia. In P. Anderson-Gerfaud (ed.) *The Prehistory of Agriculture: New Experimental and Ethnographic Approaches*. Monographie du Centre de Recherches Archéologiques. Paris: CNRS..

Geneste, J.-M. and Plisson, H. 1985. Le Solutréen de la grotte de Combe-Saunière 1 (Dordogne): première approche paléthnologique. *Gallia Préhistoire* 29: 9-27.

Hayden, B. 1977. Stone tool functions in the Western Desert. In R.V. Wright (ed.) *Stone Tools as Cultural Markers: Change, Evolution and Complexity*. Canberra: Australian Institute of Aboriginal Studies.

Hayden, B. 1979a. *Palaeolithic Reflections*. New Jersey: Humanities Press.

Hayden, B. 1979b. Snap, shatter and superfractures: use-wear of stone skin scrapers. In B. Hayden (ed.) *Lithic Use-Wear Analysis*. New York: Academic Press: 207-229.

Juel-Jensen, H. 1988. Functional analysis of prehistoric flint tools by high power microscopy: a review of Western European research. *Journal of World Prehistory* 2: 53-88.

Kamminga, J. 1979. The nature of use-polish and abrasive smoothing on stone tools. In B. Hayden (ed.) *Lithic Use-Wear Analysis*. New York: Academic Press: 143-158.

Keeley, L.H. 1980. *Experimental Determination of Stone Tool Uses: A Microwear Analysis*. Chicago: University of Chicago Press.

Keeley, L.H. and Toth, N. 1981. Microwear polishes on early stone tools from Koobi-Fora, Kenya. *Nature* 293: 464-465.

Keeley, L.H. 1982. Hafting and retooling: effects on the archeological record. *American Antiquity* 47: 798-809.

Keeley, L.H. 1987. Hafting and retooling at Verberie. In D. Stordeur (ed.) *La main et l'Outil: Manches et Emmanchements Préhistoriques*. Lyon: Travaux de la Maison de l'Orient 15: 89-96.

Kimball, L. 1989. *Functional Variability in the Upper Périgordian: Microwear Analysis of a Sample of Tools from Level 7, Le Flageolet I (Dordogne)*. Unpublished Ph.D. Dissertation, Department of Anthropology, Northwestern University.

Le Ribault, L. 1971. Présence d'une pellicule de silice amorphe à la surface de cristaux de quartz des formations sableuses. *Comptes Rendus de l'Académie des Sciences (Série D)* 272: 1933-1936.

Leroi-Gourhan, A. 1983. Une tête de sagaie à armature de lamelles de silex à Pincevent (Seine-et-Marne). *Bulletin de la Société Préhistorique Française* 80: 154-156.

Mansur-Franchomme, E. 1987. Outils ethnographiques de la Patagonie: emmanchement et traces d'utilisation. In D. Stordeur (ed.) *La Main et l'Outil: Manches et Emmanchements Préhistoriques*. Lyon: Travaux de la Maison de l'Orient 15: 297-307.

Matsutani, A. 1979. Microscopic study of amorphous silica in sediments from Douara Cave. *The Palaeolithic Site of Douara Cave in Syria*. Tokyo:University Museum: 127-131.

Mellars, P.A. 1969. The chronology of Mousterian industries in the Périgord region of southwest France. *Proceedings of the Prehistoric Society* 35: 134-171.

Mellars, P.A. 1970. Some comments on the notion of 'functional variability' in stone tool assemblages. *World Archaeology* 2: 74-89.

Miles, C. 1963. *Indian and Eskimo Artifacts of North America.* New York: Bonanza Books.

Moss, E.H. 1983. *The Functional Analysis of Flint Implements: Pincevent and Pont-d'Ambon, Two Case Studies from the French Final Paleolithic.* Oxford: British Archaeological Reports International Series S177.

Moss, E.H. 1987. Polish G and the question of hafting. In D. Stordeur (ed.) *La Main et l'Outil: Manches et Emmanchements Préhistoriques.* Lyon: Travaux de la Maison de l'Orient 15: 97-102.

Moss, E.H. (in press). Function and Spatial Distribution of Flint Artifacts from Pincevent Section 36, Level IV 40. *Oxford Journal of Archaeology.* In Press.

Moss E.H. and Newcomer, M.H. 1982. Reconstruction of tool use at Pincevent: microwear and experiments. *Studia Praehistorica Belgica* 2: 289-312.

Perlès, C. 1977. *La Préhistoire du Feu.* Paris: Masson.

Otte, M. and Caspar, J.-P. 1987. Les 'pointes' de la Font-Robert: outils emmanchés? In D. Stordeur (ed.) *La Main et l'Outil: Manches et Emmanchements Préhistoriques.* Lyon: Travaux de la Maison de l'Orient 15: 65-74.

Piperno, D.R. 1988. *Phytolith Analysis: an Archaeological and Geological Perspective.* London: Academic Press.

Plisson, H. 1985. *Etude Fonctionnelle d'Outillages Lithiques Préhistoriques par l'Analyse des Micro-Usures: Recherche Méthodologique et Archéologique.* Doctoral thesis. University of Paris I.

Plisson, H. 1987. L'emmanchement dans l'habitation no. 1 de Pincevent. In D. Stordeur (ed.) *La Main et l'Outil: Manches et Emmanchements Préhistoriques.* Lyon: Travaux de la Maison de l'Orient 15: 75-88.

Plisson, H. 1988. Technologie et tracéologie des outils lithiques moustériens en Union Sovietique: les travaux de V.E. Schelinskii. In M. Otte (ed.) *L'Homme de Néandertal.* Vol. 4: *La Technique.* Liège: Etudes et Recherches Archéologiques de l'Université de Liège (ERAUL 31): 121-168.

Plisson, H. and Mauger, M. 1988. Chemical and mechanical alteration of microwear polishes: an experimental approach. *Helenium* 28: 3-16.

Rolland, N. 1981. The interpretation of Middle Palaeolithic variability. *Man* 16: 15-42.

Rottlander R. 1983. Investigations chimiques sur les graisses en archéologie. *Nouvelles de l'Archéologie* 11: 38-42.

Rovner, I. 1983. Plant opal phytolith analysis: major advances in archeobotanical research. In M. Schiffer (ed.) *Advances in Archaeological Method and Theory* Vol. 6. New York: Academic Press: 225-266.

Semenov, S. 1964. *Prehistoric Technology.* Bradford-on-Avon: Moonraker.

Shafer H.J. and Holloway, R.G. 1979. Organic residue analysis in determining stone tool function. In B. Hayden (ed.) *Lithic Use-Wear Analysis.* New York: Academic Press: 385-399.

Shchelinskii V.E. 1974. *Proizvodstvo I Founktsii Moust'erskih Oroudii (po Dannym Eksperimental'nogo i Trasologitcheskogo Izoutcheniia). (Manufacture and Function of Mousterian Tools [According to Results from Experimental and Traceological Analysis]).* Doctoral Thesis. Leningrad.

Stordeur, D. 1981. L'outil d'os pendant la préhistoire. *La Recherche* 12: 452-465.

Stordeur, D. and Anderson-Gerfaud, P. 1985. Les omoplates encochées néolithiques de Ganj-Dareh (Iran): étude morphologique et fonctionnelle. *Cahiers de l'Euphrate* 4: 289-313.

Trinkaus, E. (in press). Bodies, brawn, brains and noses: the implications of Pliocene and Pleistocene human anatomy for the evolution of human predation. In M.H. Nitecki (ed.) *The Evolution of Human Hunting*. Oxford: Oxford University Press. In Press.

Vaughan P. 1985. *Use-Wear Analysis of Flaked Stone Tools*. Tucson: University of Arizona Press.

Vaughan, P.C. 1987. Positive and negative evidence for hafting on flint tools from various periods (Magdalenian through Bronze Age). In D. Stordeur (ed.) *La Main et l'Outil: Manches et Emmanchements Préhistoriques*. Lyon: Travaux de la Maison de l'Orient 15: 135-144.

Vincent, A. 1987. Outillage osseux du Paléolithique moyen à Bois-Roche (Cherves-Richemont, Charente): étude préliminaire. In B. Vandermeersch (ed.) *Préistoire de Poitou-Charentes: Problèmes Actuels*. Paris: Editions du Comité des Travaux Historiques et Scientifiques: 27-36.

White, R. 1989. Production complexity and standardisation in early Aurignacian bead and pendant manufacture: evolutionary implications. In P. Mellars and C. Stringer (eds) *The Human Revolution: Behavioural and Biological Perspectives on the Origins of Modern Humans*. Edinburgh: Edinburgh University Press: 366-390.

White, R.A. 1982. Rethinking the Middle-Upper Paleolithic transition. *Current Anthropology* 23: 169-192.

15: A Multiaspectual Approach to the Origins of the Upper Palaeolithic in Europe

JANUSZ K. KOZLOWSKI

The problem of continuity/discontinuity between the Middle and the Upper Palaeolithic has been usually discussed in terms of the technology and typology of lithic artefacts. Other aspects such as economy, settlement patterns and symbolism have been given only a general treatment without relating them to technological and morphological variations in lithic industries of the transitional period. The present paper is an attempt at correlating the synchronic and diachronic variability of various aspects of culture, economy and settlement in this period.

TECHNOLOGY

In the period preceding the Upper Palaeolithic three different technologies can be distinguished:

1. Flake technology which employed single-platform, multiplatform and discoidal cores to produce flakes.

2. 'Levalloisian' blade-flake technology distinguished by the use of blank forms pre-determined by preparation. The very concept of the Levalloisian core implies that blanks must have been obtained from the dorsal part of the core (i.e. the flaking face was limited by the edges which formed at the intersections of scars from bifacial preparation). In contrast to this, characteristic Upper Palaeolithic cores could be processed until they were wholly reduced, since the edges formed at the intersection of scars from bifacial preparation were normally removed in the initial phase of core preparation by the removal of a 'crested blade'.

Recently Boëda (1988) has distinguished two sub-types within the Levalloisian technique: the first referred to as 'classical' (denoted Type A) and the second, 'specialized' (denoted Type B). The latter technique allowed larger output from a core (since the prepared surface was more convex) and the production of more standardized blanks (see Figure 15.1).

These two types of the Levalloisian technique differed from the Upper Palaeolithic blade technique in that the extent of the flaking face was limited, and that cores like these yielded a diversity of blanks (flake blanks, blade-flake blanks and blade blanks). The qualitative features of blades obtained from the two types of Levalloisian cores are unlike those

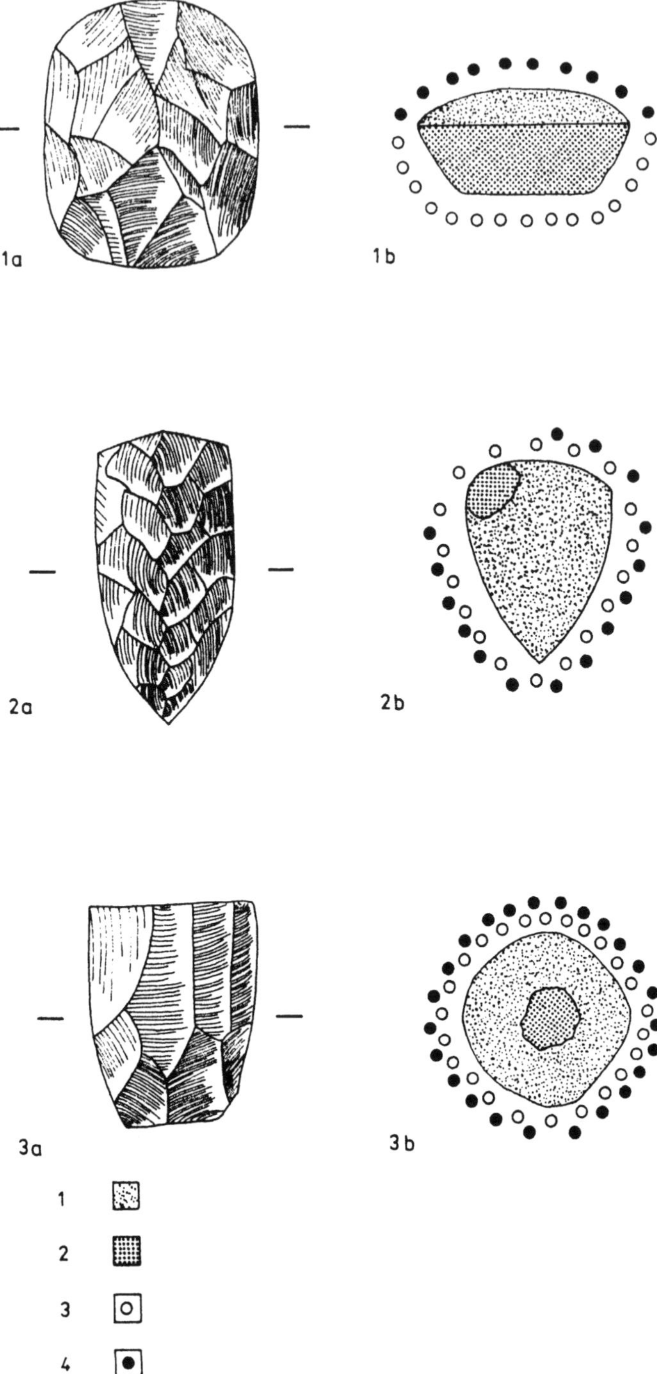

Figure 15.1. Schemes showing the differences between Levalloisian cores (no. 1) and Upper Palaeolithic cores (nos 2, 3). After E. Boëda. The symbols and shading shown in the cross-sections indicate the following: 1. reduced volume; 2. residual volume; 3. prepared surface; 4. striking surface (see key).

characterizing Upper Palaeolithic specimens, although some Type B cores could potentially yield items resembling the Upper Palaeolithic blades.

3. Non-Levalloisian blade technology, which is known to have appeared as early as the Penultimate Glaciation (e.g. in the lower layers at Piekary IIa in Poland: Morawski 1976), and in the Early Vistulian (St. Valéry-sur-Somme in Belgium: de Heinzelin and Haesaerts 1983; Rocourt in Belgium: Cahen 1984; Seclin in France: Tuffreau 1979; Rheindalen layer B1 in West Germany: Bosinski 1978, 1981; Bosinski *et al.* 1966). In this technology the core is an almost exact parallel of the Upper Palaeolithic core in that the flaking face (whether prepared or unprepared) is not delimited by lateral trimming of the edges. Standardized blanks were also obtained by this technique, but as a hard hammer was employed the quality of blanks is generally lower than that in Upper Palaeolithic assemblages (see Figure 15.2).

It is only the last-mentioned technology that can be regarded as a real precursor of the Upper Palaeolithic blade technology. The Levalloisian technique could not provide a basis for the evolution of Upper Palaeolithic blade techniques, owing to the limited number of flakes which could be struck off, the non-standardized form of the blanks, and other morphological characteristics of the blanks.

From the viewpoint of basic technology, the early Upper Palaeolithic cultural units can be divided into the following groups:

1. Those employing Levalloisian technology (i.e. the Bohunician industries);

2. Groups combining flake technique with the Upper Palaeolithic blade technology (e.g. the Szeletian and Châtelperronian industries); and

3. Groups where blade technique of fully Upper Palaeolithic type predominated (the Aurignacian industries).

The Upper Palaeolithic industries employing the Levalloisian technology are known so far only in Czechoslovakia, where the 'Bohunician' has recently been distinguished. This industry combines the features of Levalloisian technology with charateristically Upper Palaeolithic blade tools (Valoch *et al.* 1976; Oliva 1983). The percentage of Levalloisian technology in the flake group is not particularly high (5.80-20.14%). In relation to the total number of tools, on the other hand, the presence of Levalloisian blades and points is more conspicuous (25.50-46.55%). The blade index is 31.0-44.9%. Levalloisian cores amount to 13.7% of all cores. The Bohunician has been dated to the beginning of the main 'Interpleniglacial' interstadial phase, dated to around 40-43 000 BP.

The technology of the second group, represented in Central Europe by the Szeletian industries, is characterized by a blade index lower than that in the Bohunician (24.5-32.1%), and by a Levalloisian index approaching zero (0-2%). The ratio of blade cores is relatively low, even in the initial phases of core reduction. Still lower is the ratio of cores with the preparation of Upper Palaeolithic type, amounting to 6.8-14.2% (Figure 15.3: 2-4).

The third group – represented by the Aurignacian industries – is distinguished above all by a much higher ratio of blade cores (on the oldest sites such as Vedrovice II and Kupařovice this ratio constitutes 89-69% of all cores). The proportion of cores with preparation of Upper Palaeolithic type increases sharply, amounting to 15.8-19.7% of all cores. Because Aurignacian sites are functionally differentiated (depending on the relative importance of the 'workshop' element) the general blade index may fluctuate considerably. Moreover, this index also depends on the the relative intensity of core reduction on the site – as in the course of progressive reduction blade cores were transformed into flake cores. Consequently, more advanced reduction automatically produced an increase in the flake index and the drop in that of blades (Figure 15.3: 5-6).

Seen from the perspective of the evolution of technologies, the first group of industries should be regarded as a continuation of the Middle Palaeolithic 'Moustero-Levalloisian' assemblages. This trend, however, exerted no influence on the development of Upper Palaeolithic blade technologies. The second group amalgamates two diametrically differing technologies: flake techniques and blade techniques of the Upper Palaeolithic type. This group, therefore, cannot supply the missing link in the technological evolution, but is a direct continuation of the Middle Palaeolithic flake technology with the addition of Upper Palaeolithic blade technology. The formation of the Upper Palaeolithic blade technique must therefore be explained either as an external influence, or as an independent innovation. The third group mentioned above, of vital importance in the evolution of the Upper Palaeolithic techniques, is probably derived from non-Levalloisian blade techniques in the Middle Palaeolithic.

TYPOLOGY

Basic Upper Palaeolithic tool categories can be separated using as a criterion the various retouching techniques employed and the location of retouch on blanks. Steep and semi-steep retouch is characteristic of end-scrapers and backed pieces. The burin blow technique produced burins. These techniques are occasionally encountered in the Middle Palaeolithic, but in the Upper Palaeolithic their frequency increased dramatically. Despite this they are not diagnostic for *specific* Upper Palaeolithic industries.

It is the technique of flat bifacial retouch and the technique of steep, lateral blunting retouch (sporadically recorded in the Middle Palaeolithic) that constitute diagnostic elements for specific Upper Palaeolithic units. These two types of retouch can be used to divide early industries into two broad groups: those with leaf-points on the one hand, and those with backed points on the other.

Still more markedly specific are Upper Palaeolithic tool types with high lamellar transversal retouch (on carinated and nosed end-scrapers) and with lateral scalariform retouch. These are the only tool types that occur in assemblages with the highest blade indices, and they are diagnostic of

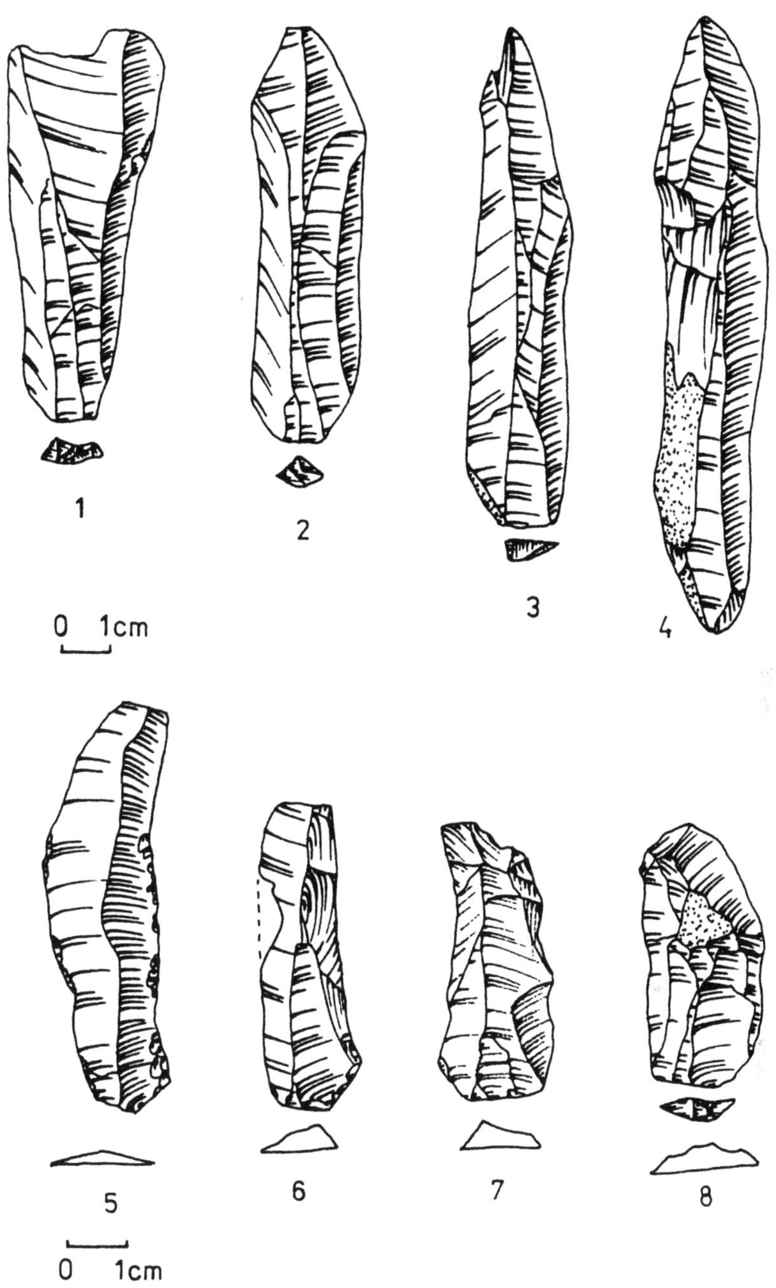

Figure 15.2. Pre-Upper Palaeolithic blade industries showing Upper Palaeolithic-type technology: 1-4. Seclin (France) (early Würm); 5-8. Rheindalen (Germany) layer B1 (Pre-Pleniglacial I). After A. Tuffreau and G. Bosinski.

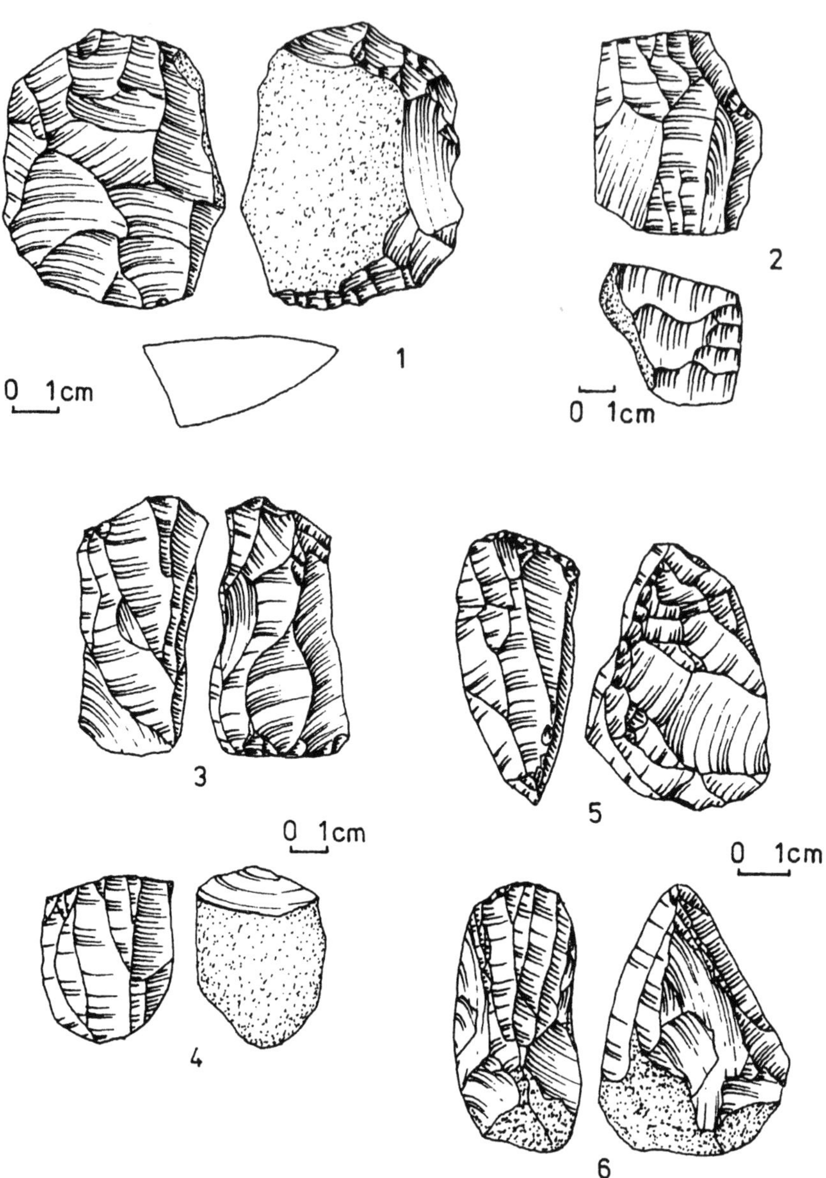

Figure 15.3. Early Upper Palaeolithic cores: 1. Bohunice (Moravia), double-platform Levalloisian core; 2-4. Dzierzyslaw (Poland), Szeletian cores for blades and flakes (2, 3. double platform; 4. single platform); 5, 6. Kupařovice site I (Moravia), Aurignacian blade cores. (Nos 1, 5, 6 after K. Valoch).

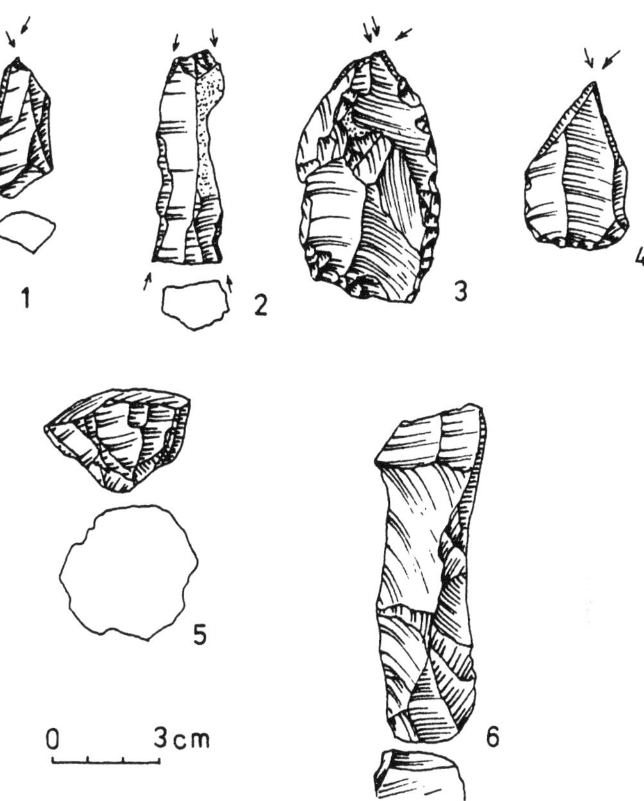

Figure 15.4. Early Upper Palaeolithic implements from Korman site IV, layer 8: 1-4. burins; 5. blade-flake conical core; 6. blade with traces of lateral core preparation.

the Aurignacian industries. So far we do not know of any Middle Palaeolithic industries that contain a similar set of tool types.

The only possible exception is the so-called 'Amudian' of the Near East, which has recently been referred to as the 'Mugharan' (Jelinek 1982), and provisionally dated to around 70 000-80 000 years BP. The assemblage from layer 15 in the rock-shelter of Jabrud 1 in Syria (Rust 1951) shows a fairly high blade component which accounts for 54.5% of all blanks; blade cores, however, comprise only 3.7%, and blade-flake cores 37.0%, of all cores. This small proportion of the blade element among the cores seem to result from the advanced level of reduction of the cores, which in the most advanced phase are transformed into blade-flake cores. Implements of Upper Palaeolithic type make up more than 75% of all retouched tools, with burins being the most numerous type (40.6%). The ratio of specifically Aurignacian end-scrapers is, however, quite low (8.4%), and that of Aurignacian retouched blades is even lower (4.3%).

For these reasons, to regard the Amudian as some kind of a 'pre-

Table 15.1. Chronological distribution of principal cultural features in the Middle

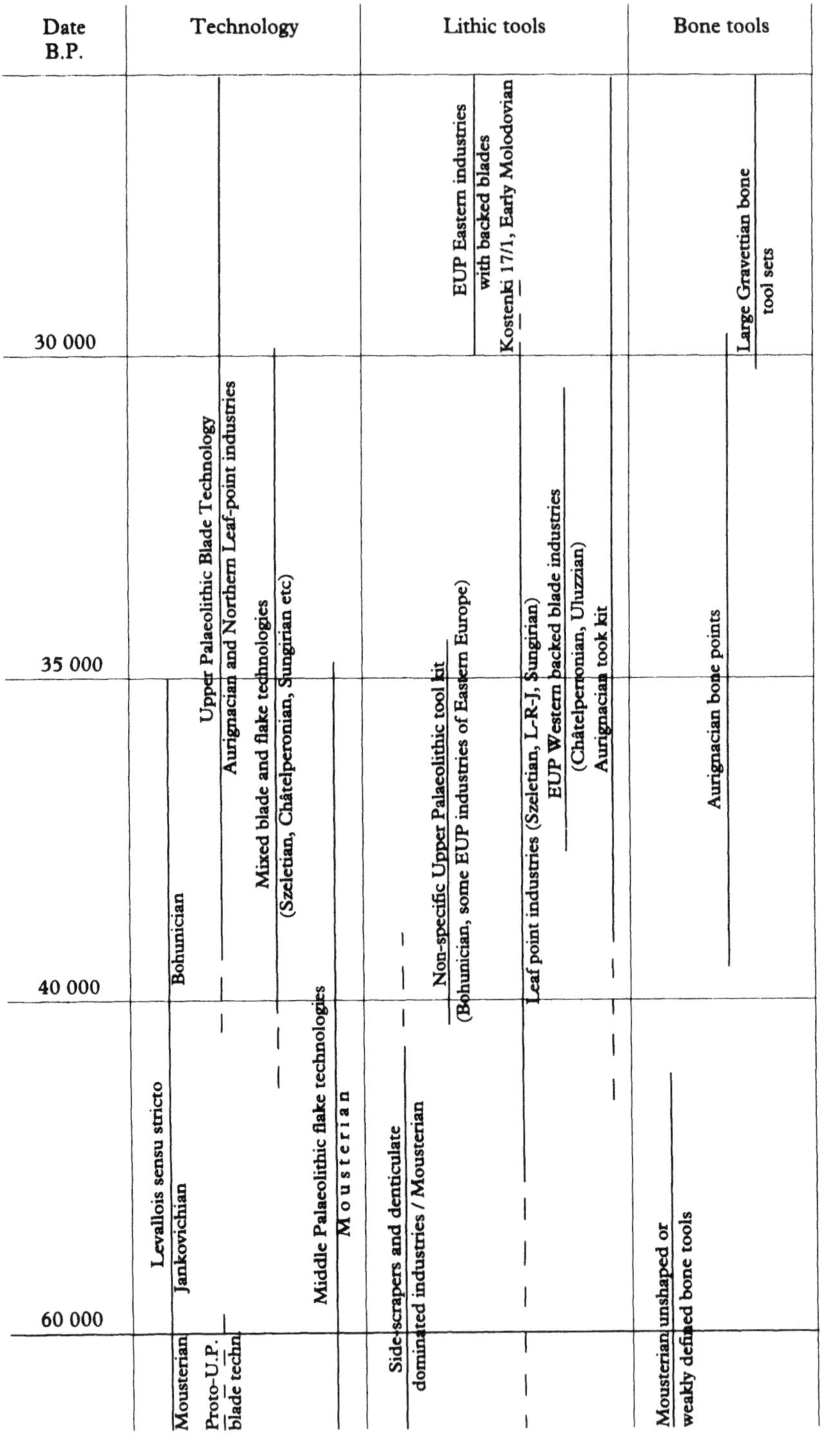

Date B.P.	Technology	Lithic tools	Bone tools
30 000 35 000 40 000 60 000	Upper Palaeolithic Blade Technology Aurignacian and Northern Leaf-point industries Mixed blade and flake technologies (Szeletian, Châtelperonian, Sungirian etc) Bohunician Levallois sensu stricto Jankovichian Mousterian Proto-U.P. blade techn. Middle Palaeolithic flake technologies M o u s t e r i a n	EUP Eastern industries with backed blades Kostenki 17/1, Early Molodovian Non-specific Upper Palaeolithic tool kit (Bohunician, some EUP industries of Eastern Europe) Leaf point industries (Szeletian, L-R-J, Sungirian) EUP Western backed blade industries (Châtelperonian, Uluzzian) Aurignacian took kit Side-scrapers and denticulate dominated industries / Mousterian	Large Gravettian bone tool sets Aurignacian bone points Mousterian unshaped or weakly defined bone tools

and Upper Palaeolithic of Europe

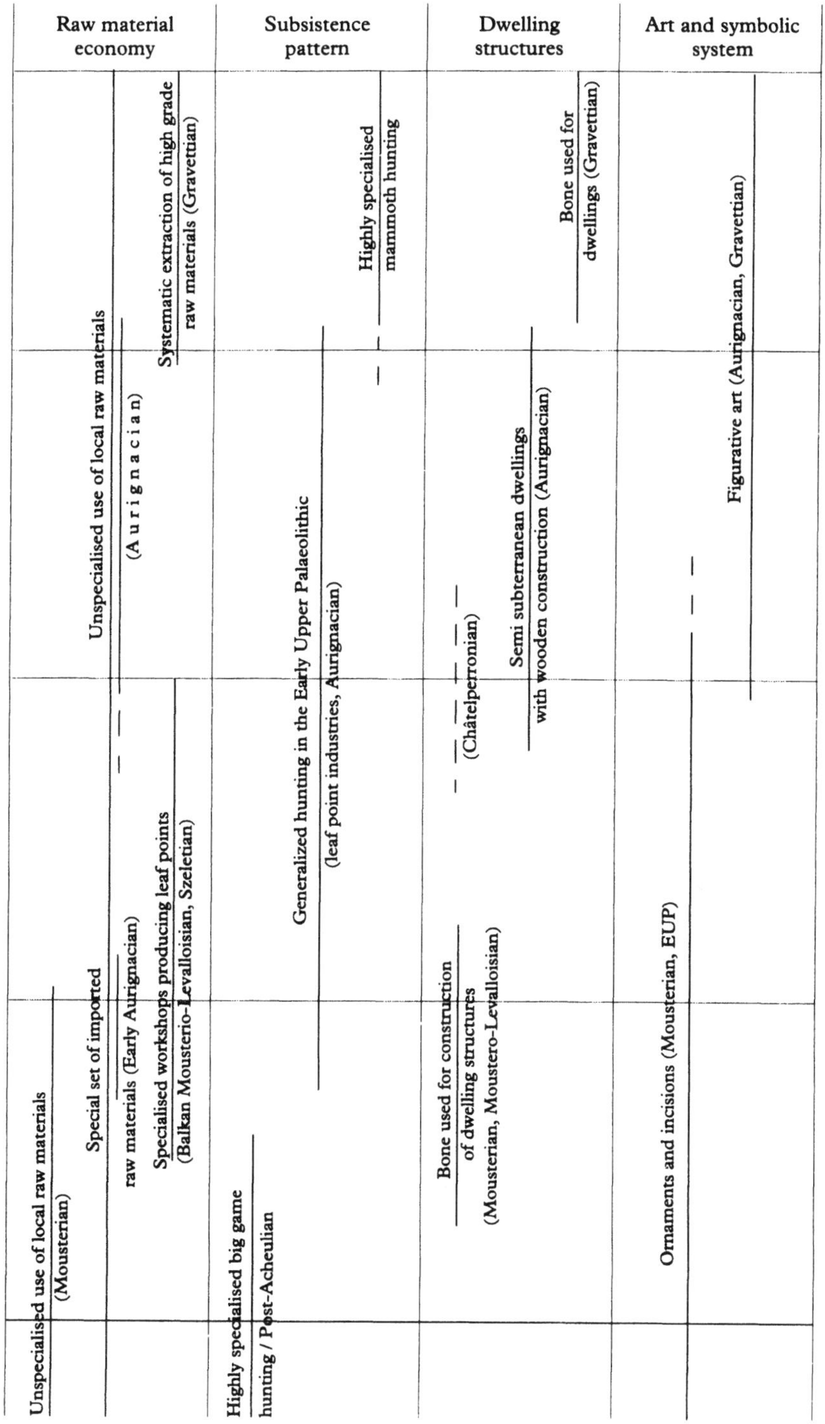

Raw material economy	Subsistence pattern	Dwelling structures	Art and symbolic system
Unspecialised use of local raw materials (Mousterian)	Highly specialised big game hunting / Post-Acheulian	Bone used for construction of dwelling structures (Mousterian, Moustero-Levalloisian)	Ornaments and incisions (Mousterian, EUP)
Special set of imported raw materials (Early Aurignacian)	Generalized hunting in the Early Upper Palaeolithic (leaf point industries, Aurignacian)	(Châtelperronian)	Figurative art (Aurignacian, Gravettian)
Specialised workshops producing leaf points (Balkan Mousterio-Levalloisian, Szeletian)	Highly specialised mammoth hunting	Semi subterranean dwellings with wooden construction (Aurignacian)	
Unspecialised use of local raw materials (A u r i g n a c i a n)		Bone used for dwellings (Gravettian)	
Systematic extraction of high grade raw materials (Gravettian)			

Aurignacian' is a risky and poorly-grounded hypothesis. The genesis of the Aurignacian set of diagnostic tools (which in its complete form first appears in the Balkans) still remains unexplained (Kozlowski 1976).

Viewed in terms of these 'morpho-technological' features, we can recognize three basic types of assemblages in the earliest phases of the European Upper Palaeolithic:

1. Assemblages with a non-specific set of retouched tools. To these belong the Bohunician industries of Czechoslovakia and some lesser-known industries from Eastern European sites (e.g. those from layer 10a at Molodova V (Tchernich 1973) and layer 7 at Korman IV (Tchernich *et al.* 1977) (Figure 15.4). As a rule these industries employed Levalloisian technology.

2. Assemblages with one specific techno-morphological group (either leaf-points or backed points) derived from specific Middle Palaeolithic assemblages. Alongside this group are found non-specific Upper Palaeolithic tool groups (such as end-scrapers, burins and truncations) and a large number of forms deriving from the earlier Middle Palaeolithic traditions (such as side-scrapers, notched and denticulated tools). Both Middle Palaeolithic flake technology and Upper Palaeolithic blade technology are used in these industries. To this assemblage group are ascribed the Szeletian of Central Europe, leaf-point industries of Northwestern Europe, the Sungirian of Eastern Europe, the Châtelperronian of France and the Uluzzian of Italy.

3. Assemblages containing a wide range of specific tool forms made on blades of Upper Palaeolithic type – notably end-scrapers, carinated and nosed scrapers, retouched blades, pointed blades and strangulated blades. Assemblages of this type make up the 'Aurignacian' technocomplex.

BONE IMPLEMENTS

Bone tools in the Middle Palaeolithic are rare. Basically, they are only partially shaped specimens (e.g. perforators with polished points or polishers with partially shaped edges) or worked like lithic implements using normal flaking techniques. The earliest bone artefacts which were finished and polished over the whole surface were bone or ivory spearpoints – with either split or unsplit bases. Whether these tools occurred in Middle Palaeolithic assemblages is highly uncertain. Vértes (1958), for example, suggested that both split- and unsplit-based points do occur in association with leaf-points in the Jankovich Cave in Hungary (where the leaf-points were subsequently ascribed to the Middle Palaeolithic: Gabori-Csank 1983), but a review of stratigraphical conditions showed that these spearpoints could not be attributed to a single assemblage. Similarly, in the Dzerava Skala Cave in western Slovakia, Prošek (1951) claimed that unsplit points co-occur with leaf-points, but, as we have already said, Valoch has stressed that such a juxtaposition is questionable.

In view of the above it seems simplest to treat split- and unsplit-based points as exclusively Aurignacian elements which are not associated with other complexes of the early Upper Palaeolithic. The only exception is a single specimen of a split-based point found by Hillebrand (1928) in layer IV of the Szeleta Cave (= lower Szeletian). Although Saad (1929) expressed some reservations about this find, Allsworth-Jones (1986) suggested that "Hillebrand's original statement is correct, and there were leaf-points in the vicinity". Similarly Allsworth-Jones (1986) maintains that a fragment of a bone point (with a broken base) discovered in the Szeleta Cave in 1947 by Saad and Nemeskeri must have been associated with leaf-points. Although there remain doubts as to the co-occurrence in the same Szeletian assemblages of bone points with split bases or points of the Mladeč type together with Szeletian leaf-points, it should be noted that Allsworth-Jones is inclined to interpret such co-occurrence as a result of external (most probably Aurignacian) influences on the cultural tradition of the Szeletian.

There is no doubt that bone and ivory points constitute a diagnostic element for the Aurignacian, whereas their possible occurrence in the Szeletian is quite unique and above all not certain. Other assemblages of the early phase of the Upper Palaeolithic have not yielded bone artefacts.

SUBSISTENCE PATTERN

Binford (1981, 1985) has recently proposed that subsistence activities in the Middle Palaeolithic were focussed primarily on scavenging rather than on the deliberate hunting of big game. But there is abundant evidence to show not only the importance of big game hunting, but also advanced specialization in the hunting of particular species. The significance of hunting has been confirmed by discoveries of killing and butchering sites – as for example at Lehringen (Veil 1985), Skaratki (Chmielewski and Kubiak 1962) and Zwolen (Schild and Sulgostowska 1988). Clear specialization in hunting activities has been documented at sites where the faunal assemblages show a high percentage of remains of one particular species (e.g. Sukhaya Metchetka on the Volga: 78% of bison; Ilskaya on the Kuban: 70% of bison; Staroselje in the Crimea: 90% of *Equus hydruntinus*; Prolom 2: 53.5% of *Saiga tatarica*). The industries from these sites have been ascribed to the Post-Acheulian ('Eastern Micoquian') tradition.

It has been suggested that the hunters of the early phase of the Upper Palaeolithic specialized in the hunting of cave bear, both in the Szeletian and in the Aurignacian. Recently, however, Allsworth-Jones (1986) has convincingly shown that the attribution of these accumulations of cave bear bones to the hunting activities of early Upper Palaeolithic groups (e.g. Vörös 1984) is highly improbable. In addition, the example of Bacho Kiro shows that, with the exception of layer 8, in all the other layers concentrations of cave bear bones are found outside the boundaries of living floors. The same remarks apply to the huge accumulation of cave bear bones found in the Istálloskö Cave. Almost certainly the accumulations of these bones were due to the natural death of very young or very old animals in the dens inhabited by bears, between periods of human occupation.

If the anthropogenic origin of accumulations of cave bear bones is rejected, then a highly diverse picture is obtained of the faunal composition in all the early Upper Palaeolithic complexes. On most sites as many as 20 different species have been recorded, most of them big game. Specialized hunting came into prominence only in the later phases of the Aurignacian when horse and reindeer remains comprise the main bulk of remains in the faunal assemblages. Subsequently, specialization is a distinctive feature of faunal assemblages belonging to the Gravettian technocomplex.

RAW MATERIALS ECONOMY

Most Mousterian industries are characterized by a varied set of local lithic raw materials, and specialized stone-working 'workshops' of this period have only rarely been identified. The only exception to this rule are the industries employing the Levalloisian technique associated with leaf-points in the Balkans and on the Middle Danube. In these regions a number of separate 'workshops' specializing in the production of leaf-points have been identified – as for example at Muselievo in Bulgaria (Haesaerts and Sirakova 1979). On some sites leaf-points were produced from imported raw materials different from those used to manufacture the rest of the inventory (e.g. at Samuilitsa II in Bulgaria (Sirakov 1981) and in some of the Bohunician sites near Brno (Oliva 1981, 1984)).

A similar pattern has been noted in the Szeletian industries in the Bükk Mountains, where leaf-points are made mainly in felsitic porphyry from the Mexiko Valley, whereas other tools are made from a variety of local raw materials. The situation in Moravia has to be reviewed afresh since Valoch (1969), in his debate with the present author, challenged the claim that separate workshops operated in southern Moravia. More recently, a wide range of new evidence has supported the existence of an organized system of lithic raw material exploitation in the Moravian Szeletian (Svoboda 1981, 1983).

The raw material economy in the Aurignacian displays a good deal of variation which is dependent on both the specific functions of the sites and their relative age. On the oldest Central European sites of the Aurignacian (Bacho Kiro layer 11: Kozlowski 1982; Vedrovice II and I: Oliva 1984) one type of raw material predominates. This raw material was often derived from fairly distant deposits. In Bacho Kiro, 52.9% of flint was imported from a distance of over 120 km; the remaining eight flint types on that site were also derived from non-local deposits, although their exact location has not yet been established. On the sites of Vedrovice I and II almost 100% of the flint comes from fairly local deposits of hornstone of the type known as 'Krumlovski Les'. In the later phases of the Aurignacian raw materials are variable, but show a general dominance of local materials (Figure 15.5). Imported raw materials do occur in these sites but only in fractional percentages (Kozlowski 1972, 1973). Only small hunting camps depart from the above model, as they may contain a higher proportion of one type of raw material (Oliva 1984).

It is only during the Gravettian technocomplex in the middle phase of

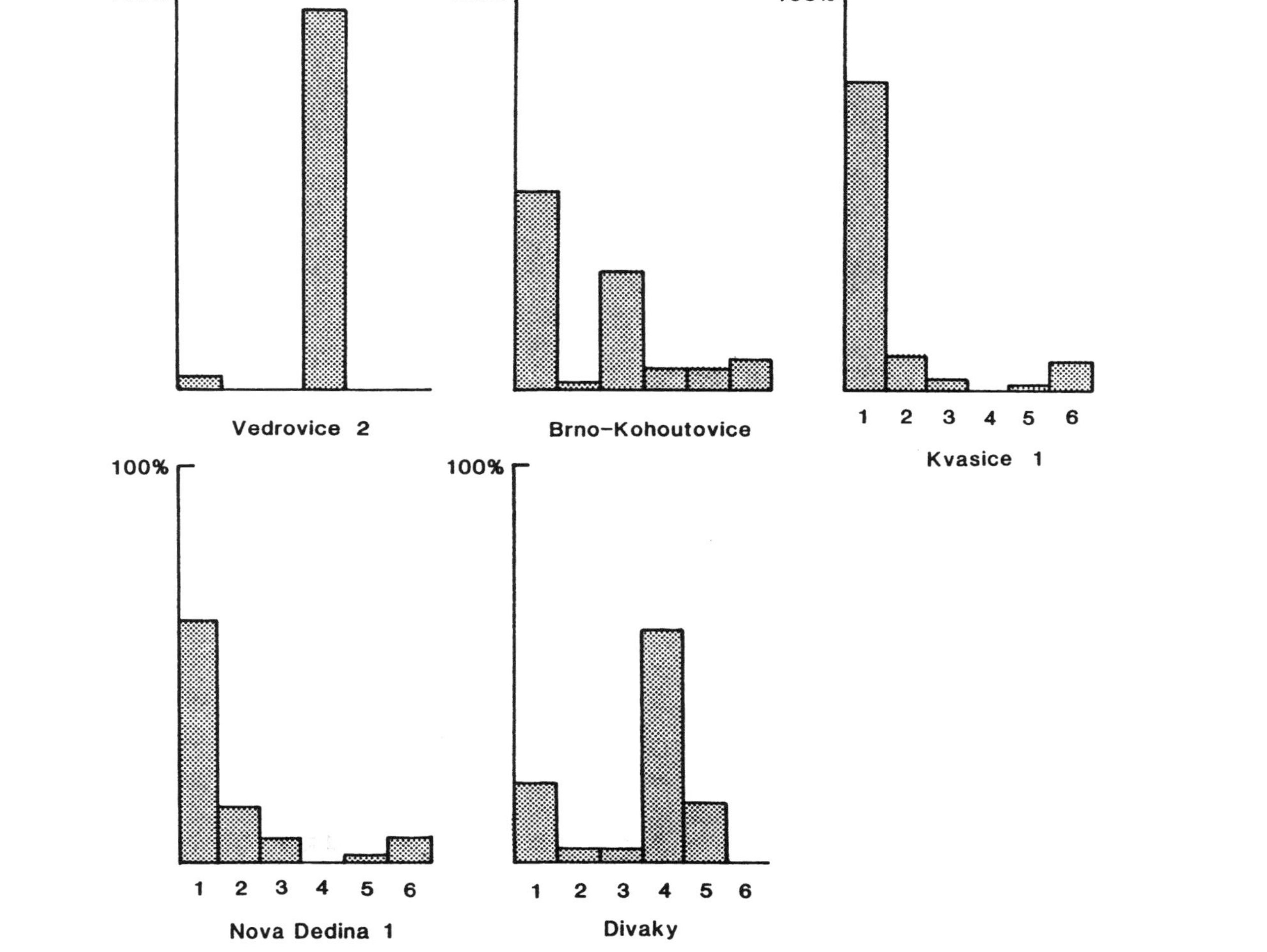

Figure 15.5. Raw materials used in Aurignacian sites in Moravia: 1. Jurassic and cretaceous flint; 2. radiolarite; 3. menilithic slate; 4. hornstone of 'Krumlovsky Les' type'; 5. cretaceous hornstone; 6. other hornstone.

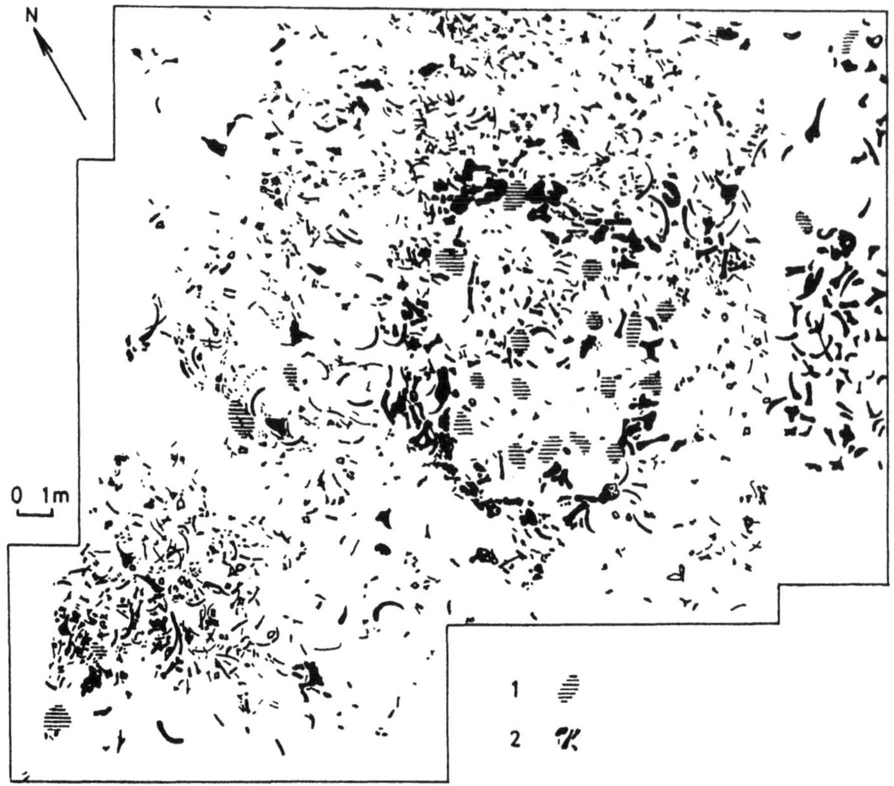

Figure 15.6. Middle Palaeolithic dwelling constructed from mammoth bones at Molodova site I, layer 2 (Russia). After O.P. Tchernich. 1. hearths; 2. bones.

the Upper Palaeolithic that another period of raw materials specialization sets in, with a stable system of procurement of high grade materials from distant deposits.

DWELLING STRUCTURES

As early as the Middle Palaeolithic we find dwelling structures built of bones (Figure 15.6). Dwellings of this kind are known from both open sites (Molodova 1 on the Dnester (Tchernich 1982) and Ripiceni Izvor in Romania (Paunescu 1965) both with circular structures of mammoth bones) and from cave sites (notably an enclosure built of reindeer antler in the Raj Cave in Poland: Kozlowski *et al.* 1972).

In the early phases of the Upper Palaeolithic in Central and Eastern Europe dwelling structures are unknown. In the leaf-point complex only one such feature is known. This is an oval structure, slightly sunk into the ground and with a foundation of large stone blocks, from the site of Dzierzyslaw 1 in Upper Silesia (Kozlowski 1964).

From the Aurignacian complex a number of habitation features are

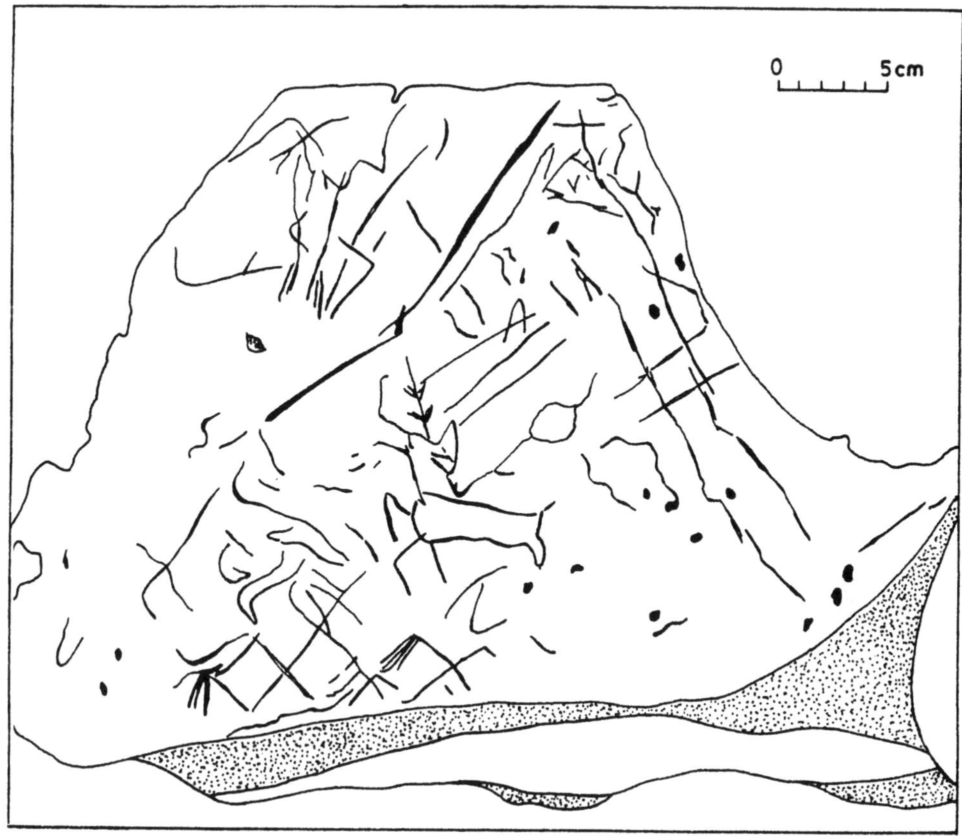

Figure 15.7. Engraving on mammoth scapula from Molodova site I, layer 2. After O.P. Tchernich.

known in both Slovakia (Barca, Tibava: Banesz 1969) and Moravia (Gottvaldov: Klima 1955). These are all semi-subterranean dwellings of the semi-dug-out type whose outlines vary. In the case of some of these features (notably Barca 1: Prošek 1951) there is some doubt as to whether they were indeed intentional structures or whether they represent simply the effects of erosional processes or other periglacial structures.

Widespread use of bones as building material is recorded only in the middle phase of the Upper Palaeolithic, commencing with the Gravettian technocomplex.

ART AND SYMBOLIC SYSTEM

On a number of Middle Palaeolithic sites objects have been uncovered, or bones with incisions, which convey symbolic information or represent notation systems. In addition, a number of drilled teeth have been reported from Mousterian sites (Marshack 1976; Stoliar 1971, 1972, 1981) which appear to reflect some form of symbolic content expressed through aesthetic preferences. As yet, however, we have no fully documented figurative art from Middle Palaeolithic sites, with the possible exception of a rather doubtful engraved representation of a deer from layer 2 at Molodova 1 on the Dnestr (Tchernich 1978) (see Figure 15.7).

Evidence of objects endowed with symbolic meaning, together with animal-tooth ornaments, occur in the early Upper Palaeolithic assemblages with leaf-points as well as in assemblages with backed points. These industries however have yielded only slight evidence of figurative art. The only exceptions are the youngest assemblages of the Moldavian Szeletian (the Bryndzeny Cave) and the late phase of the Sungirian (at Sungir itself). The Bryndzeny Cave yielded a rough schematic human figure (Ketraru 1971), whereas from Sungir there is a schematic representation of a horse (Bader *et al.* 1966). These finds, however, are very late in the early Upper Palaeolithic sequence (dating from the end of the Interpleniglacial) and are most probably contemporaneous with the Gravettian technocomplex.

The earliest examples of fully figurative art appear in the Aurignacian technocomplex, initially in Germany (Vogelherd, Geissenklösterle) and in France ('style 1' according to Leroi-Gourhan). Examples of figurative mobiliary art (such as anthropomorphic and zoomorphic figurines from southern Germany) show a high level of artistic workmanship. Indeed, these figurines resemble the art of the Gravettian technocomplex, although they are earlier by 10 000 years. The rock art of the western Aurignacian, on the other hand, is still relatively crude.

SPATIAL LIMITS OF ARCHAEOLOGICAL UNITS AND SETTLEMENT PATTERNS

Towards the end of the Middle Palaeolithic (most probably due to the increased latitudinal contrasts in ecological zones in Europe) we can observe a clear division of archaeological units into distinct environmental zones. In the south (comprising the Balkans and the middle Danube Basin) Moustero-Levalloisian industries predominate (either with or without leaf-points), whereas in the north (including the western Carpathians and Central European Uplands) we find almost exclusively industries with Acheulian-Micoquian tradition, with only sporadic occurrences of leaf-points. This situation is markedly different than that recorded earlier in the Middle Palaeolithic, when different Mousterian facies were not restricted to specific ecological provinces but occurred simultaneously in the same regions.

In the early phase of the Upper Palaeolithic the distribution of differentiated cultural units is basically exclusive, at least for the industries with leaf- and backed points. It seems that this pattern of taxonomic differentiation reflects the territorial differentiation of the preceding Middle Palaeolithic units whose continuation can be traced directly into the succeeding Szeletian, Châtelperronian and Uluzzian industries. In addition, these territorial divisions were reinforced by ecological zones. In particular, it would seem that the settlement of the northern Lowland zone during the main 'Interpleniglacial' phase created a very specific complex of

'Jerzmanowician-Lincombian-Ranisian' assemblages.

In sharp contrast to the preceding industries, the distribution of the Aurignacian shows an almost pan-European range, and was uniform over the whole territory of its distribution. Nevertheless, we are far from interpreting this phenomenon in the same way as some Mousterian facies which indeed spread as far as Asia – as for example the Charentian or the Denticulate Mousterian. Whereas in the Middle Palaeolithic the convergence in the typological structure of assemblages was determined primarily by functional factors, in the case of the Aurignacian the similarities are almost certainly a consequence of direct phyletic affiliation. Viewing the distribution of Upper Palaeolithic industries as a whole it seems that the close similarities of lithic tool kits are the result of phyletic ties rather than functional determinants or other factors such as the 'Frison' effect in the progressive resharpening of tools. In the Middle Palaeolithic it would seem that functional links predominated. It was only in the terminal stages of the Middle Palaeolithic that environmental contrasts augmented territorial patterns of differentiation between assemblages, whose typological structure then stabilized as an element of cultural tradition.

CONCLUSIONS

The preceding 'multiaspectual' analysis of the Middle-Upper Palaeolithic transition indicates that different cultural innovations did not appear simultaneously, and that the pattern of this transformation was different within the framework of each particular taxonomic unit (Table 15.1). In the Middle Palaeolithic these units can be explained either in functional terms, or as the expression of different reductional strategies in the use of tools. In the Upper Palaeolithic the main taxonomic units correspond to territorially well-delimited cultural groups. Recent data suggest that the emergence of blade technology (i.e. 'leptolithization') was not a uniform trend; different aspects of the archaeological record indicate two fundamental, and frequently opposed, adaptational strategies: the first leading from the Mousterian, through the early Upper Palaeolithic 'leaf-point' complexes, to the Gravettian; and the second representing an entirely distinct (and probably intrusive) element in the form of the Aurignacian.

REFERENCES

Allsworth-Jones, P. 1986. The Szeletian: main trends, recent results, and problems for resolution. In M. Day, R. Foley and Wu Rukang (eds) *The Pleistocene Perspective.* Precirculated papers of the World Archaeological Congress, Southampton, 1986.

Banesz, L. 1969. Paleolitické obydlia vychodného Slovenska a ich vyznam pre poznanie pervobytnej společ nosti v Evrope. *Historica Carpathica* 1: 5-66.

Boëda, E. 1988. Le concept laminaire: rupture et filiation avec le concept Levallois. In M. Otte (ed.) *L'Homme de Néanderthal.* Vol. 8: *La Mutation.* Liège: Etudes et Recherches Archéologiques de l'Université de Liège (ERAUL 35): 41-59.

Bosinski, G. 1973. Der Paläolithische Fundplatz Rheindalen. In H.J. Müller-Beck (ed.) *Neue paläolithische und mesolithische Ausgrabungen*

in der Bundesrepublik Deutschlands. Tübingen: Proceedings of IX INQUA Congress: 11-14.
Bosinski, G. 1981. Découvertes récentes de Paléolithique inférieur et moyen en Allemagne du Nord-Ouest. *Notae Praehistoricae* 1: 100-102
Bosinski, G., Brunacker K., Schütrumpf R. and Rottländer R. 1966. Der paläolithische Fundplatz Rheindalen. *Bonner Jahrbuch* 166: 318-360.
Cahen, D. 1984. Paléolithique inférieur et moyen en Belgique. In D. Cahen and P. Haesaerts (eds) *Peuples Chasseurs de la Belgique Préhistorique dans leur Cadre Naturel.* Brussels: Institut Royal des Sciences Naturelles de Belgique: 133-156.
Chmielewski, W. and Kubiak, H. 1962. The find of mammoth bones at Skaratki. *Folia Quaternaria* 9: 1-29.
Gábori-Csánk, V. 1983. La grotte Remete 'Fölsö' et le 'Szélétien de Transdanubie'. *Acta Archaeologica Academiae Scientiarum Hungaricae* 35: 249-285.
Haesaerts, P. and Sirakova, S. 1979. Le Paléolithique moyen à pointes foliacées de Mousselievo (Bulgarie). In J.K. Kozlowski (ed.) *Middle and Early Upper Palaeolithic in Balkans*. Krakow: Panstwowe Wydawnictwo Naukowe: 35-63.
Hillebrand, E. 1928. Uber eine neue Aurignacien-Lanzenspitze 'à base fendue' aus dem ungarländischen Paläolithikum. *Eiszeit und Urgeschichte* 5: 99-102.
Jelinek, A. 1982. The Tabūn Cave and Palaeolithic Man in the Levant *Science* 216: 1369-1375.
Klima, B. 1955. Nova paleoliticka stanica v Gottvaldove-Loukach. *Anthropozoikum* 5: 425-437.
Kozlowski, J.K. 1964. *Paleolit na Gornym Slasku.* Wroclaw: Ossolinem.
Kozlowski, J.K. 1972-1973. The origin of lithic raw materials used in the Palaeolithic of the Carpathian countries. *Acta Archaeologica Carpathica* 13: 5-19.
Kozlowski, J.K. 1976. L' Aurignacien dans les Balkans. In J.K. Kozlowski (ed.) *L' Aurignacien en Europe*. Etudes et Recherches Archéologiques de l'Université de Liège (ERAUL 13): 273-284.
Kozlowski, J.K. *et al.* 1972. Studies on Raj Cave near Kielce (Poland) and its deposits. *Folia Quaternaria* 41: 1-148.
Kozlowski, J.K. (ed.) 1982. *Excavation in the Bacho Kiro* Cave. Warszawa: Panstwowe Wydawnictwo Naukowe.
Marshack, A. 1976. Some implications of the Palaeolithic symbolic evidence for the origin of language. *Current Anthropology* 17: 274-282.
Morawski, W. 1976. Middle Palaeolithic flint assemblages from Piekary IIa site. *Swiatowit* 34: 139-146.
Oliva, M. 1981. Die Bohunicien-Station bei Podoli (Bez. Brno-Land) und ihre Stellung im begiwenden Jungpaläolithikum. *Časopis Moravského Muzea* 66: 7-45.
Oliva, M. 1983. Le Bohunicien, un nouveau groupe culturel en Moravie: quelques aspects psychotechniques du développement des industries paléolithiques. *L'Anthropologie* 88: 209-220.
Oliva, M. 1984. Technologie vyroby a použite suroviny štipane industrie moravského aurignacienu. *Archeologické Rozhledy* 36: 601-628.
Păunescu, A. 1965. Sur la succession des habitats paléolithiques et post-paléolithiques de Ripiceni-Izvor. *Dacia* 9: 1-32.
Prošek, F. 1951. Vyzkum jeskyne Dzerave skale v roce 1950. *Archeologické Rozhledy* 3: 293-298.

Rust, A. 1951. *Die Höhlenfunde von Jabrud (Syrien)*. Neumünster: K. Wachholtz.

Saad, A. 1929. A Bükk-hegyseben vegzett ujabb kutatasok eredmenyei. *Archaeologiai Ertesitö* 43: 237-247.

Schild, R. and Sulgostowska Z. 1988. The Middle Palaeolithic of the North European Plain at Zwolen. In M. Otte (ed.) *L'Homme de Néandertal*. Etudes et Recherches Archéologiques de l'Université de Liège (ERAUL).

Sirakov, N. 1981. Reconstruction of the Middle Palaeolithic flint assemblages from the Cave Samuilitsa II (northern Bulgaria) and their taxonomic position seen against the Palaeolithic of South-Eastern Europe. *Folia Quaternaria* 55: 1-100.

Svoboda, J. 1981. *Křemencová industrie z Ondratic. K problému počátku` mladého paleolitu*. Studie Archeologického Ustavu, Prague: Czechoslovakian Academy of Sciences IX (1).

Svoboda, J. 1983. Raw material sources in the early Upper Palaeolithic Moravia: the concept of the lithic exploitation area. *L'Anthropologie* 21: 147-158.

Stoliar, A.D. 1971. 'Naturalnoye tvortchestvo' neandertaltsev kak osnova genezisa isskustva. In R.S. Vasilevski (ed.) *Pervobytnoye Isskustvo*. Novosibirsk: Nauka.

Stoliar, A.D. 1972. O genezise izobrazitelnoy deyatelnosti i ee roli v stanovlenii soznanya. In D.A. Maksimova (ed.) *Rannye Formy Isskustva*. Moskva: Nauka.

Stoliar, A.D. 1981. Proiskhozhdenie Isskustva. Moskva: Nauka.

Tuffreau, A. 1979. Recherches récentes sur le Paléolithique inférieur et moyen de la France septentrionale. *Bulletin de la Société Royale Belge d'Anthropologie et de la Préhistoire* 90: 161-177.

Tchernich, A.P. 1973. *Paleolit i Mezolit Predniestrovia*. Moskva: Nauka.

Tchernich, A.P. (ed.) 1977. *Mnogosloynaya paleoliticheskaya stoyanka Korman IV i ee mesto v paleolite*. Moskva: Nauka.

Tchernich, A.P. 1978. O vremeni vozniknovenia paleoliticheskogo isskustva, v sviazi s issledovaniami 1976 stoyanki Molodova I. In *U istokov isskustva*. Moskva: Nauka.

Tchernich, A.P. (ed.) 1982. *Molodova I – unikalnoye musterskoye poselenie na Sredniem Dniestrie*. Moskva: Nauka.

Valoch, K. 1973. Neslovice – eine bedeutende Oberflächenfundstelle des Szeletiens in Mähren. *Casopis Moravského Muzea* 58: 5-76.

Valoch K. *et al.* 1976. Die altsteinzeitliche Fundstelle in Brno-Bohunice. Studie Archeologického Ustavu, Brno: Czechoslovakian Academy of Sciences IV (1).

Veil, S. 1985. Neue Untersuchungen zum eemzeitlichen Elefanten –Jagdplatz Lehringen Ldkr. Verden. *Die Kunde* 36: 11-58.

Vértes, L. 1965. *Az Oskör es az Atmeneti Kökör Emekei Magyarorszagon*. Budapest: Akademiai Kiado.

Vörös, I. 1984. Hunted mammals from the Aurignacian cave bear hunters' site in the Istálloskö cave. *Folia Archaeologica* 35: 7-28.

16: From the Middle to the Upper Palaeolithic: The Nature of the Transition

MARCEL OTTE

INTRODUCTION

In recent years a number of discoveries have been made which throw significant new light on the character of the transition from the Middle to the Upper Palaeolithic in Western Europe. As a prelude to a more general survey of this question, it may be useful to review, briefly, some of the evidence recovered recently from Belgian sites.

The village of Rocourt, close to Liège, has given its name to an interglacial soil contemporary with the Eemian, previously reported by Gullentops (1954). More recently, research by Haesaerts has brought to light a blade industry situated in deposits stratified at the end of this interglacial, reliably placed between St. Germain phases I and II (Haesaerts *et al.* 1981). The technology is clearly oriented towards the production of long, regular blades. The pattern of core-preparation, however, is still the same as in the Levallois technique. Facetted striking platforms and pronounced bulbs of percussion indicate the use of a hard hammer, as in the Levallois (E. Boëda, personal communication; Boëda 1986). The tool assemblage, likewise made on blades, includes only tools of an Upper Palaeolithic type – i.e. burins and backed pieces (Figure 16.1).

Interestingly, this technical facies has been found in a number of other open-air sites in Northern Europe, in each case appearing in the same chronostratigraphic position: Seclin in the north of France (Tuffreau *et al.* 1985; Leroi-Gourhan 1986), Rheindhalen in West Germany (Bosinski 1974) and Piekary in Poland (Morawski 1971).

Three aspects of these industries are of particular relevance to the present discussion:

1. This technological group should be added to the existing technological 'facies' already described by François Bordes and his students, further emphasising the regional variability of traditions within the Middle Palaeolithic.

2. Techniques of blade production were fully developed at this period, and were fully used wherever the need was appropriate (i.e. as a means of hafting, to achieve economy in the use of raw materials or to produce specialised tool forms).

3. The technique of blade production itself cannot serve as an adequate

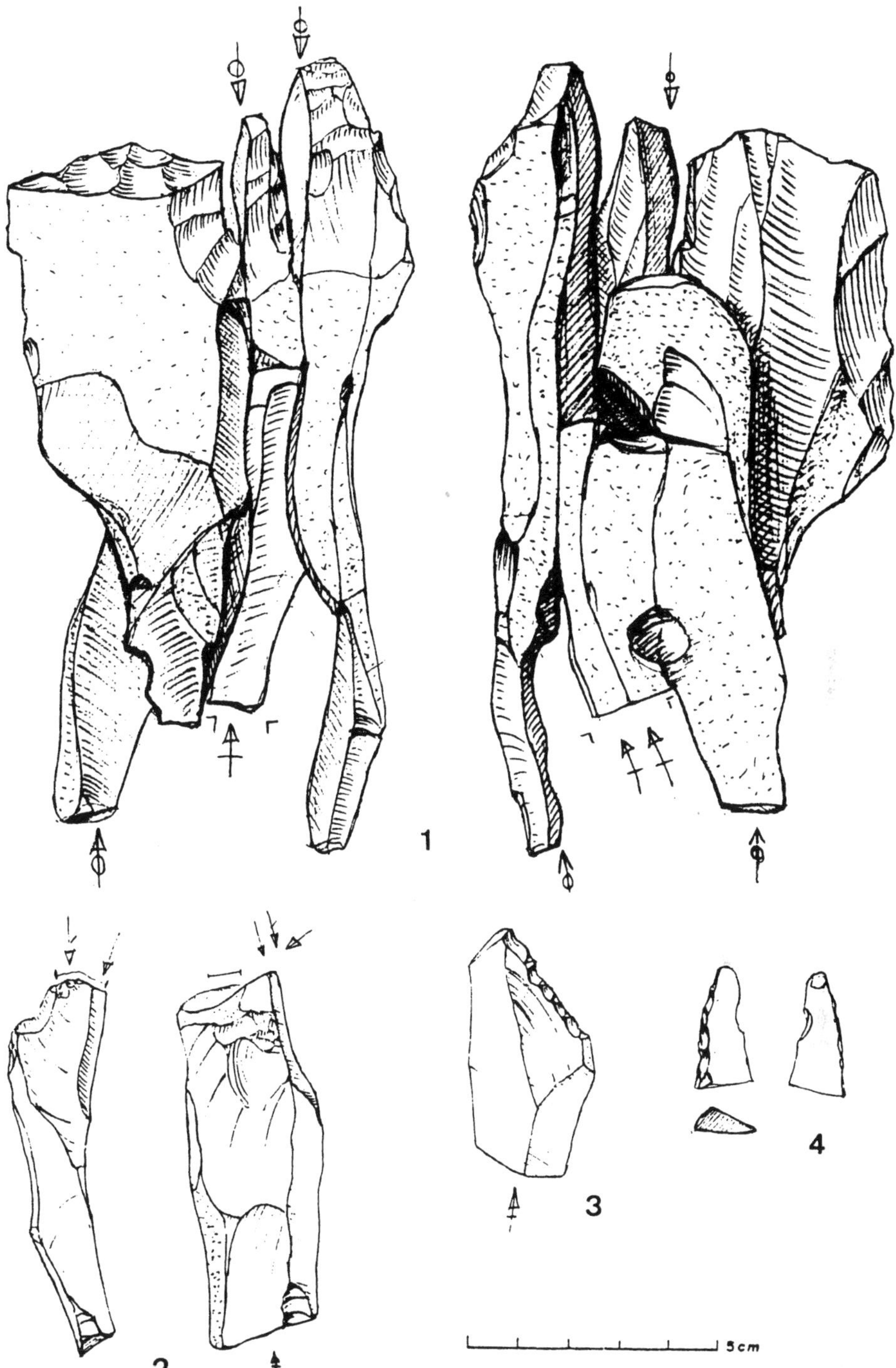

Figure 16.1. Lithic industry from Rocourt (near Liège, Belgium): 1. refitted flakes, showing blades produced by means of Levallois technique; 2-4. Upper Palaeolithic tool forms (burin, partially backed blade, and curved-back form respectively). The industry occurred in loess deposits attributed to the beginning of the last glaciation (excavation by P. Haesaerts; refitting by J. de Heinzelin; study by M. Otte).

criterion for the definition of the 'Upper Palaeolithic'; rather it is the means by which the blades were obtained, which, from the Châtelperronian onwards, seems to have involved the use of a soft hammer (Pèlegrin 1988).

THE COMPLEXITY OF MOUSTERIAN BEHAVIOUR

This ability to produce blades, apparent at Rocourt, can be added to those already documented from other Middle Palaeolithic sites under the guise of variations of the Levallois technique. It is now known that stone tools were on occasions hafted in the Mousterian, thus leading to a modification in their working angle, their manner of operation and the force involved in their use (Beyries 1987; Anderson-Gerfaud, this volume). Mousterian tools therefore reflect two distinct systems of fabrication: the one applied to the lithic material itself and the other to the organic material for the haft. This form of complexity above, therefore, cannot be used as distinguishing criterion for the Upper Palaeolithic.

Hunting methods show a similar degree of elaboration. Altuna (1988) has demonstrated clear adaptations in hunting procedures for the Cantabrian Mousterian. Chase (1988) has documented a change in the species hunted as well as the body parts brought back to the site for the Middle Palaeolithic of southwest France. Marylène Patou has noted a preferential selection towards the use of deer at Lazaret (France) and Chamois at Sclayn (Belgium) (Patou 1984). The site of Zwollen in Poland (Schild and Gautier 1988) shows a specialisation towards bison and horse. For a long time Soviet researchers have stressed the importance of ibex hunting in the Caucuses. Even though the species vary between sites, these specialisations demonstrate the complexity of economic adaptations in the Middle Palaeolithic.

Turning to the economics of raw material use, it has become clear recently that not all types of rock were manipulated by means of the same techniques; the behaviour of the knapper varied according to the relative accessibility of the available raw materials, or their particular mechanical properties. The successive stages of lithic reduction sequences are, therefore, differently represented according to the nature of the site occupied, and the source and type of raw material employed (Geneste 1985). Techniques were determined in relation to these requirements within a sort of equilibrium, balanced between the social organisation of the group, practical economics, environmental factors, and technical ability. In this respect, patterns of technology were no less elaborate than in the Upper Palaeolithic (Demars 1982) (Figure 16.2, 16.3).

Hearth deposits found in Normandy (Fosse 1987) and at the Kebara Cave in Israel (Bar Yosef *et al.* 1987) – amongst other sites – suggest that the use of space was structured from the Middle Palaeolithic onwards. Exotic items are known from Mousterian sites in southwest France (Sonneville-Bordes 1987), and a hut constructed from mammoth bones has been discovered at Molodova in Soviet Moldavia (Chernysh 1961). Ritual and symbolic behaviour seems to have been equally structured in the

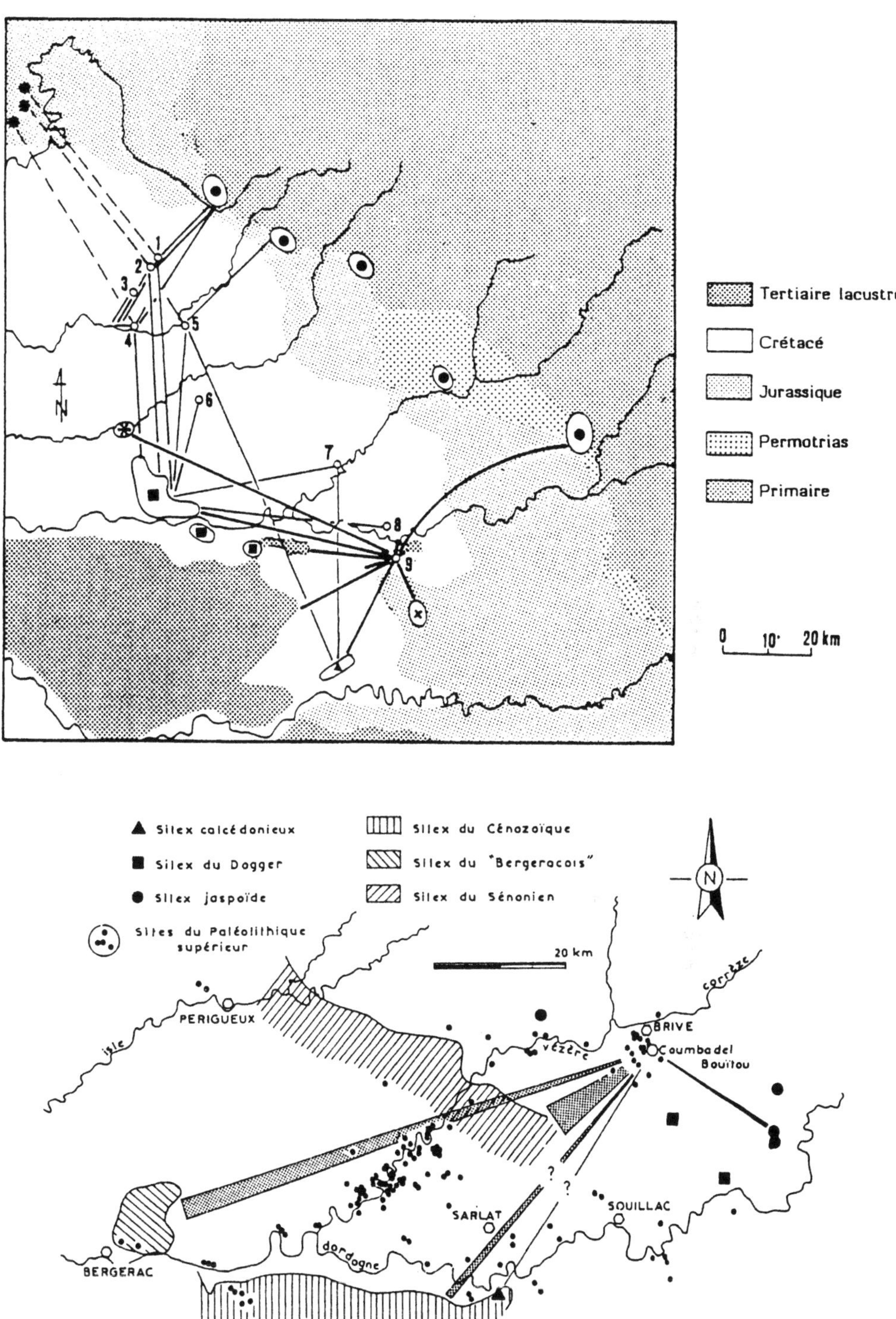

Figure 16.2. Comparison of mobility networks in the Middle Palaeolithic (upper, after Geneste 1985) and Upper Palaeolithic (lower, after DeMars 1982) based on a study of raw material sources in southwest France. From this perspective, behaviour in both periods shows equally complex patterns.

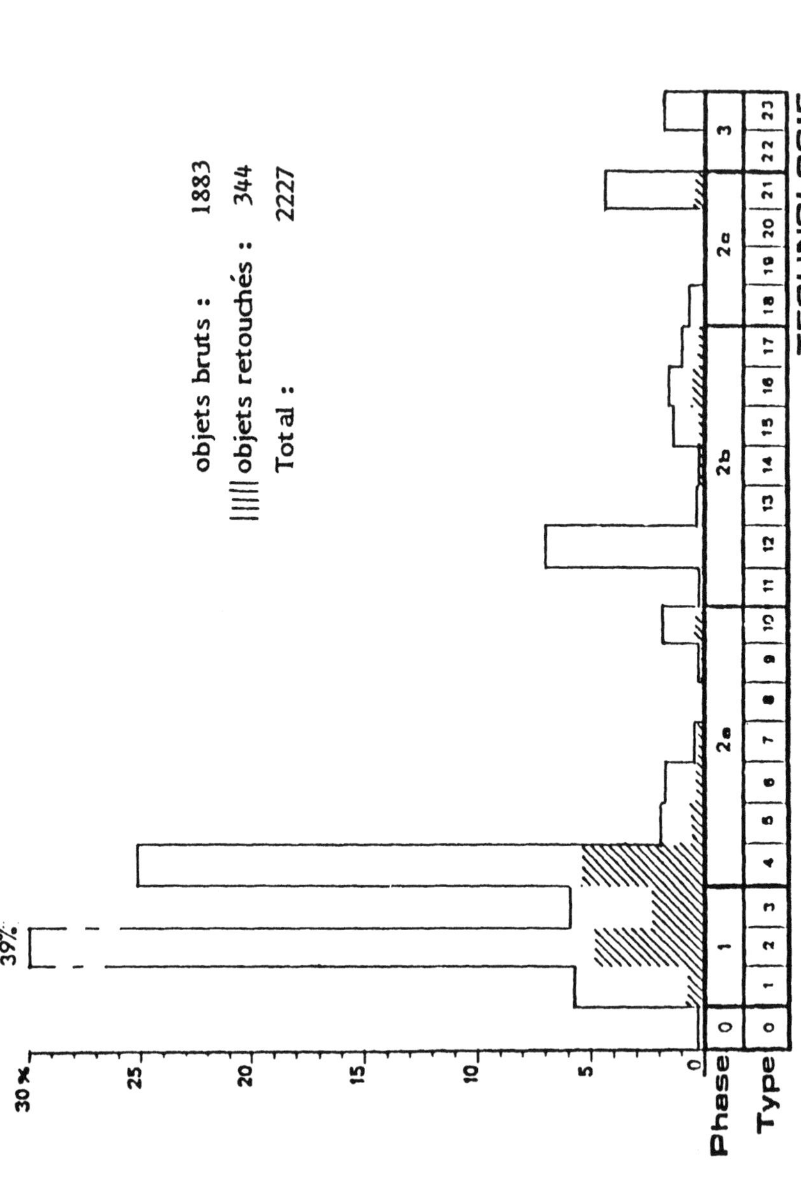

Figure 16.3. Relative proportions of different stages in the reduction sequence of lithic artefacts in the Denticulate Mousterian industry of Le Roc (Dordogne), showing representation of retouched (shaded) and unretouched (non-shaded) artifacts. After Geneste (1985).

Middle Palaeolithic, as witnessed by the burials and various transported items such as shells, fossils and minerals.

The organisation of behaviour appears most clearly when comparing the industrial groups defined by François Bordes. Their significance is probably complex and varied, but it is clear that *tradition* must have played a determining role because the facies do not have the same geographic distribution when viewed from the perspective of Europe. The functional interpretations of Binford cannot explain all the variation; if so they would imply that in some whole regions people only scraped skins whilst in others they hunted bear! The dimension of time, suggested by Mellars (1988), clearly requires careful consideration, along with the input of raw materials (Geneste 1985), seasonality patterns, ecology and the length of occupation (Rolland 1981).

All these factors must have influenced each other in a complex manner, leaving us with assemblages which are clearly structured but in a particular form which remains to be explored.

Thus it seems that the *potential* existed within the Mousterian itself for the attainment and 'creation' of the Upper Palaeolithic. But during the hundreds of millenia over which the Middle Palaeolithic was in existence, neither any stimulus nor any new development provoked this transition. As in many cases in the behavioural sciences, the evolutionary and the adaptive potentials existed, but were never fully exploited (Straus, this volume).

A DUAL ASPECT

Two apparently contradictory aspects of Mousterian behaviour can be identified in the available archaeological records:

1. Unstructured Features

These are revealed, for example, in the typology of Mousterian tools, where the tool types are not clearly defined and where one type seems to grade almost imperceptibly into another (Dibble 1988). Seemingly, there is no clear tool standardisation in the Mousterian. Reccurrent forms do not appear to correspond to a final stage in the reduction sequence (as in the Upper Palaeolithic) but rather to represent various stages of discard, during repeated phases of reworking or resharpening of the tools (Cahen 1985).

This aspect can also be seen in the functional analysis of stone tools, where there exists no simple correlation between morphology and use (Beyries 1987). In a similar manner to the fabrication of the tools themselves, their utilisation appears to be opportunistic, related to the availability of raw materials and to the form of objects, more than to any predetermined or preconceived criteria.

The distributions of activities within the areas of living space seem in general poorly defined: at one time particular activities seem to occur *within* the domestic unit and at other times between these units of occupation. Analytical methods perhaps still do not allow these differences to be made clear (Whallon 1989), or rather the migratory character of Mousterian occupation may have prevented the regular separation of activities on repeatedly occupied sites.

2. *Structured Features*

In contrast to the features noted above, other aspects of the Middle Palaeolithic demonstrate the existence of behavioural patterns which appear to have been determined by tradition or by cultural rules. The coherence and repetition of the different typological 'facies' of the Mousterian, is the clearest illustration of this, in that they show prescribed areas of geographical distribution. In addition, the technical operations applied to stone, are in themselves, structured. In their manner of reduction and the creation of systematic forms, they provide incontestable evidence of educationally-transmitted behaviour. Further reflections of this coherence can be seen of course in ritual practices (burial customs) and in the social determination of hunting methods (see above).

These two aspects of both 'structured' and 'unstructured' behaviour reveal the adaptive potential of Mousterian culture, at times approaching that of recent hunter-gatherers in their *ability*, although radically different in their *realisation*. No doubt this explains the absence, in modern primitive societies, of any state equivalent to this, either in the technical domain or in the demographic and social spheres (Whallon 1989).

THE TRANSITION TO THE RECENT PERIOD

In addition to this duality of behaviour, one can recognise during the transition from the Middle to the Upper Palaeolithic features indicating both discontinuities (i.e. 'rupture') and continuity (i.e. 'transition').

1. *Phenomena of 'rupture'*

Phenomena which show a marked change over the period of the Middle-Upper Palaeolithic transition include the development of toolkits in bone, the modification of techniques for the production of lithic blanks, the standardisation of toolkits, the development of the systems of fabrication with the production of tools destined for the fabrication of other tools (burins, perforators), and the multiplication of suspended objects such as pendants for displaying individual and group identity (White 1989).

2. *Phenomena of 'transition'*

In this category one can include the survival of Mousterian toolkits documented in the Châtelperronian levels at Arcy-sur-Cure (utilising the same raw materials as in the earlier periods: Farizy and Schmider 1985); the marked change of game animals documented in the *middle* of the Upper Palaeolithic, which does not coincide with the end of the Mousterian (Simek and Snyder 1988); and the continuity of symbolic activities marked by burials and curious, transported objects. There is also a sense of continuity with regard to the elaborate treatment of raw materials. The structure of living areas within the early Upper Palaeolithic appears to

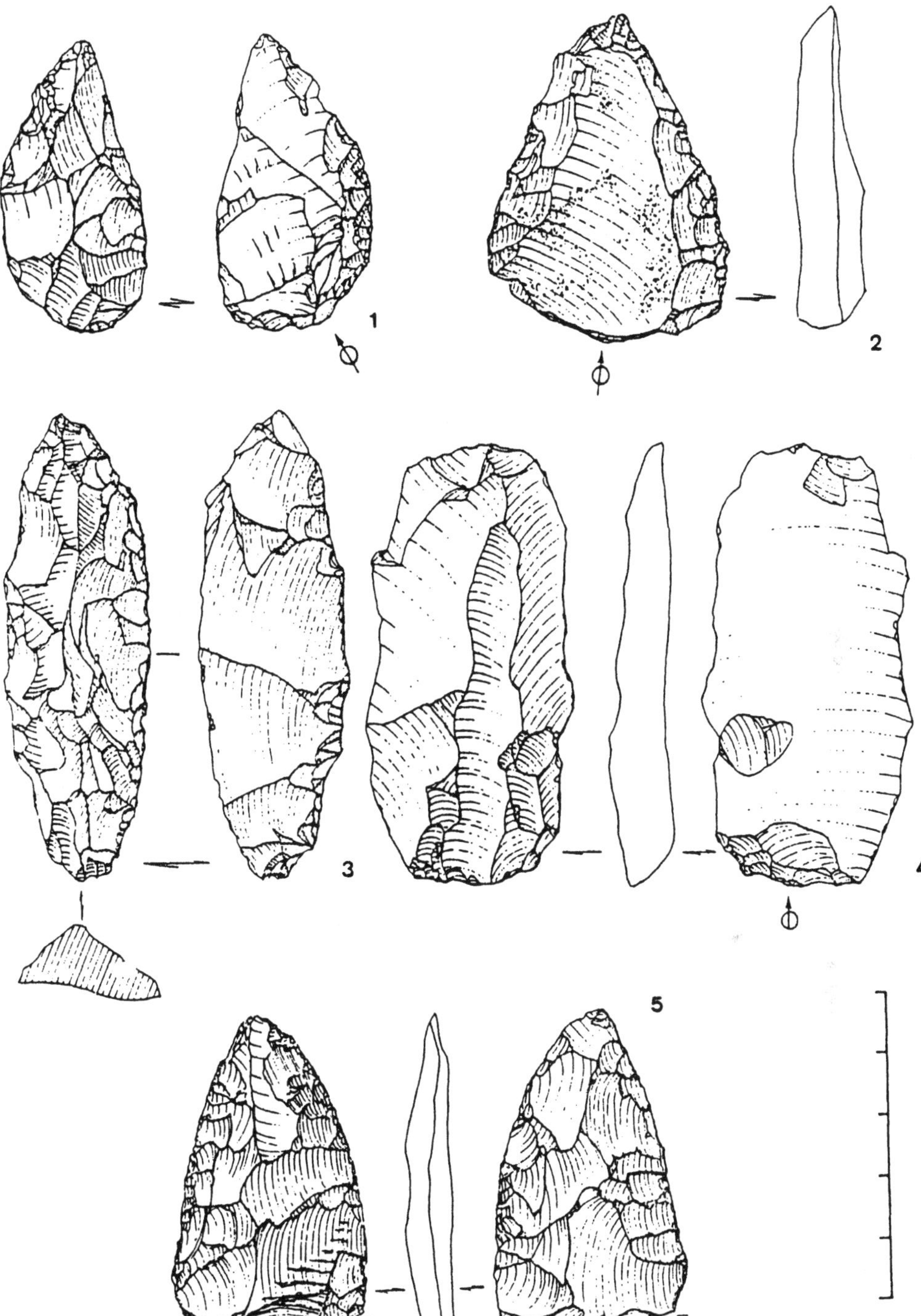

Figure 16.4. Lithic industry from Couvin (Belgium), *c.* 45 000 BP:
1. asymmetrical bifacial racloir; 2. Mousterian point; 3. leaf-point with triangular section; 4. 'Kostienki' type knife (blade with base thinned by inverse truncation, and long removals from the upper face); 5. flat, bifacial leaf-point. After Cattelain and Otte (1985).

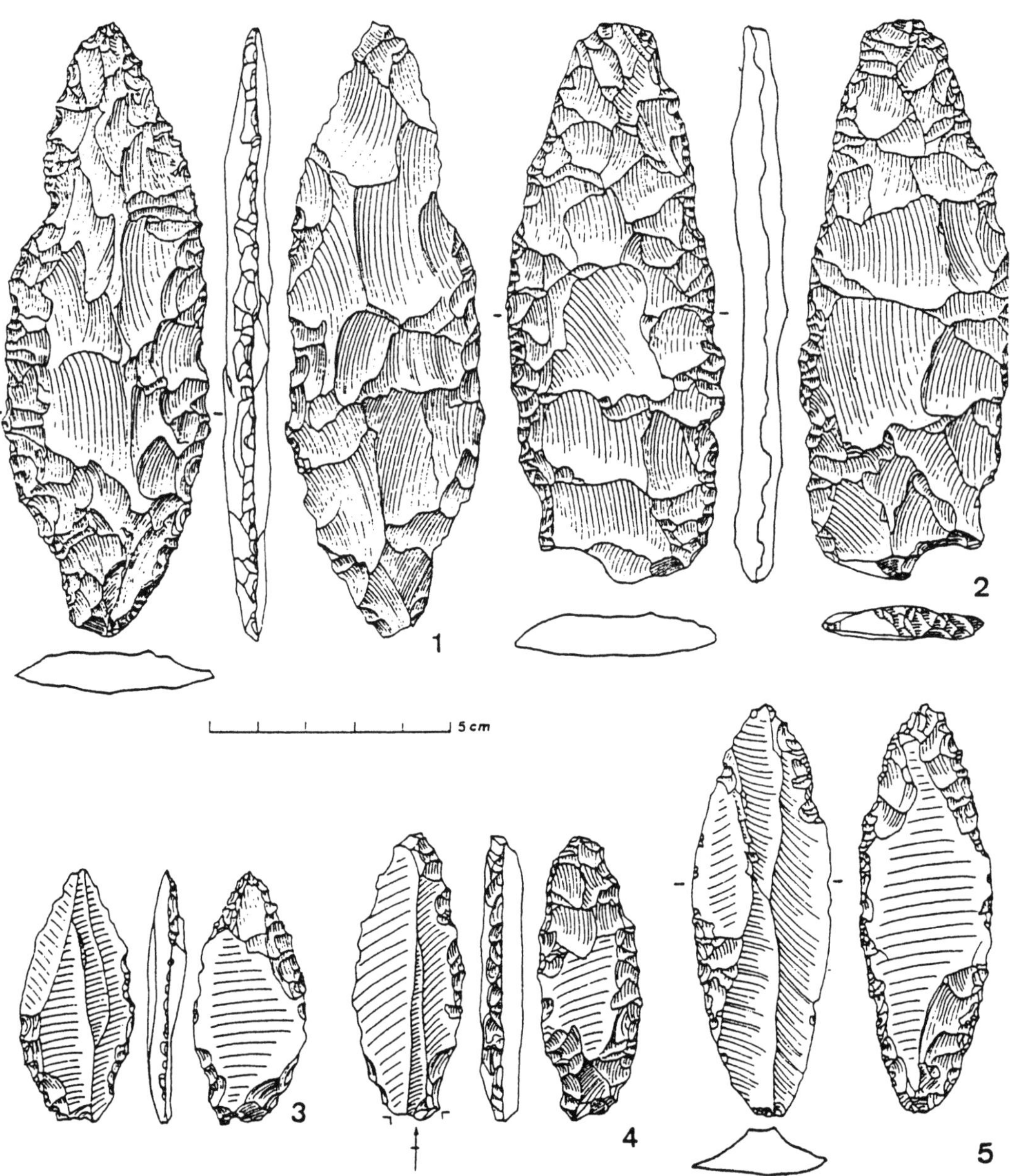

Figure 16.5. Comparison of bifacial points of the earlier phases (upper) and points made on blades with inverse retouch from the later phases (lower) of the Middle-Upper Palaeolithic transition in Northern Europe. After Otte (1985).

follow Mousterian practices (e.g. in the Châtelperronian at Arcy: Leroi-Gourhan 1982: see Figure 16.9). *Last but not least*, the discovery of human remains of clear Neanderthal character within the Châtelperronian levels at Saint-Césaire provides unquestionable proof of the transitional status of at least one Upper Palaeolithic industry (Lévêque 1988).

THE NATURE OF THE CHANGE

These characteristics of both continuity and discontinuity can be used to support two contrasting hypotheses for the Middle-Upper Palaeolithic transition: on the one hand the hypothesis of local independent evolution; and on the other, that of acculturation from some external tradition (Valoch 1984; Allsworth-Jones 1986; Harrold 1989). These two theories relate both to the state of the facts as well as to the state of current theoretical models (Gamble 1986). In reality these two theoretical processes seem, in practice, to have co-existed.

The first scenario: independent evolution

The cave of Couvin in the south of Belgium has produced an industry technically intermediate between the Middle and the Upper Palaeolithic; here, a pattern of blade production coexists with one for the production of thick triangular flakes with facetted platforms and pronounced bulbs of percussion. The tool assemblage is produced on both forms of blank, but the majority of tools are foliate points made with flat, bifacial retouch. Blades are also found with their bases thinned by the 'Kostienki knife' technique (Cattelain and Otte 1985) (Figure 16.4). Recent excavations have showed a remarkable homogeneity in the industry and dated it to *c.* 45 000 BP (^{14}C dating by E. Gilot) and a climatic period that shows a temporary oscillation according to the micro-faunal evidence (J.-M. Cordy, personal communication). This industry is, therefore, significantly older than various sites of the Middle Palaeolithic dated to c. 40 000 years BP in Belgium (Otte 1988).

This industry (which had previously been recorded in a number of disturbed contexts in several Belgian sites) is characteristic of the plains of Northwest Europe, extending from Great Britain (Jacobi 1980), through West Germany (Hülle 1977) to Poland (Chmielewski 1966) (Figures 16.5, 16.7). It shows the existence of an autonomous development of the Upper Palaeolithic much older than the Aurignacian and, in any case, largely outside of the geographical zone of the earliest stages of the Aurignacian (Otte 1985). Several evolutionary stages of this industry have been identified at Ranis (East Germany) and Jerzmanovice (Poland), illustrating the local continuity of this technical development. Only one human tooth (a deciduous molar) is as yet associated with this industry, and this is insufficient to make a distinction between a Neanderthal or a modern human.

The origin of this industry can perhaps be found in the local assemblages of Middle Palaeolithic age associated with foliate, leaf-point industries – as represented for example at Mauern in Bavaria (Von

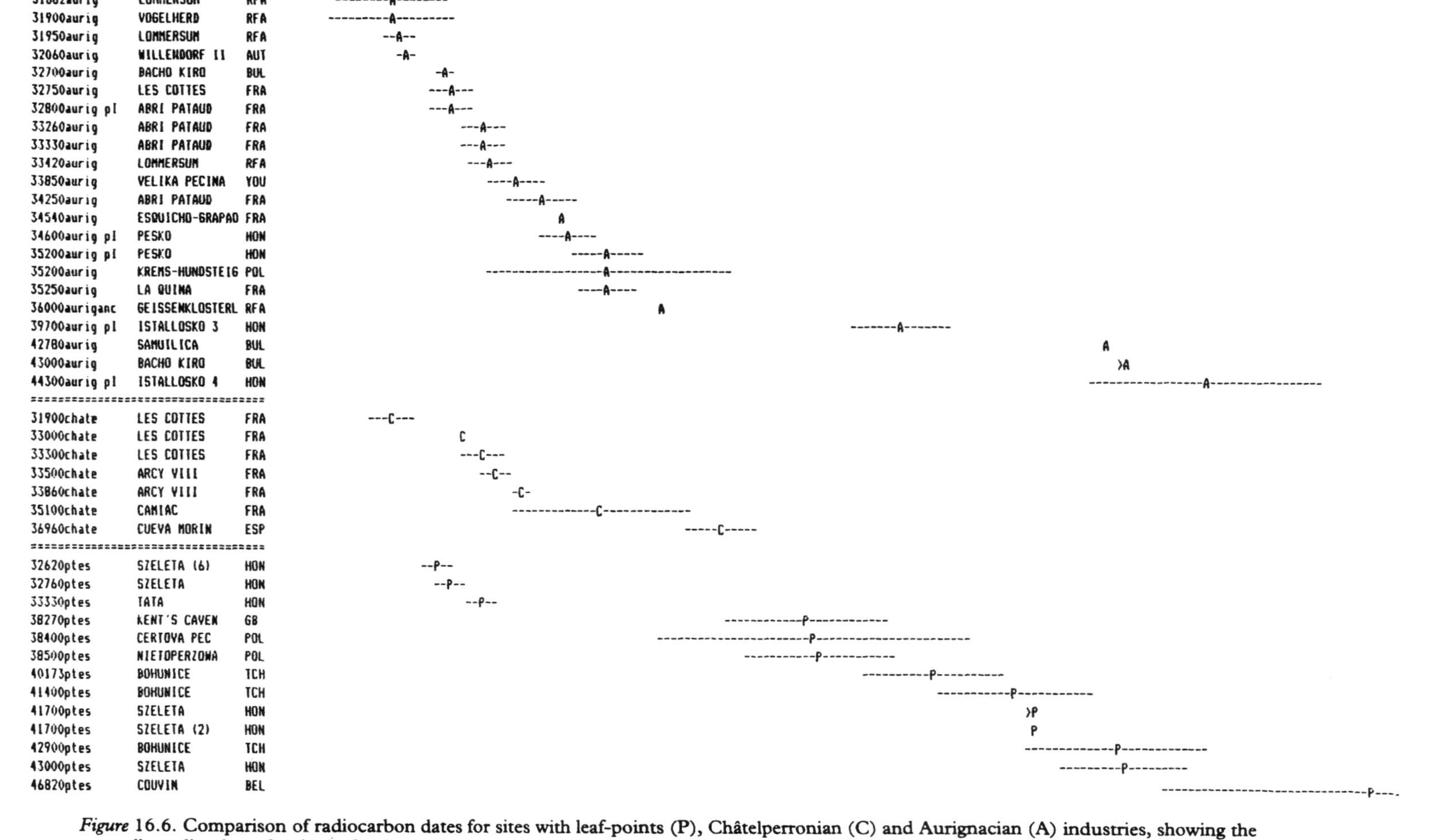

Figure 16.6. Comparison of radiocarbon dates for sites with leaf-points (P), Châtelperronian (C) and Aurignacian (A) industries, showing the generally earlier dates for the leaf-point industries (the dates for the Aurignacian industries at Istalloskö are disputed). From these dates, there is a possibility of acculturation between the Aurignacian and the Châtelperronian. In the North Europe plain, the hypothesis of an independent

Origin of the Upper Palaeolithic is more likely since the Aurignacian does not appear in this region until a relatively late stage

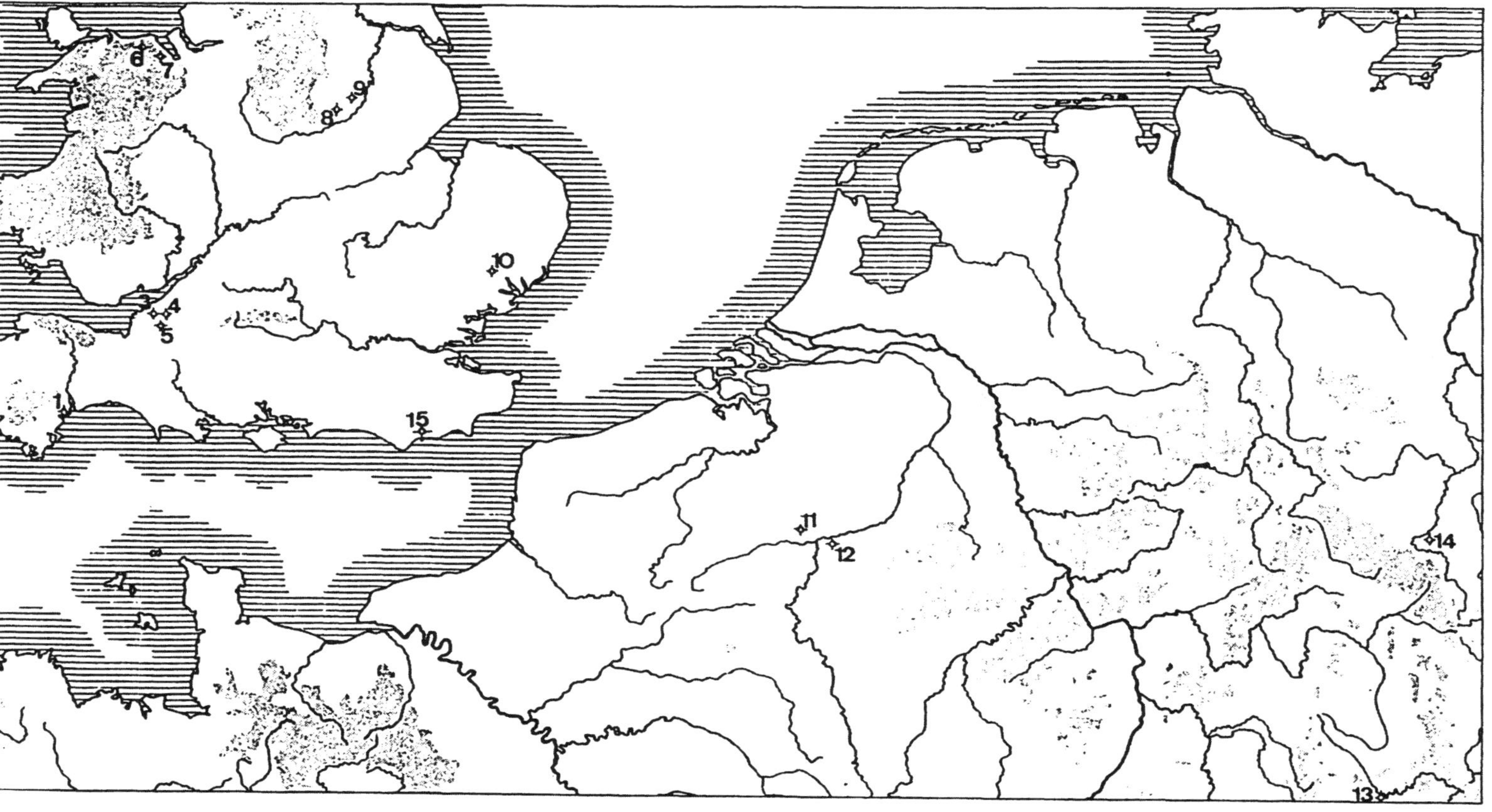

Figure 16.7. Distribution of principal early Upper Palaeolithic sites with leaf-points in Northern Europe: 1. Kents Cavern; 2. Paviland; 3. Soldiers Hole; 4. Badger Hole; 5. Hyena Den; 6. Cae Gwyn; 7. Ffynnon Beuno; 8. Pinhole; 9. Robin Hood's Cave; 10. Bramford Road; 11. Spy; 12. Goyet; 13. Mauern; 14. Ranis; 15. Pulborough. After Otte (1985).

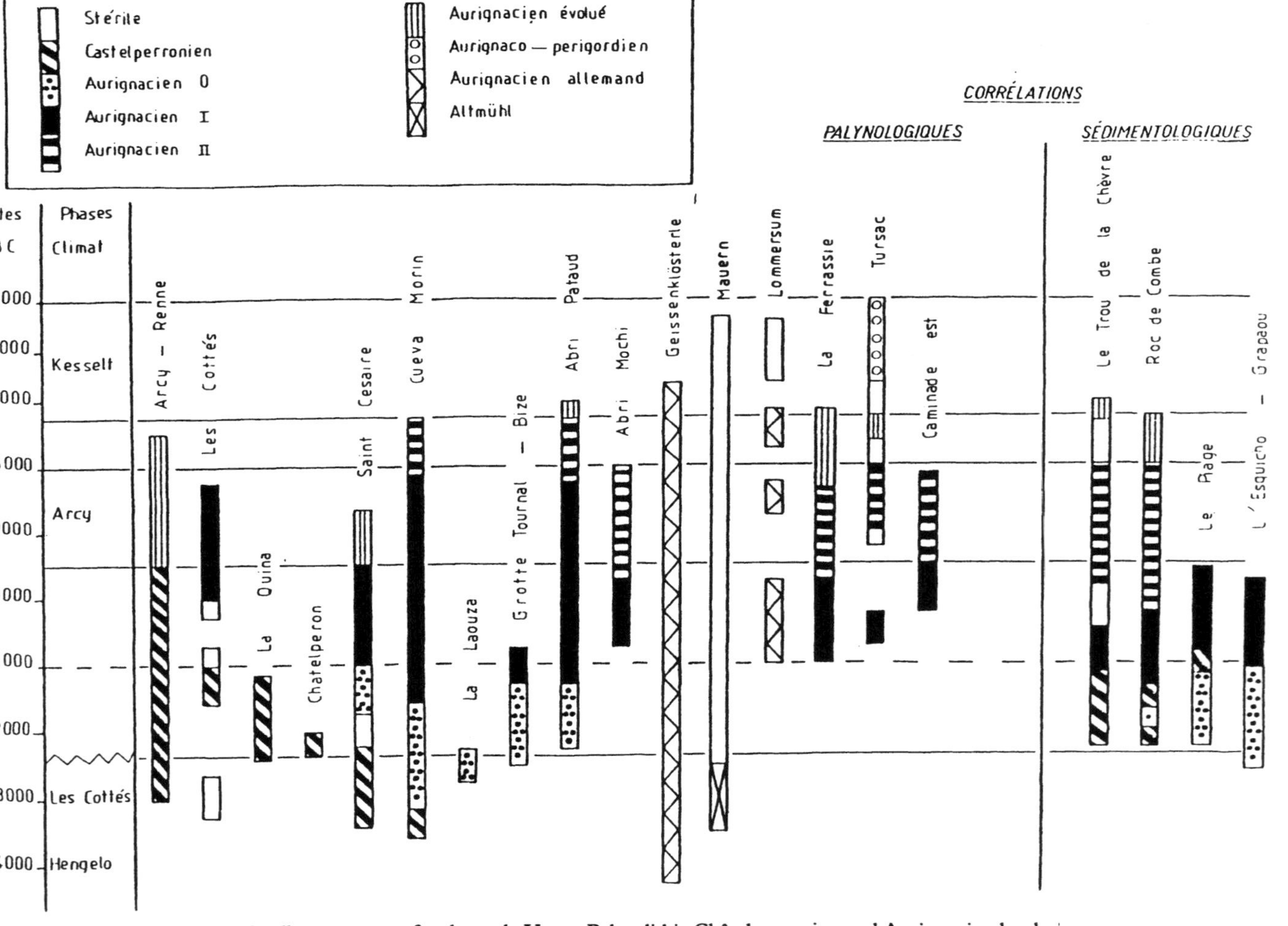

Figure 16.8. Comparison of pollen sequences for the early Upper Palaeolithic Châtelperronian and Aurignacian levels.

Koenigswald *et al.* 1974), the Doctor's Cave at Huccogne in Belgium (Ulrix-Closset 1975), and Königsaue in East Germany (Mania *et al.* 1973). Later stages of this development appear in the Gravettian industries of Central Europe, as at Petrkovice between Silesia and Moravia (Otte 1981; Klima 1966) and at Maisières in Belgium (de Heinzelin 1973; Otte 1979).

Second scenario: acculturation

In Northwest Europe, the Aurignacian appears between 32-30 000 BP in a fully developed form and without any apparent link to the local traditions. This new 'culture' seems homogeneous and distinct, not only technologically but also in economical, ritual and social terms.

The situation is clearly different from that in Southern Europe, where we can document a trail of earlier 'Proto-Aurignacian' sites extending from the Balkans, through the coasts of Mediterranean Italy and Provence, to southwestern France. In this context, the hypothesis of acculturation becomes a plausible explanation for the origins of the Châtelperronian, particularly when contrasted with the extreme duration and stability of earlier Mousterian technologies. The anatomical contrasts between the Neanderthal remains from Saint-Césaire and the 'Cro-Magnon' forms associated with the Aurignacian serves to strengthen the impression of a replacement of populations equipped with a new technology. The similarity of dates for the Châtelperronian and early Aurignacian reinforces this impression (Figure 16.6, 16.8) (cf. Leroyer 1983): at the same time as the Aurignacian moves west (Harrold 1989) there occurs the transition towards the use of the soft hammer technique within the Châtelperronian (Pèlegrin 1986). Most authors now agree with this model of a population moving westward with a new technology, diffusing the idea to the local populations without, however, completely overwhelming their own traditions. This scenario now seems to have won over more and more researchers as the most 'economical' to account for this transition (Allsworth-Jones 1986; Harrold 1989; Valoch 1984).

THE GENERAL VIEW

One is led therefore to recognise a diversity of process in the appearance of the Upper Palaeolithic in Western Europe. This complexity is apparent in the nature of the evidence, in its mode of expression, and in its articulation in both space and time (Kozlowski 1988, and this volume).

This, in turn, requires an equivalent complexity in interpretation and necessitates the existence of several associated and contemporary processes leading to the appearance of the Upper Palaeolithic along several different tracks. Each of these processes seems compatible with the current state of our knowledge.

This fundamental and irreversible change in behavioural evolution needs to be considered in its full 'depth', acknowledging on the one hand an intensity and diversity in the process of transformation, as well as the extreme variety of the cultural and geographic contexts in which they de-

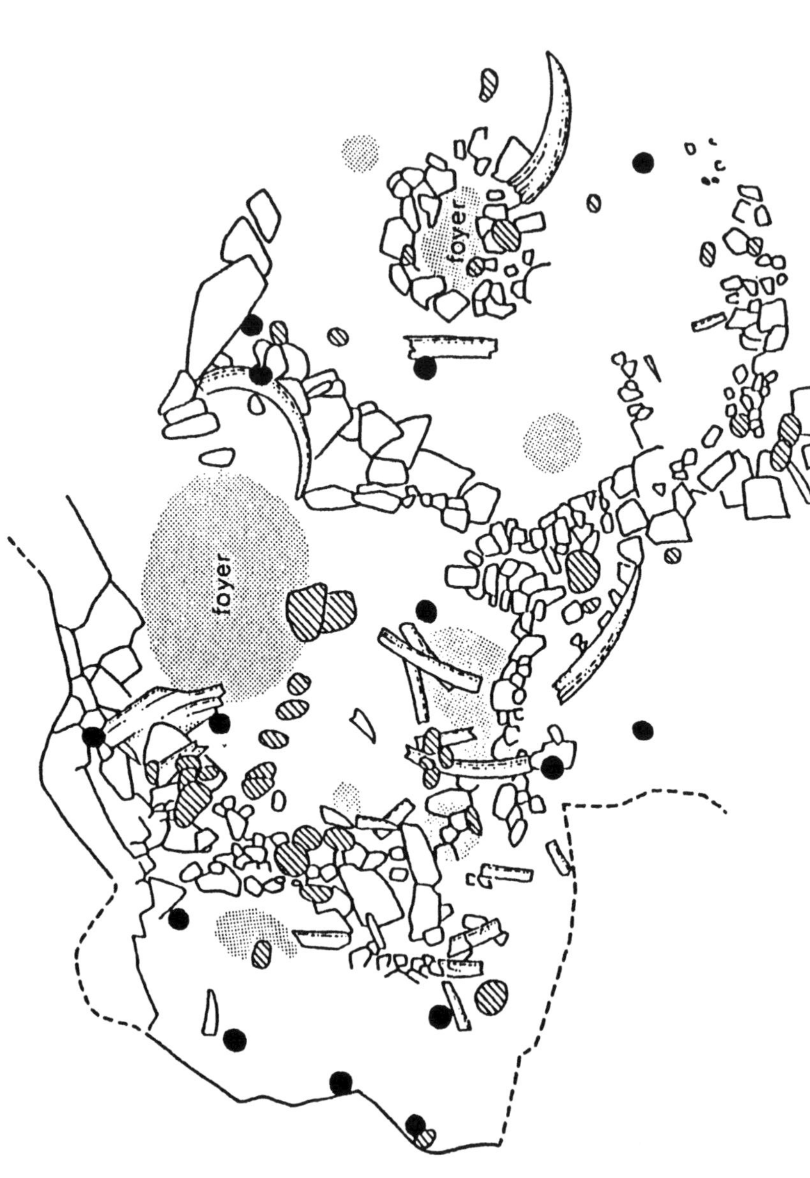

Figure 16.9. Plan of living structures recorded in the Châtelperronian levels at Arcy-sur-Cure (France) (after Leroi-Gourhan). The circular structures, partly constructed of mammoth remains, recall the later Gravettian structures of Central Europe, and continue traditions recorded earlier in the Middle Palaeolithic site of Molodova (Russia)

velop. The boundaries of this phenomenon are clear (Mousterian on the one side and Gravettian on the other) and the differences are easy to define. But the details of this process of transition still remains enigmatic in several fundamental respects (Whallon 1989; Straus, this volume; Clark and Lindly 1989).

At the end of the Middle Palaeolithic, the technical, cultural and conceptual potential of local populations was sufficient to make the step to the new technology and the new way of life. This is equally true for the Near East, the Balkans and Western Europe. It is only to be expected that on such a cultural base full of change and innovatory ability, several contemporary processes should develop, each influencing the other in the demographic explosion that the new life brought about. At the point of departure is a culture amazingly stable for hundreds of millenia. Following the 'transition' one finds civilisations of short duration, regionally highly varied, and constantly in pursuit of technical innovation or of aesthetic creation. This period of transition corresponds, therefore, to an evolutionary jump, which involved at the same time a great number of regions as well as a large number of diverse traditions. There can be neither one model for this transformation nor any single origin for this phenomena. It seems rather (paraphrasing the expression of Braidwood concerning the origin of the Neolithic) that for multiple and related reasons at the end of the Middle Palaeolithic, "culture was ready" in different places of the Eurasian continent, to make the change to the new way of life.

REFERENCES

Allsworth-Jones, P. 1986. *The Szeletian and the Transition from Middle to Upper Palaeolithic in Central Europe*. London: Clarendon Press.

Altuna, J. 1988. Subsistance d'origine animale pendant le Moustérien dans la région cantabrique (Espagne). In M. Otte (ed.) *L'Homme de Néandertal*. Vol. 6: *La Subsistence*. Liège: Etudes et Recherches Archéologiques de l'Université de Liège (ERAUL 33): 31-43.

Bar-Yosef, O., Goldberg, P. and Meignen, L. 1987. *Processus de Formation des Foyers Moustériens de Kebara*.

Beyries, S. 1987. *Variabilité de l'Industrie Lithique au Moustérien: Approche Fonctionelle sur Quelques Gisements Français*. Oxford: British Archaeological Reports International Series S328.

Boëda, E. 1986. *Approche Technologique du Concept Levallois et Evaluation de son Champ d'Application*. Unpublished Doctoral Thesis, University of Paris X.

Bosinski, G. 1973-4. Der Paläolithische Fundplatz Rheindhalen, Stadkreis Mönchengladbach (Nordrhein-Wesfale). *Archäologische Informationen* 2-3: 11-16.

Cahen, D. 1984. Paléolithique inférieur et moyen en Belgique. In D. Cahen and P. Haesaerts (eds) *Peuples Chasseurs de la Belgique Préhistorique dans leur Cadre Naturel*. Brussels: Institute Royal des Sciences Naturelles de Belgique: 133-155.

Cahen, D. 1985. Fonction, industrie et culture. In M. Otte (ed.) *La Signification Culturelle des Industries Lithiques*. Oxford: British Archaeological Reports International Series S239: 39-51.

Cattelain, P. and Otte, M. 1985. Sondage 1984 au 'Trou de l'Abîme' à Couvin, état des recherches. *Helenium* 25: 123-130.

Chase, P.G. 1988. Scavenging and hunting in the Middle Palaeolithic: the evidence from Europe. In H. Dibble and A. Montet-White (eds) *Upper Pleistocene Prehistory of Western Eurasia*. Philadelphia: University Museum, University of Pennsylvania: 225-232.

Chernysh, A.P. 1961. *Paleolithichna Stoianka Molodove* V. Kiev.

Chmielewski, W. 1961. *Civilisation de Jerzmanovice*. Inst. Hist. Kult. Nat. Polsk. Akad. Nauk, Wroklaw, Warszawa, Krakow.

Clark, G. and Lindly, J. 1989. The case for continuity: observations on the biocultural transition in Europe and Western Asia. In P. Mellars and C. Stringer (eds) *The Human Revolution: Behavioural and Biological Perspectives on the Origins of Modern Humans*. Edinburgh: Edinburgh University Press: 626-676.

Demars, P.-Y. 1982. *L'Utilisation du Silex au Paléolithique Supérieur: Chiox, Approvisionnement, Circulation. L'Exemple du Bassin de Brive*. Paris: Centre National de la Recherche Scientifique.

De Heinzelin, J. 1973. *L'Industrie du Site Paléolithique de Maisières-*Canal. Brussels: Institute Royal des Sciences Naturelles de Belgique, Mémoire 171.

Dibble, H. 1988. Typological aspects of reduction and intensity of utilization of lithic resources in the French Mousterian. In H. Dibble and A. Montet-White (eds) *Upper Pleistocene Prehistory of Western Eurasia*. Philadelphia: University Museum, University of Pennsylvania: 181-198.

Farizy, C. and Schmider, B. 1985. Contribution à l'identification culturelle du Châtelperronien: les données de l'industrie lithique de la couche X de la Grotte du Renne à Arcy-sur-Cure. In M. Otte (ed.) *La Signification Culturelle des Industries Lithiques*. Oxford: British Archaeological Reports International Series S239: 149-165.

Fosse, G. and Cliquet, D. 1987. Les structures de combustion des gisements moustériens de Saint-Vaast-la-Hogue et de Saint-Germain-des-Vaux (Manche). Precirculated papers of Colloquium *Foyers Préhistoriques*, Nemours 1987.

Gamble, C. 1986. *The Palaeolithic Settlement of Europe*. Cambridge: Cambridge University Press.

Gautier, A. and Schild, R. 1988. Le Gisement paléolithique moyen de Zwollen en Pologne. In M. Otte (ed.) *L'Homme de Néanderthal*. Vol. 6: *La Subsistance*. Liège: Etudes et Recherches Archéologiques de l'Universit; de Liège (ERAUL 33): 69-73.

Geneste, J.-M. 1985. *Analyse Lithique d'Industries Moustériennes du Périgord: Une Approche Technologique du Comportement des Groupes Humains au Paléolithique Moyen*. Unpublished Doctoral Thesis, University of Bordeaux 1.

Gullentops, F. 1954. *Contributions à la Chronologie du Pléistocene et des Formes de Reliefs en Belgique*. Louvain: Memoires de l'Institut de Géologie de l'Université de Louvain 18.

Haesaerts, P., Juvigne, E., Kuyl, O., Mucher, H., Roebroecks, W. 1981. Compte rendu de l'excursion du 13 Juin 1981, en Hesbaye et au Limbourg néerlandais, consacrée à la chronostratigraphie des loess du Pléistocène supérieur. *Annales de la Société Géologique de Belgique* 104: 223-240.

Harrold, F. 1987. The Châtelperronian and the Early Aurignacian in France. Paper presented at the Meeting of the Society for American Archaeology, Toronto, May 1987.

Harrold, F. 1989. Mousterian, Châtelperronian and Early Aurignacian: continuity or discontinuity. In P. Mellars and C. Stringer (eds) *The Human Revolution: Behavioural and Biological Perspectives on the Origins of Modern Humans*. Edinburgh: Edinburgh University Press: 677-713.

Hülle, W. 1977. *Die Ilsenhöhle unter burg Ranis/Thüringen. Eine Palöolithische Jägerstation*. Stuttgart: G. Fisher.

Jacobi, R. 1980. The Upper Palaeolithic of Britain with special reference to Wales. In J.A. Taylor (ed.) *Culture and Environment in Prehistoric Wales*. Oxford: British Archaeological Reports 76: 15-100.

Klima, B. 1966. La station paléolithique à Ostrova-Petrkovice (Moravie). In J. Filip (ed.) *Investigations Archéologiques en Tchécoslovaquie*. Prague: Czechoslovakian Academy of Sciences.

Kozlowski, J. 1988. Problems of continuity and discontinuity between the Middle and Upper Paleolithic of Central Europe. In H. Dibble and A. Montet-White (eds) *Upper Pleistocene Prehistory of Western Eurasia*. Philadelphia: University Museum, University of Pennsylvania: 349-360.

Leroi-Gourhan, A. 1982. La Grotte du Renne à Arcy-sur-Cure. In J. Combier (ed.) *Les Habitats du Paléolithique supérieur*. Precirculated papers of Colloquium held at Roanne (France), 1982.

Leroi-Gourhan, A. 1986. Les analyses palynologiques du gisement paléolithique moyen de Seclin (Nord). Precirculated papers of the Colloquium on *Cultures et Industries Paléolithiques en Milieu Loessique*, Amiens, December 1986.

Leroyer, C. 1983. *L'Aurignacio-Périgordien: Apport de la Palynologie*. Périgueux: Centre de Recherches Préhistoriques, Cahier 9.

Lévêque, F. 1988. L'homme de Saint-Césaire: sa place dans le Castelperronien de Poitou-Charente. In M. Otte (ed.) *L'Homme de Néanderthal*. Vol. 7: *L'Extinction*. Liège: Etudes et Recherches Archéologiques de l'Université de Liège (ERAUL 34): 99-108.

Mania, D. and Topfer, V. 1973. *Königsaue Gliederung, Okologie und mittelpaläolithische Funde der letzten Eiszeit*. Berlin: Deutscher Verlag der Wissenschaften.

Mellars, P. 1988. The chronology of the south-west French Mousterian: a review of the current debate. In M. Otte (ed.) *L'Homme de Néanderthal*. Vol. 4: *La Technique*. Liège: Etudes et Recherches Archéologiques de l'Université de Liège (ERAUL 31): 97-120.

Morawski, W. 1971. *Middle Paleolithic Flint Assemblages from the Piekary IIa Site*. Archaeological Research, Cracow.

Otte, M. 1979. *Le Paléolithique Supérieur Ancien en Belgique*. Monographies d'Archéologie Nationale 5. Brussels: Musées Royaux d'Art et d'Histoire.

Otte, M. 1981. *Le Gravettien en Europe Centrale*. Bruges: Dissertationes Archaeologicae Gandenses 20.

Otte, M. 1985. *Les Industries à Pointes Foliacées et à Pointes Pédonculées dans le Nord-Ouest Européen*. Artefact 2, Treignes (Belgique), CEDARC.

Otte, M. 1988. Interprétation d'un habitat au Paléolithique moyen: la grotte de Sclayn en Belgique. In H. Dibble and A. Montet-White (eds) *Upper Pleistocene Prehistory of Western Eurasia*. Philadelphia: University Museum, University of Pennsylvania: 95-124.

Patou, M. 1984. *Contribution à l'Etude des Mammifères des Couches Supérieures à la Grotte de Lazaret (Nice, Alpes-Maritimes)*. Unpublished Doctoral Thesis, University of Paris 6.

Pèlegrin, J. 1986. *Technologie Lithique: une Méthode Appliquée à l'Etude de Deux Séries du Périgordien Ancien*. Unpublished Doctoral Thesis, University of Paris X.

Rolland, N. 1981. The interpretation of Middle Palaeolithic

variability. *Man* 16: 15-42.

Simek, F. and Snyder, L.M. 1988. Changing assemblage diversity in Périgord archaeofaunas. In H. Dibble and A. Montet-White (eds) *Upper Pleistocene Prehistory of Western Eurasia*. Philadelphia: University Museum, University of Pennsylvania: 321-332.

Smith, F., Simek, J. and Harrill, M. 1989. Geographic variation in supra-orbital torus reduction during the Later Pleistocene (c. 80 000-15 000 BP). In P. Mellars and C. Stringer (eds) *The Human Revolution: Behavioural and Biological Perspectives on the Origins of Modern Humans*. Edinburgh: Edinburgh University Press: 172-193.

Sonneville-Bordes, D. de. 1987. Foyers paléolithique en Périgord. Precirculated papers of Colloquium on *Foyers Préhistoriques*, Nemours, 1987.

Tuffreau, A., Revillon, S., Sommé, J., Aitken, M.J., Huxtable, J. and Leroi-Gourhan, A. 1985. Le gisement paléolithique moyen de Seclin (Nord-France). *Archäologisches Korrespondenzblatt* 15: 131-138.

Ulrix-Closset, M. 1975. *Le Paléolithique Moyen dans le Bassin Mosan en Belgique*. Wetteren: Editions Universe.

Ulrix-Closset, M., Otte, M. and Cattelain, P. 1986. Le 'Trou de l'Abîme' à Couvin (prov. de Namur-Belgique). In M. Otte (ed.) *L'Homme de Néanderthal*. Vol. 8: *La Mutation*. Liège: Etudes et Recherches Archéologiques de l'Université de Liège (ERAUL 35): 225-239.

Valoch, K. 1984. Transition du Paléolithique moyen au Paléolithique supérieur dans l'Europe Centrale et Orientale. Salamanca: *Scripta Praehistorica, Oblata Francisco Jorda:* 439-467.

Von Koenigswald, W., Müller-Beck, H.J. and Pressmar, E. 1974. *Die Archäologie und Paläontologie in den Weinberghöhlen bei Mauern (Bayern), Grabungen* 1937-1967. Archaeologice Venatoria 3. Tübingen: Selbstverlag Institut für Urgeschichte, Tübingen University.

Whallon, R. 1989. Elements of cultural change in the Later Palaeolithic. In P. Mellars and C. Stringer (eds) *The Human Revolution: Behavioural and Biological Perspectives on the Origins of Modern Humans*. Edinburgh: Edinburgh University Press: 433-454.

Whallon, R. (in press). Unconstrained clustering for the analysis of spatial distributions in archaeology. In Press.

White, R. 1989. Production complexity and standardization in early Aurignacian bead and pendant manufacture: evolutionary implications. In P. Mellars and C. Stringer (eds) *The Human Revolution: Behavioural and Biological Perspectives on the Origins of Modern Humans*. Edinburgh: Edinburgh University Press: 366-390.

17: Early Hominid Symbol and Evolution of the Human Capacity

ALEXANDER MARSHACK

THE PROBLEM AND METHODOLOGICAL, THEORETICAL APPROACHES

Among the many problems in the present effort to understand the biological and cultural evolution and development of anatomically modern humans is the inherently skewed nature of the archaeological record and the equally skewed nature of many studies of these materials for interpretive purposes. Perhaps first among the difficulties is the persistent discussion of the so-called Mousterian-Upper Palaeolithic transition in terms of the so-called symbolic 'explosion' (Pfeiffer 1982) that occurred in the European Upper Palaeolithic, particularly in the Franco-Cantabrian area of Western Europe. Little attention has been paid to the unique regional and historical nature of the transition or to symbolic and cognitive, cultural developments in other areas of Europe and the world during the Mousterian or the period of the transition.

In an instance of Franco-Cantabrian ethnocentrism, White (1982, 1986, 1989), for instance, has used the study of a set of Aurignacian beads from the Vallon des Roches at Sergeac in the Dordogne, France, to argue for the transition from the 'palaeoculture' (Jelinek 1977, 1982) of the Mousterian to the higher and more human culture of the West European or Franco-Cantabrian Upper Palaeolithic. Binford (1981, 1983, 1985) has strongly argued for a comparative lack of culture and capacity among the Neanderthals. Chase and Dibble (1987) also argue for a comparative lack of symbolic thought or complexity in the European Mousterian, denigrating the rare examples of extant symbolism when compared to the rich evidence from the later Upper Palaeolithic. Dibble (1983, 1984, 1986, 1987, 1988) has argued as well against the typology of tool 'styles' that have traditionally been used for categorizing the Mousterian industries, suggesting that these forms were often the simple result of sequences of tool reduction occurring during the process of tool use and reuse. He has implied that the comparative lack of tool style variability in the Mousterian suggests a lack of cultural complexity and competence. Strategies of tool reduction, however, have their own styles and rules and become, themselves, aspects of culture. White, above, has argued for the sudden appearance of the concept of 'self' on the basis of the Aurignacian beads, and, as a corollary, for the development of 'social display' and its cultural consequences. According to White (1985), these were among the processes leading to language. White's suggestion, however, while referring to items of personal decoration, derives ultimately from current theories of information exchange and networking (Wobst 1977; Conkey 1978, 1980; Gamble 1980, 1982, 1983; Hodder 1982a, 1982b).

Neanderthal competence and culture and the nature of the difference between the Neanderthals and modern *Homo sapiens sapiens* have been in discussion or debate for at least a century, being most often discussed in terms of comparative morphology, technology and subsistence strategies rather than in the more difficult terms of evolving or evolved capacity or symbolic culture. There have been rare, if significant, attempts at evaluating Mousterian symbolic culture. Chase and Dibble (1987) have attempted a recent reevaluation of the Mousterian symbolic evidence and Blanc (1961), a generation earlier, attempted an interpretation and summation for his period. There have, in addition, been two major comparative studies of Mousterian and Upper Palaeolithic burial practices (S. Binford 1968; Harrold 1980), each of which suggested a lesser cultural complexity and competence for the Neanderthals. There is an inherent problem, however, in attempting to evaluate symbolic complexity or capacity in terms of a quantified comparison of the available grave goods. I will touch on the problem below.

A different type of 'cultural' comparison has been undertaken by the theoretical reconstructions of the soft tissue of the vocal tract of the Neanderthals and other early hominids (Lieberman 1955, 1989; Laitman *et al.* 1979; Laitman and Heimbuch 1982; Laitman 1983), resulting in the suggestions that the Neanderthals could 'speak' and may therefore have had 'language', but that the capacity for vocalization was not as great as that found among modern humans. Laitman *et al.* (1979) have declared that the capacity for vocalization among the Neanderthals at Monte Circeo, La Ferrassie and Saccopastore 2 was probably equal to that of a six to nine year old human child. However, it would have been an adult Neanderthal, rather than a child, that would have been using the species capacity adaptively in the culture. Besides, 6 to 9 year old human children have grammar, syntax and the capacity for human speech (Note 1). The complexity of adult Neanderthal culture and capacity cannot be derived from reconstructions of the vocal tract and descriptions of the capacity for vocalization, any more than the changes and development of Mousterian tool industries can be derived from the morphology of the hand. Vocal tract studies, while they deal with acoustic and vocal aspects of speech production, do not deal with the deeper complexities of symbolic communication and the nature of language. The capacity for language not only involves a capacity for vocalization and acoustic decoding, but is dependent as well on the capacity for the visual categorization of objects and processes, the generalized capacity to map and model or 'think' in time and space, and the capacity to solve a wide range of cognitive problems in different modes. This neurological capacity for visual mediation and planning evolved also in the development of the capacity for two-handed manipulation and problem solving. The set of the capacities necessary for language is, as we shall see, present among the symbol systems and materials of the Mousterian.

Measurable differences in anatomy during late stages in hominid evolution among groups evolved or evolving from *Homo erectus* can tell us little about unquantifiable differences in cognitive or symbolic capacity. Similarly, quantifiable differences in artifactual style or complexity, whether in stone tools or symbolic products, can tell us little about the capacity for problem solving or symbolic thought among these different late populations. The products, at most, evidence a diachronic or historical stage of cultural development or of subsistence technology. It is perhaps significant in this regard that most of the archaeological evidence for symbolic thought and productions before the Upper Palaeolithic appearance of fully modern *Homo sapiens sapiens* in Europe comes from the Mousterian period and the Eurasian area of Neanderthal habitation. Comparatively little symbolic evidence during this period comes from Africa, Australia or other parts of Asia. Even during the period of the Upper Palaeolithic, there is little evidence for symbolic production elsewhere (Belfer-Cohen 1988). We would not, for that reason suggest an absence of the capacity for symbolic thought in these other areas of *Homo sapiens sapiens* habitation or the absence of the capacity for social display or a recognition of 'self' or, for that matter, an absence of the capacity for language. There is another side to this coin. The presence of Mousterian technology among groups of early modern *Homo sapiens sapiens* is probably more the reflection of a historical stage in cultural development than a measure of biological or neurological-cognitive similarity or difference.

The difficulty of determining capacity from the products of that capacity or from rates of change among the products has never been properly addressed. It has often been assumed in discussing the late Palaeolithic cultures that these materials and rates represent a true measure of extant capacity. No such assumption would be made for cultures of the present period.

In some measure to address these problems, I have taken a somewhat different theoretical and analytical approach, arguing that the Neanderthals, as evolutionary end-products of the same three to five million year long trajectory of hominization as anatomically modern humans, had a comparable, if not equal, set of those capacities that were selected for or were 'exaptively' present or developed (Gould and Vrba 1982), but that these capacities were utilized in a historically, demographically and contextually different milieu than was present in the European Upper Palaeolithic. We are, of course, faced with the problem of describing and defining the 'human capacity', not merely in terms of the products of a culture or region, but in terms of the presumed evolutionary, neurological and cognitive level reached. I have approached the problem not only by attempting to model the evolution of the human capacity (Marshack 1976,

1984a, 1985a, 1986a, 1988a) but by conducting an exhaustive first-hand, internal analysis of all the Upper Palaeolithic symbolic traditions available in Eastern and Western Europe and Siberia, and by a first-hand analysis of all the available symbolic products of Neanderthal manufacture. There was a simultaneous study of the major available examples of symbolic products of this period in the Middle East and Africa. The comparative inquiry has necessitated a reevaluation of the traditions involved.

PERSONAL DECORATION AND 'SELF'

The earliest evidence for Aurignacian beads are the two animal tooth pendants found at Bacho Kiro, Bulgaria, in a level thousands of years earlier than the Aurignacian beads of France (Marshack 1976). Bacho Kiro stands on a possible route to the Middle East, yet there is no evidence for articles of personal decoration in the Middle East during this period (Belfer-Cohen 1988), or for that matter in any areas of the world outside of Europe. There is evidence, however, of a local transition within Central Europe from the Mousterian to the Upper Palaeolithic, both morphologically (Smith 1985; Wolpoff 1986) and technologically (Oliva 1988, Svoboda 1988). There is also evidence for a development of symbolic traditions in Europe, including the working of bone, long before the appearance of the Aurignacian beads at Bacho Kiro. In Central Europe, for instance, in the early Mousterian at Tata, Hungary, dated by uranium series to 100 000 BP by Schwarcz (personal communication), there is a carved mammoth tooth plaque (Marshack 1976). At Bacho Kiro, in a Mousterian level, Kozlowski excavated a bone fragment with an intentionally engraved accumulation of zigzag motifs (Marshack 1976). There is, in addition, apparent evidence of Micoquian beads from Germany of approximately the same age as the Tata plaque (c. 110 000 BP) at Bocksteinschmiede in Germany (Wetzel and Bosinski 1969: see below and Figure 17.7). There is a possibility, therefore, that symbolic traditions developed or assayed by the Neanderthals during the Mousterian were preparatory to traditions that were to be later developed 'explosively' in the Upper Palaeolithic.

Before beginning, however, it is necessary to address the problem of the rare and unique artifact as an instance of intentional human effort and capacity. Dibble (1987), as a specialist in statistical and quantified inquiry, has rejected the Mousterian evidence of symbolic manufacture as too sparse to be significant. The Acheulian wooden spears excavated at Clacton, England (Oakley 1961) Lehringen, Germany (Jacob-Friesen 1965) and possibly at Torralba, Spain (Howell 1966) do not document a voluminous tradition. They do, however, document an early capacity to work wood, to properly choose a hard wood such as yew and to harden a point in a fire, as at Lehringen. The Lehringen spear was found within the remains of an elephant indicating prior preparation and carriage of the spear for hunting and killing. We do not have evidence in these three spears of a major cultural tradition with regional stylistic variation. We do, however, have clear evidence for a rather complex *capacity*, involving a choice of the proper hard wood, probably in the proper season, and a use of both stone tools and fire to create a secondary tool that had a specialized use. The implied complexity lies not in any quantitative assessment of the examples, but in the cognitive contents suggested by the few examples at hand.

From the same early period we have evidence for a use of ochre. At Terra Amata, France, ochre crayons of different hue were found in the remains of a seasonal structure (Lumley 1966). In the cave of Becov, Czechoslovakia, Fridrich (1965) excavated a quartzite rubbing stone and a striated piece of ochre, together with a quantity of dispersed red ochre (Marshack 1981). These Acheulian examples for the use of ochre do not provide us, quantitatively, with evidence of cultural traditions with regional, stylistic variation, but they do document another aspect of the problem-solving *capacity* and the use of regional raw materials for nonsubsistence, probably symbolic purposes. There have been suggestions that ochre may have been used in the early Palaeolithic for tanning hides or staunching blood, but these suggest even more complex technologies and uses. The simpler suggestion, involving the obvious use of ochre for colour, adequately documents the presence of a capacity that could be used for both symbolic and a range of practical purposes. It is significant that in the use of colour, as in the use of fire to harden wood, we have early evidence of the 'exaptive' and secondary effects of hominid evolution involving selection for a broad *range* of two-handed, vision-oriented capacities. In the Acheulian natural selection would have occurred, not for an increase in the capacity to use fire, wood or ochre, but for an increase in the set of capacities being utilized. These capacities were already 'human', and any increase, in whatever part of the world, would have involved an increase in brain volume and neurological complexity. We therefore expect to find a substantial increase in these capacities among the Neanderthals.

Leroi-Gourhan and Leroi-Gourhan (1964) excavated a number of beads of different types in Châtelperronian levels at the Grotte du Renne, Arcy-sur-Cure. There are beads or pendants made of exotic fossil materials, including sea-floor creatures (a crinoid and a fossil shell), as well as beads of carved bone and animal teeth (Figure 17.1). The techniques of manufacture include boring or carving a hole and incising a horizontal groove around the top of an object to make, in each case, a hanging bead or pendant. The recent discovery of a Neanderthal skull in a Châtelperronian level at Saint-Césaire (Lévêque and Vandermeersch 1980) raises a question as to whether these objects were made by the late and terminal Neanderthals represented by that skull or by anatomically modern humans who were apparently, for a short period, in contemporaneous contact. The Châtelperronian industry has been termed Upper Palaeolithic (Harrold 1986) but with abundant Mousterian types. Whether the beads were also of Mousterian derivation (i.e., made by Neanderthals or influenced by Neanderthal tradition) is still open to question. It was the opinion of Leroi-Gourhan and Leroi-Gourhan (1964) that the Arcy beads

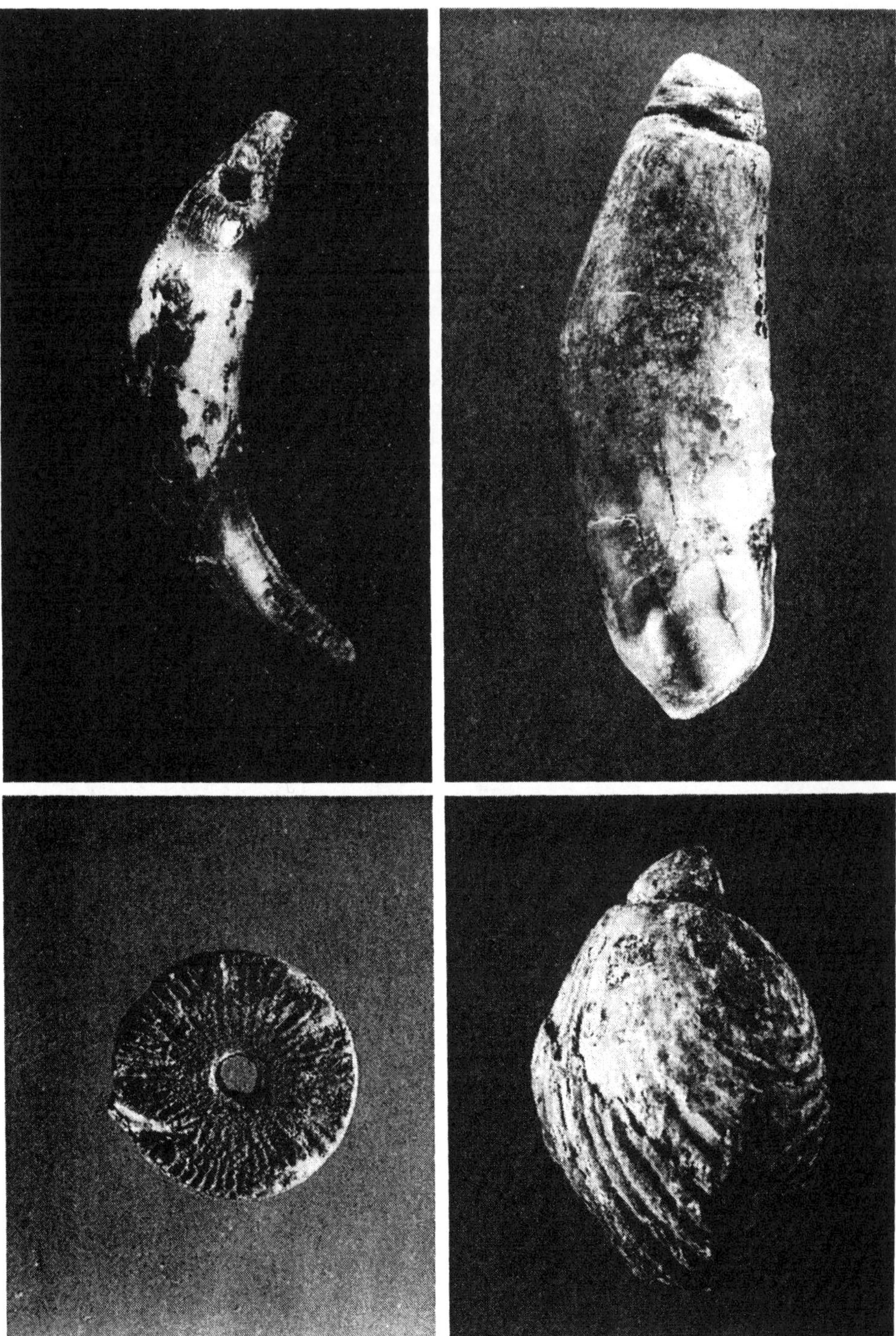

Figure 17.1. Upper: Beads made of animal teeth. One bead has a hole bored at the top, the other an incised groove circling the top. Lower: Beads made from sea fossils. One section of a crinoid with a natural hole in the centre and a sea shell with a groove incised around the top. All the specimens are from the Châtelperronian levels at Arcy-sur-Cure (Yonne, France).

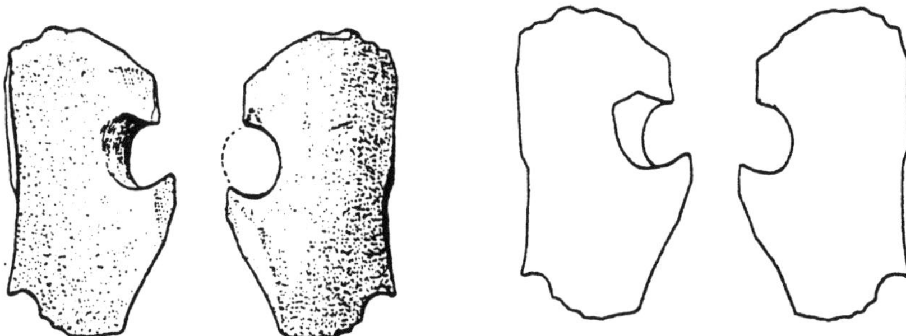

Figure 17.2. Pech de l'Azé (Dordogne, France). Bone fragment with a carved wide hole, apparently forming part of a fragmented pendant. Mousterian period (after F. Bordes).

represented the earliest known examples of personal decoration and that they were made by anatomically archaic humans.

Bordes (1969), however, found an intentionally carved fragment of bone in a Mousterian level at Pech de l'Azé which he declared to be part of a pendant (Figure 17.2). Harrold (1989) has suggested that the carved fragment might be part of a working tool, a suggestion, as we shall see, that raises an interesting set of questions of a different type. Significantly, a fragmented pendant of precisely the Pech de l'Azé type, with a similarly carved wide hole, was found in a Châtelperronian level at Arcy (Figure 17.3). Leroi-Gourhan and Leroi-Gourhan have declared that "Plusiers exemplaires fragmentés ou entiers de telles pendoloques ont été trouvés dans le Châtelperronien et dans l'Aurignacien d'Arcy" (1964: 41). One of the more famous Aurignacian pendants found at Arcy was made by carving this unique type of wide hole (Figure 17.4). The manufacture of such large holes for pendants seems to have been a regional style in the Dordogne, though the mode of carving such wide holes is common in the Aurignacian and occurs, as we shall see, in other classes of production.

The use of fossil materials for symbolic purposes at Arcy probably resulted from the discovery of fossil-bearing rock in the territory, but the fossils could as easily have been acquired in exchange or barter. Harrold (1989) mentions the presence of such exotic materials both in the Mousterian and the Châtelperronian, but he does not deal with them as materials that could also be worked and used. At Tata, Hungary, a fossil creature, a nummalite (Figure 17.5), was found in a Mousterian level, together with the well-known plaque carved from a mammoth tooth. Microscopic examination indicated that the nummalite had a natural fracture running through it. A Neanderthal had then apparently carefully incised a fine line at a right angle to the fracture to make a perfect cross, an act of delicate and accurate engraving that would be difficult for a modern human since it required a steady grip and orientation of the object between two fingers while the other hand used an engraving tool. The two-handed competence and manipulative capacity and acuity required for such an act in the Mousterian is significant because animal tooth pendants

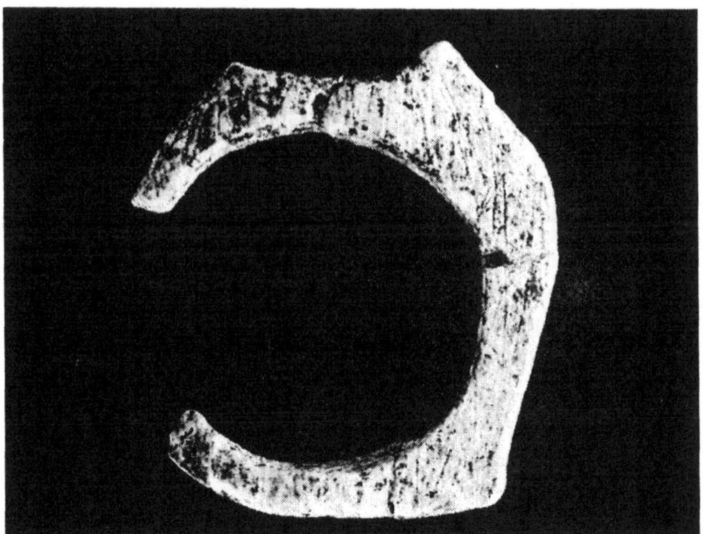

Figure 17.3. Arcy-sur-Cur (Yonne, France). Bone fragment with a carved wide hole, apparently part of a pendant or bead. Châtelperronian period.

from the Aurignacian were often begun at this scale of fine engraving and difficulty, with a multiple crossing of engraved lines, until a core was established, after which there was a gouging and widening of the hole. The process is clearly evident on the early Aurignacian beads from Bach Kiro.

It is, therefore, of importance to indicate that pendant beads with bored holes (Figure 17.7) were found in an early, pre-Mousterian, Micoquian level, c. 110 000 BP, at the site of Bocksteinschmiede (Lonetal), in Germany (Wetzel and Bosinski 1969). The beads consist of wolf-tail vertebra and a wolf footbone (metapodium). A close-up of the intentionally bored hole of the tiny vertebra (Figure 17.8) clearly indicates the gouging and funnel-like widening that is found on later Upper Palaeolithic pendants and beads. The metapodium hole is more complex. A deep natural concavity on the distal end serves as the core for the hole on one face. The other face of the metapodium has a normally high, arched and convex crest instead of a natural concavity. This has been scraped and gouged away, apparently by a tool since a long horizontal cut delimits the breakage, and a hole was then put in the resulting concavity on this face. Unfortunately the metapodium is more deteriorated than the vertebra, and close-up photography does not provide the clear results seen on the smaller vertebra. Henri Martin (1907-1910) found two pendant beads at the Mousterian site of La Quina, a reindeer phalange apparently bored through on each face in the manner of the Bocksteinschmiede vertebra, and the canine of a young fox that was fractured during the process of boring and was abandoned (Figure 17.9). The La Quina beads were for

Figure 17.4. Arcy-sur-Cur (Yonne, France). Bone pendant with a wide carved hole made in the tradition of the earlier Pech de l'Azé and Arcy-sur-Cure pendants. Aurignacian period.

almost three-quarters of a century neglected in the archaeological literature. One reason, perhaps, is that the early excavation did not adequately establish the level from which the beads had come. A more important reason was that it was universally believed that personal decoration could not possibly have existed in the Mousterian period of Neanderthal man. An added reason for their neglect was the dramatic series of discoveries of Neanderthal burials that began to be excavated in Europe and the Middle East. It was thought that these represented the incipient, primitive beginnings of symbolic thought and of a belief in life after death. No beads were found in the Neanderthal burials and it was assumed that, if beads had existed, they would have become grave goods.

Figure 17.5. A fossil nummulite from Tata (Hungary). The vertical line is a natural crack that descends through the fossil. The horizontal line was apparently engraved by a Neanderthal to make a 'cross'. Mousterian period.

The presence of these artifacts from Bocksteinschmiede, Pech de l'Azé, Arcy and La Quina, however, raises a number of additional questions. A bead or pendant requires a string for hanging or attachment, whether by strands of hair, gut or strips of hide. A bead or pendant, therefore, requires a second technology and a knowledge of different classes of materials and resources as well as different sets of skills. This multiple, hierarchical level of problem-solving, resource acquisition, planning and production is evident in many of the Upper Palaeolithic symbolic traditions, and it is present as well, as we shall see, in Mousterian traditions.

Because of the evidence for a pre-Aurignacian use of diverse materials for a range of symbolic purposes, it is necessary to raise again the question of a possible use of wood for carving small and large symbolic artifacts. The historic ethnographic record is rich in evidence for carving a wide range of symbolic products from wood. Microwear studies (Anderson-Gerfaud, this volume; Shea 1989) have documented a Neanderthal use of stone tools for working skin, wood and even grasses. The archaeological evidence of early Palaeolithic artifacts of wood is sparse but significant. There are the Acheulian spears from Europe but there are, as well, examples of decorated wooden rods, and some of bone, apparently from late Middle Stone Age levels at Border Cave, South Africa (Beaumont 1973). These have been studied by the author and they are stylistically

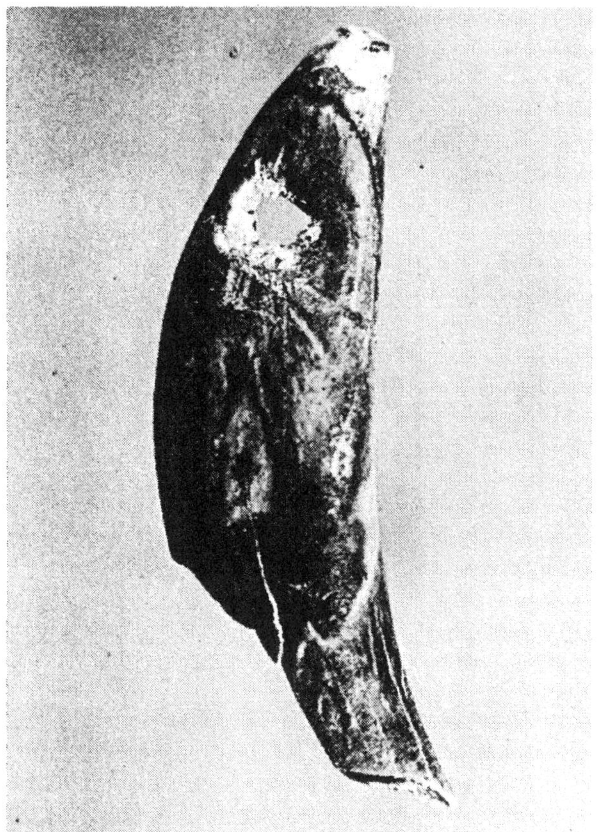

Figure 17.6. Animal tooth bead with the hole made by criss-crossing engraved lines and gouging, from Bacho Kiro (Bulgaria). Early Aurignacian period (photo by J. Kozlowski).

similar to marked and decorated points and bones from the European Upper Palaeolithic. Decorated bones from this period in South Africa have been found at other sites (Singer and Wymer 1982). The South African wood and bone decorations were apparently made by another branch of archaic *Homo sapiens sapiens.* There is nothing, therefore, in the evolving or evolved 'human' capacity or technology that would preclude a use of wood or bone for symbolic purposes at the presumed juncture separating anatomically modern humans and the Neanderthals except a *priori* theoretical assumptions of 'impossibility'.

The evidence for use of a diverse range of materials and resources for symbolic purposes in the European Middle Palaeolithic and earlier is rather complex. From the Mousterian site of Tata, Hungary, there comes the well-known, beautifully shaped non-utilitarian oval plaque (Figure 17.11) that was carved from a single lamelle of a compound mammoth molar (Figure 17.12). The rear of the plaque was bevelled back to remove

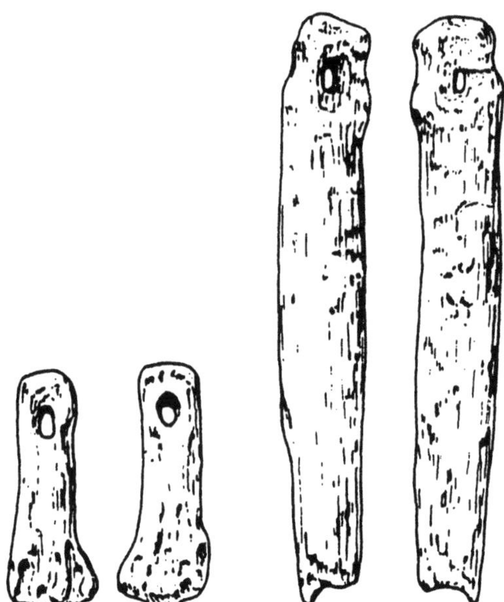

Figure 17.7. Bocksteinschmiede (Lonetal, Germany). A swan vertebra and a wolf foot bone (metapodium), with holes bored through at the top to make beads or pendants. Micoquian period. After Wetzel and Bosinki 1969.

the soft material that would have broken with persistent handling. The edge of the plaque shows the high polish of long-term handling, perhaps at times of ritual or ceremony, while the main face indicates that it was covered with red ochre. The use of ochre, of course, has been documented as far back as the Acheulian, is even more common in the Mousterian and already is abundant in the Châtelperronian. Again it is interesting that the use of ochre has been documented for the Middle Stone Age in Swaziland, South Africa (Boshier and Beaumont 1972), where a mine was in operation to secure haematite rich in specularite. It is evident from these data that the capacity to acquire and use a variety of materials for symbolic purposes does not appear at one place or time or among one group of hominids.

Apart from being carved of an 'exotic' material and being covered with ochre, the Tata plaque was apparently made to be used over a period of time. Not only was there planning involved in securing and carving the lamelle, but there was a higher order of planning, requiring a knowledge of the ritual or social context for which the plaque was intended. I have indicated in previous publications (Marshack 1975, 1979, 1984b, 1985b, 1987b, 1989b) that early symbol systems were often 'time-factored'. They involved images, signs and symbols that were often intended for long-term, continuous use or for use at particular times. They were intended to function at the proper place and time and, therefore, helped to structure and maintain the cultural fabric. It follows that the more complex a culture becomes, the more complex the set of symbolic markers and referrents

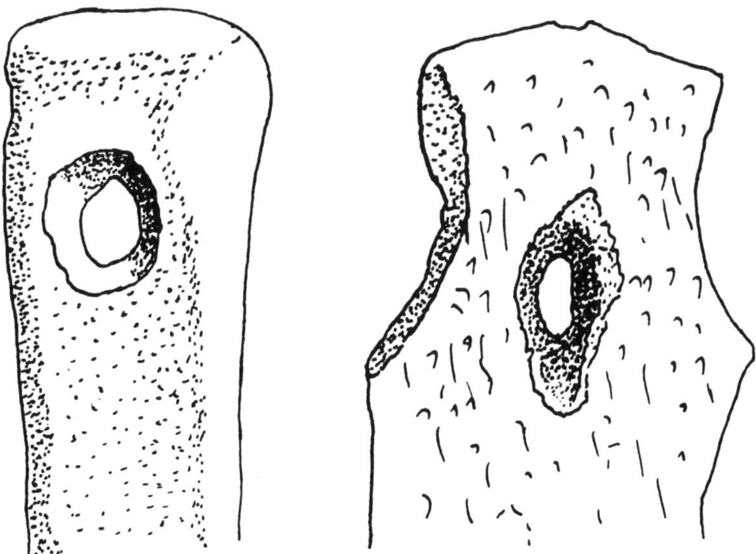

Figure 17.8. Close-up renditions of the incised and carved holes on the Bocksteinschmiede vertebra and footbone.

becomes.

The accumulating evidence for Mousterian symbol is, therefore, an indication of developing *complexity*. One does not know the context within which the Tata plaque was used, but it would seem, nevertheless, that we have an indication of a fundamental, common human symbolic mode: planning and production for later ritual or symbolic use. The suggestion, if added to a recent suggestion for Mousterian hunting of large animals and of animal drives (Jelinek *et al.* 1988), and for a developing, conceptual, perhaps seasonal modelling and mapping of the productive territory and its resources (Marks 1988a, 1988b; Geneste 1988) moves us closer to a recognition of the human or near-human capacities of the Neanderthals (Note 1).

Harrold (1989) suggests that the fragment of bone with a carved hole from Pech de l'Azé may have had a practical rather than a symbolic function. The suggestion raises interesting questions concerning the developing human capacity, questions that have not been addressed before and that have relevance for ongoing discussions concerning the origins of beads and pendants. For at least a century, there have been deepening discussions and studies of the hominid use of tools for cutting and hammering. What has not been discussed in the developing literature is the use of a far more ephemeral 'tool' and concept, the hole, a 'tool' that becomes increasingly important in the Aurignacian and that continues to develop in the later Upper Palaeolithic. In early Aurignacian levels at the Vallon des Roches whose sites (Blanchard and Castanet) have supplied us

Figure 17.9. La Quina (Charente, France). Reindeer phalange with a hole bored through both sides at the top, and the canine of a young fox with a hole that was begun but was terminated when the tooth apparently split. Mousterian period (photos by H. Martin).

with hundreds of beads (White 1989), one finds large and small holes ('*anneaux*') deliberately carved in the limestone walls and ceilings of the shelters and on limestone blocks found on the shelter floor (Figure 17.13) (Delluc and Delluc 1981). Some of the holes apparently served functionally as anchors to secure tents or hang goods, but at least one anneau from Blanchard served symbolically as a 'vulvar' hole, since it is in direct contact with an incised phallus and vulva.

From the Aurignacian also come the first 'batons', objects of antler, bone and ivory which contain large, wide holes like those carved on the Pech de l'Azé and Arcy pendant fragments. The batons were at first undecorated objects apparently made for utilitarian purposes, perhaps for straightening shafts, softening thongs, or twining cords. Aurignacian batons occur not only in the Dordogne, in France, but at Giessenklösterle in Germany (Figure 17.14). The baton from Germany has a series of holes that are threaded internally like a screw, suggesting its possible use in plaiting or twining cords. The baton is usually found with a single hole, but the baton with many holes is also found in the Magdalenian. A reindeer antler baton from the Magdalenian of Le Soucy, the Dordogne, has seven holes, reminding one of the earlier Giessenklösterle baton. It is interesting from the point of view of cultural development that the baton as a long-term, curated object eventually becomes, as well, a surface for carrying different types and classes of symbolic marking, including signs, symbols, notations, animals, and compositions that in the late Magdalenian are as complex as sanctuary cave tableaux (Marshack 1970a, 1972a, 1975). The development of the baton as a multipurpose symbolic object will be considered elsewhere. Here it is necessary to note only that in the Aurignacian it already represents another instance of the use of the hole, functionally and conceptually. It is probably from these early traditions of making and using the hole and of preparing strings for their use, that the

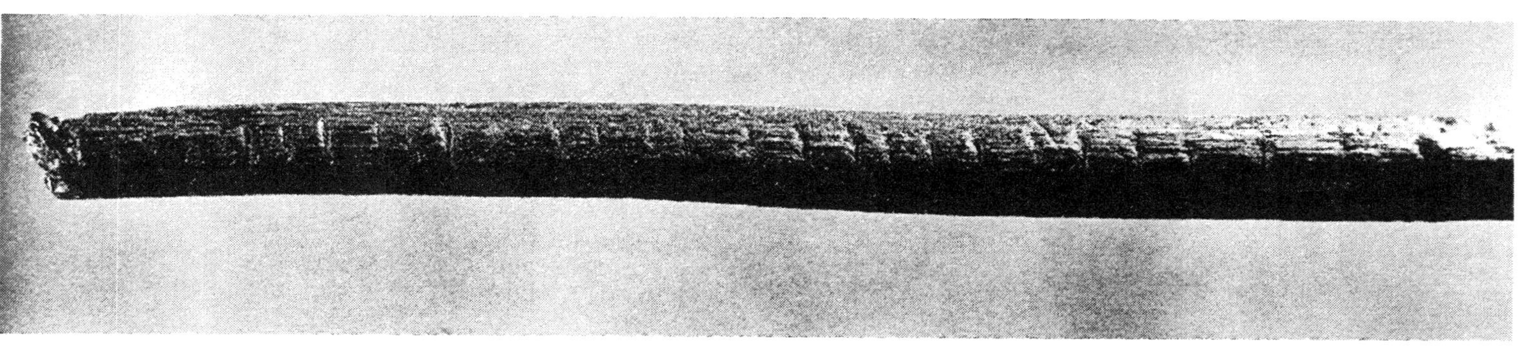

Figure 17.10. Section of a decorated wooden rod or dart, from Border Cave, South Africa. Middle Stone Age.

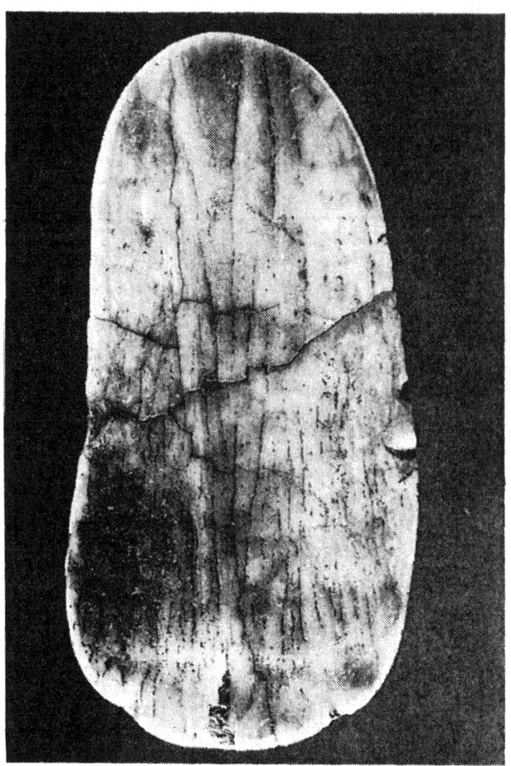

Figure 17.11. Tata (Hungary). Left: the front face of a carved oval plaque made from one section of a mammoth molar. The surface contains the remnants of red ochre, while the edges show the polish of long handling. Right: portion of the rear of the plaque indicating the bevelling of the softer material. Tool striations can be seen by microscope in the bevelled area. Mousterian period.

needle, one of the major 'inventions' of the later Upper Palaeolithic, derives. A comparison of early Aurignacian beads and their bored holes with a range of later needles clearly evidences the underlying conceptual similarity. The systematic shaping of beads and boring of holes and the extraction and shaping of bone points from bone and antler clearly prepared the way for 'invention' of the needle (Figure 17.15).

The most intriguing use of the hole in the Upper Palaeolithic, technologically and conceptually, is in the development of the many-holed flute, a 'non-utilitarian' object that was probably used in ritual and ceremony. It apparently first appears in France in the late Aurignacian and Périgordian (Saint-Périer 1950; Daleau 1963). The most beautifully made of the Upper Palaeolithic flutes, with four holes above and two underneath, comes from the Magdalenian of Pas-du-Miroir, La Roque (Peyzac, Dordogne) (Figure 17.16). The flutes of the Upper Palaeolithic, like the mammoth bone drums of Mezin, probably do not represent the 'begin-

Figure 17.12. The compound molar of a mammoth. One of the lamelles of such a tooth was used to make the Tata plaque (see Figure 17.11).

nings' of music but rather a use of a rapidly developing bone technology for the creation of diverse symbolic artifacts once made of perishable materials. Flutes and drums are still often made of reeds, wood and skins. It would be surprising if flutes and drums began to be made first in Europe of bone and later by other cultures of more accessible perishable materials.

The flute is important not only as an instance of developing cultural complexity. The neurological complexity involved in the two-handed manufacture of the instrument and in coordination of the separate actions of the two hands with the breath while playing, in order to produce a musical sequence that is determined by an acoustic evaluation of the rhythm, tone and pitch, is extraordinary. Playing the flute requires sequences of right and left hemisphere participation; it requires manipulative, acoustic, visual and breath coordination; it requires a recognition of the proper time and place for its cultural and ritual use. While music is neither personal decoration, sign, symbol or language, the cross-modal, associational, neurological capacities involved are an end-product of the same evolutionary mosaic process and exaptive effects that led to the possibility of symbolic image making and probably to aspects of language.

Here I must note that cultural developments or innovations are seldom 'invented' out of blinding inspiration and without preparation. The flute, the needle and the bead were not 'invented' by one person, at one site or at one moment. Their appearance was prepared for at many levels, biological, cultural and technological. The concept of the hole, as a 'tool' or a technology made possible by the two-handed competence, played an interesting and varied role in the late stages of hominization. It was, of

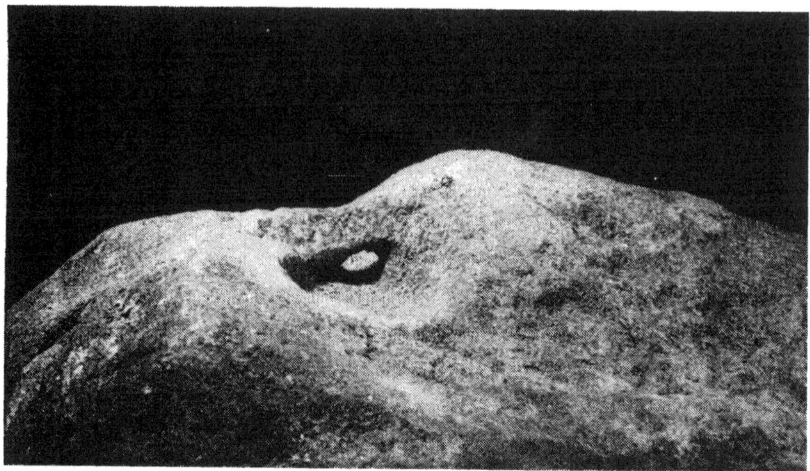

Figure 17.13. Carved and gouged *anneau* or hole in a huge limestone block, found on the floor of the site of Blanchard (Dordogne, France). Aurignacian period.

course, part of the concept of the retainer and container. It is evident in the early use of the digging stick. A digging stick made from a mammoth rib, c. 250 000 BP, has recently been found at Abri Vaufrey, Dordogne (Rigaud and Geneste, personal communication). Later in the Mousterian, the concept of the hole is involved in the problem-solving that goes into the making of graves and fire pits. It is part of the problem-solving involved in setting up poles for tents or shelters. It was probably involved in forming holes in skins or hides when making clothes, and it was present in the hafting of points into shafts or grips. The resin used in such hafting has been recovered from the German sites of Konigsaue and the Bocksteinschmiede (Bosinski 1985). At Konigsaue the resin still adhered to a retouched stone tool. It is this accumulating set of data that poses again the possibility of a Mousterian provenience for the La Quina beads or pendants, since clearly the capacity, the technology and the concepts were all present.

THE CONCEPT OF SELF AND THE OTHER

As noted above, beads, pendants and other items of personal decoration do not appear in the archaeological record outside of Europe during the period of the Mousterian, and rarely in the Upper Palaeolithic. Only a use of ochre is so far in evidence. We cannot assume for this reason that there was an absence of the capacity for personal decoration, social display, social complexity, a lack of self-awareness, or the absence of language. Such assumptions derive more from theoretical models than from the presence of data.

The capacity for a recognition of 'self', including even a use of the mirror, cosmetic powder and rouge, has been documented for chimpan-

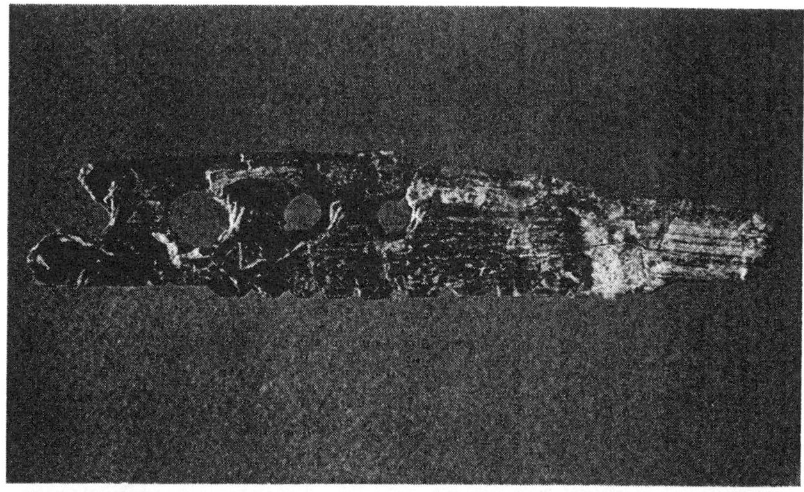

Figure 17.14. Mammoth ivory *baton* with carved holes of different sizes. The holes are carefully 'threaded' in a screw-like fashion. Vogelherd, Germany. Aurignacian period.

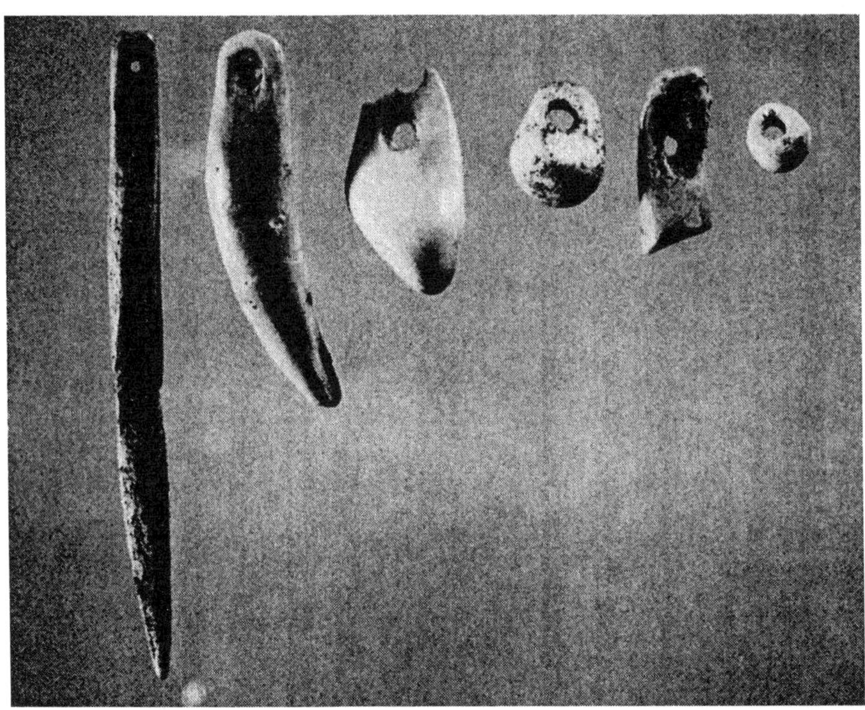

Figure 17.15. Aurignacian bone and tooth beads or pendants from the sites of Blanchard and Castanet in the valley of Castelmerle (Dordogne, France), with a Magdalenian bone needle, indicating the prevalence of the 'threaded hole' as a concept across a span of millennia. Whether worn as pendants, beads or as attachments to clothes, the early products required the technique of stringing or attachment by strings. Accumulating evidence suggests that the complex forms of problem-solving involved were present in Europe in the Châtelperronian period and far earlier.

Figure 17.16. Bone flute with four holes above and two below, from the Pas du Miroir, La Roque (Dordogne, France). Magdalenian period. A similar flute is known from the Upper Palaeolithic of the USSR. Earlier multiple-hole flutes appear in the Périgordian period..

zees in a laboratory context. In the wild, the sense of 'self' is incipient among the chimpanzees, if not consciously defined, in the playing out of age, sex, and dominance roles in social contexts. Hominization would have expanded these capacities at the same time as the generalized capacity for different types of categorization and social differentiation developed. The use of ochre indicates the early presence of a recognition and marking of self, but the capacity had strong genetic roots.

Though rare, examples of Mousterian 'personal decoration' that go beyond the much discussed categories of ochre and beads may exist. The most intriguing and suggestive evidence for such a tradition was the discovery at the site of Hortus (Valflaunes) that late Neanderthals hunted and killed leopards and other felines, apparently for their hides (Lumley and Lumley 1972). One leopard was represented by parts of the skull, footbones and tail, suggesting the presence of the full skin as a costume. Ethnography documents historic hunter-gatherer and herder practices in Africa, in which the hunter who has killed a lion or leopard and wears the skin achieves a heightened social status. In other cultures the skins of a variety of 'powerful' animals are worn in rituals. In such use, the skin or hide becomes a form of differentiation, a cultural form of marking self, and a form of social decoration. The hide, however, not only marks the individual but also represents the 'power' and 'spirit' or the myth of the animal. When the Hortus materials were first published, the concept of personal decoration in the Mousterian was widely considered to be impossible. The possibility has, I hope, been established, on both theoretical evolutionary and artifactual grounds. The possibility has been strengthened by a unique recent find from the early Aurignacian. The evidence is complex and necessitates a reassessment of early Upper Palaeolithic representational imagery.

The earliest examples of animal carving, c. 32 000 BP, come from the German Aurignacian site of Vogelherd (Lonetal), not far from Bocksteinschmiede on the Lone. The tiny mammoth ivory carvings include felines (lion and leopard), as well as food animals such as bison, horse, mammoth and reindeer. The images were at first considered to be examples of hunting magic intended to assure success in obtaining food. Lions and leopards, however, have never been regular items of diet. A microscopic examination of the Vogelherd carvings by the author (Marshack 1984b, 1988b) has revealed that the animal images, including the felines, were often marked and overmarked as though in periodic ritual. The carvings seem to have been symbols that were kept and used over a period of time. These analytical data assumed new meaning with the recent reconstruction of a carved anthropomorphic figure from Hohlenstein-Stadel (Lonetal), a site near Vogelherd (Seewald 1983).

The carved mammoth ivory figure (Figure 17.17) depicts a standing human, apparently male, with a feline head. The image looks like an animal-headed god from the late dynastic period of Egypt, but it is clearly indigenous and far earlier. If the leopard bones from Hortus suggest the possibility that leopard skins were worn by Neanderthals, perhaps to

Figure 17.17. Hohlenstein-Stadel (Lonetal, Germany). A mammoth ivory carving of an anthropomorphic figure with a lion head. Aurignacian period.

capture or encompass the feline spirit or to mark the manhood of a hunter, the Aurignacian carving from Hohlenstein-Stadel may represent the capture of that power and spirit in the more sophisticated form of a manufactured image. The carved image could perhaps now be used in rituals in a manner that was comparable to the earlier use of a leopard skin.

Data from both the Mousterian and Upper Palaeolithic suggest such a possibility. In the late Magdalenian there are images of 'sorcerers' wearing animal skins, antlers and horns in what seem to be depictions of ritual dances. It was long believed that these images represented a late Magdalenian style and development. The Hohlenstein-Stadel carving suggests that the tradition may have been tens of thousands of years older. Evidence for a comparable early tradition comes from South Africa. One of the earliest animal images found outside of Europe and made in the period of the Upper Palaeolithic comes from the Apollo 11 cave, *c.* 23 000 BP. One of the crude paintings found on a rock depicts a lion with human feet (Wendt 1976). David Lewis-Williams, a specialist in the rock art and ethnography of South Africa, reports that the medicine-men of the San bushmen during a dance trance are often overcome by the lion 'spirit' and assume the personality of a lion. Lewis-Williams reports, as well, that important symbolic animals are both danced and painted in the rock art. San dancers, like dancers of other primitive peoples, frequently wear parts of the animals they dance, including the ears of the spiritually powerful eland. The Hortus and Hohlenstein-Stadel evidence suggests the extremely early presence of a common symbolic mode of animal image manufacture and use. The difference between the Mousterian and Upper Palaeolithic modes of animal image use may at first have been more technological than conceptual.

The Aurignacian period introduced a new lithic technology and new skills for working bone. The Vogelherd and Hohlenstein-Stadel carvings may, therefore, have represented a qualitative step forward, ultimately derived from earlier traditions involving the symbolic use of animal skins and parts, instead of the sudden conceptual leap forward into the 'invention' of animal imagery and symbolism that the Vogelherd and Hohlenstein images seem to suggest (Note 2). As we have seen, there was apparently a similar prior incipience and preparation for the Aurignacian pendants and beads. Animal parts, in fact, continued to play a major role through the West European Upper Palaeolithic. In the Magdalenian there are carved and engraved amulets, pendants and engraved images that represent animal parts: a fish tail, a horse hoof, a bison foreleg, a reindeer antler, the rear flippers of a seal, an ibex head, the eye of a cervid, a horse skull, an animal jaw, etc. These animal parts were all apparently symbolically relevant images. The Hortus and Hohlenstein-Stadel material, therefore, raises questions concerning the possible beginnings of animal art and the early symbolic use of animals and animal parts. The use of animal parts is, of course, profusely documented among hunter-gatherers in the historic period, particularly in shamanistic use. The possibility of an early use of animals and animal parts is suggested, for instance, in an apparent Neanderthal 'ritual' described by Solecki (1982). In the Mousterian cave shelter of Nahr Ibrahim in Lebanon the bones of a fallow deer (*Dama mesopotamia*) were gathered in a pile and topped by the skull cap. Many of the bones were unbroken and still articulated. Around the animal were bits of red ochre. While red ochre was common in the area and so may have been introduced inadvertently, the arrangement of the largely unbroken bones suggests a ritual use of parts of the animal. There is other evidence that animal bones and parts were used symbolically in the Mousterian. In the burial of two anatomically modern humans at the Israeli sites of Qafzeh and Skhūl, the mandible of a wild boar was placed in the hands of one individual and the antler of a fallow deer in the hands of a child. The tradition may have come from an earlier Mousterian culture. The possibility that the Upper Palaeolithic development of animal art is ultimately referrable to earlier symbolic usage opposes a century of theories concerning the origins of art. The use of animal bones, animal teeth and even sea shells as items of personal decoration may also have some incipience in these earlier traditions.

The Abbé Breuil, the central figure in the study of Upper Palaeolithic art in the first half of the century, suggested that art probably began with simple meandering doodles within which images of animals were accidentally recognized (Breuil *et al.* 1915). Leroi-Gourhan (1965) later suggested that animal art in the Franco-Cantabrian area began with the simple crude animal outlines found in the Aurignacian period. Both theories are contradicted by the Hohlenstein-Stadel and Vogelherd materials and their implications. In this regard, it is important to note that in both instances, the ivory carvings are surprisingly sophisticated and seem to represent the end-product of an ancient tradition rather than the archaic and primitive beginnings of something new.

We have taken a wide and circuitous route in time and space to indicate that the effort to draw profound theoretical conclusions from the presence of Upper Palaeolithic beads and by reference to current information theory is inadequate when dealing with the complexity of the Palaeolithic materials and the problem of the evolving human symbolic capacity.

SYMBOLIC MARKING

Contemporary discussions concerning the origins of self, of social complexity, information exchange, intergroup networking and 'style' as information, while insightful in dealing with the general theoretical problems involved in the use of symbols, have not dealt methodologically or analytically with the Upper Palaeolithic traditions or materials. I have for years argued that, before undertaking sweeping generalizations about the origins, meanings and uses of art and symbol, it is necessary to undertake methodological and analytical studies of the widely dispersed and varied Upper Palaeolithic materials at first hand and to do comparative and chronological studies of the developing traditions (Marshack 1984b, 1985b, 1986b, 1986c, 1987a, 1987b, 1988b).

The European Upper Palaeolithic involves one of the most complex

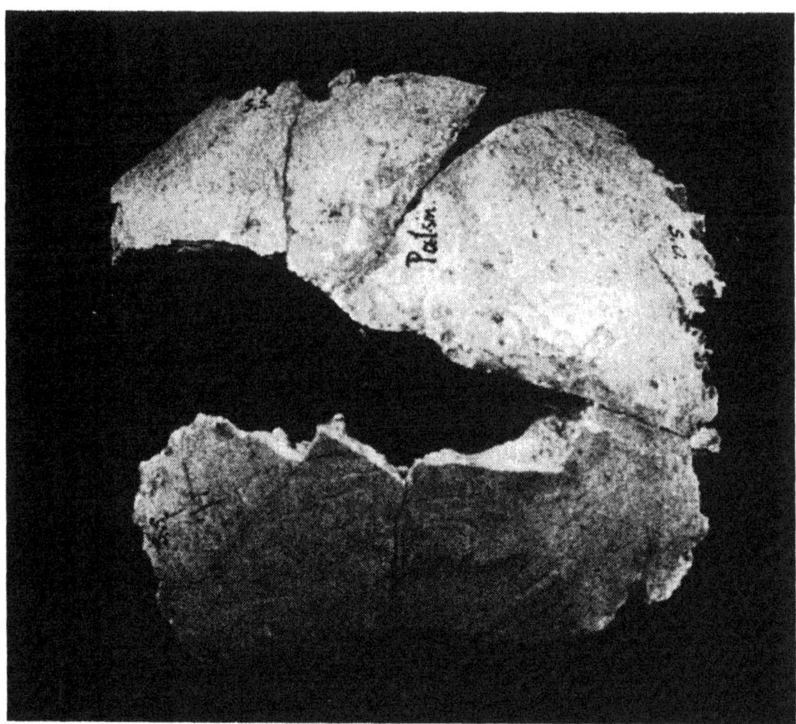

Figure 17.18. Krapina (Yugoslavia). Fragment of a Neanderthal skull cap from a multiple grave. It was originally claimed that the skull had been scraped by a flint tool long after the flesh had decomposed, in order to remove the remnants of dried adhering flesh. (See p. 488). Photo and drawing by Mary Russell.

cultural developments of a hunter-gatherer society in the record of anatomically modern man. In the last two decades the presence of more than two dozen symbol systems, each with its own iconography and mode of use, has been reported and the author has already published the analysis of many of these systems. The need to discuss systems and traditions rather than objects and style may be clarified by noting that in a modern industrial society cultural 'literacy' involves the ability to write, tell time, read tables, charts, maps, measuring devices, use the computer, and read all types of symbols (religious, political, national, technological, popular), as well as traffic signals, road signs, forms of dress, etc. (Marshack 1986a). In a non-literate society such as that found among the Australian aboriginal, the cultural products are different but the symbolic complexity is probably as great.

Attempts to categorize classes of imagery or to interpret the meaning and use of Upper Palaeolithic imagery on the basis of how these images appear to us, visually, and without investigating how different classes of imagery and symbolic artifacts were made and used, has led in the last century to a host of overly simple interpretations. One of the significant attempts to determine a system and set of relationships among the images was undertaken by Leroi-Gourhan (1965) in his study of the Franco-Cantabrian caves. Unfortunately, these innovative and profoundly influential studies were largely based on surface appearances, on how the images appeared to a modern Western viewer. There was no attempt at the internal analysis of any one image or composition. The problem is being addressed by Leroi-Gourhan's students and followers who have begun the methodological internal analysis of different classes of imagery (Delluc and Delluc 1978; Vialou 1981). The problem of overly-simple visual analysis and interpretation of the imagery, however, still exists.

White (1989: 377), for instance, states that

> At Abri Blanchard, Abri Castanet and La Souquette meandering rows of punctuations were used to decorate several bone and ivory objects, including pendants. This pattern seems to mimic that found on Atlantic sea shells from the same level. Both the pattern and the probability that it was inspired by the natural punctuations on exotic sea shells has seemingly been ignored by Marshack (1972) in arguing for lunar notation....This appropriation of a natural pattern is further verified by the fact that these same shells, complete with meandering punctuate pattern, were replicated in ivory at La Souquette and worn as pendants.

There is a failure here to differentiate between the categories and classes of objects being discussed and a lack of familiarity with the Upper Palaeolithic traditions of marking and decoration (Marshack 1987b). The Blanchard plaque was a working tool, a fine retoucher, and not a pendant, and such classes of curated objects were marked differently in the culture. The plaque was not intended to represent a sea shell any more than the Blanchard seal pendant, with its rows of dots, was intended to represent a sea shell. The mature grey seal may at times have a brown and black

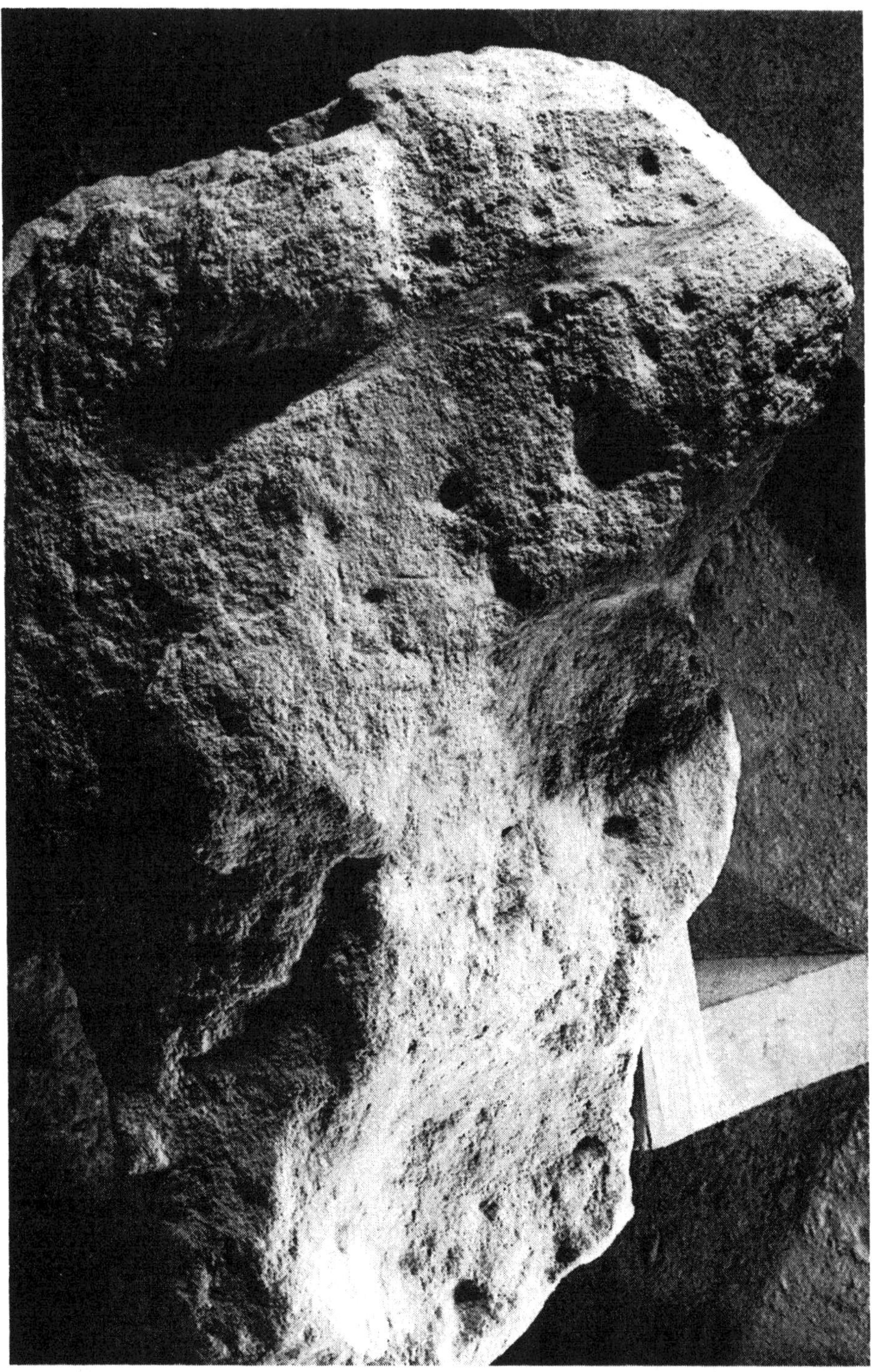

Figure 17.19. La Ferrassie (Dordogne, France). Large limestone block that covered a buried Neanderthal child, with randomly-placed cupules, most of which are made in sets of two. Mousterian period.

spotted pelt. The representation of species characteristics and animal pelage was common in the Franco-Cantabrian Upper Palaeolithic (Marshack 1988b). For instance, the spots on a certain species of sea shell, the spotted pelts or skins of the female reindeer, the trout, the salmon, the frog, the salamander, the serpent, etc. were all clearly depicted, as was the seasonal moulting of certain species.

The use of dots as 'decoration' or as symbolic marking occurs throughout the European Upper Palaeolithic, even on ivory objects from Moravia and Siberia. They occur profusely in the sanctuary caves where rows of painted dots, some made by a finger dipped in paint, others by a pad or by blowing paint, occur (Pech Merle, Combel, Trois Frères, Lascaux, Niaux, Castillo, etc.). There are meandering rows of dots gouged into limestone blocks in the Aurignacian shelters of the Dordogne, which, like the cave dots, suggest a form of ritual marking. The earliest marking of dots in this area of the Dordogne may occur in a Mousterian burial at La Ferrassie (Figure 17.19). White is, therefore, arguing not from a knowledge of the traditions involved, but from the few images he has had in hand.

The marking found in the engraved notations of the Upper Palaeolithic are analytically entirely different from the sets of dots mentioned above. They are accumulations that are not made on items of personal decoration or in difficult-to-reach areas within a cave. They are accumulations that are made on tools or intentionally formed bits of scrap bone that were kept or curated for long term use. Analysis of this unique class of artifact tells us as much about the developing social complexity of the Upper Palaeolithic as the beads or representational imagery. Microscopic examination of the Blanchard plaque, for instance, revealed that it was a fine retoucher of flint tools. The back edge was highly polished where it sat in the hand against the palm, and the front end was broken back by persistent use and pressure. The notches along the edge were intended for gripping and were highly polished and worn towards the rear where the plaque was held, but became increasingly less polished as one approached the point. The plaque was clearly a curated functional object that may have been carried about in a pouch for a considerable period to be used when and if needed. It was during this period of curation that sets of marks were carefully accumulated in a structured, linear, serpentine manner. The microscope reveals that these sets were made by different tools and styles of marking, unlike the uniform dotting on the beads or the seal pendant (Marshack 1970b, 1972a, 1972b, 1972c, 1975). The notational mode is documented on other Aurignacian bone artifacts from the Dordogne and with increasing and developing complexity in the Franco-Cantabrian Upper Périgordian, Solutrian and Magdalenian. The last known example, from the terminal Magdalenian (Magdalenian VI) of France, c. 9000 BP, found at the Grotte du Taï (Drôme), is the most complex of the Upper Palaeolithic notations, but is made on a small scrap of bone in a manner that derives from the Blanchard plaque (Marshack 1973, 1984a, 1985a).

At the Middle Magdalenian site of La Marche (Marshack 1972a, 1972b) an antler baton had been incised on one face with notational sets of marks made by different points and engraved in different directions. The baton had broken, apparently during use. The shaft had then been reshaped to become a pressure flaker and, during this second use, an entirely different set of notational marks and symbols had been engraved on the second face. The published microscopic evidence is so powerful that I have repeatedly asked that they be reexamined to verify the findings.

The multiple and variable use of artifacts and the reshaping and reworking of artifacts is well documented among hunter-gatherer societies. Discussion of the mode among the Upper Palaeolithic symbolic materials has not been common. The ongoing programme of analysis has documented the reuse and reshaping even of engraved and painted signs in the sanctuary caves, sometimes completely altering the original shape and design. Such processual studies go far beyond the categorizations of 'style', theories of information exchange, or the mere comparison of motifs.

DEATH AND BURIAL: SYMBOLS FOR THE LIVING

Of the few symbol systems that have been discussed in comparison of the Mousterian/Upper Palaeolithic cultures and transition, none has had longer and more intense scrutiny than the burials and treatment of the dead (Note 3). It is significant that the lack of personal adornment in Mousterian burials has often been used to argue for an absence of social complexity, status and rank and, by implication, for a comparative lack of symbolic capacity (S. Binford 1968; Harrold 1980; Chase and Dibble 1987). Soffer (1985), in a study of the Upper Palaeolithic cultures of the Russian plain, noted that the early burials at Kostienki and Sungir are rich in grave goods, including personal decoration, but that the later burials lack them. The symbolic complexity in these later cultures was apparently played out in other modes (Marshack 1979, 1987b). A large part of the argument concerning the 'poverty' of Neanderthal burials was initiated in 1868 when anatomically modern skeletons were found in an Aurignacian burial in the cave of Cro-Magnon, Les Eyzies (Dordogne), together with large numbers of sea-shell beads, some of which came from the Atlantic and the Mediterranean. These beads and their presence in an early Upper Palaeolithic burial became a standard against which other burials were compared. But there is a problem with such comparisons. The Upper Palaeolithic represents a temporal and regional European development; beads are not well documented in that period outside of Europe, while many of the Mousterian burials, like other instances of symboling, precede or are more complex than those found elsewhere in the world during that period. Clearly the problem of the supposed poverty of Neanderthal burials and the implications of lesser symbolic complexity and capacity in the Mousterian burials needs reevaluation.

Discussions in the field of archaeology, as in other fields, always have a historical basis, with changing areas of central contemporary concern. When certain aspects of Neanderthal behaviour began to be discussed earlier in the century, there was much made of the 'savage' custom of cannibalism as apparently evidenced by the tool-marked Neanderthal bones at Krapina, Yugoslavia, and the Neanderthal skull at Monte Circeo, Italy, which had an enlarged foramen magnum and a circle of stones apparently ritually placed around it (Blanc 1961). In addition, the seeming presence of a Neanderthal 'cave bear cult' with a ritual hunting of the cave bear was much discussed, based on the apparent intentional arrangement of bear skulls and bones in the cave of Drachenloch in the Swiss Alps (Bachler 1921, 1923). It was believed at that time that the Neanderthal burials and rituals represented the early, often mute beginnings and glimmerings of religion and an awareness of life-after-death.

In the second half of this century, when quantitative, materialist-social and analytical questions began to be asked, the forms of inquiry shifted. The grave goods found in Neanderthal burials were compared statistically to those in Upper Palaeolithic burials (S. Binford 1968; Harrold 1980), with the result that a lesser social and symbolic complexity was asserted for the earlier period, with little apparent evidence for a differentiation of status and rank, but a possible preferential treatment for males. A reexamination of the bear bones from Drachenloch revealed no evidence of human cutting or breakage (Jéquier 1975; Chase and Dibble 1987), and this apparently terminated the concept of a 'bear cult' and the idea of a systematic ritual hunting of the bear. It was suggested that natural forces, including underground streams, could have arranged the Drachenloch bones. There are alternative possibilities, however, that have not been discussed. In the Upper Magdalenian cave of Tuc d'Audoubert, there is a cave bear skull lying on the clay floor: next to it are the knee prints of a Palaeolithic visitor who had kneeled to remove one of the bear canines. The tooth may have become a powerful amulet containing the spirit of a species that had long since disappeared. When I was working in the cave of Gargas with ultraviolet light, an unseen cave bear canine suddenly fluoresced at the bottom of a pool of water. Though I was there for research, I nevertheless had a powerful feeling that this was a 'symbol' of the original inhabitants or 'owners' of the cave. It even seemed that the single tooth may have been tossed into the pool by a Palaeolithic visitor. I left the canine in the pool. A similar sense of the presence of the 'spirit' of the cave bear comes from viewing the claw marks on the limestone walls, some, as those in Cougnac, high above one' s head, suggesting an awesome, huge animal. The Neanderthals could have 'honoured' the original inhabitants of the caves, whose bones and claw marks may have been present, and have arranged them ritually without a cult of hunting and killing. This would not preclude a ritual killing of bears or even bear hunting. A Magdalenian engraving at Mas d'Azil seems to depict a ritual bear killing or baiting, and images of wounded and killed bears occur in the caves and among the mobiliary materials. The clay 'body' of a bear on the floor of the cave of Montespan was once thought to have been struck through by spears. Apparently the skull of a bear cub had been placed in front of the form and it may have been covered with a bear cub skin. Whether the clay form had been ritually stabbed or not (and Graziosi has cast doubt on the early suggestion) it is significant that the evidence suggests that there existed various forms of the symbolic use of animals and animal parts in the Palaeolithic, forms of use that apparently had a long history.

The question of Neanderthal cannibalism has recently been investigated by a reexamination of the human bones. Trinkaus (1985) studied the broken bones at Krapina and determined that they had not been broken to extract marrow, but had been crushed by the fall and pressure of overlying rock. Russell (1987) then reexamined the cranial and post-cranial bones from Krapina and seemed to find that the tool marks had been incised long after the flesh had decomposed. This suggested that the tool marks were the result of an effort to clean the bones of their last adhering remnants of flesh, with the cleaned bones being given a secondary burial (see also Le Mort 1988). T. White (personal communication) subsequently reexamined the Krapina bones and determined that the supposed 'cut' marks were due to natural causes and not to intentional later scraping (Figure 17.18). The suggestions of 'cannibalism', of 'secondary burial', of 'bear cults', etc. among the Neanderthals raise profound questions concerning the manner in which archaeologists and anthropologists tend to read contemporary meanings into the often fragmentary archaeological record. These interpretations almost always follow from the presence or absence of evidence in the record and usually entail some degree of a *priori* assumption. Chase and Dibble (1987), as noted earlier, have argued against a significant Neanderthal symbolic capacity on the grounds that the evidence in the record is *statistically* insignficant. On these grounds the presence in the European archaeological record of two or three late Acheulian wooden spears would be considered statistically insignificant, instead of being, as they clearly are, extremely significant because of the cognitive, problem-solving complexities implied in the choice of the best hard wood, the fire-hardening of the Lehringen spear, etc. The rare and idiosyncratic presence of the Tata plaque, the Beçov ochre, the Bocksteinschmiede beads, the Mousterian resin of hafting, or the Hohlenstein anthropomorph are, of course, *statistically* insignificant, but they nevertheless indicate a high order of cognitive or symbolic capacity both in their preparation and use. Their relevance, clearly, cannot be judged statistically or quantitatively. The idiosyncratic and rare example of the Aurignacian Hohlenstein lion-headed carving, while statistically insignificant, may, in fact, be one of the most important of all the early Upper Palaeolithic symbolic artifacts, requiring a fundamental reevaluation of most theories concerning the origins of symbol and art. There is a huge difference between the presence of an evolved and evolving biological, neurological capacity, as implied in even rare significant artifacts, and the historical, temporal development or *quantitative* presence of well-developed traditions involving cultural complexity and 'style'.

The trend to the quantification of Mousterian grave goods with an attribution of statistical 'poverty' in the Neanderthal burials similarly evades the problem of the cognitive complexity and symbolic capacity that may be evident in these traditions. There is evidence, for instance, that the Monte Circeo skull, with its enlarged foramen magnum, may not indicate the practice of cannibalism. Recent studies have shown that the foramen magnum represents a part of the skull that may deteriorate more rapidly than other parts and so present the appearance of intentional enlargement. The Monte Circeo skull, however, with its evidence of a severe blow to the right temporal region, and its apparent placement within an accumulation of stones, does suggest the specialized ritual treatment of one individual, who may have been killed in any of a number of ways, in a conflict with a neighbouring group, in an accident, or ritually. The ultimate significance of the skull may reside in this evidence of the specialized ritual treatment given to a single individual. Its uniqueness may, in fact, be its significance.

It is important in this regard to note that a large proportion of Neanderthal burials differ, despite their seeming simplicity, as though regional, individual and contextual variations were possible within the general tradition. These variations are probably as significant for an understanding of Neanderthal capacity and culture as the fact of burial itself. The most famous of the Neanderthal burials is that of a skeleton with an accidentally crushed skull at the back of the Shanidar cave (Solecki 1963; Stewart 1963). In 1968 Arlette Leroi-Gourhan examined samples of soil taken from the burial and determined that the body had been placed on a bed of pine boughs and had been covered with flowers (Arl. Leroi-Gourhan 1968, 1975). Specialized analysis of the soil from an apparently simple burial had uncovered what may have been a complex symbolic and participatory ritual act. The flowers provided a clue to the season of burial, but the reason for the unique use of flowers remains unknown. Was it due to the rank of the individual or the nature of his death? Did the flowers have symbolic healing powers, as suggested by Leroi-Gourhan? Would earlier analyses of other Neanderthal burials have indicated a burial, perhaps, in animal skins or the presence of other forms of perishable grave goods? While such questions are interesting they cannot at this time be answered. However, the range and variability in the Neanderthal burial tradition should begin to be recognized since it is this range, in conjunction with the range of symbolic artifacts, that will probably be of ultimate significance in the search for the Neanderthal symboling capacity.

At La Chapelle-aux-Saints (Corrèze, France) the leg of a bison was found with the skeleton. At Monte Circeo a ring or collection of stones surrounded the skull. In a number of burials the skeletons were tightly flexed and sometimes bound, either to save space as some have suggested, or to restrain the wandering 'spirit', a practice known among historic cultures. In some burials the skeletons were apparently aligned east to west as though in recognition of the direction of the rising and setting sun, a practice also known historically. In one Middle Eastern burial a simple stone was deposited. Among a number of Middle Eastern peoples in historic times, there is a practice of placing simple stones on a grave as an act of remembrance that will last longer than flowers.

These data suggest that many of the practices and concepts found in Mousterian burials may be in some ways comparable to those found later in post-Mousterian human cultures. Despite the suggestion that the quantitative 'poverty' in Neanderthal grave goods indicates an absence of status or rank and therefore of social complexity, not all burials of anatomically modern humans provide us with evidence of status or rank. Often status and rank, or the nature of a burial, are indicated in the elaborate rituals preceding incarceration, or in the ceremony following it. Besides, overt material displays of status and rank are themselves aspects of regional historical development, an aspect of culture rather than of innate cognitive capacity. The symbolic complexity of a burial tradition cannot always be read from the sophistication of the grave goods remaining in the ground. Some of the most elaborate historical rituals for disposal of the dead destroy both the body and the grave goods contributed to the ritual. As Soffer (1985) has pointed out, the early Upper Palaeolithic burials on the Russian plain are richer in grave goods and personal decoration than later burials, while I have indicated the extraordinary complexity of the other symbol systems in these later Russian cultures (Marshack 1979).

The apparent simplicity of the Neanderthal burials may, in fact, mask their true complexity. At the rock-shelter of La Ferrassie (Dordogne) a Neanderthal child was buried in an area containing five other burials. The child was covered with a huge limestone block that had cupules or 'dots', made in sets or pairs, gouged into the stone in a random, non-decorative manner (Note 4). These cupules have often been referred to as early examples of sign or symbol. They may be something entirely different, yet equally important. I have indicated in numerous publications that traditions of participatory ritual marking were present both in home sites and the caves during the Upper Palaeolithic. During the early Aurignacian in the Dordogne, for instance, there are limestone blocks on the floors of some shelters that have sequences of dots gouged into them, suggesting a form of participatory ritual marking. Some of these dots, as on one limestone block from La Ferrassie, are associated with incised vulvas. Since the dots, as we see them, are 'images', they have been referred to in the literature as signs and symbols. If, however, at the earlier burial of the Neanderthal child at La Ferrassie, each of the participants had gouged a set of marks on the stone, the act may have been intended as a gesture of participation and the marks may have been intended to last as long as the burial itself. Placing a stone or bone in a burial, placing a set of marks on a stone in a burial, or placing flowers in a burial, may have been conceptually related acts of symbolic participation, even though the end product in each case is totally different. With the La Ferrassie example, we are once again faced with the possible variability in Neanderthal symbolic behaviour despite the seeming simplicity and archaic nature of the evidence.

Though each instance of Neanderthal symbolic behaviour is unique, they do, as a group, document a potential variable capacity that is clearly related to the range of symbolic capacity found among the anatomically modern humans who followed them (Marshack 1988b, 1988c). This raises an evolutionary question that cannot be fully addressed in the present paper. It may be that evolution by natural selection screened for an increase in these conceptual capacities during the later stages of hominization. Clearly, aspects of these capacities were present among both groups, the Neanderthals and anatomically modern *Homo sapiens sapiens*. These suggestions concerning symboling capacity have relevance for the problem of possible Neanderthal speech and language.

THE NEANDERTHAL CAPACITY FOR LANGUAGE

The question as to whether the Neanderthals had 'language' and could speak has been under discussion for almost a century. When it was assumed that the group had a lesser capacity for symbolic thought, culture, social complexity and long-range planning, it was also assumed that they lacked a capacity for speech. It was the Upper Palaeolithic that provided the cultural and symbolic materials for such a comparison. As I have indicated, however, the European Upper Palaeolithic materials provide us with an essentially skewed historical and cultural regional development. Nothing quite comparable occurred during that period in other areas inhabited by anatomically modern humans.

Language is primarily a referential mode that operates in the vocal and auditory channels. It is a mode, however, that is almost totally dependent on *visual* referrents. It is an aspect of the human capacity for differentiating, categorizing and communicating information concerning those processes, objects, species, behaviours, relations and feelings that are of concern and relevance to human cultures. It is this set of observational and categorical capacities, and the ability to abstract and generalize from and about them, that is the true deep structure and foundation of language. All modern human symbol systems, whether supported by language, by imagery or by enactive forms of behaviour, mark and differentiate the diversity of the categories recognized as relevant in a culture. The question to be addressed, therefore, is whether the variability and complexity of the Neanderthal symbolic data, the increasing evidence for Mousterian cooperative hunting, the evidence for the functional and temporal mapping and modelling of a territory, the capacity to extract flint by mining, the capacity to haft points and work a range of materials, mean that there had to be language adequate to communicate information concerning that developing cultural variability and complexity?

Was evolutionary selection during hominization, apart from adaptive changes in morphology, involved in a process that tended to increase the capacity for practical problem-solving, conceptual modelling, and the ability to map and differentiate the developing cultural complexity in time and space? If so, to what degree was there a *qualitative* difference, *if any*, in these capacities between the Neanderthals and the contemporary anatomically modern *Homo sapiens*? Clearly the Neanderthals had these capacities, as well as aspects of developed material and symbolic culture. To what degree was the Upper Palaeolithic European cultural 'explosion' due to the incipient and preparatory traditions developed or assayed within Europe during the Mousterian? What relation do these data and questions have for contemporary models and debates concerning the origin and dispersal of anatomically modern humans (Smith 1985; Wolpoff 1986; Cann *et al.* 1987; Stringer 1982, 1985, 1987, 1989)? The recently suggested date of c. 92 000 BP for Qafzeh and proposed models for a South African origin for modern *Homo sapiens sapiens* leave unanswered questions concerning the relevance of the indigenous Eurasian development of Mousterian culture and the comparative lack of symbolic evidence in this period within the areas of supposed African origin and dispersal of anatomically modern humans.

NOTES

1. Dibble (personal communication) has indicated that while the plateau above La Quina and the blind drop-off to the valley would have been perfect for animal drives, the faunal remains do not yet confirm such behaviour.

2. Marshack (1989) reports on the evidence for diverse items of personal decoration in the Upper Palaeolithic made of perishable materials, including bracelets, arm bands, anklets, collars, body bands and belts.

3. Gargett (1989) has discussed the many difficulties with interpretations of Neanderthal burial. The associated comments were equally divided in supporting and opposing his analyses and interpretations.

4. Arensburg *et al.* (1989) report the presence of the hyoid bone in the Kebara, Israel, burial of a Neanderthal, suggesting that it indicates the possibility of speech. Laitman (personal communication), who believes that the Neanderthals could 'speak' (Laitman 1983; Laitman *et al.* 1989), nevertheless disputes the fact that the mere presence of the hyoid bone can indicate a capacity of speech. One would have to know the position of the hyoid in the larynx. Since the hyoid in not articulated to other bones that, apparently, is not possible. Laitman and Lieberman discuss the problem in a response to Arensburg *et al.* to appear in *Nature*.

REFERENCES

Arensburg, B., Vandermeersch, A.M.B., Duday, H., Schepartz, L.A. and Rak, Y. 1989. A Middle Palaeolithic human hyoid bone. *Nature* 338: 758-760.

Bächler, E. 1921. Das Drachenloch ob Vätis im Taminatal, 2445 m U.M. und seine Bedeutung als paläontologische Fundstätte und prähistorische Niederlasung aus der Altsteinzeit (Paläolithikum) im Schweizerlande. *Jahrbuch der St. Gallischen Naturwissenshaftlichen Gessellschaft* 57: 1-144.

Bächler, E. 1923. Die Forschungsergebnisse im Drachenloch ob Vätis im Taminatale 2445 m u.M. Nachtrag und Zusammenfassung. *Jahrbuch der St. Gallischen Naturwissenshaftlichen Gesellschaft* 59: 79-118

Beaumont, P.B. 1973. Border Cave: a progress report. *South African Journal of Science* 69: 41-46.

Belfer-Cohen, A. 1986. The evolution of symbolic expression through the Upper Pleistocene in the Levant as compared to Western Europe. In M. Otte (ed.) *L'Homme de Néandertal*. Vol. 5: *La*

Pensée. Liège: Etudes et Recherches Archéologiques de l'Université de Liège (ERAUL 32): 25-29.

Binford, L.R. 1981. *Bones: Ancient Men and Modern Myths*. New York: Academic Press.

Binford, L.R. 1983. Comment on White: 'Rethinking the Middle/ Upper Paleolithic transition'. *Current Anthropology* 23: 177-181.

Binford, L.R. 1985a. Ancestral lifeways: the faunal record. *AnthroQuest: L.S.B. Leakey Foundation News* Summer: 1, 15-20.

Binford, L.R. 1985b. Human ancestors: changing views of their behavior. *Journal of Anthropological Archaeology* 4: 292-327.

Binford, S.R. 1968. A structural comparison of disposal of the dead in the Mousterian and Upper Paleolithic. *Southwestern Journal of Anthropology* 24: 139-151.

Blanc, A.C. 1961. Some evidence for the ideologies of Early Man. In S.L. Washburn (ed.) *Social Life of Early Man*. Chicago: Aldine: 119-136.

Bordes, F. 1969. Os percé moustérien et os gravé acheuléen du Pech de l'Azé II. *Quaternaria*. 11: 1-5.

Boshier, A. and Beaumont, P. 1972. Mining in Southern Africa and the emergence of Modern Man. *Optima: Quarterly of the Anglo-American Corporation* 22: 2-12.

Bosinski, G. 1985. *Der Neandertaler und seine Zeit*. Cologne: Rheinland Verlag.

Breuil, H., Obermaier, H. and Verner, W. 1915. *La Pileta a Benaojan (Malaga)*. Monaco: Institut de Paléontologie Humaine.

Cann, R.L., Stoneking, M. and Wilson, A.C. 1987. Mitochondrial DNA and human evolution. *Nature* 325: 31-36.

Chase, P.G. and Dibble, H.L. 1987. Middle Paleolithic symbolism: a review of current evidence and interpretations. *Journal of Anthropological Archaeology* 6: 263-296.

Conkey, M. 1978. Style and information in cultural evolution: toward a predictive model for the Paleolithic. In C.L. Redman *et al.* (eds) *Social Archaeology: Beyond Subsistence and Dating*. New York: Academic Press: 61-85.

Conkey, M. 1980. Context, structure, and efficacy in Paleolithic art and design. In M.L. Foster and S.H. Brandes (eds) *Symbols as Sense*. New York: Academic Press: 225-248.

Daleau, F. 1963. *La Caverne de Pair-non-Pair*. Bordeaux: La Société Archéologique de Bordeaux.

Delluc, B. and Delluc, G. 1978. Les anneaux aurignaciens des abris Blanchard et Castanet à Sergeac. *Bulletin de la Société Historique et Archéologique du Périgord* 105: 248-263.

Delluc, B. and Delluc, G. 1981. Les anneaux aurignaciens des abris Blanchard et Castanet. *Le Périgord: les Anciennes Industries d'Aquitaine*. Périgeux: 30ème Congrès d'Etudes Régionales (1978): 171-192.

Dibble, H.L. 1983. Variability and change in the Middle Paleolithic of Western Europe. In E. Trinkaus (ed.) *The Mousterian Legacy: Human Biocultural Change in the Upper Pleistocene*. Oxford: British Archaeological Reports International Series S164: 53-71.

Dibble, H.L. 1984. Interpreting typological variation of Middle Paleolithic scrapers: function, style or sequence of reduction? *Journal of Field Archaeology* 11: 431-436.

Dibble, H.L. 1987. The interpretation of Middle Paleolithic scraper morphology. *American Antiquity* 52: 109-117.

Dibble, H.L. 1988. The interpretation of Middle Paleolithic scraper reduction patterns. In M. Otte (ed.) *L'Homme de Néandertal*. Vol.

4: *La Technique*. Liège: Etudes et Recherches Archéologiques de l'Université de Liège (ERAUL 31): 49-58.

Fridrich, J. 1976. Prispevek k problematice pocatku umeleckeho a estetick-ehno citeni u paleantropu: ein Beitrag zur Frage nach den Anfangen des kunstlerischen und aesthetischen Sinns der Urmenschens (Vor-Neandertaler, Neandertaler). *Pamatky Archeologickee* 68: 5-27.

Gábori-Csánk, V. 1988. Une mine de silex paléolithique à Budapest, Hongrie. In H. Dibble and Montet-White (eds) *Upper Pleistocene Prehistory of Western Eurasia*. Philadephia: University Museum, University of Pennsylvania: 141-143.

Gamble, C. 1980. Information exchange in the Palaeolithic. *Nature* 287: 522-523.

Gamble, C. 1982. Interaction and alliance in Palaeolithic society. *Man* 17: 92-107.

Gamble, C. 1983. Culture and society in the Upper Palaeolithic of Europe. In G. Bailey (ed.) *Hunter Gatherer Economy in Prehistory: a European Perspective*: Cambridge: Cambridge University Press: 201-211.

Gargett, R.H. 1989. Grave shortcomings: the evidence for Neandethal burial. *Current Anthropology* 30: 157-190.

Geneste, J.-M. 1988. Systèmes d'approvisionment en matières premières au Paléolithique moyen et au Paléolithique supérieur en Aquitaine. In M. Otte (ed.) *L'Homme de Néandertal*. Vol. 8: *La Mutation*. Liège: Etudes et Recherches Archéologiques de l'Université de Liège (ERAUL 35): 61-70.

Gould, S.J. and Vrba, E.S. 1982. Exaption: a missing term in the science of form. *Paleobiology* 8: 4-15.

Harrold, F.B. 1980. A comparative analysis of Eurasian Palaeolithic burials. *World Archaeology* 12: 195-211.

Harrold, F.B. 1986. Une réévaluation du Châtelperronien. *Préhistoire Ariègoise* 41: 151-169.

Harrold, F.B. 1989. Mousterian, Châtelperronian, and Early Aurignacian: Continuity and Discontinuity. In P. Mellars and C. Stringer (eds) *The Human Revolution: Behavioural and Biological Perspectives on the Origins of Modern Humans*. Edinburgh: Edinburgh University Press: 677-713.

Hodder, I. 1982a. Theoretical archaeology: a reactionary view. In I. Hodder (ed.) *Symbolic and Structural Archaeology*. Cambridge: Cambridge University Press: 1-16.

Hodder, I. 1982b. *Symbols in Action*. Cambridge: Cambridge University Press.

Howell, F.C. 1966. Observations on the earlier phases of the European Lower Paleolithic (Torralba-Ambrona). *In Recent Studies in Paleoanthropology. American Anthropologist* 68: 11-32.

Jacob-Friesen, K.H. 1956. Eiszeithliche Elefantenjäger in der Lüneburger Heide. *Jahrbuch der Romisch-Germanischen Zentralmuseums, Mainz* 3: 1-22.

Jelinek, A.J. 1977. The Lower Paleolithic: current evidence and interpretations. *Annual Review of Anthropology* 6: 11-32.

Jelinek, A.J. 1982. The Middle Palaeolithic in the Southern Levant, with comments on the appearance of modern *Homo sapiens*. In A. Ronen (ed.) *The Transition from Lower to Middle Paleolithic and the Origin of Modern Man*. Oxford: British Archaeological Reports International Series S151: 327-328.

Jelinek, A.J., Debénath, A. and Dibble, H.L. 1988. A preliminary report on evidence related to the interpretation of economic and social activities of Neanderthals at the Site of La Quina (Charente), France. In M. Otte (ed.) *L'Homme de Néandertal*. Vol. 6: *La Subsistence*. Liège: Etudes et Recherches Archéologiques de l'Université de Liège (ERAUL 33): 99-106.

Jéquier, J.P. 1975. *Le Moustérien Alpin: Eburodunum* 2. Yverdon: Institut d'Archéologie Yvrdonoise.

Laitman, J.T. 1983. The evolution of the hominid upper respiratory system and implications for the origin of speech. In E. de Grolier (ed.) *Glossogenetics: the Origin and Evolution of Language*. Paris: Harwood Academic Press: 63-90.

Laitman, J.T., Heimbuch, R.C. and Crelin, E.S. 1979. The basicranium of fossil hominids as an indicator of the upper respiratory systems. *American Journal of Physical Anthropology* 51: 15-33.

Laitman, J.T. and Heimbuch, R.C. 1982. The basicranium of Plio-Pleistocene hominids as an indicator of their upper respiratory systems 2. *American Journal of Physical Anthropology* 59: 323-343.

Le Mort, F. 1988. Cannibalisme ou rite funéraire? *Dossiers d'Histoire et d'Archéologie* 125: 46-49.

Leroi-Gourhan, A. 1965. *Préhistoire de l'Art Occidental*. Paris: Mazenod.

Leroi-Gourhan, Arl. 1968. Le Néanderthalien IV de Shanidar. *Bulletin de la Société Préhistorique Française* 65: 79-83.

Leroi-Gourhan, Arl. 1975. The flowers found with Shanidar IV, a Neanderthal burial in Iraq. *Science* 190: 562-565.

Leroi-Gourhan, Arl. and Leroi-Gourhan, A. 1964. Chronologie des Grottes d'Arcy-sur-Cure (Yonne). *Gallia Préhistoire* 17: 1-64.

Lévêque, F. and Vandermeersch, B. 1980. Découverte de restes humains dans un niveau castelperronien à Saint-Césaire (Charente-Maritime). *Comptes-Rendus de l'Academie des Sciences de Paris (series II)*: 187-189.

Lieberman, P. 1985. On the evolution of human syntactic ability: its preadaptive bases – motor control and speech. *Journal of Human Evolution* 14: 657-668.

Lieberman, P. 1989. The origins of some aspects of human language and cognition. In P. Mellars and C. Stringer (eds) *The Human Revolution: Behavioural and Biological Perspectives on the Origins of Modern Humans*. Edinburgh: Edinburgh University Press: 391-414.

Lumley, H. de. 1966. *Les Fouilles de Terra Amata à Nice*. Bulletin de la Musée d'Anthropologie de Monaco 13.

Lumley, M.-A. and Lumley, H. de. 1972. *La Grotte de l'Hortus (Valflaunes Herault)*. Etudes Quaternaire 1. Marseilles: Laboratoire de Paléontologie Humaine et de Préhistoire.

McBurney, C. 1967. *The Haua Fteah (Cyrenaica) and the Stone Age of the South-East Mediterranean*. Cambridge: Cambridge University Press.

Marks, A.E. 1988a. Early Mousterian settlement patterns in the Central Negev, Israel. In M. Otte (ed.) *L'Homme de Néandertal*. Vol. 6: *La Subsistence*. Liège: Etudes et Recherches Archéologiques de l'Université de Liège (ERAUL 33): 115-126.

Marks, A.E. 1988b. The Middle to Upper Paleolithic transition in the Southern Levant: technological change as an adaptation to increasing mobility. In M. Otte (ed.) *L'Homme de Néandertal*. Vol. 8: La *Mutation*. Liège: Etudes et Recherches Archéologiques de l'Université de Liège (ERAUL 35): 109-124.

Marshack, A. 1970a. Le baton du Montgaudier (Charente): Réexamen au microscope et interpretation nouvelle. *L'Anthropologie* 74: 321-352.

Marshack, A. 1970b. *Notation dans les Gravures du Paléolithique Supérieur*. Bordeaux: Institut de Préhistoire de l'Université de Bordeaux, Mémoire 8.

Marshack, A. 1972a. *The Roots of Civilization*. New York: McGraw Hill.

Marshack, A. 1972b. Upper Paleolithic notation and symbol. *Science* 178: 817-332.

Marshack, A. 1972c. Cognitive aspects of Upper Paleolithic engraving. Current Anthropology 13: 445-477. (Also discussion in *Current Anthropology* 15: 327-332.)

Marshack, A. 1973 Preliminary analysis of an Azilian notational engraving from the Grotte du Taï (St. Nazaire-en-Royans, Drôme). *Etudes Préhistoriques de la Société Préhistorique de l'Ardèche* 4: 13-17.

Marshack, A. 1975. Exploring the mind of Ice Age Man. *National Geographic* 147: 62-89.

Marshack, A. 1976. Some implications of the Paleolithic symbolic evidence for the origins of language. *Current Anthropology* 17: 274-282.

Marshack, A. 1979. Upper Paleolithic symbol systems of the Russian Plain: cognitive and comparative analysis of complex ritual marking. *Current Anthropology* 20: 271-311. (Also discussion in Current Anthropology 20: 604-608.)

Marshack, A. 1981. On Paleolithic ochre and the the early uses of color and symbol. *Current Anthropology* 22: 188-191.

Marshack, A. 1984a. The ecology and brain of two-handed bipedalism: an analytic, cognitive and evolutionary assessment. In H.L. Roitblat, T.G. Bever and H.S. Terrace (eds) *Animal Cognition.* Hillsdale (N.J.): Erlbaum: 491-511.

Marshack, A. 1984b. Concepts théoriques conduisant à de nouvelles méthodes analytiques, de nouveaux procédés de recherche et categories de données. *L'Anthropologie* 88: 573-586.

Marshack, A. 1985a. *Hierarchical Evolution of the Human Capacity: the Paleolithic Evidence*. 54th James Arthur Lecture on "The Evolution of the Human Brain", 1984. New York: American Museum of Natural History.

Marshack, A. 1985b. Theoretical concepts that lead to new analytical methods, modes of inquiry and classes of data. *Rock Art Research* 2: 95-111.

Marshack, A. 1986a. Reading before writing. *New York Sunday Times Book Review* April 6: 1, 40-41.

Marshack, A. 1986b. Comments and response to article. *Rock Art Research* 3: 62-82.

Marshack, A. 1986c. Comment on Davis: 'The origins of image making'. *Current Anthropology* 27: 205-206.

Marshack, A. 1987. The eye is not as clever as it thinks it is. Comment on Bahn: 'No sex, please, we're Aurignacians'. *Rock Art Research* 3: 111-116.

Marshack, A. 1988a. The species-specific evolution and contexts of the creative mind: thinking in time. In C.S. Findlay and C.J. Lumsden (eds) *The Creative Mind: Towards an Evolutionary Theory of Discovery and Innovation.* Special issue of the *Journal of Social and Biological Sciences.*

Marshack, A. 1988b. La pensée symbolique et l'art. *Dossiers d'Histoire et d'Archéologie* 124: 80-90.

Marshack, A. 1988c. The Neanderthals and the human capacity for symbolic thought: cognitive and problem solving aspects of

Mousterian symbol. In M. Otte (ed.) *L'Homme de Néandertal.* Vol. 5: *La Pensée.* Liège: Etudes et Recherches Archéologiques de l'Université de Liège (ERAUL 32): 57-91.

Marshack, A. 1989a. Methodology in the analysis and interpretation of Upper Paleolithic image: theory *versus* contextual analysis. *Rock Art Research* 6: 17-53.

Marshack, A. 1989b. The Upper Palaeolithic female image: a study in style and image use. Keynote address presented at the Prehistoric Society symposium on *Palaeolithic Art.* Oxford University, November 1989.

Oakley, K.P. 1961. *Man the Tool-Maker.* Chicago: Chicago University Press.

Oliva, M. 1986. Pointes foliacées et la technique Levallois dans le passage Paléolithique moyen/Paléolithique supérieur en Europe centrale. In M. Otte (ed.) *L'Homme de Néandertal.* Vol. 8: *La Mutation.* Liège: Etudes et Recherches Archéologiques de l'Université de Liège (ERAUL 35): 125-131.

Pfeiffer, J. 1982. *The Creative Explosion: an Inquiry into the Origins of Art and Religion.* New York: Harper and Row.

Russell, M. 1987. Mortuary practices at the Krapina Neanderthal site. *American Journal of Physical Anthropology* 72: 381-397.

Saint-Périer, R. and Saint-Périer, S. 1950. *La Grotte d'Isturitz*: Vol. 3. Paris: Archives de l'Institut de Paléontologie Humaine 25.

Seewald, C. 1983. Prähistorische Sammlungen Ulm, Menschliche Figur mit Löwenkopf. *Ulmer Statdgeschichte.* Beilage zum Gesschaf tbericht der Ulmer Volksbank: 1-5.

Shea, J.J. 1989. Tool use and human evolution in the Late Pleistocene of Israel. In P. Mellars and C. Stringer (eds) *The Human Revolution: Behavioural and Biological Perspectives on the Origins of Modern Humans.* Edinburgh: Edinburgh University Press: 611-625.

Singer, R. and Wymer, J. 1982. *The Middle Stone Age at Klasies River Mouth in South Africa.* Chicago: Chicago University Press.

Smith, F.H. 1985. Continuity and change in the origins of modern Homo sapiens. *Zeitschrift für Morphologie und Anthropologie* 75: 197-222.

Soffer, O. 1985. *The Upper Paleolithic of the Russian Plain.* New York: Academic Press.

Solecki, R. 1963. Prehistory in Shanidar Valley, northern Iraq. *Science* 129: 179-193.

Solecki, R. 1971. *Shanidar, the First Flower People.* New York: Knopf.

Solecki, R. 1975. Shanidar IV: a Neanderthal flower burial in northern Iraq. *Science* 190: 880-881.

Solecki, R. 1982. A ritual Middle Paleolithic deer burial at Nahr Ibrahim Cave, Lebanon. *Archéologie au Levant, Recueil R. Saidah.* Lyon: Collection de la Maison de l'Orient Mediterranéen 12 (Série Archéologie 9): 47-56.

Stewart, T.D. 1963. Shanidar skeletons IV and VI. Sumer 19: 8-26.

Stringer, C.B. 1982. Towards a solution to the Neanderthal problem. *Journal of Human Evolution* 11: 431-438.

Stringer, C.B. 1985. Middle Pleistocene hominid variability and the origins of Late Pleistocene humans. In E. Delson (ed.) *Human Ancestors: the Hard Evidence.* New York: Alan R. Liss: 289-195.

Stringer, C.B. 1989. The origins of early modern humans: a comparison of the European and non-European evidence. In P. Mellars and C. Stringer (eds) *The Human Revolution: Behavioural and Biological Perspectives on the Origins of Modern Humans.* Edinburgh: Edinburgh University Press: 232-244.

Svoboda, J. 1988. Early Upper Paleolithic industries in Moravia: a review of recent evidence. In M. Otte (ed.) *L'Homme de Néandertal.* Vol. 8: *La Mutation.* Liège: Etudes et Recherches Archéologiques de l'Université de Liège (ERAUL 35): 169-192.

Trinkaus, E. 1985. Cannibalism and burial at Krapina. *Journal of Human Evolution* 14: 203-216.

Vialou, D. 1981. *L'Art Pariétal en Ariège Magdalénien* (3 Vols). Laboratoire de Paléontologie Humaine et de Préhistoire 13. Paris: Musée National d'Histoire Naturelle.

Wendt. W.E. 1976. 'Art Mobilier' from the Apollo 11 Cave, South West Africa: Africa's oldest dated works of art. *South African Archaeological Bulletin* 31: 5-11.

Wetzel, R. and Bosinski, G. 1969. *Die Bocksteinschmiede im Lonetal.* Veröffentlichungen der Staatlichen Amtes für Denkmalpflege Stuttgart, Reihe A.

White, R. 1982. Rethinking the Middle/Upper Paleolithic transition. *Current Anthropology* 23: 169-192.

White, R. 1985. Thoughts on social relationships in hominid evolution. *Journal of Social and Personal Relationships* 2: 95-115.

White, R. 1986. Toward a contextual understanding of the earliest body ornaments. Paper presented at the conference on *The Origins of Modern Human Adaptations.* School of American Research Advanced Seminar, Albuquerque, New Mexico.

White, R. 1989. Production complexity and standardization in early Aurigancian bead and pendant manufacture: evolutionary implications. In P. Mellars and C. Stringer (eds) *The Human Revolution: Behavioural and Biological Perspectives on the Origins of Modern Humans.* Edinburgh: Edinburgh University Press: 366-390.

Wobst, H. 1977. Stylistic behavior and information exchange. In C. Cleland (ed.) *Papers for the Director: Research Essays in Honor of James B. Griffin.* Anthropological Papers 61. University of Michigan Museum of Anthropology: 317-342.

Wolpoff, M.H. 1986. The place of the Neanderthals in human evolution. Paper presented at the conference on *The Origins of Modern Human Adaptations.* School of American Research Advanced Seminar, Albuquerque, New Mexico.

18: Human Cognitive Changes at the Middle to Upper Palaeolithic Transition: the Evidence of Boker Tachtit

SHELDON KLEIN

INTRODUCTION

One of the main problems in understanding the development of human culture is to explain the ways in which enormously complex patterns of information can be stored and processed in the human mind. Computer models of human cognition seem to fail when applied to large-scale knowledge systems because the time it takes to compute human behaviour by rules can increase exponentially, or even combinatorially, with the size of the data base. If the human brain is like a computer, then the problem of computing behaviour fast enough for ordinary social interactions becomes extreme, for electrical impulses travel a million times slower in the brain than in computers (see Note 1).

In earlier papers I have suggested that the processing of very large quantities of information has been made possible by a basic development, at some point in human evolution, of a new pattern of cognition called 'ATO systems' (Klein 1983, 1986, 1988). These served to constrain the structure of the information and rules of behaviour in such a way that the processing time increased only linearly with the size of the data base. ATO systems can be characterized as a computational invention consisting of an analogical reasoning method in combination with a global classification scheme used in such a way that made it possible to calculate behaviour with a processing time that increases only linearly with the size of the data base. For a full description and discussion of how ATOs ('Appositional Transformation Operators') work, the reader is referred to my recent paper in *Current Anthropology* (Klein 1983: 152-4). I have suggested elsewhere (Klein 1983, 1986, 1988) that this invention can provide insights for many topics in anthropology, including divination systems, magic, shamanism, myth systems, the relation between language and culture, and the structuralism of Lévi-Strauss. The aim of the present paper is to suggest that this pattern of cognition is present throughout the Upper Palaeolithic period, that it was one of the critical developments that made possible the elaboration and rapid expansion of Upper Palaeolithic culture, and that we can identify clear indications of this development in the changes in lithic technology that were associated with the Middle-Upper Palaeolithic transition.

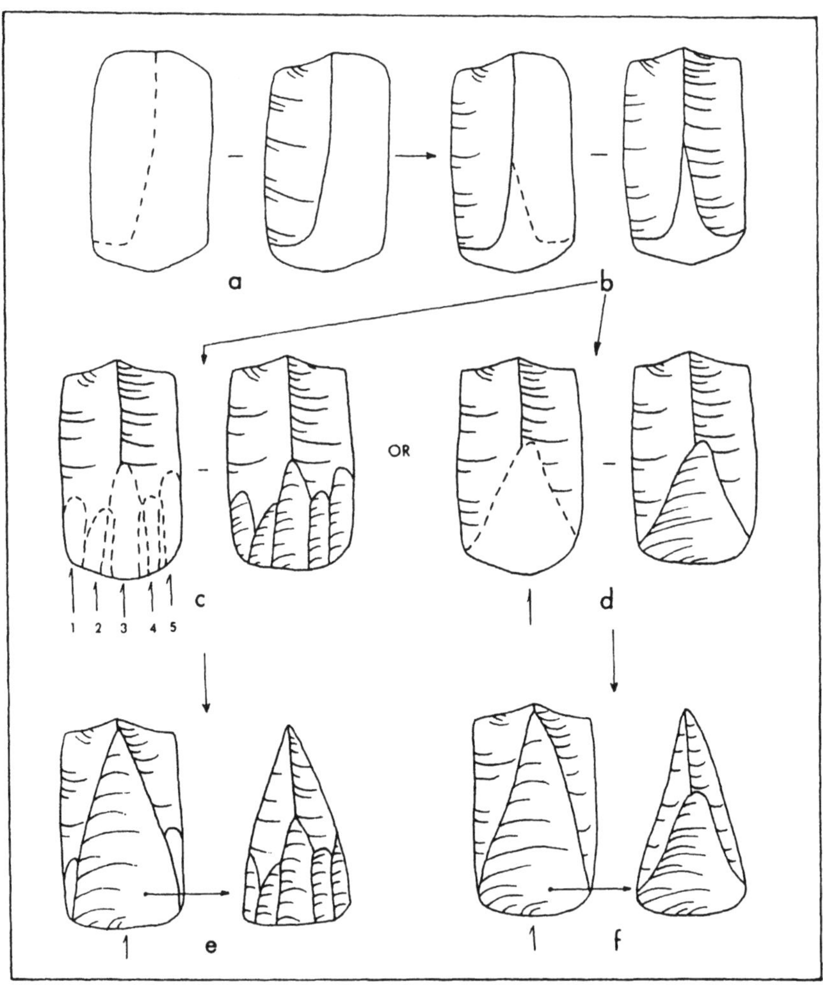

Figure 18.1. Illustration reproduced from Volkman (1983: Figure 6-14), showing two slight variations of opposed-platform Levallois Point formation at Boker Tachtit. a, b = predetermination of the form of the distal end of the point; c, e = utilization of five removals to form the butt of the point; d, f = utilization of one removal to form the butt of the point.

In the discussion which follows, I shall first examine some of the hard evidence for the patterns of technological change associated with the conventional Middle-Upper Palaeolithic transition, and then go on to discuss how this might be related to basic changes in the structure of human cognition in terms of the principles of ATO systems.

LITHIC TECHNOLOGY AT THE MIDDLE-UPPER PALAEOLITHIC TRANSITION: THE EVIDENCE FROM BOKER TACHTIT

Some of the most detailed and explicit evidence for the character of changes in lithic technology associated with the Middle-Upper Palaeolithic transition comes from the site of Boker Tachtit in the central Negev Desert of southern Israel. The site was excavated by A. Marks in the 1970s, and revealed a sequence of four superimposed occupation levels, separated by intervening sterile deposits (see Marks 1981, 1983). The results of both uranium-thorium and radiocarbon dating indicate that the occupations took place between *c.* 47 000 and 38 000 BP, placing the site clearly within the conventional time-range of the Middle-Upper Palaeolithic transition in the Near East. As Marks points out, one of the most valuable aspects of this sequence is that it represents a series of relatively short-lived episodes of human occupation and tool manufacture in a clear chronological succession (Marks 1983: 68): '...each occupation surface appears to have been lived on only briefly, as shown by the spatial distributions of reconstructable cores (Hietala and Marks 1981), and, therefore, each assemblage was produced during only a minute portion of that time. The assemblages taken together, however, should reflect accurately technological and typological patterns at specific intervals during this time span'.

The detailed core reconstructions carried out by Volkman for these levels illustrate the procedures by which typical 'Levallois point' forms were produced by means of various sequences of preparatory blade removals detached from opposed-platform cores (Volkman 1983; Marks and Volkman 1987). The procedures have been described by Volkman (1983) in the following terms:

> A blade was removed from the core face [see Figure 18.1a]. Rotating the core slightly, a second blade was removed, using the same platform. The second removal was adjacent to the first on the core face, and it slightly overlapped the first removal from the point of impact to the approximate midpoint of the core face [Figure 18.1b]. The point of impact for both of these removals occurred near the center of the width of the platform from which they were struck. One was slightly directed toward the right edge of the core face, while the second was directed toward the left edge. Thus, the scars left by these two removals created an inverted Y-shaped ridge down what would be the approximate center of the point, forming the lateral edges of the pointed end of the point.
>
> To form the butt of the point, the core was rotated 180 degrees to the opposite platform. One to five blades were struck off using this platform. These removals were shorter in length than the blades removed from the distal end. Moving from one side of the core face to the other along the platform width, these removals were struck off in succession, slightly rotating the core after each removal. Thus, the

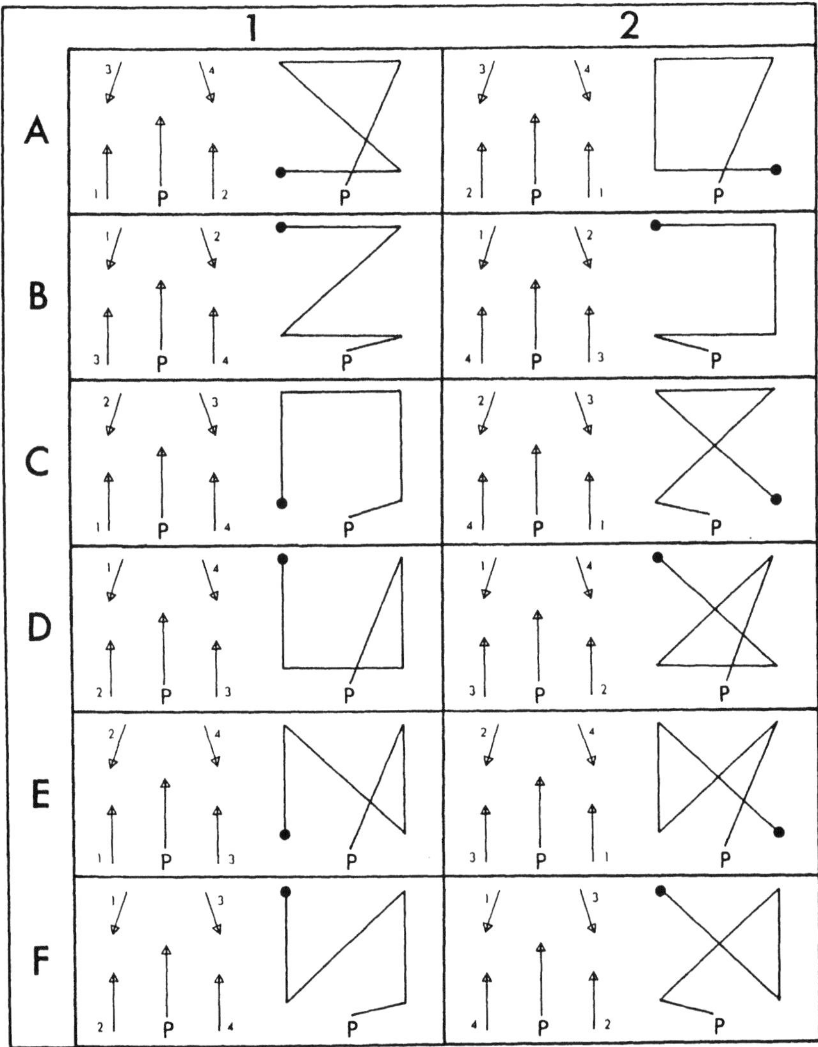

Figure 18.2. Illustration reproduced from Volkman 1983: Figure 6-25), to show schematic representation of the blade-removal sequence variations that predetermine the removal of the opposed-platform Levallois Points at Boker Tachtit.

next removal in succession was adjacent to the one before it and overlapped it [see Figure 18.1c]. If only one removal was struck from the butt end, the approximate center of the platform width was selected for striking this single removal, completing the Y-shaped ridge pattern on the dorsal face of the soon-to-be-struck point [Figure 18.1d]. To remove the point, the center of this succession of removals along the platform was selected, and the point was removed [Figure 18.1e, f]....

> The possible specific reduction sequences for producing the opposed platform Levallois points in Levels 1 through 3 at Boker Tachtit can be schematically represented as in [Figure 18.2]. The mirror images of these drawings were also represented on some points. The arrows represent the direction of the removals, depicting the opposed platform pattern, while the numbers next to each arrow present the chronological sequence of the removals. The schematics have been simplified to present only four removals in a sequence before a point was struck, but more than four removals in a sequence was common. The accompanying line drawings present a visual geometric pattern of the point production sequence. Each row presents slight variations of essentially the same type of reduction sequence (Volkman 1983: 167-70).

Volkman's illustration (Figure 18.2) combines the variants that involve from two to five removals from the butt end into classes which reflect simply *left side* removal or *right side* removal, whether or not more than one removal is made from each side of the butt. A central butt removal seems to be ignored where it occurs, and the cases where a single butt removal occurs as in Volkman's illustration (Figure 18.1d, f) are not included.

I have reformulated the data of Volkman's illustration (Figure 18.2) for an ATO analysis which appears in Table 18.1. The *types* of removals are listed in a fixed sequence, with downward pointing slanted arrows for left and right side distal removals, and upward pointing vertical arrows for left and right side butt removals. The *order* of the blows is indicated by enumerating them with binary numbers: '00' for the 1st, '01' for the 2nd, '10' for the 3rd, and '11' for the 4th. Thus in Table 18.1, the left distal removal is 3rd (10), the left base removal is 1st (00), the right distal removal is 4th (11), and the right butt removal is 2nd (01). The point removal is unnumbered and last, in exact correspondence with Volkman's illustration (Figure 18.2). Volkman classifies the six rows into three groups (1983: 167-70), and their analogical unity is corroborated by the ATO patterns I derive in Table 18.1. Rows related by identical ATOs are analogically related to each other, and it can be seen that rows A1 and B1 as well as A2 and B2 are related by the same ATOs (*A1B1 = *A2B2 = 01 01 01 01) . Also, rows A1 and A2 as well as B1 and B2 are related by the same ATOs (*A1A2 = *B1B2 = 11 10 11 10). Similar criteria apply to the other groupings. However, rows C1 and D1, C2 and D2, E1 and Fl, and E2 and F2 are all related by the same ATO (*C1D1 = *C2D2 = *E1F1 = *E2F2 = 10 10 10 10) but *C1C2 and *D1D2 (11 00 11 00) differ from *E1E2 and *F1F2 (11 01 11 01). This suggests a possible refinement of the classification in which the C1 and D1 cases are merged with E1 and F1, and in which C2 and D2 are merged with E2 and F2. Volkman's discussion of the criteria for group rows A and B are accurate (1983: 167-70): 'In rows A and B, the production sequence is divided into core ends. One end of the core and point is prepared completely before the opposite platform is utilized to prepare the other end of the point. In row A, the butt end of the point was formed first. This was followed by the preparation of

Table 18.1. An ATO analysis of the data in Volkman's figures 6-25 (see Figure 18.2) in which the types of blows are listed in a fixed sequence, with downward pointing slanted arrows for left and right-side distal removals, and updard pointing vertical arrows for left and right-side butt removals. The order of the blows is indicated by enumerating them with binary numbers: '00' for the 1st, '01' for the 2nd, '10' for the 3rd, and '11' for the 4th.

	L ↙	↑	P ↘	R ↑
A1	10	00	11	01
B1	00	10	01	11
C1	01	00	10	11
D1	00	01	11	10
E1	01	00	11	10
F1	00	01	10	11

*A1B1	01	01	01	0
*C1D1	10	10	10	10
*E1F1	10	10	10	10

	L ↙	↑	P ↘	R ↑
A2	10	01	11	00
B2	00	11	01	10
C2	01	11	10	00
D2	00	10	11	01
E2	01	10	11	00
F2	00	11	10	01

*A2B2	01	01	01	01
*C2D2	10	10	10	10
*E2F2	10	10	10	10

*A1A2	11	10	11	10
*B1B2	11	10	11	10
*C1C2	11	00	11	00
*D1D2	11	00	11	00
*E1E2	11	01	11	01
*F1F2	11	01	11	01

the distal end of the point. Subsequently, the point is struck off. In row B, the distal end of the point is prepared first before moving to the opposite platform to shape the butt of the point'. Volkman continues with a discussion that recognizes the possibility of further subgrouping, but his exact line of reasoning seems to be based on a mislabelling of some of the cases (1983: 170): 'In rows E and F, the core is divided into right side *versus* left, rather than butt end *versus* distal as in rows A and B. One side of the point is shaped completely before the other side is completed. In E1 and F1, the left side is shaped first, while the right side is prepared second. Rows C and D intermix the side and end distinctions. One side and one end are completely prepared before the other side and end are completed. More economically, however, the Cl and Dl sequences could be consid-

Table 18.2. Data reproduced from Volkman (1983) Table 6-3, showing frequencies of Levallois Points produced in the reduction sequence types shown in Figure 18.2.

Sequence Type	A1	A2	B1	B2	C1	C2	D1	D2	E1	E2	F1	F2	Totals
Level 1	1		1	1	1	3	1	2			2		12
Level 2			6	3	3	3	3		4	2	3	1	28
Emireh, Level 2			1	1	1	1	1	1					5
Level 3			1										1
Sub-total	1	–	9	4	5	7	5	3	4	2	5	1	
Totals	1		13		12		8		6		6		46

ered variations of the E/F type, and the C2 and D2 sequences variation of the A/B type'. Actually, both E1 and F1 *and* C1 and D1 prepare the left side first; also, both E2 and F2 *and* C2 and D2 intermix the side end distinctions.

Many alternative analogical groupings of the sequence patterns are possible, and for purposes of this analysis it is best to ignore Volkman's groupings and deal with individual cases, as Volkman himself does in his tabulation of reduction sequence type occurrences in here in Table 18.2.

The data shown in Table 18.2 must be treated with caution. He states that mirror images of the patterns occur, but it is not clear if they are included with their mirrored counterparts nor, if so, *all* of the mirrored counterparts were actually found. Also, Emireh Level 2 row contains a count error, as does the B2 column, and there is no data for reduction type A2 which is presumed to occur because of its appearance in his illustration (Figure 18.2).

There are 4! or 24 real reduction sequences. The binary numbering scheme used to indicate the order of the removals permits a specification of 2^8 or 256 possible sequences that include the 'real' ones in which each removal occurs only once, as well as 'surreal' sequences in which some or all removals may have the same sequence number and appear to occur 'simultaneously'. The AT0s of Table 18.1 are examples of the latter type; real sequences may also function as ATOs in some analogies.

Sequence types E1, E2 and F2 appear only in Level 2 (Table 18.2). The sequences present in Level 1 provide a basis for deriving the the missing sequences by ATO analogies. F2 appears as the ATO, *A1C1 which might emerge in any analogies involving relations between sequences A1 and C1. E2 can be derived from the analogy,

$$\frac{\text{C1} = 01\ 00\ 10\ 11}{\text{D1} = 00\ 01\ 11\ 10} :: \frac{\text{F2} = 00\ 11\ 10\ 01}{?}$$

'?' = *F2(*C1D1) = *(00 11 10 01)(10 10 10 10) = 01 10 11 00 = E2.

Similarly, E1 can be derived from the analogy,

Table 18.3.
An ATO analysis of the data in Volkman's figure 6-25 (see Figure 18.2) in which the order of the removals is listed in a fixed sequence and binary numbers are assigned to the types of removals. Two binary opposition sets, left = 0/ Right = 1, and distal 0/ base = 1, are used to determine the number names of the removals so that a left-side distal removal = '00', a left-side base removal = '01', a right-side distal removal = '10', and a right-side base removal = '11'.

						L		P	R
						1	2	3	4
					A1	01	11	00	10
*A1B1	10	10	10	10					
					B1	00	10	01	11
					C1	01	00	01	11
*C1D1	10	10	10	10					
					D1	00	01	11	10
					E1	01	00	11	10
*E1F1	10	10	10	10					
					F1	00	01	10	11

	L		P	R					
	1	2	3	4					
A2	11	01	00	10					
					*A2B2	00	00	00	00
B2	00	10	11	01					
C2	11	00	10	01					
					*C2D2	00	00	00	00
D2	00	11	01	10					
E2	11	00	01	10					
					*E2F2	00	00	00	00
F2	00	11	10	01					

*A1	01	01	11	11
*B1B2	11	11	01	01
*C1C2	01	11	01	01
*D1D2	11	01	01	11
*E1E2	01	11	01	11
*F1F2	11	01	11	01

$$\frac{C2}{D2} :: \frac{F1}{?}$$

The reader may verify the ATO calculations using the data of Table 18.1. Accordingly, the Level 2 patterns E1, E2 and F2 can be derived by ATO analogy from patterns observed in Level 1. It is also possible to reformulate the data of Volkman's illustration (Figure 18.2) in a different way. Instead of listing the *types of removals in a fixed sequence* and indicating the *order of removals with binary integers*, as was done in Table 18.1, one may list the *order of the removals* in a fixed sequence, and assign binary numbers to the *types of removals*. If this is done, one may choose the binary number 'names' of the removals so that they reflect componential semantic features or categories. For example, if one selects two binary opposition sets, left = 0/right = 1 and distal = 0/base = 1, then a left-side distal removal would be coded or named '00' a left-side base removal '01', a right-side distal removal '10', and a right-side base removal '11'. The results appear in Table 18.3. Volkman's classification is no longer directly apparent. The sequence pairs A1 and B1, C1 and D1, and E1 and F1 form one grouping, each pair related by the ATO, '10 10 10 10'. Similarly, A2 and B2, C2 and D2, and E2 and F2 form another grouping, each pair related by the ATO '00 00 00 00'. Some of the number patterns are altered, and yet many are identical in the representations of both versions. It is still possible to derive the Level 2 sequences E1, E2 and F2 analogically from the Level 1 sequences in the Table 18.3 version, but the analogies and ATOs used are different: *A1C1 = E2, *F1(*A1B1) = E1, *E2(*A2B2) = F2.

It may, of course, be the case that E1, E2 and F2 also existed at the time of Level 1, but were not reflected in the cores that could be reconstructed. This would not affect the value of the Boker Tachtit evidence for ATO theory. Each final reduction sequence may have involved an event of less than 15 minutes duration. Of the 24 real sequences that would yield Levallois points, Volkman portrays 12 and indicates that some or all of the mirror images of these were present. There must have been a first reduction sequence, and a second, and a third, etc. that developed by analogical extension, even if all the possibilities were derived in a *single day* of Level 1 time.

SIGNIFICANCE OF THE EVIDENCE

The 256 possible reduction sequences, including the 24 real ones, can be characterized by the concept of mathematical groups. The group concept is essential to characterizations of models in the structuralism of both Lévi-Strauss and Piaget (Gardner 1972; Klein 1977, 1982, 1983, 1986, 1988; Klein *et al.* 1981; Hage 1979; Hage and Harary 1983a, 1983b; Wynn 1979, 1981, 1985; Piaget 1960, 1970a, 1970b, 1972). Essentially, it reflects the mathematics of permutations and combinatorics. ATO logic permits one to calculate by analogy with a processing time that increases only *linearly* with increases in the number of elements, in contrast with combinatoric calculations for which the processing time increases factorially. The preceding analysis of the Boker Tachtit evidence demonstrates that the analogical reasoning process integral to ATO theory *must* have been part of the cognitive repertoire of some *Homo sapiens* at least as early as the Middle-Upper Palaeolithic transition in the Negev. Semantic categories related by the group concept and ATOs were all present, implicit in the technology of the blade removal sequence variations that predetermine the removal of opposed platform Levallois points at Boker Tachtit.

It is certainly possible to view the development of the reasoning process as biological and to evaluate stone tool production techniques developmentally, within the framework of Piaget, as does Wynn (1979, 1981, 1985). But Wynn states (1985: 41): 'It was between 1.5 million and 300 000 years ago that modern intelligence evolved. Indeed, the geometry of later Acheulian handaxes requires a stage of intelligence that is typical of fully modern adults. No subsequent developments in stone tool technology require a more sophisticated intelligence'.

Wynn (1985) bases his analysis on the mental processes necessary to produce Late Acheulian pointed handaxes:

> ...with bilateral symmetry that was achieved by a minimal amount of trimming. Furthermore, the trimming is not contiguous but is concentrated in four distinct regions. The knapper must have had an intended result and, more importantly, a plan for achieving that result with a minimum of flaking. This may seem a simple task but it is not. The knapper had to have conceived of the shape as resulting from an arrangement of parts (potential trimming) and to choose those parts that were essential. Pre-correction of errors was necessary in order to choose which trimming was essential and which superfluous. Trial and error trimming can achieve an imagined shape only by trimming and checking until the imagined shape is approximated, usually by contiguous flaking and rarely through a minimum amount of work. In sum, the shapes of later Acheulean handaxes and the means for achieving these shapes required the organizational abilities of operational intelligence. As we have seen, operational intelligence is typical of modern adults (Wynn 1985: 39).

Certainly the four distinct areas of secondary flaking in Wynn's example *might* have been knapped in more than one order, and the production process, in principle, might reflect the same group concept as in the Boker Tachtit data. There is, however, no evidence of sequence variations and, no matter how sophisticated Wynn attempts to make the cognitive processes appear in the pre-correction of errors, the *ordering* in any particular knapping sequence may *always* have been predetermined by random factors.

It is still possible, nevertheless, that the group concept may also have been implicit in the production of late Acheulian handaxes. There is, however, a flaw in Wynn's arguments, and it lies in his oversimplification of Piaget's concept of operational intelligence:

> Operational thought actually consists of two stages, *concrete operations* and *propositional operations*. While both kinds of operations employ the same organizational principles, reversibility for example, they differ in their realms of application. Concrete operations, which are achieved earlier, organize physical entities-objects, peoples, etc. – while propositional operations organize propositions and hypothetical entities – the square root of minus one, for example. There is, however, considerable controversy about the validity of formal operations as a separate stage. It may be that they represent operational concepts applied in abstract areas of thinking rather than a true

successor to concrete operations that is logically necessary (Dasen 1977). Moreover, propositional operations would be invisible archaeologically, at least until the introduction of writing. Because of these problems, and because both concrete and propositional operations employ the same organizational principles, I will downplay the distinction in the following analysis (Wynn 1985: 34-5).

Wynn's analysis of late Acheulian handaxes yields evidence only of *concrete* operations. Wynn's claim that *propositional* operations are invisible archaeologically can be challenged. The *process of inventing* the Boker Tachtit technique for the production of Levallois points must have involved propositional operations, even though the *manufacturing* process involved only concrete ones, for both a specific tool (a point) and a set of intermediate units (the blades) are produced in the same cycle, which may, itself, be repeated. The concept is suggestive of the notion of 'embedding rules' in phrase structure grammar, which is discussed in the next section of this paper. Accordingly, within the framework of a Piaget model, there would seem to be evidence of a significant change in human cognition. Wynn considered only the cognitive aspects of the *techniques of production* rather than the cognitive requirements for *inventing* those techniques. The difference is comparable to that in the cognitive functions of an engineer who invents a production technique and a factory assembly worker who uses it.

ATOS AND LANGUAGE

Phrase structure grammar works on ATO principles. This can be demonstrated by creating a categorial grammar in which appropriately chosen binary integers are used to represent grammatical categories, together with information about their right and left combining properties. Consider a simple phrase structure type grammar rule of the form,

$$S = A\ B$$

If we represent categories A and B by binary numbers say A = 1010 and B = 0101, plus information that A combines with an element to the right and that B combines with an element to the left, then

$$S = 1010R\ \ 0101L$$

If we apply the * operation to A and B (*1010R 0101L) we compute a binary integer name for S, 0000. Accordingly, a grammar could be constructed consisting of a table of all binary codes that represent grammatical categories in the language – in effect a list of all valid phrase structure units. One could then parse sentences by taking the * product of adjacent sequences of the binary integer codes that represent the grammatical categories of the elements that make up a sentence in the language in an iterative and recursive way. If the result of any * operation is in the list of valid phrase structure units, one may accept it and continue the parsing process until one obtains a single category, representing a sentence. For natural languages, each element might have have more than one code, resulting in multiple parses, some of which might be incomplete because they cannot reduce to a single category. Recursive rules can be fitted in the scheme. Consider a rule of the type,

B = A B

In this case it is necessary to assign the recursive element A the code 1111R. If we assign B the code 0000L then,

0000L = 1111R 0000L

(As mentioned earlier, ambiguity is usually present in natural language phrase structure grammars, and elements may have more than one code.) The sentence, A A B might be parsed as follows:

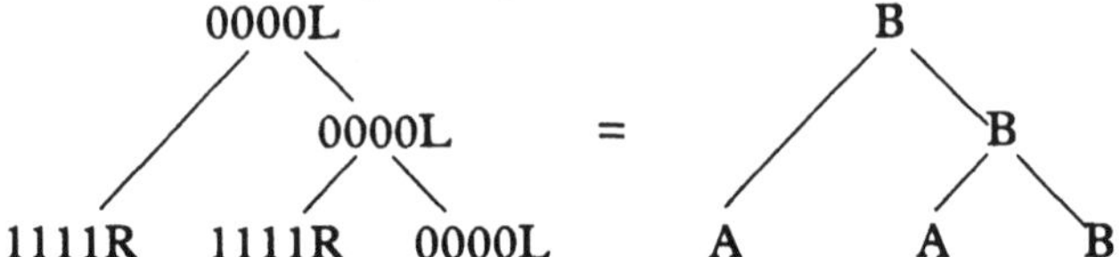

Recursive embedding rules for structures of the type A^nB^n (a string of n A's followed by a string of n B's) require a pair of rules of the type,

S = A S B

S = A B

These rules might be represented in an ATO categorial grammar by assigning A the code 1010R, B the code 0101L, and S, *two* codes, 0000LR and 1111LR. An 'LR' at the end of a code means that the * operator is to apply twice, combining the LR unit and its left partner first, then combining the result with its right partner. The sentence A A A B B B would yield the phrase,

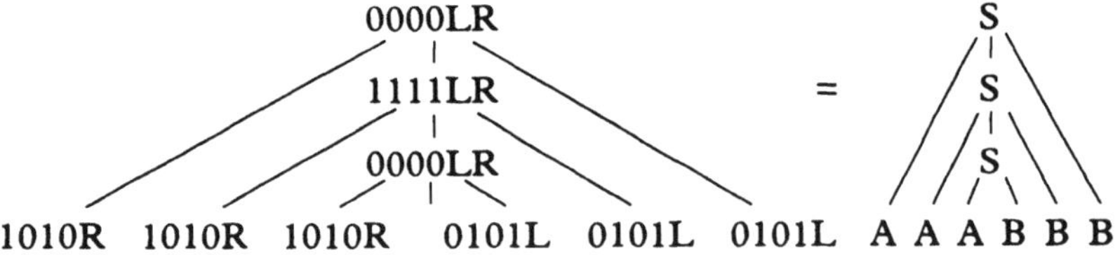

By demonstrating that ATO logic can model both phrase structure grammar rules and Boker Tachtit opposed platform Levallois point production techniques, I have provided a mechanism for connecting the development of phrase structure grammar with the emergence of new stone tool technology during the Middle-Upper Palaeolithic transition, as suggested by Mellars (1989). Let us now consider some implications for the development of other kinds of behaviour.

ATOS AND THE GROWTH OF WORLD KNOWLEDGE SYSTEMS

The representation of elements of a conceptual universe by bundles of semantic features has some interesting consequences. The number, n, of semantic features limits the size of that universe to 2^n named elements. If there are n semantic features, each unique element in that universe can be represented by a unique n-place binary integer. From a developmental point of view, this means that each time a new semantic feature is added, the size of that universe doubles. A system with 12 features implies a universe containing 4096 elements, and a system with 15 features implies 32,768 elements. 20 features would imply a possible 1,048,576 elements. A relatively small number of semantic features, which may have been used to categorize a rather small, observed subset of the real world, could imply a vastly larger universe that is essentially unknown and unexplored. ATO systems provide a means of exploration by analogy and with a computational processing time that increases only linearly with the number of semantic features. This is in sharp contrast with an exponentially or even combinatorially (f(n!)) increasing computation time associated with propositional logic methods.

The use of ATO logic in combination with the invention of global classification schemes solves a major problem of calculating human behaviour with 'frames' or 'scripts' ('frame/scripts'). The concept is characterized by Lakoff (1987: 116) as follows:

> Minsky's *frames* (1975), like Schank and Abelson's scripts (1977) and Rumelhart's *schemas* (1975), are akin to Fillmore's *frames* and to what we have called *propositional models*. They are all network structures with labelled branches than can code propositional information. In fact, Rumelhart's schemas, which are widely used in computational approaches to cognitive psychology, were developed from Fillmore's earlier work on case frames (Fillmore 1968). Frames, scripts and schemas are all attempts to provide a format for representing human knowledge in computational models of the mind. They attempt to do so by providing conventional propositional structures in terms of which situations can be understood. The structures contain empty slots, which can be filled by the individuals occurring in a given situation that is to be structured.

The same concept has been analysed by myself in the following terms (Klein 1988: 479):

> Major computational problems arise when frames or scripts are used to parse human behavior sequences. They can be used well with rather small knowledge domains, but as the size and heterogeneity of a knowledge system increases, the computation time necessary to determine which frame/scripts apply and which individuals match which slots can increase combinatorially. The problem becomes acute when one attempts to recognize metaphorical relations. To do so, it is necessary to discover mappings between the 'role' or slot fillers in two or even three frame/scripts, each of which is drawn from a different semantic domain (Klein, Kaufer and Neuwirth 1980; Gentner 1983). The computational problems associated with such matchings are generally recognized in the field of artificial intelligence, and the general problem of determining consistent selection of role or slot fillers across sequences of frame/scripts can be characterized as one of logical quantification. First order predicate calculus quantification and, arguably, higher order quantification are involved.
>
> Planning in frame/script propositional models can also involve

combinatoric calculations. The planning process requires a determination of the preconditions of a desired goal state, the preconditions of those preconditions, and so forth, for as many intermediate states as may exist between an initial situation and a desired goal.

In contrast to propositional frame/script models, the ATO model permits planning by analogy with a computation time that increases only linearly:

'If a sequence of events, A, B, C, D, occurs, then

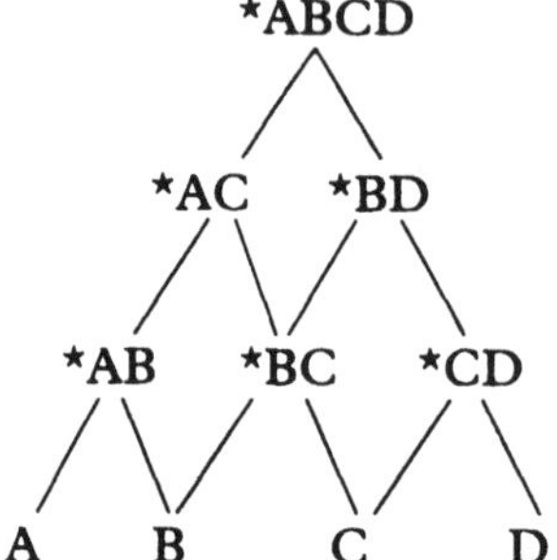

= patterns behind patterns behind patterns behind events

= patterns behind patterns behind events

=ATO patterns behind events

= event sequence

If we wish to obtain a state E instead of D, *without changing* any of the ATOs, we derive by analogy, a sequence leading to E by replacing A, B, C, respectively, with *A(*DE), *B(*DE), *C(*DE). If we wish to make a plan that specifies more than one goal state in the event sequence, we must alter some ATOs.

The meaning of 'culturally defined behavior' is that members of a society plan in a way that minimizes the level and numbers of ATOs affected. It follows that deviant behavior may be interpreted as behavior that violates acceptable levels and numbers of ATOs. ATO patterns are part of the knowledge acquired by children. They are encoded in multiple media of expression, both material and symbolic, and are the source of metaphor. It is this encoding that gives form to a culture, and it is their widely distributed presence in the environment that makes calculation of social behavior computationally feasible for the human mind' (Klein 1983: 154)

A DEVELOPMENTAL SEQUENCE FOR THE GROWTH OF LANGUAGE AND CULTURE ON STRUCTURALIST PRINCIPLES DURING THE UPPER PALAEOLITHIC

The implications of the present study for the developmental sequence of language and culture during the Upper Palaeolithic – viewed in terms of structural principles – may be summarized as follows:

1. At some point during the Middle-Upper Palaeolithic transition, ATO analogical reasoning emerged in conjunction with the development of new types of stone tool technology, as suggested by the Boker Tachtit evidence presented earlier. A change in brain structure may have been involved, but the discovery of the group property may have emerged from

the physical constraints of stone tool production.

2. The ATO logic manifest in the opposed-platform Levallois point core reduction strategy subsequently extended to other domains of human behaviour, for it may have seemed a 'law of nature', and its extension to activities that involved stone tools may have seemed 'natural'. The generality of the extension may have followed from the fact that the group property was associated with the production of blades, which were potentially transformable into tools for a variety of activities at a later time. The extension to language structures and semantic categories made language a medium for analogical reasoning.

3. Global classification schemes were invented at least as early as the Upper Palaeolithic. Their invention made it possible to extend frame/scripts from old domains to new ones. All that was required was to replace their slot-fillers in old contexts with their classificatory counterparts in new ones. The frame/scripts consisted, for the most part, of situation descriptions related to ATOs rather than propositional logic structures.

4. As a result of the extension of these ATO frame/scripts to new areas of behaviour, a limited number of the same ATOs began to apply in an increasing number of frame/scripts in different domains. New ATOs were derived analogically from previously existing ones. The process consisted of generating new ATOs from old ones, through the use of a limited number of second-order ATOs. The process may have occurred recursively, yielding a hierarchy of higher-order ATOs that, together with a global classification scheme, defined the concept of 'social structure'.

5. The emergence of art, which may have involved a biological change, may also have been associated with visual encoding of higher-level ATOs in surrealistic imagery. At this stage ATOs could be represented in exactly the same media as realistic situation descriptions. Realistic art and surrealistic imagery encoding ATO operators may have appeared simultaneously.

6. *At the same time*, encoding of higher-level ATOs in myth structures and in the form of spirits and deities may have marked the development and elaboration of religious phenomena. As I have discussed elsewhere:

> 'The emergence of a canonical hierarchy of ATOs, applicable to multiple domains of social reality through the mediation of a global classification scheme, would be a natural consequence of organizing social life on the basis of ATO logic. If we make the assumption that the human mind encodes ATOs in iconic imagery, we may also suggest that such imagery is given metaphysical interpretation. A hierarchic ATO system may be interpreted by the human mind as a hierarchy of spiritual beings, and the spirit journey of a Shaman seeking to resolve problems in a spirit realm can be interpreted as ... a kind of ATO manipulation... Magic spells and rituals would appear as devices for inserting desirable ATOs in given situations, and it might be possible to predict their form and general content from the global classification scheme' (Klein: in press).

Divination systems based on ATO principles may also have appeared at this time, and the use of methods involving analysis of randomly obtained patterns of cracks in fire-heated bones may have been motivated by a recognition that ATO logic is a 'law of nature' in both tool technology and its extension to social behaviour.

DISCUSSION

The developmental sequence just presented, although highly speculative, offers a strong functionalist rationale for the thesis that the structural principles discerned by Lévi-Strauss reflect principles of organization of human behaviour that were operative in the development and evolution of human behavioural systems at least as early as the Upper Palaeolithic. Structuralism appears as an invention, rooted perhaps in a change in brain structure that facilitated computation by analogy, that made it possible for humans to develop complex social systems, the behavioural requirements for which could be computed in real time. The theory offers a mechanism for the views of Durkheim and Mauss (1963) about the relation of spirits, deities and divination systems to classification schemes, mechanism for the view of Marx that religion is emergent from social structure, and mechanism for the structuralism of Lévi-Strauss. It also offers an explanation of how innovations in stone tool technology could lead to the elaboration of language structures and the development of complex social and cultural behaviour. Certainly, behavioural systems existed in relatively simple forms before the Upper Palaeolithic, but it might have been the development of ATO systems that enabled the exponential expansion of the human conceptual universe that *began* in the Upper Palaeolithic.

NOTE

1. An assumption that the brain is a massively parallel computer does not mitigate the problem. The addition of n parallel processors can reduce the computation time by a factor of n, but the problem domain involves a processing time that can increase combinatorially with the size of the data base. If an additional computer processor is added for *each* new item in the data base, the processing time may increase at a rate of $n!/n = (n-1)!$. A connectionist brain model presents an analogous difficulty; the need for combinatorially increasing processing time is replaced by a need for combinatorially increasing connectivity.

ACKNOWLEDGEMENTS

This research was conducted while I was a Visiting Fellow at Clare Hall and the Department of Archaeology at the University of Cambridge, England, January-June 1988, and was supported by a grant from the Wisconsin Alumni Research Foundation. I am grateful for advice given to me by Ian Hodder, Paul Mellars and Tony Sinclair.

REFERENCES

Dasen, P.R. 1977. *Piagetian Psychology: Cross-Cultural Contributions.* New York: Garden Press.

Fillmore, C. 1968. The case for Case. In E. Bach and R. Harms (eds) *Universals in Linguistic Theory*. New York: Holt, Rinehart and Winston: 1-90.

Gardner, H. 1972. *The Quest for Mind: Piaget, Lévi-Strauss, and the Structuralist Movement*. London: Coventure.

Gentner, D. 1983. Structure-mapping: a theoretical framework for analogy. *Cognitive Science* 7: 155-170.

Hage, P. 1979. Symbolic culinary mediation: a group model. *Man* 14: 81-92.

Hage, P. and Harary, F. 1983a. *Structural Models in Anthropology*. Cambridge: Cambridge University Press.

Hage, P. and Harary, F. 1983b. Arapesh sexual symbolism, primitive thought and Boolean Groups. *L'Homme* 23: 55-77.

Hietala, H.J., and Marks, A.E. 1981. Changes in spatial organization at the Middle to Upper Paleolithic transitional site of Boker Tachtit, Central Negev, Israel. In J. Chauvin and P. Sanlaville (eds) *Préhistoire du Levant*. Paris: Centre National de la Recherche Scientifique: 305-318.

Klein, S. 1977. Whorf transforms and a computer model for propositional/appositional reasoning. Paper presented at the Applied Mathematics Colloquium, University of Bielefeld, at the Computer Science Colloquium, University of Paris-Orsay, December 1987, and at a joint colloquium of the Anthropology and Computer Science Departments, University of California, Irvine, March 1978.

Klein, S. 1982. Culture, mysticism, and social structure and the calculation of behavior. *Proceedings of the 1982 European Conference on Artificial Intelligence, July 12-14*. Paris: Society for the Study of Artificial Intelligence and the Simulation of Behavior: 141-146.

Klein, S. 1983. Analogy and mysticism and the structure of culture. *Current Anthropology* 24: 151-180.

Klein, S. 1988. Reply to S.D. Siemens' critique of Klein (1983). *Current Anthropology* 29: 478-483.

Klein, S. (in press). The invention of computationally plausible knowledge systems in the Upper Paleolithic. In R. Foley (ed.) *The Origins of Human Behaviour*. London: Allen and Unwin. In Press.

Klein, S., Kaufer, D.S., and Neuwirth, C.M. 1980. The locus of metaphor in frame-driven text grammars. In W.C. McCormack and H.J. Izzo (eds) *The Sixth LACUS Forum 1979*. Columbia (SC): Hornbeam Press: 53-67.

Klein, S., Ross, D.A., Manasse, M.S., Danos, J., Bickford, M.S., Burt, W.A. and Jensen, K.L. 1981. Surrealistic imagery and the calculation of behavior. In R. Wilensky (ed.) *Proceedings of the 3rd Annual Meeting of the Cognitive Science Society, University of California, Berkeley, August 19-21*. Hillsdale (N.J.): Lawrence Erlbaum Associates: 307-309.

Lakoff, G. 1987. *Women, Fire and Dangerous Things: What Categories Reveal about the Mind*. Chicago: University of Chicago Press.

Lévi-Strauss, C. 1962. *La Pensée Sauvage*. Paris: Plon.

Lévi-Strauss, C. 1964-71. *Mythologiques*. Paris: Plon. 4 Volumes.

Marks, A.E. 1981. The Middle Paleolithic of the Negev. In J. Chauvin and P. Sanlaville (eds) *La Préhistoire du Levant*. Paris: Centre National de la Recherche Scientifique: 287-298.

Marks, A.E. 1983. The Middle to Upper Paleolithic Transition in the Levant. In F. Wendorf and A.E. Close *Advances in World Archaeology* Vol. 2. New York: Academic Press: 51-98.

Marks, A.E. and P. Volkman. 1987. Technological variability and change seen through core reconstruction. In G. de Sieveking and M.H. Newcomer (eds) *The Human Uses of Flint and Chert*. Cambridge: Cambridge University Press: 11-20.

Mellars, P. 1989. Technological changes at the Middle-Upper Paleolithic transition: technological, social and cognitive perspectives. In P. Mellars and C. Stringer (eds) *The Human Revolution: Behavioural and Biological Perspectives on the Origins of Modern Humans.* Edinburgh: Edinburgh University Press: 338-365.

Minsky, M. 1975. A framework for representing knowledge. In P.H. Winston (ed) *The Psychology of Computer Vision*. New York: McGraw-Hill: 211-277.

Piaget, J. 1960. *The Psychology of Intelligence* (Translated by M. Piercy and D.E. Berlyne). Totowa: Littlefield, Adams.

Piaget, J. 1970a. *Genetic Epistemology* (Translated by E. Duckworth). New York: Viking.

Piaget, J. 1970b. *Structuralism* (Translated by C. Maschler). New York: Kegan Paul.

Piaget, J. 1972. *The Principles of Genetic Epistemology* (Translated by W. Mays). London: Kegan Paul.

Rumelhart, D. 1975. Notes on a schema for stories. In D.G. Bobrow and A.M. Collins (eds) *Representations and Understanding: Studies in Cognitive Science*. New York: Academic Press: 211-236.

Schank, R.C. and R.P. Abelson. 1977. *Scripts, Plans, Goals, and Understanding*. Hillsdale (NJ): Lawrence Erlbaum Associates.

Volkman, P. 1983. Boker Tachtit: core reconstructions. In A. Marks (ed.) *Prehistory and Paleoenvironments in the Central Negev, Israel.* Vol. 3: *The Avdat/Aqev Area* (*Part 3*). Dallas: Southern Methodist University Press: 127-190.

Wynn, T. 1979. The intelligence of later Acheulean hominids. *Man* 14: 371-391.

Wynn, T. 1981. The intelligence of Oldowan hominids. *Journal of Human Evolution*: 10: 529-541.

Wynn, T. 1985. Piaget, stone tools and the evolution of human intelligence. *World Archaeology* 17: 33-43.

19: Symbolic Origins and Transitions in the Palaeolithic

MARY LECRON FOSTER

INTRODUCTION

For archaeology, the Palaeolithic is most notable as the period in which culture emerged and evolved. If archaeologists are to understand the first stirrings of culture, as well as subsequent prehistoric changes and innovations, it is important for them to be aware of the structure of culture. Both cultural anthropology and linguistics can provide many insights into cultural structure – insights that to date have been little exploited both within archaeology (cf. Conkey 1980) and within physical anthropology (Note 1). Among other things, this paper is a plea for greater theoretical integration of the various subdisciplines of anthropology. As knowledge increases, so does specialization, and today anthropology is in danger of serious and damaging fragmentation. In what follows I will bring insights from cultural anthropology and linguistics to bear on the problem of cultural origins – problems which are usually addressed only by archaeologists and physical anthropologists.

It is highly unlikely that culture arose through a single massive change, or behavioural rubicon, on the earlier side of which was non-culture, or pre-culture – the socially shared activities and understandings of pre-human primates – and, on the later side, culture – or the socially shared activities and understandings of the genus *Homo*. (See also Conkey 1980: 227; also for gradualist views of cognitive evolution from the standpoint of neuro-biology, see Lamendella 1980: 147-74, and from that of cognitive ecology, see Lock 1982: 112-7.)

If culture is unique to the human species and came into being gradually, what are the characteristics, or the abilities, that distinguish non-cultural mammalian practices from the practices of culture-bearing groups or individuals? In my opinion this question can only be successfully answered by means of examination of the *structural* characteristics of culture, determined by comparative analysis of the cultural systems of human groups. It is not enough to explore single systems, such as language, tool-making, living arrangements, etc. What is crucial is to explore the structure of each human activity and then to make cross-modal comparisons in order to discover the commonalities of cognitive structure that have made it possible for each cultural structure to come into being.

The only way to discover cognitive structure is through examination of behaviour. For archaeologists, prehistoric behaviour is available only through human products, their structures and distributions.

Cultural comparison as a means to reconstruct past stages is not a well-developed methodology except in comparative linguistics where, in the mid-nineteenth century, a scientific method was developed that permitted the reconstruction of past stages in the history of particular language families. Recent reconstruction of stages of far greater time depths, across, rather than within, families, but still using the strengths of the comparative method, provides clues to the cognitive structure of language at very remote time periods (Foster 1978, 1980, 1981, 1983, 1986, In Press). When these linguistic structures are compared with other cultural structures available through archaeological research, observation of the behaviour of other animal species, and the behaviour of language-acquiring children, much progress can be made in determining cognitive stages in the evolution of culture.

RATIONALITY

According to popular belief, the distinguishing characteristic of human beings is their ability to reason, or, as my dictionary defines the term, 'to think coherently and logically; draw inferences or conclusions from facts known or assumed'. However, a moment's thought will show that without the ability to reason, no animal could survive for more than a moment. All animals must make judgements as to what is food and what non-food, what is prey to be hunted and effective means to hunt it, what is reasonably safe territory in which to rear the young, which animals are to be trusted, and what predators are to be feared. Thus, rationality cannot constitute the difference between human and non-human mammals, for the degree to which any given animal can make judgements of this kind is the key to its survival. Rational strategies are learned through experience, as psychologists experimenting with maze-running and other learning devices have amply demonstrated.

It has often been argued that there are two human forms of thought, sometimes designated as 'logical' and 'prelogical'. Prelogical thought is conceived of as non-realistic, or fanciful, while logical thought is based on premises of cause and effect. Since human beings are often touted as supremely rational, the assumption has seemed to follow that logical thought was preceded prehistorically by fanciful (metaphoric) thought. It would seem that human beings have, indeed, made progress in the complexity of their potential for rational decision-making, but a reverse temporal ordering of the sequence would seem to be indicated since associational, analogical (or symbolic) thinking quite clearly is the basis for culture.

Logic, as a formalized, and highly valued, system of thought began with Aristotle, who based his thinking on the kinds of propositions that could be made with language. From this it seemed to follow that language was a logical system, but what Aristotle failed to recognize was that this only seemed so because of certain uses to which language could be put, rather than because of the actual structure of language, which is associational rather than logical. Some later philosophers, such as Cassirer (1953) recognized that symbolism was the basis of language and culture, but empirically oriented anthropologists, and perhaps this includes most archaeologists, still view symbolic analysis with distrust. This is partly a heritage of the Aristotelian fallacy, reinforced by the belief that symbolism implies fantasy, and held despite the fact that whatever the meaning of particular symbols, when examined structurally they are not found to be rooted in idealism but to be solidly grounded in the phenomenological world.

SYMBOLIZATION AND CULTURE

According to Langer (1942: 32), the basic need of human beings alone, is '*the need of symbolization.* The symbol-making function is one of man's primary activities, like eating, looking, or moving about. It is the fundamental process of his mind, and goes on all the time'. Langer then quotes from Ritchie's *The Natural History of the Mind*: 'As far as thought is concerned, and at all levels of thought, it is a symbolic process... The essential act of thought is symbolization'.

There are two essential characteristics of symbolization. The first is that no symbol exists in isolation. Culture is a system, and *system implies classification* (Note 2). Classification, in turn, implies that *different things are perceived as like.* The second characteristic then, is that a symbol is defined by likeness. Everything that plays any role in human culture is a cultural symbol, whether a product of nature or of human activity, as long as it has taken on some social use and consensual meaning. Because of classification, symbolism is analogically motivated.

Words have, however, been said to be 'arbitrary', for, according to Saussure (1959), the father of modern linguistics, the meaningful elements of language are signs, because they are arbitrary, while symbols are non-arbitrary, or metaphoric (Note 3). This judgement is due partly to the fact that in popular usage language is generally conceived of as 'naming', or the assigning of referential meaning to a vocalized or written configuration that is in every way dissimilar to any object designated by that 'name', unless onomatopoeia is involved. Hence, words are usually not directly iconic with whatever they signify.

There are problems with this judgement. The first is that language is actually *not* naming. There is no direct connection between a word and some object or action which can be referred to by that word. Every word is general in scope rather than particular, or concrete. The word *chair* does not imply *this* chair, but any or every object with the general characteristics of chair-ness. According to Langer (1942: 49), 'symbols are not proxy for their objects, but are vehicles for *conception of objects...it is the conceptions not the things that symbols directly mean*'.

What symbolic analysis must always begin with, and refer back to is the concrete evidence: in linguistic terminology, the syntagma (Note 4). Only the syntagmatic dimension of language (or of any other cultural system) expresses relationships which are actually available, since they occur distributionally, or spatio-temporally. Syntagmatic units may also be referred to as paradigmatic units, if what is in question is their relationship to other segments that can fill the same syntagmatic slot or slots. The units themselves are the same, it is only their relationship to other units that differs. A unit that is paradigmatic does not *co-occur* syntagmatically with other units of the same paradigmatic class, whereas a syntagma occurs in sequential or otherwise spatial relationship to other syntagmata.

This can be illustrated in a simple way with language: If I say, 'I am the mother of a ...', the empty syntagmatic slot may be filled either with 'daughter', or with 'son'. The two together constitute a paradigmatic class. Class membership can be determined both by the fact that they are potential fillers of the same slot, or by the fact that they have the same hierarchical relationship to the same cover-term, 'child'. Thus, 'I am the parent of a child', contains members of two obviously hierarchical paradigmatic classes, *parent* and *child* that are related to one another in exactly the same way – parent:child :: mother (or father):daughter (or son). It should be apparent that the analytic establishment of classes and their hierarchy of membership is a very complex semantic task, as is the establishment of rules governing the syntagmatic (spatio-temporal) relationships between members of paradigms.

In this, language is no different from any other cultural system. Cultural meanings, like linguistic meanings are established through class membership as revealed in their syntagmatic relationships. No symbol has meaning in a cultural vacuum. All symbols and their meanings form part of a complex system, a network of interconnected meanings by means of which human beings communicate with one another. Although the major system of communication is language, since other systems are generally translatable into linguistic terms other systems can be analyzed by the same *structural* method. This method was available to, and formed the analytic strength of, linguistics, from the early decades of this century, a method of which Saussure's lectures (1959; first published in 1916) are an enduring model.

In popular usage, a 'symbol' is usually construed as an ideologically charged artifact. In the parlance of anthropologists concerned with symbolism, such an artifact is called variously a 'key symbol', a 'dominant symbol', a 'central symbol', an 'elaborating symbol', or a 'condensed symbol', because a great deal of affect has come to be attached to it. However it is not the affect which makes a symbol a symbol but the fact that it is connected with other symbols in a meaningful way. Every symbol is so connected, but not every symbol is similarly charged with affect.

What makes a symbol a symbol is, then, not the fact of its potential to arouse human emotion but the fact of its organization within meaningful but covert (because paradigmatic organization is necessarily covert) symbolic networks.

The ability to operate in terms of *gross* classification is apparent from the behaviour of even the very lowest creatures in the ability to distinguish food from non-food, and, with the advent of sexual dimorphism, to distinguish potential mate from non-mate, both in terms of gender and in terms of species. Somewhat higher on the evolutionary cognitive scale comes the ability to distinguish parent from non-parent and child from non-child. Thus, pre-symbols only become symbols and are processed as symbols because they are organized according to some standard of likeness.

SYMBOLISM IN ARCHAEOLOGY

According to the symbolic interpretation of culture , all of the materials that Palaeolithic archaeologists discover which provide evidence of human activity are symbols, whether it be a chopper, a handaxe, a spear-thrower, a hearth fire, a dwelling, or an enigmatic marking on bone or stone – all are symbols belonging to an undeciphered, or only very partially deciphered, human culture. Symbols always have concrete manifestation. They are *things*. Since archaeologists deal in the discovery of past things, it is the business of archaeology, just as it is the business of ethnology, which deals in present things, to decipher culture on the basis of its products and their interrelationships. The only difference is that some of the things that ethnologists deal in are observable behavioural sequences. Otherwise, deciphering techniques can, or should, be identical. It is the relationships between *things* which are crucial to interpretation because symbols have no meaning except in their structure of interconnectedness.

Relationships to be explored in a structural approach are always spatial and temporal. Since symbols are always physically manifested, it is to be supposed that some very early symbols will have survived. Those that have been discovered are found with associative material (i.e. in syntagmatic interconnections) that suggests something of their use. Stratigraphic associations, while also spatial in nature, provide a temporal dimension, so that it is possible to trace temporal changes in shapes of artifacts and in their associations with other symbolic material. Much natural material is not very durable, and early human actions can only be determined rather crudely by their enduring products. For much of the Pleistocene, the most durable symbols were stone tools.

Isaac (1976) developed a chronology of early tool styles as evidence of symbolic change in the Pleistocene, implying (correctly in my view) that the cognitive processes in all cultural domains are similarly structured. In interaction with Isaac I had developed an hypothesis that the organization of early tools and other artifacts revealed cognitive patterns that were reflected in language and other systems of symbolic activity. The symmetry of handaxes and the binarism of bifaces seemed to have parallels in symmetries and binarisms of phonology. Composite tools, which began with hafting, struck me as cognitively similar to the composition of phonemes into simple words, words into phrases, or morphemes bound together in multi-morpheme words. Cultural meanings that I have reconstructed for early language in a long-term comparative project seem also to be reflected in early technology.

Some of the tool types discussed by Isaac (1976) occur at the same time periods as other cultural forms. Thus, symmetry and regularity of design are found in the Middle Pleistocene, as are shelters and huts and the domestic use of fire. There was not only regularity of design in tools but also in huts with rule-governed structure, the placement of the hearth and of areas used for distinguishable purposes. Fine discrimination of form is also found in this period with the Levallois method of predetermining flakes. In the late Acheulian and early Mousterian, forms prepared for hafting are found, as is organized asymmetry, including backing. In both the Acheulian and Mousterian the first engraved pieces with deliberate design motifs are found, and in the Mousterian, the first burials and the first associated grave goods. It is apparent that by the Mousterian, the ability to exploit relationships to achieve cultural meaning had greatly expanded, and by the Upper Palaeolithic, not only had complex relationships begun to be explored in tool-making, with such devices as composite armatures, the atlatl, the bow and arrow, and the needle, but as is indicated by the painting and engraving on bone and stone that occurred in this period, symbolism had burgeoned and taken on new meanings through signal expansion of the capacity to discover and exploit likeness. If we look at the earliest known literary productions, those of the Sumerians, we find a flourishing use of polysemic metaphor. It looks very much as if the polysemy of words was already assuming the elaboration of metaphor in the culture of the Upper Palaeolithic, and metaphoric extension of meaning was certainly very much in place in the art of the Neolithic (Gimbutas 1982).

Isaac (1976) failed to mention the extraordinary variety of Oldowan tools that was discussed by Montagu (1976: 266-7) in the conference where both of their papers were presented. Montagu believes that the kind of cognitive processes needed to make, and to transmit the art of making, so many different tools – five types of chopper, polyhedrons, discoids, light duty scrapers, burins, other heavy duty tools, artificially flaked and abraded bone tools – 'implies the existence of some form of speech, however rudimentary'. Montagu relates the advent of rudimentary language to the development of small animal hunting and butchering and thinks it probably initially involved only names for things and words for processes relating to tool-making and the hunt.

THE INCEPTION OF LANGUAGE

Correlating these tool developments with a possible schema for the development of language, we can assume rather gross attempts at conveying meaning by single but paradigmatically contrastive sounds in the Oldowan period. The hypothesized contrastiveness of sounds correlates with the contrastiveness of tool types as dictated by the uses to which they were put (i.e. their syntagmatic usage) as well as with other activities than hunting or tool-making. Tanner and Zihlman (1976) stress the predominance of visual and tactile communication for monkeys and apes, and the extent to which this is integrated into human behaviour, suggesting that, 'However speech developed, it clearly evolved in a matrix of extensive nonverbal communication' (1976: 474) in which face-to-face contact was important. According to these authors (1976: 475), 'Quite possibly the context for communicatory control and the development of arbitrary symbols for expressing meaning is to be sought in leisure or in food-processing activities back in camp where many individuals were associated, rather than in seeking food away from camp' (Note 5). In this context, it is interesting that Wright (1978) demonstrated by experimenting with an orangutan that mimicry rather than speech was all that was necessary for the learning of striking flakes from a core with the goal of cutting a cord in order to retrieve food.

Since young apes, like young humans, spend a great deal of time in face to face interaction with mothers and other caregivers, it would seem that it is primarily in this context that control of facial expression and mouth movements would have become important.

To understand the advent of symbol-use during the Pleistocene it is useful to look at the potential of animals to exploit analogy. Any form of reasoning involves at least some measure of classification: many things are fearsome and elicit fear, many things are edible and evoke hunger and eating. This is elementary use of the analogical potential. How and why did the analogic faculty develop beyond this for the human species?

A more advanced use of analogy that can be traced to lower mammal behaviour is mimicry. Imitation of what others do plays a major role in learning for all mammals, but mimicry is especially characteristic of primates. Mimicry is behavioural use of analogy, but without the systemic classification which is the hallmark of symbolism.

One thing that is apparent from observation of the behaviour of young mammals and young children is that both derive pleasure from movement for its own sake and from experimentation with new movements involving others or involving inanimate objects. Much is learned by all mammals from manipulative activity of many kinds. Since tool-making involves manipulation in accordance with methods learned through imitation and in expectation of future rather than present use of the tool, it is quite different from even chimpanzee preparation of termiting twigs, or even Wright's (1978) orangutan which learned to strike flakes from a stone for immediate use, when provided with the materials.

Normally, a primate will use a finger or tongue for extraction. Termite holes are finer and deeper than a finger or tongue can either penetrate or reach. Twigs are substitutions for fingers or tongues, but used similarly, based on a recognition that different kinds of insertion need spatially different implements. Tools for insertion are perceived as both bodily and non-bodily and indicate a dawning aptitude for classification in the fact that finger, tongue and twig come to occupy the same paradigmatic class in the syntagma of extraction of food from a hole. The experimental teaching of 'language' to great apes also shows that they possess a rudimentary capacity for classification.

On the pre-hominid level, classification was minimal but began to increase as animals organized themselves into social groupings with hierarchical status rankings. Lawick-Goodall (1971: 113-4) describes the strategy of the male chimpanzee, Mike, in raising his status through manipulation of man-made articles. This strategy involved adding a new item to the class of objects used for gaining and maintaining status through brandishing display. Addition of man-made items to display paraphernalia which heretofore only included trees and branches, or perhaps also rocks, not only increased the paradigm but apparently gave it a hierarchical dimension in that Mike tried to add other man-made items to the ones used initially, and as a result was able to become dominant male through effective displays. Rules for syntagmatic displays apparently included brandishing, throwing and hooting, followed by submissive grooming of the defeated animal by the now dominant challenger.

We can assume that in early hominoid society there were already patterned (rule-governed) ways of behaving. In addition to status preserving behaviour, males and females probably began to have differentiated social and economic roles. Isaac (1976) assumed a division of labour in the Plio-Pleistocene, with male hunting and female gathering of vegetal food and return of both to a 'home base' for food sharing. It is probable that any rudimentary 'language' used at that time was associated in rule-governed syntagmatic ways with particular conventionalized behavioural stretches of this kind.

For additional insight into the question of the development of meaning, studies of child language acquisition are useful. An article by Bloom (1976) contains the following:

> First, children talk about ways in which objects relate to one another; they do not simply tell other people what objects are or the names for objects. The relational information that children talk about has to do with the existence, disappearance, nonexistence, or recurrence of an object (the relation of an object to itself); the actions of persons that affect objects, the locations and changes in location of objects and persons; the objects that are possessions or otherwise within the domains of other persons. Children talk about what they do to objects and what they do with objects, but they do not mention objects as instruments of actions on other objects (the importance of toolmaking in the evolution of the species notwithstanding)...
>
> Movement is critically important in early child language among the first names of objects that children learn are the names of objects that move (e.g. persons, pets, balls, etc.), and semantic-syntactic structures are used to encode action events before they are used to encode stative events. With respect to movement, children talk overwhelmingly of their own intentions to act, e.g. 'eat cookie' as the child reaches for the cookie, or their desire for another to act, e.g. 'Mommy open' as the child gives Mommy a box to open. Children talk far less frequently about what they have already done or what they have just seen others do. It is entirely reasonable to speculate

that each of these generalizations about early child language has its analog in the origins of language in the human species: talking and understanding about movement, intentions to act, and the relations among objects. It is relevant that observations of the creation of 'new' languages – pidgins and Creoles – have revealed the dominance of encoding dynamic aspects involving movement in events over the static aspects of events in the origins of such languages.

The major point is that children talk about spatial relationships first in terms of *movement*. They name movement rather than the object that is moving, but relate this to themselves (as moving or movers) and others (as moving or movers). Hence movement, and the manner in which movement through space is conducted is the kind of meaning that is primary in early language. Temporally, the immediate present, or immediate future in terms of command, is the only recognized time sequence. This would seem to tally with Montagu's judgement of the kinds of meanings conveyed by early language, and with the meanings in early language that I have derived from comparative reconstruction.

THE COMPARATIVE METHOD OF LINGUISTIC RECONSTRUCTION

Traditionally, linguistic reconstruction has taken place within groups of languages which inspection has shown to exhibit considerable word similarity, indicating genetic relationship. Reconstruction is used to confirm and to explicate common origin.

Reconstruction proceeds very much in the manner of scientific experimentation in which one variable is changed while others are kept constant. On the diachronic level of reconstruction, linguists make comparisons between meaningful segments or whole words in two or more languages in the attempt to find regularities in sound change. Thus, if English is compared with Latin and Greek, we find that the p, t, and k of the latter two languages are always reflected as f, th, and h respectively in English. Thus, the Proto-Indo-European (PIE) root that is reflected in Greek as pet- 'to fly' and in Latin as pet- 'to dash against', is found in the English word *feather*, with the root as feath- rather than as pet-. Similarly, the English word *haft*, with the root haf-, is a reflex of the prototype form for Greek kapt- 'to snatch', and Latin capi- 'to take'. The English word *have* also reflects the same PIE root. The PIE roots *pet- and *kap- (Note 6), as adapted in these three languages, have meanings that can be abstracted from the varied meanings found in the range of meanings in daughter languages as 'to fly' and 'to grasp' respectively. The meanings of verbal roots are rarely semantically identical from language to language within a single family. These examples give a fairly representative range, which is enlarged if more cognate languages are added in. Judgement of root meaning is arrived at by a process of abstraction across languages, by factoring out the common semantic ground from reflexes in the various daughter languages.

Regularity of sound change is the rock upon which the comparative method is founded. If, in some words, we should find that English did not have f, th, or h in positions corresponding to those of p, t, and k in Latin, we would either have to suspect borrowing from a p/t/k language, or to discover some phonologically environmental factor that had interfered with the expected changes (Note 7). Because /p/, /t/ and /k/ as reflexes of these proto-phonemes are much more common across languages than any other reflexes, the proto-phonemes are assigned those values.

Even in cases where simple inspection indicates that languages clearly belong to the same family, reconstruction is often very incomplete because not all of the segments of words that seem to be cognate reconstruct to one another in ways that can be determined. The most extensively reconstructed language is PIE, from which most of the modern languages of Europe derive. It is very rare that every IE language contains a whole word that is cognate with every other, or even with the majority. Meanings can shift, so that, for example, German *Hund* means 'dog' in English, while the cognate form, *hound* has a somewhat altered meaning. This is a mild form of semantic alteration – often alteration is extreme. However, PIE is far and away the most thoroughly reconstructed of the world's linguistic stocks, so that it is usually fairly easy to discover the root-origin of words with semantic change.

THE COMPARATIVE RECONSTRUCTION OF PRIMORDIAL LANGUAGE

In order to get beyond the relatively short time spans of differentiation that the comparative method has traditionally disclosed (up to *c*. 5000 years), it is possible to recover a great deal of prehistoric language history by means of comparison of languages which simple inspection has not revealed to be genetically related – i.e. comparison and reconstruction across linguistic families and stocks. For this purpose it is necessary to have a rather complete dictionary for each language rather than lists of reconstructed forms, even where these exist, because, unless they are relatively complete (which is only true of PIE) they prove inadequate to the task, and also contain a preponderance of cultural terms which would have had no applicability at early stages. In cases in which reconstruction has taken place, and sound regularities have been established, these should be taken into account. The inadequacy of incompletely reconstructed lexicons is principally due to the fact that they rarely give enough information to allow one to differentiate roots from affixes. For this purpose one needs to have evidence of root recurrence across words within the language.

Early in my reconstructive efforts it became clear to me that the linguistic segments that were reconstructible were not whole words but roots, and also even that many words that at the present time appeared to be made up of single roots were, on earlier levels, made up of still earlier roots plus one or more affixes. In other words, at the level at which most language families had separated from one another, historically attested grammar had not yet come into being, and what appeared as grammatical segments in one language, or language family were reconstructible with roots in another. In making reconstructions across language families and stocks, it is rare that one cannot find a PIE form which does not have a root cognate with a root, or with a segment that proves to have originally been a root, of a word from any other of the world's languages.

In attempting reconstructions across languages, I also discovered very early that nouns were not reconstructible unless they shared roots with verbs. Every noun does ultimately share such a primordial root but it is often so overlain with later accretions that it may require a great deal of analysis to isolate the root.

The variation in word or root meaning across languages representing different language families does not reach a much greater spread than that within a family. In fact, in order to be sure that reconstruction is correct, it is safest to maintain a fairly strict meaning correlation across languages, at least until reflex regularities are well established.

Reconstructible primordial segments that constitute 'roots' in historical languages invariably turn out to be complex, because some phonemes become lost over time and others are hidden within fused phoneme innovations. For example, Indoeuropean (IE) languages have lost all of the primordial fricatives. Sounds like f, th, s and h in English do not represent primordial fricatives, but instead are derived from the voiceless stops, p, t, c and k, respectively. Also in IE, labialized velars and palatalized consonants have developed from *kw and *ky respectively. Post-consonantal falling diphthongs *wa and *ya were common in PL and account for vowel apophony in PIE and other languages.

Examination of ancient Egyptian and the Semitic languages suggested this reason for the virtually ubiquitous vowel apophony found in historical languages. Occurrence of a semivowel after a root consonant and before the phememically meaningless vowel *a tended to affect the root vowel so that it became converted to a backed and rounded vowel, usually [o] in the case of a vanished *w, or [e] in the case of a vanished *y. Thus, the rounding of [w] became transferred from the consonant to the adjacent vowel, and the fronting and raising of the tongue of [y] was transferred to the adjacent vowel, which became raised and fronted instead. In the case of PIE, if a fricative was the pre-vocalic consonant, the consonant was eliminated as well as the semivowel. Thus, the celebrated 'laryngeal' or 'laryngeals' hypothesized as the cause of vowel apophony in the IE languages prove instead to be post-consonantal falling diphthongs, *ya and *wa, transformed into [e] and [o] respectively. Other changes, such as lengthening, also attributed in PIE reconstruction to lost laryngeals, are instead caused by the loss of some one of a full range of fricatives. Among IE languages alone, these are phonemically reflected in Hittite.

Examples of sound loss using the PIE reconstructions, *pet- 'to fly' and *kap- 'to grasp', illustrated above, show both consonant loss and vowel refashioning. Thus *pet-, in PIE occurring in words with meanings of flying or birds, proves cognate in Egyptian (E) with pʔyt 'feathered fowl, birds'. E /ʔ/ represented either PL *ʔ or PL *h, or perhaps both (Note 8); phememes which have vanished from IE languages. In any event, PIE and

Table 19.1. Primordial language phenemes.

	Labial (external)	Dental (internal)	Alveo-palatal (axial)	Velar	Glottal
Glides	w encircling		y extending		
Sonarants	m coactive	n internal	l weakening	r rising	
Stops	p protruding	t intruding	c extruding	k descending	ʔ stopping
Fricatives	f overarching	θ joining	s expanding	x opposing	h starting

E can be seen to derive from an earlier form, either *phyat, or *pʔyat (depending on the value given to the glottal phememe), because of the prevocalic *y yielding *e vocalization in PIE.

The Egyptian word, in turn, is seen to derive from a shorter word or root, illustrated by Egyptian words pʔ, pʔy and py, all of them translated as 'to fly', and all probably representing a form with varied patterns of vowel insertion or vocalization of the semivowel: *pʔay, *paʔy, *pʔy, *paʔay. In Yana, a California Indian language, bahbil- means 'to fly', and bahbingu- means 'to flutter about'. The root is clearly reduplicated in this language from an original PL *pʔy (or *phy), first with, then without, a vowel nucleus between consonants.

Even from within PIE it can be seen that *kap-, 'to grasp', is a compound form because of *ap- and *ep-, 'to grasp, take, reach', which in Hittite is eip-, or ep-, 'to take'. In E we find xp 'grasp, fist, palm of the hand as a measure', and in Arabic, (A) kafq 'to beat, to tread, to flutter, flap the wings'. And again in E, hp 'to bind, regulate', hp-t 'something seized or snatched' and in A hafw 'to snatch at something with, to flutter, to fly'. A /k/ is equivalent to E x, both from PL *x, and both E and A /h/ derive from PL *xy when it occurs before a vowel. Thus, once more, PIE interconsonantal *e is seen to derive from the PL postconsonantal sequence *ya. The reflex of PL *xyap- in Yana is haba(a)- 'long object moves, do with a long object', and of PL *xyp is hiba(a) 'to pull, stretch', and hibʔzil- 'to tie up'. The Yana translation of 'a long object to move or do', is a global meaning, similar to those of PL, which encompasses movement of arm, foot, wing, as it emerges variously in these reflexes.

While I assumed monogenesis for analytic purposes (see also Wescott 1976 for a similar recommendation) and expected to be able to make reconstructions of primordial phonemes and phoneme sequences, I did not expect meaning to emerge from the phonology of these sequences. However, in the course of what proved to be a successful monogenetic reconstruction, it suddenly occurred to me that substitution of consonants in reconstructed roots signalled semantic changes of a regular nature, much as addition of -s to English nouns signals plurality. Thus, a reconstructed *p, as in the reconstructions discussed above, has the meaning of 'projection', or 'extending outward from', and *t the meaning of 'striking against or through'. *l regularly signaled a meaning of looseness, weakness, relaxation of tension or the like. In English, without reconstructing more of their phonology than the initial phoneme, we find words such as loose, lax, lazy, lethargic, limp, lose, let, as legacies of a primordial initial *l, and such words as fail, fall, spill, kill, dull, mill, as legacies of a primordial final *-l. The relative lack of tension in the touching of tongue to alveolar ridge and releasing it provides iconic explanation for the meaning. The underlying meaning of *l is so transparent in virtually every language that a search for *l words with a meaning of attenuation provides a quick way of predicting monogenetic relationship. We find the same semantics in words with reflexes of a primordial *p, *t or *l in other languages as well. The fact that this and other phememes, each with its uniquely iconic meaning, are easily found, seems to be clear indication of monogenesis. To date I have found no languages which do not Pit this model.

It is standard linguistic practise to designate contrastive sounds (i.e, sounds which will, if substituted for others, change the meaning of a word) as *phonemes*. However, phonemes are iconically arbitrary, whereas these primordial contrastive sounds are not, therefore I have named them *phememes*, meaning-bearing minimal units of primordial speech. The phememes of primordial language, as I currently analyze them, are presented systematically in Table 19.1.

Each word is, of course, the product of its own history, and grammatical additions at any point in that history can modify original meanings. For this reason, not every word with an /l/ in English will, on casual inspection, show the primordial meaning as clearly as do these examples, but usually painstaking analysis of the phememic prototypes reflected in the phonemic constituents of the forms provides a semantic rationale for the meaning of the whole.

Most verb roots, in most languages, are analyzable as two-consonant sequences with or without a vowel nucleus. By comparing reconstructed sequences of this kind it has been possible to discover a reasonable facsimile of the original meaning of what have become over time modern phonemes – 'arbitrary' in the sense that there remains no iconic relationship between sound and meaning. I have made some revisions, usually fairly minor, in my early assessments of semantic values (see Tables in Foster 1978, 1980, and 1983 for earlier charts of these values as well as examples of reconstructed roots with examples taken from modern languages, or from reconstructed languages such as PIE) (Note 9).

ORAL ICONICITY AND CULTURAL EVOLUTION

What is of interest in analysis of the evolution of culture is that in the early days of language the smallest units of oral articulation, phememes, bore meaning, but later became phonemes, which did not. What is of equal interest, is that the meanings that they conveyed were not specific, or of the type that we view as 'naming', but were oral gestural analogs of spatial relationships of a very general nature, with a focus on movement. It becomes apparent from the iconic nature of phememes: that for early hominids, like the language-learning children described by Bloom (1976), the salient aspect of nature was movement in all of its spatial-relational variety rather than the objects that did the moving. Semantics were global rather than specific. The particularity of movement was abstracted from the particularity of its nominal moorings and selected for expression. We can assume with Montagu (1976) that such iconically derived express ions of global movement were applicable to the manufacture of tools in a very general way consistent with the unrefined nature of Oldowan tool assemblages.

From the standpoint of primordial language (PL) it makes sense that meaningful segments were made up of single meaningful sounds because nothing else in culture was as yet complex or compounded. For Oldowan and early Middle Pleistocene periods it would seem that roots were constituted of single phonemes, and that compounding of roots into two or more phonemes came much later, along with hafting, for example, at a period when greater cognitive complexity arose.

It can perhaps be assumed that the globally iconic, oral gesture that became language was originally derived from whole body mimicry. Monkeys and apes are good mimics and because of this faculty are better observational learners than other non-humans. Great apes can learn novel motor patterns by observation of other apes (Lancaster 1958: 61) or even another species (Wright 1978). I have not found anything in the literature on ape behaviour that indicates that apes are capable of mimicking movements 'out there' that are not produced by a similar creature. The metaphoric extension of the iconicity of early primordial phememes indicates an early development of this capacity, since it would seem that almost from the start, the movements imitated were conceived as global rather than concrete – applicable to any similar spatial-relational movement.

It is almost never possible to find single morphemes representing single phememes in historical languages, except very occasionally as affixes or grammatical particles such as prepositions. Verbally (and derivatively in other form classes) they have been concatenated into longer sequences, even longer than seems to be the case if we look at modern two phoneme roots, as illustrated above.

STAGES IN LANGUAGE EVOLUTION

Language comparison has led me to the conclusion that not every phememe assignable to primordial language was in place in the very earliest stages. In all historical languages some phonemes are of far greater frequency than others, and, by the same phememic token, certain phoneme related meanings are of much greater frequency than others.

Frequencies relate both to overall occurrence in vocabularies and to frequency of roots with these phonemes that occur in longer constructions. It can be hypothesized that there is a relationship between frequency and time depth. Analysis of language universals bears on this point. For example, it has been found that every language has at least one nasal consonant (Ferguson 1963: 44). Primordial language reconstruction indicates that in some cases *m and *n have fallen together, and in others that *m has become /n/ and *n has become /m/.

According to the frequency hypothesis (Note 10), the earliest primordial phememes were *m, *w and *y. These consonants and their phememic meanings were probably present in the speech of early Oldowan hominids. I would speculate that *m, *w and *y were probably used before *n because they were visual as well as auditory gestures. In the production of all three, lip shaping is crucial. [y] is not thought of as labial consonant, but to produce [y], or its vocalic allophone [i], lip shaping is in distinct contrast with that of [w]. For [y], lips must be stretched lineally and horizontally, whereas for [w] they must be rounded.

In order to understand the contrast, I suggest that the reader pronounce in sequence the English words *two* and *tea*. The first contains the phone [u] and the second the phone [i]. When *two* is uttered, the lips must be brought into a rounded position; when *tea* is uttered, the lips are elongated into a horizontal line. The same mouth positions are effected when the semivocalic allophones [w] and [y] are pronounced: contrasting the words *wet* and *yet* we see that the lips are rounded at the onset of the first, and elongated at the onset of the second.

By virtue of this shaping, PL *w acquired the meaning of 'rounding', 'circularity', or 'encircling', while PL *y acquired the meaning of 'extending', 'elongating, or 'lineality'.

The meaning of *m as iconic with oral gesture is equally clear. It derives from the bringing together of the two surfaces of the lips and underlies words expressing movement of mouth, eye, and hand. It probably originally derives from such gestures as grasping, biting, blinking – movements in which two surfaces were brought into contact, or into contact around another object. It is found in all sorts of words expressing interaction, and in this became the token of social intercourse *par excellence* – one reason for its extraordinary frequency.

In Table 19.2, I present a hypothesis of stages of phememic evolution and the ultimate conversion of phememes into phonemes (i.e. with loss of iconicity) that seems consistent with prehistoric stages of artifact structure available from the archaeological record, and with linguistic ontogeny.

Stage one in the evolutionary language sequence only finds contrasts between phememes in one articulatory position, in contrast to the symmetrically organized positions of stage two, caused by addition of the phememe *n to the system. The two articulatory positions were probably not very well defined during this stage, but only contrasted as front versus back, or, better perhaps, outer *versus* inner. Here, the raising of the blade of the tongue toward the hard palate in pronouncing [y], or [i], contrasts

Table 19.2. Stages in language evolution.

A. *Early Oldowan*

Front
w around
m coactive
y along

B. *Early Middle Pleistocene*

Front	Back	
w around	y along	oral
m coactive	n internal	Nasal

C. *Middle Pleistocene*

Front	Back
w around	y along
m coactive	n internal
p protruding	t intruding

D. *Acheulean-Mousterian*

Labial	Dental	Alveo-Palatal	Velar
w around		y along	
m coactive	n internal	l weak	r up
p protruding	t intruding	c exiting	k down

E. *Early Upper Palaeolithic*

Labial	Dental	Alveo-palatal	Velar
w around		y along	
m coactive	n internal	l weak	r upward
p protruding	t intruding	c extruding	k downward
f overarching	θ touching	s expanding	x opposing

F. *Upper Palaeolithic*

Labial	Dental	Palato-Alveolar	Velar	Glottal	
w around		y along			Semi- vocalic
m coactive	n internal	l weak	r up		Sonorant
p protruding	t intruding	c extruding	k down	ʔ negating	Obstruent
f overarching	θ joining	s expanding	x opposing	h initiating	Fricative

G *Mesolithic* (or perhaps late Upper Palaeolithic)
Phememes as in PL (Upper Palaeolithic Primordial Language); binary roots; birth of derivational grammar; beginning of consonant fusion, consonant loss and semivocalic vowel coloration; beginning of dialect branching with loss of mutual intelligibility; transition from phememe to phoneme.

H. *Neolithic*
Beginning of polysynthesis; birth of inflectional grammar; increasing dialect branching with increasing mutual unintelligibility to create language families; dialect variation within families.

---

with the raising of the back of the tongue toward the velum in pronouncing [w] or [u]. Tongue movement is now included in articulation in addition to lip movement. *n as 'inner', or 'acted upon', contrasts with *m as 'outer', or 'interactive'.

The point at which phememes began to be organized paradigmatically into a system of binary oppositions – i.e. symmetrically – simple tool symmetry would most likely have made its debut. Symmetrical tools, handaxes, are first seen in the Acheulian. These designs would seem not to have been 'arbitrary' (as Isaac 1976, would have it) but highly motivated, with the symmetry of both tools and phememe organization derived from the symmetry found in nature.

An interesting corollary of stages in the addition of phememes is that the earlier that a phememe came into operation, the more transparent its meaning. In other words, the farther toward the front of the mouth the articulation takes place, the easier the identification of meaning is. Velar and glottal phememes are the most difficult to decipher.

Later stages in development of the system of phememes continued to increase oral symmetry in point and manner of articulation, but each added phememe tended also to unbalance the system so that it required reorganization to maintain symmetricality. Reorganization through addition of new phememes continued for about the length of time needed for major articulatory positions to be easily occupied. After that, the refinement took the course of refinement of meaning through concatenation of phememes – first binarily, and later, multiply.

Expression of meaning from the combining of two phememes probably arose with hafting in the Late Acheulian. These were probably the earliest manifestations of what we would consider to be words. In most languages they, or expansions of them, now serve as roots. Ordering of the two phememes in these 'words' seems not to have been rigid. One can still find roots with reversed ordering and unchanged meaning. However, at some point during the Acheulian syntagmatic involvement increased. We can see this in the spatial organization of shelters, in the advent of engraving on bone and stone in the Acheulian and Mousterian and in the decorative use of red ochre, at least in the Mousterian (Marshack 1976).

As concatenation made meanings increasingly concrete, more and more phememes were added to the concatenations. During the Upper Palaeolithic, artifacts and their uses became more complex, with additions of such tools as the atlatl, the bow and arrow, and the advent of paintings and engravings on cave walls with highly developed syntagmatic interrelationships.

To recapitulate, If we look at the paradigms of primordial language as they must have developed over time, and are hypothesized in Table 19.2, we find a steady increase in the complexity of paradigmatic organization. Syntagmatic organization, when it arose, probably involved only two juxtaposed items, but by the Upper Palaeolithic concatenation of phememes was well underway. I would assign development of grammatical derivation (a form of transformation) to the late Upper Palaeolithic, or the Mesolithic, and development of grammatical inflection to the Neolithic, at which time the greatest degree of differentiation between groups would seem to have taken place. Grammatical particles or affixes did not spring into being arbitrarily, but were manufactured in particular dialects either from single phememes, or from short concatenations of phememes that serve as roots in other languages. Adaptation of sequences for grammatical purposes tended to eliminate use of the same concatenated phememes as roots in those languages.

TRANSFORMATION

For culture as we know it to have been operative, certain universal cultural features would have to have been in place. These are (1) paradigms, (2) syntagmata, (3) realization rules governing syntagmata, (4) transformational rules governing repositioning of paradigms.

Paradigms are based upon systemic organization of units into analogically motivated classes. Syntagmata are composed of members of paradigms organized spatially and temporally. Realization rules operate on paradigms to ensure the ordering of their classes into behavioural sequences and their spatially distributed products. Transformational rules reorder paradigms within syntagmata.

I stated above that human burial, first known from the Mousterian, is the first clear indication of transformation, in that the enclosure of gestation becomes equated with enclosure in death. Cognitively, what has taken place is that, by means of syntagmatic reorganization, one paradigm has been incorporated into the other in such a way that they have become, in some sense, one.

Lévi-Strauss, in 1955 (1963) first introduced the notion of transformation into anthropology, although without use of the term. Lévi-Strauss (1953: 224) says that mythical thought always progresses from the awareness of oppositions toward their resolution. This takes place by means of a successive series of 'permutations'. A law governing the entire permutation sequence reconciles the oppositions introduced in the course of the myth by reordering elements of mythic structure in such a way that previous oppositions become cancelled or resolved through syntagmatic association of paradigmatic elements (1953: 228). In any single myth, syntagmata undergo successive transformations, and as such a myth changes over time and across space, further transformation is effected, such that, unless the successive steps can be reconstituted, it is not apparent that one myth is related to another. It is only through patient reconstruction of such stages that semantic foundations for myth can be uncovered.

Lévi-Strauss' reconstruction methodology is very similar in concept to that of comparative linguistics, whereby earlier stages of language, as embodied in words, can be reconstructed by careful comparison of known (written or modern) stages across languages and dialects. To discover mythic transformations he compares the syntagmatic organization from segment to segment of a single myth and discovers successive shifts in paradigmatic realization such that, by the time the myth is complete, there has been a dramatic reversal in the terms of the original mythic statement. The successive transformations build in such a way that the original meaning embodies something approaching a total semantic reversal, allowing us to see that everything has the germ of its opposite in its structure.

Reversals are commonplace in culture. Leach (1976: 27) gives as a simple example, 'the Christian and European customs by which brides are veiled and dressed in white and widows are veiled and dressed in black are both part of the same message. A bride is entering marriage, a widow is leaving it. The two customs are *logically* related. The reason we do not ordinarily see that they are logically related is because they are normally widely separated in time'. Similarly, birth is preceded by enclosure, and in burial, death is followed by enclosure. Mousterian burial with flower petals emphasizes the semantics of life, now transferred to death. The modern custom of floral funeral tributes says something of the same nature, as well as do more ancient customs of burying the dead with goods that they enjoyed during life.

It seems clear that transformations cannot be enacted without some syntagmatic reorganization. Linguistically, until elements were strung together in rule-governed ways, transformation would hardly be possible. Syntagmatic organization seems to have begun in a limited way in the late Acheulian, but was probably not very complex until the Upper Palaeolithic – the period when culture came into its own and change suddenly began to take place with unprecedented and ever-accelerating speed.

I feel that the potential of reversal, introduced somewhat prior to, and perfected during, the Upper Palaeolithic, gave a great impetus to the enormous cultural advances that the archaeology of that period shows.

The possibility of reversal does a great deal to explain both the anomalies of culture and the extraordinary dynamism of culture. Where other animal societies have gone on behaving in much the same ways for millennia, human society has been changing with ever-increasing rapidity.

CONCLUSIONS

Two competing anthropological approaches to the study of culture have been called *materialist* and *idealist*. 'Materialists' think of culture change as determined primarily by strategies for survival. 'Idealists' think of culture change as fostered by processes of interaction between symbols and symbol systems. The terminology is unfortunate because the term *idealist* is basically pejorative – referring to a behavioural mode that is governed by ideals rather than by reason or logic, considered to be the preferred way of responding to circumstance. I believe that the terms might better be exchanged for *survivalist* and *symbolist*. Materialist is just as inept for the one as idealist is for the other. Symbols are material in nature and the manipulation of symbols is the very stuff of culture. The symbolic mode, although governed by analogic classification rather than logic has proved extraordinarily adaptive, else why has *Homo sapiens* become material master of the world? It would seem obvious that a prehistoric marriage of the analogical and the logical must be assumed in order to account for the signal success of the human species. Since all animals can be seen to operate in accordance with the dictates of reason in order to survive, then the new ingredient that accounts for culture must certainly be a steady development of the analogic faculty which substantially increased the adaptive odds over what was available through the simple functioning of reason. Reconstructive discovery of the iconism of early language dramatically confirms this.

The *sense* produced by symbolism is not the conscious sense of reason but the more subtle sense of unconscious association between things that are somehow assumed to be like. It is more obvious if the likeness is iconic, as it was in the earliest stirrings of language, but assumed, or relational, likenesses can be distinguished if one sets out to discover them.

To my mind, one of the best discussions of symbolism is that of Cohen (1974). He stresses that the essential *ambiguity* of symbolism is due to the fact that symbols embody a multiplicity of disparate meanings. It is apparent from the gross, or abstract, quality of Pleistocene phememes that symbolic polysemy was a characteristic of symbols from the very start. Cohen (1974: 8) also says, 'symbols are essential for the development and maintenance of social order. To do their job efficiently their social functions must remain largely unconscious and unintended by the actors. Once these functions become known to the actors, the symbols lose a great deal of their efficacy'.

One wonders how unconscious symbolism was at the outset, when it was so patently iconic. It may well be that its steady burrowing into the unconscious was what made it become so powerfully adaptive. It is one thing to understand the workings of symbolism in the abstract and quite another to grasp the situational operation of the symbolic networks in which one is oneself enmeshed.

In any event, if we are to understand the origin and evolution of culture, the only avenue is to attempt to unravel the cognitive advances of our ancestors through investigation of their symbolic products.

NOTES

1. Washburn and McCown (1978) cast the blame on social anthropology, because of an apparent unfamiliarity with various seminal developments in cultural and linguistic anthropology and linguistics proper. According to these critics, 'During a period of remarkable advances in astronomy, physics, and biology, the social sciences have undergone no revolution' (1978: 286). The authors seemed to be lamentably unaware of revolutionary developments in linguistics with publication of Chomsky's

Syntactic Structures in 1957, and in cultural and linguistic anthropology with publication of Lévi-Strauss' work in the 1960s, and in various works of Goodenough, Conklin, Lounsbury and Frake from the late 1950s onward. These seminal works introduced a strong involvement with meaning in cultural anthropology, and through the efforts of generative revisionists, in linguistics as well. It is largely the fault of biological and archaeological evolutionists that these advances have not been exploited, rather than that of cultural anthropologists as Washburn and McCown would have it. If these authors are correct (which I seriously doubt) in assuming that 'biology is much more complex than social systems' (1978: 287), then the fault is all the more theirs for failure to assimilate this material.

2. I use the term culture as it is employed, either implicitly or explicitly, by a growing number of cultural and social anthropologists (e.g. Geertz 1973; Leach 1976; Turner 1967), as a shared system of symbols. However, anthropologists discussing symbolism differ in the way that they define 'symbol'.

3. This is, of course, the reverse of the semiotic usage whereby a sign is considered to be motivated, while a symbol is not.

4. A syntagma has also been called a unit of process (Hjelmslev 1953: 24) or a unit of structure (Halliday 1961: 254-9).

5. The assumption of arbitrariness in early linguistic symbols is a product of Saussure's (1959) proclamation of the arbitrariness of linguistic signs.

6. Reconstructed forms are starred.

7. English *have*, with /v/ where /f/ would be expected, is an example that must be explained, as it can be through study of the history of consonant voicing in relation to accent shifts – a topic too complex to discuss here.

8. I have not yet found sure indications for a contrast between a reconstructible *? and *h. There was at least one glottal phememe, and perhaps two. I suspect the latter because, for example, words with /?/ in Egyptian seem often to contrast through positive *versus* negative polarization. This has made it difficult to ascertain the meaning of this (or these) phememe, which I tend to think of as either creation or extinction. The phememe or phememes have been lost in most languages.

9. Since these earlier charts were published, by using material from additional languages I have succeeded in eliminating four of the original sound-meaning units by finding them to result from later compositions of two phememes. I have also added two additional phememes to the system, and slightly modified some of the original meanings that I had assigned to the units.

10. I admit to not having made a frequency count. My assessment is based largely on impressionistic assessment of the relative length of dictionary entries with initial phonemes that reflect particular PL phememes, and relative numbers of reflexes of particular root final phememes – taking 'root' as the two phoneme sequences characteristic of roots in most historical languages.

REFERENCES

Bloom, L. 1976. Child language and the origins of language. In S.R. Harnad, H.D. Steklis and J. Lancaster (eds) *Origins and Evolution of Language and Speech.* (Annals of the New York Academy of Sciences 280). New York: New York Academy of Sciences: 170-172.

Cassirer, E. 1953. *The Philosophy of Symbolic Forms.* New Haven: Yale University Press.

Chomsky, N. 1957. *Syntactic Structures.* The Hague: Mouton.

Cohen, A. 1974. *Two-Dimensional Man: an Essay on the Anthropology of Power.* Berkeley: University of California Press.

Conkey, M.W. 1980. Context, structure, and efficacy in Paleolithic art and design. In M.L. Foster and S.H. Brandes (eds) *Symbol as Sense: New Approaches to the Analysis of Meaning.* New York: Academic Press: 225-248.

Ferguson, C.A. 1963. Assumptions about nasals: a sample study in phonological universals. In J.H. Greenberg (ed.) *Universals of Language.* Cambridge (Mass): MIT Press: 42-47.

Foster, M.L. 1978. The symbolic structure of primordial language. In S.L. Washburn and E.R. McCown (eds) *Human Evolution Biosocial Perspectives.* (Perspectives on Human Evolution 4). Menlo Park (CA): Benjamin/Cummings: 77-121.

Foster, M.L. 1980. The growth of symbolism in culture. In M.L. Foster and S.H. Brandes (eds) *Symbol as Sense: New Approaches to the Analysis of Meaning.* New York: Academic Press 371-398.

Foster, M.L. 1981. Culture as metaphor: a new look at language and culture in the Pleistocene. *Kroeber Anthropological Society Papers* 59, 60: 1-12.

Foster, M.L. 1983. Solving the insoluble: language genetics today. In E. de Grolier (ed.) *Glossogenetics: the Origin and Evolution of Language.* Paris: Harwood Academic Press: 455-480.

Foster, M.L. 1986. Iconic abstraction in prehistoric language and signs. Paper presented at the annual meeting of the Language Origins Society. Oxford, 1986.

Foster, M.L. (in press). Reconstruction of the evolution of language. In A. Lock and C. Peters (eds) *Handbook of Symbolic Evolution.* Oxford: Oxford University Press. In Press.

Geertz, C. 1973. *The Interpretation of Cultures.* New York: Basic Books.

Gimbutas, M. 1982. *The Goddesses and Gods of Old Europe.* Berkeley: University of California Press.

Halliday, M.A.K. 1961. Categories of the theory of grammar. *Word* 17: 241-292.

Hjelmslev, L. 1953. *Prolegomena to a Theory of Language.* (International Journal of American Linguistics Memoir 7). Baltimore: Waverly Press.

Isaac, G.L. 1976. Stages of cultural elaboration in the Pleistocene: possible archaeological indicators of the development of language capabilities. In S.R. Harnad, H.D. Steklis and J. Lancaster (eds) *Origins and Evolution of Language and Speech.* (Annals of the New York Academy of Sciences 280). New York: New York Academy of Sciences: 275-287..

Lamendella, J. 1980. Neurofunctional foundations of symbolic communication. In M.L. Foster and S.H. Brandes (eds) *Symbol as Sense: New Approaches to the Analysis of Meaning.* New York: Academic Press: 147-174.

Lancaster, J. 1968. On the evolution of tool-using behavior. *American Anthropologist* 70:56-66.
Langer, S.K. 1942. *Philosophy in a New Key: a Study in the Symbolism of Reason, Rite and Art*. New York: Mentor.
Lawick-Goodall, J. van. 1971. *In the Shadow of Man*. New York: Delta.
Leach, E. 1976. *Culture and Communication*. Cambridge: Cambridge University Press.
Lévi-Strauss, C. 1963. *Structural Anthropology*. New York: Basic Books.
Lock, A.J. 1982. A note on the ecology of meaning. *Quaderni di Semantica* 3: 112-117.
Marshack, A. 1976. Some implications of the Paleolithic symbolic evidence for the origin of language. In S.R. Harnad, H.D. Steklis and J. Lancaster (eds) *Origins and Evolution of Language and Speech*. (Annals of the New York Academy of Sciences 280). New York: New York Academy of Sciences: 289-311.
Montagu, A. 1976. Toolmaking, hunting, and the origin of language. In S.R. Harnad, H.D. Steklis and J. Lancaster (eds) *Origins and Evolution of Language and Speech*. (Annals of the New York Academy of Sciences 280). New York: New York Academy of Sciences: 266-274.
Saussure, F. de. 1959. *Course in General Linguistics*. New York: Philosophical Library.
Tanner, N. and Zihlman, A. 1976. The evolution of human communication: what can primates tell us? In S.R. Harnad, H.D. Steklis and J. Lancaster (eds) *Origins and Evolution of Language and Speech*. (Annals of the New York Academy of Sciences 280). New York: New York Academy of Sciences: 467-480.
Turner, V. 1967. *The Forest of Symbols: Aspects of Ndembu Ritual*. Ithaca: Cornell University Press.
Washburn, S.L. and McCown, E.R. 1978. *Human Evolution: Biosocial Perspectives*. (Perspectives on Human Evolution 4). Menlo Park (CA): Benjamin/Cummings.
Wescott, R. 1976. Protolinguistics: the study of protolanguages as an aid to glossogonic research. In S.R. Harnad, H.D. Steklis and J. Lancaster (eds) *Origins and Evolution of Language and Speech*. (Annals of the New York Academy of Sciences 280). New York: New York Academy of Sciences: 104-116.
Wright, R.V.S. 1978. In S.L. Washburn and E.R. McCown (eds) *Human Evolution: Biosocial Perspectives*. Menlo-Park: Cummings: 215-236.

Index of Authors

Abelson, R.P., with R.C. Schank, 511
Adamenko, O.M., 175, 176
Airvaux, J., 370
Alimen, H., 58
Allsworth-Jones, P., 161, 163, 164, 167, 168, 170, 171, 173, 174, 179, 180, 183, 184, 185, 186, 187, 188, 190, 194, 196-7, 198, 201, 202, 204, 205, 206, 207, 208, 209, 210, 211, 212, 213, 215, 216, 219, 221, 223, 224, 225, 226, 228, 230-1, 234, 235, 429, 447, 451
Almagro Gorbea, M., 277(fig.)
Altuna, J., 277, 277(+fig.), 284-5(fig.), 286, 440
Ambrose, S.H., 16, 349
Amirkhanov, Kh.A., 217, 221, 222, 227
Ammerman, A.J., 352
Andersen, H.H., 395
Anderson-Gerfaud, P.C., 389, 391, 394, 395, 398, 400, 401, 402, 404, 405, 406, 407, 410, 413, 440, 466; with D. Stordeur, 391, 413
Anikovich, M.V., 223, 226-9; with A.N. Rogachev, 221, 227
Anisyutkin, N.K., 228
Apsimon, A.M., 207-8
Arambourg, C., 374
Arambourou, R., 276, 281, 283, 293
Arensburg, B., 492
Aristotle, 518-19
Arkell, W.J., with K.S. Sandford, 139
Armstrong, A.L., 209
Audouze, F., 390, 399
Avery, D. M., 42-3, 44-6(+fig.), 47, 52
Azoury, I., 58, 68, 69, 92, 94, 96, 97-100, 104, 112, 114, 125, 129, 133

Bächler, E., 487
Bácskay, E., 169
Bada, J.L., 37
Bader, O.N., 352, 434
Baffier, D., 320
Bahn, P., 287, 288(fig.), 289, 293, 276, 277(fig.)
Bailey, G.N., 349, 369
Bailey, R.C., 12
Balch, H.E., 208-9
Banesz, L., 433
Bar-Yosef, O., 57, 61, 62, 63, 65, 66, 67, 72-3, 114, 156, 440; with A. Belfer-Cohen, 73; with B. Vandermeersch, 62
Barandiarán, I., 294-5; with J. González Echegaray, 277(fig.)
Barbetti, M., with R.H. Pearce, 335, 338
Barker, G.W., 353(fig.), 355
Bárta, J., 232-3
Barth, F., 361
Bartholomew, G.A., 352
Basler, D., 212
Bate, D.M.A., with D.A.E. Garrod, 60, 62, 73, 81-2
Baumhoff, M., with R.L. Bettinger, 3, 8
Bayer, J., 191, 193, 195-6
Beaumont, P.B., 7, 24, 25, 34, 39, 41, 43(fig.), 45, 46, 47, 49, 50, 51, 52, 466; with A. Boshier, 468; with J.C. Vogel, 5; with K.W. Butzer, 42(fig.)
Beaune, S. de, 376
Behm-Blancke, G., 163, 353(fig.), 354
Belfer, A., with O. Bar-Yosef, 156
Belfer-Cohen, A., 73, 459, 460
Bellwood, P., 336
Bender, B., 376
Beregovaya, N.A., 228
Bergman, C.A., 70, 72, 73, 74, 91, 94, 95, 111, 112, 114, 125, 126, 129, 132, 133, 413; with K. Ohnuma, 101
Bernaldo de Quirós, F., 276, 292
Besançon, J., 134
Bettinger, R.L., 3, 8
Bevacqua, R., 339
Beyries, S., 390, 398, 399, 400, 402, 404, 405, 411, 414, 440, 443
Bicchieri, M.G., 11
Binford, L.R., 5, 9, 16, 28, 243, 257, 267-8, 350, 354, 356, 361, 362, 367, 372, 377, 389, 395, 398, 405, 429, 457

Binford, S.R., 458, 486, 487; with L.R. Binford, 267-8, 362, 395, 405
Birdsell, J.B., 327-9, 352; with G.A. Bartholomew, 352
Biró, K.T., 170, 171, 172, 173, 174, 175, 236
Bischoff, J., 280; with V. Cabrera, 280
Bitiri, M., 215, 216-17, 219
Blackburn, R.H., 16
Blackwell, B., 359
Blanc, A.C., 376, 458, 487
Bloom, L., 524-5, 531
Blundell, V.J., 339
Boas, F., 375
Boëda, E., 268, 269, 419, 420(fig.), 438
Bofinger, E., 277(fig.)
Bonifay, E., 357, 359
Bordes, F., 4, 57, 83, 85, 97, 101, 104, 112, 132, 246, 263, 265-6, 267-8, 270, 271, 272, 335, 350, 353(fig.), 358, 359, 360-1, 362, 363, 365, 377, 389, 394, 395, 408-9(fig.), 413, 414, 443, 463(+fig.); with M. Sireix, 352-4
Borziyak, I.A., 221-2, 223-4, 228
Boshier, A., 468
Bosinski, G., 358, 359, 421, 423(fig.), 438, 474; with R. Wetzel, 460, 464, 468(fig.)
Böszörmenyi, Z., 236
Bottet, B., with H. de Lumley, 357
Bouchud, J., 353(fig.), 357, 368(fig.); with R. Van Campo, 368(fig.)
Bouyssonie, A., 365
Bowdler, S., 327, 329, 330, 332, 333, 336
Bowler, J.M., 327, 336
Boyd-Dawkins, W., 207
Braidwood, R.J., 452
Brandt, S.A., 16
Bräuer, G., 34, 50
Breuil, H., 91, 168, 293, 294, 295, 365, 481
Bricker, H., 259
Brink, J., 408-9(fig.)
Briuer, F.L., 391, 413
Brodar, M., 195, 196; with S. Brodar, 191, 193, 194-5, 196, 197
Brodar, S., 191, 193, 194-5, 196, 197
Brooks, P.M., 46
Brown, J.A., with T.D. Price, 351, 374, 376
Brown, J.L., 3
Buikstra, J.E., with D.K. Charles, 16
Butzer, K.W., 3, 6, 7, 34, 36, 38, 39-42(+figs), 43, 45, 46, 47, 48, 50, 57, 277(fig.), 280
Buzy, D., 56

Cabrera, V., 277, 280, 284-5(fig.), 286, 287, 294
Cacho, C., 276, 281, 283, 290, 292
Cahen, D., 421, 443
Caldwell, J.R., 16
Callow, P., 358, 365
Campbell, J.B., 204, 208, 209
Cann, R.L., 50, 243, 492
Cârciumaru, M., 215-16, 217, 219, 220; with M. Bitiri, 215, 216-17, 219
Carter, P.L., 5, 24-5, 48, 49, 52
Caspar, J.-P., with M. Otte, 390, 399, 404
Cassirer, E., 519
Caton-Thompson, G., 58, 67, 139, 145
Cattelain, P., 206, 445(fig.), 447
Cazeneuve, J., 357
Champagne, F., 243
Chapman, R.W., 16
Chappell, J., 39
Chard, C.S., 352
Charles, D.K., 16
Chase, P.G., 349, 354, 398, 440, 457, 458, 486, 487, 488
Chavaillon, J., 365
Chernysh, A.P., 440; with I.K. Ivanova, 357
Childe, V.G., 350, 351, 356, 357, 363
Chmielewski, W., 66, 163, 429, 447
Chollet, A., with J. Airvaux, 370
Chomsky, N., 536-7
Christopher, C., 70
Clark, G.A., 61, 287, 290, 291, 453; with L. Straus, 277(fig.), 290, 294
Clark, J.D., 4, 19, 64, 204, 376
Clark, J.G.D., 347, 350, 351, 374
Clarke, D.L., 350, 372
Cliquet, D., with G. Fosse, 440
Close, A.E., 141, 156; with W.L. Singleton, 66, 144
Clottes, J., 276, 283, 293
Clutton-Brock, T.H., 13
Cohen, A., 536
Coinman, N., 65
Colman, S.M., 6
Combier, J., 351, 353(fig.), 358, 359
Commont, V., 363
Conkey, M.W., 244, 293, 372, 458, 517
Cook, J., 163, 191
Cooke, H.B.S., 50
Copeland, L., 57, 58, 59, 60, 61, 64, 67, 68, 69, 91, 92, 114, 132; with C.A. Bergman, 91
Cordy, J.-M., 206, 447
Crabtree, D., with F. Bordes, 4, 97, 104
Crew, H., 147
Croisset, E. de, 325

Daleau, F., 472
Dasen, P.R., 509
David, A.I., 221, 222
David, F., 319, 325
Davidson, I., 289(fig.), 290; with E. Bofinger, 277(fig.)

Davis, H., with H. Harpending, 8, 9-11
Davis, S., 354
Day, A.E., 91
de Beaune, S., 376
de Heinzelin, J., 57, 141, 421, 439(fig.), 451
de Puydt, M., 208
de Villiers, H., with P.B. Beaumont, 43(fig.)
Deacon, H.J., 4, 5, 6, 7, 9, 16, 17, 21, 24, 26, 34, 36, 37, 39, 47
Deacon, Janette, 3, 7, 16, 17, 21, 22(+fig.), 24, 26, 36-7, 42, 49, 51, 52
Debénath, A., 359
Deems, L., with J.L. Bada, 37
Del Bene, T., 413
Delluc, B., 470, 483
Delluc, G., with B. Delluc, 470, 483
Delporte, H., 277
Demars, P.-Y., 267, 440, 441(fig.)
Dennell, R., 244, 263, 266, 352, 356, 374
Deraprahamian, G., 402-3(fig.), 414
Dibble, H.L., 65, 271, 357, 363, 365, 412, 414, 443, 457, 460, 492; with P.G. Chase, 457, 458, 486, 487, 488
Dobosi, V.T., 165, 173, 174, 175, 192
Dombek, G., 231
Dortch, C.E., 130, 331
Drury, W.H., 370
Dumond, D.E., 352
Duport, L., 359
Durkheim, E., 514
Dyson-Hudson, R., 3, 8, 9(fig.), 11, 13, 14, 15(fig.), 16, 17, 18, 28
Dzambazov, N.S., 211, 212

Eastham, A., 290
Eisenberg, J.F., 16
Espitalié, R., with F. Champagne, 243
Ewing, J.F., 67, 91, 92, 132

Farizy, C., *see* Girard-Farizy, C.
Farrand, W., 57
Ferguson, C.A., 531
Fernández-Miranda, M., with M. Almagro Gorbea, 277(fig.)
Ferring, C.R., 73; with A.E. Marks, 70
Feustel, R., 351
Fillmore, C., 511
Fish, P.R., 370
Flannery, K.V., 8; with F. Hole, 58
Flood, J., 338
Foley, R., 18, 349, 372
Fortea, J., 277(+fig.), 294, 295
Fosse, G., 440
Foster, M.L., 518, 529
Fourie, D., 52
Frayer, D.W., 235
Freeman, L.G., 277(+fig.), 290, 291, 292, 296, 349, 355, 356, 358, 359, 371; with J. González Echegaray, 277, 292, 294
Freidel, D.A., with A.E. Marks, 349, 358, 371
Frelih, M., 236
Freund, G., 161
Freundlich, J.C., 49
Fridrich, J., 461
Fuentes, C., 284-5(fig.), 286
Fullager, R., 413
Fullola Pericot, J., 276, 277(+fig.), 283, 289
Fülöp, J., 173

Gábori, M., 164, 196, 212, 352, 354
Gábori-Csánk, V., 164, 165, 173, 192, 356, 357, 372, 428
Gálffy, I., 168
Gamble, C., 8, 16, 17, 21, 26, 293, 372, 376, 447, 458
Garcia-Soto, E., with J. Moure, 277(fig.)
Gardner, H., 507
Gargett, R.H., 492
Garrod, D.A.E., 58, 59, 60, 61, 62, 63, 67, 68, 73, 81-2, 83, 84-5, 87, 89
Gaughwin, D., 335
Gaussen, J., 297
Gautier, A., 440
Gearing, F., 362
Geertz, C., 537
Geist, V. van, 16
Geneste, J.-M., 266, 267, 270, 353(fig.), 390, 399, 440, 441(fig.), 442(fig.), 443, 469
Gentner, D., 511
Gilead, I., 61, 72, 73, 132, 135(+fig.)
Gilot, E., 447
Gimbutas, M., 522
Girard-Farizy, C., 236, 319, 322, 353(fig.), 358, 359, 373, 444
Gladilin, V.N., 175-6
Glen, E., 235
Golson, J., 333, 340
González Echegaray, J., 277(+fig.), 292, 294; with L. Freeman, 296
Goodwin, A.J.H., 49
Gordon, B.C., 348
Gould, R.A., 17, 19, 357
Gould, S.J., 459
Gowlett, J.A.J., 207, 208, 209
Grayson, D.K., 7, 245, 248
Graziosi, P., 487-8
Grigoriev, G.P., 226
Grigorieva, G.V., 221, 224, 225
Groube, L., 338
Guichard, G., with J. Guichard, 57, 66, 141-2(+fig.), 143
Guichard, J., 57, 66, 141-2(+fig.), 143
Guillien, Y., 367
Gullentops, F., 438

Habgood, P.J., 337
Haesaerts, P., 231, 232, 430, 438, 439(fig.); with J. de Heinzelin, 421
Hage, P., 507
Hahn, J., 203, 232; with H. Laville, 231, 232
Hall, M., 47
Hallam, S.J., 331-2
Haller, J., 67
Halliday, M.A.K., 537
Hamilton, W.J. III, 12
Hansen, C., with K.W. Butzer, 57
Harako, R., 16
Harary, F., with P. Hage, 507
Harpending, H., 8, 9-11; with J. Yellen, 16, 361
Harrold, F.B., 243, 247, 250, 259, 291, 447, 451, 458, 461, 463, 469, 486, 487
Hassan, F., 352
Hayden, B., 13, 16, 17, 18, 20, 26, 351, 372, 399, 402, 404, 405
Heimbuch, R.C., with J.T. Laitman, 458
Heinzelin, J. de, 57, 141, 421, 439(fig.), 451
Hellebrandt, M., 166, 167, 169
Helmer, D., 402-3(fig.), 408-9(fig.), 414; with P.C. Anderson-Gerfaud, 389, 400, 404, 406, 407, 410, 413
Hendey, Q.B., 6, 37
Henri-Martin, B.L., 464, 470(fig.)
Henri-Martin, G., 359; with Y. Guillien, 367
Henry, D.O., 61, 65
Herman, O., 168
Hester, J., 67
Hietala, H.J., 71, 501
Higgs, E.S., 355
Hillebrand, E., 168, 170, 191, 193, 429
Hjelmslev, L., 537
Hodder, I., 362, 458, 514
Hodson, F.R., with I. Azoury, 100
Hoebler, P., with J. Hester, 67
Hole, F., 58
Holloway, R.G., with H.J. Shafer, 391, 413
Holloway, R.L., 377
Honea, K., 215, 216, 218, 219, 220(+fig.), 236
Horn, H.S., 3, 14
Horowitz, A., 57, 64
Horton, D.R., 331
Houghton, G., 339
Hours, F., 59, 64, 114
Howell, F.C., 375, 460
Hülle, W., 192, 204, 206-7, 447
Humbert, R., 325
Humphreys, A.B.J., 26
Huzayyin, S., 58
Inizan, M.-L., with J. Tixier, 91, 94, 114-17, 131
Irwin, H.T., 154-5
Isaac, G.L., 377, 521-2, 524, 533
Itani, J., 12
Ivanova, I.K., 357
Ivanova, M.A., 227
Izawa, K., 12

Jacob-Friesen, K.H., 460
Jacobi, R.M., 210, 236, 447
Jánossy, D., 198
Jarman, P.J., 16
Ječmínek, J., 186
Jelinek, A.J., 60, 61, 63, 64, 68, 83, 84, 88, 354, 362, 363, 367, 371, 425, 457, 469
Jenkinson, R.D.S., 191, 209
Jéquier, J.-P., 356, 487
Jewell, P.A., 355
Jochim, M., 291, 293
Johnson, G.A., 376
Jolly, K., 49
Jones, G., with R. Leonard, 248
Jones, M., 72
Jones, P.R., 19
Jones, R., 330, 336; with M. Lorblanchet, 339
Jorda, F., with J. Fortea, 277(+fig.)
Juel-Jensen, H., 391, 399, 405, 406
Julien, M., 320

Kaczanowski, K., 235
Kadić, O., 174, 191, 192-3
Kaminská, L., 162
Kamminga, J., 395
Kaplan, J., 48, 49, 52
Karásek, J., 189
Karlin, C., with M.H. Newcomer, 94
Kaufman, D.: with A.E. Marks, 72; with P. Volkman, 59
Keeley, L.H., 389, 390, 398, 399, 401, 404, 410, 411, 413
Keith, A., with T. McCown, 61
Keller, C.M., 5, 26, 47-8, 49
Kelly, R.L., 9
Kershaw, A.P., 335
Ketraru, N.A., 221, 222, 225, 228, 434; with I.A. Borziyak, 222
Kiernan, K., 338
Kimball, L., 390, 399, 414
Kintigh, K., 245, 248, 250, 259
Kirkbride, D., with D.A.E. Garrod, 82, 89
Klein, R.G., 4, 7, 8, 24, 26, 27, 44, 45-6, 47, 52, 353(fig.), 367
Klein, S., 499, 507, 511-12, 513
Klíma, B., 184, 185, 188, 433, 451
Kobayashi, T., 268
Kopper, J.S., 358
Kordos, L., with M. Hellebrandt, 166, 167
Korek, J., 169, 170
Kowalski, K., 357

Kozlowski, J.K., 169, 173, 187, 198, 203, 204-5, 206, 211, 232, 233, 235, 428, 430, 432, 451, 467(fig.)
Kozlowski, S.K., 204-5, 206
Kretzoi, M., 165
Kroeber, A.L., 376
Kronfeld, J., 37
Kruuk, H., 13
Kubiak, H., with W. Chmielewski, 429
Kurtén, B., 197, 348

Laitman, J.T., 377, 458, 492
Lakoff, G., 511
Lamendella, J., 517
Lampert, R.J., 329, 335
Lancaster, J., 530
Lancaster, N., with J. Deacon, 42
Langer, S.K., 519
Lantier, R., with H. Breuil, 365
Laplace, G., 186, 277
Larson, P.A., 349, 356, 358, 371
Laughlin, W.S., 17, 357
Laville, H., 231, 232, 245, 248, 249, 252, 256, 266, 353(fig.)
Lawick-Goodall, J. van, 524
Le Mort, F., 488
Le Ribault, L., 395
Le Tensorer, J.M., 265
Leach, E., 376, 535, 537
Leclerc, J., with C. Girard-Farizy, 353(fig.)
Lee, R.B., 9, 17, 372
Leonard, R., 248
Leroi-Gourhan, André, 303, 308, 319, 322, 324, 350, 357, 358, 359, 363, 366, 373, 376, 377, 399, 434, 438, 447, 452(fig.), 481, 483; with Arlette Leroi-Gourhan, 320, 461-3
Leroi-Gourhan, Arlette, 162, 164, 233, 303, 320, 461-3, 489; with André Leroi-Gourhan, 303; with C. Leroyer, 232, 233
Leroyer, C., 232, 233, 450(fig.), 451; with Arlette Leroi-Gourhan, 303
Lévêque, F., 243, 270, 447, 461
Lévi-Strauss, C., 376, 499, 507, 514, 534-5, 537
Levkovskaya, G.M., 224
Lewis, H.T., 357
Lewis-Williams, David, 480
Lieberman, P., 458, 492
Lindly, J.M., 61; with G.A. Clark, 61, 453
Lindner, K., 349, 354
Lock, A.J., 517
Lohest, M., 208
Lorblanchet, M., 339
Lorenz, K. G., with S.H. Ambrose, 349
Lourandos, H., 333, 376
Lovejoy, O., 352
Loy, T.H., 349, 413
Lumley, H. de, 357, 358, 372, 461; with M.-A. Lumley, 478
Lumley, M.-A., 478
Lynch, T.F., 162

McBurney, C.B.M., 67, 156, 204, 205, 350
McCown, E. R., with S. L. Washburn, 536-7
McCown, T., 61
Macdonald, I.A.W.: with H.K. Watson, 46; with P.M. Brooks, 46
McGhee, R., 350
Malan, B.D., 49
Mania, D., 451
Mansur-Franchomme, E., 410
Marean, C.W., 5
Mariezkurrena, K., with J. Altuna, 284-5(fig.), 286
Marks, A.E., 57, 58, 59, 61, 65, 67-8, 69, 70, 72, 73, 104, 111, 112, 114, 134, 135, 142-3, 145, 150(fig.), 151, 152, 154-5, 156, 281, 349, 358, 371, 469, 501; with H.J. Hietala, 71, 501
Maroto, J., with N. Soler, 277
Marquardt, W.H., 373
Marshack, A., 433, 459-60, 461, 468, 470, 478, 481, 483, 485, 486, 490, 492, 533
Marx, K., 514
Mason, R., 49
Matsutani, A., 394, 401
Mauger, M., with H. Plisson, 398, 412
Mauss, M., 356, 514
Maynard, L., 339
Mazel, A., 52
Meehan, B., 332, 333
Meggers, B., 355
Meignen, L., 60, 353(fig.), 365
Mellars, P., 6, 135, 243, 244, 245, 262, 263, 264-5, 271, 298, 340, 352, 355, 357, 359, 362, 373, 395, 443, 510, 514
Mentis, M.T., 46
Miles, C., 408-9(fig.), 410
Mills, Glen, 52
Minsky, M., 511
Moguşanu, F., 216, 218
Montagu, A., 522, 525, 530
Montet-White, A., 212, 214, 373
Mook, W.G., 219, 232
Moore, J.A., 16
Morawski, W., 421, 438
Morse, K., 338, 340
Moss, E.H., 390, 399, 407, 410, 413
Mottl, M., 198
Moure, J., 277(fig.)
Movius, H.L., 373
Müller-Beck, H.J., 192
Murphy, J.W., 91, 94
Murray, P., 359
Musil, R., 192

Neal, R., 338
Nemeskéri, J., 429
Neter, J., 249
Neuville, R., 63
Newcomer, M.H., 70, 73, 92, 94, 96, 101, 104, 114, 125, 126, 129, 134, 268; with C.A. Bergman, 413; with E.H. Moss, 390, 399, 413
Nicolaescu-Plopşor, C.S., 164, 215, 218, 219
Nie, N.H., 95-6

Oakley, K.P., 64, 460
Obermaier, H., 296
O'Connell, J.F., 359; with J.P. White, 330, 331
O'Connor, S., 339, 340
Ohnuma, K., 70, 72, 95, 101, 133
Okladnikov, A.P., 352
Oliva, M., 176, 179, 182, 184, 185, 189, 190, 201-4, 233, 421, 430, 460
Orians, G.H., 3; with J.L. Brown, 3
Otte, M., 203, 204, 205, 207, 208, 210-11, 224, 351, 390, 399, 404, 439(fig.), 446(fig.), 447, 449(fig.), 451; with P. Cattelain, 206, 445(fig.), 447; with P. Haesaerts, 231, 232
Owen, L.R., with J. Hahn, 203

Palma di Cesnola, A., 233-4
Pálosi, M., with K.T. Biró, 173, 174, 175
Papp, K., 171
Paquereau, M.M., 370
Parkington, J.E., 3, 5, 18, 22, 48; with C.A. Poggenpoel, 48
Parry, R.F., 207
Partridge, T.C., 6
Patou, M., 440
Patte, E., 353(fig.)
Patterson, L.W., 347
Paulissen, E., 141, 144, 152, 155
Păunescu, A., 357, 432
Payne, S., 354, 355
Pearce, R.H., 335, 338
Pearson, N., with G. Orians, 3
Pèlegrin, J., 440, 451; with E. Boëda, 268, 269
Pelísek, J., 179
Perez Ripoll, M., 354
Perkins, D., 353(fig.)
Perlès, C., 357, 399
Perrot, J., 292; with D. de Sonneville-Bordes, 100, 246, 263
Peyrony, D., 265, 292
Pfeiffer, J., 457
Phillips, J.L., 156; with O. Bar-Yosef, 57, 72-3
Piaget, J., 507-9
Pielou, E.C., 245, 246
Piperno, D.R., 394, 401
Piperno, M., 359
Piveteau, J., 374
Plisson, H., 389, 390, 398, 399, 404, 410, 412, 413, 414; with J.-M. Geneste, 390, 399; with P.C. Anderson-Gerfaud, 410
Poggenpoel, C.A., 48
Porter, R.N., with A. Whateley, 46
Pradel, J.H., with L. Pradel, 356, 358, 359
Pradel, L., 356, 358, 359
Praslov, N.D., 228
Prat, F., with F. Bordes, 353(fig.)
Price, T.D., 351, 374, 376
Prošek, F., 161-2, 192, 428, 433
Puydt, M. de, 208

Radin, P., 376
Ranov, V.A., 352
Ravasz-Baranyai, L., 174
Reynolds, T.E.G., 266
Rhoades, R.E., 362
Rigaud, J.-P., 243, 244, 353(fig.), 355, 373
Rightmire, G.P., 34, 50, 52
Ringer, A., 168, 169, 170, 171-3
Ripoll, E., 290
Roche, J., 290
Rogachev, A.N., 221, 227
Rolland, N., 355, 356, 358, 363, 365, 366(fig.), 367, 370, 395, 443
Ronen, A., 60, 73, 358
Rottlander, R., 413
Rouse, I., 349
Rovner, I., 394, 401
Rowell, T.E., 12
Rumelhart, D., 511
Runnels, C., 372
Russell, M., 482(fig.), 488
Rust, A., 81, 82, 89, 425

Saád, A., 170, 429
Sacchi, D., 276, 277(+fig.), 283, 289, 293, 295, 296
Saggers, S., with R.A. Gould, 19
Said, R., 57, 141
Saint-Périer, R. de, 287, 288(fig.), 289, 293, 296, 472
Saint-Périer, S. de, with R. de Saint-Périer, 287, 288(fig.), 289, 293, 296, 472
Sampson, C.G., 3, 4, 5, 25
Sandford, K.S., 139
Sanlaville, P., 61, 63
Sauer, C.O., 357
Saussure, F. de, 519, 520, 537
Sauvet, G., 282
Schank, R.C., 511
Schiffer, M.B., 16, 17, 18
Schild, R., 57, 141, 144, 429; with A. Gautier, 440; with F. Wendorf, 57,

141, 143, 144, 145, 146, 150(fig.), 152, 154, 155
Schmider, B., with C. Girard-Farizy, 444
Schoeninger, M.J., 355
Schönweiss, W., 207
Schwarcz, H., 61, 165
Schwede, M.L., 338
Scott, L., with J. Deacon, 42
Scott, K., 353(fig.), 354
Seewald, C., 478
Semenov, S., 389
Servello, A.F., with D.O. Henry, 61
Severin, T., 196
Shackleton, J.C., with T.H. Van Andel, 370
Shackleton, N.J., 6, 37-8, 39; with J. Chappell, 39
Shafer, H.J., 391, 413
Shchelinskii, V.E., 389, 398, 402, 404, 405, 411, 412, 413, 414
Shea, J.J., 466
Shennan, S.J., 362
Sherwood, J., 335
Silberbauer, G.B., 18
Simán, K., 168-71, 172, 174, 175
Simek, J.F., 243-4, 246, 248, 257, 258, 355, 358, 444
Sinclair, A., 514
Singer, R., 3, 4, 5, 6, 22(fig.), 25, 49, 50(fig.), 467
Singh, G., 335
Singleton, W.L., 66, 144
Sirakov, N., 211-12, 213(fig.), 221, 430
Sirakova, S., with P. Haesaerts, 430
Sireix, M., 352-4
Skinner, J.H., 350
Smith, E.A., 8; with R. Dyson-Hudson, 3, 8, 9(fig.), 11, 13, 14, 15(fig.), 16, 17, 18, 28
Smith, F.H., 204, 234, 243, 244, 460, 492
Smith, M.A., 339
Smith, P.E.L., 374
Smolíková, L., 188
Snyder, L.M., with J.F. Simek, 243-4, 246, 248, 258, 355, 444
Soergel, W., 354
Soffer, O., 27-8, 227, 357, 486, 490
Soldatenko, L.V., 175, 176
Solecki, R.S., 359, 376, 481, 489
Soler, N., 277
Sollas, W.J., 208
Soltész, B., 198, 199
Sonneville-Bordes, D. de, 100, 246, 263, 272, 292, 440; with F. Bordes, 395
Spencer, F., with F.H. Smith, 243, 244
Štělcl, J., 179
Stevens, D., with H. Hietala, 71
Steward, J.H., 8, 11
Stewart, T.D., 489
Stock, E., with R. Neal, 338
Stoliar, A.D., 433
Stordeur, D., 391, 399, 404, 413
Straus, L.G., 276, 277(fig.), 282, 283, 284-5(fig.), 286, 287, 290, 291, 293, 294, 352, 443, 453; with G. Clark, 287, 290, 291
Stringer, C.B., 50, 62, 67, 208, 209, 243, 492
Strong, W.D., 375
Sulgostowska, Z., with R. Schild, 429
Suzuki, A., 12; with J. Itani, 12
Svoboda, H., with J. Svoboda, 176, 178, 182, 183, 185, 186-7
Svoboda, J., 176, 178, 179-84, 185, 186-7, 189-90, 201, 202-3, 229, 230, 234, 236, 430, 460

Taborin, Y., 320
Tankersley, K.B., 5
Tanner, N., 522-3
Tasnádi-Kubacska, A., 198, 199
Taylor, R.E., 37
Tchernich, A.P., 428, 432(+fig.), 433(fig.), 434
Tchernov, E., 61
Tensorer, J.M. Le, 265
Testart, A., 372, 374, 376
Thackeray, J.F., 52; with H. J. Deacon, 4, 5, 9, 24, 26, 34, 36
Thomson, D.F., 361
Thorne, A.G., 327, 337
Tixier, J., 91, 92(+fig.), 94, 97, 114-17, 131; with P. Mellars, 135
Tobias, P.V., 374, 377
Topfer, V., with D. Mania, 451
Torrence, R., 13
Toth, L., 170, 174; with M. Hellebrandt, 166, 167
Tóth, N., 351; with L.H. Keeley, 398, 411
Trinkaus, E., 62, 230, 243, 291, 352, 356, 389, 488
Troilett, G., 339
Tuffreau, A., 421, 423(fig.), 438
Turner, V.W., 376, 537
Turq, A., 370
Turville-Petre, F., 81

Ulrix-Closset, M., 205, 206, 366, 451

Valladas, H., 264
Valoch, K., 161, 163, 176, 179, 182, 183, 184, 185, 186, 187-9, 190, 200-1, 202, 204, 207, 217, 227, 421, 424(fig.), 430, 447, 451
Van Andel, T.H., 370
Van Campo, R., 353(fig.), 368(fig.)
Van Geist, V., 16
Van Peer, P., 66, 67, 144
Vandermeersch, B., 62, 63, 372; with F. Lévêque, 243, 270, 461; with L.

Duport, 359; with O. Bar-Yosef, 61, 67
Vaughan, P.C., 390, 405, 406, 413
Veil, S., 429
Vencl, S., 210
Vereschagin, N.K., 353(fig.), 354
Vermeersch, P.M., 66, 67, 143, 154, 155; with E. Paulissen, 141, 144, 155
Vértes, L., 161, 162, 164, 165, 166, 168, 169, 170, 172, 174, 191, 193, 196, 197, 198-9, 215, 428
Veth, P.M., 339, 340
Veyrier, M.E., 358
Vézian, J. and J., 289
Vialou, D., 483
Villa, P., 325
Villiers, H. de, with P.B. Beaumont, 43(fig.)
Vincent, A., 404
Vogel, J.C., 5, 48, 52, 92; with J. Kronfeld, 37; with K.W. Butzer, 42(fig.); with P.B. Beaumont, 43(fig.); with P.L. Carter, 48, 49
Volkman, P., 59, 68, 69, 500(fig.), 501-7(+figs); with A.E. Marks, 59, 65, 67-8, 69, 114, 134, 501
Volman, T.P., 4, 5, 7, 25, 34, 47-8, 49, 52, 259; with Q.B. Hendey, 6, 37
Von Koenigswald, W., 447-51
Vörös, I., 192, 198-9, 200(fig.), 429
Vrba, E.S., with S.J. Gould, 459

Wadley, L., 52
Waechter, J., 114, 281
Wagner, P., 362
Wainscoat, J.S., 50
Washburn, S.L., 536-7
Waterbolk, H.T., with J.C. Vogel, 92
Watson, H.K., 46
Watson, J., with M.H. Newcomer, 73
Webbley, D.P., with E.S. Higgs, 355
Wendorf, F., 57, 66, 139, 141, 143, 144, 145, 146, 150(fig.), 152, 154-5, 156; with R. Schild, 141
Wendt, W.E., 6, 480
Weniger, G.-C., 349
Werner, H.J., with W. Schönweiss, 207
Wescott, R., 528
Wettstein, O., 193
Wetzel, R., 460, 464, 468(fig.)
Whallon, R., 28, 358, 444, 453
Whateley, A., 46
White, J.P., 330, 331
White, L.A., 356
White, R., 64, 243, 244, 245, 250, 262, 356, 395, 404, 444, 457-8, 470, 483
White, T., 488
Whitlow, H.J., with H.H. Andersen, 395
Willey, G.R., 349
Williams, S., 339
Wilmsen, E.N., 3, 8, 14, 15(fig.), 17, 18, 20, 26
Wissler, C., 357
Wiszniowska, T., 198
Wobst, H., 458
Wójcik, M., 197
Wolpoff, M.H., 337, 357, 460, 492
Woodburn, J., 17, 372
Wright, R.V.S., 335, 523, 530
Wylie, M.A., 349
Wymer, J., with R. Singer, 3, 4, 5, 6, 22(fig.), 25, 49, 50(fig.), 467
Wynn, T., 507-9

Yates, R., 52
Yellen, J., 16, 17, 18, 361

Zar, J.H., 249
Ziegler, A.C., 349
Zihlman, A., with N. Tanner, 522-3
Zilhão, J., 276, 281, 283, 290, 295
Zotz, L., 161
Zvelebil, M., 356, 374

Index of Sites and Locations

Les Abeilles, 279(fig.)
Abou Sif, 112
Abri Agut, 355, 371
Abri Blanchard, 469-70, 474(fig.), 476(fig.), 483-5
Abri Castanet, 469-70, 476(fig.), 483
Abri Lartet, 359
Abri Mochi, 450(fig.)
Abri Pataud, 297, 448(fig.), 450(fig.)
Abri Vaufrey, 266, 390, 474
Abri Zumoffen, 81, 82, 85, 88, 89
Abric Romani, 280
Agenais, 281
Agut, Abri, 355, 371
Aitzbitarte, 279(fig.), 284-5(fig.)
L'Aldène, 357, 359
Alsószentgyörgy, 169-70
Altamira, 295
Amalda, 277, 279(fig.)
Ambrona, 354
Ambrosio, 281, 283, 290
Amúd, 63
Anosovka, 226
Apollo XI, 5, 35(fig.), 49, 480
Arad, 216
L'Arbreda, 277(+fig.), 279(fig.), 280, 281, 283, 289
Arcy-sur-Cure, 303-25(+figs), 359, 390, 444, 447, 448(fig.), 450(fig.), 452(fig.), 461-3(+fig.), 464(fig.), 465(fig.), 466, 470
Argolid, 372
Ariège, 281
Arnero, 279(fig.)
Asprochaliko, 367
Asturias, 281, 282, 292
Atapuerca, 282
Atxurra, 279(fig.)
Aurignac, 279(fig.), 281, 288(fig.), 289
Auvergne, 280
Avas, 162, 166, 167-70, 173, 174-5
Avastetö, 169

Bacho Kiro, 161, 190, 197-8, 200, 203, 211, 232, 235, 429, 430, 448(fig.), 460, 464, 467(fig.)
Badger Hole, 209, 449(fig.)
Baia de Fier, 224
Bakers Hole, 269
Ballavölgy, 164
Bambatan, 4
Bañolas, 296
Barca, 433
Barkly Tableland, 328(fig.), 338
Barranc Blanc, 281, 283
Basaharc, 165
Bass Point, 328(fig.), 329
Basté, 279(fig.)
Baume-Bonne, 357
Beçov, 461, 488
Bédeilhac, 293
Belvis, 293
Bénesse, 279(fig.)
Biache-St-Vaast, 374, 390, 405, 410, 414
Biarritz, 283
Bidache, 279(fig.)
Bidart, 279(fig.)
Bison, Grotte du, *see* Arcy-sur-Cure
Bistricioara-Lutărie, 215, 219-20(+fig.), 231
Bize, 279(fig.), 450(fig.)
Blanchard, Abri, 469-70, 474(fig.), 476(fig.), 483-5
Bobrava valley, 187
Bobuleshty, 222-3, 225
Bockstein-Törle, 202
Bocksteinschmiede, 460, 464, 466, 468-9(figs), 474, 488
Bodo, 376
Bohunice, 160, 163, 164, 176-8(+fig.), 179, 182, 183, 185-6, 188, 200, 232, 424(fig.), 448(fig.)
Boinești, 216
Bois-Roche, 404
Boker Tachtit, 59, 67-8, 69-71, 72, 134, 500(fig.), 501-7(+figs), 508, 509, 510, 512
Bolinkoba, 279(fig.), 282, 284-5(fig.), 287

Boomplaas, 5, 6, 7, 25, 35(fig.), 44(fig.), 49
Border Cave, 5, 6, 7, 25, 34, 35(fig.), 39-47(+figs), 49 , 50-1, 466-7, 471(fig.)
Boskovice, 185
Boussens, 279(fig.)
Bouzin, 279(fig.)
Bramford Road Pit, 210, 449(fig.)
Brassempouy, 277, 279(fig.), 281, 283
Brno, 430
Brynzeny, 222-3, 225, 226, 434
Buccaneer Archipelago, 339
Büdöspest, 164
Buffelskloof, 35(fig.)
Bükk mountains, 160, 162, 164, 168, 170, 171, 173, 174, 186, 192, 430
Burrill Lake, 328(fig.), 329
Bushman Rock Shelter, 35(fig.)
Buteshty, 221
Buzduzhany, 222
Býčí Skála, 187, 189, 201
Byneskranskop, 35(fig.), 44(fig.)

Cae Gwyn, 449(fig.)
Calcoix, 279(fig.)
Las Caldas, 296
Caldeirão, Gruta do, 290
Camargo, 279(fig.), 296
Camiac, 448(fig.)
Caminade, 450(fig.)
Can Crispins, 279(fig.)
Canecaude, 277(+fig.), 279(fig.), 281, 283, 289
Cantabria, 355, 356
Cape Peninsula, 49
La Carane, 279(fig.)
Carigüela, 277(fig.), 296
Carmel, Mount, 358
Cassegros, 406, 410
Castanet, Abri, 469-70, 476(fig.), 483
Castelcivita, Grotta de, 234
Castelmerle, 476(fig.); *see also* Blanchard *and* Castanet
El Castillo, 277, 278(fig.), 279(fig.), 280, 282, 284-5(fig.), 286, 287, 294, 295, 296, 485
Les Cauneilles-basses, 279(fig.)
Cavallo, Grotta del, 233-4
Cave Bay Cave, 329, 338
Cave of Hearths, 25, 49
Ceahlaû basin, 215, 217, 219
Ceahlaû-Cetățica, 221, 223
Ceahlaû-Dîrțu, 215, 219, 220(fig.), 231
Čertova Díra, 175
Čertova Pec, 162, 164, 448(fig.)
Chabiague, 279(fig.)
La Chaise, 359, 374
Chalosse, 283
La Chapelle-aux-Saints, 365, 368(fig.), 489
Charentes, 280, 352
Châtelperron, 450(fig.)
Chinchon, 361
El Chorro, 281
Chuntu, 222-3
El Cierro, 279(fig.)
Cioarei de la Boroșteni, 219, 220(fig.)
Clacton, 460-1
Cladova, 216
Cleland Hills, 328(fig.), 339
Colless Creek, 338
Columbeira, Gruta Nova da, 296
Combe-Capelle, 360
Combe-Grenal, 265-6, 269, 323, 353(fig.), 359, 360, 361, 365, 370(fig.), 390, 392-3(fig.), 394-5, 396-7(fig.), 400
Combe-Saunière, 390, 399
Combel, 485
El Conde, 277, 279(fig.), 284-5(fig.), 294
Conop, 216
Corbehem, 390
Corbiac, 389-90, 392-3(fig.), 394, 396-7(fig.), 400, 402-3(fig.), 405, 406-13(+figs)
Cotte de St-Brélade, 353(fig.), 354, 358, 365, 376
Les Cottés, 206, 368(fig.), 448(fig.), 450(fig.)
Cougnac, 487
Coupe Gorge, 279(fig.)
Couteret, 279(fig.)
Couvin, 205, 206-7, 230, 445(fig.), 447, 448(fig.)
Cova Negra, 296, 354
Creswell Crags, 209
Crkvina, 212, 214
Cro-Magnon, 486
Crouzade, 279(fig.), 281, 289, 293, 296
El Cudón, 279(fig.)
Cueto de la Mina, 279(fig.), 282, 296
Cueva Ambrosio, 281, 283, 290
Cueva del Conde, 277, 279(fig.), 294
Cueva Millán, 278(fig.), 280
Cueva Morín, 277, 278(fig.), 279(fig.), 280, 282, 284-5(fig.), 287, 291, 294, 296, 359, 448(fig.), 450(fig.)
Cueva Oscura de Perán, 279(fig.)
Cure Valley, 358; *see also* Arcy-sur-Cure
Cyrenaica, 156

Devil's Lair, 328(fig.), 331
Devil's Tower, 278(fig.), 291, 296
Diepkloof, 35(fig.), 41, 48, 49
Diósgyör-Tapolca, 165, 166-7(+fig.)
Djebel Qafzeh, 375
Dniester-Carpathian, 221, 222, 223, 229
Docteur, Grotte du, 205, 451
Dolní Kounice, 202

Dolní Věstonice, 178, 188, 233
Don, 227, 229
Dorog, 175
Douara Cave, 354, 401
Drachenloch, 487
Drahany Plateau, 179
Dufaure, 296, 297
Duruthy, 293, 296, 297
Dzeravá Skála (Pálffy), 162-3, 164, 192, 194, 235, 428
Dzierzysław, 424(fig.), 432

Ebro, 281
Eger, 162
Eger-Köporostetö, 170, 174
Ehringsdorf, 163-4, 353(fig.), 354, 360, 367, 374
Ein Aqev, 73
Ekain, 278(fig.), 279(fig.), 280, 284-5(fig.), 296
El Kilh, 140(fig.), 146, 147, 149-51(+figs), 154, 155
El Kowm, 64
El Wad, 67
Elands Bay Cave, 35(fig.)
Elis, 356, 365
Els Ermitons, 277(fig.)
Emireh, 505(+fig.)
Enlène, 293, 295, 296
Epirus, 367
Érd, 196, 216
Erevan, 402, 404, 414
Erralla, 296
Escoural, 295
Espalungue, 293
Espélugues, 293
Esquicho-Grapaou, 448(fig.), 450(fig.)
Estremadura, 281
Eyres, 279(fig.)
Les Eyzies, 486

Fabbrica, Grotta de la, 234
Le Facteur, 297
Felsöszentgyörgy, 169, 170
La Ferrassie, 265, 450(fig.), 458, 484(fig.), 485, 490
Ffynnon Beuno, 208, 449(fig.)
Le Flageolet, 390, 399
La Flecha, 278(fig.), 280
Font Yves, 132
Fontéchevade, 374
Fontmaure, 356, 359
Forbes Quarry, 296
Forty Acres Pit, 210
Franklin River, 328(fig.), 337-8, 340

Gahuzère, 279(fig.)
Gánovce, 234
Gargas, 279(fig.), 281, 283, 288(fig.), 289, 293-4, 295, 487
Gatzarría, 277, 279(fig.), 281, 282, 283, 289
Gaujacq, 279(fig.)
Gazel, Grotte, 295
Gebel Maghara, 156
Das Geissenklösterle, 190, 203, 230, 231, 232, 434, 448(fig.), 450(fig.), 470
George, Lake, 335
Glentyre Cave, 45
Gönnersdorf, 297
Gordineshty, 223, 225, 226, 228
Gorham's Cave, 278(+fig.), 281, 290, 291
Gottwaldov, 433
Gottwaldov-Louky, 163, 185
La Gouba, 402, 414
Gourdan, 293
Goyet, 204, 208, 210, 449(fig.)
Le Grand Pastou, 297
La Griega, 282
Grimaldi, 376
Grotta de Castelcivita, 234
Grotta del Cavallo, 233-4
Grotta de la Fabbrica, 234
Grotta Guattari, 359, 376
Grotta del Poggio, 353(fig.)
Grotte du Bison, *see* Arcy-sur-Cure
Grotte du Docteur, 205, 451
Grotte Gazel, 295
Grotte de l'Hyène, *see* Arcy-sur-Cure
Grotte du Loup, *see* Arcy-sur-Cure
Grotte du Pape, *see* Brassempouy
Grotte du Renne, *see* Arcy-sur-Cure
Grotte du Taï, 485
Grotte Tournal, 450(fig.)
Gruta do Caldeirão, 290
Gruta Nova da Columbeira, 296
Guattari, Grotta, 359, 376
Guipuzcoa, 281
Gura Cheii Rîşnov, 219, 220(fig.)

Ha Soloja, 35(fig.), 48, 49
Halfan, 58
Haréguy, 279(fig.)
Haua Fteah, 156
Haute-Garonne, 281
Hauteroche, 359, 368(fig.)
Hearths, Cave of, 25, 49
Helena River, 338
Herman Ottó Cave, 162
L'Hermitage, 210
Heuningneskrans, 35(fig.)
Highlands, 35(fig.)
El Higueron, 281
Hluhluwe, 46
Höhlenstein-Stadel, 478-81(+fig.), 488
Hont-Babat, 162
Hont-Csitár, 162
L'Hôpital, 279(fig.)
Hornos de la Peña, 279(fig.), 294-5
Hortus, 478-81

Hostějov, 178(+fig.), 184
Howieson's Poort, 6, 35(fig.)
Huccogne, 451
Hunter Island, 327, 328(fig.), 329
Huon Peninsula, 39, 328(fig.), 338
Hyena Den, 449(fig.)
l'Hyène, Grotte de, *see* Arcy-sur-Cure

Il'skaya, 228, 367, 429
Ioton, 353(fig.), 365
Isa, Mount, 327, 328(fig.)
Isle Valley, 297
Isna, 140(fig.), 152-4, 155
Istállóskő, 190, 191-2, 193(+fig.), 194, 196, 197, 198-200(+fig.), 224, 225, 232, 429, 448(fig.)
Isturitz, 279(fig.), 281, 282, 283, 287-9(+fig.), 291, 293, 294, 295, 296

Jabrud, 81, 82, 88, 89, 425
Jankovich, 162, 165, 175, 191-2, 193(fig.), 194, 196, 428
Jerzmanowice, 447
Jezeřany, 173, 190, 202
Jihlava, River, 189
Jordan, 61, 64, 65, 69
El Juyo, 296

Kaán-Károly, 174
Kadar, 212-13
Kamen, 212, 214
Kánástető, 171, 172-3(+fig.)
Kangaroo Island, 328(fig.), 335, 338
Kangkara, 35(fig.)
Kebara, 62, 64, 354, 375, 440, 492
Kecskésgalya, 164
Keilor, 335
Die Kelders, 35(fig.), 44(fig.)
Kent's Cavern, 163, 208, 210, 448(fig.), 449(fig.)
El Kilh, 140(fig.), 146, 147, 149-51(+figs), 154, 155
Kimberley, 338-9
Klasies River Mouth, 4, 5, 6-7, 22-3(fig.), 24, 25, 26, 35(fig.), 37-9(+fig.), 41, 46, 49, 50(fig.), 51
Klausennische, 205, 207
Klimautsy, 222-3
Klipfonteinrand, 35(fig.), 49
Kokkinopilos, 164
Königsaue, 207, 451, 474
Koobi Fora, 411
Koolan Island, 328(fig.), 339
Korlát-Ravaszlyuktető, 170-1, 173
Korman, 425(fig.), 428
Korolevo, 175-8, 230
Korpach, 223, 224-5, 226(fig.), 233
Korpach-mys, 223-4, 225
Kösten, 164, 173
Kostienki, 226-9(+fig.), 486
Kostienki-Tel'manskaya, 163
Kow Swamp, 327, 328(fig.), 337
El Kowm, 64
Koziarnia, 205
Krapina, 214, 234, 354, 374, 482(fig.), 487, 488
Krems-Hundsteig, 448(fig.)
Křepice, 184
Křížova, 162-3, 191-2, 193(fig.), 196
Krómĕvříž, 184
Krumlovskýles, 179, 180, 185
Ksar Akil, 60, 65, 67, 68, 69-71, 72, 73, 74, 91-136(+figs)
Kubsice, 202
Kůlna, 163, 175, 230
Kupařovice, 160, 178, 180-1(fig.), 187, 189-90, 201-2, 230, 422, 424(fig.)
Kurtzía, 279(fig.)

Labastide, 279(fig.), 293
Lagama, 72-3
Lake George, 335
Lake Mungo, 327, 328(fig.), 330, 335, 337, 340
Lambrecht Kálmán cave, 166
Les Landes de Gascogne, 281
Langmannersdorf, 202
La Laouza, 450(fig.)
Lapa de Rainha, 290
Lartet, Abri, 359
Lascaux, 485
Lazaret, 440
Lehringen, 354, 429, 460-1, 488
Lesotho, 48, 52
Lezetxiki, 278(fig.), 279(fig.), 280, 282, 284-5(fig.), 286, 296
Lezía, 279(fig.), 289
Lindenmier, 20
Lion Cavern, 35(fig.)
Líšeň, 176-8(+fig.), 183-4, 185-6
Líšeň-Čtvrtě, 185
Líšeň-Lepiny, *see* Podolí
Lombard, 390
Lömmersum, 448(fig.), 450(fig.)
Londža, 212, 214
Lorthet, 293
Loup, Grotte du, *see* Arcy-sur-Cure
Lovas, 192
Lumentxa, 279(fig.)
Lunel-Viel, 359
Lusčvić, 212, 214, 231

Maas River, 366
Mackay, 338
La Madeleine, 405
Maine d'Euche, 355
Maisières, 204, 205, 210, 451
Makhadma, 140(fig.), 144
Malarnaud, 296
Les Mallaetes, 277(+fig.), 281, 283,

289(fig.), 290, 294
Mal'ta, 376
Maltravieso, 282
Mályi-Öreghegy, 171-3(+fig.), 174
Mamaia, 164
Mamutowa, 205
Mandu Mandu Creek, 338
La Marche, 485-6
Máriaremete, 161, 164-5, 192, 235
Marillac, 390
Marsoulas, 279(fig.), 289, 295
Maršovice, 202
Mas d'Azil, 279(fig.), 293, 487
Mas-des-Caves, 359
Mauern, 164, 191-3(+fig.), 196, 206-7, 230, 447-51(+figs)
Mauran, 353(fig.), 355
Mazouco, 282
Melca, Roc de la, 277(fig.), 279(fig.)
Melikane, 35(fig.), 49
Melkhoutboom, 35(fig.)
Meseta del Norte, 281
Mexiko Valley, 430
La Micoque, 353(fig.), 360
Miskolc, 168, 169, 171-2
Miskolc-Kánástetö, *see* Kánástetö
Miskolc-Szabadkatetö, *see* Szabadkateto
Mitoc, 221-2
Mitoc Malu Galben, 219, 220-1(+fig.), 231
Mitoc-Valea Izvorului, 216-18(+fig.), 221, 223, 229
Mladeč, 234-5
Mochi, Abri, 450(fig.)
Mokriška jama, 196
Moldavia, 160-1
Molochnyi Kamen', 224
Molodova, 221, 223, 224, 353(fig.), 357, 358-9, 428, 432(+fig.), 433(fig.), 434, 440, 452(fig.)
Montagu Cave, 5, 6, 26, 35(fig.), 47-8, 49
Montaut, 279(fig.)
Monte Circeo, 359, 458, 487, 489
Montespan, 487-8
Montgaudier, 359
Montmaurin, 279(fig.)
Morava, River, 184
Los Morceguillos, 281, 290
Morín, 277, 278(fig.), 279(fig.), 280, 282, 284-5(fig.), 287, 291, 294, 296, 359, 448(fig.), 450(fig.)
Moshebi's Shelter, 5, 35(fig.), 49
Mouligna, 279(fig.)
Mount Carmel, 358
Mount Isa, 327, 328(fig.)
Mount Newman, 328(fig.), 339
Le Moustier, 264, 360
Mungo, Lake, 327, 328(fig.), 330, 335, 337, 340
Muselievo, 163, 430
Naáme, 61, 63, 65
Nad Kačákem, 163
Nahr Ibrahim, 60, 481
Nandru-Spurcatâ, 215-16
Napajedla, 184
Nazlet Khater, 140(fig.), 144, 147, 148(fig.), 149(fig.), 154, 155, 156
Negev, 61, 65, 69
Nelson Bay Cave, 6, 7, 22-4(+fig.), 25, 35(fig.), 37
Newman, Mount, 328(fig.), 339
Niaux, 295, 485
Nietoperzowa, 163, 164, 191-2, 193(fig.), 196, 197, 204, 207, 448(fig.)
Nile Valley, 57, 58, 66-7, 139-56(+figs)
Normandy, 440
Northwest Cape, 328(fig.)
Nos, 35(fig.)
Nosovo, 414

Oaş, 216, 229
Ohaba Ponor, 219, 220(fig.)
Ojo Guareña, 282
Ondratice, 163, 177(fig.), 178, 179-81, 186, 187, 203, 230
Orange River Valley, 25
Ořechov, 187
Otaslavice, 163
El Otero, 279(fig.), 282, 294
Les Ourteix, 353(fig.)

Las Palomas, 281
Parpalló, 277(fig.), 281, 283, 290, 293, 294
Pas-du-Miroir, 472-3, 476(fig.) Patagonia, 410
Pataud, Abri, 297, 448(fig.), 450(fig.)
Paviland, 163, 208, 210, 449(fig.)
Pavlov, 178, 233
Pech de l'Azé, 359, 361, 370(fig.), 389-90, 396-7(fig.), 400, 404-13(+figs), 463(+fig.), 465(fig.), 466, 469, 470
Pech Merle, 485
Peers Cave, 5, 35(fig.), 49
Penches, 282
El Pendo, 277, 279(fig.), 282, 284-5(fig.), 287, 291, 294
Perán, Cueva Oscura de, 279(fig.)
Périgord, 243-4, 245-59, 263, 280, 352, 354, 356, 389-90, 391
Las Perneras, 281
Pesko, 448(fig.)
Peştera Cioarei de la Boroşteni, 219, 220(fig.)
Petřkovice, 451
Petrovany, 162
Pi, 390
Le Piage, 233, 319, 450(fig.)
Picken's Hole, 207-8, 230
Piekary, 421, 438

Pietersburg, 4
Pilbara, 339, 340
La Pileta, 294
Pin Hole, 209, 210, 449(fig.)
Pincevent, 399
Pockenbank, 35(fig.)
Pod Hradem, 162-3, 192, 197, 232, 234
Pod Tureckom, 232-3
Podolí, 176-8(+fig.), 179, 182, 185
Podstránká, 186, 187
Poggio, Grotta del, 353(fig.)
Pont d'Ambon, 410
Le Portel, 279(fig.), 283, 289, 295, 296, 368(fig.)
Posada de Llanes, 282
Potočka Zijalka, 160, 191-2, 193-6(+figs), 197
Pouillon, 279(fig.)
Předmostí, 163, 178-9
Pulborough, 210, 449(fig.)

Qafzeh, 62, 63, 65, 297, 481, 492
Quercy, 280, 281
La Quina, 324, 353(fig.), 354, 359, 365, 367, 368(fig.), 448(fig.), 450(fig.), 464-6, 470(fig.), 474, 492
Quinçay, 324

Rachat, 279(fig.)
Raj Cave, 357, 359, 432
Ranis, 164, 191-2, 193(fig.), 196, 204, 206-7, 210, 447, 449(fig.)
El Rascaño, 278(fig.), 279(fig.), 284-5(fig.)
Reclau Viver, 279(fig.), 281, 283, 289
Remete cave, 165
Remetea-Şomoş, 216
Renne, Grotte du, *see* Arcy-sur-Cure
Rheindalen, 359, 421, 423(fig.), 438
Rhône Valley, 358
Les Rideaux, 279(fig.), 283, 289
La Riera, 278(fig.), 279(fig.), 280, 284-5(fig.), 294, 296
Le Rigabe, 360
Ripiceni-Izvor, 215, 216, 217, 219-20(+fig.), 221, 223, 357, 358-9, 432
Robin Hood's Cave, 209, 449(fig.)
Le Roc, 353(fig.), 355, 442(fig.)
Roc de Combe, 233, 319, 450(fig.)
Roc-de-Marsal, 353(fig.), 368(fig.)
Roc de la Melca, 277(fig.), 279(fig.)
Roc-en-Pail, 357, 359, 368(fig.)
Rocky Cape, 328(fig.)
Rocourt, 421, 438, 439(fig.)
Romani, 280
La Roque, 472-3, 476(fig.)
Roquecourbère, 279(fig.)
Rörshain, 163
Rose Cottage Cave, 5, 25, 35(fig.)
Rosh Ein Mor, 71
Ruprechtov, 179
Rytířská, 162-3

Saccopastore, 374, 458
Sahara, 143, 144
Sahulland, 329
Saint-Césaire, 233, 234, 297, 375, 447, 450(fig.), 451, 461
Saint-Jean-de-Verges, 277(fig.), 279(fig.), 281, 283, 288(fig.), 289
St Lon, 279(fig.)
St Valéry-sur-Somme, 421
Sajóbábony, 166, 173, 174
Sajóbábony-Kóvesoldal, 171, 172-3(+fig.)
Sajóbábony-Méhésztetó, 171, 172-3(+fig.)
Sajószentpéter-Nagykorcsolás, 172
Salat-Volp-Garonne, 283
Salemas, 290
Saliès-de-Béarn, 279(fig.)
El Salitre, 279(fig.)
Samuilitsa, 164, 211-12, 213(fig.), 219, 430, 448(fig.)
Sandougne, 353(fig.)
Santander, 281, 282, 296
Santimamiñe, 279(fig.)
Schambach, 205, 207
Sclayn, 440
Sde Divshon, 73
Seclin, 421, 423(fig.), 438
Sehonghong, 5, 35(fig.), 48, 49
Sergeac, 457
Serrón, 281
Shanidar, 352, 353(fig.), 376, 489
Shongweni, 35(fig.)
Shukwikhat, 140(fig.), 152, 153(fig.), 154, 155
Sinai, 56, 156
Šipka, 175, 230
Skaratki, 429
Skhul, 61, 62, 64, 297, 481
Soldier's Hole, 207-8, 230, 449(fig.)
Sorde, 279(fig.)
Le Soucy, 470
Soulabé, 296
La Souquette, 483
Southern Cape, 25
Spitsynskaya, 227, 228
Sprimont, 208
Spy, 163, 204, 210, 449(fig.)
Starosel'ye, 163, 211, 228, 353(fig.), 354, 429
Stinka, 217, 222
Stradbroke Island, 328(fig.), 338
Stránská Skála, 160, 176-8(+fig.), 179, 182-3, 185-6, 187, 200, 231, 232
Seletskaya, 226, 227-9
Stříbrnice, 178, 184
Subalyuk, 164, 167, 170, 173, 175, 219, 223, 234

Sukhaya Mechetka, 367, 429
Sümeg-Mogyorósdomb, 169
Sundalland, 329
Sungir, 227, 434, 486
Švédův Stůl, 175, 230
Swabian Alb, 21
Swan Bridge, Upper, 328(fig.), 331, 335, 338, 340
Swaziland, 468
Szabadkatetö, 171, 172-3(+fig.)
Szeleta, 160, 162, 164, 170, 191-3(+fig.), 196, 197, 223, 227, 233, 428, 448(fig.)
Szinva valley, 171

Tabún, 60-2, 63, 64, 68, 81-8(+figs), 89
Tahta, *see* Nazlet Khater
Taï, Grotte du, 485
Tambourets, 279(fig.)
Tarté, 279(fig.), 283, 289
Tata, 165, 354, 448(fig.), 460, 463, 466(fig.), 467-9, 472-3(figs), 488
Tata-Kálváriadomb, 169
Taubach, 353(fig.), 354
Tenaghi Philippon, 213
Téoulé, 279(fig.)
Tercis, 279(fig.)
Terra Amata, 461
Thessaly, 356
Tibava, 433
Tito Bustillo, 295, 296
Torralba, 354, 460-1
Tournal, Grotte, 450(fig.)
Toutifaut, 361
Trencianske Bohuslavice, 232-3
Trenčín, 162
Trinka, 228-9
Les Trois Frères, 294, 295, 485
Trou de la Chèvre, 450(fig.)
Trou Magrite, 210
Trou du Renard, 208
Troubky, 178(+fig.), 184
Le Tuc d'Audoubert, 295, 487
Tursac, 208, 450(fig.)
La Tuto de Camalhot, 279(fig.), 283
Tüzköves, 169, 170

Uherske Hradiste, 184
Umfolozi, 46
Umhlatuzana, 35(fig.), 48, 49, 52
Usategi, 279(fig.)

La Vache, 296
Vallon des Roches, 457, 469-70; *see also* Blanchard *and* Castanet
Vaufrey, Abri, 266, 390, 474
Vedrovice, 160, 176-8(+fig.), 179, 180-1(fig.), 187-90, 200, 201-2, 230, 232, 422, 430
Velika Pecina, 448(fig.)
Velký Šariš, 162
Verberie, 399, 401, 404
Vergisson, 353(fig.), 359
Veternica, 214
Villefranue, 279(fig.)
La Viña, 279(fig.), 294
Vindija, 214, 234
Visoko Brdo, 212, 214
Vizcaya, 281
Vogelherd, 434, 448(fig.), 475(fig.), 478, 480, 481
Volkovskaya, 226, 227-8(+fig.)

El Wad, 67
Wadi Amud, 82
Wadi Halfa, 140(fig.), 142-3, 145, 146, 147, 149-52(+figs), 155
Wadi Hasa, 64
Wadi Kubbaniya, 140(fig.), 144-5
Wadi el-Mughara, 81
Wadi Skifta, 81
Warrnambool, 328(fig.), 335, 337
Western Desert (Australia), 399, 402-4
Willandra Lakes, 330, 336
Willendorf, 231, 232, 448(fig.)
Wookey Hole Hyaena Den, 207

Zâbrani, 216
Zajara, 281
Zdislavice, 178, 184
Zebrarivier, 35(fig.)
Zeitlarn, 207
Zelešice, 187
Zobište, 212, 214
Zumoffen, Abri, 81, 82, 85, 88, 89
Zuttiyeh, 63, 81, 82
Zwierzyniec, 163, 205, 225
Zwollen, 429, 440